MODELLING TRANSPORT

MODELLING TRANSPORT

Fourth Edition

Juan de Dios Ortúzar

Department of Transport Engineering and Logistics
Pontificia Universidad Católica de Chile
Santiago
Chile

Luis G. Willumsen

Luis Willumsen Consultancy
and University College London
London
UK

A John Wiley and Sons, Ltd., Publication

This edition published 2011
© 2011 John Wiley & Sons, Ltd

Previous editions published 1990, 1994, 2001 © John Wiley & Sons, Ltd

Registered office
John Wiley & Sons Ltd, The Atrium, Southern Gate, Chichester, West Sussex, PO19 8SQ, United Kingdom

For details of our global editorial offices, for customer services and for information about how to apply for
permission to reuse the copyright material in this book please see our website at www.wiley.com.

Library of Congress Cataloguing-in-Publication Data

Ortuzar, Juan de Dios (Ortuzar Salas), 1949-
  Modelling Transport / Juan de Dios Ortuzar, Luis G. Willumsen. – Fourth edition.
    p. cm
  Includes bibliographical references and index.
  ISBN 978-0-470-76039-0 (hardback)
  1. Transportation–Mathematical models.   2. Choice of transportation–Mathematical models.
3. Trip generation–Mathematical models.   I. Willumsen, Luis G.   II. Title.
  HE147.7.O77 2011
  388.01'5118–dc22

                                                                        2010050373

A catalogue record for this book is available from the British Library.

Print ISBN: 9780470760390
E-Pdf ISBN: 9781119993315
O-book ISBN: 9781119993308
E-Pub ISBN: 9781119993520
Mobi ISBN: 9781119993537

Typeset in 9/11pt Times by Aptara Inc., New Delhi, India.

Contents

About the Authors

Juan de Dios Ortúzar is Professor of Transport Engineering at Pontificia Universidad Católica de Chile, his alma mater and where he has worked since 1972. From this remote setting he has managed to form generations of young researchers in Latin America and Europe. Outside academia, he has been advisor to governments and international agencies, and has directed several transport studies involving the application of advanced demand modelling techniques and the collection of large-scale travel survey data. A keen golfer he also enjoys playing guitar and singing with friends.

Luis (Pilo) Willumsen has some 40 years of experience as a consultant, transport planner and researcher with a distinguished academic career. He studied Engineering in Chile and has been based in Britain since 1975: he was a researcher and lecturer at Leeds University and then at University College London. He was a Board Director of Steer Davies Gleave having joined it full-time in 1989 with a special responsibility for technical and international development. He left that company in 2009 and is now Director of the Luis Willumsen Consultancy and Visiting Professor in the Department of Civil, Environmental and Geomatic Engineering at University College London.

Preface

This book is a result of nearly 40 years of collaboration, sometimes at a distance and sometimes working together in Britain and in Chile. Throughout these years we discussed many times what we thought were the strong and weak aspects of transport modelling and planning. We speculated, researched and tested in practice some new and some not so new ideas. We have agreed and disagreed on topics like the level of detail required for modelling or the value of disaggregate or activity based models in forecasting; we took advantage of a period when our views converged to put them in writing; here they are.

We wish to present the most important (in our view) transport modelling techniques in a form accessible to students and practitioners alike. We attempt this giving particular emphasis to key topics in contemporary modelling and planning:

- the practical importance of theoretical consistency in transport modelling;
- the issues of data and specification errors in modelling, their relative importance and methods to handle them;
- the key role played by the decision-making context in the choice of the most appropriate modelling tool;
- how uncertainty and risk influence the choice of the most appropriate modelling tool;
- the advantages of variable resolution modelling; a simplified background model coupled with a much more detailed one addressing the decision questions in hand;
- the need for a monitoring function relying on regular data collection and updating of forecasts and models so that courses of action can be adapted to a changing environment.

We have approached the subject from the point of view of a modelling exercise, discussing the role of theory, data, model specification in its widest sense, model estimation, validation and forecasting. Our aim in writing this book was to create both a text for a diploma or Master's course in transport and a reference volume for practitioners; however, the material is presented in such a way as to be useful for undergraduate courses in civil engineering, geography and town planning. The book is based on our lecture notes prepared and improved over several years of teaching at undergraduate and graduate levels; we have also used them to teach practitioners both through in-house training programmes and short skills-updating courses. We have extended and enhanced our lecture notes to cover additional material and to help the reader tackling the book without the support of a supervisor.

Chapters 3 to 9, 12 and 15 provide all the elements necessary to run a good 30 sessions course on transport demand modelling; in fact, such a course – with different emphasis on certain subjects – has been taught by us at undergraduate level in Chile, and at postgraduate level in Australia, Britain, Colombia, Italy, Mexico, Portugal and Spain; the addition of material from Chapters 10 and 11 would make it a transport modelling course. Chapters 4 to 6 and 10 to 12 provide the basic core for a course on network modelling and equilibrium in transport; a course on transport supply modelling would require more material, particularly relating to important aspects of public transport supply which we do not discuss in enough detail. Chapters 13, 14 and 16 cover material which is getting more important as time goes by, in particular as the shift in interest in the profession is moving from passenger issues to freight and logistics, and to the role models play not only in social evaluation but also in the analysis of private projects. Chapter 1 provides an introduction to transport planning issues and outlines our view on the relationship between planning and modelling. Chapter 2 is there mainly for the benefit of those wishing to brush up their analytical and statistical skills and to make the volume sufficiently self-contained.

During our professional life we have been fortunate to be able to combine teaching with research and consultancy practice. We have learnt from papers, research, experimentation and mistakes. We are happy to say the latter have not been too expensive in terms of inaccurate advice. This is not just luck; a conscientious analyst pays for mistakes by having to work harder and longer to sort out alternative ways of dealing with a difficult modelling task. We have learnt the importance of choosing appropriate techniques and technologies for each task in hand; the ability to tailor modelling approaches to decision problems is a key skill in our profession. Throughout the book we examine the practical constraints to transport modelling for planning and policy making in general, particularly in view of the limitations of current formal analytical techniques, and the nature and quality of the data likely to be available.

We have avoided the intricate mathematical detail of every model to concentrate instead on their basic principles, the identification of their strengths and limitations, and a discussion of their use. The level of theory supplied by this book is, we believe, sufficient to select and use the models in practice. We have tried to bridge the gap between the more theoretical publications and the too pragmatic 'recipe' books; we do not believe the profession would have been served well by a simplistic 'how to' book offering a blueprint to each modelling problem. In this latest edition we have also marked, with a shaded box, material which is more advanced and/or still under development but important enough to be mentioned. There are no single solutions to transport modelling and planning. A recurring theme in the book is the dependence of modelling on context and theory. Our aim is to provide enough information and guidance so that readers can actually go and use each technique in the field; to this end we have striven to look into practical questions about the application of each methodology. Wherever the subject area is still under development we have striven to make extensive references to more theoretical papers and books which the interested reader can consult as necessary. In respect of other, more settled modelling approaches, we have kept the references to those essential for understanding the evolution of the topic or serving as entry points to further research.

We believe that nobody can aspire to become a qualified practitioner in any area without doing real work in a laboratory or in the field. Therefore, we have gone beyond the sole description of the techniques and have accompanied them with various application examples. These are there to illustrate some of the theoretical or practical issues related to particular models. We provide a few exercises at the end of key chapters; these can be solved with the help of a scientific pocket (or better still, a spreadsheet) calculator and should assist the understanding of the models discussed.

Although the book is ambitious, in the sense that it covers quite a number of themes, it must be made clear from the outset that we do not intend (nor believe it possible) to be up-to-the-minute in every topic. The book is a good reflection of the state of the art but for leading-edge research the reader should use the references provided as signposts for further investigation.

We wrote most of the first edition during a sabbatical visit by the first of us to University College London in 1988–89. This was possible thanks to support provided by the UK Science and Engineering Research Council, The Royal Society, Fundación Andes (Chile), The British Council and The Chartered Institute of Transport. We thank them for their support as we acknowledge the funding provided for our research by many institutions and agencies over the past 30 years. The third and this fourth edition benefited greatly from further sabbatical stays at University College London in 1998–99 and 2009; these were possible thanks to the support provided by the UK Engineering and Physical Sciences Research Council. We also wish to acknowledge the support to our research provided by the Chilean Fund for

Developing Scientific and Technical Research (FONDECYT) and the Millennium Institute on Complex Engineering Systems (ICM: P05-004F; FONDECYT: FB016). Steer Davies Gleave also allowed the second author to spend time updating the second and third editions.

We have managed to maintain an equal intellectual contribution to the contents of this book but in writing and researching material for it we have benefited from numerous discussions with friends and colleagues. Richard Allsop taught us a good deal about methodology and rigour. Huw Williams's ideas are behind many of the theoretical contributions in Chapter 7; Andrew Daly and Hugh Gunn have helped to clarify many issues in Chapters 3, 7–9 and 15. Dirck Van Vliet's emphasis in explaining assignment and equilibrium in simple but rigorous terms inspired Chapters 10 and 11. Tony Fowkes made valuable comments on car ownership forecasting and stated-preference methods. Jim Steer provided a constant reference to practical issues and the need to develop improved approaches to address them.

Many parts of the first edition of the book also benefited from a free, and sometimes very enthusiastic, exchange of ideas with our colleagues J. Enrique Fernández and Joaquin de Cea at the Pontificia Universidad Católica de Chile, Sergio Jara-Díaz and Jaime Gibson at the Universidad de Chile, Marc Gaudry at the Université de Montréal, Roger Mackett at University College London, Dennis Gilbert and Mike Bell at Imperial College. Many others also contributed, without knowing, to our thoughts.

Subsequent editions of the book have benefited from comments from a number of friends and readers, apart from those above, who have helped to identify errors and areas for improvement. Among them we should mention Michel Bierlaire from the Ecole Polytechnique Fédérale de Lausanne, Patrick Bonnel from the French Laboratoire d'Economie des Transports, David Boyce at the University of Illinois, Victor Cantillo from Universidad del Norte, Barranquilla, Elisabetta Cherchi from University of Cagliari, Michael Florian from Université de Montréal, Rodrigo Garrido, Luis I. Rizzi and Francisca Yañez from Pontificia Universidad Católica de Chile, Cristián Guevara now at Universidad de Los Andes in Chile, Stephane Hess at Leeds University, Ben Heydecker from University College London, Frank Koppelman from Northwestern University, Mariëtte Kraan at the University of Twente, Francisco J. Martínez and Marcela Munizaga at the Universidad de Chile, Piotr Olszewski from Warsaw University of Technology, Joan L. Walker from University of California at Berkeley, and Sofia Athanassiou, Gloria Hutt, Neil Chadwick, John Swanson, Yaron Hollander and Serbjeet Kohli at Steer Davies Gleave. Special thanks are due to John M. Rose at ITLS, University of Sydney, for his contributions to Chapter 3.

Our final thanks go to our graduate and undergraduate students in Australia, Britain, Chile, Colombia, México, Italy, Portugal and Spain; they are always sharp critics and provided the challenge to put our money (time) where our mouth was.

We have not taken on board all suggestions as we felt some required changing the approach and style of the text; we are satisfied future books will continue to clarify issues and provide greater rigour to many of the topics discussed here; transport is indeed a very dynamic subject. Despite this generous assistance, we are, as before, solely responsible for the errors remaining in this latest edition. We genuinely value the opportunity to learn from our mistakes.

Juan de Dios Ortúzar and Luis G. Willumsen

1

Introduction

1.1 Transport Planning and Modelling

1.1.1 Background

The world, including transport, is changing fast. We still encounter many of the same transport problems of the past: congestion, pollution, accidents, financial deficits and pockets of poor access. We are increasingly becoming money rich and time poor. However, we have learnt a good deal from long periods of weak transport planning, limited investment, emphasis on the short term and mistrust in strategic transport modelling and decision making. We have learnt, for example, that old problems do not fade away under the pressure of attempts to reduce them through better traffic management; old problems reappear in new guises with even greater vigour, pervading wider areas, and in their new forms they seem more complex and difficult to handle.

We now have greater confidence in technical solutions than in the previous century. This is not the earlier confidence in technology as the magic solution to economic and social problems; we have also learnt that this is a mirage. However, Information Technology has advanced enough to make possible new conceptions of transport infrastructure (e.g. road transport informatics), movement systems (e.g. automated driverless trains) and electronic payment (e.g. smartcards, video tolling). Mobile phones and GPS services are changing the way to deliver useful traveller information, facilitating payment and charging for the use of transport facilities. Of particular interest to the subject of this book is the advent of low-cost and high-speed computing; this has practically eliminated computing power as a bottleneck in transport modelling. The main limitations are now human and technical: contemporary transport planning requires skilled and experienced professionals plus, as we will argue below, theoretically sound modelling techniques with competent implementations in software.

Emerging countries are becoming more significant in the world stage but they suffer serious transport problems as well. These are no longer just the lack of roads to connect distant rural areas with markets. Indeed, the new transport problems bear some similarities with those prevalent in the post-industrialised world: congestion, pollution, and so on. However, they have a number of very distinctive features deserving a specific treatment: relatively low incomes, fast urbanisation and change, high demand for public transport, scarcity of resources including capital, sound data and skilled personnel.

The birth of the twenty-first century was dominated by two powerful trends affecting most aspects of life and economic progress. The stronger trend is *globalisation*, supported and encouraged by the other trend, cheap and high-capacity *telecommunications*. The combination of the two is changing the way we perceive and tackle many modern issues; their influence in transport planning is starting to be

Modelling Transport, Fourth Edition. Juan de Dios Ortúzar and Luis G. Willumsen.

felt. Some of these influences are the role of good transport infrastructure in enhancing the economic competitiveness of modern economies; a wider acceptance of the advantages of involving the private sector more closely in transport supply and operations; the possible role of telecommunications in reducing the need to travel.

Important technical developments in transport modelling have taken place since the mid-1970s, in particular at major research centres; these developments have been improved and implemented by a small group of resourceful consultants. However, many of these innovations and applications have received limited attention outside the more academic journals. After these years of experimentation there is now a better recognition of the role of modelling in supporting transport planning. This book attempts a review of the best of current practice in transport modelling; in most areas it covers the 'state of the art' but we have selected those aspects which have already been implemented successfully in practice. The book does not represent the leading edge of research into modelling. It tries, rather, to provide a survival tool-kit for those interested in improving transport modelling and planning, a kind of bridge or entry-point to the more theoretical papers that will form the basis of transport modelling in the future.

Transport modelling is not transport planning; it can only support planning, and in a few cases it may have the most important role in the process. We have known many good professionals who have developed sophisticated transport models but are frustrated because their work has apparently been ignored in many key planning decisions. In truth, planning and implementation have the power to change the world and transport modelling can only assist in this if adopted as an effective aid to decision making. This requires wise planners and, above all, better modellers.

1.1.2 Models and their Role

A *model* is a simplified representation of a part of the real world–the system of interest–which focuses on certain elements considered important from a particular point of view. Models are, therefore, problem and viewpoint specific. Such a broad definition allows us to incorporate both physical and abstract models. In the first category we find, for example, those used in architecture or in fluid mechanics which are basically aimed at design. In the latter, the range spans from the mental models all of us use in our daily interactions with the world, to formal and abstract (typically analytical) representations of some theory about the system of interest and how it works. Mental models play an important role in understanding and interpreting the real world and our analytical models. They are enhanced through discussions, training and, above all, experience. Mental models are, however, difficult to communicate and to discuss.

In this book we are concerned mainly with an important class of abstract models: mathematical models. These models attempt to replicate the system of interest and its behaviour by means of mathematical equations based on certain theoretical statements about it. Although they are still simplified representations, these models may be very complex and often require large amounts of data to be used. However, they are invaluable in offering a 'common ground' for discussing policy and examining the inevitable compromises required in practice with a level of objectivity. Another important advantage of mathematical models is that during their formulation, calibration and use the planner can learn much, through experimentation, about the behaviour and internal workings of the system under scrutiny. In this way, we also enrich our mental models thus permitting more intelligent management of the transport system.

A model is only realistic from a particular perspective or point of view. It may be reasonable to use a knife and fork on a table to model the position of cars before a collision but not to represent their mechanical features, or their route choice patterns. The same is true of analytical models: their value is limited to a range of problems under specific conditions. The appropriateness of a model is, as discussed in the rest of this chapter, dependent on the context where it will be used. The ability to choose and adapt models for particular contexts is one of the most important elements in the complete planner's tool-kit.

This book is concerned with the contribution transport modelling can make to improved decision making and planning in the transport field. It is argued that the use of models is inevitable and that of formal models highly desirable. However, transport modelling is only one element in transport planning: administrative practices, an institutional framework, skilled professionals and good levels of communication with decision makers, the media and the public are some of the other requisites for an effective planning system. Moreover, transport modelling and decision making can be combined in different ways depending on local experience, traditions and expertise. However, before we discuss how to choose a modelling and planning approach it is worth outlining some of the main characteristics of transport systems and their associated problems. We will also discuss some very important modelling issues which will find application in other chapters of this book.

1.2   Characteristics of Transport Problems

Transport problems have become more widespread and severe than ever in both industrialised and developing countries alike. Fuel shortages are (temporarily) not a problem but the general increase in road traffic and transport demand has resulted in congestion, delays, accidents and environmental problems well beyond what has been considered acceptable so far. These problems have not been restricted to roads and car traffic alone. Economic growth seems to have generated levels of demand exceeding the capacity of most transport facilities. Long periods of under-investment in some modes and regions have resulted in fragile supply systems which seem to break down whenever something differs slightly from average conditions.

These problems are not likely to disappear in the near future. Sufficient time has passed with poor or no transportation planning to ensure that a major effort in improving most forms of transport, in urban and inter-urban contexts, is necessary. Given that resources are not unlimited, this effort will benefit from careful and considered decisions oriented towards maximising the advantages of new transport provision while minimising their money costs and undesirable side-effects.

1.2.1   Characteristics of Transport Demand

The demand for transport is *derived*, it is not an end in itself. With the possible exception of sight-seeing, people travel in order to satisfy a need (work, leisure, health) undertaking an *activity* at particular locations. This is equally significant for goods movements. In order to understand the demand for transport, we must understand the way in which these activities are distributed over space, in both urban and regional contexts. A good transport system widens the opportunities to satisfy these needs; a heavily congested or poorly connected system restricts options and *limits* economic and social development.

The demand for transport services is highly *qualitative* and *differentiated*. There is a whole range of specific demands for transport which are differentiated by time of day, day of week, journey purpose, type of cargo, importance of speed and frequency, and so on. A transport service without the attributes matching this differentiated demand may well be useless. This characteristic makes it more difficult to analyse and forecast the demand for transport services: tonne and passenger kilometres are extremely coarse units of performance hiding an immense range of requirements and services.

Transport demand takes place over *space*. This seems a trivial statement but it is the distribution of activities over space which makes for transport demand. There are a few transport problems that may be treated, albeit at a very aggregate level, without explicitly considering space. However, in the vast majority of cases, the explicit treatment of space is unavoidable and highly desirable. The most common approach to treat space is to divide study areas into zones and to code them, together with transport networks, in a form suitable for processing with the aid of computer programs. In some cases, study areas can be simplified assuming that the zones of interest form a corridor which can be collapsed into a linear form. However, different methods for treating distance and for allocating origins and destinations (and their attributes) over space are an essential element in transport analysis.

The spatiality of demand often leads to problems of lack of coordination which may strongly affect the equilibrium between transport supply and demand. For example, a taxi service may be demanded unsuccessfully in a part of a city while in other areas various taxis may be plying for passengers. On the other hand, the concentration of population and economic activity on well-defined corridors may lead to the economic justification of a high-quality mass transit system which would not be viable in a sparser area.

Finally, transport demand and supply have very strong *dynamic* elements. A good deal of the demand for transport is concentrated on a few hours of a day, in particular in urban areas where most of the congestion takes place during specific peak periods. This time-variable character of transport demand makes it more difficult–and interesting–to analyse and forecast. It may well be that a transport system could cope well with the *average* demand for travel in an area but that it breaks down during peak periods. A number of techniques exist to try to spread the peak and average the load on the system: flexible working hours, staggering working times, premium pricing, and so on. However, peak and off-peak variations in demand remain a central, and fascinating, problem in transport modelling and planning.

1.2.2   Characteristics of Transport Supply

The first distinctive characteristic of transport supply is that it is a *service* and not a good. Therefore, it is not possible to stock it, for example, to use it in times of higher demand. A transport service must be consumed when and where it is produced, otherwise its benefit is lost. For this reason it is very important to estimate demand with as much accuracy as possible in order to save resources by tailoring the supply of transport services to it.

Many of the characteristics of transport systems derive from their nature as a service. In very broad terms a transport system requires a number of fixed assets, the *infrastructure*, and a number of mobile units, the *vehicles*. It is the combination of these, together with a set of rules for their operation, that makes possible the movement of people and goods.

It is often the case that infrastructure and vehicles are not owned nor operated by the same group or company. This is certainly the case of most transport modes, with the notable exception of many rail systems. This separation between supplier of infrastructure and provider of the final transport service generates a rather complex set of interactions between government authorities (central or local), construction companies, developers, transport operators, travellers and shippers, and the general public. The latter plays several roles in the supply of transport services: it represents the residents affected by a new scheme, or the unemployed in an area seeking improved accessibility to foster economic growth; it may even be car owners wishing to travel unhindered through somebody else's residential area.

The provision of transport infrastructure is particularly important from a supply point of view. Transport infrastructure is 'lumpy', one cannot provide half a runway or one-third of a railway station. In certain cases, there may be scope for providing a gradual build-up of infrastructure to match growing demand. For example, one can start providing an unpaved road, upgrade it later to one or two lanes with surface treatment; at a later stage a well-constructed single and dual carriageway road can be built, to culminate perhaps with motorway standards. In this way, the provision of infrastructure can be adjusted to demand and avoid unnecessary early investment in expensive facilities. This is more difficult in other areas such as airports, metro lines, and so on.

Investments in transport infrastructure are not only lumpy but also take a long time to be carried out. These are usually large projects. The construction of a major facility may take from 5 to 15 years from planning to full implementation. This is even more critical in urban areas where a good deal of disruption is also required to build them. This disruption involves additional costs to users and non-users alike.

Moreover, transport investment has an important political role. For example, politicians in developing countries often consider a road project a safe bet: it shows they care and is difficult to prove wrong or uneconomic by the popular press. In industrialised nations, transport projects usually carry the risk of alienating large numbers of residents affected by them or travellers suffering from congestion and delay in overcrowded facilities. Political judgement is essential in choices of this kind but when not supported by planning, analysis and research, these decisions result in responses to major problems and crises only; in the case of transport this is, inevitably, too late. Forethought and planning are essential.

The separation of providers of infrastructure and suppliers of services introduces economic complexities too. For a start, it is not always clear that all travellers and shippers actually perceive the total costs incurred in providing the services they use. The charging for road space, for example, is seldom carried out directly and when it happens the price does not include congestion costs or other external effects, perhaps the nearest approximation to this being toll roads and modern road-pricing schemes. The use of taxes on vehicles and fuels is only a rough approximation to charging for the provision of infrastructure.

But, why should this matter? Is it not the case that other goods and services like public parks, libraries and the police are often provided without a direct charge for them? What is wrong with providing free road space? According to elementary economic theory it does matter. In a perfect market a good allocation of resources to satisfy human needs is only achieved when the marginal costs of the goods equal their marginal utility. This is why it is often advocated that the price of goods and services, i.e. their perceived cost, should be set at their marginal cost. Of course real markets are not perfect and ability to pay is not a good indication of need; however, this general framework provides the basis for contrasting other ways of arranging pricing systems and their impact on resource allocation.

Transport is a very important element in the welfare of nations and the well-being of urban and rural dwellers. If those who make use of transport facilities do not perceive the resource implications of their choices, they are likely to generate a balance between supply and demand that is inherently inefficient. Underpriced scarce resources will be squandered whilst other abundant but priced resources may not be used. The fact that overall some sectors of the economy (typically car owners) more than pay for the cost of the road space provided, is not a guarantee of more rational allocation of resources. Car owners probably see these annual taxes as fixed, *sunk*, costs which at most affect the decision of buying a car but not that of using it.

An additional element of distortion is provided by the number of concomitant- or *side-effects* associated with the production of transport services: accidents, pollution and environmental degradation in general. These effects are seldom *internalised*; the user of the transport service rarely perceives nor pays for the costs of cleaning the environment or looking after the injured in transport related accidents. Internalising these costs could also help to make better decisions and to improve the allocation of demand to alternative modes.

One of the most important features of transport supply is *congestion*. This is a term which is difficult to define as we all believe we know exactly what it means. However, most practitioners do know that what is considered congestion in Leeds or Lampang is often accepted as normal in London or Lagos. Congestion arises when demand levels approach the capacity of a facility and the time required to use it (travel through it) increases well above the average under low demand conditions. In the case of transport infrastructure the inclusion of an additional vehicle generates supplementary delay to all other users as well, see for example Figure 1.1. Note that the contribution an additional car makes to the delay of all users is greater at high flows than at low flow levels.

This is the external effect of congestion, perceived by others but not by the driver originating it. This is a cost which schemes such as electronic road pricing attempt to internalise to help more reasoned decision making by the individual.

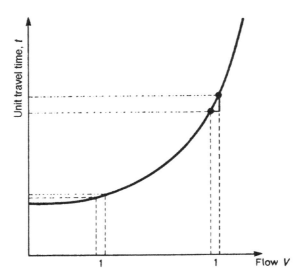

Figure 1.1   Congestion and its external effects

1.2.3   Equilibration of Supply and Demand

In general terms the role of transport planning is to ensure the satisfaction of a certain demand **D** for person and goods movements with different trip purposes, at different times of the day and the year, using various modes, given a transport system with a certain operating capacity. The transport system itself can be seen as made up of:

- an infrastructure (e.g. a road network);
- a management system (i.e. a set of rules, for example driving on the right, and control strategies, for example at traffic signals);
- a set of transport modes and their operators.

Consider a set of volumes on a network **V**, a corresponding set of speeds **S**, and an operating capacity **Q**, under a management system **M**. In very general terms the speed on the network can be represented by:

$$\mathbf{S} = f\{\mathbf{Q}, \mathbf{V}, \mathbf{M}\} \tag{1.1}$$

The speed can be taken as an initial proxy for a more general indicator of the *level of service* (LOS) provided by the transport system. In more general terms a LOS would be specified by a combination of speeds or travel times, waiting and walking times and price effects; we shall expand on these in subsequent chapters. The management system **M** may include traffic management schemes, area traffic control and regulations applying to each mode. The capacity **Q** would depend on the management system **M** and on the levels of investment **I** over the years, thus:

$$\mathbf{Q} = f\{\mathbf{I}, \mathbf{M}\} \tag{1.2}$$

The management system may also be used to redistribute capacity giving priority to certain types of users over others, either on efficiency (public-transport users, cyclists), environmental (electric vehicles) or equity grounds (pedestrians).

As in the case of most goods and services, one would expect the level of demand **D** to be dependent on the level of service provided by the transport system and also on the allocation of activities **A** over space:

$$\mathbf{D} = f\{\mathbf{S}, \mathbf{A}\} \qquad (1.3)$$

Combining equations (1.1) and (1.3) for a fixed activity system one would find the set of equilibrium points between supply and demand for transport. But then again, the activity system itself would probably change as levels of service change over space and time. Therefore one would have two different sets of equilibrium points: short-term and long-term ones. The task of transport planning is to forecast and manage the evolution of these equilibrium points over time so that social welfare is maximised. This is, of course, not a simple task: modelling these equilibrium points should help to understand this evolution better and assist in the development and implementation of management strategies **M** and investment programmes **I**.

Sometimes very simple cause-effect relationships can be depicted graphically to help understand the nature of some transport problems. A typical example is the car/public-transport vicious circle depicted in Figure 1.2.

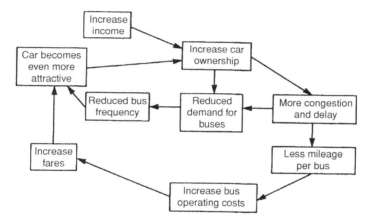

Figure 1.2 Car and public-transport vicious circle

Economic growth provides the first impetus to increase car ownership. More car owners means more people wanting to transfer from public transport to car; this in turn means fewer public-transport passengers, to which operators may respond by increasing the fares, reducing the frequency (level of service) or both. These measures make the use of the car even more attractive than before and induce more people to buy cars, thus accelerating the vicious circle. After a few cycles (years) car drivers are facing increased levels of congestion; buses are delayed, are becoming increasingly more expensive and running less frequently; the accumulation of sensible individual decisions results in a final state in which almost everybody is worse off than originally.

Moreover, there is a more insidious effect in the long term, not depicted in Figure 1.2, as car owners choose their place of work and residence without considering the availability (or otherwise) of public transport. This generates urban sprawl, low density developments that are more difficult and expensive to serve by more efficient public transport modes. This is the 'development trap' that leads to further congestion and a higher proportion of our time spent in slow moving cars.

This simple representation can also help to identify what can be done to slow down or reverse this vicious circle. These ideas are summarised in Figure 1.3. Physical measures like bus lanes or other bus-priority schemes are particularly attractive as they also result in a more efficient allocation of road space. Public transport subsidies have strong advocates and detractors; they may reduce the need for fare increases, at least in the short term, but tend to generate large deficits and to protect poor management from the consequences of their own inefficiency. Car restraint, and in particular congestion charging, can help to internalise externalities and generate a revenue stream that can be distributed to other areas of need in transportation.

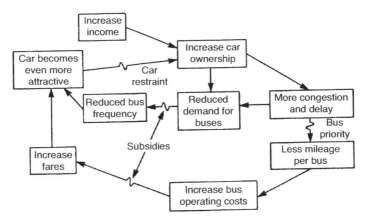

Figure 1.3   Breaking the car/public-transport vicious circle

The type of model behind Figures 1.2 and 1.3 is sometimes called a *structural model*, as discussed in Chapter 12; these are simple but powerful constructs, in particular because they permit the discussion of key issues in a fairly parsimonious form. However, they are not exempt from dangers when applied to different contexts. Think, for example, of the vicious circle model in the context of developing countries. Population growth will maintain demand for public transport much longer than in industrialised countries. Indeed, some of the bus flows currently experienced in emerging countries are extremely high, reaching 400 to 600 buses per hour one-way along some corridors. The context is also relevant when looking for solutions; it has been argued that one of the main objectives of introducing bus-priority schemes in emerging countries is not to protect buses from car-generated congestion but to organise bus movements (Gibson *et al.* 1989). High bus volumes often implement a *de facto* priority, and interference between buses may become a greater source of delay than car-generated congestion. To be of value, the vicious circle model must be revised in this new context.

It should be clear that it is not possible to characterise all transport problems in a unique, universal form. Transport problems are context dependent and so should be the ways of tackling them. Models can offer a contribution in terms of making the identification of problems and selection of ways of addressing them more solidly based.

1.3   Modelling and Decision Making

1.3.1   Decision-making Styles

Before choosing a modelling framework one needs to identify the general decision-making approach adopted in the country, government or decision unit. It must be recognised that there are several decision-making styles in practice and that not all of them use modelling as a basic building block. Previous editions of this text have characterised decision-making styles following the ideas of Nutt (1981); in practice, no decision-making style fits any of these categories exactly. This time, we would just like to distinguish two different paradigms: 'substantive rationality' and 'muddling through', following the lines of the very important book by Kay (2010).

The *substantive rationality* view of the world assumes that we know what our objectives are and we can envisage all alternative ways of achieving them and, with some luck, quantify the costs and benefits associated to each approach. This would apply to important decisions like choosing a place to live and less important ones like choosing a place to eat. This is the rational or normative decision-making approach implicit in most textbooks about transport planning. It is sometimes referred to as the 'systems approach' to planning. Here, quantification is essential. The decision problem is seen as one of choosing options from a complete set of alternatives and scenarios, with estimates on their probability of occurrence; the *utility* of each alternative is quantified in terms of benefits and costs and other criteria like environmental protection, safety, and so on.

In some cases it may even be possible to cast a decision problem into a mathematical programming framework. This means that the objective function is well understood and specified, and that the same applies to the constraints defining a solution space. However, for most real problems some elements of the objective function or constraints may be difficult to quantify or to convert into common units of measurement, say money or time. It may also be difficult to include some of the probabilistic elements in each case, but a good deal about the problem is learnt in the process. Modelling is at the core of this approach. The evaluation of plans or projects using Cost Benefit Analysis or a Multi-Criteria Framework is also based on this view of reality.

Some of the problems of applying normative decision theory are:

- Difficulties in actually specifying what the objectives are beyond generalities like reducing congestion or improving accessibility; as soon as we develop a measure or indicator for that objective, we find that it is actually misleading in respect of the things we want to achieve.
- The accusation of insensitivity to the aspirations of the public; people do not actually care about 'optimised' systems, they just want to see progress that is sustained along lines that are difficult to identify: they ask for speed but when it is delivered they are dissatisfied with the associated noise and emissions.
- Its high costs; substantive rationality is expensive to implement, requires advanced models and many runs for alternative arrangements and sensitivity analyses; efforts to apply this approach often overrun in time and budget; and
- The alienation of decision makers who may not understand, nor accept, the analytical treatment of the problem. This is a common complaint in our profession; the recurrent requisite to demonstrate the usefulness of our simulations may be irritating but reflects a real need to make our models and results relevant and communicable.

Moreover, there is very limited evidence that countries or organisations that do not follow this approach fare worse that those who do. Kay (2010) argues that many of the companies that were once hailed as paragons of good rational decision making failed spectacularly a few years later; there seem to be plenty of examples of this.

The main alternative approach to substantive rationality is what Lindblom (1959) called *muddling through*. The name, misleadingly self-deprecating, is not meant to imply that intuitive and unstructured decision making is desirable. On the contrary, in Lindblom's eyes, muddling through is a disciplined process but not one based on the substantive rational handling of defined objectives. The approach uses a combination of high-level (often unquantifiable) objectives, intermediate goals and immediate actions or experiments. Muddling through, or what Kay calls 'oblique or indirect approach', is characterised by:

- The use of high level objectives that are only loosely defined with no attempt to quantify them.
- Abandoning any clear distinction between objective, goals and actions; we learn about high-level objectives by adopting goals and implementing actions.
- Recognising that the environment is uncertain and that we cannot even know the range of events that might take place in the future, and
- Accepting that we can never identify, nor describe, all the range of options available; we can only deal with a limited set without aspiring to exhaust the search.

The following table, adapted from Kay's ideas, identifies additional contrasts between the two basic approaches:

Substantive rationality	Issue	Indirect approach
Interactions with others are limited and their response depend on our actions alone	Interactions	The outcome of interactions with others depend on context and their interpretation of our intentions
The relationships between objectives, states, goals and actions are understandable	Complexity	Our understanding of the relationships between objectives, states, goals and actions is imperfect but can be improved by experience
The problem and context can be described by a well specified and estimated analytical model	Abstraction	Appropriate simplification of complex problems must rely on judgement and understanding of context
What happens is what we intended to happen	Intentionality	What happens is the result of complex processes whose totality nobody fully understands
Decisions are made on the basis of the fullest possible information	Information	Decisions are recommended and made acknowledging that only limited knowledge is or can be available
The best outcome is achieved through a conscious process of maximisation	Adaptation	Good results are obtained through continual adaptation to constantly changing conditions
Rules and guidelines can be defined that allow people to find the correct solutions	Expertise	Experts can do things that others cannot – and can only learn with difficulty

In practice, no organisation relies on (attempts to) substantive rationality alone. Most apply an eclectic mixture of approaches using models, narratives, political context and sources of evidence. Modelling plays an important role in each of these approaches and the professional modeller should be ready to offer flexibility and capacity for adaptation, including new variables as required and responding quickly in the analysis of innovative policies and designs.

1.3.2   Choosing Modelling Approaches

This book assumes that the decision style adopted involves the use of models but it does not advocate a single (i.e. a normative) decision-making approach. The acceptability of modelling, or a particular modelling approach, within a decision style is very important. Models which end up being ignored by decision makers not only represent wasted resources and effort, but result in frustrated analysts and planners. It is further proposed that there are several features of transport problems and models which must be taken into account when specifying an analytical approach:

1. **Precision and accuracy required**. These concepts are sometimes confused. *Accuracy* is the degree to which a measurement or model result matches true or accepted values. Accuracy is an issue pertaining

to the quality of data and model. The level of accuracy required for particular applications varies greatly. It is often the case that the accuracy required is just that necessary to discriminate between a good scheme and a less good one. In some cases the best scheme may be quite obvious, thus requiring less accurate modelling. Remember, however, that common sense has been blamed for some very poor transport decisions in the past.

Precision refers to the level or units of measurement used to collect data and deliver model outputs. One may measure travel times between two points in fractions of a second, but individuals may estimate and state the same much less precisely in five minute intervals. Precision is not accuracy and it is often misleading. Reporting estimates with high precision is often interpreted as confidence in their accuracy, whereas transport modellers often use precise numbers to report uncertain estimates. There is a difference between stating that 'traffic on link X was measured as 2347 vehicles between 8:00 and 9:00 AM yesterday' and saying that 'traffic on link X between 8:00 and 9:00 AM in five years time will be 3148 vehicles': the first statement may be both precise and accurate where the second is equally precise but certainly inaccurate. It is less misleading to report the second figure as 3150. As in the quote attributed to John Maynard Keynes 'it is much better to be roughly right than precisely wrong'.

2. **The decision-making context**. This involves the adoption of a particular *perspective* and a choice of a *scope* or coverage of the system of interest. The choice of perspective defines the type of decisions that will be considered: strategic issues or schemes, tactical (transport management) schemes, or even specific operational problems. The choice of scope involves specifying the level of analysis: is it just transport or does it involve activity location too? In terms of the transport system, are we interested in just demand or also on the supply side at different levels: system or suppliers' performance, cost minimisation issues within suppliers, and so on? The question of how many options need to be considered to satisfy different interest groups or to develop a single best scheme is also crucial. The decision-making context, therefore, will also help define requirements on the models to be used, the variables to be included in the model, or considered given or exogenous.

3. **Level of detail required**. The level of resolution of a model system can be described along four main dimensions: geography, unit of analysis, behavioural responses and the handling of time.

   Space is very important and it can be handled in an aggregate way, as a few zones with area-wide speed flow curves, or at the detailed level of the individual addresses for trips with links described in detail. There is a wide range of options in this field and the choice will depend on the application in hand: if the issue is a detailed design for traffic in a small area, highly disaggregated zones with an accurate account of the physical characteristics of links would be appropriate in a microsimulation model. Strategic planning may call for a more aggregate zoning system with links described in terms of their speed-flow relationships alone.

   The unit of interest for modelling may be the same zone with trips emanating and ending there or, at the other end of the spectrum, sampled or synthesised individuals; somewhere in between there will be different household or person strata as representative of the travelling population.

   The behavioural responses included may vary from fairly simple route choice actions in a traffic model to changes in time of travel, mode, destination, tour frequency and even land use and economic activity impacts.

   Time, in turn, can be treated either as a discrete or a continuous variable. In the first case the model may cover a full day (as in many national models), a peak period or a smaller time interval: all relevant responses will take place in that period although there may be interactions with other periods. Alternatively, time may be considered as a continuous variable which allows for more dynamic handling of traffic and behavioural responses like the choice of time of travel. Considering discrete time slices is a common option as treating time as a continuous variable is much more demanding.

4. **The availability of suitable data**, their stability and the difficulties involved in forecasting their future values. In some cases very little data may be available; in others, there may be reasons to

suspect the information, or to have less confidence in future forecasts for key planning variables as the system is not sufficiently stable. In many cases the data available will be the key factor in deciding the modelling approach.

5. **The state of the art in modelling** for a particular type of intervention in the transport system. This in turn can be subdivided into:
   - behavioural richness;
   - mathematical and computer tractability;
   - availability of good solution algorithms.

   It has to be borne in mind that in practice all models assume that some variables are exogenous to it. Moreover, many other variables are omitted from the modelling framework on the grounds of not being relevant to the task in hand, too difficult to forecast or expected to change little and not influence the system of interest. An explicit consideration of what has been left out of the model may help to decide on its appropriateness for a given problem.

6. **Resources available for the study**. These include money, data, computer hardware and software, technical skills, and so on. Two types of resource are, however, worth highlighting here: time and level of communication with decision makers and the public. *Time* is probably the most crucial one: if little time is available to make a choice between schemes, shortcuts will be needed to provide timely advice. Decision makers are prone to setting up absurdly short timescales for the assessment of projects which will take years to process through multiple decision instances, years to implement and many more years to be confirmed as right or wrong. On the other hand, a good level of communication with decision makers and the public will alleviate this problem: fewer unrealistic expectations about our ability to accurately model transport schemes will arise, and a better understanding of the advantages and limitations of modelling will moderate the extremes of blind acceptance or total rejection of study recommendations.

7. **Data processing requirements**. This aspect used to be interpreted as something like 'how big a computer do you need?' The answer to that question today is 'not very big', as a good microcomputer will do the trick in most cases. The real bottleneck in data processing is the human ability to collect, code, input the data, run the programs and interpret the output. The greater the level of detail, the more difficult all these human tasks will be. The use of computer-assisted data collection and graphics for input–output of programs reduces the burden somewhat.

8. **Levels of training and skills of the analysts**. Training costs are usually quite high; so much so that it is sometimes better to use an existing model or software that is well understood, than to embark on acquiring and learning to use a slightly more advanced one. This looks, of course, like a recipe for stifling innovation and progress; however, it should always be possible to spend some time building up strengths in new advanced techniques without rejecting the experience gained with earlier models.

9. **Modelling perspective and scope**. Florian *et al*. (1988) formalise decision-making contexts using a two-dimensional framework: the *level of analysis* and the *perspective*. The levels of analysis may include six different groups of *procedures*, where a procedure centres on one or more models and their specific solution algorithms. These are:
   - *activity location* procedures **L**;
   - *demand procedures* **D**;
   - *transport system performance* procedures **P**, which produce as output levels of service, expenditure and practical capacities, and depend on demand levels and on transport supply conditions;
   - *supply actions* procedures **S**, which determine the actions taken by suppliers of transport services and infrastructure; these depend on their objectives (profit maximisation, social welfare), institutional environment, their costs and estimates of future states of the system;
   - *cost minimisation* procedures **CM**;
   - *production* procedures **PR**.

The last two have more to do with the microeconomic issues affecting the suppliers in their choice of input combinations to minimise costs.

The perspectives dimension considers the six level procedures **L, D, P, S, CM, PR** and three perspectives: a *strategic* perspective **STR**, a *tactical* perspective **TAC** and an *operational* perspective **OPE**. These are, of course, related to the planning horizons and the levels of investment; however, in this context they must be seen as generic concepts dealing with the capacity:

* to visualise the levels **L, D, P, S, CM, PR** in their true and relative importance;
* to choose, at any level, what is to be regarded as fixed and what as variable.

Figure 1.4 summarises the way in which different perspectives and levels usually interact. The largest and most aggregate is, of course, the strategic level; analysis and choice at this level have major system-wide and long-term impacts, and usually involve resource acquisition and network design. Tactical issues have a narrower perspective and concern questions like making the best use of existing facilities and infrastructure. The narrowest perspective, the operational one, is concerned with the short-term problems of suppliers of transport services which fall outside the scope of this book; nevertheless, the actual decisions on, for example, levels of service or vehicle size, are important exogenous input to some of the models discussed in this book, and this is depicted in Figure 1.4.

Figure 1.4   The two-dimensional conceptual framework

This is, of course, a rather abstract and idealised way of visualising planning problems. However, it helps to clarify the choices the analyst must face in developing a transport modelling approach. In this book we are mainly concerned with strategic and tactical issues at the demand and performance procedure levels. Nevertheless, some of the models discussed here sometimes find application outside these levels and perspectives.

1.4   Issues in Transport Modelling

We have already identified the interactions between transport problems, decision-making styles and modelling approaches. We need to discuss now some of the critical modelling issues which are relevant to the choice of model. These issues cover some general points like the roles of theory and data, model specification and calibration. But perhaps the most critical choices are those between the uses of aggregate or disaggregate approaches, cross-section or time-series models, and revealed or stated preference techniques.

1.4.1   General Modelling Issues

Wilson (1974) provides an interesting list of questions to be answered by any would-be modeller; they range from broad issues such as the *purpose* behind the model-building exercise, to detailed aspects such as *what techniques* are available for building the model. We will discuss some of these below, together with other modelling issues which are particularly relevant to the development of this book.

1.4.1.1   The Roles of Theory and Data

Many people tend to associate the word 'theory' with endless series of formulae and algebraic manipulations. In the urban transport modelling field this association has been largely correct: it is difficult to understand and replicate the complex interactions between human beings which are an inevitable feature of transport systems.

Some theoretical developments attempting to overcome these difficulties have resulted in models lacking adequate data and/or computational software for their practical implementation. This has led to the view, held strongly by some practitioners, that the gap between theory and practice is continually widening; this is something we have tried to redress in this book.

An important consideration on judging the contribution of a new theory is whether it places any meaningful restrictions on, for example, the form of a demand function. There is at least one documented case of a 'practical' transport planning study, lasting several years and costing several million dollars, which relied on 'pragmatic' demand models with a faulty structure (i.e. some of its elasticities had a wrong sign; see Williams and Senior 1977). Although this could have been diagnosed *ex ante* by the pragmatic practitioners, had they not despised theory, it was only discovered *post hoc* by theoreticians.

Unfortunately (or perhaps fortunately, a pragmatist would say), it is sometimes possible to derive similar functional forms from different theoretical perspectives (this, the *equifinality issue*, is considered in more detail in Chapter 8). The interpretation of the model output, however, is heavily dependent on the theoretical framework adopted. For example, the same functional form of the gravity model can be derived from analogy with physics, from entropy maximisation and from maximum utility formalisms. The interpretation of the output, however, may depend on the theory adopted. If one is just interested in flows on links it may not matter which theoretical framework underpins the analytical model function. However, if an evaluation measure is required, the situation changes, as only an economically based theory of human behaviour will be helpful in this task. In other cases, phrases like: 'the gravitational pull of this destination will increase', or 'this is the most probable arrangement of trips' or 'the most likely trip matrix consistent with our information about the system' will be used; these provide no help in devising evaluation measures but assist in the interpretation of the nature of the solution found. The theoretical framework will also lend some credence to the ability of the model to forecast future behaviour. In this sense it is interesting to reflect on the influence practice and theory may have on each other. For example, it has been noted that models or analytical forms used in practice have had traditionally a guiding influence on the assumptions employed in the development of subsequent theoretical frameworks. It is also well known that widely implemented forms, like the gravity-logit model we will discuss in Chapters 6 and 7, have been the subject of strong *post hoc* rationalisation:

> theoretical advances are especially welcome when they fortify existing practice which might be deemed to lack a particularly convincing rationale (Williams and Ortúzar, 1982b).

The two classical approaches to the development of theory are known as *deductive* (building a model and testing its predictions against observations) and *inductive* (starting with data and attempting to infer general laws). The deductive approach has been found more productive in the pure sciences and the inductive approach has been preferred in the analytical social sciences. It is interesting to note that data are central to both; in fact, it is well known that data availability usually leaves little room for negotiation and compromise in the trade-off between modelling *relevance* and modelling *complexity*. Indeed, in very many cases the nature of the data restricts the choice of model to a single option.

The question of data is closely connected with issues such as the type of variables to be represented in the model and this is, of course, closely linked again to questions about theory. Models predict a number of dependent (or endogenous) variables given other independent (or explanatory) variables. To test a model we would normally need data about each variable. Of particular interest are the *policy variables*, which are those assumed to be under the control of the decision maker, e.g. those the analyst may vary in order to test the value of alternative policies or schemes.

Another important issue in this context is that of aggregation:

- How many population strata or types of people do we need to achieve a good representation and understanding of a problem?
- In how much detail do we need to measure certain variables to replicate a given phenomenon?
- Space is crucial in transport; at what level of detail do we need to code the origin and destination of travellers to model their trip making behaviour?

1.4.1.2   Model Specification

In its widest and more interesting sense this issue considers the following themes.

Model Structure   Is it possible to replicate the system to be modelled with a simple structure which assumes, for example, that all alternatives are independent? Or is it necessary to build more complex models which proceed, for example, to calculate probabilities of choice conditional on previous selections? Disaggregate models, such as those discussed in Chapters 7 to 9, usually have parameters which represent aspects of model structure and the extensions to methodology achieved by the mid-1980s have allowed the estimation of more and more general model forms. However, as Daly (1982b) has remarked, although it might be supposed that ultimately all issues concerned with model form could be resolved by empirical testing, such resolution is neither possible nor appropriate.

Functional Form   Is it possible to use linear forms or does the problem require postulating more complex non-linear functions? The latter may represent the system of interest more accurately, but certainly will be more demanding in terms of resources and techniques for model calibration and use. Although theoretical considerations may play a big role in settling this question, it is also possible to examine it in an inductive fashion by means of 'laboratory simulations', for example in stated intentions/preferences experiments.

Variable Specification This is the more usual meaning attached to the specification issue; which variables to use and how (which form) they should enter a given model. For example, if income is assumed to influence individual choice, should the variable enter the model as such or deflating a cost variable? Methods to advance on this question range from the deductive ('constructive') use of theory, to the inductive statistical analysis of the data using transformations.

1.4.1.3   Model Calibration, Validation and Use

A model can be simply represented as a mathematical function of variables X and parameters θ, such as:

$$Y = f(\mathbf{X}, \boldsymbol{\theta}) \tag{1.4}$$

It is interesting to mention that the twin concepts of *model calibration* and *model estimation* have taken traditionally a different meaning in the transport field. Calibrating a model requires choosing its parameters, assumed to have a non-null value, in order to optimise one or more *goodness-of-fit* measures which are a function of the observed data. This procedure has been associated with the physicists and engineers responsible for early aggregate transport models who did not worry unduly about the statistical properties of these indices, e.g. how large any calibration errors could be.

Estimation involves finding the values of the parameters which make the observed data more likely under the model specification; in this case one or more parameters can be judged *non-significant* and left out of the model. Estimation also considers the possibility of examining empirically certain specification issues; for example, structural and/or functional form parameters may be estimated. This procedure has tended to be associated with the engineers and econometricians responsible for disaggregate models, who placed much importance on the statistical testing possibilities offered by their methods. However, in essence both procedures are the same because the way to decide which parameter values are better is by examining certain previously defined goodness-of-fit measures. The difference is that these measures generally have well-known statistical properties which in turn allow confidence limits to be built around the estimated values and model predictions.

Because the large majority of transport models have been built on the basis of *cross-sectional* data, there has been a tendency to interpret model *validation* exclusively in terms of the goodness-of-fit achieved between observed behaviour and base-year predictions. Although this is a *necessary*, it is by no means a *sufficient* condition for model validation; this has been demonstrated by a number of cases which have been able to compare model predictions with observed results in *before-and-after* studies (see the discussion in Williams and Ortúzar, 1982a). Validation requires comparing the model predictions with information *not used* during the process of model estimation. This obviously puts a more stringent test on the model and requires further information or more resources.

One of the first tasks a modeller faces is to decide which variables are going to be predicted by the model and which are possibly required as inputs to it. Some will not be included at all, either because the modeller lacks control over them or simply because the theory behind the model ignores them (see Figure 1.5). This implies immediately a certain degree of error and uncertainty (we will come back to this problem in Chapter 3) which of course gets compounded by other errors which are also inherent to modelling; for example, sampling errors and, more important, errors due to the unavoidable simplifications of reality the model demands in order to be practical (see Figure 1.5).

Thus, the main use of models in practice is for *conditional forecasting*: the model will produce estimates of the dependent variables given a set of independent variables. In fact, typical forecasts are conditional in two ways (Wilson 1974):

• in relation to the values assigned to the policy variables in the plan, the impact of which is being tested with the model;
• in relation to the assumed values of other variables.

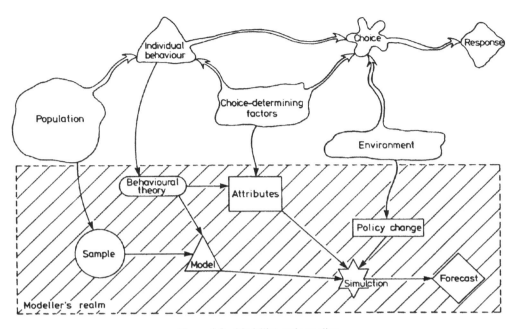

Figure 1.5   Modelling and sampling

A model is normally used to test a range of alternative plans for a range of possible assumptions about the future value of the other variables (e.g. low- and high-income scenarios). This means that it might be 'run' many times in the context of examining a particular problem. For this reason it may be of crucial importance that its specification allows for quick turn-around time in a computer; this is not an easy task in the case of a full-scale transportation model which involves complex processes of equilibration between supply and demand, as we will discuss in Chapter 11.

1.4.1.4   Modelling, Forecasting and Judgement

There is a subtle difference between modelling and forecasting. Modelling focuses on building and applying appropriate tools that are sensitive to the choices of interest and respond logically to changes in key policy instruments. The successful modeller will provide useful and timely advice to the decision-making process, even if the data and timescales are limited. In this case, it is important that the model produces consistent results for all expected interventions, policies and projects, such that they can be ranked fairly, even if the correspondence to reality is not perfect.

Forecasting is an attempt to envision and quantify future conditions. It normally involves estimating future travel demand and the resulting multimodal flows and costs over time. In the case of private sector projects, see Chapter 16, these projections are usually accompanied by revenue forecasts and investors will take considered risks based on these forecasts. Forecasting is usually based on formal models, but they alone cannot provide the full picture; it is necessary to incorporate other analyses and assumptions. Given the uncertainty about the future, several complementary approaches might be used in forecasting. For example a formal model may be supported by consideration about the main economic drivers of future travel activity in a region; in that way it is made clear how forecasts are dependent on the future of these activities. The success of forecasts can only be objectively measured through before and after studies.

The importance of formal models increases as the interventions under consideration diverge further from what is on the ground and known today. For example, when introducing a mode not currently available in a city, the model will often have to rely on stated preference data, information from other regions, or rational decision making theory. The same is true when evaluating any sort of policy not currently in existence (congestion charging) or when considering fuel prices or congestion conditions radically different than at present. In general, good advice on these issues cannot be given only on the basis of good modelling, however excellent. This requires intelligent consideration of other factors and assumptions, in particular about the limitations of any modelling approach.

Given the nature of analytical models, interpretation of their output is essential. Interpretation requires good judgement and this is only acquired with experience and a thorough understanding of the theories underpinning models and their limitations. For instance, most of the models described in this text are supported by random utility theory (see Chapter 7) that in turn assumes rational decision making on the part of travellers. However, there is an increasingly solid body of evidence, provided mostly by Behavioural Economics and Psychology, that humans are neither entirely rational nor consistent in their choices. This evidence (see Ariely 2009) punctures the theory underpinning our models–even the most advanced activity based approaches–and makes the application of judgement in the interpretation of model outputs even more important.

1.4.2   Aggregate and Disaggregate Modelling

The level of aggregation selected for the measurement of data is an important issue in the general design of a transportation planning study. Of central interest is the aggregation of exogenous data, that is, information about items other than the behaviour of travellers which is assumed endogenous (i.e. the model attempts to replicate it). For example, throughout the years it has been a cause for concern whether a given data item represents an average over a group of travellers rather than being collected specifically for a single individual. When the model at base aims at representing the behaviour of more than one individual (e.g. a population segment like car owners living in a zone), such as in the case of the *aggregate* models we will examine in Chapters 5 and 6, a certain degree of aggregation of the exogenous data is inevitable. But when the model at base attempts to represent the behaviour of individuals, such as in the case of the *disaggregate* models we will study in Chapters 7 to 9, it is conceivable that exogenous information can be obtained and used separately for each traveller. An important issue is then whether, as is often the case, it might be preferable on cost or other grounds to use less detailed data (see Daly and Ortúzar 1990).

Forecasting future demand is a crucial element of the majority of transport planning studies. Being able to predict the likely usage of new facilities is an essential precursor to rational decision making about the advantages or otherwise of providing such facilities. It may also be important to have an idea about the sensitivities of demand to important variables under the control of the analyst (e.g. the price charged for its use). In most cases the forecasts and sensitivity estimates must be provided at the aggregate level, that is, they must represent the behaviour of an entire population of interest. Therefore, the analyst using disaggregate models must find a sound method for aggregating model results to provide these indicators.

Aggregate models were used almost without exception in transportation studies up to the late 1970s; they became familiar, demanded relatively few skills on the part of the analyst (but required arcane computer knowledge) and had the property of offering a 'recipe' for the complete modelling process, from data collection through the provision of forecasts at the level of links in a network. The output of these models, perhaps because they were generated by obscure computer programs, were often considered more accurate than intended, for example predicting turning movement flows 15 years in the future. Aggregate models have been severely (and sometimes justifiably) criticised for their inflexibility, inaccuracy and cost. Unfortunately, many disaggregate approaches which have adopted sophisticated treatments of the choices and constraints faced by individual travellers have failed to take the process through to the production of forecasts, sometimes because they require data which cannot reasonably be forecast.

Disaggregate models, which became increasingly popular during the 1980s, offer substantial advantages over the traditional methods while remaining practical in many application studies. However, one important problem in practice is that they demand from the analyst quite a high level of statistical and econometric skills for their use (in particular for the interpretation of results), certainly much higher than in the case of aggregate models. Moreover, the differences between aggregate and disaggregate model systems have often been overstated. For example, the disaggregate models were first marketed as a radical departure from classical methods, a 'revolution' in the field, while eventually it became clear that an 'evolutionary' view was more adequate (see Williams and Ortúzar 1982b). In fact, in many cases there is complete equivalence between the forms of the forecasting models (Daly 1982a). The essential difference lies in the treatment of the description of behaviour, particularly during the model development process; in many instances the disaggregate approach is clearly superior to the grouping of behaviour by zones and by predefined segments.

Attempts to clarify the issue of whether disaggregate or aggregate approaches were to be preferred, and in what circumstances, have basically concluded that there is no such thing as a definitive approach appropriate to all situations (see Daly and Ortúzar 1990). These attempts have also established the need for guidelines to help the despairing practitioner to choose the most appropriate model tools to apply in a particular context. We have striven to answer that call in this book.

1.4.3   Cross-section and Time Series

The vast majority of transport planning studies up to the late 1980s relied on information about trip patterns revealed by a cross-section of individuals at a single point in time. Indeed, the traditional use of the cross-sectional approach transcended the differences between aggregate and disaggregate models.

A fundamental assumption of the cross-sectional approach is that a measure of the response to incremental change may simply be found by computing the derivatives of a demand function with respect to the policy variables in question. This makes explicit the assumption that a realistic *stimulus-response* relation may be derived from model parameters estimated from observations at one point in time. This would be reasonable if there were always enough people changing their choices, say of mode or destination, in *both* directions and without habit or time-lag effects.

However, the cross-sectional assumption has two potentially serious drawbacks. First, a given cross-sectional data set may correspond to a particular 'history' of changes in the values of certain key variables influencing choice. For example, changes in mode or location in time may have been triggered by a series of different stimuli (petrol prices, life-cycle effects, etc.) and the extent to which a system is considered to be in *disequilibrium* (because of, say, inertia) will depend on these. The trouble is that it can be shown (see Chapter 7) that the response of groups with exactly the same current characteristics, but having undergone a different path of changes, may be very different indeed. Second, data collected at only one point in time will usually fail to discriminate between alternative model formulations, even between some arising from totally different theoretical postulates. It is always possible to find 'best-fit' parameters from base-year data even if the model suffers severe mis-specification problems; the trouble is, of course, that these do not guarantee good response properties for a future situation. As we saw in section 1.4.1, a good base-year fit is not a sufficient condition for model validation.

Thus, in general it is not possible to discriminate between the large variety of possible sources of dispersion within a cross-sectional data set (i.e. preference dispersion, habit effects, constraints, and so on). Real progress in understanding and assessing the effectiveness of forecasting models, however, can only be made if information is available on response over time. From a theoretical point of view, it is also desirable that appropriate frameworks for analysis are designed which allow the eventual refutation of hypotheses relating to response. Until this is achieved, a general problem of potential misrepresentation will continue to cast doubts on the validity of cross-sectional studies.

The discussion above has led many people to believe that, where possible, longitudinal or time-series data should be used to construct more dependable forecasting models. This type of data incorporates information on response by design. Thus, in principle, it may offer the means to directly test and even perhaps reject hypotheses relating to response.

Longitudinal data can take the form of *panels* or more simply *before-and-after* information. Unfortunately, models built on this type of data have severe technical problems of their own; in fact, up to the end of the 1990s progress in this area had been limited. We will discuss some of the issues involved in the collection and use of this type of information in Chapters 3 and 7.

1.4.4 Revealed and Stated Preferences

The development of good and robust models is quite difficult if the analyst cannot set up experiments to observe the behaviour of the system under a wide range of conditions. Experimentation of this kind is neither practical nor viable in transport and the analyst is restricted, like an astronomer, to make observations on events and choices they do not control. Up to the mid-1980s it was almost axiomatic that modelling transport demand should be based on information about observed choices and decisions, i.e. *revealed-preference* data. Within this approach, project evaluation requires expressing policies in terms of changes in attributes which 'map onto' those considered to influence current behaviour. However, this has practical limitations basically associated with survey costs and the difficulty of distinguishing the effects of attributes which are not easy to observe, e.g. those related to notions such as quality or convenience. Another practical embarrassment has been traditionally the 'new option' problem, whereby it is required to forecast the likely usage of a facility not available at present and perhaps even radically different to all existing ones.

Stated-preference/intentions techniques, borrowed from the field of market research, were put forward by the end of the 1970s as offering a way of experimenting with transport-related choices, thus solving some of the problems outlined above. Stated-preference techniques base demand estimates on an analysis of the response to *hypothetical choices*; these, of course, can cover a wider range of attributes and conditions than the real system. However, these techniques were severely discredited at the start because it was not known how to discount for the over enthusiasm of certain respondents, e.g. not even half of the individuals stating they would take a given course of action actually did so when the opportunity eventually arose.

It took a whole decade for the situation to change, but by the end of the 1980s stated-preference methods were perceived by many to offer a real chance to solve the above-mentioned difficulties. Moreover, it has been found that, in appropriate cases, revealed-and stated-preference data and methods may be employed in complementary senses with the strengths of both approaches recognised and combined. In particular, they are considered to offer an invaluable tool for assisting the modelling of completely new alternatives. We will examine data-collection aspects of stated-preference methods in Chapter 3 and modelling issues in Chapter 8.

1.5   The Structure of the Classic Transport Model

Years of experimentation and development have resulted in a general structure which has been called the classic transport model. This structure is, in effect, a result from practice in the 1960s but has remained more or less unaltered despite major improvements in modelling techniques since then.

The general form of the model is depicted in Figure 1.6. The approach starts by considering a zoning and network system, and the collection and coding of planning, calibration and validation data. These data would include base-year levels for population of different types in each zone of the study area as well as levels of economic activity including employment, shopping space, educational and recreational facilities. These data are then used to estimate a model of the total number of trips generated and attracted by each zone of the study area (*trip generation*). The next step is the allocation of these trips to particular destinations, in other words their *distribution* over space, thus producing a trip matrix. The following stage normally involves modelling the choice of mode and this results in *modal split*, i.e. the allocation of trips in the matrix to different modes. Finally, the last stage in the classic model requires the *assignment* of the trips by each mode to their corresponding networks: typically private and public transport.

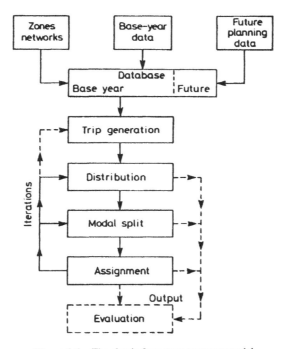

Figure 1.6   The classic four-stage transport model

The classic model is presented as a sequence of four sub-models: trip generation, distribution, modal split and assignment. It is generally recognised that travel decisions are not actually taken in this type of sequence; a contemporary view is that the 'location' of each sub-model depends on the form of the utility function assumed to govern all these travel choices (see Williams 1977). Moreover, the four-stage model is seen as concentrating attention on only a limited range of travellers' responses. Current thinking requires an analysis of a wider range of responses to transport problems and schemes. For example, when faced with increased congestion a trip maker can respond with a range of simple changes to:

- the **route** followed to avoid congestion or take advantage of new links; this includes choice of parking place or combination of services in the case of public transport;
- the **mode** used to get to the destination;
- the **time** of departure to avoid the most congested part of the peak;

- the **destination** of the trip to a less congested area;
- the **frequency** of journeys by undertaking the trip at another day, perhaps combining it with other activities.

Furthermore, other more complex responses take place in the longer term, for example changes in jobs, residential location, choice of shopping areas and so on; all of these will respond, at least partially, to changes in the accessibility provided by the transport system.

Despite these comments, the four-stage sequential model provides a point of reference to contrast alternative methods. For example, some contemporary approaches attempt to treat simultaneously the choices of trip frequency (trips per week), destination and mode of travel thus collapsing trip generation, distribution and mode choice in one single model. Other approaches emphasise the role of household activities and the travel choices they entail; concepts like sojourns, circuits, and time and money budgets are used in this context to model travel decisions and constraints. These modelling strategies are more difficult to cast in terms of the four main decisions or sub-models above. However, the improved understanding of travel behaviour these activity based models provide is likely to enhance more conventional modelling approaches, see Chapter 14.

The trip generation–distribution–modal split and assignment sequence is the most common but not the only possible one. Some past studies have put modal split before trip distribution and immediately after (or with) trip generation. This permits a greater emphasis on decision variables depending on the trip generation unit, the individual or the household. However, forcing modal split before the destination is known requires "averaging" the attributes of the journey and modes in the model. This detracts policy relevance from the modal-split model. Another approach is to perform distribution and mode choice simultaneously, as discussed in Chapter 6. Note also that the classic model makes trip generation inelastic, that is, independent of the level of service provided in the transport system. This is probably unrealistic but only recently techniques have been developed which can take systematic account of these effects.

Once the model has been calibrated and validated for base-year conditions it must be applied to one or more planning horizons. In order to do this it is necessary to develop *scenarios* and plans describing the relevant characteristics of the transport system and planning variables under alternative futures. The preparation of realistic and consistent scenarios is not a simple task as it is very easy to fall into the trap of constructing futures which are neither financially viable nor realistic in the context of the likely evolution of land use and activities in the study area. Despite these difficulties, scenario writing is still more of an art than a technique and requires a good deal of engineering expertise combined with sound political judgement; unfortunately these are scarce resources seldom found together in planning teams.

Having prepared realistic scenarios and plans for testing, the same sequence of models is run again to simulate their performance. A comparison is then made between the costs and benefits, however measured, of different schemes under different scenarios; the idea is to choose the most attractive programme of investment and transport policies which satisfies the demand for movement in the study area.

An important issue in the classic four-stage model is the consistent use of variables affecting demand. For example, at the end of the traffic assignment stage new flow levels, and therefore new travel times, will be obtained. These are unlikely to be the same travel times assumed when the distribution and mode choice models were run, at least when the models are used in the forecasting mode. This seems to call for the re-run of the distribution and modal-split models based now on the new travel times. The subsequent application of the assignment model may well result in a new set of travel times; it will be seen that in general the naive feed-back of the model does not lead to a stable set of distribution, modal split and assignment models with consistent travel times. This problem will be treated in some detail in Chapter 11; its particular relevance is in the risk of choosing the wrong plan depending on how many cycles one is prepared to undertake.

1.6   Continuous Transport Planning

Transport planning models on their own do not solve transport problems. To be useful they must be utilised within a decision process adapted to the chosen decision-making style. The classic transport model was originally developed for an idealised normative decision-making approach. Its role in transport planning can be presented as contributing to the key steps in a 'rational' decision-making framework as in Figure 1.7:

Figure 1.7   A framework for rational decision making with models

1. **Formulation of the problem**. A problem can be defined as a mismatch between expectations and perceived reality. The formal definition of a transport problem requires reference to objectives, standards and constraints. The first reflect the values implicit in the decision-making process, a definition of an ideal but achievable future state. Standards are provided in order to compare, at any one time, whether minimum performance is being achieved at different levels of interest. For example, the fact that many signalised junctions in a city operate at more than 90% degree of saturation can be taken to indicate an overloaded network. Constraints can be of many types, financial, temporal, geographical, technical or simply certain areas or types of building that should not be threatened by new proposals.

2. **Collection of data** about the present state of the system of interest in order to support the development of the analytical model. Of course, data collection is not independent from model development, as the latter defines which types of data are needed: data collection and model development are closely interrelated.

3. **Construction of an analytical model** of the system of interest. The tool-set provided in this book can be used to build transport models including demand and system performance procedures from a

tactical and strategic perspective. In general, one would select the simplest modelling approach which makes possible a choice between schemes on a sound basis. The construction of an analytical model involves specifying it, estimating or calibrating its parameters and validating its performance with data not used during calibration.

4. **Generation of solutions** for testing. This can be achieved in a number of ways, from tapping the experience and creativity of local transport planners and interested parties, to the construction of a large-scale design model, perhaps using optimisation techniques. This involves supply- and cost-minimisation procedures falling outside the scope of this book.

5. In order to test the solutions or schemes proposed in the previous step it is necessary to **forecast the future values of the planning variables** which are used as inputs to the model. This requires the preparation of consistent quantified descriptions, or scenarios, about the future of the area of interest, normally using forecasts from other sectors and planning units. We will come back to this issue in Chapter 15.

6. **Testing the model and solution**. The performance of the model is tested under different scenarios to confirm its reasonableness; the model is also used to simulate different solutions and estimate their performance in terms of a range of suitable indicators. These must be consistent with the identification of objectives and problem definition above.

7. **Evaluation of solutions** and recommendation of a plan/strategy/policy. This involves operational, economic, financial and social assessment of alternative courses of action on the basis of the indicators produced by the models. A combination of skills is required here, from economic analysis to political judgement.

8. **Implementation of the solution** and search for another problem to tackle; this requires recycling through this framework starting again at point (1).

Although based on the idea of a normative decision theory approach, this framework could also be used within behavioural decision-theory styles, to formulate master plans or to provide ammunition in the bargaining involved in adaptive decision making. It implicitly assumes that the problem can be fully specified, the constraints and decision space can be defined and the objective function identified, even if not necessarily completely quantified.

However, one of the main arguments of this book is that real transport systems do not obey the restrictions above: objective functions and constraints are often difficult to define. With hindsight these definitions often turn out to be blinkered: by narrowing a transport problem we may gain the illusion of being able to solve it; however, transport problems have the habit of 'biting back', of reappearing in different places and under different guises; new features and perspectives are added as our understanding of the transport system progresses; changes in the external factors and planning variables throw our detailed transport plans off course. A strong but fixed normative decision-making framework may be suitable for simpler, well-defined and constrained problems but it hardly helps to deal with richer, more complex, many-featured and multi-dimensional transport issues.

How can we improve this general approach to cope with an ever-changing world? It seems essential to recognise that the future is much more tenuous than our forecasting models would lead us to believe. If this is the case, master plans need revising at regular intervals and other decision-making strategies need supporting with the inclusion of fresh information regularly collected to check progress and correct course where necessary. Adaptive or mixed-mode decision-making styles seem more flexible and appropriate to the characteristics of transport problems. They recognise the need to continually redefine problems, arenas and goals as we understand them better, identify new solution strategies, respond to political and technological changes and enhance our modelling capabilities through training, research and experience.

The introduction of a monitoring function is an important addition to the scheme in Figure 1.7. A monitoring system is not restricted to regular data collection; it should also facilitate all other stages in the decision-making framework, as highlighted in Figure 1.8. There are two key roles for a monitoring system. First, it should provide data to identify departures from the estimated behaviour of the transport system and of exogenous key variables such as population and economic growth. Second, the data collected should be valuable in further validating and enhancing the modelling approach followed in preparing the plans.

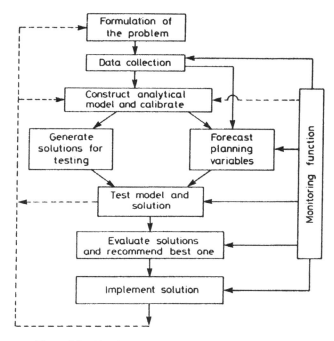

Figure 1.8   Planning and monitoring with the help of models

A good monitoring system should also facilitate learning by the planning team and provide ideas on how to improve and modify models. In this sense, major disruptions to the transport system, like public-transport strikes, short-term fuel shortages or major roadworks which may temporarily change the network structure and its characteristics, should provide a major source of information on transport behaviour to contrast with model predictions. These unplanned experiments should enable analysts to test and enhance their models. A monitoring system fits very well with the idea of a regular or continuous planning approach in transport. If the monitoring system is not in place, it should be established as part of any transportation study.

Monitoring the performance of a transport system and plans is such an important function that it deserves to influence the choice of transport models used to support planning and policy making. The use of models which can be re-run and updated using low-cost and easy-to-collect data, seems particularly appropriate to this task. As we shall see in subsequent chapters, these simpler models cannot provide all the behavioural richness of other more detailed approaches. However, there is scope for combining the two techniques, applying the tool with the highest resolution to the critical parts of the problem and using coarser tools that are easier to update to monitor progress and identify where and when a new detailed modelling effort is needed. We have made an attempt to identify the scope for trade-offs of this kind in the remainder of this book.

The adoption of a monitoring function enables the implementation of a continuous planning process. This is in contrast to the conventional approach of spending considerable resources over a period of one or two years to undertake a large-scale transport study. This burst of activity may be followed by a much longer period of limited effort in planning and updating of plans. Soon the reports and master plans become obsolete or simply forgotten, and nobody capable of running the models again is left in the planning unit. Some years later a new major planning and modelling effort is embarked upon and the cycle is repeated. This style of planning with the help of models in fits and starts is wasteful of resources, does not encourage learning and adaptation as a planning skill, and alienates analysts from real problems. This approach is particularly painful in developing countries: they do not have resources to waste and the rapid change experienced there speeds up plan and data obsolescence. The use of models that are simpler and easier to update is advocated in Chapter 12 to help the implementation of a sound but low-cost monitoring function.

1.7   Theoretical Basis Versus Expedience

One of the recurring themes of transport modelling practice is the distance, and some would say mistrust, between theoreticians and practitioners. The practitioner would often refer to the need to choose between a theoretically sound but difficult to implement set of models, and a more pragmatic modelling approach reflecting the limitations of the data, time and resources available for a study. The implication is that the 'pragmatic' method can deliver the answers needed in the time period available for the study, even if shortcuts must be taken.

The authors have nothing against pragmatic approaches provided they deliver the answers needed to make sound decisions. There is no point in using sophisticated and expensive (but presumably theoretically sound) models for the sake of winning some credit in the academic fraternity. However, there are several reasons to prefer a model based on a sound theoretical background:

1. To guarantee stable results. The recommendations from a study should not depend on how many iterations of a model were run. Prescriptions like 'always start from free flow costs' or 'iterate twice only' are not good enough reasons to assume stable results: next time somebody will suggest running a couple more iterations or a different, and quite justifiable, starting point; this should not be able to change the recommendations for or against a particular scheme.
2. To guarantee consistency. One should be careful about using a particular model of travellers' choice in one part of a model system and a different one in another. Pragmatic models sometimes fail to pass this test. Model consistency is necessary to pass the test of 'reasonableness' and public scrutiny.
3. To give confidence in forecasting. It is almost always possible to fit a model to an existing situation. However, there are plenty of examples of well-fitting models that make no sense, perhaps because they are based on correlated variables. Variables which are correlated today may not be so tomorrow; for example, a strong correlation between banana production and car ownership in a particular country may disappear once oil is discovered there. Therefore models should be backed by some theory of travel behaviour so that one can interpret them consistently and have some confidence that they will remain valid in the future.
4. To understand model properties and develop improved algorithms for their solution. When one is able to cast a problem in mathematical programming or maximum likelihood terms, to mention two popular approaches to model generation, one has a wealth of technical tools to assist in the development of good solution algorithms. These have been developed over the years by researchers working in many areas besides transport.
5. To understand better what can be assumed constant and what must be accepted as variable for a particular decision context and level of analysis. The identification of exogenous and endogenous

variables and those which may be assumed to remain constant is a key issue in modelling economics. For example, for some short-term tactical studies the assumption of a fixed trip matrix may be reasonable as in many traffic management schemes. However, even in the short term, if the policies to be tested involve significant price changes or changes to accessibility, this assumption no longer holds valid.

On the other hand practitioners have often abandoned the effort to use theoretically better models; some of the reasons for this are as follows:

1. They are too complex. This implies that heuristic approaches, rules of thumb, and *ad hoc* procedures are easier to understand and therefore preferable. This is a reasonable point; we do not advocate the use of models as 'black boxes'; quite the contrary. Model output needs interpretation and this is only possible if a reasonable understanding of the basis for such a model is available. Without ignoring the important role of academic literature in advancing the state of the art, there is a case for more publications explaining the basis of models without recourse to difficult notation or obscure (to the practitioner) concepts. Most models are not that complex, even if some of the statistics and computer implementations needed may be quite sophisticated. Good publications bridging the gap between the practitioner and the academic are an urgent need.

2. Theoretical models require data which are not available and are expensive to collect. This is often not entirely correct; many advanced models make much better use of small-sample data than some of the most pragmatic approaches. Improvements in data-collection methods have also reduced these costs and improved the accuracy of the data.

3. It is better to work with 'real' matrices than with models of trip making behaviour. This is equivalent to saying that it is better to work with fixed trip matrices, even if they have to be grossed up for the planning horizon. We will see that sampling and other data-collection errors cast doubts on the accuracy of such 'real' matrices; moreover, they cannot possibly respond to most policies (e.g. improvements in accessibility, new services, and price changes) nor be reasonable for oversaturated do-minimum future conditions. Use of observations alone may lead to 'blinkered' decision making, to a false sense of accuracy and to underestimating the scope for change.

4. Theoretical models cannot be calibrated to the level of detail needed to analyse some schemes. There may be some truth in this statement, at least in some cases where the limitations of the data and time available make it necessary to compromise in detail if one wishes to use a better model. However, it may be preferable to err in this way than to work with the illusion of sufficient detail but undermined by potentially pathological (predictions of the wrong sign or direction) or insensitive results from *ad hoc* procedures.

5. It is better to use the same model (or software) for most problems because this ensures consistency in the evaluation methods. This is, in principle, correct provided the model remains appropriate to these problems. It has the advantage of consistent approach, ease of use and interpretation, and reduced training costs. However, this strategy breaks down when the problems are not of the same nature. Assumptions of fixed trip matrices, or insensitivity to mode choice or pricing policies, may be reasonable in some cases but fail to be acceptable in others. The use of the same model with the same assumptions may be appropriate in one case and completely misleading in another.

The importance of these criteria depends, of course, on the decision context and the levels of analysis involved in the study. What we argue in this book is for the use of the appropriate level of resolution to the problem in hand. Our own preference is for striving to use good, sound models as far as possible even if some level of detail has to be sacrificed. One has to find the best balance between theoretical consistency and expedience in each particular case and decision-making context. We have striven to provide material to assist in this choice.

2

Mathematical Prerequisites

2.1 Introduction

This book is aimed at practitioners and students in transport modelling and planning. Some of these may have a sound mathematical background; they may skip this chapter without loss of continuity. Other readers may have a weaker mathematical background or may simply welcome the opportunity to refresh ideas and notation. This chapter is addressed to these readers. It aims only to outline the most important mathematical prerequisites needed to benefit from this book.

Most of the mathematical prerequisites, however, are not that demanding; the reader can get by with little more than school algebra and some calculus. We introduce first the idea of functions and some specialised notation together with the idea of plotting functions in Cartesian (orthogonal) coordinates. After introducing the concept of series we treat the very important topic of matrix algebra; this is particularly important in transport as we often deal with trip and other matrices. Elements of calculus come next, including differentiation and integration. Logarithmic and exponential functions deserve some special attention as we will find them often in transport models. Finding maxima and minima of functions plays an important role in model development and the generation of solution algorithms. Finally, a few statistical concepts are introduced in the last section of this chapter. Statistics play a key part in contemporary transport modelling techniques and this section provides only an elementary entry point to the subject. A few other statistical concepts and techniques will be introduced in subsequent chapters as needed.

There are several books available as reference works for the more informed reader and as first textbooks for readers needing to brush up their mathematical background. These include those by Morley (1972), Stone (1966), Wilson and Kirby (1980) and Wonnacott and Wonnacott (1990) for the statistical elements discussed. We have seen transport modelling practice moving steadily away from expedience through shortcuts and 'fudge factors', and increasingly adopting models with sounder theoretical backing. This trend results from the need to provide consistent advice to decision makers; this advice should not depend on an arbitrarily chosen number of iterations or starting points, or on models likely to produce pathological results when used to forecast completely new options. This increased rigour will rely on better mathematical and statistical representations of problems and therefore requires further reading in these areas.

Modelling Transport, Fourth Edition. Juan de Dios Ortúzar and Luis G. Willumsen.
© 2011 John Wiley & Sons, Ltd. Published 2011 by John Wiley & Sons, Ltd.

2.2   Algebra and Functions

2.2.1   Introduction

Elementary algebra consists of forming expressions using the four basic operations of ordinary mathematics on letters which stand for numbers. It is useful to distinguish between *variables* (generally denoted by letters such as x, y and z), which represent measured quantities, and *constants* or *parameters* (generally denoted by letters such as a, b, c, ..., k, m, n, ..., or by letters from the Greek alphabet). The value of a constant is supposed to remain invariant for the particular situation examined.

Variables, and constants, are related through equations such as:

$$y = a + bx \tag{2.1}$$

and if we were interested in x, we could 'solve' (2.1) for x, obtaining:

$$x = (y - a)/b \tag{2.2}$$

The variables x and y in (2.1) and (2.2) are related by the '=' sign. However, in algebra we may also have *inequalities* of the following four types:

$<$ which means *less than*

$\leq$ which means *less than or equal to*

$>$ which means *greater than*, and

$\geq$ which means *greater than or equal to*

and which are used to constrain variables, for example:

$$x + 2y \leq 5 \tag{2.3}$$

This expression, unlike an equation, cannot be 'solved' for x or y, but note that both variables can take only a restricted range of values. For example, if we restrict them further to be positive integers, it can easily be seen that x cannot be greater than 3 and y cannot be greater than 2.

It is possible to manipulate inequalities in much the same way as equations, thus:

- we can add or subtract the same quantity to/from each side;
- we can also multiply or divide each side by the same quantity, but if the number which is being multiplied or divided is negative, the inequality is reversed.

Example 2.1   If we subtract 5 on both sides of (2.3) we get

$$x + 2y - 5 \leq 0$$

which is certainly the same constraint. However, if we multiply it by -2 we obtain:

$$-2x - 4y \geq -10$$

which can be checked by the reader to provide the same constraint as (2.3).

The use of different letters to denote each variable is only convenient up to a certain point. Soon it becomes necessary to use indices (e.g. subscripts or superscripts) to define additional variables, as in x_1, x_2, x_3, ..., x_n, which we can conveniently summarise as x_i, $i = 1, 2, ..., n$; it does not matter if we use another letter for the index if it has the same numerical range. For example, we could have defined also $x_k, k = 1, 2, \ldots, n$.

The use of indices facilitates a very convenient notation for summations and products:

$$\sum_{i=1}^{n} x_i = x_1 + x_2 + x_3 + \ldots + x_n \tag{2.4}$$

or

$$\prod_{j=1}^{m} y_j = y_1 y_2 y_3 \ldots y_m \tag{2.5}$$

In certain cases a single index is not enough and two or more may be used. For example we could define the following six variables, $T_{11}, T_{12}, T_{21}, T_{22}, T_{31}, T_{32}$ as $T_{ij}, i = 1, 2, 3$, and $j = 1, 2$. With two-subscript variables we can have double summations or double products, as in:

$$\sum_{i=1}^{3} \sum_{j=1}^{2} T_{ij} = \sum_{i=1}^{3} (T_{i1} + T_{i2}) = T_{11} + T_{12} + T_{21} + T_{22} + T_{31} + T_{32} \tag{2.6}$$

2.2.2 Functions and Graphs

We have already referred to variables as being related by equations and inequalities; in general these can be called functional relations. A particular function is some specific kind of relationship between two or more variables. For example, the power function:

$$y = \phi x^n \tag{2.7}$$

yields values of the *dependent* variable y, given values of the parameters ϕ and n, and of the *independent* variable x; a function requires that for every value of x in some range, a corresponding value of y is specified. Often we do not wish to refer to a particular function, but only to state that y is 'some function of x' or vice versa; this can be written as:

$$y = f(x) \tag{2.8}$$

A large range of functions exists and readers should familiarise themselves with these as they arise. It is usually convenient to plot functions graphically on a Cartesian co-ordinate system (see Figure 2.1).

A dependent variable may be a function of several independent variables, for example:

$$y = f(x_1, x_2, \ldots, x_n) \tag{2.9}$$

but this would require $n + 1$ dimensions to represent it (n for the independent variables and 1 for y). Cartesian coordinates can be used in three or more dimensions, in the case of three dimensions the orientation of the third axis is out of this side of the paper in Figure 2.1. More than three dimensions cannot be easily visualised physically but are dealt with algebraically in just the same way as one, two and three dimensions. For example, in the case of $n = 2$ the function can be represented by a surface over the relevant part of the (x_1, x_2) plane.

Generally, any equation for an unknown quantity x can be put in the form $f(x) = 0$; for example, the linear equation:

$$ax = b$$

is equivalent to

$$ax - b = 0$$

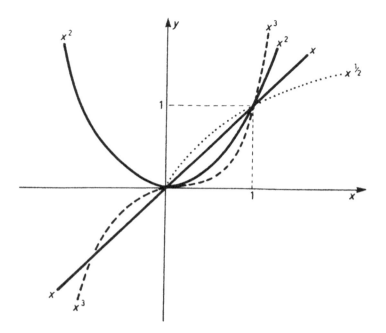

Figure 2.1   Plot of various power functions

where $f(x) = ax - b$. Solving the equation is therefore equivalent to finding the points on the curve $y = f(x)$ which meet the x axis. These points are called *real solutions* or *zeros* of $f(x)$; for example, x_1 and x_2 in Figure 2.2.

We are sometimes interested in what happens to the value of a function $f(x)$, as x increases indefinitely ($x \rightarrow \infty$); it can easily be seen that the possibilities are only the following:

- tend to infinity (e.g. when $f(x) = x^2$)
- tend to minus infinity (e.g. when $f(x) = -x$)

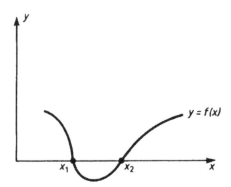

Figure 2.2   Real solutions of a general function

- oscillate infinitely (e.g. when $f(x) = (-1)^x x^2$)
- tend to a finite limit (e.g. $f(x) = 1 + 1/x$).

For more complex functions some ingenuity may be required to find out if they tend to a finite limit when $x \to \infty$.

We may also be interested in finding the *limit* when x approaches a finite value. For example, if $f(x) = 1/(x + 3)$, it can easily be seen that the limit when $x \to 0$ is $1/3$. If for some value α we have that $f(x) \to \infty$ as $x \to \alpha$, the curve $y = f(x)$ is said to have an asymptote $x = \alpha$ (see Figure 2.3).

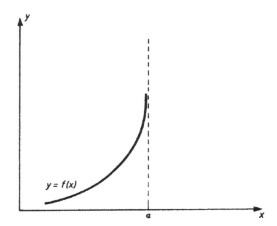

Figure 2.3   General function with asymptote at α

One of the most important functions is the *straight line*, shown in Figure 2.4 and whose general equation is (2.1). It can easily be seen that b is the value of y when $x = 0$; this is usually called the *intercept* on the y axis. The constant a is called the *gradient* and it can be shown to be given by:

$$a = \frac{y_2 - y_1}{x_2 - x_1} \tag{2.10}$$

where (x_1, y_1) and (x_2, y_2) are any two points on the line (see Figure 2.4a). Although a straight line has by definition a constant gradient, this can be either positive or negative as shown in the figure.

Unless two straight lines are parallel they will intersect at one point; this can be represented either graphically (as in Figure 2.5) or algebraically as a system of equations as follows:

$$y = x + 2 \tag{2.11a}$$

$$y = -x + 4 \tag{2.11b}$$

Solving for x in (2.11b) and replacing this value (i.e. $-y + 4$) in (2.11a) we get that the solution is point P with coordinates ($x = 1, y = 3$).

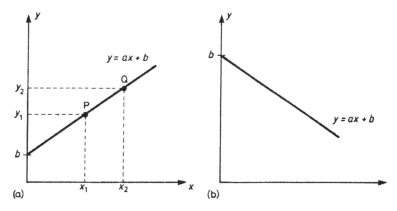

Figure 2.4   Two straight lines $y = ax + b$: (a) positive gradient, (b) negative gradient

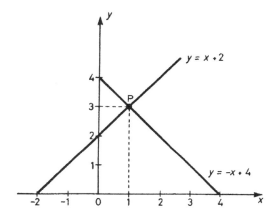

Figure 2.5   Intersection of two straight lines

2.2.3   Sums of Series

A series is simply defined as a sequence of numbers u_n, $n = 1, 2, \ldots, N$. In many cases it may be interesting to find its sum given by:

$$S_N = u_1 + u_2 + \ldots + u_N = \sum_n u_n \qquad (2.12)$$

In some cases, such as the *arithmetic progression* given by:

$$u_n = u_{n-1} + d \qquad (2.13)$$

it can be shown that the series has a sum to N terms. For example, if the first term is b the sum can be shown to be:

$$S_N = Nb + N(N - 1)d/2 \qquad (2.14)$$

The *geometric progression* (2.15), formed by multiplying successive terms by a constant factor r, also has an expression for the sum of N terms. If b is again the first term, the sum can be shown to be given by (2.16) if r is different from 1:

$$u_n = r u_{n-1} \tag{2.15}$$

$$S_N = \frac{b(1 - r^n)}{1 - r} \tag{2.16}$$

In other cases the series may have a simple expression for its sum, such as in:

$u_n = n$, where the sum is given by $S_N = N(N + 1)/2$,
or $u_n = x^n$, where it is given by $S_N = x(1 - x^N)/(1 - x)$ for x different from 1;

but still *diverge* (i.e. S_N keeps increasing indefinitely when N tends to infinity). That happens to $u_n = n$; it also happens to $u_n = x^n$ if $x > 1$ above; however, the latter converges to $S_N = x/(1 - x)$ for the range $0 < x < 1$.

2.3 Matrix Algebra

2.3.1 Introduction

Any variable with two subscripts can be called a *matrix*. We will denote matrices by the notation $\mathbf{B} = \{B_{ij}\}$, where the variables B_{ij}, $i = 1, 2, \ldots, N$; $j = 1, 2, \ldots, M$ are the elements of $\mathbf{B}$. This can be written as follows:

$$\mathbf{B} = \begin{pmatrix} B_{11} & B_{12} & B_{13} & \ldots & B_{1M} \\ B_{21} & B_{22} & B_{23} & \ldots & B_{2M} \\ \vdots & & & & \\ B_{N1} & B_{N2} & B_{N3} & \ldots & B_{NM} \end{pmatrix} \tag{2.17}$$

As can be seen, the matrix has N rows and M columns; for this reason it is known as a $N \times M$ matrix. A *vector* is an important special case, being a one-dimensional array or a $N \times 1$ matrix. In these cases the second index is redundant, so we write:

$$\mathbf{V} = \{V_i\} = \begin{pmatrix} V_1 \\ V_2 \\ V_3 \\ \vdots \\ V_N \end{pmatrix} \tag{2.18}$$

Formally, a non-indexed variable, or even a constant, can be thought of as a 1×1 matrix and it is known as a *scalar*.

If we interchange rows and columns we obtain an $M \times N$ matrix known as the *transpose* $\mathbf{B}^T$ of $\mathbf{B}$, which is given by:

$$\mathbf{B}^T = \begin{pmatrix} B_{11} & B_{21} & B_{31} & \ldots & B_{N1} \\ B_{12} & B_{22} & B_{32} & \ldots & B_{N2} \\ \vdots & & & & \\ B_{1M} & B_{2M} & B_{3M} & \ldots & B_{NM} \end{pmatrix} \tag{2.19}$$

Similarly, the transpose of an $N \times 1$ vector (also known as a *column* vector) is a *row* vector:

$$\mathbf{V}^T = [V_1 \, V_2 \, V_3 \dots V_N] \tag{2.20}$$

A *square* matrix $\mathbf{S}$ is one where $N = M$; a square matrix such that $\mathbf{S} = \mathbf{S}^T$ is called symmetric. A *diagonal* matrix $\mathbf{D} = \{D_{ij}\}$ is one where $D_{ij} = 0$ unless $i = j$. The *unit* matrix is a square diagonal matrix with each diagonal element equal to 1, that is:

$$\mathbf{I} = \begin{pmatrix} 1 & 0 & 0 & \dots & 0 \\ 0 & 1 & 0 & \dots & 0 \\ 0 & 0 & 1 & \dots & 0 \\ \vdots & & & & \\ 0 & 0 & 0 & \dots & 1 \end{pmatrix} \tag{2.21}$$

2.3.2 Basic Operations of Matrix Algebra

We will define the operations between two matrices $\mathbf{A}$ and $\mathbf{B}$ by setting a new matrix $\mathbf{C}$ which will represent the combination required. First matrix *addition*:

$$\mathbf{C} = \mathbf{A} + \mathbf{B} = \mathbf{B} + \mathbf{A} \tag{2.22}$$

is defined by $C_{ij} = A_{ij} + B_{ij}$ and requires that both matrices being combined are of the same size, say both $N \times M$ matrices; then $\mathbf{C}$ is also an $N \times M$ matrix. This is also a requirement for matrix subtraction:

$$\mathbf{C} = \mathbf{A} - \mathbf{B} \tag{2.23}$$

similarly defined as $C_{ij} = A_{ij} - B_{ij}$. An operation which is unique to matrix algebra is *multiplication by a scalar*:

$$\mathbf{C} = k\mathbf{A} \tag{2.24}$$

defined by $C_{ij} = k \, A_{ij}$, where obviously the new 'grossed up' matrix has the same size as the old one.

Matrix *multiplication* is more complex, as:

$$\mathbf{C} = \mathbf{AB} \tag{2.25}$$

is defined by $C_{ij} = \sum_k A_{ik} B_{kj}$, where $\mathbf{A}$ is an $N \times M$ matrix and $\mathbf{B}$ is any $M \times L$ matrix (i.e. the number of columns in $\mathbf{A}$ must equal the number of rows in $\mathbf{B}$ but there are no other restrictions). In this case $\mathbf{C}$ is an $N \times L$ matrix.

It is easy to see that in general $\mathbf{AB}$ is not equal to $\mathbf{BA}$, i.e. the operation is non-commutative, as opposed to elementary algebra. However, this is not the case with the unit matrix $\mathbf{I}$; in fact, it can easily be checked that:

$$\mathbf{IA} = \mathbf{AI} = \mathbf{A} \tag{2.26}$$

Thus, although it is possible to define the product of any number of matrices, order must always be preserved. In fact we refer to *pre-multiplication* of $\mathbf{A}$ by $\mathbf{B}$ to form the product $\mathbf{BA}$, and to *post-multiplication* to form $\mathbf{AB}$.

To define *division* it is convenient to use the concept of inverse of a matrix. Unfortunately this only exists for square matrices and then not always. If the inverse exists, it is denoted as $\mathbf{B}^{-1}$ and is the matrix that satisfies:

$$\mathbf{B}^{-1}\mathbf{B} = \mathbf{BB}^{-1} = \mathbf{I} \tag{2.27}$$

In this case **B** is said to be *non-singular*. Another related interesting concept is that of a *positive definite* matrix. A real symmetric matrix **M** is positive definite if $\mathbf{w}^T \cdot \mathbf{M} \cdot \mathbf{w} > 0$ for all non-zero vectors **w** with real entries. We will not give a procedure for the calculation of the elements of the inverse matrix as it is fairly complicated. It is sufficient to know that under suitable conditions it exists. Division is then just pre- or post-multiplication by $\mathbf{B}^{-1}$.

In this book matrices and vectors are mostly used to provide a shorthand notation for such things as sets of simultaneous equations and for obtaining their solution in terms of the inverse matrix.

2.4   Elements of Calculus

The two main branches of calculus are differentiation and integration; their basic nature can be intuitively identified by reference to the function $y = f(x)$ depicted in Figure 2.6. Consider the points P and Q and the straight line (*chord*) connecting them. Differentiation is concerned with the calculation of the gradient of a curve at a point. To do this, it is useful to consider Q approaching P; in the limit the chord PQ becomes the *tangent* to the curve at P = Q (i.e. when their horizontal 'distance' h is 0) and by definition its gradient is equal to that of the curve.

Integration, on the other hand, is concerned with calculating the area under a curve, say the shaded area in Figure 2.6; as we will see below these two operations are closely related.

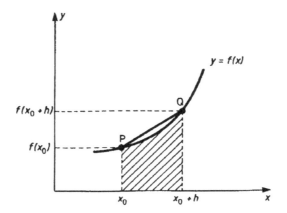

Figure 2.6   Gradient at a point and area under a curve

2.4.1   Differentiation

Using (2.10) the gradient of the chord PQ in Figure 2.6 can be written as:

$$\delta(x) = [f(x_0 + h) - f(x_0)] / h$$

If the limit of $\delta(x)$ when $h \to 0$ exists and is the same whether h tends to zero from above or below, it is called the *derivative* of y, or of $f(x)$, with respect to x at x_0 and it is often written as $f'(x_0)$ or $dy/dx|x_0$. The process of finding the derivative is called differentiation.

If $f(x)$ is given as an expression in x, it is usually not difficult to find $f'(x)$ as a function of x using the results in Table 2.1, plus others we will give below.

Table 2.1   Common derivatives

Function $f(x)$		Derivative $f'(x)$
k	(k constant)	0
x^b	(b constant, $x > 0$)	bx^{b-1}
$ku(x)$	(k constant)	$ku'(x)$
$u(x) + v(x)$		$u'(x) + v'(x)$
$u(x)v(x)$		$u'(x)v(x) + u(x)v'(x)$
$u[v(x)]$		$u'[v(x)]v'(x)$

Since derivatives are themselves functions of x, we can also define second-and higher-order derivatives (i.e. $f_0''(x)$ or $d^2 y/dx^2$ and so on). For example, if we differentiate the first derivative of $y = x^b$ in Table 2.1, we get:

$$\frac{d^2 y}{dx^2} = b(b-1)x^{b-2} \tag{2.28}$$

2.4.2   Integration

This is the reverse of differentiation; if we know the gradient of some curve at every point then the equation of the curve itself is known as the *integral*. For example, if $g = g(x)$ is the gradient, the equation of the curve is written

$$y = \int_x g(x)\, dx$$

and this result is always arbitrary up to an additive constant; for example, if $g = bx^{b-1}$ we know from Table 2.1 that the *indefinite* integral of $g(x)$ is given by:

$$y = G(x) = \int_x bx^{b-1} dx = x^b + C \tag{2.29}$$

where C is an arbitrary constant of integration (i.e. the derivative of $x^b + C$ is bx^{b-1} no matter the value of C). The most practical elementary use of integration is to obtain the *area under a curve* as the *definite* integral, as shown in Figure 2.7a.

$$\text{Area abcd} = [F(x)]_a^b = F(b) - F(a) = \int_a^b y\, dx = \int_a^b f(x)\, dx \tag{2.30}$$

For example, if we take the simple case of a straight line parallel to the x axis at height h and want to integrate between the values a and b (see Figure 2.7b), we get:

$$y = f(x) = h$$

and

$$F(x) = hx + C$$

then

$$\text{Area} = F(b) - F(a) = h(b - a)$$

which is indeed the area of the shaded rectangle in the figure.

Table 2.1 can be used in reverse to help to find indefinite integrals. In particular, if

$$\int u(x)dx = U(x) + C_1 \text{ and } \int v(x)dx = V(x) + C_2$$

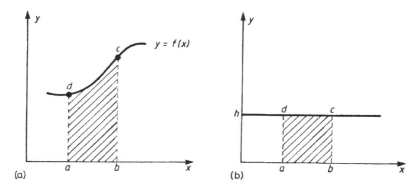

Figure 2.7 Areas under a curve: (a) general case, (b) line parallel to x axis

then

$$\int u[v(x)]v'(x)\,\mathrm{d}x = U[v(x)] + C_3$$

and

$$\int U(x)v(x)\,\mathrm{d}x = U(x)V(x) - \int u(x)V(x)\,\mathrm{d}x$$

Of course not all functions, even some that are deceptively simple in appearance, have indefinite integrals which are similarly simple expressions. However, for those that do not it is still possible to evaluate definite integrals numerically.

2.4.3   The Logarithmic and Exponential Functions

Among the functions we have considered so far, the simplest one with no indefinite integral is the inverse function $f(x) = 1/x$, depicted in Figure 2.8.

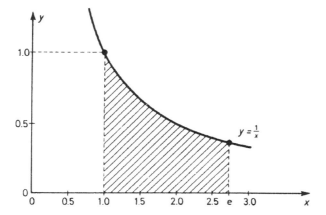

Figure 2.8 Inverse function and Nepper's constant

The integral of this function has been defined as the *natural logarithm* of x, or $\log_e (x)$, where e is Nepper's constant. Its value of approximately 2.7183 corresponds to the point on the x axis of Figure 2.8 such that the shaded area is 1, i.e. $\log_e (e) = 1$. As in this book we will only use natural logarithms, we will drop the base e from our notation.

In common with any other logarithm, $\log (x)$ has the following properties:

$$\log(1) = 0;$$
$$\text{As } t \to \infty, \ \log (t) \to \infty;$$
$$\text{As } t \to 0, \ \log (t) \to -\infty;$$
$$\log (uv) = \ \log (u) + \log (v).$$

Another useful related function is the *exponential function* exp (x) or e^x for short, defined as the number w such that $\log (w) = x$. Then, as expected of a power function, we have:

$$e^{(x+y)} = e^x e^y;$$

moreover,

$$e^{\log(x)} = x$$

Both functions $\log (x)$ and $\exp (x)$ are easy to differentiate; by definition:

$$\frac{d}{dx} \log (x) = \frac{1}{x} \tag{2.31}$$

and it is not difficult to show that:

$$\frac{d}{dx} (e^x) = e^x \tag{2.32}$$

Thus the exponential is the function which remains unaltered under differentiation.

2.4.4   Finding Maximum and Minimum Values of Functions

This is one important use of differentiation. Consider Figure 2.9 for example; the function shown has a *maximum* at x_1 and a *minimum* at x_2. Both are characterised by the gradient of the curve being zero at those points, so the first step in finding them is to solve the equation $f'(x) = 0$.

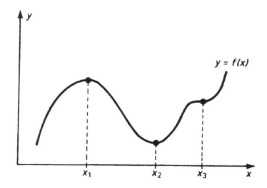

Figure 2.9   Maximum, minimum and point of inflexion

It is important to note, however, that not all zeros are maxima or minima; an example of one that is not (called a *point of inflexion*) is x_3 in Figure 2.9. To find out more precisely what a zero gradient stands for, it is necessary to evaluate $f''(x)$ at each zero of $f'(x)$. Thus, for a maximum we require:

$$f''(x) < 0 \tag{2.33}$$

For a minimum we need:

$$f''(x) > 0 \tag{2.34}$$

and for a point of inflexion,

$$f''(x) = 0 \tag{2.35}$$

These cases are illustrated in Figure 2.10, which suggests a good mnemonic. Consider the function as a cup of water; if it is facing downwards as in the case of the maximum, the liquid will drop (i.e. a minus sign). Conversely if it is facing upwards (e.g. a minimum) the liquid will stay (i.e. a plus sign).

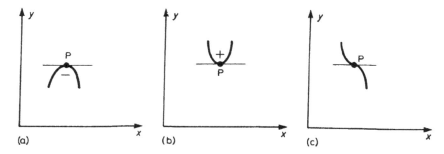

Figure 2.10   Stationary points: (a) maximum, (b) minimum, (c) point of inflexion

In order to develop a theory directed toward characterising global, rather than local, minimum (or maximum) points mathematicians have found it necessary to introduce the complementary notions of *convexity* and *concavity*. These result not only in a more powerful (although more restrictive) theory, but also provide an interesting geometric interpretation of the second-order conditions (2.33) to (2.35).

Figure 2.11 presents some examples of convex and non-convex functions. Geometrically, a function is convex if the line joining two points on its graph lies nowhere below the graph, as shown in Figure 2.11a; in two dimensions, a convex function would have a bowl-shaped graph. Similarly and simply, a function g is said to be concave if the function $f = -g$ is convex. A nice property of convex functions is that the sum of two such functions is also convex.

2.4.5   Functions of More Than One Variable

It is useful to consider the application of differential and integral calculus to this kind of function. Suppose that we have:

$$y = f(x_1, x_2, \ldots, x_n) \tag{2.36}$$

Then the derivative of y with respect to one of these variables may be calculated assuming the other variables remain constant during the operation. This is known as a *partial* derivative and is written $\partial y/\partial x_i$.

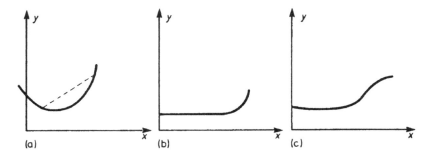

Figure 2.11   Convex and nonconvex functions: (a) convex, (b) convex, (c) nonconvex

Example 2.2 Consider the following function:

$$y = 2x_1 + x_2^3 x_3$$

then the partial derivatives are given by:

$$\frac{\partial y}{\partial x_1} = 2$$

$$\frac{\partial y}{\partial x_2} = 3x_2^2 x_3$$

$$\frac{\partial y}{\partial x_3} = x_2^3$$

It can be shown that maxima and minima of a function such as (2.36) can be found by setting all the partial derivatives to zero:

$$\frac{\partial y}{\partial x_i} = 0, \qquad i = 1, 2, \ldots, n \tag{2.37}$$

which gives a set of simultaneous equations to solve. A particularly interesting case is that of the restricted maximum or minimum. Assume we wish to maximise (2.36) subject to the following restrictions:

$$\begin{aligned} r_1(x_1, x_2, \ldots, x_n) &= b_1 \\ r_2(x_1, x_2, \ldots, x_n) &= b_2 \\ &\vdots \\ r_K(x_1, x_2, \ldots, x_n) &= b_K \end{aligned} \tag{2.38}$$

This can be done by defining *Lagrangian multipliers* $\lambda_1, \lambda_2, \ldots, \lambda_K$ for each of the equations (2.38) in turn, and maximising

$$L = f(x_1, x_2, \ldots, x_n) + \sum_k \lambda_k [r_k(x_1, \ldots, x_n) - b_k] \tag{2.39}$$

as a function of $x_1, x_2, \ldots, x_n$ and $\lambda_1, \lambda_2, \ldots, \lambda_K$. Thus, we solve:

$$\frac{\partial L}{\partial x_i} = 0, \quad i = 1, 2, \ldots, n \tag{2.40}$$

and

$$\frac{\partial L}{\partial \lambda_k} = 0, \qquad k = 1, 2, \ldots, K \tag{2.41}$$

The equations (2.41) are simply the restrictions (2.38) in another form; the device of introducing the multipliers as additional variables enables the restricted maximum to be found.

The Hessian Matrix This is the matrix of second-order partial derivates of a function. Given the real value function: $y = f(x_1, x_2, \ldots, x_n)$ if all second partial derivatives of f exist, then the ij element of its Hessian matrix is $h_{ij} = \frac{\partial^2 y}{\partial x_i \partial x_j}$; thus, the matrix is given by:

$$H = \begin{bmatrix} \dfrac{\partial^2 y}{\partial x_1 \partial x_1} & \dfrac{\partial^2 y}{\partial x_1 \partial x_2} & \cdots & \dfrac{\partial^2 y}{\partial x_1 \partial x_n} \\ \dfrac{\partial^2 y}{\partial x_2 \partial x_1} & \dfrac{\partial^2 y}{\partial x_2 \partial x_2} & \cdots & \dfrac{\partial^2 y}{\partial x_2 \partial x_n} \\ \cdot & \cdot & \cdots & \cdot \\ \cdot & \cdot & \cdots & \cdot \\ \cdot & \cdot & \cdots & \cdot \\ \dfrac{\partial^2 y}{\partial x_n \partial x_1} & \dfrac{\partial^2 y}{\partial x_n \partial x_2} & \cdots & \dfrac{\partial^2 y}{\partial x_n \partial x_n} \end{bmatrix}$$

and it is often used in optimisation problems as we will see in Chapters 7 and 8.

2.4.6 Multiple Integration

In the case of integration, multiple integrals can be defined. For example, given (2.36) we might have:

$$V = \int \int \cdots \int f(x_1, x_2, \ldots, x_n) \, dx_1 \, dx_2, \ldots dx_n \tag{2.42}$$

with n integral signs. In order to get an intuitive feeling of its meaning it is useful to consider the two-dimensional case. The function

$$S = f(x_1, x_2) \tag{2.43}$$

may be considered as defining a surface in a three-dimensional Cartesian system. Therefore,

$$V = \int \int f(x_1, x_2) \, dx_1 \, dx_2 \tag{2.44}$$

measures a volume under this surface, in a similar way to the single variable measuring an area under a curve.

2.4.7 Elasticities

The elasticity of a dependent variable y with respect to another variable x_i in a function such as (2.9) is given by the expression:

$$E(y, x_i) = \frac{\partial y}{\partial x_i} \frac{x_i}{y} \tag{2.45}$$

and can be interpreted as the percentage change in the dependent variable with respect to a given percentage change in the relevant independent variable.

In econometrics we will often be interested in the elasticities of a given demand function with respect to changes in the values of some explanatory variables or *attributes*. We will generally distinguish between *direct-* and *cross-elasticities*; the first relate to attributes of the service or good under consideration and the second to attributes of competing options or goods. For example, it is often stated that the elasticity of public transport demand to fares is around -0.33; this means that if we increase fares by 1% we should expect patronage to decrease by approximately 0.3%.

2.4.8 Series Expansions

It is sometimes necessary to estimate the values of a function $f(x)$ in the neighbourhood of a particular value x_0 of x, in terms of the values of the function and its derivatives at this value. For suitable functions this can be done by means of *Taylor's series* expansion; first we require to define the concept of a *factorial number* $(n!)$ which applies to non-negative integers:

$$n! = n(n-1)(n-2)\ldots 3 \cdot 2 \cdot 1$$
$$0! = 1 \tag{2.46}$$

A Taylor's series expansion is defined as:

$$f(x_0 + h) = f(x_0) + hf'(x_0) + (h^2/2!)f''(x_0) + (h^3/3!)f'''(x_0) + \ldots \tag{2.47}$$

and it is most useful when h is small enough for the higher-order terms to become rapidly smaller so that a good approximation is obtained by stopping the summation after just a few terms – even just after two terms.

The special case when $x_0 = 0$ is known as *Maclaurin's series*, which upon setting h to x in the left-hand side of (2.47) yields:

$$f(x) = f(0) + hf'(0) + (h^2/2!)f''(0) + (h^3/3!)f'''(0) + \ldots \tag{2.48}$$

This provides a method of expressing certain functions as power series, for example:

$$e^x = 1 + x + x^2/2! + x^3/3! + \ldots$$

which allows us very easily to see why expression (2.32) holds.

2.5 Elementary Mathematical Statistics

In this section we provide only a basic review of the more fundamental statistical concepts. In the rest of the book we take for granted that the reader is not only aware of the most important distributions (e.g. binomial, normal, Student, chi-squared and Fisher) but also has some knowledge about basic statistical inference (e.g. estimators, confidence intervals and tests of hypotheses). As there are very good textbooks about this subject, the reader is strongly advised to consult them for further reference. In particular we recommend Wonnacott and Wonnacott (1990) and Chapter 7 of Wilson and Kirby (1980).

Certain specialised subjects, such as basic sampling theory and linear regression analysis are presented at greater length in the relevant chapters (i.e. 3 and 4 respectively).

2.5.1 Probabilities

The most intuitive definition of the probability that a certain result will occur (e.g. obtaining a six by rolling a dice) is given by the limit of its *relative frequency*, that is:

$$P(e_i) = p_i = \lim_{n \to \infty} \frac{n_i}{n} \tag{2.49}$$

where e_i is the desired result, n is the number of times the experiment is repeated and n_i the number of times e_i occurs. Expression (2.49) allows deducing certain basic properties of probabilities:

$$0 \leq p_i \leq 1 \tag{2.50}$$

as n_i can take both the values 0 and n, and

$$\sum_i p_i = 1 \tag{2.51}$$

as $n_1 + n_2 + \ldots = n$. An alternative view of the expected probability of the result can be expressed in terms of a fair bet. If a person regards as fair a bet in which they win \$35 if e_i happens and loses \$$x$ if it does not, then their estimate of p_i is $x/(x + 35)$. This is so because they have solved the following equation which makes their expected gains or losses equal to zero, i.e. a fair bet:

$$35p_i - x(1 - p_i) = 0$$

On many occasions the probabilities of certain experiments are not simple to calculate. It is convenient to define an event (E) as a subset of the set of possible results of an experiment, $E = \{e_1, \ldots, e_i\}$. The probability of an event is the sum of the probabilities of the results it is composed of,

$$P(E) = \sum_i p_i, \qquad e_i \in E$$

Example 2.3 The event E: {to obtain at least two heads in three throws of a coin} includes (the first) four results out of the eight possible ones: (H, H, H), (H, H, T), (H, T, H), (T, H, H), (T, T, H), (T, H, T), (H, T, T) and (T, T, T). As each of these results has a probability of 1/8 (if the probabilities of getting heads and tails are equal), the probability of the event is 1/2.

For *combinations* of events (i.e. two heads but such that not all throws give the same result) it becomes necessary to work with the concepts of *union* ($\cup$) and *intersection* ($\cap$) of set theory as presented in Figure 2.12. The rectangle in the figure represents the event space and A and B are events within it.

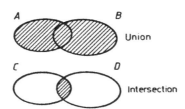

Figure 2.12   Venn diagram for events and probabilities

In general, it is true that

$$P(A \cup B) = P(A) + P(B) - P(A \cap B) \tag{2.52}$$

and if A and B are mutually exclusive,

$$P(A \cup B) = P(A) + P(B) \tag{2.53}$$

The *conditional probability* $P(A/B)$, of A happening given that B is true, is:

$$P(A/B) = P(A \cap B)/P(B) \tag{2.54}$$

An event F is statistically independent of another event E, *if and only if* (iff) $P(F|E)$ is equal to $P(F)$. Therefore, for independent events we have:

$$P(E \cap F) = P(E)P(F) \tag{2.55}$$

which we applied intuitively when estimating event probability in Example 2.1.

2.5.2   Random Variables

These can be defined as those which take values following a certain probability *distribution* (see Figure 2.13).

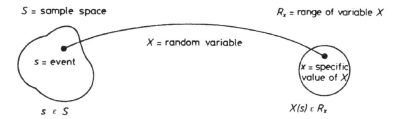

Figure 2.13   Random variable mapping from sample space

Example 2.4   The experiment 'spinning a coin twice', can yield only the following results (sample space): $S = \{$HH, HT, TH, TT$\}$. If we define the random variable $X =$ number of heads, it is easy to see that it can only take the following three values: $X(\text{HH}) = 2$, $X(\text{HT}) = X(\text{TH}) = 1$ and $X(\text{TT}) = 0$. Therefore, an advantage of the random variable concept becomes immediately apparent: the set of results (sample space) is reduced to a smaller, more convenient, numerical set (the range of the variable). The probabilities of X are as follows:

$$P(X = 1) = P(\text{HT} \cup \text{TH}) = P(\text{HT}) + P(\text{TH}) = 1/2$$
$$P(X = 2) = P(X = 0) = 1/4.$$

Random variables may be *discrete* or *continuous*. In the former case they can take values from a finite set with probabilities belonging to a set $P(X)$ which satisfy (2.51) and $p(x_i) \geq 0$. In the latter case it is necessary to define a probability *density* function $f(x)$, such that:

$$\int_x f(x)\mathrm{d}x = 1$$
$$f(x) \geq 0, \qquad \forall x \tag{2.56}$$

2.5.3 Moments around Zero

2.5.3.1 Expected Value

If X is a random variable, then its *expected value* $E(X)$ is the function obtained by taking the weighted average of the x_i values times their probabilities, thus:

$$E(X) = \sum_i x_i \, p_i(x_i), \qquad \text{discrete case}$$
$$E(X) = \int_a^b x \, f(x) \, dx, \qquad \text{continuous case} \tag{2.57}$$

where $f(x)$ is defined for the range ($a \leq x \leq b$). The expected value corresponds to the concept of *mean* ($\bar{X}$) of a sample in descriptive statistics and is normally found by direct application of the *expectation operator* to the random variable X. It can be applied also to functions of random variables; the operator has the important property of *linearity*, whereby for any random variables X and Y, and constants a, b and c we have:

$$E(a + bX + cY) = a + bE(X) + cE(Y) \tag{2.58}$$

When dealing with statistical data, summary information may be provided conveniently by specifying certain key features rather than the whole of a distribution. For example, the distribution of a random variable might be described with reference to its mean value and the dispersion around it. These descriptive statistics can be used to make simple comparisons between distributions without going into full details. More interestingly, certain standard distributions can be completely specified by just a few descriptive statistics.

Another two usual descriptive statistics that attempt to indicate the 'middle' of a distribution (i.e. measures of *central tendency*) are the *mode* X^*, which is the value of X which maximises $p_i(x_i)$, and the *median* $X_{0.5}$ is the value of X below which lies half of the distribution, that is:

$$P(X_{0.5}) = \sum_{x=1}^{X_{0.5}} P(X) = 0.5 \qquad \text{discrete case}$$
$$P(x < X_{0.5}) = \int_a^{X_{0.5}} f(x) \, dx = 0.5 \quad \text{continuous case} \tag{2.59}$$

Neither the median nor the mode can be found by direct calculation, but need the solution of a problem.

2.5.3.2 Variance

The *variance* of a random variable X is defined as:

$$Var(X) = E\{[X - \mu]^2\} = E(x^2) - E^2(x) \tag{2.60}$$

where μ denotes the population mean. Unlike expectation, the variance is not a linear operator, so:

- $Var\,(a + bX) = b^2\,Var\,(X)$, i.e. adding a constant does not affect the spread of the distribution.
- $Var\,(aX + bY) = a^2\,Var\,(X) + b^2\,Var\,(Y) + 2ab\,Cov\,(X, Y)$, where the *covariance* of X and Y is given by:

$$Cov\,(X, Y) = E((X - \mu_x) \cdot (Y - \mu_y)) = E(XY) - E(X)E(Y) \tag{2.61}$$

Thus, the variance is a special case of covariance, for $X = Y$; it is also easy to see that the covariance of two mutually independent random variables is 0. In the continuous case the variance is written as:

$$Var(X) = \int (x - \mu)^2 f(x)\, dx$$

On the other hand, the population variance is usually denoted as σ^2 and is given by:

$$\sigma^2 = \sum_i (x_i - \mu)^2 \cdot p_i(x_i) = \sum_i x_i^2 \cdot p_i(x_i) - \mu^2$$

An important concept for the rest of this text is the *covariance matrix* which has the general form:

$$\underline{\underline{\Sigma}} = \begin{pmatrix} \sigma_1^2 & \rho_{1,2}\sigma_1\sigma_2 & \rho_{1,3}\sigma_1\sigma_3 & \cdots \\ \rho_{1,2}\sigma_1\sigma_2 & \sigma_2^2 & \rho_{2,3}\sigma_2\sigma_3 & \cdots \\ \vdots & \vdots & \ddots & \cdots \\ \vdots & \vdots & \vdots & \sigma_n^2 \end{pmatrix} \tag{2.62}$$

where the variances are in the diagonal, ρ_{ij} is the *coefficient of correlation* and $\rho_{ij}\sigma_i\sigma_j$ denotes the covariance between variables i and j; the coefficient of correlation lies between -1 and 1 and when it is zero indicates that the variables are independent. In descriptive statistics we compute the *sample dispersion* s^2 that is given by:

$$s^2 = \frac{\sum_i (x_i - \bar{X})^2}{n - 1} \tag{2.63}$$

and the reason for $(n - 1)$ in the denominator is due to the fact that we lost one *degree of freedom* when we calculated $\bar{X}$ from the x_i.

The *standard deviation* se (x) is the square root of the variance. This, in contrast with the variance, has the same dimensions as the random variable X and the measures of central tendency. Finally, the *coefficient of variation* CV, is the ratio of the standard deviation to the mean, and constitutes a useful dimensionless measure of spread.

2.5.4 More Advanced Statistical Concepts

In this section we will present a few more advanced statistical concepts that will be heavily used throughout the book. Most of them refer to the Normal distribution and extensions that can be made using it as a starting point. For this reason, we will first present this important distribution (also called Gauss distribution) in some detail.

2.5.4.1 The Normal or Gauss Distribution

This is undoubtedly the more useful function in statistics. Not only do many random processes have a Normal distribution, but also many other useful probability distributions can be approximated (e.g. the Binomial distribution) or are related to the Normal (i.e. the Student t distribution, the Fisher F distribution). The probability density function of a *standard* Normal random variable Z is defined as:

$$f(Z) = \frac{1}{\sqrt{2\pi}} \exp\left(-\frac{1}{2}Z^2\right)$$

where the constant $\frac{1}{\sqrt{2\pi}}$ appears so that condition (2.56) is fulfilled in this case.

It is easy to show that the mean μ_z of $f(Z)$ is equal to zero and that its variance σ_z^2 is equal to one; for this reason the standard Normal is also known as N(0, 1). In general a variable X distributes $N(\mu, \sigma^2)$ if its density function is given by:

$$f(Z) = \frac{1}{\sqrt{2\pi}\sigma} \exp\left(-\frac{1}{2}\left(\frac{x-\mu}{\sigma}\right)^2\right) \tag{2.64}$$

and from here it is easy to see that X can be 'standardised' by applying the transformation $Z = \frac{X-\mu}{\sigma}$. The advantage of standardising is that one can use tabulated values of the function (see Figure 2.14). Note that, conversely, if one wants to generate Normal values with mean b and variance s^2 these are given as $X = b + s \cdot Z$; this will come in handy when we need to generate draws of Normal (and other) distributions for various simulation procedures in Chapters 3, 8, 10 and 14.

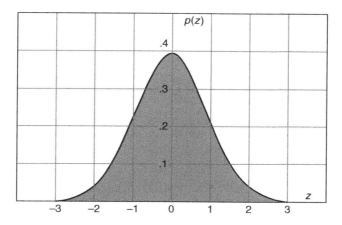

Figure 2.14   Standard Normal curve

Some useful properties of this well-known bell-shaped function are that:

$$\text{between} \begin{cases} \mu \pm \sigma \\ \mu \pm 2\sigma \\ \mu \pm 3\sigma \end{cases} \text{we find} \begin{cases} 68.20\% \text{ of the distribution} \\ 95.44\% \text{ of the distribution} \\ 99.73\% \text{ of the distribution.} \end{cases}$$

and we will see later that precisely 95% of the distribution lies in the range $\mu \pm 1.96\,\sigma$. Another interesting property is that if we have n variables $\mathbf{x}$ that distribute with any distribution with finite variance, according to the Central Limit Theorem (see Wonnacott and Wonnacott 1990) it can be shown that:

$$\frac{\overline{x} - \mu}{\sigma/\sqrt{n}} \sim N(0, 1) \qquad \text{if } n \geq 30. \tag{2.65}$$

Finally, it is important to mention that the Normal distribution is *closed* to algebraic summation, i.e. the sum and the difference (and indeed any linear combination) of Normal variables is also Normal distributed. This will come in very handy in several parts of the text.

Defining the Quadratic Form   The portion $\left(\frac{x-\mu}{\sigma}\right)^2$ in (2.64) is known as *quadratic form* (*QF*) and distributes Chi-squared with one degree of freedom (χ_1^2).

For the bivariate Normal distribution we have:

$$\mathbf{x} = (x_1, x_2)^\mathsf{T} \sim \mathrm{N}\left[\bar{\mathbf{x}} = \left(\bar{\mathbf{x}}_1 \ \bar{\mathbf{x}}_2\right)^\mathsf{T}, \Sigma_x\right]$$

where Σ_x is the covariance matrix. In this case, the quadratic form is given by:

$$QF_2 = (\mathbf{x} - \bar{\mathbf{x}})^\mathsf{T} \Sigma_x^{-1} (\mathbf{x} - \bar{\mathbf{x}})$$

and this distributes χ_2^2 (i.e. with two degrees of freedom). In this case the density function is given by:

$$f(x_1, x_2) = \frac{|\Sigma_x|^{1/2}}{2\pi^{2/2}} \exp\left(-\frac{1}{2}QF_2\right)$$

where $|\Sigma_x|$ is the determinant of the covariance matrix (note that if it was a p-variate Normal the denominator above would be $2\pi^{p/2}$). The covariance matrix in this case is:

$$\Sigma_x = \begin{pmatrix} \sigma_1^2 & \rho\sigma_1\sigma_2 \\ \rho\sigma_1\sigma_2 & \sigma_1^2 \end{pmatrix}$$

and the quadratic form can for the last time be written in extended (rather than matrix) form:

$$QF_2 = \frac{1}{1-\rho^2}\left[\left(\frac{x_1 - \bar{x}_1}{\sigma_1}\right)^2 - \frac{2\rho\,(x_1 - \bar{x}_1)\,(x_2 - \bar{x}_2)}{\sigma_1\sigma_2} + \left(\frac{x_2 - \bar{x}_2}{\sigma_2}\right)^2\right]$$

Finally note that the following theorem holds for quadratic forms in multivariate Normal distributions. If $\mathbf{X} = (x_1, \ldots, x_n)$ distributes multivariate Normal with mean $\bar{\mathbf{x}}$ and non-singular covariance matrix Σ (i.e. $|\Sigma| \neq 0$), then the random scalar variable QF_n, defined by $QF_n = (\mathbf{x} - \bar{\mathbf{x}})\,\Sigma^{-1}(\mathbf{x} - \bar{\mathbf{x}})^\mathsf{T}$ distributes χ^2 with n degrees of freedom.

Choleski Decomposition for the Multivariate Normal As described above, a univariate Normal variable with mean b and variance s^2 is obtained as $x = b + s \cdot z$ where z is a standard Normal. An analogous procedure can be used to take draws from a multivariate Normal distribution.

Let $\mathbf{x}$ be a vector with n elements distributed N($\mathbf{b}, \Sigma$). A Choleski transformation (or factorisation) of the matrix Σ is defined as a lower triangular matrix $\mathbf{L}$ such that $\mathbf{L} \cdot \mathbf{L}^\mathsf{T} = \Sigma$ (see Daganzo 1979). It can also be called generalised square root of Σ or generalised standard deviation of $\mathbf{x}$; in fact, when $n = 1$ the Choleski factor is precisely s (see Train 2009). Nowadays, most statistical packages have routines to calculate a Choleski factorisation for any positive definite symmetric matrix.

2.5.4.2 The Extreme Value Type I (Gumbel or Weibull) Distribution

This is another distribution that will feature heavily in this book, as it is the basis of the famous family of Logit models (as the Normal distribution is the basis for the Probit model), which have found use in all models associated with transport systems.

The cumulative distribution function of the EV1 distribution (as it is now more generally called) is given by the following expression:

$$F(\varepsilon) = \exp\left[-\exp\left(-\lambda\,(\varepsilon - \eta)\right)\right]$$

and deriving it we get the following density function:

$$f(\varepsilon) = \lambda \exp\left[-\lambda\,(\varepsilon - \eta)\right] \exp\left[-\exp\left(-\lambda\,(\varepsilon - \eta)\right)\right] \tag{2.66}$$

where η is the mode of the function and λ is a scale factor; these two parameters allow to represent the EV1 function completely, so it is generally said that $\varepsilon \sim \mathrm{EV1}(\eta, \lambda)$.

The mean of the distribution is at $\eta + \gamma/\lambda$ where γ is Euler's constant ($\gamma \approx 0,577$), and the variance is given by $\pi^2/6\lambda^2$. The shape of the distribution is shown in Figure 2.15 for conditions that allow the mean to be zero.

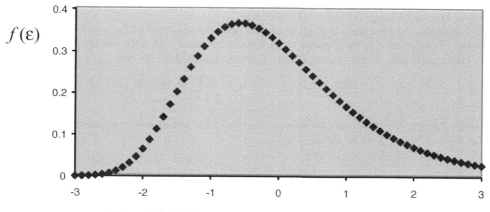

Figure 2.15   EV1 density function for $\lambda = 1$ and $\eta = -0,577$.

An important characteristic of this distribution is that it is closed to maximisation; this will come in very handy for deriving the most popular discrete choice models in Chapter 7. Also, the difference of two variables independent and identically distributed (i.e. with the same variance) EV1 follows the logistic distribution.

2.5.4.3   Some Notions about Statistical Inference

The goal of statistical inference is to deduce certain characteristics of a population from a sample taken from it. The population parameters are usually indicated by Greek letters (i.e. μ and σ) and are unknown; the sample parameters are denoted by Latin letters ($\bar{x}$ and s) and must estimated from the data. If we denote by $\hat{\theta}$ an estimator of the population characteristic θ, it is easy to see that this must be a function of the sampled data:

$$\hat{\theta} = f(x_1, x_2, \ldots, x_n)$$

For example, it is well-known that $\hat{\mu} = \bar{x}$ and $\hat{\sigma} = s$. There are several methods for estimating parameters, but probably the most popular is *maximum likelihood* (and we will use it extensively throughout the text). The main hypotheses of this method are:

- The sample is random, all draws x_i ($i = 1, \ldots, n$) taken from the population are independent of each other and the whole sample corresponds to the same population.
- The distribution function is known in the population, with the exception of the parameter θ.

If every value x_i is assumed to have a density function $f(x_i, \theta)$, as they are independent the joint density function for all **x** can be written as:

$$g(x_1, x_2, \ldots, x_n, \theta) = \prod_n f(x_i, \theta)$$

Note that the usual interpretation of this density function is with **x** as unknown variables and θ fixed. Inverting the process, the previous equation can be interpreted as a likelihood function $L(\theta)$; if we maximise it with respect to θ, the result $\hat{\theta}$ is called maximum likelihood estimate, because it corresponds to the parameter value which has the greatest probability of having generated the observed sample. Of course the idea may be extended to several parameters; for example, in *Linear Regression* models (see section 4.2.1) it can be shown that the least squares coefficients are in fact maximum likelihood estimates. We will come back to these in section 8.4.1 and others.

In calculating the maximum, it is easier to work with the logarithm of $L(\theta)$, which is called *log-likelihood* function $l(\theta)$; as the logarithm of a function increases with $\mathbf{x}$ the maximisation procedure yields the same results. In this latter case then, we would maximise the function:

$$l(\theta) = \ln g(x_1, x_2, \ldots, x_n, \theta) = \sum_n \ln f(x_n, \theta)$$

Example 2.5 We wish to estimate the mean μ of a $N(\mu, \sigma^2)$ distribution from a random sample of size n. If each $x_i \sim N(\mu, \sigma^2)$, we have that:

$$f(x_i, \mu) = \frac{1}{\sqrt{2\pi}\sigma} e^{-\frac{1}{2}\left(\frac{x-\mu}{\sigma}\right)^2}$$

then

$$l(\mu) = g(x_1, x_2, \ldots, x_n, \mu) = \left(\frac{1}{\sqrt{2\pi}\sigma}\right)^n \exp\left(-\frac{1}{2}\sum_{i=1}^{n}\left(\frac{x-\mu}{\sigma}\right)^2\right)$$

Taking logarithm, we get:

$$l(\mu) = \ln\left(\frac{1}{\sqrt{2\pi}\sigma}\right)^n - \frac{1}{2}\sum_{i=1}^{n}\left(\frac{x_i-\mu}{\sigma}\right)^2$$

deriving, and equalising to zero, we get:

$$\frac{\partial l}{\partial \mu} = 0 = \sum x_i - n\hat{\mu}$$

so finally: $\hat{\mu} = \frac{\sum x_i}{n} = \bar{x}$ as we had commented above.

Properties of a Good Estimator   It is usually accepted that a good estimator should first be *unbiased*. That means that $E(\hat{\theta}) = \theta$. If this is not possible, at least it should be *asymptotically unbiased*; this means: $\lim_{n\to\infty} E(\hat{\theta}) = \theta$, i.e. any possible bias should tend to zero as sample size increases. A second useful property is that the estimator should be *efficient*, that is, it should have minimum variance. Finally, a good estimator should be *consistent*. This happens when:

$$\lim_{n\to\infty} E(\hat{\theta} - \theta)^2 = 0$$

and this can be shown to mean that both any bias and the variance will tend to zero when the sample tends to infinity.

A Note on Hypothesis Testing   A test of hypothesis has the objective of verifying, using a sample, if a certain property of a process is actually taking place. For this the analyst expresses a *null hypothesis* (H_0) and an *alternative hypothesis* (H_1), where between them the universe of possible values of the parameter θ has to be covered. The test is a rule that allows accepting or rejecting H_0 on the basis of the sample values and the confidence intervals that can be established for each parameter.

Table 2.2 shows the possible results of this test:

Table 2.2 Type I and type II errors

	Accept H$_0$	Reject H$_0$
H$_0$ is true	Desirable result ($P = 1-\alpha$, "confidence level")	Type I error ($P = \alpha$, "significance level")
H$_0$ is false	Type II error ($P = \beta$)	Desirable result ($P = 1 - \beta$, "power")

It is ideal to have a low probability of incurring in both types of error. Unfortunately, looking at Figure 2.16 it can be seen that if α decreases β increases and vice versa. The figure incorporates an acceptance region C$_0$ defined as the confidence interval $\theta^{\mathrm{inf}} \leq \theta_0 \leq \theta^{\mathrm{sup}}$, where θ_0 is the value of the parameter consistent with the null hypothesis and θ_1 an inconsistent value (which has also associated a certain confidence interval) In fact it is only possible to diminish both types of errors by increasing the sample size n (as $\sigma_{\bar{x}} = \sigma/\sqrt{n}$, see 2.65) but this has associated higher costs.

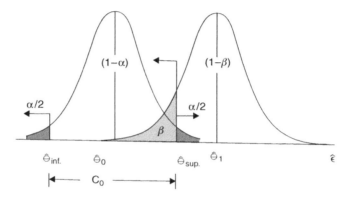

Figure 2.16 The relation between α and β

3

Data and Space

This chapter is devoted to issues in data collection and their representation for use in transport modelling. We present here a wide range of data collection methods but this is by no means complete. The nature of the data to be collected depends, of course, on the models chosen for a particular study. Moreover, advances in telecommunications are changing travel data collection with more general use of personal GPS units that offer specific advantages in tracking movement over longer periods of time. The treatment here is general. We will consider five subjects which are a prerequisite for other subjects treated in the rest of the book. Firstly, we will provide a brief introduction to statistical sampling theory, which will complement in part the elementary concepts discussed in section 2.5. Interested readers are advised that there is a complete book on the subject (Stopher and Meyburg 1979) which may be consulted for more details. In section 3.2 we will discuss the nature and importance of errors which can arise both during model estimation and when forecasting with the aid of models; the interesting question of data accuracy versus model complexity and cost is also addressed.

In section 3.3 we will consider various types of surveys used in applied transport planning; we will be particularly interested in problems such as the correction, expansion and validation of survey data, and we will also discuss issues involved in the collection of longitudinal (e.g. panel) data, and travel time data. Section 3.4 gives a fairly complete treatment of the most important issues involved in the experimental design and collection of stated preference data. Finally, section 3.5 considers the important practical problems of network representation and zoning design; this is where the 'spatial capabilities' of the model are actually decided. Poor network representations or too coarse zoning systems may invalidate the results of even the most theoretically appealing model.

3.1 Basic Sampling Theory

3.1.1 Statistical Considerations

Statistics may be defined as the science concerned with gathering, analysing and interpreting data in order to obtain the maximum quantity of useful information. It may also be described as one of the disciplines concerned with decision making under uncertainty; its goal would be in this case to help determine the level of uncertainty associated with measured data in order to support better decisions.

Data usually consist of a sample of observations taken from a certain population of interest which is not economically (or perhaps even technically) feasible to observe in its entirety. These observations are made about one or more attributes (say income) of each member of the population. Inferences can be

Modelling Transport, Fourth Edition. Juan de Dios Ortúzar and Luis G. Willumsen.
© 2011 John Wiley & Sons, Ltd. Published 2011 by John Wiley & Sons, Ltd.

made then about the mean value of these attributes, often called parameters of the population. Sample design aims at ensuring that the data to be examined provide the greatest amount of useful information about the population of interest at the lowest possible cost; the problem remains of how to use the data (i.e. expand the values in the sample) in order to make correct inferences about this population. Thus two difficulties exist:

- how to ensure a *representative* sample; and
- how to extract valid conclusions from a sample satisfying the above condition.

Neither of these would constitute a problem if there was no variability in the population. To solve the second difficulty, a well-established procedure exists which does not present major problems if certain conditions and assumptions hold. The identification of a representative sample, however, may be a more delicate task in certain cases, as we shall see below.

3.1.1.1 Basic Definitions

Sample   The sample is defined as a collection of units which has been especially selected to represent a larger population with certain attributes of interest (i.e. height, age, income). Three aspects of this definition have particular importance: first, which population the sample seeks to represent; second, how large the sample should be; and third, what is meant by 'especially selected'.

Population of Interest   This is the complete group about which information is sought; in many cases its definition stems directly from the study objectives. The population of interest is composed of individual elements; however, the sample is usually selected on the basis of sampling units which may not be equivalent to these individual elements as aggregation of the latter is often deemed necessary. For example, a frequently used sampling unit is the household while the elements of interest are individuals residing in it.

Sampling Method   Most of the acceptable methods are based on a form of random sampling. The key issue in these cases is that the selection of each unit is carried out independently, with each unit having the same probability of being included in the sample. The more interesting methods are:

- *Simple random sampling*, which is not only the simplest method but constitutes the basis of all the rest. It consists in first associating an identifier (number) to each unit in the population and then selecting these numbers at random to obtain the sample; the problem is that far too large samples may be required to ensure sufficient data about minority options of particular interest. For example, it may well be that sampling households at random in a developing country would provide little information on multiple car ownership.
- *Stratified random sampling*, where *a priori* information is first used to subdivide the population into homogeneous strata (with respect to the stratifying variable) and then simple random sampling is conducted inside each stratum using the same sampling rate. The method allows the correct proportions of each stratum in the sample to be obtained; thus it may be important in those cases where there are relatively small subgroups in the population as they could lack representation in a simple random sample.

It is also possible to stratify with respect to more than one variable, thus creating an n-dimensional matrix of group cells. However, care must be taken with the number of cells created as it increases geometrically with the number of strata; large figures may imply that the average number of sampling units per cell is too small. Nevertheless, even stratified sampling does not help when data are needed about options with a low probability of choice in the population; in these cases a third method called *choice-based sampling*, actually a subset of the previous one, is required. The method consists in stratifying the population based on the result of the choice process under consideration. This method is fairly common in transport studies, as we will see in section 3.3. A major advantage is that data may be produced at a much lower cost than with the other sampling methods; its main drawback is that the sample thus formed may not be random and therefore the risk of bias in the expanded values is greater.

Sampling Error and Sampling Bias   These are the two types of error that might occur when taking a sample; combined, they contribute to the measurement error of the data. The first is simply due to the fact that we are dealing with a sample and not with the total population, i.e. it will always be present due to random effects. This type of error does not affect the expected values of the means of the estimated parameters; it only affects the variability around them, thus determining the degree of confidence that may be associated with the means; it is basically a function of sample size and of the inherent variability of the parameter under investigation.

The sampling bias, on the other hand, is caused by mistakes made either when defining the population of interest, or when selecting the sampling method, the data collection technique or any other part of the process. It differs from the sampling error in two important respects:

- it can affect not only the variability around the mean of the estimated parameters but the values themselves; therefore it implies a more severe distortion of the survey results;
- while the sampling error may not be avoided (it can only be reduced by increasing sample size), the sampling bias may be virtually eliminated by taking extra care during the various stages of sampling design and data collection.

Sample Size   Unfortunately, there are no straightforward and objective answers to the calculation of sample size in every situation. This happens, in spite of the fact that sample size calculations are based on precise statistical formulae, because many of their inputs are relatively subjective and uncertain; therefore they must be produced by the analyst after careful consideration of the problem in hand.

Determining sample size is a problem of trade-offs, as:

- too large a sample may imply a data-collection and analysis process which is too expensive given the study objective and its required degree of accuracy; but
- too small a sample may imply results which are subject to an unacceptably high degree of variability reducing the value of the whole exercise.

Somewhere between these two extremes lies the most efficient (in cost terms) sample size given the study objective. In what follows it will be assumed that this consists in estimating certain population parameters by means of statistics calculated from sample data; as any sample statistics are subject to sampling error, it is also necessary to include an estimate of the accuracy that may be associated with its value.

3.1.1.2   Sample Size to Estimate Population Parameters

This depends on three main factors: variability of the parameters in the population under study, degree of accuracy required for each, and population size. Without doubt the first two are the most important; this may appear surprising at first sight because, to many, it seems intuitively necessary to take bigger samples in bigger populations in order to maintain the accuracy of the estimates. However, as will be shown below, the size of the population does not significantly affect sample size except in the case of very small populations.

The Central Limit Theorem, which is at the heart of the sample size estimation problem, postulates that the estimates of the mean from a sample tend to become distributed Normal as the sample size (n) increases. This holds for any population distribution if n is greater than or equal to 30; the theorem holds even in the case of smaller samples, if the original population has a Normal-like distribution.

Consider a population of size N and a specific property which is distributed with mean μ and variance σ^2. The Central Limit Theorem states that the distribution of the mean ($\bar{x}$) from successive samples is distributed Normal with mean μ and standard deviation se ($\bar{x}$), known as the standard error of the mean, and given by:

$$\text{se}(\bar{x}) = \sqrt{(N-n)\sigma^2/[n(N-1)]} \tag{3.1}$$

If only one sample is considered, the best estimate of μ is $\bar{x}$ and the best estimate of σ^2 is s^2 (the sample variance); in this case the standard error of the mean can be estimated as:

$$\text{se}(\bar{x}) = \sqrt{(N-n)s^2/nN} \tag{3.2}$$

and, as mentioned above, it is a function of three factors: the parameter variability (s^2), the sample size (n) and the size of the population (N). However, for large populations and small sample sizes (the most frequent case) the factor $(N-n)/N$ is very close to 1 and equation (3.2) reduces to:

$$\text{se}(\bar{x}) = \frac{s}{\sqrt{n}} \tag{3.3}$$

Thus, for example, quadrupling sample size will only halve the standard error, i.e. it is a typical case of diminishing returns of scale. The required sample size may be estimated solving equation (3.2) for n and this is usually simpler to do in two stages, first calculating n from equation (3.3) such that:

$$n' = \frac{s^2}{\text{se}(\bar{x})^2} \tag{3.4}$$

and then correcting for finite population size, if necessary, by:

$$n = \frac{n'}{1 + \dfrac{n'}{N}} \tag{3.5}$$

Although the above procedure appears to be both objective and relatively trivial it has two important problems that impair its application: estimating the sample variance s^2 and choosing an acceptable standard error for the mean. The first one is obvious: s^2 can only be calculated once the sample has been taken, so it has to be estimated from other sources. The second one is related with the desired degree of confidence to be associated with the use of the sample mean as an estimate of the population mean; normal practice does not specify a single standard error value, but an interval around the mean for a given confidence level. Thus, two judgements are needed to calculate an acceptable standard error:

- First, a confidence level for the interval must be chosen; this expresses how frequently the analyst is prepared to make a mistake by accepting the sample mean as a measure of the true mean (e.g. the typical 95% level implies an acceptance to err in 5% of cases).
- Second, it is necessary to specify the limits of the confidence interval around the mean, either in absolute or relative terms; as the interval is expressed as a proportion of the mean in the latter case, an estimate of this is required to calculate the absolute values of the interval. A useful option considers expressing sample size as a function of the expected coefficient of variation ($CV = \sigma/\mu$) of the data.

For example, if a Normal distribution is assumed and a 95% confidence level is specified, this means that a maximum value of 1.96 se ($\bar{x}$) would be accepted for the confidence interval (i.e. $\mu \pm 1.96\sigma$

contains 95% of the Normal probability distribution); if a 10% error is specified we would get the interval ($\mu \pm 0.1\mu$) and it may be seen that:

$$se(\bar{x}) = 0.1\mu/1.96 = 0.051\mu$$

and replacing this value in (3.4) we get:

$$n' = (s/0.051\mu)^2 = 384CV^2 \tag{3.6}$$

Note that if the interval is specified as ($\mu \pm 0.05\mu$), i.e. with half the error, n' would increase fourfold to 1536 CV^2.

To complete this point it is important to emphasise that the above exercise is relatively subjective; thus, more important parameters may be assigned smaller confidence intervals and/or higher levels of confidence. However, each of these actions will result in smaller acceptable standard errors and, consequently, bigger samples and costs. If multiple parameters need to be estimated the sample may be chosen based on that requiring a larger sample size.

3.1.1.3   Obtaining the Sample

The last stage of the sampling process is the extraction of the sample itself. In some cases the procedure may be easily automated, either on site or at the desk (in which case care must be taken that it is actually followed on the field), but it must always be conducted with reference to a random process. Although the only truly random processes are those of a physical nature (i.e. roll of a dice or flip of a coin), they are generally too time consuming to be useful in sample selection. For this reason pseudo-random processes, capable of generating easily and quickly a set of suitable random-like numbers, are usually employed in sampling.

Example 3.1 Consider a certain area the population of which may be classified in groups according to: automobile ownership (with and without a car); and household size (up to four and more than four residents).

Let us assume that m observations are required by cell in order to guarantee a 95% confidence level in the estimation of, say, trip rates; assume also that the population can be considered to have approximately the following distribution (i.e. from historic data).

Car ownership	Household size	% of population
With car	Four or less	9
	More than four	16
Without car	Four or less	25
	More than four	50

There are two possible ways to proceed:

1. Achieve a sample with m observations by cell by means of a random sample. In this case it is necessary to select a sample size which guarantees this for each cell, including that with the smallest proportion of the population. Therefore, the sample size would be:

$$n = 100m/9 = 11.1m$$

2. Alternatively, one can undertake first a preliminary random survey of $11.1m$ households where only cell membership is asked for; this low-cost survey can be used to obtain the addresses of m households even in the smallest group. Subsequently, as only m observations are needed by cell, it would suffice to randomly select a (stratified) sample of $3m$ households from the other groups to be interviewed in detail (together with the m already detected for the most restrictive cell).

As can be seen, a much higher sample is obtained in the first case; its cost (approximately three times more interviews) must be weighed against the cost of the preliminary survey.

3.1.2   Conceptualisation of the Sampling Problem

In this part we will assume that the final objective of taking the sample is to calibrate a choice model for the whole population. Following Lerman and Manski (1976) we will denote by P and f population and sample characteristics respectively. We will also assume that each sampled observation may be described on the basis of the following two variables:

i = observed choice of the sample individual (e.g. took a bus);

$\mathbf{X}$ = vector of characteristics (attributes) of the individual (age, sex, income, car ownership) and of the alternatives in his choice set (walking, waiting and travel times, cost)

We will finally assume that the underlying choice process in the population may be represented by a model with parameters θ; in this case, the joint distribution of i and $\mathbf{X}$ is given by:

$$P(i, \ \mathbf{X}/\theta)$$

and the probability of choosing alternative i among a set of options with attributes $\mathbf{X}$ is:

$$P(i/\mathbf{X}, \ \theta)$$

Depending on the form in which each observation is extracted, the sample will have its own joint distribution of i's and $\mathbf{X}$'s which we will denote by $f(i, \ \mathbf{X}/\theta)$. On the basis of this notation the sampling problem may be formalised as follows (Lerman and Manski 1979).

3.1.2.1   Random Sample

In this case the distribution of i and $\mathbf{X}$ in the sample and population should be identical, that is:

$$f(i, \ \mathbf{X}/\theta) = P(i, \ \mathbf{X}/\theta) \tag{3.7}$$

3.1.2.2   Stratified or Exogenous Sample

In this case the sample is not random with respect to certain independent variables of the choice model (e.g. a sample with 50% low-income households and 50% high-income households is stratified if and only if a random sample is taken inside each stratum). The sampling process is defined by a function $f(\mathbf{X})$, giving the probability of finding an observation with characteristics $\mathbf{X}$; in the population this probability is of course $P(\mathbf{X})$. The distribution of i and $\mathbf{X}$ in the sample is thus given by:

$$f(i, \ \mathbf{X}/\theta) = f(\mathbf{X})P(i/\mathbf{X}, \ \theta) \tag{3.8}$$

It is simple to show that a random sample is just a special case of stratified sample where $f(\mathbf{X}) = P(\mathbf{X})$, because:

$$f(i, \mathbf{X}/\theta) = P(\mathbf{X})P(i/\mathbf{X}, \theta) = P(i, \mathbf{X}/\theta) \tag{3.9}$$

3.1.2.3 Choice-based Sample

In this case the sampling procedure is defined by a function $f(i)$, giving the probability of finding an observation that chooses option i (i.e. it is stratified according to the choice). Now the distribution of i and $\mathbf{X}$ in the sample is given by:

$$f(i, \mathbf{X}/\theta) = f(i)P(\mathbf{X}/i, \theta) \tag{3.10}$$

We had not defined the rightmost probability in (3.10), but we may obviate it on the basis of a Bayes theorem stating:

$$P(\mathbf{X}/i, \theta) = P(i/\mathbf{X}, \theta)P(\mathbf{X})/P(i/\theta) \tag{3.11}$$

The expression in the denominator, which had not been defined either, may be obtained assuming discrete $\mathbf{X}$ from:

$$P(i/\theta) = \sum_X P(i/\mathbf{X}, \theta)P(\mathbf{X}) \tag{3.12}$$

Therefore the final expression for the joint probability of i and $\mathbf{X}$ for a choice-based sample is clearly more complex:

$$f(i, \mathbf{X}/\theta) = f(i)P(i/\mathbf{X}, \theta)P(\mathbf{X})/\sum_X P(i/\mathbf{X}, \theta)P(\mathbf{X}) \tag{3.13}$$

and it serves to illustrate not only that choice-based sampling is intuitively more problematic than the other two approaches, but also that it has higher bias potential in what really concerns us: choice.

Thus, each sampling method yields a different distribution of choices and characteristics in the sample, and there are no *a priori* reasons to expect that a single parameter estimation method would be applicable in all cases.

Example 3.2 Assume that for the purposes of a transport study the population of a certain area has been classified according to two income categories, and that there are only two modes of transport available (car and bus) for the journey to work. Let us also assume that the population distribution is given by:

	Low income	High income	Total
Bus user	0.45	0.15	0.60
Car user	0.20	0.20	0.40
Total	0.65	0.35	1.00

1. Random sample. If a random sample is taken, it is clear that the same population distribution would be obtained.

2. Exogenous sample. Consider a sample with 75% low income (LI) and 25% high income (HI) travellers. From the previous table it is possible to calculate the probability of a low-income traveller using bus, as:

$$P(\text{Bus/LI}) = \frac{P(\text{LI and Bus})}{P(\text{LI and Bus}) + P(\text{LI and Car})} = \frac{0.45}{0.45 + 0.20} = 0.692$$

Now, given the fact that the exogenous sample has 75% of individuals with low income, the probability of finding a bus user with low income in the sample is: $0.75 \times 0.692 = 0.519$. Doing this for the rest of the cells, the following table of probabilities for the stratified sample may be built:

	Low income	High income	Total
Bus user	0.519	0.107	0.626
Car user	0.231	0.143	0.374
Total	0.750	0.250	1.000

3. Choice-based sample. Let us assume now that we take a sample of 75% bus users and 25% car users. In this case the probability of a bus user having low income may be calculated as:

$$P(\text{LI/Bus}) = \frac{P(\text{LI and Bus})}{P(\text{LI and Bus}) + P(\text{HI and Bus})} = \frac{0.45}{0.45 + 0.15} = 0.75$$

Therefore, the probability of finding a low-income traveller choosing bus in the sample is 0.75 times 0.75, or 0.563. Proceeding analogously, the following table of probabilities for the choice-based sample may be built:

	Low income	High income	Total
Bus user	0.563	0.187	0.750
Car user	0.125	0.125	0.250
Total	0.688	0.312	1.000

As was obviously expected, each sampling method produces in general a different distribution in the sample. The importance of the above example will increase when we consider what is involved in the estimation of models using the various samples. For this it is necessary to acquire an intuitive understanding of what calibration programs do; they simply search for the 'best' values of the model coefficients associated with a set of explanatory variables; in this case best consists in replicating the observed choices more accurately.

For the population as a whole the probability of actually observing a given data set may be found, conceptually, simply by calculating the probabilities of choosing the observed option by different types of traveller (with given attributes and choice sets). For example, in the first table in Example 3.2 (simple random sample) the probability that a high-income traveller selects car is given by the ratio between the probability of him having high income and using car, and the probability of him having high income, that is:

$$\frac{0.20}{0.15 + 0.20} = 0.572$$

If we consider the second table (exogenous sample), the same probability is now given by:

$$\frac{0.143}{0.107 + 0.143} = 0.572$$

This is no coincidence; in fact it was one of the most important findings of an interesting piece of research by Lerman *et al.* (1976) in the USA. In practice it means that standard software may be used to estimate models with data obtained from an exogenous sample.

It is also important to note that this is not the case for choice-based samples. To prove this, consider calculating the same probability but using information from the third table:

$$\frac{0.125}{0.187 + 0.125} = 0.400$$

As can be seen, the result is completely different. To end this theme it is interesting to mention that Lerman *et al.* (1976) did also propose a method to use data from choice-based samples in model estimation avoiding bias at the expense only of requiring knowledge of the actual market shares. This involves weighting the observations by factors calculated as:

$$\frac{\text{Prob (select the option in a random sample)}}{\text{Prob (select the option in a choice based sample)}}$$

Thus, in our example the weighting factor for bus-based observation should be:

$$\frac{0.45 + 0.15}{0.563 + 0.187} = 0.8$$

and for car users:

$$\frac{0.20 + 0.20}{0.125 + 0.125} = 1.6$$

Note that it is necessary to have data about choices on each alternative, i.e. it would not be possible to calibrate a model for car and bus, based on data for the latter mode only. We will come back to this problem in section 8.4.2.

3.1.3   Practical Considerations in Sampling

3.1.3.1   The Implementation Problem

Stratified (and choice-based) sampling requires random sampling inside each stratum; to do so it is first necessary to isolate the relevant group and this may be difficult in some cases. Consider for example a case where the population of interest consists of all potential travellers in a city. Thus if we stratify by area of residence, it may be relatively simple to isolate the subpopulation of residents inside the city (e.g. using data from a previous survey); the problem is that it is extremely difficult to isolate and take a sample of the rest, i.e. those living outside the city.

An additional problem is that in certain cases even if it is possible to isolate all subpopulations and conforming strata, it may still be difficult to ensure a random sample inside each stratum. For example, if we are interested in taking a mode choice-based sample of travellers in a city we will need to interview bus users and for this it is first necessary to decide which routes will be included in the sample. The problem is that certain routes might have, say, higher than average proportions of students and/or old age pensioners, and this would introduce bias (Lerman and Manski 1979).

3.1.3.2 Finding the Size of Each Subpopulation

This is a key element in determining how many people will be surveyed. Given certain stratification, there are several methods available to find out the size of each subpopulation, such as:

1. Direct measurement. This is possible in certain cases. Consider a mode choice-based sample of journey-to-work trips; the number of bus and metro tickets sold, plus traffic counts during the peak hour in an urban corridor, may yield an adequate measure (although imperfect as not all trips during the peak are to work) of the number of people choosing each mode. If we have a geographical (i.e. zonal) stratification, on the other hand, the last census may be used to estimate the number of inhabitants in each zone.
2. Estimation from a random sample. If a random sample is taken, the proportion of observations corresponding to each stratum is a consistent estimator of the fraction of the total corresponding to each subpopulation. It is important to note that the cost of this method is low as the only information sought is that necessary to establish the stratum to which the respondent belongs.
3. Solution of a system of simultaneous equations. Assume we are interested in stratifying by chosen mode and that we have data about certain population characteristics (e.g. mean income and car ownership). Taking a small on-mode sample we can obtain modal average values of these variables and postulate a system of equations which has the subpopulation fractions as unknowns.

Finally, the 'failure rate' of different types of surveys must be considered when designing sampling frameworks. The sample size discussed above corresponds to the number of successful and valid responses to the data-collection effort. Some survey procedures are known to generate low valid response rates (e.g. some postal surveys), but they may still be used because of their low cost (Richardson *et al.* 1995).

Example 3.3 Assume the following information is available:

Average income of population (I): 33 600 $/year

Average car ownership (CO): 0.44 cars/household

Assume also that small on-mode surveys yield the following:

Mode	I ($ / year)	CO (cars/household)
Car	78 000	1.15
Bus	14 400	0.05
Metro	38 400	0.85

If F_i denotes the subpopulation fraction of the total, the following system of simultaneous equations holds:

$$33\,600 = 78\,000 F_1 + 14\,400 F_2 + 38\,400 F_3$$
$$0.44 = 1.15 F_1 + 0.05 F_2 + 0.85 F_3$$
$$1 = F_1 + F_2 + F_3$$

the solution of which is:

$$F_1 = 0.2451$$
$$F_2 = 0.6044$$
$$F_3 = 0.1505$$

This means that if the total population of the area was 180 000 inhabitants, there would be approximately 44 100 car users 108 800 bus users and 27 100 metro users.

3.2   Errors in Modelling and Forecasting

The statistical procedures normally used in (travel demand) modelling assume not only that the correct functional specification of the model is known *a priori*, but also that the data used to estimate the model parameters have no errors. In practice, however, these conditions are often violated; furthermore, even if they were satisfied, model forecasts are usually subject to errors due to inaccuracies in the values assumed for the explanatory variables in the design year.

The ultimate goal of modelling is often forecasting (i.e. the number of people choosing given options); an important problem model designers face is to find which combination of model complexity and data accuracy fits best the required forecasting precision and study budget. To this end, it is important to distinguish between different types of errors, in particular:

- those that could cause even correct models to yield incorrect forecasts, e.g. errors in the prediction of the explanatory variables, transference and aggregation errors; and
- those that actually cause incorrect models to be estimated, e.g. measurement, sampling and specification errors.

In the next section consideration is given first to the types of errors that may arise with the broad effects they may cause; then the trade-off between model complexity and data accuracy is examined with particular emphasis on the role of simplified models in certain contexts.

3.2.1   Different Types of Error

Consider the following list of errors that may arise during the processes of building, calibrating and forecasting with models.

3.2.1.1   Measurement Errors

These occur due to the inaccuracies inherent in the process of actually measuring the data in the base year, such as: questions badly registered by the interviewee, answers badly interpreted by the interviewer, network measurement errors, coding and digitising errors, and so on. These errors tend to be higher in less developed countries but they can always be reduced by improving the data-collection effort (e.g. by appropriate use of computerised interview support) or simply by allocating more resources to data quality control; however, both of these cost money.

Measurement error, as defined here, should be distinguished from the difficulty of defining the variables that ought to be measured. The complexity that may arise in this area is indicated in Figure 3.1. Regretfully, modeller and traveller use different 'units' to express variables like time and distance. The modeller works in seconds and metres where travellers perceive something they find it difficult to convert in precise minutes and kilometres. Modellers just hope that measurements in our own units reflect, with some unknown degree of error, the travellers' perceptions that influence their choices. Ideally, modelling should be based on the information perceived by individual travellers but whilst reported data may give some insight into perception, its use raises the difficult question of how to forecast what users are going to perceive in the future. So it appears inevitable that models will be endowed with perception errors which tend to be greater for non-chosen alternatives due to the existence of *self-selectivity* bias (i.e. the attributes of the chosen option are perceived as better and those of the rejected option as worse than they are, such as to reinforce the rationality of the choice made).

Most models are used in order to forecast future conditions. This poses an interesting problem in the choice of variables to be included. Let us assume that we can fit a model very well to a set of variables with current data but that it may be difficult to forecast the future value of at least some of these. If

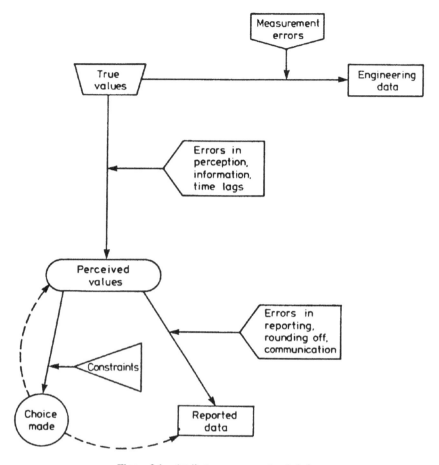

Figure 3.1   Attribute measurement and choice

this is the case, although we can get accurate values for these independent variables during the survey, their future values are only known with great uncertainty (a wide confidence interval). This problem can take a simple form if the difficult variable is, for example, future fuel prices that are difficult to forecast with any accuracy. But it may take a more complex form if the problem is to estimate the number of individuals with specific characteristics of age, gender, income, employment type, marriage status and number of children that will reside in a particular zone or location ten years from now.

3.2.1.2   Sampling Errors

These arise because the models must be estimated using finite data sets. Sampling errors are approximately inversely proportional to the square root of the number of observations (i.e. to halve them it is necessary to quadruple sample size); thus, reducing them may be costly. Daganzo (1980) has examined the problem of defining optimal sampling strategies in the sense of refining estimation accuracy.

3.2.1.3   Computational Errors

These arise because models are generally based on iterative procedures for which the exact solution, if it exists, has not been found for reasons of computational costs. These errors are typically small in comparison with other errors, except for cases such as assignment to congested networks and problems of equilibration between supply and demand in complete model systems, where they can be very large (see De Cea *et al.* 2005).

3.2.1.4   Specification Errors

These arise either because the phenomenon being modelled is not well understood or because it needs to be simplified for whatever reason. Important subclasses of this type of error are the following:

- Inclusion of an irrelevant variable (i.e. one which does not affect the modelled choice process). This error will not bias the model (or its forecasts) if the parameters appear in linear form, but it will tend to increase sampling error; in a non-linear model, however, bias may be caused (see Tardiff 1979).
- Omission of a relevant variable; perhaps the most common specification error. Interestingly, models incorporating a random error term (such as many of those we will examine in Chapters 4 and 7) are designed to accommodate this error; however, problems can arise when the excluded variable is correlated with variables in the model or when its distribution in the relevant population is different from its distribution in the sample used for model estimation (see Horowitz 1981).
- Not allowing for *taste variations* on the part of the individuals will generally produce biased models, as shown in Chapter 8. Unfortunately this is the case in many practical models of choice; exceptions are the less yielding Multinomial Probit and Mixed Logit models, which we discuss in Chapters 7 and 8.
- Other specification errors, in particular the use of model forms which are not appropriate, such as linear functions to represent non-linear effects, *compensatory* models to represent behaviour that might be *non-compensatory* (see the discussion in Chapter 8), or the omission of effects such as *habit* or *inertia* (see Cantillo *et al.* 2007). A full discussion of these forms of error is given by Williams and Ortúzar (1982a).

All specification errors can be reduced in principle simply by increasing model complexity; however, the total costs of doing this are not easy to estimate as they relate to model operation, but may induce other types of errors which might be costly or impossible to eliminate (e.g. when forecasting more variables and at a higher level of disaggregation). Moreover, removal of some specification errors may require-extensive behavioural research and it must simply be conceded that such errors may be present in all feasible models.

3.2.1.5   Transfer Errors

These occur when a model developed in one context (time and/or place) is applied in a different one. Although adjustments may be made to compensate for the transfer, ultimately the fact must be faced that behaviour might just be different in different contexts. In the case of spatial transfers, the errors can be reduced or eliminated by partial or complete re-estimation of the model to the new context (although the latter would imply discarding the substantial cost savings obtainable from transfer). However, in the case of temporal transfer (i.e. forecasting), this re-estimation is not possible and any potential errors must just be accepted (see the discussion in Chapter 9). This type of error will be greater for long-range planning applications as time will reduce the validity of the model as attitudes and preferences change over time, and perhaps more importantly, the accuracy of the planning variables used as input.

3.2.1.6 Aggregation Errors

These arise basically out of the need to make forecasts for groups of people while modelling often needs to be done at the level of the individual in order to capture behaviour better. The following are important subclasses of aggregation error:

- Data aggregation. In most practical studies the data used to define the choice situation of individual travellers is aggregated in some form or another; even when travellers are asked to report the characteristics of their available options, they can only have based their choice on the expected values of these characteristics. When network models are used there is aggregation over routes, departure times and even zones; this means that the values thus obtained for the explanatory variables are, at best, averages for groups of travellers rather than exact values for any particular individual. Models estimated with aggregate data will suffer from some form of specification error (see Daly and Ortúzar 1990). Reducing this type of aggregation error implies making measurements under many more sets of circumstances: more zones, more departure times, more routes, more socio-economic categories; this costs time and money and increases model complexity.
- Aggregation of alternatives. Again due to practical considerations it may just not be feasible to attempt to consider the whole range of options available to each traveller; even in relatively simpler cases such as the choice of mode, aggregation is present as the large variety of services encompassing a bus option, say (e.g. one-man operated single-decker, two-man operated double-decker, mini-buses, express services), are seldom treated as separate choices.
- Model aggregation. This can cause severe difficulties to the analyst except in the case of linear models where it is a trivial problem. Aggregate quantities such as flows on links are a basic modelling result in transportation planning, but methods to obtain them are subject to aggregation errors which are often impossible to eliminate. We will examine this problem in some detail in Chapters 4 and 9.

3.2.2   The Model Complexity/Data Accuracy Trade-off

Given the difficulties discussed above, it is reasonable to consider the dual problem of how to optimise the return of investing in increasing data accuracy, given a fixed study budget and a certain level of model complexity, to achieve a reasonable level of precision in forecasts. In order to tackle this problem we must understand first how errors in the input variables influence the accuracy of the model we use.

Consider a set of observed variables $\mathbf{x}$ with associated errors $\mathbf{e}_x$ (i.e. standard deviation); to find the output error derived from the propagation of input errors in a function such as:

$$z = f(x_1, x_2, \ldots, x_n)$$

the following formula may be used (Alonso 1968):

$$e_z^2 = \sum_i \left(\frac{\partial f}{\partial x_i} \right)^2 e_{x_i}^2 + \sum_i \sum_{j \neq i} \frac{\partial f}{\partial x_i} \frac{\partial f}{\partial x_j} e_{x_i} e_{x_j} r_{ij} \tag{3.14}$$

where r_{ij} is the coefficient of correlation between x_i and x_j; the formula is exact for linear functions and a reasonable approximation in other cases. Alonso (1968) used it to derive some simple rules to be followed during model building in order to prevent large output errors; for example, an obvious one is to avoid using correlated variables, thus reducing the second term of the right-hand side of equation (3.14) to zero.

If we take the partial derivative of e_z with respect to e_{x_i} and ignore the correlation term, we get:

$$\frac{\partial e_z}{\partial e_{x_i}} = \left(\frac{\partial f}{\partial x_i} \right)^2 \frac{e_{x_i}}{e_z} \qquad (3.15)$$

Using these marginal improvement rates and an estimation of the marginal costs of enhancing data accuracy it should be possible, in principle, to determine an optimum improvement budget; in practice this problem is not easy though, not least because the law of diminishing returns (i.e. each further percentage reduction in the error of a variable will tend to cost proportionately more) might operate, leading to a complex iterative procedure. However, equation (3.15) serves to deduce two logical rules (Alonso 1968):

- concentrate the improvement effort on those variables with a large error; and
- concentrate the effort on the most relevant variables, i.e. those with the largest value of $(\partial f / \partial x_i)$ as they have the largest effect on the dependent variable.

Example 3.4 Consider the model $z = xy + w$, and the following measurement of the independent variables:

$$x = 100 \pm 10; \quad y = 50 \pm 5; \quad w = 200 \pm 50$$

Assume also that the marginal cost of improving each measurement is the following:

$$\text{Marginal cost of improving } x \ (\text{to } 100 \pm 9) = \$5.00$$
$$\text{Marginal cost of improving } y \ (\text{to } 50 \pm 4) = \$6.00$$
$$\text{Marginal cost of improving } w \ (\text{to } 200 \pm 49) = \$0.02$$

Applying equation (3.14) we get:

$$e_z^2 = y^2 e_x^2 + x^2 e_y^2 + e_w^2 = 502\,500$$

then $e_z = 708.87$; proceeding analogously, values of improved e_z in the cases of improving x, y or w may be found to be 674.54, 642.26 and 708.08 respectively. Also from (3.15) we get:

$$\frac{\partial e_z}{\partial e_x} = \frac{10y^2}{708.87} = 35.2; \quad \frac{\partial e_z}{\partial e_y} = 70.5; \quad \frac{\partial e_z}{\partial e_w} = 0.0705$$

These last three values are the marginal improvement rates corresponding to each variable. To work out the cost of the marginal improvements in e_z we must divide the marginal costs of improving each variable by their respective marginal rates of improvement. Therefore we get the following marginal costs of improving e_z arising from the various variable improvements:

$$\text{Marginal improvement in } x = 5/35.2 = \$0.142$$
$$\text{Marginal improvement in } y = 6/70.05 = \$0.085$$
$$\text{Marginal improvement in } w = 0.02/0.705 = \$0.284$$

Therefore it would be decided to improve the measurement accuracy of variable y if the marginal reduction in e_z was worth at least $0.085.

Let us now define complexity as an increase in the number of variables of a model and/or an increase in the number of algebraic operations with the variables (Alonso 1968). It is obvious that in order to reduce specification error (e_s) complexity must be increased; however, it is also clear that as there are more variables to be measured and/or greater problems for their measurement, data measurement error (e_m) will probably increase as well.

If total modelling error is defined as $E = \sqrt{(e_s{}^2 + e_m{}^2)}$, it is easy to see that the minimum of E does not necessarily lie at the point of maximum complexity (i.e. maximum realism). Figure 3.2 shows not only that this is intuitively true, but also that as measurement error increases, the optimum value can only be attained at decreasing levels of model complexity.

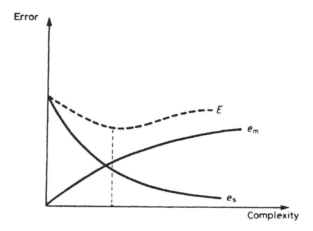

Figure 3.2   Variation of error with complexity

Example 3.5 Consider the case of having to make a choice between an extremely simple model, which is known to produce a total error of 30% in forecasts, and a new model which has a perfect specification (i.e. $e_s = 0$) given by:

$$z = x_1 x_2 x_3 x_4 x_5$$

where the x_i are independent variables measured with a 10% error (i.e. $e_m = 0.1\ x_i$). To decide which model is more convenient we will apply equation (3.14):

$$e_z^2 = 0.01[x_1^2(x_2 x_3 x_4 x_5)^2 + x_2^2(x_1 x_3 x_4 x_5)^2 + \ldots + x_5^2(x_1 x_2 x_3 x_4)^2]$$
$$e_z^2 = 0.05[x_1 x_2 x_3 x_4 x_5]^2 = 0.05z^2$$

that is, $e_z = 0.22z$ or a 22% error, in which case we would select the second model.

The interested reader can check to see that if it is assumed that the x_i variables can only be measured with 20% error, the total error of the second model comes out as 44.5% (i.e. we would now select the first model even if its total error increased up to 44%).

Figure 3.3 serves to illustrate this point, which may be summarized as follows: if the data are not of a very good quality it might be safer to predict with simpler and more robust models (Alonso 1968). However, to learn about and understand the phenomenon, a better-specified model will always be preferable. Moreover, most models will be used in a forecasting mode where the values of the planning

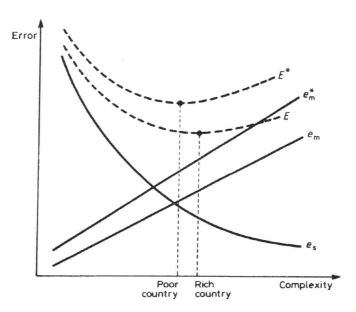

Figure 3.3 Influence of the measurement error

variables x_i will not be observed but forecast. We know that some planning variables are easier to forecast than others and that disaggregation makes predicting their future values an even less certain task. Therefore, in choosing a model for forecasting purposes preference should be given to those using planning variables which can be forecast, in turn, with greater confidence.

Consider, for example, that x_1 is fuel price. An accuracy of 10 % can be expected in gasoline costs over the year and areas where data was collected. However, the accuracy of the estimate of fuel prices in ten years from now will decrease, probably to something like 40 %. If the errors in the other variables remain as stated, it will be enough for the error in X_1 to increase to 25 % to make the total error 32 % and the simpler model preferable.

3.3   Basic Data-Collection Methods

3.3.1   Practical Considerations

The selection of the most appropriate data collection methods will depend significantly on the type of models that will be used in the study; they will define what type of data is needed and therefore what data collection methods are more appropriate. However, practical limitations will also have a strong influence in determining the most appropriate type of survey for a given situation. The specification of the desired model system and the design of a survey plan is not a simple matter and require considerable skill and experience. For basic information on recruiting, training, questionnaire design, supervision and quality control, the reader is referred to the ever popular book by Moser and Kalton (1985). Information about survey procedures with a particular transport planning flavour may also be found in Stopher and Meyburg (1979) and Richardson *et al.* (1995). In what follows we briefly discuss some of the most typical practical constraints in transport studies.

3.3.1.1   Length of the Study

This obviously has great importance because it determines indirectly how much time and effort it is possible to devote to the data-collection stage. It is very important to achieve a balanced study (in terms of its various stages) avoiding the all too frequent problem of eventually finding that the largest part of the study budget (and time) was spent in data collection, analysis and validation (see Boyce *et al.* 1970).

3.3.1.2   Study Horizon

There are two types of situation worth considering in this respect:

- If the design year is too close, as in a tactical transport study, there will not be much time to conduct the study; this will probably imply the need to use a particular type of analysis tool, perhaps requiring data of a special kind.
- In strategic transport studies, on the other hand, the usual study horizon is 20 or more years into the future. Although in principle this allows time to employ almost any type of analytical tool (with their associated surveys), it also means that errors in forecasting will only be known in 20 or more years time. This calls for flexibility and adaptation if a successful process of monitoring and re-evaluation is to be achieved.

3.3.1.3   Limits of the Study Area

Here it is important to ignore formal political boundaries (i.e. of county or district) and concentrate on the whole area of interest. It is also necessary to distinguish between this and the study area as defined in the project brief; the former is normally larger as we would expect the latter to develop in a period of, say 20 years. The definition of the area of interest depends again on the type of policies examined and decisions to be made; we will come back to this issue below.

3.3.1.4   Study Resources

It is necessary to know, as clearly and in as much detail as possible, how many personnel and of what level will be available for the study; it is also important to know what type of computing facilities will be available and what restrictions to their use will exist. In general, the time available and study resources should be commensurate with the importance of the decisions to be taken as a result. The greater the cost of a wrong decision, the more resources should be devoted to getting it right.

There are many other possible restrictions, ranging from physical (i.e. sheer size and topography of the locality) to social and environmental (e.g. known reluctance of the population to answer certain types of questions), which need to be taken into account and will influence sample design.

A general practical consideration is that travellers are often reluctant to answer 'yet another' questionnaire. Responding to questions takes time and may sometimes be seen as a violation of privacy. This may result in either flatly refusing to answer or in the provision of simplistic but credible responses, which is actually worse. In many countries it is necessary to obtain permission from the authorities before embarking on any traffic survey involving disruptions to travellers.

Modern technology offers a number of methods to collect information about trips, tours, destination, modes or route choice, without requiring the active participation of the traveller, for example tracking mobile phone locations. However, these present privacy issues which need to be handled carefully, with sensitivity and, of course, according to the law which at the time of writing is not particularly clear in most places. Moreover, these tracking methods do not offer much insight into the underlying behavioural intentions although some may be inferable from locations and timings. This is another fertile area for further research.

3.3.2 Types of Surveys

Up to the mid-1970s a large number of household origin–destination (O–D) surveys, using a simple random sample technique, were undertaken in urban areas of industrialised countries and also in many important cities in developing countries. This large effort was very expensive and demanded enormous quantities of time (a problem with collecting too much information is that a lot of time and money must also be spent analysing it); in fact, as we have commented already, the data-collection effort has traditionally absorbed a vital part of the resources available to conduct these large studies leaving, in many cases, little time and money for the crucial tasks of preparing and evaluating plans.

In many urban areas, and particularly in large metropolitan areas, there is an important role for travel survey data. In some situations this kind of data is used almost entirely for its richness in portraying the existing situation and thus helping the analyst to identify problems related to the transport system. In others, and such is our main interest, data is collected primarily for use in strategic transport modelling and hence forecasting, but it still may be used for both purposes (Battellino and Peachman 2003).

Understanding the use of the data is one of the key steps in determining the survey methodology for any travel survey. For example, activity models (Beckman *et al.* 1995) require large amounts of data not only about the activities people perform, but also on the activity 'infrastructure' (e.g. opening times of shops). However, the usual needs of travel survey data are to provide the basis for accurate predictions, typically by a strategic transport planning model. In this case the key data elements are trips between origins and destinations, rather than the underlying behavioural determinants, hence the term, 'origin–destination' study.

Stopher and Jones (2003) provide a rigorous, complete and useful guide of the elements a state-of-the-art survey should consider. Here we will only concentrate on certain key elements required to enhance the usefulness of the data as an aid to calibrating a contemporary supply-demand equilibration strategic transport planning model. In that case, current best practice suggests that the data set would be likely to have the following characteristics (Ampt and Ortúzar 2004):

- Consideration of stage-based trip data, ensuring that analyses can relate specific modes to specific locations/times of day/trip lengths, etc.
- Inclusion of all modes of travel, including non-motorised trips.
- Measurements of highly disaggregated levels of trip purposes.
- Coverage of the broadest possible time period, e.g. 24 hours a day, seven days a week, and perhaps 365 days a year (to cover all seasons).
- Data from *all* members of the household.
- High-quality information robust enough to be used even at a disaggregate level (Daly and Ortúzar 1990).
- Be part of an integrated data collection system incorporating household interviews as well as origin–destination data from other sources such as cordon surveys.

Unfortunately, collecting data at this level of precision is not an easy task and is often precluded by the sheer difficulty of convincing a sufficiently large sample of individuals to participate in such a strenuous effort. Then, depending on the modelling objectives (i.e. strategic analysis versus detailed tactical studies), the analyst may need to ease the burden on the respondents and settle for less detailed information.

3.3.2.1   Survey Scope

Figure 3.4 is useful to describe the scope of a study to capture all trips affecting a metropolitan area. It is first necessary to define the study's area of interest. Its external boundary is known as the *external cordon*. Once this is defined, the area is divided into zones (we will look at some basic zoning rules in section 3.4) in order to have a clear and spatially disaggregated idea of the origin and destination of trips, and so we can spatially quantify variables such as population and employment.

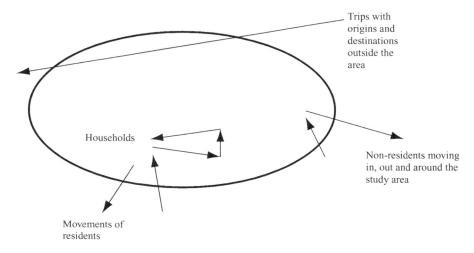

Figure 3.4   Scope of data collection needed for a metropolitan O–D survey

The area outside the external cordon is also divided into zones but at a lesser level of detail (larger zones). Inside the study area there can also be other *internal* cordons, as well as *screen lines* (i.e. an artificial divide following a natural or artificial boundary with few crossings, such as a river or a railway line), the purposes of which are discussed below. There are no hard and fast rules for deciding the location of the external cordon and hence which areas will be considered external to the study; it depends on the scope and decision levels adopted for the study, i.e. it is a very contextual problem.

Figure 3.4 implies that the following data are needed:

- *Household survey*: trips made by all household members by all modes of transport both within the study area and leaving/arriving to the area during the survey period; this survey should include socio-economic information (income, car ownership, family size and structure, etc.). This information is very efficient at generating data that permits the estimation of trip generation and mode split models; furthermore, data on household travel provides good information on the distribution of trip lengths in the city, an important element in the estimation of trip distribution models.
- *Intercept surveys, external cordon*: data on people crossing the study area border, particularly non-residents of the study area. This data can also be used to check and amplify the household data on study area crossings, since there is usually only a small amount of data collected, even in a very large survey. These are shorter surveys, carried out at points that intercept trips arriving and departing the study area: off-kerb surveys, on board public transport vehicles or at mode interchange points (i.e. airports).

- *Intercept surveys, internal cordons and screen lines*: these are required to measure trips by non-residents, and again to verify household data to some extent. They are important inputs to other models.
- *Traffic and person counts*: they are low cost and are required for calibration, validation and for further checks to other surveys. The integration of this data into the survey methodology is discussed below.
- *Travel time surveys*: these are required to calibrate and validate most models and may be needed for both car and public transport travel.
- *Other related data*: to create robust forecasting models as needed in large metropolitan areas, it is also important to have a survey methodology which allows integration of related data items that influence travel behaviour (Richardson *et al.* 1995). Here we include:
    - Land-use inventory; residential zones (housing density), commercial and industrial zones (by type of establishment), parking spaces, etc.; these are particularly useful for trip generation models.
    - Infrastructure and existing services inventories (public and private transport networks, fares, frequency, etc.; traffic signal location and timings); these are essential for model calibration, especially distribution and assignment models.
    - Information from special surveys on attitudes and elasticity of demand (e.g. stated preference and other methods).

Each of the above survey components requires a detailed design together with a carefully selected sampling strategy. In what follows we give insights into the total methodological design and clues to the necessary measures to integrate the diverse survey components.

3.3.2.2 Home Interview Travel Surveys

Home Interview or Household Travel Surveys are the most expensive and difficult type of survey but offer a rich and useful data set. However, on many occasions interest will not be centred on gathering data for the complete model system, but only for parts of it: the most typical case is that of mode choice and assignment in short-term studies.

An interesting method, particularly suitable for corridor-based journey-to-work studies and which has proved very efficient in practice, is the use of workplace interviews (see Dunphy 1979; Ortúzar and Donoso 1983). These involve the local authority asking a sample of institutions (employers) in, for example, the Central Business District (CBD) permission to interview a sample of their employees; in certain cases it has been found efficient to ask for the sample to be distributed by residence of the employee (e.g. those living in a certain corridor). It must be noted, however, that contrary to the case of random household surveys, the data obtained in this case are choice based in terms of destination; nevertheless it is mostly random with respect to mode.

Although we will be referring mainly to household surveys, most aspects of the general discussion and indeed those about the design of measurement instrument are equally applicable to any other type of origin-destination (O–D) survey.

General Considerations   Both the procedures and measurement instruments used to collect information on site have a direct and profound influence on the results derived from any data-collection effort. This is why it has been recommended to include the measurement procedure as yet another element to be considered explicitly in the design of any project requiring empirical data for its development. Wermuth (1981), for example, has even proposed a categorisation of all the stages comprising a measurement procedure. In this part we will refer to only two of these categories: the *development* and the *use* of instruments designed to measure activity patterns outside the household.

We have already mentioned that the empirical measurement of travel behaviour is one of the main inputs to the decision-making process in urban transport planning; in fact, it provides the basis for the formulation and estimation of models to explain and predict future travel activities. For this reason, methodological deficiencies at this stage will have direct repercussions in all subsequent stages of the transport planning process.

Frequent criticisms about household or workplace travel surveys have included:

- the surveys only measured average rather than actual travel behaviour of individuals;
- only part of the individual's movements could be investigated;
- level-of-service information (for example about travel times) is poorly estimated by the respondent.

In fact, it has been found that variable measurements derived from traditional O–D surveys – for example related to times, distances and costs of travel – have proved inadequate when compared with values measured objectively for the same variables. That is, the reported characteristics have tended to differ substantially from reality in spite of the fact that the individuals responding to the survey experience the actual values of these level-of-service variables twice per day. It has also been concluded that the bias has a systematic nature and is apparently related with user attitudes with respect to each mode; for example in the case of public transport, access, waiting and transfer times (which are rather bothersome) tend to be severely overestimated. It is interesting to note that from a conceptual point of view these results would indicate that the subjective perception of level-of-service variables constitutes an important determinant in modal choice (see the discussion in Ortúzar *et al.* 1983).

A methodological analysis of these criticisms leads to two conclusions (Brög and Ampt 1982). First, travel behaviour information should not be sought in general terms (i.e. average values) but with reference to a concrete temporal point of reference (e.g. a pre-assigned travel day). Second, it is not recommended to examine the various activities in isolation, but rather to take the complete activity pattern as the basis for analysis; for example, it can be shown that asking for starting and ending times of a trip yields more accurate results than asking for its total duration. Thus, contemporary travel surveys employ an activity recall framework (Ampt and Ortúzar 2004).

An Ongoing Data Collection Process   The best approach to household travel data collection postulates that information should be gathered for *each day* of the week *throughout the year* and over several years (Richardson *et al.* 1995). In order to allow the use of data at any level of aggregation and to move away from the need for a standard zoning system, the methodology also recommends geocoding all origin and destination information.

Collecting data for each day of a given year allows capturing seasonal variations, as well as weekend/weekday differences. Therefore, the approach has numerous advantages:

- It permits measuring changes in demand over time, and in particular, it allows the correlation of these changes to changes in the supply system.
- Since respondents only report data for one or two days, it makes their task easy and reliable, at the same time giving data over a longer period.
- The spread of the study over a year also results in lower operational costs.
- It allows for better quality control.

On the other hand, in this new approach there are several issues that need to be addressed on each specific circumstance:

- It is necessary to wait for up to a year before there is sufficient data to meet the purposes for which the study was designed (typically calibrate a full-scale model).
- If interviewers are used, it is necessary to keep them motivated over a longer period.
- It is necessary to develop weighting processes which take account of seasonal variations.

- It is necessary to develop special methods for post-weighting annual data if it is combined with ongoing survey data (as we discuss below).

Periodic Update of Matrices and Models   Matrices and models to match the ongoing data collection system should be updated periodically in order to maximise the benefit of the continuous information. Notwithstanding, although it is possible to consider the preparation of partial trip matrices given specific requirements; trip tables and models for the whole study area should not be updated more frequently than every 12–18 months, depending on the type of city under study.

Implications for Data Collection   Periodic updating of models and matrices is likely to have an effect on the data collected. For example, which information is most sensitive to updating? In this context there may be several elements that are worth periodic updating:

- Trip generation and attraction models.
- Travel matrices, reflecting the differential growth in different parts of the study area.
- Modal split, including non-motorised modes, reflecting the possible impacts of different transport policies.
- Traffic levels in different parts of the network, allowing identification of differential growth in the primary, secondary, access and local networks.
- Car ownership and household formation trends in various city boroughs.

The priority to apply in the case of each of these indicators will depend on the type of transport policies being considered, the need to monitor their performance, and the general modelling needs. It will also depend on their expected rates of change, their importance, and the costs associated with collecting data for updating, including the social cost of bothering users of the transport system (see DICTUC 1998 for a more detailed discussion).

It is also important to note the convenience of allowing estimation of other types of models likely to be needed in the future, such as time-of-day choice models and dynamic models. On the other hand, the availability of data collected on a continual basis allows monitoring user behaviour with respect to radical interventions in the transport system. Examples are environmental emergencies where CBD car-entry restrictions are increased, main road works, bus strikes, or changes in petrol prices, bus fares or parking charges. The response to such policies (predictable or otherwise) provides basic information about users' behavioural thresholds and creates a temporal database which should facilitate the development of more sophisticated models.

Questionnaire Format and Design   Since one of the aims of a survey is to achieve the highest possible response rate to minimise non-response bias, it is recommended that mixed methods (i.e. based on self-completion and personal interviews) are used to collect the data (Goldenberg 1996). In particular, a self-completion system seems more appropriate in districts where people are used to 'filling in forms' (with personal interview validation follow-up) or where households cannot be accessed other than by remote security bell systems, where attempts at personal interviews result in low response rates. This combination capitalises on the cost-effectiveness and efficiency of high-quality self-completion designs and ensures minimum response burden for all participants (Richardson et al. 1995).

Telephone-based surveys, although widely used in North America for their relative cost-effectiveness, are not recommended for several reasons (Ampt and Ortúzar 2004):

- Even where prompts are sent to households in advance (e.g. in the form of mini-diaries), they tend to suffer from extensive proxy reporting (i.e. one person reporting on behalf of others) leading to significant under-reporting of trips.

- Although phone ownership is very high in countries where phone lists are used as sampling frame, there are often up to 40% of unlisted households (Ampt 2003); this means that at least 40% (plus those households without phones) of the population cannot give data if a single method approach was used.
- Where random digit dialling is used to overcome this difficulty, the problem can be exacerbated by the ire of people with unlisted numbers receiving calls.
- An increasing number of people in many countries have now only mobile phones; this means that, even if mobile phone numbers are available, there is a mixture of household-based and person-based sampling, which would need considerable effort if the weighting stage is to be effective.

In terms of layout, the order of the questions normally seeks to minimise the respondent's resistance to answering them, so difficult questions (e.g. relating to income) are usually put at the end. The survey instrument (and any personal interviews) should try to satisfy the following criteria:

- The questions should be simple and direct.
- Make sure each question serves a specific purpose; an excessive number of questions degrades the response rate and increases trip omissions.
- The number of open questions should be minimised.
- Travel information must include the purpose of the trip. It is interesting to acquire *stage-based trip data* (i.e. all movements on a public street) to ensure that analyses can relate specific modes to specific locations, times of day, etc.
- Collect information so that complete tours can be re-constructed during analysis.
- Seek information about *all modes of travel*, including non-motorised travel.
- All people in the household should be included in the survey, including non-family members, like maids in developing countries.
- To facilitate the respondent's task of recording all travel, an activity-recall framework is recommended, whereby people record travel in the context of activities they have undertaken rather than simply trips they have made; this has been shown to result in much more accurate travel measurement (Stopher 1998).
- Since people have difficulty recalling infrequent and discretionary activities, even when they are recent, a travel day or days should be assigned to each household in advance. Respondents should be given a brief diary in advance of these days; the information in the diary may then be transferred to the self-completion form or reported to the interviewer at the end of the day (or as soon as possible).
- Finally, all data should be collected at the maximum level of disaggregation (x-y co-ordinate level) based on a geographical information system (GIS).

The survey instrument needs to be designed for minimum respondent burden (Ampt 2003), maximum response rate (CASRO 1982) and hence greatest robustness of the data:

- Self-completion designs need to focus on overall layout since they are the researchers' only contact with the respondent. The layout needs to be clear and concise, and in general it should lead respondents onto the next question. Layouts should usually be designed to encourage every respondent to reply, whether or not they are used to filling out forms (i.e. be user friendly, nicely presented and using simple language).
- The strength of personal interviews lies in the ease of response for survey participants, so the focus needs to be on training interviewers to understand the context of the survey and making sure the survey designs are easy for them to administer.

For either type of household survey, it is recommended that the survey be divided into two parts: (1) personal and household characteristics and identification and (2) trip data. We will briefly review the information sought in each part:

- *Personal and household characteristics and identification*: this part includes questions designed to classify the household members according to their relation to the head of the household (e.g. wife, son), sex, age, possession of a driving licence, educational level and occupation. In order to reduce the possibility of a subjective classification, it is important to define a complete set of occupations (non-household surveys are usually concerned only with the person being interviewed; however, the relevant question are the same or very similar). This part also includes questions designed to obtain socio-economic data about the household, such as characteristics of the house, identification of household vehicles (including a code to identify their usual user), house ownership and family income.
- *Trip data*: this part of the survey aims at detecting and characterising all trips made by the household members identified in the first part. A trip is now defined as any movement outside a building or premises with a given purpose; but the information sought considers trips by stages, where a stage is defined by a change of mode (including walking). Each stage is characterised on the basis of variables such as origin and destination (normally expressed by their nearest road junction or full postcode, if known), purpose, start and ending times, mode used, amount of money paid for the trip, and so on. Ideally, analysis should be able to link trips in a logical way to re-construct tours and to generate Productions and Attractions by household.

Definition of the Sampling Framework   The scope of mobility surveys usually includes all travellers in an area (Figure 3.4). Thus, it not only includes residents, but also visitors to households, in hotels, other people in non-private dwellings (such as hospitals) and travellers that pass through the area on the survey days.

Once the scope has been defined, the sampling frame needs to be determined. In other words, what type of list will provide information on all residents, visitors and people who pass through the area, in order to choose a sample of those people and trips. Although there are various options, the household sample frame, while complex, is usually the most straightforward. If a census has been conducted recently and information on all dwellings is available, this can be ideal. Alternatively, a block list of the whole region (prepared for any reason, e.g. for a utility company or for a previous survey) could be used, but a key issue is that it should be very up-to-date. Although Census data is only available every ten years in most places, in certain countries the government possesses a list of all dwellings officially registered for paying property taxes and this can be a useful starting point. If such lists are not available, several other methods can be used, the most typical one in industrialised nations being telephone listings (Stopher and Metcalf 1996) complemented by other methods if telephone ownership or listings are not universal. If no 'official' frame is available, it is always possible to simply sample blocks at random, enumerate the households in the block, and randomly sample from these.

Choosing the sampling frame for travel by non-residents is more complicated. It is recommended that this be done in the following way:

- Obtain a list of all non-private dwellings and select a sample (possibly stratified by size or by type of visitor).
- Obtain a list of public transport interchanges where people are likely to arrive and leave the metropolitan region (e.g. airports, train stations, and long-distance bus stations). Ideally this will produce a sampling of travellers at each intercept point, although in some cases it will be necessary to sample sites.
- Obtain a list of all road crossing points of the area's external cordon. As with public transport interchanges, ideally all cross-points should be included, although in some cases they will need to be sampled.

However, the above procedure does not guarantee a perfect sampling frame. Fortunately, with some clear exceptions, the importance of trips made by visitors is generally much smaller than that of residents in any given study area.

Sample Size   Travel surveys are always based on some type of sampling. Even if it were possible to survey all travellers on a specific service on a given day, this would only be a sample of travellers making trips in a given week, month or year. The challenge in sampling design is to identify sampling strategies and sizes that allow reasonable conclusions, and reliable and unbiased models, without spending excessive resources on data collection. Often there is more than one way of obtaining the relevant information. For some data needs it may be possible to gather the information either through household surveys or through intercept surveys. In these situations it is best to use the method that delivers the most precise data at the lowest cost (DICTUC 1998).

There are well-documented procedures for estimating the sample size of household surveys so that it is possible to satisfy different objectives; for example, estimation of trip rates, and trip generation by categories, levels of car ownership and even of mode choice variables for different income strata (Stopher 1982). Given reasonable budget limitations, the analyst faces the question of whether it is possible to achieve all these objectives with a given sample of households in a certain metropolis (see for example, Purvis 1989). In general, these methods require knowledge about the variables to be estimated, their coefficients of variation, and the desired accuracy of measurement together with the level of significance associated with it.

The first requirement, although both obvious and fundamental, has been ignored many times in the past. The majority of household O–D surveys have been designed on the basis of vague objectives, such as 'to reproduce the travel patterns in the area'. What is the meaning of this? Is it the elements of the O–D matrix which are required, and if this is the case, are they required by purpose, mode and time of day, or is it just the flow trends between large zones which are of interest?

The second element (coefficient of variation of the variable to be measured) was an unknown in the past, but now it may be estimated using information from the large number of household O–D surveys which have been conducted since the 1970s. Finally, the accuracy level (percentage error acceptable to the analyst) and its confidence level are context-dependent matters to be decided by the analyst on the basis of personal experience. Any sample may become too large if the level of accuracy required is too strict. It can be said that this aspect is where the 'art' of sample size determination lies.

Once these three factors are known, the sample size (n) may be computed using the following formula (M.E. Smith 1979):

$$n = \frac{CV^2 \, Z_\alpha^2}{E^2} \tag{3.16}$$

where CV is the coefficient of variation, E is the level of accuracy (expressed as a proportion) and Z_α is the standard normal value for the confidence level (α) required.

Example 3.6   Assume that we want to measure the number of trips per household in a certain area, and that we have data about the coefficient of variation of this variable for various locations in the USA as follows:

Area	CV
Average for U.S.A. (1969)	0.87
Pennsylvania (1967)	0.86
New Hampshire (1964)	1.07
Baltimore (1962)	1.05

As all the values are near to one, we can choose this figure for convenience. As mentioned above, the decision about accuracy and confidence level is the most difficult; equation (3.16) shows that if we postulate levels which are too strict, sample size increases exponentially. On the other hand, it is convenient to fix strict levels in this case because the number of trips per household is a crucial variable (i.e. if this number is badly wrong, the accuracy of subsequent models will be severely compromised). In this example we ask for 0.05 level of accuracy at a 95% level.

For $\alpha = 95\%$ the value of Z_α is 1.645, therefore we get:

$$n = 1.0(1.645)^2/(0.05)^2 = 1084$$

that is, it would suffice to take a sample of approximately 1100 observations to ensure trip rates with a 5% tolerance 95% of the time. The interested reader may consult M.E. Smith (1979) for other examples of this approach.

The situation changes, however, if it is necessary to estimate origin-destination (O–D) matrices. For example, M.E. Smith (1979) argues that a sample size of 4% of all trips in a given study area would be needed to estimate levels higher than 1100 trips between O–D pairs at the 90% confidence level with a standard error of 25%. This effectively means that if there are less than 1100 trips between two zones, a sample size of less than 4% would not be sufficient to detect them.

Example 3.7 Trips by O–D cell in Santiago, at the municipality level (e.g. just 34 zones), were analysed using data from the 1991 Household survey (Ortúzar *et al.* 1993). It was observed that only 58% of the O–D cells contained more than 1000 trips. Thus, it would seem necessary to postulate a sample size of 4% of trips (and by deduction, 4% of households) to estimate an O–D matrix at the municipality level with a 25% standard error and 90% confidence limits. However, if the effect of response rates is considered (even if they were as high as 75%), as there were about 1 400 000 households in the city this would imply an initial sample size of nearly 75 000 households. It is doubtful that such a large sample size (and the associated costs and levels of effectiveness) are justified to accomplish such a meagre objective.

Clearly, the driving force behind large sample sizes is the need to obtain trip matrices at the zone level. It has also been shown that it is very difficult to reduce the measurement error to an acceptable level in areas with more than say 100 zones, since the sample size required is close to that of the population (M.E. Smith 1979). Hence, if the objective of the study includes estimating an O–D matrix, it is necessary to use a combination of survey methods, including both household and intercept surveys, to take advantage of their greater efficiencies for different data objectives.

Optimisation Strategies for Sample Design To achieve a sampling design that yields a smaller sample size, it is necessary to devise strategies that estimate, say, trip generation rates by socio-economic status. One approach is to use a multi-stage stratified random sampling heuristic which produces better results than the classic method devised by M.E. Smith (1979); unfortunately, it requires a lot of effort by the analyst and does not guarantee a unique solution (DICTUC 1998). The heuristic begins by ordering the socio-demographic classes according to the degree to which they are represented in the population. Next the zones in the study area are allocated a class based on the most frequently occurring socio-economic group in them. Then a random sample of zones of each socio-economic type is selected (i.e. of the order of 1% of all households). After that the remaining zones are categorised in priority order and are chosen as necessary to reach the difference between the sample already selected and the minimum required for each new class. The procedure is repeated until all classes have the minimum sample size required (say 30 or 50 observations).

This procedure was applied in Santiago for the 264-zone system defined in the 1991 O–D survey (Ortúzar *et. al.* 1993) and for a stratification of 14 classes based on income and car ownership. The final solution achieved was a sample size of 1312 households located in only 15 zones, which guaranteed a minimum of 30 observations per class. However, in certain zones, notably those containing high-income people, the solution implied somewhat unreasonable sample sizes (i.e. around 20% of the zonal population). A better solution was found by solving the following optimisation problem (Ampt and Ortúzar 2004):

Minimise

$$\sum_{i \in \{classes\}} \sum_{j \in \{zones\}} \alpha_j \eta_{ij}$$

subject to

$$0 \le \alpha_j \le \delta$$

$$\sum_{j \in \{zones\}} \alpha_j \eta_{ij} \ge \mu_i$$

where α_j is the proportion of households to interview in zone j and δ a reasonable limit (e.g. a maximum of 5%), η_{ij} is the number of households of class i in zone j, and μ_i is the minimum acceptable sample size for each class i (i.e. 30 or 50 observations).

Using the same information as in the previous case, it was found that the problem could be optimally solved yielding a sample of just 482 households. More interestingly, for a stratification with 26 classes (i.e. adding household size as stratifying variable), it was found that an optimum sample size of 1372 households, guaranteeing a minimum of 30 observations in each of the specified classes, would be possible by collecting data in only 17 of the 264 zones.

However, as no limit was enforced for δ, some values of α_j were again near 20% of the zone population. So, by applying the restriction of δ being less than 5%, a sample of 1683 households in only 27 zones was finally obtained. Note that the method permits segmentations other than by socio-economic criteria. For example, it is also possible to identify spatial differences in terms of physical area (i.e. distance from the CBD) or access to the public transport network, and to increase the number of classes considered for the optimisation (Ortúzar et al. 1998).

Finally, remember that the design can also be improved by allowing for different response rates between different groups. In principle it is possible to estimate the number of households required in a gross sample (μ_i) to achieve a given minimum number of responses for each class, thereby ensuring a design that will yield even higher-quality trip generation data.

Sample Size for a Continuous Survey   A final challenge consists of designing a sampling strategy for a continuous survey. If a sample of say 15 000 households is required in year 1 to fulfil the initial modelling requirements of a metropolitan area, an ongoing survey would probably have the following form or something similar:

Year 1	Year 2	Year 3	Year 4	Year 5
15 000	5000	5000	5000	5000

This method requires smaller ongoing input after the first year, which offers several advantages:

- A smaller well-trained field force and administrative procedures which are likely to ensure very high quality data with minimal effort in subsequent years.
- The appropriate authorities make a financial commitment for four years in year 1, reducing the risk of difficulties over repeat funding in say year 4.

But it does require the development of an annual weighting and integration system to ensure the data is readily usable for modelling, and this system needs to be robust and easy to use. We need to ensure that *all* the data at the end of year 2 is representative of year 2, that *all* the data at the end of year 3 is representative of that year, and so on. Such a procedure provides an up-to-date representation of existing travel behaviour for modelling and other purposes. In developing cities (i.e. where rapid changes occur in car ownership, land-use spread and distribution), this would mean a more accurate modelling capability than has ever been possible in the past. It would also provide a larger sample size for use in the second and subsequent years, enabling more detailed questions to be asked of the data in them. Furthermore, if it is assumed that the data will be used for other purposes as well as modelling, the annual data collection method will provide essential time series data. Here are some examples:

• Changes in travel patterns (by mode) related to changes in car ownership levels and distribution, pollution levels or land-use patterns.
• Changes in choice of mode related to changes in supply patterns, e.g. improvements for pedestrians, expansion of the public transport network.

The way in which data from the second and subsequent years should be integrated and combined with data from the first year has to occur at four levels: household, vehicle, person and trip. In this sense it is important to consider three things:

• Careful sample selection and high response rates to ensure the 15 000 households in year 1 are representative of the city; then weighting and expansion procedures should be applied as described below.
• Make sure the 5000 households in year 2 are representative of the city (i.e. spatially and on all other parameters used for the first year of the sample selection); again, weighting procedures need to be applied.
• At the end of year 2, the database will consist of 20 000 households but it will contain the raw data and the weighting factors only.

In smaller-sized cities, or in areas where there is little change in size and structure, it may not be necessary to have such a complicated sampling strategy, but it still depends on the uses of the data. For example, an equal sample for each of the years in a five-year period could be appropriate.

3.3.2.3 Other Important Types of Surveys

Roadside Interviews These provide useful information about trips not registered in household surveys (e.g. external–external trips in a cordon survey). They are often a better method for estimating trip matrices than home interviews as larger samples are possible. For this reason, the data collected are also useful in validating and extending the household-based information.

Roadside interviews involve asking a sample of drivers and passengers of vehicles (e.g. cars, public transport, goods vehicles) crossing a roadside station, a limited set of questions; these include at least origin, destination and trip purpose. Other information such as age, sex and income is also desirable but seldom asked due to time limitations; however, well trained interviewers can easily add at least part of these data from simple observation of the vehicle and occupants (with obvious difficulties in the case of public transport).

The conduct of these interviews requires a good deal of organisation and planning to avoid unnecessary delays, ensure safety and deliver quality results. The identification of suitable sites, co-ordination with the police and arrangements for lighting and supervision are important elements in the success of these surveys. We shall concentrate here on issues of sample size and accuracy.

Example 3.8 Let us assume a control point where N cars cross and we wish to take a sample of n vehicles to survey. Let us also assume that of these n, X_1 cars travel between the origin–destination pair O–D$_1$. In this case it can be shown that X_1 has a hyper geometric distribution $H(N, N_1, n)$, where N_1 is the total number of travellers between pair O–D$_1$, and that its expected value and variance are given by:

$$E(X_1) = np \text{ with } p = N_1/N$$
$$V(X_1) = np(1 - p)(1 - n/N)$$

Using a Normal approximation (based on the Central Limit theorem) the distribution of X_1 is:

$$X_1 \sim N(np, np(1 - p)(1 - n/N))$$

and an estimator for p is:

$$\hat{p} = \frac{X_1}{n}$$

Therefore

$$\hat{p} \sim N\left(p, \frac{p(1 - p)(1 - n/N)}{n}\right)$$

and an approximate $100(1 - \alpha)\%$ confidence interval for p is given by:

$$\left[\hat{p} - z\sqrt{\frac{p(1 - p)(1 - n/N)}{n}}, \hat{p} + z\sqrt{\frac{p(1 - p)(1 - n/N)}{n}}\right]$$

where z is the standard Normal value for the required confidence level (1.96 for the 95% level). We typically require that the absolute error e associated with $\hat{p}$ does not exceed a pre-specified value (usually 0.1), that is:

$$E = z\sqrt{\frac{p(1 - p)(1 - n/N)}{n}} \leq e$$

Working algebraically on this expression we get:

$$n \geq \frac{p(1 - p)(1 - n/N)}{(e/z)^2}$$

or equivalently:

$$n \geq \frac{p(1 - p)}{(e/z)^2 + p(1 - p)/N} \tag{3.17}$$

It can be seen that, for a given N, e and z, the value $p = 0.5$ yields the highest (i.e. most conservative) value for n in (3.17). Taking this value and considering $e = 0.1$ (i.e. a maximum error of 10%) and z equal 1.96, we obtain the values in Table 3.1.

Table 3.1 Variation of sample size with observed flow

N (passengers/period)	n (passengers/period)	100 n/N (%)
100	49	49.0
200	65	32.5
300	73	24.3
500	81	16.2
700	85	12.1
900	87	9.7
1100	89	8.1

Example 3.9 An examination of historical data during preparatory work for a roadside interview revealed that flows across the survey station varied greatly throughout the day. Given this, it was considered too complex to try to implement the strategy of Table 3.1 in the field. Therefore, the following simplified table was developed:

Estimated observed flow (passengers/period)	Sample size (%)
900 or more	10.0 (1 in 10)
700 to 899	12.5 (1 in 8)
500 to 699	16.6 (1 in 6)
300 to 499	25.0 (1 in 4)
200 to 299	33.3 (1 in 3)
1 to 199	50.0 (1 in 2)

The fieldwork procedure requires stopping at random the corresponding number of vehicles, interviewing all their passengers and asking origin, destination and trip purpose. In the case of public-transport trips, given the practical difficulties associated with stopping vehicles for the time required to interview all passengers, the survey may be conducted with the vehicles in motion. For this it is necessary to define road sections rather than stations and the number of interviewers to be used depends on the observed vehicle-occupancy factors at the section. However, even this approach may be unworkable if the vehicles are overloaded.

Cordon Surveys   These provide useful information about external–external and external–internal trips. Their objective is to determine the number of trips that enter, leave and/or cross the cordoned area, thus helping to complete the information coming from the household O–D survey. The main one is taken at the external cordon, although surveys may be conducted at internal cordons as well. In order to reduce delay they sometimes involve stopping a sample of the vehicles passing a control station (usually with police help), to which a short mail-return questionnaire is given. In some Dutch studies a sample of licence plates is registered at the control station and the questionnaires are sent to the corresponding addresses stored in the Incomes and Excise computer. An important problem here is that return-mail surveys are known to produce biased results: this is because less than 50% of questionnaires are usually returned and it has been shown that the type of person who returns them is different to those that do not (see Brög and Meyburg 1980). This is why in many countries roadside surveys often ask a rather limited number of questions (i.e. occupation, purpose, origin, destination and modes available) to encourage better response rates.

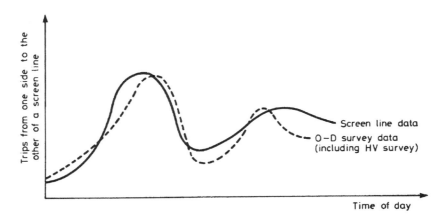

Figure 3.5   Household survey data consistency check

Screen-line Surveys   Screen lines divide the area into large natural zones (e.g. at both sides of a river or motorway), with few crossing points between them. The procedure is analogous to that of cordon surveys and the data also serve to fill gaps in and validate (see Figure 3.5) the information coming from the household and cordon surveys. Care has to be taken when aiming to correct the household survey data in this way, because it might not be easy to conduct the comparison without introducing bias.

3.3.3   Survey Data Correction, Expansion and Validation

Correction and weighting are essential in any travel survey (Stopher and Jones 2003); the following sections discuss an approach deemed appropriate for the contemporary surveys described above, which are conducted over a period of several years.

3.3.3.1   Data Correction

The need to correct survey data in order to achieve results which are not only representative of the whole population, but also reliable and valid, has been discussed at length (Brög and Erl 1982; Wermuth 1981). It is now accepted that simply expanding the sample is not appropriate, although for many years it was the most commonly practised method. Brög and Ampt (1982) identify a series of correction steps as follows.

Corrections by Household Size and Socio-Demographic Characteristics   To make corrections that guarantee that the household size, age and sex, housing type and vehicle ownership distributions of the sampled data represent that in the population (based on Census data), an iterative approach is needed, since more simplistic methods do not guarantee correct results (see the discussion by Deville *et al.* 1993). Multi-proportional fitting (see sections 5.2.3 and 5.6.2), also known as 'raking ratio' (Armoogum and Madre 1998), is probably the best approach in this case, since it guarantees convergence in few iterations. Furthermore, its application has the additional advantage of not requiring the subsequent calculation of expansion factors. Stopher and Stecher (1993) give an almost pedagogical example of this approach.

The method is particularly valid if the secondary population data has been gathered close to the time of the travel survey. However, it may not be appropriate if the travel survey is done several years after the Census as the urban population may change rather quickly, particularly in less-developed countries. In this case it would be necessary to calculate proportions of households in each group and to compute expansion factors in the more traditional form (Ortúzar *et al.* 1993; Richardson *et al.* 1995).

The multi-proportional method does not guarantee that each cell value will be identical in the Census and in the travel survey since in any matrix there is an important degree of indeterminacy (i.e. many combinations of cell values can give rise to the same totals of rows and columns). In particular, due to its multiplicative characteristics, a cell with a zero will always end up with a zero value. Furthermore, certain special matrix structures (that contain zeroes in some key positions) can lead to non-convergence of the method (see section 5.6.1 for an example).

To avoid bias in the multi-proportional correction, because we are correcting by items as diverse as, say, household size (number of persons) on the one hand, and personal characteristics (sex and age) on the other, it is better to define unique categories, thus avoiding classes that consider – for example – two to four persons, six or more persons, etc. Nevertheless, it is easy to imagine occasions on which it would be necessary to group some category because it is not represented in the sample for a given zone. In that case, it is convenient to check if it is possible to group similar zones instead of making the correction at such a disaggregate level (Stopher and Stecher 1993).

Additional Corrections in Household Surveys In addition to the corrections by household size, vehicle ownership and socio demographics, there are two other correction procedures necessary – depending on whether it is a personal interview or self-completion survey (Richardson *et al.* 1995). These procedures are noted below:

Corrections for non-reported data These are needed when certain elements of the survey have not been answered (item non-response). In self-completion surveys, interviewing a validation sample of people using personal interviews and then weighting the data accordingly (Richardson *et al.* 1995) is used to address this. This type of correction is not usually needed when personal interviews are used because interviewers must be well trained and supervised thereby decreasing the incidence of item non-response (but see the discussion in Stopher and Jones 2003).

Corrections for non-response These are needed when a household or individual does not respond, i.e. does not return the survey instrument or refuses verbally or by mail to respond to the survey (Zimowski *et al.* 1998). This can be attributed to a variety of causes, and it is important to differentiate between genuine sample loss (e.g. vacant dwellings which do not generate travel should be ineligible), and refusals (where the person could be travelling but not responding, clearly eligible). In the case of personal interviews, it has been recommended that corrections should be based on the number of visits necessary to achieve a response, since it has been shown that this is associated with strong differences in travel behaviour (Kam and Morris 1999; Keeter *et al.* 2000); however, there is also evidence suggesting that these differences might be small (Kurth *et al.* 2001).

In self-completion surveys, on the other hand, it was originally believed that corrections could be done based on the number of follow-up reminders needed to generate a household response (Richardson *et al.* 1995) but the problem is likely to be more complex than for personal interviews (see the discussions by Polak 2002, and Richardson and Meyburg 2003). Related to this, it is interesting to mention that reductions in non-response bias due to the inadequate representation of certain population strata (i.e. by income) have been reported using special factoring techniques that take into account the differences in return rates by different types of households by zones (Kim *et al.* 1993).

A final related point is how to decide when the response by a household is considered complete. The US National Travel Survey uses the 'fifty-percent' rule (at least 50% of adults over 18 years of age completed the survey), after arguments that excluding households where not everybody responded may exaggerate bias, and data are weighted to mitigate the person-level non-response in sampled households. Research on this subject allowed detecting the most likely types of households and the most likely non respondent (DRCOG 2000). Interestingly, trip rates by the sample including 50% households have been found to be not statistically different to a sample including only households with 100% of members responding (see Ampt and Ortúzar 2004).

Integration Weighting for a Continuous Survey   Integration weighting is required to unite each wave of the survey; in this case it is recommended to proceed as follows (Ampt and Ortúzar 2004):

- *Household weighting* should occur for each 'important' variable (as chosen in prior consultation), for example household size, car ownership or household income.
- *Vehicle weighting* should be done in the same way. A variable of particular importance here is the age of the vehicle, since without correct weighting it would appear as if the fleet was not ageing.
- *Person weighting.* Here factors of importance are likely to be income and education, for example.
- *Trip weighting.* Number of trips and mode are likely to be the key variables in this case – all done according to the same general principles described above.

In this way the data will be representative of the population in every year of the survey. Of course this is not perfect, but with a good sampling scheme it should be very robust. For example, in year 2 the sample will actually reflect real changes in household size (say) that may be occurring. Hence if one wanted to use years 1 and 2 to reflect the situation in year 2 (which is exactly what a government agency would like to do), it would be necessary to weight the year 1 data set to have the proper household size that is actually observed in year 2. Clearly if a given year coincides with a Census year, the weighting process can take on a whole new meaning, although this is likely to occur only about once a decade.

Example 3.10 Table 3.2 presents the number of samples gathered in the first three years of a continuous survey, stratified according to household size. If we consider households of size 1 say, we can see that they constitute 13.33% of the sample in year 1 (i.e. 2000/15 000) 17% of the sample of year 2, and if added without reweighting 14.25% of the sample for both years. However, this would be akin to the proverbial mixing of apples and pears.

Table 3.2   Weighting procedures for integration

Weighted values for years 1 and 2											
	Household size										
	1		2		3		4		5		Total
Year 1	2000	13.33%	3000	20.00%	4000	26.67%	5000	33.33%	1000	6.67%	15 000
Year 2	850	17.00%	1200	24.00%	1000	20.00%	1500	30.00%	450	9.00%	5 000
Total	2850	14.25%	4200	21.00%	5000	25.00%	6500	32.50%	1450	7.25%	20 000
Reweighting values for year 1											
	1		2		3		4		5		
Year 1	17.00/13.33	1.275	1.200		0.750		0.900		1.350		
Reweighting procedure											
	1		2		3		4		5		
Year 1	2550		3600		3000		4500		1350		15 000
Year 2	850		1200		1000		1500		450		5 000
Total	3400	17%	4800	24%	4000	20%	6000	30%	1805	9%	20 000

To integrate the data properly we need first to calculate (appropriate) weights for year 1 to ensure that both sets have the same proportions as measured in the latest year (based on the assumption that the new sample drawn each year represents the characteristics of that year's population). These weights are calculated in the next part of the table, and are equal to the ratio between the percentages (for each strata) of years 2 and 1 (i.e. 24/20 = 1.2 in the case of households of size 2). The final part of the table shows the result of adding the weighted year-1 data to the year-2 data, to achieve a final sample of 20 000 households that has the same distribution according to household size as it occurs in year 2.

3.3.3.2 Imputation Methods

Survey non-response makes identification of population parameters problematic and, normally, identification of non-response data is only possible if certain assumptions (frequently not testable) are made about the distribution of missing data. Non-response does not, however, necessarily preclude identification of the bounds on parameters. There are several state-of-practice imputation methods ranging from deductive imputation, to use of overall or class means, to hot and cold-deck imputation and so on (Armoogum and Madre 1998). In fact, the organizations conducting major surveys usually release data files that provide non-response weights or imputations to be used for estimating population parameters. Stopher and Jones (2003) recommend distinguishing between imputation and inference. The latter can be used initially and is particularly useful for certain types of variables (i.e. if a person does not indicate s/he is a worker but reports trips to work). However, there are some variables that may not be safe to infer due, for example, to changing social structures.

Imputation is defined as the substitution of values for missing data, based on certain rules or procedures. It is worth noting, however, that most imputation methods do not preserve the variance of the imputation variable (for example income), and therefore, they can produce inconsistent estimates when the variable that contains imputations is included in a model. For this reason, some researchers even believe that to impute values increases the bias in some instances, and is simply translated in makeshift data. Thus, it is recommended that the changes produced upon imputing values are registered, and if it is possible, to have their effects evaluated. Horowitz and Manski (1998) show how to bind the asymptotic bias of estimates using typical weights and imputations. They provide a thorough mathematical treatment of the subject and illustrate it with empirical examples using real data.

Another approach to solving this and other problems consists of making multiple imputations and thereafter combining the estimators of the resulting models in each case to obtain consistent values that include a consideration of the errors associated with the imputation process (Brownstone 1998).

In the Santiago 2001 O-D Survey (DICTUC 2003), 543 households out of 15 537 did not answer the family income question. Due to the strong asymmetry of the income distribution a logarithmic transformation of the data was used which allowed us to centre the distribution and achieve a better resemblance of a Normal distribution. Multiple imputations were successfully produced using a linear model based on the Student t-distribution with five degrees of freedom (Lange *et al.* 1989), estimated using Gibbs sampling (Geman and Geman 1984). Outliers were detected and removed from the estimation process; as it turned out, they were found to be wrongly coded meaning that the process had the secondary advantage of allowing for further checks on the quality of the data.

3.3.3.3 Sample Expansion

Once the data have been corrected it is necessary to expand them in order to represent the total population; to achieve this expansion factors are defined for each study zone as the ratio between the total number of addresses in the zone (A) and the number obtained as the final sample. However, often data on A are outdated leading to problems in the field. The following expression is fairly general in this sense:

$$F_i = \frac{A - A(C + CD/B)/B}{B - C - D}$$

where F_i is the expansion factor for zone i, A is the total number of addresses in the original population list, B is the total number of addresses selected as the original sample, C is the number of sampled addresses that were non-eligible in practice (e.g. demolished, non-residential), and D is the number of sampled addresses where no response was obtained. As can be seen, if A was perfect (i.e. $C = 0$) the factor would simply be $A/(B - D)$ as defined above. On the other hand, if $D = 0$ it can be seen that the formula takes care of subtracting from A the proportion of non-eligible cases, in order to avoid a bias in F_i.

3.3.3.4   Validation of Results

Data obtained from O–D surveys are normally submitted to three validation processes. The first simply considers on site checks of the completeness and coherence of the data; this is usually followed by their coding and digitising in the office. The second is a computational check of valid ranges for most variables and in general of the internal consistency of the data. Once these processes are completed, the data is assumed to be free of obvious errors.

In mobility studies the most important validation is done within the survey data itself and not with secondary data such as traffic counts at screen lines and cordons in the study area. The reason is that each method has its own particular biases which confound this task. For example, gross comparisons, such as number of trips crossing a cordon or number of trips by mode, often give relatively poor comparisons.

Although state-of-the-art survey techniques minimise these problems, the use of independent data to check figures from all components of a metropolitan O–D travel survey (see the discussion in Stopher and Jones 2003) is still recommended. Objective comparisons of these figures, taking into account the strengths and weaknesses of each survey method make it possible to detect potential biases and to take steps to amend them. Furthermore, if matrices are to be adjusted (see section 12.4.7), it is essential to reserve independent data to validate the final results. This requires good judgement and experience, since if insufficient care is given to the task it is easy to produce corrections to the O–D matrices that do not correspond to reality.

3.3.4   Longitudinal Data Collection

Most of the discussion so far has been conducted under the implicit assumption that we are dealing with cross-sectional (snap-shot) data. However, as we saw in Chapter 1, travel behaviour researchers are becoming increasingly convinced that empirical cross-sectional models have suffered from lack of recognition of the inter-temporality of most travel choices. Panel data are a good alternative to incorporate temporal effects because in this data structure a given group of individuals is interviewed at different points in time.

In this part we will attempt to provide a brief sketch of longitudinal or time-series data-collection methods and problems; we will first define various approaches and then we will concentrate on the apparently preferred one: panel data. In Chapter 8 we will consider the added problems of modelling discrete choices in this case.

We will finally examine some evidence about the likely costs of a panel data-collection exercise in comparison with the more typical cross-sectional approach.

3.3.4.1   Basic Definitions

1. Repeated cross-sectional survey. This is one which makes similar measurements on samples from an equivalent population at different points in time, without ensuring that any respondent is included in more than one round of data collection. This kind of survey provides a series of snapshots of the population at several points in time; however, inferences about the population using longitudinal models may be biased with this type of data and it may be preferable to treat observations as if they were obtained from a single cross-sectional survey (see Duncan et al. 1987).
2. Panel survey. Here, similar measurements (i.e. the panel waves), are made on the same sample at different points in time. There are several types of panel survey, for example:
   - Rotating panel survey. This is a panel survey in which some elements are kept in the panel for only a portion of the survey duration.
   - Split panel survey. This is a combination of panel and rotating panel survey.

- Cohort study. This is a panel survey based on elements from population sub-groups that have shared a similar experience (e.g. birth during a given year).

	Although the use of panel data has increased in many areas, especially since the pioneering work of Heckman (1981), in transport there are only a few examples, which can be classified into two groups:

- Long survey panels. These consist of repeating the same survey (i.e. with the same methodology and design) at 'separate' times, for example once or twice a year for a certain number of years or before-and-after an important event. Some famous examples are the Dutch Panel (Van Wissen and Meurs 1989) and the Puget Sound Transportation Panel (PSTP) in the United States (Murakami and Watterson 1990).The main problem of this kind of panel is attrition (i.e. losing respondents) between successive surveys (known as waves).

- Short survey panels: These are multi-day data where repeated measurements on the same sample of units are gathered over a 'continuous' period of time (e.g. two or more successive days), but the survey is not necessarily repeated in subsequent years. Some recent examples of this type of panel are the two-day time-use diary for the US National Panel Study of Income Dynamics and the six-week travel and activity diary data panels collected in Germany (Axhausen et al. 2002) and Switzerland (Axhausen et al. 2007). In this case attrition is not a problem, but the infrequent changes in mode choice and low data variability (both the attributes of each mode and the respondents' socioeconomic characteristics are practically fixed) are, as this may cause difficulties in estimating models, as discussed by Cherchi and Ortúzar (2008b).

If a substantive intervention is planned for a system, panels have even more significant advantages for evaluating changes (Kitamura 1990a). Indeed, Van Wissen and Meurs (1989), based on the Puget Sound Panel, described how the effects of policies could change trends; also, it is easier to capture these changes using observations of the same individuals, as part of their current behaviour may be explained by previous experiences. Although the advantages of panels seem clear, there are precious few panels built around a substantial system change that would allow modelling changes in mode choice; notable exceptions are the before and after study developed in Amsterdam around an extension of its urban motorway system (Kroes et al. 1996) and the *Santiago Panel* (Yañez et al. 2009a) built around the introduction of Transantiago, a radical change to the public transport system of Santiago, Chile (Muñoz et al. 2009).

It is important to distinguish between longitudinal survey and panel data. The former consist of periodic measurements of certain variables of interest. Finally, although in principle it is possible to obtain panel data from a cross-sectional survey, measurement considerations argue for the use of a panel survey design rather than retrospective questioning to obtain reliable panel data.

3.3.4.2 Representative Sampling

Panel designs are often criticised because they may become unrepresentative of the initial population as their samples necessarily age over time. However, this is only strictly true in cohort study designs considering an unrepresentative sample to start with; for example, if the sample consists of people with a common birth year, individuals joining the population either by birth or immigration will not be represented in the design.

In general, a panel design should attempt to maintain a representative sample of the entire population over time. So, it must cope not only with the problems of birth, immigration or individual entry by other means, but also be able to handle the incorporation of whole new families into the population (e.g. children leaving the parental home, couples getting divorced). A mechanism is needed to maintain a representative sample that allows families and individuals to enter the sample with known probabilities, but this is not simple (for details see Duncan *et al.* 1987).

3.3.4.3   Sources of Error in Panel Data

A panel design may add to (or detract from, if it is not done with care) the quality of the data. Although repeated contact and interviewing are generally accepted to lead to better-quality information, panels have typically higher rates of non-response than cross-sectional methods, and run the risk of introducing *contamination* as we discuss below.

Effects on Response Error   Respondents in long survey panels have repeated contact with interviewers and questionnaires at relatively long time intervals; this may improve the quality of the data for the following reasons:

- Repeated interviewing over time reduces the amount of time between event and interview, thus tending to improve the quality of the recalled information.
- Repeated contact increases the chances that respondents will understand the purpose of the study; also they may become more motivated to do the work required to produce more accurate answers.
- It has been found that data quality tends to improve in later waves of a panel, probably because of learning, by respondents, interviewers or both.

However, in the case of short survey panels, the quality of responses tends to decrease with the number of days considered (i.e. less trips are reported and travel by slow modes is omitted) due to fatigue.

Non-response Issues   Under the generic non-response label, there are included several important issues which have two basic sources: the loss of a unit of information (attrition) and/or the loss of an item of information. Hensher (1987) discusses in detail how to test and correct for this type of error.

   The non-response problems associated with the initial wave of a panel are not different to those of cross-sectional surveys, so very little can be done to adjust for their possible effects. In contrast, plenty of data have been gathered about non- respondents in subsequent waves; this can be used to determine their main characteristics, enabling non-response to be modelled as part of the more general behaviour of interest (see Kitamura and Bovy 1987).

   Typical large panel designs spend a great amount of effort attending to the 'care and feeding' of respondents: this involves instructing interviewers to contact respondents many times and writing letters of encouragement specifically tailored to the source of respondents' reluctance. This 'maintenance policies' are often considered important by panel administrators, as are the use of incentives to encourage cooperation (Yañez *et al.* 2009a).

Response Contamination   Evidence has been reported that initial-wave responses in panel studies may differ from those of subsequent waves; for this reason in some panel surveys the initial interviews are not used for comparative purposes. A crucial question is whether behaviour itself, or just its reporting, is being affected by panel membership. Evidence about this is not conclusive, but it seems to depend on the type of behaviour measured. For example, Traugott and Katosh (1979) found that participants in a panel about voting behaviour increased their voting (i.e. changed behaviour) as time went by; however, it was also found that this was caused partly by greater awareness of the political process and partly by the fact that individuals who were less politically motivated tended to drop out of the panel.

Treatment of Repeated Observations   Another problem, which is more specific to short survey panels, relates to the presence of repeated observations. It is normal to expect that individuals, in different days, may repeat exactly the same trips (typical cases are the systematic trips to work that are often made every day with the same characteristics: time, cost, purpose, mode, and so on). So, especially when these data are used for model estimation, a crucial question here is which should be the optimum length of the short survey panel as the way in which repeated information is treated may affect the estimation results (Cherchi *et al.* 2009; Yañez *et al.* 2009b).

3.3.4.4 Relative Costs of Longitudinal Surveys

Questions about the relative costs of longitudinal studies cannot be answered without reference to the alternatives to them. One obvious comparison is between a single cross-sectional survey, with questions about a previous period, and a two-wave panel. However, if the longitudinal study is designed to keep its basic sample representative each year and if enough resources are devoted to the task, it can also serve as an (annual) source of representative cross-sectional data and thus ought to be compared with a series of such surveys rather than just a single one.

Duncan *et al.* (1987) have made rough calculations on these lines, concluding that in the first case the longitudinal survey would cost between 20 to 25% more than the cross-sectional survey with retrospective questions. However, they also conclude that in the second case the field costs of each successive wave of the cross-sectional study would be between 30 and 70% higher than additional waves of the panel survey, depending on the length of the interview.

Other costs are caused by the need to contact and persuade respondents in the case of panels and by the need to sample again with each fresh cross-section in the other case. Finally, there are other data processing costs associated with panels but these must be weighed against the greater opportunity to check for inconsistencies, analysis of non-response, consideration of inertia effects in modelling and so forth.

3.3.5 Travel Time Surveys

The requirement for detailed and accurate travel time, vehicle speed and delay data is important for the calibration and validation of model systems. Travel times are a key determinant of travel costs therefore it is important to ensure the model correctly represents delays in the network of interest. In principle, one would expect traffic on the road network to be subject to variability in their travel time. Nevertheless, travel times on buses, and even metro, can also be affected by congestion and disruption. The focus here is mostly on vehicle travel times on congested networks but the principles are applicable to other modes. Travel times can be divided into:

- Running times, whilst the vehicle is moving.
- Delays, when the vehicle is stopped because of congestion or traffic control measures (traffic lights, stop sign, etc.).

Travel times can be very variable as a result of traffic control measures and simply congestion levels. For short links measurement errors will also be significant. As the variability over a single link would be too high for most models, it is preferable to observe travel times on segments covering several junctions to reduce it and make results more representative of the type of model used. For strategic models segments should include at least five links or be at least 1 km long (whatever is the longest). For smaller scale models one may use shorter segments but observations will have to be repeated more to obtain a reliable estimate despite local variability in travel times.

The most common technique for travel time measurements is known as the 'moving observer method'. In this case, a probe car is driven at the average speed of the traffic stream and times are recorded for stretches of road. Maintaining an average speed is difficult and the normal requirement is for the driver to overtake as many vehicles (in the relevant class) as vehicles overtake him. An observer in the car (or a

GPS based instrument), record times at regular intervals or when passing identifiable locations (i.e. key junctions, a particular bridge or building).

The design of a travel time survey requires:

- Specifying the level of accuracy required.
- Identification of one or more circuits to be surveyed.
- Identification of the road sections of interest.
- Selection of a method for data collection: observer, GPS or other.
- Selection of the days and times when the surveys will be conducted.
- Number of runs that will be needed for each circuit and survey times.

The accuracy required will depend on the objective of the model. For large scale strategic models it is desirable to have an accuracy of some 5 to 8 km/hr (around ± 10%). For operational studies, a better accuracy of 2 to 5 km/hr is desirable. In the case of disaggregate mode choice models, as discussed in Chapters 7 to 9, it has been found that the level of accuracy should be very high indeed (i.e. average travel times for the peak period will not adequately represent those experienced by travellers within the peak, and it has been recommended to group individuals according to departure time in, at most, 15 min intervals, see Daly and Ortúzar 1990).

The circuits to be surveyed should be representative of the study area of interest. They should cover roads and streets of different types and flow levels, with emphasis on those types considered most important. The length of the road sections should be chosen to reduce the variability encountered at junctions, especially if signal controlled. For dense urban areas, sections should contain between 7 and 10 signal controlled junctions.

The sample size and the number of runs to be undertaken, will depend also on the variability observed on different types of roads under different conditions. Equation (3.16) above can be used to estimate more accurately both the number of segments (links between junctions) in a road section and the number of runs. Ideally, the coefficient of variation CV should be estimated from observations. Typical values for the CV would be between 9 and 15 for roads with low and high variability. If we require 90% confidence to be within a 10% error, this results in three to seven runs for this range. It is often recommended that at least five runs are undertaken to ensure any special circumstance does not unduly affect the results.

3.4 Stated Preference Surveys

3.4.1 Introduction

The previous discussion has been conducted under the implicit assumption that any choice data corresponded to *revealed preference* (RP) information; this means data about actual or observed choices made by individuals. It is interesting to note that we are seldom in a position to actually observe choice; normally we just manage to obtain data on what people report they do (or more often, what they have done on the previous day or, better, in the pre-assigned travel day).

In terms of understanding travel behaviour, RP data have limitations:

- Observations of actual choices may not provide sufficient variability for constructing good models for evaluation and forecasting. For example, the trade-offs between alternatives may be difficult to distinguish so the attribute level combinations may be poor in terms of statistical efficiency.
- Observed behaviour may be dominated by a few factors making it difficult to detect the relative importance of other variables. This is particularly true with secondary qualitative variables (e.g. public-transport information services, security, décor) which may also cost money and we would like to find out how much do travellers value them before allocating resources among them.

- The difficulties in collecting responses for policies which are entirely new, for example a completely new mode (perhaps a people mover) or cost-recovery system (e.g. electronic road pricing).

These limitations would be surmounted if we could undertake real-life controlled experiments within cities or transport systems, but the opportunities for doing this in practice are very limited. Thus, where data from real markets is not available for predicting behaviour or eliciting reliable preference functions, researchers have had to turn to *stated preference* (SP) methods. These cover a range of techniques, which have in common the collection of data about respondent's intentions in hypothetical settings as opposed to their actual actions as observed in real markets. The three most common SP methods have been *contingent valuation* (CV), *conjoint analysis* (CA) and *stated choice* (SC) techniques. In transport, SC techniques have tended to dominate (see some examples in Ortúzar 2000) and for this reason, we will focus on this method providing only a brief description of the CV and CA survey approaches. Note also that in the transport arena the SP label has not embraced contingent valuation, as in fields such as marketing or environmental economics; further, in transport practice the SP label has generally referred to either CA or SC without a formal distinction (see the discussion in Ortúzar and Garrido 1994b).

3.4.1.1 Contingent Valuation and Conjoint Analysis

As a coherent technique, CV primarily deals solely with eliciting *willingness-to-pay* (WTP) information for various policy or product options (Mitchell and Carson 1989). In this case, the policy (e.g. a way to reduce accident risk) is presented to respondents who are then asked how much they are willing to pay for having it. Four types of CV questions are typically used in practice; direct questioning, biding games, payment options and referendum choices. In CV studies, the policy or product is kept static and the outcome, in the form of WTP, is for the entire product or policy. As such, CV questions cannot be used to disentangle the WTP for individual characteristics or attributes of the product or policy under study. We will come back to this technique in section 15.4.

Unlike CV, traditional conjoint analysis allows the researcher to examine the preferences, and even WTP if a price or cost attribute is included, not only for the entire policy or product, but also of the individual characteristics of the object(s) under study. In CA, respondents are presented with a number of alternative policies or products and are asked to either rate or rank them (see Figure 3.6). The levels of the characteristics or attributes of the various policies or products are systematically varied and become the independent variables which are regressed against the ratings or rankings data. The parameter weights for each attribute reflect the marginal preference or 'part-worth' for that attribute. Thus, if a cost or price attribute is included as part of the product or policy presented to respondents, then the ratio of any non-price parameter to the price or cost parameter reflects the marginal WTP for the associated non-price attribute (Gaudry *et al.* 1989). The special difficulties associated with estimating WTP when flexible discrete choice functions, such as those we will discuss in section 8.6 are used to model the situation in hand, are discussed by Sillano and Ortúzar (2005).

Traditional CA has had limited acceptance in transport studies due to a number of criticisms that have been levelled against the method over the years (Louviere and Lancsar 2009). Firstly, it has been argued that the statistical methods primarily used to analyse CA data are inappropriate, in that the dependent variable of a linear regression model should be, at a minimum, interval scaled. As such, using ranking data as a dependent variable certainly violates this assumption, although some argue that even ratings data also is not interval level data, given how respondents psychologically use the ratings metric. A second criticism lies not in how the data is analysed but with the very use of ratings or rankings data as measurement metric. Respondents in real life do not rate or rank alternatives and even if they did different people would approach such scales in psychologically different manners. As such, it has been argued that outputs of CA surveys have no psychologically meaningful interpretation (Louviere and Lancsar 2009). So, SC methods have tended to dominate transport studies.

Fare	Interchange	Time on bus	Walk time
70 p	No change	15 mins	10 mins

Fare	Interchange	Time on bus	Walk time
70 p	No change	20 mins	8 mins

Fare	Interchange	Time on bus	Walk time
85 p	No change	15 mins	10 mins

Fare	Interchange	Time on bus	Walk time
85 p	1 change	15 mins	8 mins

Figure 3.6   Example of stated-preference ranking exercise

3.4.1.2   Stated Choice Methods

Stated choice studies are similar to CA methods insofar as respondents are presented with a number of hypothetical alternatives; however, the two methods differ in terms of the response metric. Whereas CA asks respondents to rank or rate the alternatives (with all alternatives shown to respondents at the same time), respondents undertaking a SC survey are asked to choose their preferred alternative from amongst a subset of the total number of hypothetical alternatives constructed by the analyst. In asking respondents to make a choice, rather than a rating or ranking, the two criticisms levelled at CA are avoided. Firstly, the analysis of discrete choice data requires a different set of econometric models specifically developed to analyse such data; thus, the choice metric is consistent with the statistical model applied to it. Secondly, the selection of the single preferred alternative is psychologically consistent across respondents and a task that is common to individuals in real markets. A further distinction between the two methods is that CA tasks typically present respondents with a relatively large number of alternatives, simultaneously, to rate or rank, whereas SC methods typically present only a few alternatives at a time (and in most cases only two), changing them and having respondents repeat the choice task.

On the other hand, the primary distinction between RP and SC surveys is that in the latter case individuals are asked about what they would choose to do (or how would they rank/rate certain options) in one or more hypothetical situations. The degree of artificiality of these situations may vary, according to the needs and rigour of the exercise:

- The *decision context* may be a hypothetical or a real one; in other words, the respondent may be asked to consider an actual journey or one that she might consider undertaking in the future.
- Some of the *alternatives* offered may be hypothetical although it is recommended that one of them be an existing one, for example the mode just chosen by the respondent including all its attributes.

A crucial problem with stated preference data collection in general, is how much faith we can put on individuals actually doing what they stated they would do when the case arises (for example, after introducing a new option). In fact, experience in the 1970s was not good in this sense, with large differences between predicted and actual choice (e.g. only half the people doing what they said they would) found in many studies (see Ortúzar 1980a).

The situation improved considerably in the 1980s and good agreement with reality was reported from models estimated using SC data (Louviere 1988a). However, this occurred because data-collection methods improved enormously and became very demanding, not only in terms of survey design expertise but also in their requirements for trained survey staff and quality-assurance procedures. The interested reader can consult the excellent book by Louviere *et al.* (2000).

The main features of an SC survey may be summarised as follows:

(a) It is based on the elicitation of respondents' statements of how they would respond to different hypothetical (travel) alternatives.
(b) Each option is represented as a 'package' of different attributes like travel time, price, headway, reliability and so on.
(c) The analyst constructs these hypothetical alternatives so that the individual effect of each attribute can be estimated; this is achieved using *experimental design* techniques that ensure the parameters of the chosen attributes are estimated with the smallest standard errors. In reality, an experimental design is nothing more than a matrix of numbers used to assign values to the attributes of each alternative. By using experimental design theory, the assignment of these values occurs in some non-random manner, and by systematically varying the design attributes, the analysts are able to control as many factors as possible influencing the observed choices. In creating the design in a specific and precise manner, the analyst seeks to ensure the ability to obtain reliable parameter estimates with minimal *confoundment* with the other parameter estimates.
(d) The researcher has to make sure that respondents are given hypothetical alternatives they can understand, appear plausible and realistic, and relate to their current level of experience.
(e) The responses given by individuals are analysed to provide quantitative measures of the relative importance of each attribute; for this choice models are estimated as discussed in detail in Chapter 8.

However, the process of constructing effective SP surveys is far from simple and quite time consuming if done correctly. Extensive qualitative and secondary research is advised to determine the relevant set of alternatives, attributes and attribute levels that will be used to make up the hypothetical alternatives. In what follows we give advice based on useful discussions and comments by Dr. John M. Rose, Institute of Transport and Logistics Studies, University of Sydney, one of the leading experts in this subject.

In preparing a SP survey, the analyst will need to address at least the following issues:

- Will the experiment be *labelled* (i.e. the names of the alternatives have substantive meaning beyond their ordering; see Figure 3.7) or *unlabelled* (see Figure 3.8) and will a *non purchase* or *status quo* alternative be presented (see Figure 3.7b)? We will come back to this last issue in section 3.4.2.6.
- In deciding what attributes to use, we need to determine what factors best represent those influencing choices between the various alternatives. Note that other external criteria may also influence this task;

	Train	Bus		Car
Fare	$3.00	$4.00	Petrol Costs	$1.00
			Toll Cost	$3.00
			Parking Cost	$8.00
Access Time	5 mins	10 mins		
In-vehicle Time	35 mins	25 mins	In-vehicle Time	15 mins
Egress Time	15 mins	10 mins	Egress Time	5 mins
I would choose	○	○	or	○

(a) Standard design

	Train	Bus		Car	None
Fare	$3.00	$4.00	Petrol Costs	$1.00	
			Toll Cost	$3.00	
			Parking Cost	$8.00	
Access Time	5 mins	10 mins			
In-vehicle Time	35 mins	25 mins	In-vehicle Time	15 mins	
Egress Time	15 mins	10 mins	Egress Time	5 mins	
I would choose	○	○	or	○	○

(b) Design including a non-purchase option

Figure 3.7 Example of labelled mode SC tasks

for example, if the outputs from the study will be used as inputs into, say a network model, the attributes should accommodate the constraints or needs of the latter (e.g. if a network model does not allow for a comfort attribute, the analyst will need to determine whether it is worthwhile including comfort in the SC study); we will come back to this issue below.

	Route A	Route B	Route C
Petrol Costs	$1.50	$2.00	$1.00
Toll Cost	$2.00	$4.00	$0.00
Prob. of arriving late	0.3	0.5	0.1
Prob. of arriving early	0.1	0.2	0.3
Free Flow Time	15 mins	10 mins	20 mins
Congested Time	10 mins	15 mins	20 mins
Egress Time	15 mins	10 mins	5 mins
Please rank in order of preference (1 = best)	○	○	○

Figure 3.8 Example of unlabelled route SC task

• With regards to attribute levels, the analyst needs to define values for each one, including specific quantitative values (e.g. $5, $10 and $20) or qualitative labels ('low', 'medium' and 'high'). Once the above have been defined, further pre-testing and piloting is also recommended. This may result in further refinements of the survey instrument. Only once the analyst is satisfied with the survey, should the SC study be put out to field.

3.4.2   The Survey Process

In setting up a SP survey, analysts should aim to follow the five stages illustrated in Figure 3.9. The first stage requires that the study objectives be clearly defined and clarified. This involves identifying the population of interest as well as refining the experimental objects, or alternatives that will be studied. Definitions and descriptions of new alternatives should also be defined and tested.

The second stage requires outlining the set of assumptions reflecting our overall beliefs as to what qualities are important for an experimental design to display. These assumptions will dictate the statistical properties of the design generated in Stage 3 of the process. As there exist many different possible

Figure 3.9   Steps in designing a stated preference experiment

experimental designs for any given problem (each with different statistical properties), specifying the assumptions and outlining the properties that the analyst deems important is critical to generate the design. Unfortunately, in the vast majority of SP studies, this second stage is generally ignored with researchers generating designs without fully appreciating what assumptions led to them, or whether the generated designs are appropriate for meeting the needs of the study.

The actual method for constructing a design in Stage 3 is dependent on the assumptions made in Stage 2, with different assumptions requiring different design generation methods. Thus, even if the analyst skips Stage 2, implicit assumptions are still being made in generating the design.

Stage 4 represents an ideal stage in the process rather than a necessary one; unfortunately, as with Stage 2, it is often ignored in practice. In this stage, the analyst performs tests, usually in the form of simulations, in order to determine how the design is likely to perform in practice. This type of tests may allow the analyst to correct any issues with the design before going to field.

The final stage of the design generation process involves taking the design and using it to construct the questionnaire that will be given to respondents.

3.4.2.1   Clarifying Study Objectives and Defining Objects of Interest

This stage involves the analyst gaining an understanding of the specific context or problem under study, the population of interest, as well as the types of choices that sampled respondents will be asked to make.

Typically, the choice context (experimental conditions) is an input that is not under the analyst's control, being supplied by an external client or determined by the study objectives. Nevertheless, understanding the context of the study is crucial to the success of SP studies and in special circumstances in-depth interviews and seeking specialist knowledge may be vital in this task (Ortúzar and Palma 1992). Armed with this knowledge, the analyst will then need to determine what behavioural outputs are of direct interest, such as determining the subjective value of time (SVT), WTP for risk reductions, or just estimating a generalised cost of travel formulation.

After gaining a full understanding of the problem under study, the analyst is next required to indentify and understand the population of interest. This involves determining who the sample respondents are, where they are likely to be located, how they will be sampled and how will they be surveyed. Understanding such questions at this stage is important, as they will influence the type of questionnaire that will be used and this will likely influence the type of experimental design generated.

For example, if respondents are located in a geographically dispersed pattern and have limited access to internet, mail-back paper and pencil surveys may be the only option. In such cases, the experimental design will be more difficult to adapt to individual specific circumstances. Where respondents may be surveyed using a computer or over the internet, the experiment may adapt to each individual's reported circumstances (e.g. if a respondent does not have access to a car, then the car alternative may be removed from the survey for that respondent). The type of survey used will also have implications in terms of the data collected, which will determine whether the assumptions made in Stage 2 of the survey design process transfer from the design over to the data finally gathered. As well as having an impact upon survey design, understanding the population of interest will also provide further insights in the sampling required for the study (see Stage 3). Finally, in understanding the population of interest, the analyst may determine for example, whether different segments should be sampled, and hence whether more than one experimental design or survey questionnaire is necessary.

Understanding the population of interest as well as the overall study objectives will provide insights into the number of and diversity of alternatives applicable to various sampled individuals when making decisions within the study context. Such knowledge will assist in constructing the choice tasks that will be used in the SP survey. Figure 3.7 showed two different choice tasks that might be considered for a mode choice study. That in Figure 3.7a requires respondents to choose between train, bus and car alternatives, whereas the choice task in Figure 3.7b allows the respondent to also select none of these alternatives. If the objective of the study was to model commuter choice, then the first choice task can be applied without problems for respondents who cannot choose not to travel to work; however, it might not be adequate for respondents who can telecommute (i.e. work from home), and in that case the second choice task should be preferable. Similarly, for non-commuting trips the second choice task might be more appropriate given that most non-commuting trips can be considered discretionary in nature (at least for non-teenagers).

Attributes and Alternatives   The construction of realistic, or technologically feasible, alternatives requires the following four distinct tasks:

a) The range of options is usually given by the objective of the exercise; however, one should not omit realistic alternatives a user might consider in practice. For example, in studying potential responses of car drivers to new road-pricing initiatives it may not be sensible to consider only alternative modes of travel; changes to departure time or to alternative destinations (to avoid the most expensive road charges) may be very relevant responses. By ignoring them one places the respondent in a more artificial (less realistic) context, perhaps triggering inappropriate or unrealistic responses (we will discuss this type of issue further in section 3.4.2.6).

b) The set and nature of the attributes should also be chosen to ensure realistic responses. The most important attributes must be present and they should be sufficient to describe the technologically feasible alternatives. Care must be applied here as particular combinations of attributes (e.g. a high-quality, high-frequency, low-cost alternative) may not be seen as realistic by respondents thus reducing the value of the whole exercise. Care must be taken also if the number of attributes is deemed excessive (say higher than six); Carson *et al.* (1994) found that fatigue effects make respondents simplify their choices by focusing on a smaller number of attributes or simply answering at random or in lexicographic fashion (Sælensminde 1999). In this sense, recent work has shown that there may be limits, which are culturally affected, on the number of choice tasks, alternatives, attributes and even their range of variation, that are acceptable in a given study (Caussade *et al.* 2005; Rose *et al.* 2009a).

c) To ensure that the right attributes are included and that the options are described in an easy-to-understand manner, it is advantageous to undertake a small number of group discussions (e.g. focus groups) with a representative sample of individuals. A trained moderator will make sure all relevant questions regarding perception of alternatives, identification of key attributes and the way in which they are described and perceived by subjects, and the key elements establishing the context of the exercise are all discussed and reported. Focus groups cost money and in many cases the researcher will be tempted to skip them believing a good understanding of the problem and context already exists. In that case, it will be even more essential to undertake a carefully monitored pilot survey where any issues of attribute description and alternative presentation can be explored.

d) The selection of the metric for most attributes is relatively straightforward. However, there are some situations that may require more careful consideration, in particular with respect to qualitative attributes like 'comfort' or 'reliability'. For example travel time reliability can be presented as a distribution of journey times on different days of a working week, or as the probability of being delayed by more of a certain time. For more on this issue see the discussion of stimulus presentation below.

e) Finally, in relation to the number of levels that each attribute can take, it is important to bear in mind that Wittink *et al.* (1982) found evidence that variables with more levels could be perceived as more important by respondents; we will come back to this issue in relation to another topic below.

3.4.2.2   Defining Experimental Assumptions

For any SC study, there exist many potential experimental designs that can be constructed. The analyst's aim is to choose a particular design construction method and generate the design. This will depend upon many different considerations, most of which reflect the personal beliefs of the analyst as to what are important properties the design must possess. However, some decisions do not reflect the personal biases or beliefs of the analyst but, rather, are influenced by the problem being studied.

Labelled or Unlabelled Experiments   In many instances the decision to treat an experiment as either *labelled* or *unlabelled* will depend upon the problem under study. In particular, mode choice studies will generally require a labelled experiment, whereas route choice problems are in general amenable to unlabelled SC experiments. Nevertheless, the decision as to whether either type of experiment is used is crucial, as it typically impacts upon the number and type of parameters that will be estimated as part of the study.

Generally, unlabelled experiments require only the estimation of generic parameters whereas labelled experiments may require the estimation of either alternative specific or generic parameters, or combinations of both. Prior knowledge of the number of likely (design related) parameter estimates is important as each one represents an additional degree of freedom required for estimation purposes. General experimental design theory posits that the Fischer Information (or Hessian) matrix ($\mathbf{I}$) will be singular if the number of choice observations (each one equivalent to a choice task) is smaller than the number of parameters (see Goos 2002). As such, the minimum number of choice tasks required for an experimental design is equal to or greater than the number of (design related) parameters to be estimated. Note that the inclusion of a *status quo* alternative does not impact upon the minimum number of choice tasks required for a design, as it does not require the estimation of any attribute related parameter estimates. The decision to use a labelled rather than an unlabelled choice experiment may also impact upon the generation of *orthogonal designs*, as discussed in section 3.4.2.3.

Imposing Attribute Level Balance   This is another consideration in generating designs. Attribute level balance occurs when each attribute level appears an equal number of times, within each attribute, over the entire design. This is generally considered a desirable property, although it may impact upon the statistical efficiency of the design (see section 3.4.2.3). If present, it ensures that each point in preference space is equally represented, so that parameters can be estimated equally well on the whole range of levels, instead of having more or less data points at only some of the attribute levels (which may affect how the design performs in practice). Nevertheless, it is worth noting that attribute level balance may require larger designs than dictated by the number of parameter estimates requirement.

Example 3.11   Consider a design with four attributes, where two have two levels, one has three levels and the last has four levels. In the classical jargon in this field we would refer to this as a $2^2\, 3^1\, 4^1$ factorial design; note that the product of levels to the power of attributes (48 in this case) represents the total number of choice tasks needed to recover all effects (i.e. main or linear effects and all interactions), i.e. a full factorial design (more about this below).

Assuming each attribute will produce a unique parameter estimate (i.e. main effects only), the smallest design would require just four choice tasks based on the number of parameters criterion; however, to maintain attribute level balance, the smallest possible design would require 12 choice tasks (12 being divisible without remainder by 2, 3 and 4).

Number of Attribute Levels   This should reflect the researchers' belief as to the relationship each level has to the overall contribution to utility and whether the relationship is expected to be linear or nonlinear from one level to the next. If nonlinear effects are expected for a certain attribute and the analyst suspects that the attribute will be, say, *dummy coded* (see Example 3.11) prior to analysis, then more than two levels will be needed to model appropriately the suspected nonlinearities. Where dummy coded (or *effects* and/or *orthogonal* coded) attributes are included, the number of levels for these attributes is predetermined. However, the more levels used, the higher the potential number of choice tasks required due to additional parameters being estimated. Also, mixing the number of attribute levels for different attributes may yield a higher number of choice tasks (due to attribute level balance).

Varying the Range of Attributes   Research into the impact of this suggests that using a wide range (e.g. \$0–\$30) is statistically preferable to using a narrow range (e.g., \$0–\$10) as this will theoretically lead to parameter estimates with a smaller standard error; however, using too wide a range may also be problematic (see Bliemer and Rose, 2008). In fact, having too wide an attribute level range may result in choice tasks with dominated alternatives; whereas having too narrower a range may result in alternatives for which the respondent will have trouble distinguishing between (see Cantillo *et al.* 2006). However, such considerations are purely statistical in nature and analysts should also consider practical limitations upon the possible range that the attribute levels can take; that is, the attribute levels shown to respondents must make sense to them (must be realistic). Hence there will be often a trade-off between the statistical preference for a wider attribute level range and practical considerations that may limit this range.

Inclusion of Interaction Effects   These are important when the effects of two variables are not additive (see Figure 3.10); including interactions will impact upon the number of choice tasks required of a design. This is because each interaction effect will have a corresponding parameter estimate and hence it requires an additional degree of freedom, and in turn, an additional choice task. As such, Rose and Bliemer (2009) suggest starting the design generation process by specifying the 'worst case' utility specification (i.e. in terms of all the effects that might be tested, along with any non-linear parameterisation that may be estimated). Generating a design with too few choice tasks will likely preclude the estimation of potentially valid utility specifications at a later stage, whilst generating a design with more than the minimum number of choice tasks does not preclude the estimation of simpler model forms.

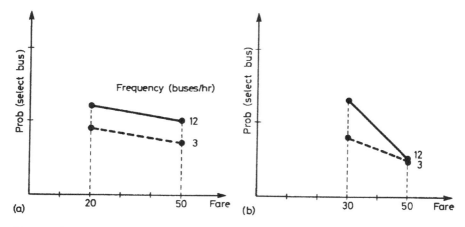

Figure 3.10   Presence and absence of attribute interaction: (a) without interaction, (b) with interaction

Once decisions for each of the above have been made, several different experimental *design generation procedures* can be considered. The easiest method is to employ a *full factorial* (FF) design, i.e. one consisting of all possible choice tasks. One benefit of using a FF design is that all main effects and interaction effects will be orthogonal. Unfortunately, the number of choice tasks in a FF design will typically be too large and many of the choice tasks will have dominated or unrealistic alternatives.

Fractional Factorial Designs   Due to the practical impossibility of dealing with FF designs, many analysts rely on the so-called *fractional factorial* designs, which consist of a subset of choice tasks from the full factorial. To construct a fractional factorial design, one could randomly select choice tasks from the FF; however, more intelligent strategies are possible. Numerous methods have been explored within the literature as to how to select choice tasks in a structured manner, so that the best possible data from the SC experiment will be produced for estimating models. The most widely known fractional factorial design type is the *orthogonal design*, which is produced so as to have zero correlations between the attributes within the SC experiment (and thus it is excellent for estimating linear models, see Rose and Bliemer 2009). Although there are several types of orthogonal designs, we will consider here only the most popular one, consisting of constructing a simple orthogonal array.

Example 3.12 Consider a situation with five attributes, two at two levels and the rest at three levels (i.e. a $2^2\,3^3$ design). In this case, depending on the number of interactions to be tested, the number of options required would vary as follows in a classical orthogonal design:

- 108 to consider all effects (i.e. a full factorial design);
- 81 to consider principal effects and all interactions between pairs of attributes, ignoring effects of a higher order;
- 27 to consider principal effects and interactions between one attribute and all the rest;
- 16 only if no interactions are considered.

More recently, several researchers have suggested other types of fractional factorial designs such as *D-optimal* or *D-efficient* designs (Rose and Bliemer 2008). In generating these types of designs, researchers define its efficiency in terms of variances (the roots of which are the standard errors, as we will see in Chapter 8) and covariances of the parameter estimates; the lower these (co)variances, the more efficient the experimental design. As such, the objective in generating this type of design is to choose attribute level combinations that will result in the smallest possible parameter (co)variances. In order to do so, the analyst must make a number of assumptions about the model to be estimated as well as the parameter estimates that will be obtained. This enables the expected asymptotic covariance matrix ($\mathbf{S}^2$) of the design to be calculated, from which the (co)variances are derived; note that it is calculated as the negative inverse of $\mathbf{I}$ (see section 8.4.1), the Fisher information or Hessian matrix. Understanding what model will be estimated is important, as $\mathbf{S}^2$ for a given design will be different for different econometric model specifications.

Nevertheless, two competing schools of thought have emerged within the literature as to what parameter priors are appropriate to use in generating experimental designs for SC studies. The first creates designs under the so-called *null hypothesis*, namely zero valued parameter priors (Street *et al.* 2005), whilst the competing school assumes non-zero valued parameter priors. Clearly, in the latter case, one has to decide what these non-zero valued parameter priors are, typically leading to more efficient designs if the population parameter estimates are truly non-zero. However, this comes at the expense of the effort to obtain parameter priors (Rose and Bliemer 2008).

Within the first school of thought, whilst not necessary, further assumptions are typically made in generating SC designs. Firstly, it is generally assumed that designs that are orthogonal within alternatives and which maximise the differences in attribute levels between alternatives (i.e. will be correlated between alternatives) will be optimal (see Burgess and Street 2003; 2005; Street and Burgess 2004; 2007). This is because under the assumption that the parameters are all zero, any discrete choice model will collapse to a linear model and therefore an orthogonal design will be optimal. This also acknowledges the fact that discrete choice models are really difference in the utilities models (see section 7.3.1).

A second assumption which has been less well communicated for this class of designs, is that $\mathbf{S}^2$ is usually constructed under the assumption that the analyst will apply orthogonal codes to models estimated using data collected with the design. As such, the efficiency of the design may be reduced if a different coding system is used in practice, which is typically the case. Finally, optimal designs for this type of design have only been produced under the assumption of estimating Multinomial Logit (MNL) models, which are the simplest discrete choice models (Domencich and McFadden 1975). Nevertheless, the appeal of this approach is two-fold:

- Respondents are forced to make trade-offs on each and every attribute of the design, as no two attributes in any given choice situation will, where possible, take the same value.
- The approach does not require *a priori* knowledge of the parameter estimates, and this may be particularly suitable for designs which have mainly qualitative attributes.

The second design school of thought utilise non-zero valued parameter priors, assuming some prior knowledge about their values. Original research into generating *efficient* designs assumed that researchers had *exact* knowledge about the expected parameter estimates (e.g. $\theta_1 = -0.2$; $\theta_2 = 1.0$). Designs generated under such an assumption are known as *locally optimal designs* as their (co)variances will be minimised only at the precise values assumed for the prior parameters (see Rose and Bliemer 2005; Scarpa and Rose 2008).

More recently, researchers have produced *Bayesian efficient designs* that do not assume precise knowledge of the parameter estimates. Such designs allow for the true population to fall within some distribution of possible parameter estimates, such that the analyst optimises over the distribution of possible priors (e.g. $\theta_1 \sim N(-0.8, 0.2)$). In this approach, we generally let go of the principle of orthogonality and generate designs in a manner that can be expected to minimise the elements of S^2 associated with the (non-linear) discrete choice model estimated on the data (Bliemer and Rose 2006; Bliemer *et al.* 2009; Carlsson and Martinsson 2002; Ferrini and Scarpa 2007; Fowkes 2000; Huber and Zwerina 1996; Kanninen 2002; Kessels *et al.* 2006; Rose and Bliemer 2008; Sándor and Wedel 2001; 2002; 2005; Scarpa and Rose 2008; Toner *et al.* 1998; Watson *et al.* 2000).

The main advantage of this approach is that the generated design is directly related to the expected outcome of the modelling process. Besides, it can be optimised for any model type, not just MNL, or for a range of model types (Rose *et al.* 2009b). Further, the approach can assume any data structure (i.e. it is not limited to assuming orthogonal coding). However, while the first school can prove optimality of their designs (under the null hypothesis), the second school generally cannot (under the non-null hypothesis). Therefore these designs are typically called *efficient* and not *optimal* designs.

A Note on Dummy, Effects and Orthogonal Coding Data   If the marginal impact upon utility is believed to be non-linear from one attribute level to another, the analyst may wish to test this by transforming the data using dummy, effects or orthogonal codings. Within the literature, the former remains the preferred method, although effects and orthogonal codings offer a number of advantages over it.

In all cases, the analyst creates $D = L - 1$ new variables in the data, where L is the total number of levels for the attribute being transformed. For *dummy coding* transformations, the researcher uses a series of zeros and ones to map the original levels to the newly created D variables. To create the mapping, each newly dummy variable corresponds to the first $L - 1$ levels of the original attribute. To create the dummy codes, every time level l appears for the original attribute, the corresponding newly created dummy variable takes the value 1; otherwise, it takes the value zero. This occurs for all but the last attribute level (the base level) which does not have a corresponding dummy variable. In that case, the last attribute level will take the value zero for all dummy coded variables (see Table 3.3). *Effects coding* use the same pattern of mapping creating $E = L - 1$ effects coded variables; however, the base level takes the value -1 for all the E effects coded variables (see Table 3.3).

Finally, like dummy and effects coding, *orthogonal coding* requires the creation of $O = L - 1$ new variables. However, unlike dummy and effects coding here we use orthogonal polynomial contrasts to populate the O new variables (see Table 3.3). Thus, each successive orthogonal coded variable corresponds to a higher order polynomial effect for the pre-transformed attribute (i.e. $o_{k1} \rightarrow x_k$ (linear effect), $o_{k2} \rightarrow x_k^2$ (quadratic effect), $o_{k3} \rightarrow x_k^3$ (cubic effect), etc.). Rather than directly taking the polynomial power of the original variable and using these directly (which will induce correlation in

(continued)

the data), orthogonal coding uses orthogonal polynomial contrasts which retain orthogonality within each attribute (see Chihara 1978).

Table 3.3 Example of dummy, effects and orthogonal coding

Attribute Levels		Dummy Coding					Effects Coding					Orthogonal Coding				
		D1	D2	D3	D4	D5	E1	E2	E3	E4	E.5	O1	O2	O3	O4	O5
2	1	1	–	–	–	–	1	–	–	–	–	1	–	–	–	–
	2	0	–	–	–	–	–1	–	–	–	–	–1	–	–	–	–
3	1	1	0	–	–	–	1	0	–	–	–	–1	1	–	–	–
	2	0	1	–	–	–	0	1	–	–	–	0	–2	–	–	–
	3	0	0	–	–	–	–1	–1	–	–	–	1	1	–	–	–
4	1	1	0	0	–	–	1	0	0	–	–	–3	1	–1	–	–
	2	0	1	0	–	–	0	1	0	–	–	–1	–1	3	–	–
	3	0	0	1	–	–	0	0	1	–	–	1	–1	–3	–	–
	4	0	0	0	–	–	–1	–1	–1	–	–	3	1	1	–	–
5	1	1	0	0	0	–	1	0	0	0	–	–2	2	–1	1	–
	2	0	1	0	0	–	0	1	0	0	–	–1	–1	2	–4	–
	3	0	0	1	0	–	0	0	1	0	–	0	–2	0	6	–
	4	0	0	0	1	–	0	0	0	1	–	1	–1	–2	–4	–
	5	0	0	0	0	–	–1	–1	–1	–1	–	2	2	1	1	–
6	1	1	0	0	0	0	1	0	0	0	0	–5	5	–5	1	–1
	2	0	1	0	0	0	0	1	0	0	0	–3	–1	7	–3	5
	3	0	0	1	0	0	0	0	1	0	0	–1	–4	4	2	–10
	4	0	0	0	1	0	0	0	0	1	0	1	–4	–4	2	10
	5	0	0	0	0	1	0	0	0	0	1	3	–1	–7	–3	–5
	6	0	0	0	0	0	–1	–1	–1	–1	–1	5	5	5	1	1

Example 3.13 Consider the orthogonal design for two attributes, A and B, and four choice tasks shown in Table 3.4. The rows represent choice tasks and the columns have the attribute level values that would be shown in each choice task. In this case the correlation structure (i.e. the correlation coefficients for the variables in each column) is orthogonal by construction (see Wonnacott and Wonnacott 1990).

Table 3.4 Original orthogonal design

Choice task	A	B
1	1	3
2	2	1
3	3	4
4	4	2
Correlation Structure		
	A	B
A	1	0
B	0	1

Table 3.5 demonstrates dummy, effects and orthogonal coding transformations for this design. The correlation structure is also given at the base of the table for each coding type (it can be simply computed using the *data analysis tool* in Excel). As can be seen, both dummy and effects codings

induce correlation within the data, even if the original design from which they were created was orthogonal (uncorrelated). Orthogonal coding, despite being largely ignored within the literature, avoids this problem.

Table 3.5 Dummy, effect and orthogonal code comparison

Choice task	Dummy Codes						Effects Codes						Orthogonal Codes					
	A_{D1}	A_{D2}	A_{D3}	B_{D1}	B_{D2}	B_{D3}	A_{E1}	A_{E2}	A_{E3}	B_{E1}	B_{E2}	B_{E3}	A_{O1}	A_{O2}	A_{O3}	B_{O1}	B_{O2}	B_{O3}
1	1	0	0	0	0	1	1	0	0	0	0	1	−3	1	−1	1	−1	−3
2	0	1	0	1	0	0	0	1	0	1	0	0	−1	−1	3	−3	1	−1
3	0	0	1	0	0	0	0	0	1	−1	−1	−1	1	−1	−3	3	1	1
4	0	0	0	0	1	0	−1	−1	−1	0	1	0	3	1	1	−1	−1	3
	Correlation Structure																	
	A_{l1}	A_{l2}	A_{l3}	B_{l1}	B_{l2}	B_{l3}	A_{l1}	A_{l2}	A_{l3}	B_{l1}	B_{l2}	B_{l3}	A_{l1}	A_{l2}	A_{l3}	B_{l1}	B_{l2}	B_{l3}
A_{l1}	1	0	−0.3	−0.3	−0.3	−0.3	1	0.5	0.5	0	−0.5	0.5	1	0	0	0	0	1
A_{l2}	−0.3	1	−0.3	1	−0.3	−0.3	0.5	1	0.5	0.5	−0.5	0	0	1	0	0	−1	0
A_{l3}	−0.3	−0.3	1	−0.3	−0.3	−0.3	0.5	0.5	1	−0.5	−1	−0.5	0	0	1	−1	0	0
B_{l1}	−0.3	1	−0.3	1	−0.3	−0.3	0	0.5	−0.5	1	0.5	0.5	0	0	−1	1	0	0
B_{l2}	−0.3	−0.3	−0.3	−0.3	1	−0.3	−0.5	−0.5	−1	0.5	1	0.5	0	−1	0	0	1	0
B_{l3}	1	−0.3	−0.3	−0.3	−0.3	1	0.5	0	−0.5	0.5	0.5	1	1	0	0	0	0	1

Aside from the issue of correlation, a further reason for preferring to use effects and orthogonal codes over dummy codes, is that the base level of dummy coded variables will be perfectly confounded with the model constants and hence indistinguishable from each other. By using non-zero base level codes, effects and orthogonal codings avoid such confoundment and allow for independent estimates of the base level.

3.4.2.3 Generating the Experimental Design

In practice, there are several different approaches that one might employ to generate a workable experimental design, all of which reflect the analyst own beliefs about what are the most important properties for a SC experimental design to display.

Traditional Orthogonal Designs Methods These have been historically the most common experimental design types. Orthogonality is related to the correlation structure of the design attributes. By forcing them to have zero correlations, each attribute is independent of all others. Several methods for constructing orthogonal designs exist in practice, including but not limited to methods such as generating balanced incomplete blocked designs (BIBD), Latin Squares designs, orthogonal in the differences fractional factorial designs, and fold-over designs; all these have been discussed extensively elsewhere (Bunch *et al.* 1996; Fowkes and Wardman 1988; Louviere *et al.* 2000; Rose and Bliemer 2008).

We will only consider the most widely applied orthogonal design type: the L^{KJ} orthogonal fractional factorial design (where L is the number of levels, K the number of attributes and J the number of alternatives); two types of L^{KJ} designs have been described in the past. The first involves attributes that are uncorrelated both within and between alternatives; such designs are termed *simultaneous* orthogonal designs, as all alternatives are generated at the same time. The second type involves

(continued)

first locating an orthogonal design for the first alternative, and using the same design to construct subsequent alternatives by re-arranging the rows of the design (Louviere *et al.* 2000); such designs are known as *sequentially generated L^{KJ}* orthogonal fractional factorial designs.

In generating an orthogonal design sequentially, the analyst needs only locate an orthogonal design for a single alternative, whereas the simultaneous design approach requires the generation of an orthogonal design considering the correlation structure of all attributes, irrespective of which alternative they belong to. For this reason, the sequential design approach will generally result in designs with smaller numbers of choice tasks, as the theoretical minimum number of choice tasks required for a design does not necessarily guarantee that an orthogonal design may be located.

Example 3.14 Consider a design with three alternatives, each described by seven attributes with three attribute levels. The smallest simultaneous fractional factorial orthogonal design that can be constructed with 21 design attributes (7 attributes across 3 alternatives) has 72 choice tasks (see Hahn and Shapiro, 1966 or the websites mentioned below).

In comparison, the smallest sequential orthogonal design (where it is only necessary to locate an orthogonal design that is uncorrelated for 7 attributes) has only 12 choice tasks.

One limitation of the sequential design process is that such designs are appropriate only for unlabelled SC experiments. Note also that designs generated under the null hypothesis of zero prior parameters, as described above, are sequentially generated L^{KJ} orthogonal fractional factorial designs.

Independent of the actual process used, a number of useful websites and software are available for obtaining orthogonal designs (see Hedayat *et al.* 1999). Further, several software packages such as SPSS (www.spss.com), SAS (www.sas.com) and Ngene 1.0 (www.choice-metrics.com) are also able to generate a range of orthogonal designs.

D-Optimal Design Method Under the Null Hypothesis   Traditionally, an analyst would construct a sequential orthogonal design by simply assigning choice tasks randomly from the first alternative to make up the second and subsequent alternatives. More recently, a new optimality criterion has been developed to construct optimal orthogonal SC designs specifically generated for MNL models using orthogonal codes. These designs maintain (within alternative) orthogonality, whilst also minimizing S^2 under the assumption that the parameters will be zero and the attributes will be orthogonal coded.

In practice, this typically results in the attribute levels across alternatives being made as different as possible. As such, these designs will generally increase the trade-offs that respondents are forced to make across all attributes maximising the information obtained in terms of the importance that each attribute plays on choice (Burgess and Street 2005; Street and Burgess 2004). Street and Burgess (2007), Street *et al.* (2005) and Rose and Bliemer (2008), provide detailed discussions of the exact procedures used in generating this class of design. Finally, Ngene 1.0 and Burgess (http://crsu.science.uts. edu.au/choice/choice.html) provide computing capabilities for generating such designs.

D-Efficient Design Methods Under the Non-Null Hypothesis   An alternative approach to generating SC experiments involves selecting a design that is likely to provide an S^2 matrix containing values which are as small as possible, under the assumption that the parameters will not be zero. Given that their asymptotic standard errors obtained from discrete choice models are simply the square roots of the leading diagonal of this matrix, the smaller the matrix elements (or at a minimum, it's diagonal elements), the smaller the asymptotic standard errors for each parameter. However, these *efficient* designs are unlikely to be orthogonal.

Efficient designs constructed under the non-null hypothesis differ to those generated under the null hypothesis in that they attempt to mimic the performance of the model to be estimated post data collection. If after extensive pre-design research, including focus groups, in-depth interviews and pilot studies, one expects that the selected attributes have no impact upon choice (equivalent to assuming that the parameters will be statistically equal to zero), then one could question why the study is being conducted at all. In this way, the objective function defining optimality in generating a non-null prior parameter efficient design may be considered a practical one, as the design seeks to minimise the standard errors one is expecting to obtain in practice.

Nevertheless, efficient designs constructed under the non-null parameter prior hypothesis require a number of strong assumptions:

- The analyst must first decide what model type is likely to be estimated once the data has been collected, in order to decide what matrix will be specifically used in generating the design. This is because S^2 for one discrete choice model will differ from that of any other discrete choice model; for example, that corresponding to the MNL model is different to that of a Nested Logit (NL) or Mixed Logit (ML) model (see Bliemer and Rose 2008; 2009; Rose *et al.* 2009b). But recall that optimal designs generated under the null-hypothesis assumption can only assume a MNL model specification.
- The analyst must also assume what the population parameter estimates will be in order to predict S^2 for a design. The reason is that for any Logit model S^2 is analytically equal to the negative inverse of the second derivatives of the model's log-likelihood function (as we will see in section 8.3) and these are, in turn, a function of the model probabilities. But the model probabilities are a function of the utilities, which are in turn a function of the design attributes and the parameter estimates.

As discussed above, the analyst can assume prior parameter estimates in a Bayesian-like fashion when constructing a design. The assumption of prior parameters does not need to be too restrictive. Precise prior parameter values need not be provided (though such designs have been generated in the past; see for example, Carlsson and Martinsson 2002). Rather, prior parameter distributions that (hopefully) contain the true population parameter values can be used. Such designs are then optimised over a range of possible parameter values, without the analyst having to know the precise population values in advance (see Sándor and Wedel 2001; Kessels *et al.* 2006). This, however, increases substantially the computing time required to generate the design. Rose and Bliemer (2008) outline the precise steps used to generate this type of design, whilst Bliemer and Rose (2008), Bliemer *et al.* (2009) and Rose *et al.* (2009b) provide details of the analytical second derivatives (needed to compute S^2) for a range of different Logit models.

Measuring Statistical Efficiency   Rather than attempting to minimise the elements of S^2 of a given design directly, a number of measures of the statistical efficiency of a design have been proposed and can be used instead. The most common is the *D-error*, which uses the scaled determinant (i.e. raised to the power $1/K$ to account for the number of parameters to be estimated) of S^2 to measure efficiency. Another, less popular measure, is *A-error* that is based on the trace of S^2.

The determinant of a matrix is a single summary statistic of the magnitude of the elements contained within the matrix. The smaller the determinant, the smaller, on average, the values contained within the matrix will be. In the case of designs generated under the null hypothesis, assuming orthogonal coding and a MNL model structure, the *D-error measure* is converted to a *D-optimality* statistic, which is a percentage value of the design's overall efficiency. Typically designs with values of around 90 percent or higher, are said to represent desirable designs of this class. For all other efficient designs, the objective is to minimise the *D-error*.

Another measure of statistical efficiency, proposed by Bliemer and Rose (2009), is *S-error*. McFadden (1974) described a direct relationship between the matrix S^2 of Logit models and the sample size required to locate statistically significant parameter estimates. Bliemer and Rose (2009) proposed exploiting this relationship to calculate the sample size requirements for SC experiments. The *S-error* of a design provides the theoretically minimum sample size required to obtain asymptotically significant parameter estimates from it. As with *D-error*, the objective is to find a design that minimises *S-error*. In order to calculate the *S-error* of a design, the analyst must also construct its matrix $\mathbf{S}^2$.

Independent of the precise efficiency measure used, minimizing the elements of $\mathbf{S}^2$ for a design also reduces the expected asymptotic standard errors (i.e. the square roots of the diagonals of the matrix). As such, for any given sample size, smaller asymptotic standard errors mean smaller confidence intervals around the parameters estimates, as well as larger asymptotic *t*-ratios for each parameter. Hence, efficient designs are constructed specifically for the purpose of producing more reliable study results.

Alternatively, efficient designs may produce the same asymptotic standard errors as other designs given smaller sample sizes. This is because the $\mathbf{S}^2$ of all discrete choice models are divisible by N, the sample size, and as such, the asymptotic standard errors are also divisible by the square root of N. The result of this is that there exists an inescapable diminishing return in terms of the statistical significance of the parameter estimates obtained from each additional respondent added to a survey.

Example 3.15 Let the Fisher information matrix with N respondents be denoted by $\mathbf{I}_N(\theta)$. Since $\mathbf{I}_N(\theta) = N \cdot \mathbf{I}_1(\theta)$, it holds that $\mathbf{S}_N^2 = (\mathbf{I}_N(\theta))^{-1} = \frac{1}{N}(\mathbf{I}_1(\theta))^{-1} = \frac{1}{N}\mathbf{S}_1^2$, such that:

$$se_N(\theta) = \frac{se_1(\theta)}{\sqrt{N}}. \tag{3.18}$$

Hence, it is clear that the asymptotic standard errors provide diminishing improvements (decreases) for larger sample sizes.

Methods for Generating Designs Under the Non-Null Hypothesis   A number of algorithms have been implemented for generating efficient designs under the non-null hypothesis; these tend to be either row or column-based algorithms. *Row-based algorithms* (e.g. Modified Federov algorithm, see Cook and Nachtsheim 1980) typically start by creating a set of candidate choice tasks (either generating a full factorial or a fractional factorial design) and then select choice tasks either randomly or based on some form of efficiency criterion.

Column-based algorithms start with a random design and change the attribute levels within each attribute of the design. Row-based algorithms have the benefit of being able to remove dominated choice tasks from the candidate set; however, such algorithms typically struggle with maintaining attribute level balance. Column-based algorithms, on the other hand, typically have little difficulty in maintaining attribute level balance but can often result in dominated choice tasks.

The most popular algorithm appears to be the RSC (relabelling, swapping and cycling) algorithm (Huber and Zwerina 1996; Sándor and Wedel 2001). It has three separate operations; however, some may be omitted if required. The algorithm begins with a randomly constructed design after which the columns and rows are changed using *relabelling*, *swapping*, and *cycling* techniques, or combinations thereof. Relabelling occurs where all attribute levels of an attribute are switched (e.g. the combination $\{1, 2, 1, 3, 2, 3\}$ might be relabelled $\{3, 2, 3, 1, 2, 1\}$). The *swapping* operation involves switching the levels of only a few attribute levels at a time rather than all attribute levels (e.g. the attribute combination $\{1, 2, 1, \mathbf{3}, 2, 3\}$ might become $\{\mathbf{3}, 2, 1, \mathbf{1}, 2, 3\}$). Finally, *cycling* replaces all attribute levels in each choice task simultaneously, by replacing the first attribute level with the second level, the second level with the third, etc. As such, cycling can only be performed if all attributes have exactly the same set of feasible levels.

A Note on Blocking of Designs   Often the total number of choice tasks generated from a design may be too large for any one respondent to handle. In such cases, the researcher may turn to *blocking* the design.

One way of doing this is by means of modular algebra deciding one effect that will be confounded (see Galilea and Ortúzar, 2005). For orthogonal designs, blocking involves finding an additional 'blocking' column which may be used to allocate subsets of the generated choice tasks to different respondents. By using an orthogonal blocking column, the allocation of the choice tasks to respondents will be independent of the attribute levels shown to each (i.e. one respondent will not view choice tasks with only high prices and another choice tasks with low prices). For non-orthogonal efficient designs, it is unlikely that an orthogonal blocking column may be located. In such a case, the analyst may produce a near orthogonal blocking column by minimising the largest correlation between the blocking column and the design attributes. This approach will be demonstrated in section 3.4.3.

A Note on the Need for Prior Information in Generating Designs One of the main criticisms of generating SC experiments is the requirement for prior knowledge about parameter priors and model structure. With regards to the first issue, researchers have shown that where no prior information about likely parameter values is known, a traditional orthogonal design will most likely generate good results as it is actually generated under the assumption of no priors. However, where prior information is available it is generally possible to obtain greater statistical efficiency by relaxing the orthogonality constraint (see Rose and Bliemer 2009 for a review of this literature). Thus, even if the only information that the analyst has is that a parameter will take a particular sign (e.g. a cost parameter will be negative), a Bayesian uniform distribution may be used to generate a D-efficient design under the non-null hypothesis with advantage.

The issue of requiring advanced knowledge about the model structure is far more complicated. Our case study above assumed a MNL model. Unfortunately, different discrete choice models have different S^2 matrices as each one produces more or less parameter estimates. As such, optimising a design for a MNL model does not ensure that it will be optimal for other model forms.

Equations for constructing the S^2 matrix for other model structures have been reported; for example, Bliemer *et al.* (2009) compare designs optimised for the Nested Logit (NL) model with those generated for the MNL. Sándor and Wedel (2002) examined the cross-sectional formulation of the random parameters Mixed Logit (ML) model, and Bliemer and Rose (2008) explored the panel formulation of this same model. It is worth noting, however, that these researchers have found that designs optimised for the MNL model typically perform well when analysed using other model forms, with the exception of the cross sectional version of the random parameters ML model. In any case and to overcome this criticism, Rose *et al.* (2009b) introduced a form of model averaging, where the S^2 matrix for different model structures can be computed and a weighted efficiency measure generated.

A Note on Interaction Effects and SC Designs Quite often, analysts are interested in estimating interaction effects in addition to main effects (see Figure 3.10). Using the traditional design method of constructing orthogonal designs, this meant that the attributes of the design were allocated to particular columns so that not only were the effects orthogonal with each other, but so too were some or all of the interaction columns (formed by multiplying two or more main effects columns together). Recall that orthogonal designs minimise the elements contained within the S^2 matrix of linear models. Indeed, they tend to produce zero covariances suggesting that the parameter estimates of the effects of interest are independent of one another.

Given that the experimental design literature originated with the examination of linear models (i.e. ANOVA and regression models), such designs were deemed important. Nevertheless, discrete choice models are not linear models (although they do collapse to linear models when all parameter estimates are simultaneously zero). As such, the same problems that exist for main effects exist for interaction terms when it comes to estimating discrete choice models. In fact, a design that is capable of detecting independent interaction effects under the null hypothesis will produce non-zero covariances when the parameters are no longer zero, suggesting that there exist correlation between the parameters of interest. Thus, in order to minimise the standard errors and covariances of any interaction effects of interest, prior parameter estimates are required for these also.

3.4.2.4 Conduct Post Design Generation Testing

Once a design has been generated, it is possible to test how it might be expected to perform in practice. When conducted, such tests have typically taken one of two forms. Given a design, fix it and change the parameter priors to test its efficiency under the new set of assumed parameters (see Rose and Bliemer 2008). In taking this approach, the researcher is able to determine the robustness of the design to misspecification of the prior parameters. Some analysts have also employed Monte Carlo simulation to test whether the correct data generation process has been used in generating the design as well as how accurate the parameter estimates will be for various sample sizes (see Kessels *et al.* 2006; Ferrini and Scarpa 2007).

A number of different statistical measures have been used to compare the prior parameters to those obtained from the Monte Carlo simulation process. The two most popular are the mean square error (MSE) and the relative absolute error (RAE), defined as follows:

$$MSE = \frac{1}{R} \sum_{r=1}^{R} \left(\theta_k^{(r)} - \bar{\theta} \right)^2 \tag{3.19}$$

$$RAE = \frac{1}{R} \sum_{r=1}^{R} \left(\theta_k^{(r)} - \bar{\theta} \right) / \bar{\theta} \tag{3.20}$$

where $\theta_k^{(r)}$ is the parameter estimate for attribute k obtained at sample iteration r, and $\bar{\theta}$ is the known prior parameter estimate used in constructing the Monte Carlo simulation.

Another popular statistic often used in these tasks is the expected mean square error of the parameter estimates (EMSE). Unlike the MSE and RAE, this measure provides a single summary statistic of the overall bias and variance across all parameter estimates, rather than for individual parameter estimates:

$$EMSE = \frac{1}{R} \sum_{r=1}^{R} \left(\theta_k^{(r)} - \bar{\theta} \right)^T \left(\theta_k^{(r)} - \bar{\theta} \right). \tag{3.21}$$

3.4.2.5 Conduct Questionnaire

Once the experimental design has been generated, the next stage is to construct the questionnaire. Given that the design is simply nothing more than a matrix of values, the analyst needs to convert this matrix into something that respondents can meaningfully interpret and respond to. The task for the analyst is therefore to convert each row of the design into a choice similar to that shown in Figures 3.7 and 3.8. This may call, for example, for the use of high-quality graphic material to convey an impression of what new rolling stock might be like. The researcher must be careful to avoid any implicit bias in the illustrative material used. Graphic illustrations are often preferred to photographs because of the higher control afforded in respect of the details included in them.

Example 3.16 In a study of the role of train frequency over demand for intercity travel (Steer and Willumsen 1983) it was found that although different people perceived the key variable (frequency) in different ways, almost nobody thought about it in terms of trains per hour or per day. Therefore the SC survey started by ascertaining how was frequency (i.e. the analyst's concept) viewed by the traveller, for instance:

- 'I took the last train that puts me in Newcastle before 11 a.m.; it was the 7:50 from Kings Cross', or
- 'I just turned up at the station and found that the next train to Newcastle was due in 15 minutes'.

The interviewer then converted the different frequency attributes of the experimental design into the same terms, for example: 'To get to Newcastle before 11 a.m. you must now take the 7:30 train' in a low-frequency option, or ' . . . the 8:00 train', in a high frequency one. Alternatively, ' . . . the next train to Newcastle was in 30 . . . ' or ' . . . 10 minutes', for each option. Travellers were then asked to choose among alternatives described in terms they were familiar with and which affected their current journey choices thus increasing realism and relevance.

In constructing the questionnaire, it is advised that the analyst randomise the order of the choice tasks shown to different respondents in order to minimise possible order effects. That is, respondents may use the first few choice tasks to learn what it is they are being asked to do, whereas they might suffer fatigue for the last few choice tasks. Randomising the choice tasks over respondents should reduce any interaction of these biases with the specific choice tasks of the design that might otherwise occur. Although less common, it is also advised that the order of the alternatives and attributes be randomised between respondents.

3.4.2.6   Nothing is Important

It has been recommended to add a null option to the experimental design, also known as a *non-purchase option*. The reason is that if two or more options are presented to an individual who finds them all unacceptable, and has no opportunity to reject the lot, it is possible that this will trigger a secondary decision-making mechanism that could bias the results of the experiment. This important problem has been ignored far too often in practice.

Example 3.17 Olsen and Swait (1998) studied the veracity of the following propositions for the case of buying a product the sale of which is subject to strict prerequisites (e.g. concentrated orange juice, where it is expected that many consumers would require it to be unsweetened):

- If a non-purchase option (NPO) is not present, the attribute weights will differ from those observed when an NPO is offered in the design.
- If an NPO is included in the experimental design, the analyst should be able to identify more non-linear preference structures than if the NPO were missing.
- Models based on data without an NPO may show low predictive capacity for choice situations including an NPO, whereas models based on data including an NPO will present good predictive capacity in any scenario.

Olsen and Swait (1998) used an experimental design with three brands, two levels for orange quality, two levels of sweetness, two types of packing (single and in lots of four) and two levels of price per unit. They also added a cheap option (with the supermarket brand name), consisting of a sweet orange juice made from low-quality oranges. They postulated a factorial design allowing them to estimate the main effects and all interactions between pairs of attributes.

Equal-sized samples (70 individuals) were presented with 16 situations involving three options, for the designs with and without NPO. They also asked if consumers would veto the sale of a product if one of its attributes had an unacceptable level.

They found that the parameters of the estimated models not only differed in magnitude but, as expected, the model with NPO presented significant non-linear (interaction) effects. These results were confirmed by an analysis of the responses to the question about vetoing a product depending on its characteristics (i.e. 63% of the sample found the sweetened juice unacceptable; 57% of it found unacceptable the requirement to buy packages of four units). Table 3.6 shows the percentage error in the predictions

Table 3.6 Cross errors of prediction in market shares

	Prediction error (%)	
Alternative	with NPO → without NPO	without NPO → with NPO
1	3.8	24.8
2	−1.9	21.6
3	−2.8	37.9
NPO	−	−47.8

for each data set using the parameters estimated with the other set. There is no doubt that their initial hypotheses were confirmed; so it may be concluded that nothing *is* indeed important.

3.4.2.7 Realism and Complexity

A key element in the success of SP surveys is the degree of realism achieved in the responses. Realism must be preserved in the *context* of the exercise, the *options* that are presented and the *responses* that are allowed. This can be achieved in a number of ways:

- Focusing on *specific* rather than general behaviour; for example, respondents should be asked how they would respond to an alternative on a given occasion, rather than in general; the more abstract the question the less reliable the response.
- Using a realistic choice context, in particular one the respondents have had recent personal experience of (i.e. a *pivot* design).
- Retaining the constraints on choice required to make the context realistic; this usually means asking respondents to express preferences in respect of a very recent journey without relaxing any of its constraints: e.g. 'if today you would prefer to use the car to visit your dentist in the evening directly from work, then retain this restriction in your choices'. Easing these constraints will just produce unrealistically elastic responses.
- Using existing (perceived) levels of attributes so that the options are built around existing experience.
- Using respondents' perceptions of what is possible to limit the attribute values in the exercise. For example, in considering improved rail services, do not offer options where the station is closer to home than feasible.
- Ensuring that all relevant attributes are included in the presentation; this is especially important if developing travel choice models and not just measuring the relative importance of different attributes.
- Keeping the choice experiments as simple as possible, without overloading the respondent. Remember we respond to very complex choices in practice but we do so over a long period of time, acquiring experience about alternatives at our own pace and selecting the best for us. In an SP exercise these choices are compressed on a very short period of time and must, therefore, be suitably simplified.
- Allowing respondents to opt for a response outside the set of experimental alternatives. For example, in a mode choice exercise if all options become too unattractive the respondent may decide to change destination, time of travel or not to travel at all; allow her a 'will do something else' alternative. If a computer-based interview is used it could be programmed to branch then to another exercise exploring precisely these other options.
- Making sure that all the options are clearly and unambiguously defined. This could be quite difficult when dealing with qualitative attributes like security or comfort; for example, do not express

alternatives as 'poor' or 'improved' as this is too vague and prone to different interpretations by respondents. Describe instead what measures or facilities are involved in improving security or ride comfort (closed circuit TV in all stations/attendants present at all times, . . ., air-conditioning in all coaches as in InterCity trains).

3.4.2.8   Use of Computers in SP Surveys

Computers have been used now for several years in the conduct of surveys of many kinds, including SC surveys. Computers do offer very significant advantages over 'paper and pen' methods but they have, given present technology, a few limitations. Let us consider them first.

In the case of SC surveys one is most likely to use portable, preferably notebook size, microcomputers. In the past, their main limitations were battery life and weight but modern machines have practically overcome these problems. The second restriction is screen size and quality. Contemporary portable computers offer a reasonable screen size and high-resolution colour screens permit the display of more information, at a price. Also, a computer screen is perfectly suited to paired choices in their pure and generalised (i.e. with an associated rating scale, see Ortúzar and Garrido 1994a) form. It is also possible to display not only the attributes that vary as part of the SC experiment but also other features that remain fixed, such as destination, clock times, or indeed anything else that may be relevant or useful to the respondent.

What makes computer-based interviewing most attractive is, however, the task of tailoring the experiment to the subject. Most stated preference interviews will include a questionnaire in which information about the respondent and a recent journey (or purchase, etc.) is collected and used to build a subsequent experiment (i.e. pivot design). This questionnaire can be reproduced in software with the added advantages of automatic entry validation and automatic routing (see Figure 3.11). With a computerised system the responses to this initial questionnaire can be used to generate the SC experiments and options automatically for each subject, following a specified design. Automatic routeing can be used to select the appropriate experiment for each individual depending on her circumstances. Furthermore, range

Computerised Interviews by Steer Davies Gleave	
How long did it take to walk from the parking place to your destination?	5 minutes
How long will the car be parked there?	8h 0m
How much do you pay for a gallon of petrol ? (press F1 for £/litre)	1 Pounds 89 Pence
How many miles per gallon do you get from your car? (press F1 for kms/ltre)	35.0
	Get a lift from someone else Bus Train Taxi Walk Cycle
If you had not been able to go by car, what would you have done ?	Travel another way Not travel

Figure 3.11   Example of computerised questionnaire

and logic checks on the responses and pop-up help screens or look-up information windows (e.g. for timetables) can be incorporated to improve the quality of the interview.

Computers also allow for experimenting with adaptive designs, that is, modifying the experimental design in the light of the responses of the subject (Holden *et al.* 1992; Toubia *et al.* 2007); although there can be gains by adapting the design in a Bayesian sense, care must be exercised not to lose the desirable properties of the sample and general design. In fact, Bradley and Daly (2000) caution against the use of adaptive designs as they may lead to bias; also, the methods to implement this in the case of efficient designs for more complex discrete choice models are very complex.

The use of computers for SC surveys also makes it possible to design more complex interviews than might be attempted manually, although this complexity may never be apparent to the respondent, or even to the interviewer. Moreover, good software permits randomisation of the order in which the options are offered to each individual thus removing a further potential source of bias in the responses. Finally, as all responses are stored directly on disk there are no data entry costs nor errors and data are available immediately for processing.

A number of software packages offer excellent facilities for designing and coding very complex interviews with a minimum of understanding of computing itself; among the best known are ACA (Sawtooth Software), ALASTAIR (Steer Davies Gleave), MINT (Hague Consulting Group) and Ngene 1.0 (Choice Metrics). In summary, the practical advantages of computer-based SP interviews are:

- An interesting format that is consistent across interviews and respondents.
- Automatic question branching, prompting and response validation.
- Automatic data coding and storage.
- The ease with which the SP exercise can be tailored to each individual.
- The reduction in interview time achieved because the interviewer does not have to calculate and prepare written options.
- Reduced training and briefing costs.
- The statistical advantages of randomising the sequence of choices.

On the debit side one has the initial cost of investing in hardware, software, insurance and the requirement to provide some back-up services (disks, spare battery packs, modems, technical advice to interviewers and supervisors, etc.) on location.

Another, everyday more attractive, possibility is conducting the interview remotely via a Web page survey distributed through the Internet (see for example, Iragüen and Ortúzar 2004; Hojman *et al.* 2005). In this case the sampling frame is an important issue as well as the even more careful design of the survey instrument; this has to follow the already noted special recommendations for mail-back surveys.

3.4.2.9   Quality Issues in Stated Preference Surveys

Stated-preference (SP) techniques have proved to be a powerful instrument in research and model development in transport and other fields. Their value depends on the careful application of the guidelines developed so far and discussed in the preceding pages. A key element in this is restricting the artificiality of the exercise to the minimum required. The more the analyst is interested in predicting future behaviour the more important it is to make sure the *decision context* is specific (an actual journey, not a hypothetical one) and the *response space* is behavioural.

But one of the dangers of these techniques is that it is relatively easy to cut corners in order to reduce costs. For example, one can allow the decision context to become less specific and more generic; this makes the sampling easier, the questionnaire simpler and, not surprisingly, the resulting models quite believable as they reflect 'ideal' rather than constrained behaviour. The value of goodness-of-fit indicators in stated preference surveys is entirely dependent on the quality and realism of the experiment. The problem is that the models resulting from 'cheaper' studies will only be found to be flawed much later.

The same is true of the analysis techniques discussed in Chapter 8. Good analysis will often require combining SP and RP data to make sure the resulting models are well anchored (scaled) in the restrictions and noise of real behaviour.

SP surveys can be a cost-effective way of refining and improving modelling tools but too much emphasis on low cost, at the expense of quality assurance and sound analysis, is likely to lead to disappointments and poor decision support.

3.4.3   Case Study Example

Consider a simple hypothetical transport evaluation study in which respondents will be asked to choose between three different hypothetical routes. The study objective is to determine the role that prices (i.e. petrol and toll costs), and travel times (i.e. travelling in free flow and congested traffic conditions), have upon route choice. The study also requires determining how travel time reliability may influence the choice of route.

Assume that secondary and qualitative research confirmed the above as the relevant set of attributes influencing choice; the same research further identified the attribute levels shown in Table 3.7 as being relevant to the study. Note that each attribute has three levels in this example but this need not be the case, as different attributes are allowed to take different numbers of attribute levels. This was only done here for the sake of simplicity.

Table 3.7   Attribute and attribute levels

Travel costs	
Petrol	$1.00, $1.50, $2.00
Toll	$0.00, $2.00, $4.00
Travel times (minutes)	
Free flow time	10, 15, 20
Congested time	10, 15, 20
Egress time	5, 10, 15
Travel time reliability	
Probability of arriving early	0.1, 0.2, 0.3
Probability of arriving late	0.1, 0.3, 0.5

Further, given that we have chosen a route choice problem as our case study, an unlabelled SC experiment is the most appropriate experimental design approach to consider. Recall, however, that the processes and principles in constructing unlabelled SC surveys are a little different to those for generating labelled SC surveys. As such, where differences do exist, we will make a special note.

Now armed with the appropriate set of alternatives (3), attributes (7) and attribute levels (3 for each attribute), the goal becomes to generate an experimental design that can be used to capture data on the behavioural responses of individuals that will assist in answering the identified study objectives.

(continued)

The first step is to write out the most likely set of representative utility functions (V) that will be estimated for the study. Equation (3.22) shows the expected utility functions to be used once data has been collected. In writing out the equations, as shown below, it is easy to see that we anticipate estimating generic parameters for the main effects only (i.e. the parameters are the same across alternatives and there are no interactions) and no alternative specific constants (ASC); note that if we had assumed a labelled choice experiment, the utility functions should reflect the mix of expected alternative specific and generic parameters to be estimated. If ASC or interaction effects were expected, these should be included in the utility specification. Likewise, any dummy, effects, or orthogonal coded variables should also be included.

$$
\begin{aligned}
V(A) &= \theta_1 x_{Pet_A\{1.00,1.50,2.00\}} + \theta_2 x_{Toll_A\{0,2,4\}} + \theta_3 x_{FFT_A\{10,15,20\}} + \theta_4 x_{CongT_A\{10,15,20\}} \\
&\quad + \theta_5 x_{EgT_A\{10,15,20\}} + \theta_6 x_{\mathrm{Prearl}_A\{0.1,0.2,0.3\}} + \theta_7 x_{\mathrm{Prearl}_A\{0.1,0.3,0.5\}}, \\
V(B) &= \theta_1 x_{Pet_B\{1.00,1.50,2.00\}} + \theta_2 x_{Toll_B\{0,2,4\}} + \theta_3 x_{FFT_B\{10,15,20\}} \\
&\quad + \theta_4 x_{CongT_B\{10,15,20\}} + \theta_5 x_{EgT_B\{10,15,20\}} + \theta_6 x_{\mathrm{Prearl}_B\{0.1,0.2,0.3\}} + \theta_7 x_{\mathrm{Prearl}_B\{0.1,0.3,0.5\}}, \\
V(C) &= \theta_1 x_{Pet_C\{1.00,1.50,2.00\}} + \theta_2 x_{Toll_C\{0,2,4\}} + \theta_3 x_{FFT_C\{10,15,20\}} \\
&\quad + \theta_4 x_{CongT_C\{10,15,20\}} + \theta_5 x_{EgT_C\{10,15,20\}} + \theta_6 x_{\mathrm{Prearl}_C\{0.1,0.2,0.3\}} + \theta_7 x_{\mathrm{Prearl}_C\{0.1,0.3,0.5\}}.
\end{aligned}
\tag{3.22}
$$

At the same time as the utility specification is considered, the most likely model structure to be estimated should also be decided. This is because different model structures will have more or less parameter estimates (as we will see in Chapter 7). For example, if a Mixed Logit model with random parameters is to be estimated (see section 7.6.2), then more than one parameter may be associated with each attribute. Similarly, the Nested Logit model will require the estimation of additional scale related parameters (see section 7.4.3) than the simpler MNL.

Given the model structure and expected utility specification, it is then possible to determine the smallest number of choice tasks required in generating the design. For the present case study, assume that we would like to estimate a simple MNL model with utility specification (3.22) on the data; in that case, seven parameters need to be estimated. As such, the design is required to have seven choice tasks at a minimum. As we are uncertain as to whether interaction effects might be present, and we would like to allow for the possibility of a more advanced econometric model being estimated, we may wish to produce a design with more than seven choice tasks.

On the other hand, assuming attribute level balance is required the final design should aim to have more than seven choice tasks with their total number being divisible by three (i.e. as all attributes have three levels, the number of choice tasks must be divisible by this number). Given this, we decided to generate a design with 12 choice tasks, but we will block it so that each respondent faces only six of them (i.e. assume that initial qualitative research indicated that respondents could answer only this number of tasks comfortably).

We have constructed three different designs in our example (although in practice, it would be usual to construct only one): an orthogonal design, a D-optimal design under the null-hypothesis, and a D-efficient design under the non-null hypothesis. In generating the latter, information is required as to the expected values of the parameter estimates. Now, prior parameter values may be established from a number of sources. For example, the analyst may have some prior expectation as to the likely sign, such as a cost parameter should be negative in a utility function. Alternatively, previous research may also provide evidence as to what values the parameter estimates might take, or a pilot study may provide an indication as to reasonably likely values.

For the current case study, assume that a review of the literature established that the set of parameter estimates in Table 3.8 could act as good priors in setting up the experiment (see Zi *et al.* 2009). Note that our reference did not include an *Egress time* attribute, and hence the prior parameter for this attribute was taken as zero.

Table 3.8 Parameter priors and prior standard errors

Travel costs ($)	
Petrol	−0.479 (0.0311)
Toll	−0.426 (0.0362)
Travel times (min)	
Free flow time	−0.098 (0.0087)
Congested time	−0.147 (0.0108)
Egress time	0.0 (0.0)
Travel time reliability	
Probability of arriving early	−0.120 (0.0827)
Probability of arriving late	−0.305 (0.032)

Note that should the exact parameter estimates reported in Table 3.8 be used, with no additional information, then a locally optimal design would result. In fact, the parameter estimates for the current study are unlikely to match exactly those obtained in our reference study; for this reason we will assume Bayesian prior parameter distributions in generating the design. To do this, we will take draws from Bayesian multivariate Normal distributions using the reported parameter estimates as the means of the distributions and their standard errors as the standard deviations (e.g. the Bayesian prior parameter distribution for the petrol attribute is $\theta_1 \sim N(-0.479, 0.0311)$). Although we have chosen a multivariate Normal distribution here, we could have just as easily assumed Uniform distributions or any other distributional assumption in generating the design (Rose and Bliemer 2009).

The three designs generated are shown in Table 3.9 and their correlation structures are reported in Table 3.10. All designs were generated using Ngene 1.0. The first one, the traditional orthogonal design, was constructed using the sequential design process. Thus, an orthogonal design was first built for the first alternative; then its choice tasks were randomly re-arranged to build the second and third alternatives (e.g. the first choice task in alternative 1 was randomly selected to be choice task 11 in alternative 2 and choice task 8 in alternative 3, and so on). As can be seen in Table 3.11, this results in a design where the *within alternative* correlations are zero for the design attributes, but the *between alternative* correlation structure may be non-zero. We will return to discuss the second design after discussing the third.

Design 3 was constructed using the prior parameter estimates mentioned above. Here we require estimating the design's $\mathbf{S}^2$ matrix, and hence the Fisher information matrix $\mathbf{I}_N(\theta)$. Re-arranging the design so that each row represents an alternative, and hence several rows combine to form a choice task, $\mathbf{I}_N(\theta)$ may be calculated using simple matrix algebra (see Rose and Bliemer 2006), by means of equations (3.23) and (3.24):

$$\mathbf{I}_N(\theta) = (\mathbf{Z}^T \mathbf{Z}) = \sum_{c=1}^{C} \sum_{j=1}^{J_c} \mathbf{z}_{jc}^T \mathbf{z}_{jc} \tag{3.23}$$

where

$$\mathbf{z}_{jc} = \left(x_{jkc} - \sum_{i=1}^{J_c} x_{ikc} P_{ic} \right) \sqrt{P_{ic}} \tag{3.24}$$

Here j and i are used to denote alternatives, c a choice task, k a particular attribute and x_{jkc} the attribute level for the kth attribute of alternative j in choice task c. Finally, P_{ic} represents the choice probability of alternative i being chosen in choice task c.

(continued)

Table 3.9 Experimental designs

Orthogonal Design

Choice task	Alternative A								Alternative B								Alternative C							
	Petrol	Toll	Pr. Late	Pr. Early	F.F. Time	Cong. Time	Eg. Time		Petrol	Toll	Pr. Late	Pr. Early	F.F. Time	Cong. Time	Eg. Time		Petrol	Toll	Pr. Late	Pr. Early	F.F. Time	Cong. Time	Eg. Time	Block
1	1.5	2	0.3	0.1	15	10	15		1.5	2	0.5	0.2	10	10	15		2	0	0.1	0.1	20	15	10	1
2	2	0	0.5	0.3	20	15	10		1	0	0.5	0.2	15	20	5		1	4	0.5	0.3	20	15	10	1
3	2	4	0.3	0.1	10	20	5		1	4	0.1	0.3	20	15	10		2	0	0.5	0.3	15	15	10	1
4	1	0	0.1	0.2	10	10	5		1	4	0.5	0.1	20	20	10		2	4	0.3	0.3	15	10	5	1
5	1	4	0.1	0.3	20	15	10		1.5	2	0.1	0.2	15	10	15		1	4	0.5	0.1	20	15	10	1
6	1.5	2	0.1	0.2	15	20	15		1	2	0.1	0.2	10	10	5		1.5	2	0.3	0.3	10	20	15	1
7	2	4	0.3	0.3	15	10	5		2	0	0.1	0.1	20	15	10		1	0	0.5	0.2	15	15	5	2
8	1	0	0.5	0.2	15	20	5		2	4	0.3	0.1	15	10	5		1.5	2	0.3	0.1	15	10	15	2
9	2	0	0.1	0.1	20	15	10		1.5	2	0.3	0.3	10	20	15		1.5	2	0.1	0.2	15	20	15	2
10	1	4	0.5	0.1	20	15	10		2	4	0.3	0.1	10	20	10		1.5	2	0.5	0.2	10	10	15	2
11	1.5	2	0.3	0.3	10	20	15		1.5	2	0.3	0.1	15	10	15		1	0	0.1	0.2	10	10	5	2
12	1.5	2	0.5	0.2	10	10	15		2	0	0.5	0.3	20	15	10		2	4	0.3	0.1	10	20	5	2

Efficient Design under Null Hypothesis Assumption

Choice task	Alternative A								Alternative B								Alternative C							
	Petrol	Toll	Pr. Late	Pr. Early	F.F. Time	Cong. Time	Eg. Time		Petrol	Toll	Pr. Late	Pr. Early	F.F. Time	Cong. Time	Eg. Time		Petrol	Toll	Pr. Late	Pr. Early	F.F. Time	Cong. Time	Eg. Time	Block
1	1.5	2	0.3	0.1	15	10	15		2	4	0.5	0.2	10	15	10		1	0	0.1	0.3	20	20	5	1
2	1	0	0.5	0.2	20	20	5		1.5	2	0.1	0.3	10	10	15		2	4	0.3	0.1	20	15	10	1
3	2	0	0.1	0.1	20	15	10		1	2	0.3	0.2	15	20	5		1.5	4	0.5	0.3	10	10	15	1
4	1.5	2	0.5	0.2	10	10	15		2	4	0.1	0.3	20	15	10		1	0	0.3	0.1	15	20	5	1
5	2	4	0.3	0.1	10	20	5		1	4	0.5	0.2	20	20	5		2	2	0.1	0.3	15	15	5	1
6	1	4	0.1	0.3	20	15	10		1.5	0	0.3	0.1	15	20	5		1.5	2	0.5	0.2	10	10	15	1
7	2	4	0.3	0.3	15	10	5		1	0	0.5	0.1	10	15	15		1.5	2	0.1	0.2	20	20	10	2
8	2	0	0.5	0.3	20	15	10		1	2	0.1	0.1	15	20	5		1.5	4	0.3	0.2	10	10	15	2
9	1	4	0.5	0.1	20	15	10		1.5	2	0.1	0.2	15	20	10		2	2	0.1	0.3	10	10	15	2
10	1.5	2	0.3	0.3	10	20	15		2	4	0.5	0.1	20	10	10		1	0	0.1	0.2	15	15	5	2
11	1	0	0.1	0.2	10	10	5		1	0	0.3	0.3	20	15	15		2	4	0.5	0.1	15	20	10	2
12	1.5	2	0.1	0.2	15	20	5		2	4	0.3	0.3	10	10	10		1	0	0.5	0.1	20	15	5	2

Table 3.9 (*Continued*)

Efficient Design under Non-Null Hypothesis Assumption

Choice task	Alternative A							Alternative B							Alternative C							
	Petrol	Toll	Pr. Late	Pr. Early	F.F. Time	Cong. Time	Eg. Time	Petrol	Toll	Pr. Late	Pr. Early	F.F. Time	Cong. Time	Eg. Time	Petrol	Toll	Pr. Late	Pr. Early	F.F. Time	Cong. Time	Eg. Time	Block
1	2	2	0.1	0.2	15	15	10	1.5	2	0.1	0.2	10	10	5	1	0	0.5	0.2	15	15	15	1
2	1	4	0.1	0.1	10	10	15	2	0	0.5	0.3	20	10	5	1.5	4	0.3	0.2	20	15	10	1
3	1.5	2	0.5	0.1	20	10	5	1.5	0	0.1	0.3	10	20	10	1.5	4	0.3	0.1	15	20	10	1
4	1	0	0.3	0.3	20	10	15	1.5	4	0.3	0.1	10	15	10	2	2	0.5	0.2	10	15	10	1
5	1.5	4	0.3	0.2	20	15	5	2	0	0.5	0.1	15	20	15	1	2	0.1	0.3	15	10	5	1
6	1	0	0.5	0.3	10	20	10	1	2	0.3	0.2	15	20	5	2	0	0.1	0.1	15	15	15	1
7	2	4	0.5	0.3	10	10	15	2	2	0.3	0.2	20	20	10	1	0	0.5	0.1	20	20	5	2
8	1	2	0.1	0.2	20	15	5	1.5	4	0.1	0.1	15	15	15	2	2	0.3	0.3	10	20	10	2
9	2	0	0.1	0.1	15	20	10	1	4	0.1	0.3	15	10	15	1.5	2	0.5	0.2	20	20	15	2
10	1.5	4	0.3	0.3	15	20	10	2	0	0.5	0.3	20	15	10	1	4	0.5	0.1	10	20	5	2
11	2	2	0.3	0.2	15	20	10	1	2	0.5	0.1	10	20	5	1.5	2	0.1	0.3	20	10	15	2
12	1.5	0	0.5	0.1	10	20	15	1	4	0.3	0.2	20	10	10	2	4	0.3	0.3	10	10	5	2

(continued)

Table 3.10   Design correlation structures

								Traditional Orthogonal Design														
	Petrol	Toll	Pr. Late	Pr. Early	F.F. Time	Cong. Time	Eg. Time	Petrol	Toll	Pr. Late	Pr. Early	F.F. Time	Cong. Time	Eg. Time	Petrol	Toll	Pr. Late	Pr. Early	F.F. Time	Cong. Time	Eg. Time	Block
Petrol	1.00																					
Toll	0.00	1.00																				
Pr. Late	0.00	0.00	1.00																			
Pr. Early	0.00	0.00	0.00	1.00																		
F.F. Time	0.00	0.00	0.00	0.00	1.00																	
Cong. Time	0.00	0.00	0.00	0.00	0.00	1.00																
Eg. Time	0.00	0.00	0.00	0.00	0.00	0.00	1.00															
Petrol	-0.25	0.25	0.50	0.00	0.13	-0.25	0.00	1.00														
Toll	-0.50	0.00	0.00	-0.50	-0.13	0.25	-0.50	0.00	1.00													
Pr. Late	-0.13	-0.63	0.38	-0.13	-0.13	-0.50	0.25	0.00	0.00	1.00												
Pr. Early	0.25	-0.25	0.13	-0.38	0.00	0.25	-0.25	0.00	0.00	0.00	1.00											
F.F. Time	0.13	0.13	0.13	0.38	-0.63	-0.25	0.00	0.00	0.00	0.00	0.00	1.00										
Cong. Time	0.13	0.13	0.00	0.00	0.63	-0.25	-0.25	0.00	0.00	0.00	0.00	0.00	1.00									
Eg. Time	0.13	0.13	-0.50	0.00	-0.13	-0.25	0.25	0.00	0.00	0.00	0.00	0.00	0.00	1.00								
Petrol	-0.13	-0.13	0.00	-0.75	-0.50	-0.25	0.00	-0.13	0.38	0.38	0.13	0.38	-0.25	0.00	1.00							
Toll	-0.38	-0.38	0.00	0.25	0.25	0.00	0.00	-0.13	-0.13	0.38	-0.13	0.13	0.50	-0.25	0.00	1.00						
Pr. Late	-0.25	0.75	0.00	0.00	0.00	-0.50	0.25	0.25	0.25	-0.63	-0.25	0.13	0.25	-0.38	0.00	0.00	1.00					
Pr. Early	0.38	-0.13	-0.25	0.00	-0.13	-0.25	0.00	-0.75	0.00	-0.13	-0.25	0.13	0.13	-0.38	0.00	0.00	0.00	1.00				
F.F. Time	0.25	0.00	-0.13	0.00	0.25	-0.13	-0.38	-0.38	0.13	0.00	0.25	0.13	0.25	0.25	0.00	0.00	0.00	0.00	1.00			
Cong. Time	0.63	0.13	-0.25	0.00	0.13	-0.25	0.25	0.00	-0.75	-0.25	0.38	0.00	0.13	0.13	0.00	0.00	0.00	0.00	0.00	1.00		
Eg. Time	-0.13	-0.13	0.00	-0.50	0.63	0.50	0.00	0.00	0.25	-0.25	0.38	-0.75	0.13	-0.38	-0.20	0.00	0.00	0.00	0.00	0.00	1.00	
Block	0.00	0.00	0.41	0.00	0.00	0.00	0.00	0.82	0.00	0.00	0.00	0.00	0.00	0.00	-0.20	-0.20	0.00	-0.41	-0.61	0.00	0.00	1.00

Table 3.10 (Continued)

D-optimal Design Under the Null Hypothesis

	Petrol	Toll	Pr. Late	Pr. Early	F.F. Time	Cong. Time	Eg. Time	Petrol	Toll	Pr. Late	Pr. Early	F.F. Time	Cong. Time	Eg. Time	Block
Petrol	1.00														
Toll	0.00	1.00													
Pr. Late	0.00	0.00	1.00												
Pr. Early	0.00	0.00	0.00	1.00											
F.F. Time	0.00	0.00	0.00	0.00	1.00										
Cong. Time	0.00	0.00	0.00	0.00	0.00	1.00									
Eg. Time	0.00	0.00	0.00	0.00	0.75	0.75	1.00								
Petrol	-0.50	0.00	0.00	0.00	0.38	0.00	0.00	1.00							
Toll	0.00	-0.50	0.00	0.00	-0.38	0.00	0.00	0.75	1.00						
Pr. Late	0.38	0.38	-0.50	0.00	-0.38	0.00	0.00	0.38	0.38	1.00					
Pr. Early	-0.38	-0.38	0.00	-0.50	-0.38	0.00	0.00	0.38	0.00	-0.38	1.00				
F.F. Time	0.00	0.00	0.00	0.00	-0.50	0.00	0.00	0.00	-0.38	-0.38	-0.38	1.00			
Cong. Time	0.00	0.00	0.00	0.00	0.00	-0.50	0.00	-0.38	-0.38	-0.38	-0.38	0.75	1.00		
Eg. Time	0.00	0.00	0.00	0.00	0.00	0.00	-0.50	0.75	0.75	0.75	-0.38	-0.50	-0.75	1.00	
Block	0.00	0.00	0.00	0.41	0.00	0.00	0.00	0.00	0.00	0.00	-0.20	0.00	0.00	0.00	1.00

(continued)

Table 3.10 *(Continued)*

D-efficient Design Under the Non-Null Hypothesis

	Petrol	Toll	Pr. Late	Pr. Early	F.F. Time	Cong. Time	Eg. Time	Petrol	Toll	Pr. Late	Pr. Early	F.F. Time	Cong. Time	Eg. Time	Petrol	Toll	Pr. Late	Pr. Early	F.F. Time	Cong. Time	Eg. Time	Block
Petrol	1.00																					
Toll	0.13	1.00																				
Pr. Late	0.00	-0.13	1.00																			
Pr. Early	-0.13	0.13	0.25	1.00																		
F.F. Time	-0.13	0.00	-0.25	-0.25	1.00																	
Cong. Time	0.13	-0.25	0.13	0.13	-0.25	1.00																
Eg. Time	-0.13	0.00	0.25	0.25	-0.63	-0.13	1.00															
Petrol	-0.13	0.88	-0.13	0.25	0.13	-0.50	0.13	1.00														
Toll	0.00	-0.75	-0.13	0.00	0.00	0.13	0.13	0.00	1.00													
Pr. Late	0.13	0.00	-0.13	-0.25	-0.25	0.00	0.13	-0.13	0.00	1.00												
Pr. Early	0.13	0.13	0.00	-0.38	-0.38	-0.13	0.00	0.13	0.13	-0.13	1.00											
F.F. Time	-0.13	0.38	0.13	0.00	-0.63	0.13	0.38	0.38	-0.13	0.13	0.38	1.00										
Cong. Time	0.00	0.13	0.63	0.50	0.13	0.00	-0.13	0.00	-0.25	-0.13	-0.25	-0.25	1.00									
Eg. Time	0.00	0.13	-0.13	0.00	0.50	0.00	-0.50	0.25	0.13	-0.13	0.00	0.25	-0.13	1.00								
Petrol	-0.63	-0.75	0.13	-0.13	0.00	0.13	0.13	-0.63	0.63	0.00	-0.25	-0.13	0.00	-0.13	1.00							
Toll	-0.13	0.13	0.13	-0.50	0.00	0.00	0.13	0.13	-0.38	0.13	0.50	0.25	-0.25	0.13	0.00	1.00						
Pr. Late	-0.25	-0.13	-0.50	0.00	0.38	-0.13	0.00	0.13	0.25	-0.63	0.13	-0.13	-0.50	0.25	-0.25	0.13	1.00					
Pr. Early	0.00	-0.13	-0.38	-0.38	0.25	0.25	-0.13	-0.25	0.38	0.38	-0.63	-0.13	-0.38	0.13	0.50	0.13	0.00	1.00				
F.F. Time	0.50	0.25	-0.13	-0.25	-0.38	-0.25	0.00	0.00	-0.25	0.63	0.25	0.00	0.13	-0.38	-0.13	0.13	-0.63	-0.13	1.00			
Cong. Time	0.13	0.13	-0.25	-0.25	0.38	-0.75	-0.38	0.25	0.13	-0.13	0.13	-0.13	0.00	0.38	-0.13	-0.13	-0.13	-0.13	0.25	1.00		
Eg. Time	0.13	-0.50	-0.38	-0.13	0.00	0.13	-0.25	-0.63	0.25	0.00	0.00	-0.63	0.00	-0.50	-0.63	0.25	-0.63	0.00	0.38	0.00	1.00	
Block	0.41	0.00	0.00	0.00	-0.20	0.41	0.00	-0.20	0.41	0.00	0.00	0.41	0.00	0.41	0.00	0.25	0.00	0.20	0.00	0.00	-0.20	1.00

Table 3.11 $\mathbf{I}_N(\Theta)$ and $\mathbf{S}^2$ for the third design

	$\mathbf{I}_N(\Theta)$						
	Var(θ₁)	**Var(θ₂)**	**Var(θ₃)**	**Var(θ₄)**	**Var(θ₅)**	**Var(θ₆)**	**Var(θ₇)**
Var(θ₁)	**2.088**	−1.974	0.026	0.045	−0.807	−1.525	2.761
Var(θ₂)	−1.974	**27.323**	0.065	−0.077	−20.218	−30.855	−5.141
Var(θ₃)	0.026	0.065	**0.376**	0.002	−2.105	0.608	0.362
Var(θ₄)	0.045	−0.077	0.002	**0.091**	0.374	−0.330	0.507
Var(θ₅)	−0.807	−20.218	−2.105	0.374	**188.628**	−36.823	3.175
Var(θ₆)	−1.525	−30.855	0.608	−0.330	−36.823	**159.857**	−14.522
Var(θ₇)	2.761	−5.141	0.362	0.507	3.175	−14.522	**213.312**
	$\mathbf{S}^2$						
	Var(θ₁)	**Var(θ₂)**	**Var(θ₃)**	**Var(θ₄)**	**Var(θ₅)**	**Var(θ₆)**	**Var(θ₇)**
Var(θ₁)	**0.588**	0.081	0.002	−0.179	0.016	0.024	−0.004
Var(θ₂)	0.081	**0.071**	0.016	0.027	0.011	0.017	0.002
Var(θ₃)	0.002	0.016	**2.852**	−0.161	0.034	−0.001	−0.005
Var(θ₄)	−0.179	0.027	−0.161	**11.441**	−0.018	0.022	−0.022
Var(θ₅)	0.016	0.011	0.034	−0.018	**0.008**	0.004	0.000
Var(θ₆)	0.024	0.017	−0.001	0.022	0.004	**0.011**	0.001
Var(θ₇)	−0.004	0.002	−0.005	−0.022	0.000	0.001	**0.005**

For the third design, Figure 3.12 shows the calculations used to construct $\mathbf{I}_N(\Theta)$ assuming the mean of the Bayesian parameter distributions given in Table 3.9. For known design values $\mathbf{x}$, the first problem is to calculate the expected probabilities for each choice task given the assumed model and utility functions. Once the choice probabilities have been calculated, it is possible to construct the $\mathbf{Z}$ matrix in equation (23) as well as $\mathbf{I}_N(\Theta)$. Taking the inverse of $\mathbf{I}_N(\Theta)$ produces $\mathbf{S}^2$ for the third design, under the parameter priors assumed; these two matrices are shown in Table 3.11.

The D-error value for this design is then calculated as

$$D\text{-error} = \det\left(\mathbf{S}^2(\mathbf{x}, \theta)\right)^{1/K} \qquad (3.25)$$

and the value computed for the matrix $\mathbf{S}^2$ in Table 3.12 is 0.109162.

However, in generating this design we will not simply assume the parameter estimates shown in the above figures and tables. Rather, we will use simulation methods to take draws from the assumed multivariate Bayesian prior parameter distributions, let us call these $\mathbf{\Omega}$ in general (i.e. recall they can be Normal or any other distribution), and calculate $\mathbf{S}^2$ for each draw taken. The Bayesian $D_b\text{-error}$ can therefore be calculated as:

$$D_b\text{-error} = \int_\theta \det\left(\mathbf{S}^2(\mathbf{x}, \theta)\right)^{1/K} \phi(\theta|\mathbf{\Omega})d\theta \qquad (3.26)$$

By fixing the simulated draws and changing the attribute level combinations of the design using the algorithms discussed in section 3.4.2.3, the Bayesian D_b-error can be calculated for each newly generated design. That producing the lowest value is retained and used. In particular, the Bayesian D_b-error for Design 3 using 100 Halton draws (see Bhat 2001) in the simulation was 0.109749. Note that the actual D-error or Bayesian D_b-error value is actually meaningless, and can only be used to compare designs constructed under the same set of assumptions.

(continued)

s	J	θ=	-0.48 Petrol	-0.43 Toll	-0.31 Late	-0.12 Early	-0.10 FF	-0.15 CongT	0.00 EgT
1	1		2	2	0.1	0.2	15	15	10
1	2		1.5	2	0.1	0.2	10	10	5
1	3		1	0	0.5	0.2	15	15	15
2	1		1	4	0.1	0.1	10	10	15
2	2		2	0	0.5	0.3	20	10	5
2	3		1.5	4	0.3	0.2	20	15	10
3	1		1	4	0.1	0.1	10	10	15
3	2		2	0	0.5	0.3	20	10	5
3	3		1.5	4	0.3	0.2	20	15	10
4	1		1.5	2	0.5	0.1	20	10	5
4	2		1.5	0	0.1	0.3	10	20	10
4	3		1.5	4	0.3	0.1	15	20	10
5	1		1.5	2	0.5	0.1	20	10	5
5	2		1.5	0	0.1	0.3	10	20	10
5	3		1.5	4	0.3	0.1	15	20	10
6	1		1.5	2	0.5	0.1	20	10	5
6	2		1.5	0	0.1	0.3	10	20	10
6	3		1.5	4	0.3	0.1	15	20	10
11	1		1.5	4	0.3	0.2	20	15	5
11	2		2	0	0.5	0.1	15	15	15
11	3		2	2	0.1	0.3	15	15	5
12	1		1	0	0.5	0.3	10	20	10
12	2		1	2	0.3	0.2	15	20	5
12	3		2	0	0.1	0.1	15	10	15

Middle (V, P):

s	J	V	P
1	1	-5.54	0.12
1	2	-4.07	0.50
1	3	-4.33	0.39
2	1	-4.67	0.45
2	2	-4.58	0.49
2	3	-6.70	0.06
3	1	-4.67	0.45
3	2	-4.58	0.49
3	3	-6.70	0.06
4	1	-5.16	0.36
4	2	-4.71	0.57
4	3	-6.93	0.06
5	1	-5.16	0.36
5	2	-4.71	0.57
5	3	-6.93	0.06
6	1	-5.16	0.36
6	2	-4.71	0.57
6	3	-6.93	0.06
11	1	-6.70	0.08
11	2	-4.80	0.52
11	3	-5.07	0.40
12	1	-4.59	0.31
12	2	-5.85	0.09
12	3	-3.94	0.60

Right table (Z matrix):

s	J	Petrol	Toll	Late	Early	FF	CongT	EgT
1	1	0.22	0.26	-0.05	0.00	0.85	0.85	0.19
1	2	0.10	0.54	-0.11	0.00	-1.77	-1.77	-3.13
1	3	-0.23	-0.76	0.15	0.00	1.55	1.55	3.46
2	1	-0.35	1.32	-0.14	-0.07	-3.69	-0.20	3.50
2	2	0.34	-1.42	0.13	0.07	3.15	-0.21	-3.35
2	3	-0.01	0.48	0.00	0.00	1.09	1.14	0.05
3	1	-0.35	1.32	-0.14	-0.07	-3.69	-0.20	3.50
3	2	0.34	-1.42	0.13	0.07	3.15	-0.21	-3.35
3	3	-0.01	0.48	0.00	0.00	1.09	1.14	0.05
4	1	0.00	0.62	0.15	-0.07	3.65	-3.84	-1.92
4	2	0.00	-0.74	-0.12	0.06	-2.99	2.75	1.38
4	3	0.00	0.75	0.01	-0.03	0.26	0.90	0.45
5	1	0.00	0.62	0.15	-0.07	3.65	-3.84	-1.92
5	2	0.00	-0.74	-0.12	0.06	-2.99	2.75	1.38
5	3	0.00	0.75	0.01	-0.03	0.26	0.90	0.45
6	1	0.00	0.62	0.15	-0.07	3.65	-3.84	-1.92
6	2	0.00	-0.74	-0.12	0.06	-2.99	2.75	1.38
6	3	0.00	0.75	0.01	-0.03	0.26	0.90	0.45
11	1	-0.02	0.81	-0.01	0.00	1.29	0.00	-1.46
11	2	0.32	-0.80	0.13	-0.06	-0.28	0.00	3.45
11	3	-0.36	0.56	-0.14	0.07	-0.25	0.00	-3.30
12	1	-0.33	-0.10	0.14	0.07	-1.92	3.35	-1.43
12	2	-0.18	0.54	0.02	0.01	0.47	1.78	-2.24
12	3	0.31	-0.14	-0.11	-0.06	1.21	-3.11	1.89

Figure 3.12   Calculating the Z matrix

Table 3.12   Prior parameter misspecification test

% variation	θ_1	Design 1	Design 2	Design 3
−100	0	0.195474	0.177806	0.119139
−80	−0.0958	0.196038	0.17958	0.117869
−60	−0.1916	0.196801	0.181486	0.116844
−40	−0.2874	0.197772	0.183523	0.116066
−20	−0.3832	0.198958	0.18569	0.11532
0	**−0.479**	**0.200367**	**0.187988**	**0.115242**
20	−0.5748	0.202007	0.190418	0.115196
40	−0.6706	0.203888	0.192982	0.115396
60	−0.7664	0.20617	0.195684	0.11584
80	−0.8622	0.208405	0.19528	0.11653
100	−0.958	0.21106	0.201519	0.117468

Although we do not show it here, a S^2 matrix can also be calculated for the other two designs. The interested reader can check that under the same assumptions used to generate the D-efficient design for the non-null hypothesis (i.e. the same utility functions and prior parameter estimates), the traditional orthogonal design would produce a D-error of 0.200367 and a Bayesian D_b-error of 0.201293, whilst the D-optimal design under the null hypothesis would produce a D-error of 0.187988 and a Bayesian D_b-error of 0.1897077.

Coming back to the second design now (i.e. under the null-hypothesis), recall that it was generated assuming a MNL model structure with parameter priors equal to zero and attribute levels recoded into orthogonal codes. We do not show the precise design generation process here and the interested reader is referred to Rose and Bliemer (2008) or Street *et al.* (2005) for the methods employed to come up with the exact attribute level combinations for generating it. However, the design can also be generated using similar methods as those shown for constructing the D-efficient design under the non-null hypothesis.

In taking this approach, the design matrix would need to be transformed to allow for orthogonal codes rather than the actual levels shown to respondents. Next, the parameter priors would be set to zero and $\mathbf{I}_N(\theta)$ calculated. Unlike the D-efficient design under the non-null hypothesis, Street and Burgess (2005) have derived a set of equations that allow the analyst to determine precisely how efficient this class of design is. Rather than use the D-error value, Street and Burgess (2005) maximise the D-efficiency of the design. Note that for our example, the D-efficiency is 72.07 percent, which suggests that there could possibly be a better (i.e. nearer to the optimum) design under the assumptions used in its construction.

Examining the design itself (Table 3.9), it is clear that under the assumptions used to generate it, a typical outcome is that the attribute level differences will be maximised between the alternatives. This can be seen, for example, by examining the petrol price attribute which never takes the same level across alternatives. Similar differences exist for all other attributes. So, as we had mentioned, this particular design generation process tends to maximise the trade-offs that respondents are asked to make in choosing amongst their choice alternatives.

Once the design has been generated, various tests of how it might be expected to perform in practice can be conducted. A simple one is to fix the design and systematically vary the parameter priors, observing the impact of each change on the efficiency of the design. Table 3.12 shows the D-error calculations when the first parameter prior is varied over a range of values. By examining

(continued)

the impact upon efficiency given such parameter changes, the analyst may gain an understanding of how robust the design will be over varying degrees of prior parameter misspecification.

Even though not explicitly stated, traditional orthogonal designs are generated under the null hypothesis. As such, testing for prior parameter misspecification is not limited to D-efficient designs constructed under the non-null hypothesis. Indeed, although rarely done in practice, it can be applied to any generated design, including those generated under the null hypothesis assumption.

Once the design has been generated the survey can be finalised. To construct the choice tasks we need to translate the generated design into a format that respondents can meaningfully interpret and respond to. An examination of Figure 3.8 will reveal that the choice task shown there corresponds to the first choice task taken from the D-optimal design under the null hypothesis in Table 3.10.

Sample Size and SC Experiments   Equation (3.18) provides clues as to the sample size requirements for SC experiments. Bliemer and Rose (2009a; b) demonstrate how the sample size requirement to obtain asymptotically statistically significant parameter estimates can be derived from this equation. The asymptotic t-ratio value for a given parameter θ_k may be calculated as:

$$t_N(\theta_k) = \theta_k / \left(\frac{se_k(\theta_k)}{\sqrt{N_k}} \right) \tag{3.27}$$

where N_k is the sample size that would be derived from the calculation for attribute k. Re-arranging it we get:

$$N_k = \frac{t_N(\theta_k)^2 se_k(\theta_k)^2}{\theta_k^2} \tag{3.28}$$

Thus, for any desirable asymptotic t-ratio value, say that for a 95% confidence level, 1.96, it is possible to calculate the sample size requirement to achieve that asymptotic t-ratio value for a design under various prior parameter assumptions.

For the case study above, taking the means of the prior parameter distributions as the true population parameters, and obtaining the parameter standard errors from Table 3.11, the sample size requirements according to each parameter, in order, would be: 11.12, 1.74, 125.27, 3096.82, 3.25 and 2.30 (the reader can easily check this). Note that no sample size can be calculated for *Egress time* given that it was assumed a zero prior parameter. Taking the largest sample size as the critical one, we could say that the design requires at least 3097 respondents for all parameters to be statistically significant. However, note that in making such calculations, Bliemer and Rose (2009a; b) suggest that these sample sizes represent a theoretical minimum and that a larger sample size should probably be adopted.

3.5   Network and Zoning Systems

One of the most important early choices facing the transport modeller is that of the level of detail (resolution) to be adopted in a study. This problem has many dimensions: it refers to the schemes to be tested, the type of behavioural variables to be included, the treatment of time, and so on. This section concentrates on design guidelines for two of these choices: zoning system and network definition.

We shall see that in these two cases, as in other key elements of transport modelling, the final choices reflect a compromise between two conflicting objectives: accuracy and cost. In principle greater accuracy could be achieved by using a more detailed zoning and network system; in the limit, this would imply recognising each individual household, its location, distance to access points to the network, and so on. With a large enough sample (100% rate over several days) the representation of the current system could be made very accurate indeed. However, the problem of stability over time weakens this vision of accuracy as one would need to forecast, at the same level of detail, changes at the individual household level that would affect transport demand. This is a very difficult and mostly unnecessary task. Lesser levels of detail therefore, are not only warranted on the grounds of economy but also on those of accuracy whenever forecasting is involved (recall our discussion in section 3.2).

3.5.1 Zoning Design

A zoning system is used to aggregate the individual households and premises into manageable chunks for modelling purposes. The main two dimensions of a zoning system are the number of zones and their size. The two are, of course, related. The greater the number of zones, the smaller they can be to cover the same study area. It has been common practice in the past to develop a zoning system specifically for each study and decision-making context. This is obviously wasteful if one performs several studies in related areas; moreover, the introduction of different zoning systems makes it difficult to use data from previous studies and to make comparisons of modelling results over time.

The first choice in establishing a zoning system is to distinguish the study area itself from the rest of the world. Some ideas may help in making this choice:

- In choosing the study area one must consider the decision-making context, the schemes to be modelled, and the nature of the trips of interest: mandatory, optional, long or short distance, and so on.
- For strategic studies one would like to define the study area so that the majority of the trips have their origin and destination inside it; however, this may not be possible for the analysis of transport problems in smaller urban areas where the majority of the trips of interest are through-trips and a bypass is to be considered.
- Similar problems arise with traffic management studies in local areas where again, most of the trips will have their origin, destination or both, clearly outside the area of interest. What matters in these cases is whether it is possible to model changes to these trips arising as a result of new schemes.
- The study area should be somewhat bigger than the specific area of interest covering the schemes to be considered. Opportunities for re-routeing, changes in destination and so on, must be allowed for; we would like to model their effects as part of the study area itself.

The region external to the study area is normally divided into a number of *external zones*. In some cases it might be enough to consider each external zone to represent 'the rest of the world' in a particular direction; the boundaries of these different slices of the rest of the world could represent the natural catchment areas of the transport links feeding into the study area. In other cases, it may be advantageous to consider external zones of increasing size with the distance to the study area. This may help in the assessment of the impacts over different types of travellers (e.g. long- and short-distance).

The study area itself is also divided into smaller *internal* zones. Their number will depend on a compromise between a series of criteria discussed below. For example, the analysis of traffic management schemes will generally call for smaller zones, often representing even car parks or major generators/attractors of trips. Strategic studies, on the other hand, will often be carried out on the basis of much larger zones. For example, strategic studies of London have been undertaken using fine zoning systems of about 1000 zones (for about 7.2 million inhabitants) and several levels of aggregation of them down to about 50 zones (at borough level). Examples of zone numbers chosen for various studies are presented in Table 3.13.

As can be seen, there is a wide variety of models and number of zones per million inhabitants. These vary depending on the nature of the model (tactical and short term planning, strategic and long term), the resources available and the particular focus or set of problems addressed.

Table 3.13 Typical zone numbers for studies

Location	Population	Number of zones	Comments
London (2006)	7.2 million	2252	Fine level subzones
		~1000	Normal zones at LTS
		~230	LTS districts
		52	Traffic boroughs
Montréal (2008)	3.4. million	1425	Normal zones
Leeds UK (2009)	0.7 million	~560	Normal zones
Santiago (2009)	5.5 million	~700	Normal zones
Dallas-Forth Worth (2004)	6.5 million	4875	Including 61 external zones
Washington DC (2008)	6.5 million	~2200	Normal zones
		463	Coarse zones
Bogotá (2000)	6.1 million	637	Normal zones
Dublin (2010)	1.7 million	~650	And some 10,000 road links
Sydney (2006)	3.6 million	2690	Normal zones

Note also that sometimes a more aggregated zoning system is used for part of the model system, for example trip generation and distribution; then a finer zoning system is often used for mode choice and assignment: these are often referred to as Traffic Assignment Zones or TAZs.

Zones are represented in the computer models as if all their attributes and properties were concentrated in a single point called the *zone centroid*. This notional spot is best thought of as floating in space and not physically on any location on a map. Centroids are attached to the network through *centroid connectors* representing the average costs (time, distance) of joining the transport system for trips with origin or destination in that zone. Nearly as important as the cost associated with each centroid connector is the node in the network it connects to. These should be close to natural access/egress points for the zone itself. Locating centroids automatically at the centre of gravity of each zone and measuring their distance to key nodes to produce centroid connectors is a quick fix valid only for the simplest 'first cut' network runs.

Centroids and centroid connectors play a key role in the quality of the rest of the models, but their definition and coding does not follow a strict and objective approach; they rely a good deal on the experience of the modeller. The centroide connector influences the route followed to load trips onto both the road and public transport networks and therefore affects the total cost of travelling from Origin to Destination and all the models that include them.

The following is a list of zoning criteria which has been compiled from experience in several practical studies:

1. Zoning size must be such that the aggregation error caused by the assumption that all activities are concentrated at the centroid is not too large. It might be convenient to start postulating a system with many small zones, as this may be aggregated in various ways later depending on the nature of the projects to be evaluated.
2. The zoning system must be compatible with other administrative divisions, particularly with census zones; this is probably the fundamental criterion and the rest should only be followed if they do not lead to inconsistencies with it.
3. Zones should be as homogeneous as possible in their land use and/or population composition; census zones with clear differences in this respect (i.e. residential sectors with vastly different income levels) should not be aggregated, even if they are very small.

4. Zone boundaries must be compatible with cordons and screen lines and with those of previous zoning systems. However, it has been found in practice that the use of main roads as zone boundaries should be avoided, because this increases considerably the difficulty of assigning trips to zones, when these originate or end at the boundary between two or more zones.

5. The shape of the zones should allow an easy determination of their centroid connectors; this is particularly important for later estimation of intra-zonal characteristics. A zone should represent the natural catchment area of the transport networks and its centroid connector(s) identified so as to represent the average costs to access them.

6. Zones do not have to be of equal size; if anything, they could be of similar dimensions in travel time units, therefore generating smaller zones in congested than in uncongested areas.

It is advantageous to develop a hierarchical zoning system, as in the London Transportation Studies, where subzones are aggregated into zones which in turn are combined into districts, traffic boroughs and finally sectors. This facilitates the analysis of different types of decisions at the appropriate level of detail. Hierarchical zoning systems benefit from an appropriate zone-numbering scheme where the first digit indicates the broad area, the first two the traffic borough, the first three the district, and so on.

3.5.2 Network Representation

The transportation network is deemed to represent a key component of the supply side of the modelling effort, i.e. what the transport system offers to satisfy the movement needs of trip makers in the study area. The description of a transport network in a computer model can be undertaken at different levels of detail and requires the specification of its structure, its properties or attributes and the relationship between those properties and traffic flows. For an early general review of network representation issues, see Lamb and Havers (1970).

3.5.2.1 Network Details

The transport network may be represented at different levels of aggregation in a model. At one extreme one has models with no specific links at all; they are based on continuous representations of transport supply (Smeed 1968). These models may provide, for example, a continuous equation of the average traffic capacity per unit of area instead of discrete elements or links. At a slightly higher level of disaggregation one can consider individual roads but include speed-flow properties taken over a much larger area; see for example Wardrop (1968).

Normal practice, however, is to model the network as a *directed graph*, i.e. a system of nodes and links joining them (see Larson and Odoni 1981), where most nodes are taken to represent junctions and the links stand for homogeneous stretches of road between junctions. Links are characterised by several attributes such as length, speed, number of lanes and so on, and are normally unidirectional; even if during input a single two-way link is specified for simplicity, it will be converted into two one-way links in the internal computer representation of the network. A subset of the nodes is associated with zone centroids, and a subset of the links to centroid connectors. Currently, the principal source of network data would be one of the many digital maps available for most cities. One should not assume, however, that they are error free. They will need checking, updating, pruning (to focus on the network of interest) and complementing with observations on items like on-street parking, pedestrian friction, bus lanes and other features that may affect their performance. A very simple configuration of this type is presented in Figure 3.13.

A problem with this scheme is that 'at-node' connectivity is offered to each link joining it at no cost. In practice, some turning movements at junctions may be much more difficult to perform than others;

Figure 3.13   A road network coded as nodes and links

indeed, some turning movements may not be allowed at all. In order to represent these features of real road networks better, it is possible to penalise and/or ban some turning movements. This can be done manually by expanding the junction providing separate (sometimes called dummy) links for each turning movement and associating a different cost to each. Alternatively, some commercial computer programs are capable of performing this expansion in a semi-automatic way, following simple instructions from the user about difficult or banned movements.

The level of disaggregation can be increased further when detailed traffic simulation models are used. In these cases additional links are used at complex junctions to account for the performance of reserved lanes, give-way lines, and so on.

Sometimes networks are subsets of larger systems; they may be cordoned off from them thus defining access or cordon points where the network of interest is connected to the rest of the world. These points are sometimes called 'gateways' and dummy links may be used to connect them to external zones.

A key decision in setting up a network is how many levels to include in the road hierarchy. If more roads are included, the representation of reality should be better; however, there is again a problem of economy versus realism which forces the modeller to select some links for exclusion. Moreover, it does not make much sense to include a large number of roads in the network and then make coarse assumptions about turning movements and delays at junctions. It is not sensible either to use a very detailed network with a coarse zoning system as then *spatial aggregation* errors (i.e. in terms of centroid connections to the network) will reduce the value of the modelling process. This is particularly important in the case of public-transport networks, as we will see in Chapter 11. What matters is to make route choice and flows as realistic as possible within the limitations of the study.

Jansen and Bovy (1982) investigated the influence of network definition and detail over road assignment accuracy. Their conclusion was that the largest errors were obtained at the lower levels in the hierarchy of roads. Therefore, one should include in the network at least one level below the links of interest: for example, in a study of A (trunk) roads one should also include B (secondary) roads.

In the case of public-transport networks an additional level of detail is required. The modeller must specify the network structure corresponding to the services offered. These will be coded as a sequence of nodes visited by the service (bus, rail), normally with each node representing a suitable stop or station. Junctions without bus stops can, therefore, be excluded from the public-transport network. Two types of extra links are often added to public-transport networks. These are walk links, representing the parts of a journey using public transport made on foot, and links to model the additional costs associated with transferring from one service (or mode) to another.

3.5.2.2   Link Properties

The level of detail provided about the attributes of links depends on the general resolution of the network and on the type of model used. At the very minimum the data for each link should include its length, its travel speeds (either free-flow speeds or an observed value for a given flow level) and the capacity of the link, usually in passenger car equivalent units (pcu) per hour.

In addition to this a cost-flow relationship is associated with each link as discussed below. In some cases, more elaborate models are used to relate delay to traffic flow, but these require additional information about links, for example:

- Type of road (e.g. expressway, trunk road, local street).
- Road width or number of lanes, or both.
- An indication of the presence or otherwise of bus lanes, or prohibitions of use by certain vehicles (e.g. lorries).

- Banned turns, or turns to be undertaken only when suitable gaps in the opposing traffic become available, and so on.
- Type of junction and junction details including signal timings.
- Storage capacity for queues and their presence at the start of a modelling period.

Some research results have identified other attributes of routes as important to drivers, for example tolls (see Chapter 16), signposting and fuel consumption (see for example Outram and Thompson 1978 and Wootton *et al*. 1981). Work in the Netherlands has shown that (weighted) time and distance explains only about 70% of the routes actually chosen. The category of the road (motorway, A road, B road), the predictability of the time taken, scenic quality, traffic signals and capacity help to explain additional routes. As our understanding of how these attributes influence route choice improves, we will be able to develop more accurate assignment models. The counterpart of this improvement will be the need to include other features of roads, like their scenic quality, gradient, and so on.

3.5.2.3   Network Costs

Most current assignment techniques assume that drivers seek to minimise a linear combination of time, distance (associated to fuel costs) and tolls (if any), sometimes referred to as generalised cost for route choice. This is known to be a simplifying assumption as there may be differences not only in the perception of time, but also about its relative importance compared with other route features. However, the majority of network models in use today deal only with travel time, distance and tolls.

When modelling travel time as a function of flow one must distinguish two different cases. The first case is when the assumption can be made that delay on a link depends only on the flow on the link itself; this is typical of long links away from junctions and therefore it has been used in most inter-urban assignment models so far. The second case is encountered in urban areas where the delay on a link depends in an important way on flows on other links, for example for non-priority traffic at a give-way or roundabout junction.

The introduction of very general flow-delay formulations is not difficult until one faces the next issue, equilibration of demand and supply. There, the mathematical treatment of the first case (often called the separable cost function case) is simpler than the second; however, there are now techniques for balancing demand and supply in the case of link-delay models depending on flows on several links, i.e. when the effect of each link flow cannot be separated. Chapter 11 will provide a fuller discussion of cost-flow relationships.

3.5.2.4   Public Transport Networks

Public Transport networks are more complex than road networks. They require an identification of the route taken by each service as a unique sequence of links. It is also necessary to identify the locations where stops are possible and also those where interchange with other services is permissible. The frequency of the service and in some cases the actual timetable and the fare, must also be specified and included in the network description. Access to stops may be on foot or by another mode and this can be represented by centroide connectors in the simpler models and by one or more auxiliary networks of access modes in more realistic undertakings. This is why the centroide connectors for public transport are always different from those used for the road network.

The public transport network could be entirely independent of the road network as in the case of most rail and metro services; alternatively, the speed of the service may be affected by road traffic as in the case of buses and on-street running of trams, even when priority measures to support them help to reduce the impact of road congestion.

In addition to the effect of road congestion, it is sometimes necessary to account for the issue of passenger congestion: crowding on buses and trains leading to discomfort and even having to miss a service because it was full and impossible to board.

Exercises

3.1 We require estimating the population of a certain area for the year 2020 but we only have available reliable census information for 1990 and 2000, as follows:

$$P_{1990} = 240 \pm 5 \text{ and } P_{2000} = 250 \pm 2$$

To estimate the future population we have available the following model:

$$P_n = P_b t^d$$

where P_n is the population in the forecast year (n), P_b the population in the base year (b), t is the population growth rate and $d = (n - b)/10$, is the number of decades to extrapolate growth.

Assume that the data from both censuses is independent and that the model does not have any specification error; in that case,

(a) find out with what level of accuracy is it possible to forecast the population in the year 2020;

(b) you are offered the census information for 2010, but you are cautioned that its level of accuracy is worse than that of the previous two censuses:

$$P_{2010} = 265 \pm 8$$

Find out whether it is convenient to use this value.

(c) repeat the analysis assuming that the specification error of the model is proportional to d, and that it can be estimated as $12d\%$.

3.2 Consider the following modal-split model between two zones i and j (but we will omit the zone indices to alleviate notation):

$$P_1(\Delta t/\theta) = \frac{\exp(-\theta t_1)}{\exp(-\theta t_1) + \exp(-\theta t_2)} = \frac{1}{1 + \exp -\theta(t_2 - t_1)} = \frac{1}{1 + \exp(-\theta \, \Delta t)}$$

$$P_2(\Delta t/\theta) = 1 - P_1 = \frac{\exp(-\theta \, \Delta t)}{1 + \exp(-\theta \, \Delta t)}$$

where t_k is the total travel time in mode k, and θ a parameter to be estimated.

During the development of a study, travel times were calculated as the average of five measurements (observations) for each mode, at a cost of $1 per observation, and the following values were obtained:

$$t_1 = 12 \pm 2 \text{ min} \qquad t_2 = 18 \pm 3 \text{ min}$$

(a) If the estimated value for θ is 0.1, compute a confidence interval for P_1.

(b) Assume you would be prepared to pay $3 per each percentage point of reduction in the error of P_1; find out whether in that case it would be convenient for you to take 10 extra observations in each mode whereby the following values for t_k would be obtained:

$$t_1 = 12 \pm 1 \text{ min} \qquad t_2 = 17.5 \pm 1.5 \text{ min}$$

3.3 Consider an urban area where 100 000 people travel to work; assume you possess the following information about them:

(i) General information:

Mode	Average number of cars per household	Family income (1000$/year)
Car	2.40	120
Underground	1.60	60
Bus	0.20	10
Total	0.55	25

(ii) Population distribution

Family income (1000$/year)	Cars per household			
	0	1	2+	Total
Low (< 25)	63.6	15.9	0.0	79.5
Medium (25–75)	6.4	3.7	2.4	12.5
High (> 75)	0.0	2.4	5.6	8.0
Total	70.0	22.0	8.0	100.0

You are required to collect a sample of travellers to estimate a series of models (with a maximum of 8 parameters) which guarantee a negligible specification error if you have available at least 50 observations per parameter. You are also assured that if you take a 20% random sample of all travellers the error will be negligible and there will be no bias.

Your problem is to choose the most convenient sampling method (random, stratified or choice-based), and for this you have available also the following information:

Hourly cost of an interviewer ....$2 per hour
Questionnaire processing cost...0.3 per form
Time required to classify an interviewee..4 min
Time required to complete an interview..10 min

You are also given the following table containing recommended choice-based sample sizes:

Subpopulation size	% to be interviewed
< 10 000	25
10 000–15 000	20
15 000–30 000	15
30 000–60 000	10
>60 000	5

3.4 Consider the following results of having collected stratified (based on income I) and choice-based samples of a certain population:

Stratified sample

Mode	Low I	Medium I	High I
Car	3.33	18.00	20.00
Bus	33.34	7.20	4.00
Underground	3.33	4.80	6.00
Total	40.00	30.00	30.00

Choice-based sample

Mode	Low I	Medium I	High I	Total
Car	6.67	20.00	13.33	40.00
Bus	17.24	2.07	0.69	20.00
Underground	16.67	13.33	10.00	40.00

(a) If you know that the income-based proportions in the population are 60, 25 and 15% respectively for low, medium and high income, find an equivalent table for a random sample. Is it possible to validate your answer?

(b) Compute the weighting factors that would be necessary to apply to the observations in the choice-based sample in order to estimate a model for the choice between car, bus and underground using standard software (i.e. that developed for random samples).

4

Trip Generation Modelling

As we saw in Chapter 1, the trip generation stage of the classical transport model aims at predicting the total number of trips generated by (O_i) and attracted to (D_j) each zone of the study area. This can be achieved in a number of ways: starting with the trips of the individuals or households who reside in each zone or directly with some of the properties of the zones: population, employment, number of cars, etc. The subject has also been viewed as a *trip frequency* choice problem: how many shopping (or other purpose) trips will be carried out by this person type during a representative week? This is usually undertaken using discrete choice models, as discussed in Chapters 7 to 9, and it is then cast in terms like: what is the probability that this person type will undertake zero, one, two or more trips with this purpose per week?

In this chapter we concentrate on the first approach (i.e. predicting the totals O_i and D_j from data on household socioeconomic attributes), and also give a glimpse about the second which has advantages in some studies.

We will start by defining some basic concepts and will proceed to examine some of the factors affecting the generation and attraction of trips. Then we will review the main modelling approaches, starting with the simplest growth-factor technique. Before embarking on more sophisticated approaches we will present a reasonable review of linear regression modelling, which complements well the previous statistical themes presented in Chapters 2 and 3.

We will then consider zonal and household-based linear regression trip generation models, giving some emphasis to the problem of non-linearities which often arise in this case. We will also address for the first time the problem of aggregation (e.g. obtaining zonal totals), which has a trivial solution here precisely because of the linear form of the model. Then we will move to cross-classification models, where we will examine not only the classical category analysis specification but also more modern approaches including the person category analysis model. We then examine the relationship between trip generation and accessibility including a short discussion on trip frequency models. The chapter ends with two short sections: the first discusses the problem of predicting future values for the explanatory variables in the models, and the second the problems of stability and updating of trip generation parameters.

4.1 Introduction

4.1.1 Some Basic Definitions

As always, definitions play an important role in our understanding of any phenomenon. Travel is no different. The question is, what is the basic event of interest for our models, activities involving a short

Modelling Transport, Fourth Edition. Juan de Dios Ortúzar and Luis G. Willumsen.
© 2011 John Wiley & Sons, Ltd. Published 2011 by John Wiley & Sons, Ltd.

period of stay in a location (sojourn), the displacement from one location to another (trip or journey), or a sequence of such trips and sojourns that start and end at home and constitute an outing or a tour? What about tours that are not based at home? The modeller must bear all these options in mind while settling on a set from the following list to develop and apply a set of models.

Trip or Journey   This is a one-way movement from a point of origin to a point of destination. We are usually interested in all vehicular trips. Walking trips less than a certain study-defined threshold (say 300 metres or three blocks) have often been ignored as well as trips made by infants of less than five years of age. However, this emphasis is changing with greater attention being paid to non-motorised trips and, as discussed in the previous chapter, due to the requirements of the recall activity framework recommended for mobility surveys.

Home-based (HB) Trip   This is one where the home of the trip maker is either the origin or the destination of the journey. Note that for visitors from another city their Hotel acts as a temporary home in most studies.

Non-home-based (NHB) Trip   This, conversely, is one where neither end of the trip is the home of the traveller.

Trip Production   This is defined as the home end of an HB trip or as the origin of an NHB trip (see Figure 4.1).

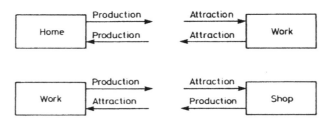

Figure 4.1   Trip productions and attractions

Trip Attraction   This is defined as the non-home end of an HB trip or the destination of an NHB trip (see Figure 4.1).

Trip Generation   This is often defined as the total number of trips generated by households in a zone, be they HB or NHB. This is what most models would produce and the task then remains to allocate NHB trips to other zones as trip productions.

Sojourn   A short period of stay in a particular location. It usually has a purpose associated with this stay: work, study, shopping, leisure, etc.

Activity   An endeavour or interest often associated with a purpose as above but not necessarily linked to a fixed location. One could choose to go shopping or to the cinema in different locations.

Tour or Trip Chain   A set of linked sojourns and trips. The last three concepts correspond better to the idea of travel as derived demand (i.e. it depends strongly on the demand for other activities) but initially were used mainly by discrete choice modellers in practice (see Daly *et al*. 1983). Contemporary models, particularly of the trip frequency type, are more typically interested in tours.

4.1.2   Characterisation of Journeys

4.1.2.1   By Purpose

It has been found in practice that a better understanding of travel and trip generation models can be obtained if journeys by different purposes are identified and modelled separately. In the case of HB trips, a number of categories have been employed:

- travel to work;
- travel to school or college (education trips);
- shopping trips;
- social and recreational journeys;
- escort trips (to accompany or collect somebody else);
- other journeys.

The first two are usually called compulsory (or mandatory) trips and all the others are called discretionary (or optional) trips. The latter category encompasses all journeys made for less routine purposes, such as health and personal business (need to obtain a passport or a certificate). Note that social and cultural contexts may change the importance of different types of trips and therefore the most appropriate classification. NHB trips are sometimes separated into 'on business' and 'other' but are often kept as a single category because they only amount to 15–20% of all total travel.

4.1.2.2   By Time of Day

Trips are sometimes classified into peak and off-peak period trips; the proportion of journeys by different purposes usually varies greatly with time of day. This type of classification, although important, gets more complicated when tours rather than trips are of interest, as a complete tour may comprise trips made at several times of the day.

Table 4.1 summarises data from the Greater Santiago 1977 Origin Destination Survey (DICTUC, 1978) as an example of good and bad traits; the morning (AM) peak period (the evening peak period is sometimes assumed to be its mirror image) occurred between 7:00 and 9:00 and the representative off-peak period was taken between 10:00 and 12:00. Some comments are in order with respect to this table. Firstly, note that although the vast majority (87.18%) of trips in the AM peak are compulsory (i.e. either to work or education), this is not the case in the off-peak period. Secondly, a typical trait of a developing country emerges from the data: the large proportion of trips for bureaucratic reasons in both

Table 4.1   Example of trip classification

Purpose	AM Peak		Off Peak	
	No.	%	No.	%
Work	465 683	52.12	39 787	12.68
Education	313 275	35.06	15 567	4.96
Shopping	13 738	1.54	35 611	11.35
Social	7 064	0.79	16 938	5.40
Health	14 354	1.60	8 596	2.74
Bureaucracy	34 735	3.89	57 592	18.35
Accompanying	18 702	2.09	6 716	2.14
Other	1 736	0.19	2 262	0.73
Return to home	24 392	2.72	130 689	41.65

periods. Thirdly, a problem caused by faulty classification, or lack of forward thinking at the data-coding stage, is also clearly revealed: the *return to home* trips (which account for 41.65% of all off-peak trips) are obviously trips with another purpose; the fact that they were returning home is not as important as to why they left home in the first place. In fact, these data needed recoding in order to obtain adequate information for trip generation modelling (see Hall *et al*. 1987). This kind of problem used to occur before the concepts of trip productions and attractions replaced concepts such as origins and destinations, which did not explicitly address the generating capacity of home-based and non-home-based activities.

4.1.2.3 By Person Type

This is another important classification, as individual travel behaviour is heavily dependent on socioe-conomic attributes. The following categories are usually employed:

- income level (e.g. three strata: low, middle and high income);
- car ownership (typically three strata: 0, 1 and 2 or more cars);
- household size and structure (e.g. six strata in the classical British studies).

It is important to note that the total number of strata can increase very rapidly (e.g. 54 in the above example) and this may have strong implications in terms of data requirements, model calibration and use. For this reason trade-offs, adjustments and aggregations are usually required (see the discussion in Daly and Ortúzar 1990).

4.1.3 Factors Affecting Trip Generation

In trip generation modelling we are typically interested not only in person trips but also in freight trips. For this reason models for four main groups (i.e. personal and freight, trip productions and attractions) have usually been required. In what follows we will briefly consider some factors which have been found important in practical studies. We will not discuss freight trip generation modelling, however (although a little had been done by the end of the century), but postpone a discussion on the general topic of freight demand modelling until Chapter 13.

4.1.3.1 Personal Trip Productions

The following factors have been proposed for consideration in many practical studies:

- income;
- car ownership;
- family size;
- household structure;
- value of land;
- residential density;
- accessibility.

The first four (income, car ownership, household structure and family size) have been considered in several household trip generation studies, while value of land and residential density are typical of zonal studies. The last one, accessibility, has rarely been used although many studies have attempted to include it. The reason is that it offers a way to make trip generation elastic (responsive) to changes in the transport system; we will come back to this issue in section 4.3.

4.1.3.2   Personal Trip Attractions

The most widely used factor has been roofed space available for industrial, commercial and other services. Another factor used has been zonal employment, and certain studies have attempted to incorporate an accessibility measure. However, it is important to note that in this case not much progress has been reported.

4.1.3.3   Freight Trip Productions and Attractions

These normally account for few vehicular trips; in fact, at most they amount to 20% of all journeys in certain areas of industrialised nations, although they can still be significant in terms of their contribution to congestion. Important variables include:

- number of employees;
- number of sales;
- roofed area of firm;
- total area of firm.

To our knowledge, neither accessibility nor types of firm have ever been considered as explanatory variables; the latter is curious because it would appear logical that different products should have different transport requirements.

4.1.4   Growth-factor Modelling

Since the early 1950s several techniques have been proposed to model trip generation. Most methods attempt to predict the number of trips produced (or attracted) by household or zone as a function of (generally linear) relations to be defined from available data. Prior to any comparison of results across areas or time, it is important to be clear about the following aspects mentioned above:

- what trips to be considered (e.g. only vehicle trips and walking trips longer than three blocks);
- what is the minimum age to be included in the analysis (i.e. five years or older).

In what follows we will briefly present a technique which may be applied to predict the future number of journeys by any of the categories mentioned above. Its basic equation is:

$$T_i = F_i t_i \tag{4.1}$$

where T_i and t_i are respectively future and current trips in zone i, and F_i is a growth factor.

The only problem of the method is the estimation of F_i, the rest is trivial. Normally the factor is related to variables such as population (P), income (I) and car ownership (C), in a function such as:

$$F_i = \frac{f(P_i^d, I_i^d, C_i^d)}{f(P_i^c, I_i^c, C_i^c)} \tag{4.2}$$

where f can even be a direct multiplicative function with no parameters, and the superscripts d and c denote the design and current years respectively.

Example 4.1 Consider a zone with 250 households with car and 250 households without car. Assuming we know the average trip generation rates of each group:

$$\text{car-owning households produce:} \quad 6.0 \text{ trips/day}$$
$$\text{non-car-owning households produce:} \quad 2.5 \text{ trips/day}$$

we can easily deduce that the current number of trips per day is:

$$t_i = 250 \times 2.5 + 250 \times 6.0 = 2125 \text{ trips/day}$$

Let us also assume that in the future all households will have a car; therefore, assuming that income and population remain constant (a safe hypothesis in the absence of other information), we could estimate a simple multiplicative growth factor as:

$$F_i = C_i^d / C_i^c = 1/0.5 = 2$$

and applying equation (4.1) we could estimate the number of future trips as:

$$T_i = 2 \times 2125 = 4250 \text{ trips/day}$$

However, the method is obviously very crude. If we use our information about average trip rates and make the assumption that these will remain constant (which is actually the main assumption behind one of the most popular forecasting methods, as we will see below), we could estimate the future number of trips as:

$$T_i = 500 \times 6 = 3000$$

which means that the growth factor method would overestimate the total number of trips by approximately 42%. This is very serious because trip generation is the first stage of the modelling process; errors here are carried through the entire process and may invalidate work on subsequent stages.

In general growth factor methods are mostly used in practice to predict the future number of *external* trips to an area; this is because they are not too many in the first place (so errors cannot be too large) and also because there are no simple ways to predict them. In some cases, they are also used, at least as a sense check, for interurban toll road studies. In the following sections we will discuss other (superior) methods which can also be used in principle to model personal and freight trip productions and attractions. However, we will just make explicit reference to the case of personal trip productions as this is the area not only where there is more practical experience, but also where the most interesting findings have been reported.

4.2   Regression Analysis

The next subsection provides a brief introduction to linear regression. The reader familiar with this subject can proceed directly to subsection 4.2.2.

4.2.1   The Linear Regression Model

4.2.1.1   Introduction

Consider an experiment consisting in observing the values that a certain variable $\mathbf{Y} = \{Y_i\}$ takes for different values of another variable $\mathbf{X}$. If the experiment is not deterministic we would observe different values of Y_i for the same value of X_i.

Let us call $f_i(Y|X)$ the probability distribution of Y_i for a given value X_i; thus, in general we could have a different function f_i for each value of $\mathbf{X}$ as shown in Figure 4.2.

However, such a completely general case is intractable; to make it more manageable certain hypotheses about population regularities are required. Let us assume that:

1. The probability distributions $f_i(Y \mid X)$ have the same variance σ^2 for all values of $\mathbf{X}$.
2. The means $\mu_i = E(Y_i)$ form a straight line known as the *true regression line* and given by:

$$E(Y_i) = a + bX_i \tag{4.3}$$

where the population parameters a and b, defining the line, must be estimated from sample data.

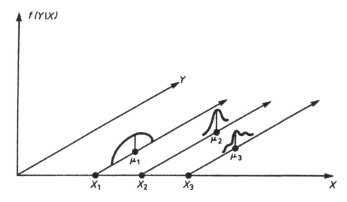

Figure 4.2   General distributions of Y given X

3. The random variables **Y** are statistically independent; this means, for example, that a large value of Y_1 does not tend to make Y_2 large.

The above *weak set of hypotheses* (see for example Wonnacott and Wonnacott 1990) may be written more concisely as:

The random variables Y_i are statistically independent with mean $a + b X_i$ and variance σ^2.

With these Figure 4.2 changes to the distribution shown in Figure 4.3.

It is sometimes convenient to describe the deviation of Y_i from its expected value as the error or disturbance term e_i, so that the model may also be written as:

$$Y_i = a + bX_i + e_i \tag{4.4}$$

Note that we are not making any assumptions yet about the shape of the distribution of **Y** (and **e**, which is identical except that their means differ) provided it has a finite variance. These will be needed later, however, in order to derive some formal tests for the model. The error term is as usual composed of measurement and specification errors (recall the discussion in Chapter 3).

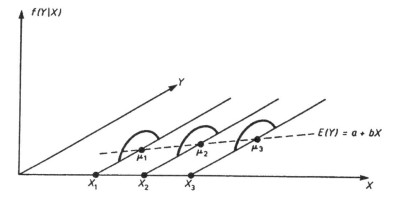

Figure 4.3   Distribution of Y assumed in linear regression

4.2.1.2   *Estimation of a and b*

Figure 4.4 can be labelled the *fundamental graph* of linear regression. It shows the true (dotted) regression line $E(Y) = a + bX$, which is of course unknown to the analyst, who must estimate it from sample data about **Y** and **X**. It also shows the estimated regression line $\hat{Y} = \hat{a} + \hat{b}X$; as is obvious, this line will not coincide with the previous one unless the analyst is extremely lucky (though he will never know it). In general the best he can hope is that the parameter estimates will be close to the target.

Figure 4.4   True and estimated regression lines

It is important to distinguish between the errors e_i, which are not known and pertain to the true regression line, and the differences ε_i, between observed (Y_i) and fitted values $(\hat{Y}_i)$. Least squares estimation, which is the most attractive line-fitting method, results from the minimization of $\sqrt{\varepsilon_i}$.

If we make the following change of variables $x_i = X_i - \overline{X}$, where $\overline{X}$ is the mean of **X**, it is easy to show that the previous regression lines keep their slopes (b and $\hat{b}$ respectively) but obviously change their intercepts (a and $\hat{a}$ respectively) in the new axes (Y, x). The change is convenient because the new variable **x** has the following property: $\sum_i x_i = 0$.

Under this transformation, the least square estimators are given by:

$$\hat{a} = \overline{Y} \tag{4.5}$$

which ensures that the fitted line goes through the *centre of gravity* $(\overline{X}, \overline{Y})$ of the sample of n observations, and

$$\hat{b} = \frac{\sum_i x_i Y_i}{\sum_i x_i^2} \tag{4.6}$$

It worth noting that if X_i can be written as a linear combination of the intercept a, that is, if X_i is equal to some constant for all i, expression (4.6) would be undefined since x_i would be equal to zero for all i, and, therefore, its denominator would be equal to zero.

These estimators have the following interesting properties:

$$E(\hat{a}) = a \quad \text{Var}(\hat{a}) = \sigma^2/n$$
$$E(\hat{b}) = b \quad \text{Var}(\hat{b}) = \sigma^2 \Big/ \sum_i x_i^2$$

In passing, the formula for the variance of $\hat{b}$ has interesting implications in terms of experimental design. First, it can be noted that the variance of both estimators decrease with the sample size n. Also, if the values $\mathbf{X}$ are too close together, as in Figure 4.5a, their deviations from the mean $\overline{X}$ will be small and consequently the sum of x_i will be small; for this reason the variance of $\hat{b}$ will be large and so $\hat{b}$ will be an unreliable estimator. In the contrary case, depicted in Figure 4.5b, even though the errors ε are of the same size as previously, $\hat{b}$ will be a reliable estimator. Therefore, the analyst can improve the quality of the estimators by increasing the sample size and by sampling more cases for which $\mathbf{X}$ takes values that are further apart from the mean $\overline{X}$.

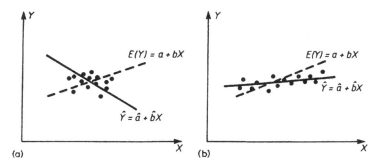

Figure 4.5 Goodness-of-fit and experimental design: (a) unreliable (X_i close), (b) reliable (X_i spread out)

If a fourth, stronger, hypothesis is considered, that the expected value of e conditional on $\mathbf{X}$ is zero, the least squares estimators (4.5) and (4.6) acquire some desirable statistical properties. In this case (i.e. when the mean of $\mathbf{e}$ weighed by the probability of occurrence of $\mathbf{X}$ is zero) the least square estimators are said to be unbiased (i.e. their expected values are equal to the true values a and b) and consistent (i.e. they can be as near as desired to the true values as the sample size goes to infinity).

This important assumption may be easily violated if a relevant variable is omitted from the model and it is correlated with the observed $\mathbf{X}$. For example, consider that the number of trips generated depends on household's income and number of cars, variables that are positively correlated since it is more likely to have a car as income grows. If the number of cars is omitted from the model, the least square estimator of the effect of income will account for both the effect of income and of the number of cars, and will therefore be larger than the true coefficient of income. Methods to test and to correct for the violation of this assumption exist in the literature; the interested reader is referred to Greene (2003) for further descriptions and to Guevara and Ben-Akiva (2006) for an application to discrete choices.

If in addition to Hypothesis 4, hypotheses 1 and 3 hold, the least squares estimators (4.5) and (4.6) are not only consistent and unbiased, but are also the Best (the most efficient, and those with the smallest variance) among all possible Linear and Unbiased Estimators (BLUE). This result is known as the Gauss-Markov theorem; methods to test and to correct for violations of hypotheses 1 and 3 can be found in the literature. In section 4.2.2 we present a practical case where the failure of Hypothesis 3 is corrected. For further examples and applications, the interested reader is again referred to Greene (2003).

4.2.1.3 Hypothesis Tests for $\hat{b}$

To carry out these hypothesis tests we need to know the distribution of $\hat{b}$ and for this, we need to consider the strong hypothesis that the variables $\mathbf{Y}$ are distributed Normal. Note that in this case the least squares estimators will not just be BLUE, but BUE (i.e. Best Unbiased Estimators) among all possible linear and non-linear estimators. This assumption may be strong when the sample is small, but as the sample size increases it will begin to hold no matter which is the true distribution thanks to the *Law of Large Numbers*.

Now, as $\hat{b}$ is just a linear combination of the Y_i, it follows that it is also distributed N $(b, \sigma^2/\sum_i x_i^2)$. This means we can standardise it in the usual way, obtaining

$$z = \frac{\hat{b} - b}{\sigma / \sqrt{(\sum_i x_i^2)}} \tag{4.7}$$

which is distributed N $(0,1)$; it is also useful to recall that z^2, the *quadratic form* (see 2.5.4.1), is distributed χ^2 with one degree of freedom. However we do not know σ^2, the variance of $\mathbf{Y}$ with respect to the true regression. A natural estimator is to use the *residual variance* s^2 around the fitted line:

$$s^2 = \frac{\sum_i (Y_i - \hat{Y}_i)^2}{n - 2}$$

We divide by $(n - 2)$ to obtain an unbiased estimator because two degrees of freedom have been used to calculate $\hat{a}$ and $\hat{b}$ which define $\hat{Y}_i$ (see Wonnacott and Wonnacott 1990).

However, if we substitute s^2 by σ^2 in (4.7) the standardised $\hat{b}$ becomes distributed Student (or t) with $(n - 2)$ degrees of freedom:

$$t = \frac{\hat{b} - b}{s / \sqrt{(\sum_i x_i^2)}} \tag{4.8}$$

The denominator of (4.8) is usually called *standard error* of $\hat{b}$ and is denoted by s_b, hence we can write (4.8) as: $t = (\hat{b} - b)/s_b$.

The t-test   A typical null hypothesis is H_0: $b = 0$; in this case (4.8) reduces to:

$$t = \hat{b}/s_b \tag{4.9}$$

and this value needs to be compared with the critical value of the Student statistics for a given significance level α and the appropriate number of degrees of freedom. One problem is that the alternative hypothesis H_1 may imply unilateral (b > 0) or bilateral (b not equal 0) tests; this can only be determined by examining the phenomenon under study.

Example 4.2   Assume we are interested in studying the effect of income (I) in the number of trips by non-car-owning households (T), and that we can use the following relation:

$$T = a + bI$$

As in theory we can conclude that any influence must be positive (i.e. higher income always means more trips) in this case we should test H_0 against the unilateral alternative hypothesis H_1: $b > 0$. If H_0 is true, the t-value from (4.9) is compared with the value $t_{\alpha:d}$, where d are the appropriate number of degrees of freedom, and the null hypothesis is rejected if $t > t_{\alpha:d}$ (see Figure 4.6).

On the other hand, if we were considering incorporating a variable the effect of which in either direction was not evident (for example, number of female workers, as these may or may not produce more trips than their male counterparts), the null hypothesis should be the bilateral H_1: $b \neq 0$, and H_0 would be rejected if 0 is not included in the appropriate confidence interval for $\hat{b}$.

The F-test for the Complete Model   Figure 4.7a shows the set of values $(\hat{a}, \hat{b})$ for which null hypotheses such as the one discussed above are accepted individually. If we were interested in testing the hypothesis that both estimators are equal to 0, for example, we could have a region such as that depicted in

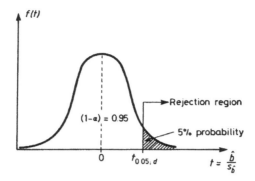

Figure 4.6   Rejection region for $\alpha = 5\%$

Figure 4.7b; i.e. accepting that each parameter is 0 individually does not necessarily mean accepting that both should be 0 together.

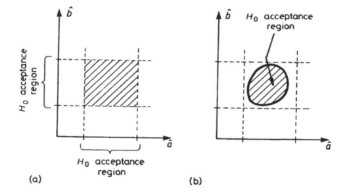

Figure 4.7   Acceptance regions for null hypothesis: (a) both parameters individually, (b) both parameters together

Now, to make a two-parameter test it is necessary to know the joint distribution of both estimators. In this case, as their marginal distributions are Normal, the joint distribution is also bivariate Normal. The F-statistic used to test the trivial null hypothesis H_0: $(a, b) = (0, 0)$, provided as one of the standards in commercial computer packages, is given by:

$$F = \left(n\,\hat{a}^2 + \sum_i x_i^2\,\hat{b}^2 \right) / 2s^2$$

H_0 is accepted if F is less than or equal to the critical value $F_\alpha (2, n-2)$. Unfortunately the test is not very powerful (i.e. it is nearly always rejected), but similar ones may be constructed for more interesting null hypotheses such as $(a, b) = (\overline{Y}, 0)$.

4.2.1.4   The Coefficient of Determination R^2

Figure 4.8 shows the regression line and some of the data points used to estimate it. If no values of **x** were available, the best prediction of Y_i would be $\overline{Y}$. However, the figure shows that for a particular

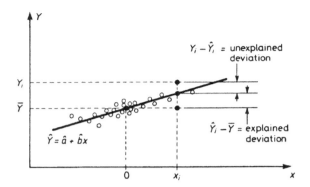

Figure 4.8   Explained and unexplained deviations

value x_i the error of this method could be high: $(Y_i - \overline{Y})$. When x_i is known, on the other hand, the best prediction for Y_i is $\hat{Y}_i$ and this reduces the error to just $(Y_i - \hat{Y}_i)$, i.e. a large part of the original error has been explained. From Figure 4.8 we have:

$$\underset{\text{total deviation}}{(Y_i - \overline{Y})} = \underset{\text{explained deviation}}{(\hat{Y}_i - \overline{Y})} + \underset{\text{unexplained deviation}}{(Y_i - \hat{Y}_i)}, \qquad \forall i$$

If we square the total deviations and sum over all values of i, we get the following:

$$\underset{\text{total variation}}{\sum_i (Y_i - \overline{Y})^2} = \underset{\text{explained variation}}{\sum_i (\hat{Y}_i - \overline{Y})^2} + \underset{\text{unexplained variation}}{\sum_i (Y_i - \hat{Y}_i)^2} \tag{4.10}$$

Now, because $(\hat{Y}_i - \overline{Y}) = \hat{b}x_i$ it is easy to see that the explained variation is a function of the estimated regression coefficient $\hat{b}$. The process of decomposing the total variation into its parts is known as *analysis of variance* of the regression, or ANOVA (note that variance is just variation divided by degrees of freedom).

The coefficient of determination is defined as the ratio of explained to total variation:

$$R^2 = \frac{\Sigma(\hat{Y}_i - \overline{Y})^2}{\Sigma(Y_i - \overline{Y})^2} \tag{4.11}$$

It has limiting values of 1 (perfect explanation) and 0 (no explanation at all); intermediate values may be interpreted as the percentage of the total variation explained by the regression. The index is trivially related to the sample correlation R, which measures the degree of association between X and Y (see Wonnacott and Wonnacott 1990).

4.2.1.5   Multiple Regression

This is an extension of the above for the case of more explanatory variables and, obviously, more *regressors* ($\hat{b}$ parameters). The solution equations are similar, although more complex, but some extra problems arise which are usually important, such as the following:

1. Multicollinearity. This occurs when there is a linear relation between the explanatory variables. Equivalent with what occurred when one explanatory variable was a linear function of the intercept in (4.6), in this case the equations for the regressors $\hat{b}$ are not independent and cannot be solved uniquely.

2. How many regressors to include. To make a decision in this case, several factors have to be taken into consideration:
   - Are there strong theoretical reasons to include a given variable, or is it important for policy testing with the model?
   - Is the estimated sign of the coefficient consistent with theory or intuition and is the variable significant (i.e. is H_0 rejected in the t-test?)?

   If in doubt, one way forward is to take out the variable in question and re-estimate the regression in order to examine the effect of its removal on the rest of the coefficients; if this is not too important the variable can be left out for *parsimony* (the model is simpler and the rest of the parameters can be estimated more accurately). Commercial software packages provide an 'automatic' procedure for tackling this issue (the *stepwise* approach); however, this may induce some problems, as we will comment below. We will come back to this general problem in section 8.4 (Table 8.1) when discussing discrete choice model specification issues.

3. Coefficient of determination. This has the same form as (4.11). However, in this case the inclusion of another regressor always increases R^2; to eliminate this problem the corrected R^2 is defined as:

$$\overline{R}^2 = [R^2 - k/(n-1)][(n-1)/(n-k-1)] \tag{4.12}$$

   where n stands for sample size as before and k is the number of regressors $\hat{b}$.

   In trip generation modelling the multiple regression method has been used both with aggregate (zonal) and disaggregate (household and personal) data. The first approach has been practically abandoned in the case of trip productions, but it is still the premier method for modelling trip attractions. In this sense, it is worth noting that expressions (4.11) and (4.12) will have values between 0 and 1 if and only if the least square model considers an intercept, that is, if model (4.4) is not forced to consider a equal to zero. Also, (4.12) is a good tool to compare models as long as the variables $\mathbf{Y}$ used for the cases under analysis are the same. For example, if the analyst wants to compare a model for the number of trips as a function of zone attributes and another one using the logarithm of the number of trips, the measures are not appropriate since the denominator in (4.11) is not the same for both models.

4. Hypothesis testing. If the analyst is interested in testing a hypothesis regarding a specific estimator, the t-test described in (4.8) may be used. However, if the hypothesis involves a linear restriction between many estimators, an F-test should be used instead. In this case we need to estimate first a *restricted* model, where the restrictions to be tested hold and calculate the Sum of Squared Residuals of the Restricted model (SSR_R) which is equal to $\Sigma \varepsilon_i^2 = \Sigma (Y_i - \hat{Y}_i)^2$ and is often an output of regression software. Second, we need to estimate an *unrestricted* model (i.e. where the restrictions are not imposed) and calculate the SSR_U. Then, the F statistic is computed as follows, where k is the number of variables in the unrestricted model, and r is the number of restrictions imposed:

$$\hat{F} = \frac{\{SSR_R - SSR_U\}}{SSR_U} \frac{(n-k)}{r} \sim F_{r,n-k}$$

This statistic follows an F distribution with r and n-k degrees of freedom. The intuition of the test is as follows: if the restrictions are true, SSR_R should be similar to SSR_U and the statistic should be near to zero. On the contrary, if the statistic is larger than $F_{r,n-k}$ the null hypothesis can be rejected for some desired confidence level.

4.2.2   Zonal-based Multiple Regression

In this case an attempt is made to find a linear relationship between the number of trips produced or attracted by zone and average socioeconomic characteristics of the households in each zone. The following are some interesting considerations:

1. Zonal models can only explain the variation in trip making behaviour between zones. For this reason they can only be successful if the inter-zonal variations adequately reflect the real reasons behind trip variability. For this to happen it would be necessary that zones not only had a homogeneous socioeconomic composition, but represented as wide as possible a range of conditions. A major problem is that the main variations in person trip data occur at the intra-zonal level.

2. Role of the intercept. One would expect the estimated regression line to pass through the origin; however, large intercept values (i.e. in comparison to the product of the average value of any variable and its coefficient) have often been obtained. If this happens the equation may be rejected; if on the contrary, the intercept is not significantly different from zero, it might be informative to re-estimate the line, forcing it to pass through the origin.

3. Null zones. It is possible that certain zones do not offer information about certain dependent variables (e.g. there can be no HB trips generated in non-residential zones). Null zones must be excluded from analysis; although their inclusion should not greatly affect the coefficient estimates (because the equations should pass through the origin), an arbitrary increment in the number of zones which do not provide useful data will tend to produce statistics which overestimate the accuracy of the estimated regression.

4. Zonal totals versus zonal means. When formulating the model the analyst appears to have a choice between using aggregate or total variables, such as trips per zone and cars per zone, or rates such as trips per household per zone and cars per household per zone. In the first case the regression model would be:

$$Y_i = \theta_0 + \theta_1 X_{1i} + \theta_2 X_{2i} + \ldots + \theta_k X_{ki} + E_i$$

whereas the model using rates would be:

$$y_i = \theta_0 + \theta_1 x_{1i} + \theta_2 x_{2i} + \ldots + \theta_k x_{ki} + e_i$$

with $y_i = Y_i/H_i$; $x_i = X_i/H_i$; $e_i = E_i/H_i$ and H_i the number of households in zone i.

Both equations are almost identical, in the sense that they seek to explain the variability of trip making behaviour between zones, and in both cases the parameters have the same meaning. Their unique and fundamental difference relates to the error-term distribution in each case; it is obvious that the constant variance condition of the model cannot hold in both cases, unless H_i was itself constant for all zones i.

Now, as the aggregate variables directly reflect the size of the zone, their use should imply that the magnitude of the error actually depends on zone size; this *heteroskedasticity* (variability of the variance) has indeed been found in practice. Using multipliers, such as $1/H_i$, allows heteroskedasticity to be reduced because the model is made independent of zone size. In this same vein, it has also been found that the aggregate variables tend to have higher intercorrelation (i.e. multicolinearity) than the rates. It is important to note that models using aggregate variables often yield higher values of R^2, but this is just a spurious effect because zone size obviously helps to explain the total number of trips (see Douglas and Lewis 1970). What is certainly unsound is the mixture of rates and aggregate variables in a single model.

To end this theme it is important to remark that even when rates are used, zonal based regression is conditioned by the nature and size of zones (i.e. the spatial aggregation problem). This is clearly exemplified by the fact that inter-zonal variability diminishes with zone size as shown in Table 4.2, constructed with data from Perth (Douglas and Lewis 1970).

Table 4.2   Inter-zonal variation of personal productions for two different zoning systems

Zoning system	Mean value of trips/household/zone	Inter-zonal variance
75 small zones	8.13	5.85
23 large zones	7.96	1.11

4.2.3   Household-based Regression

Intra-zonal variation may be reduced by decreasing zone size, especially if zones are homogeneous. However, smaller zones imply a greater number of them and this has two consequences:

- more expensive models in terms of data collection, calibration and operation;
- larger sampling errors, which are assumed non-existent by the multiple linear regression model.

For these reasons it seems logical to postulate models which are independent of zone boundaries. At the beginning of the 1970s it was believed that the most appropriate analysis unit in this case was the household (and not the individual); it was argued that a series of important interpersonal interactions inside a household could not be incorporated even implicitly in an individual model (e.g. car availability, that is, which member has use of the car). This thesis was later challenged as we will see in section 4.3.3, but with little practical success.

In a household-based application each home is taken as an input data vector in order to bring into the model all the range of observed variability about the characteristics of the household and its travel behaviour. The calibration process, as in the case of zonal models, may proceed stepwise, testing each potential explanatory variable in turn until the best model (in terms of some summary statistics for a given confidence level) is obtained. Care has to be taken with automatic stepwise computer packages because they may leave out variables which are slightly worse predictors than others left in the model, but which may prove much easier to forecast.

In actual fact, stepwise methods are not recommended; it is preferable to proceed the other way around, i.e. test a model with all the variables available and take out those which are not essential (on theoretical or policy grounds) and have low significance or an incorrect sign.

Example 4.3   Consider the variables trips per household (Y), number of workers (X_1) and number of cars (X_2). Table 4.3 presents the results of successive steps of a stepwise model estimation; the last row also shows (in parenthesis) values for the t-ratio (equation 4.9). Assuming large sample size, the appropriate number of degrees of freedom ($n - 2$) is also a large number so the t-values may be compared with the critical value 1.645 for a 95% significance level on a one-tailed test (we know the null hypothesis is unilateral in this case as Y should increase with both X_1 and X_2).

Table 4.3   Example of stepwise regression

Step	Equation	R^2
1	$Y = 2.36\,X_1$	0.203
2	$Y = 1.80\,X_1 + 1.31\,X_2$	0.325
3	$Y = 0.91 + 1.44X_1 + 1.07X_2$	
	(3.7) (8.2) (4.2)	0.384

The third model is a reasonable equation in spite of its low R^2. The intercept 0.91 is not large (compare it with 1.44 times the number of workers, for example) and the regression coefficients are significantly different from zero (H_0 is rejected in all cases). The model could probably benefit from the inclusion of further variables if they were available.

An indication of how good these models are may be obtained from comparing observed and modelled trips for some groupings of the data (see Table 4.4). This is better than comparing totals because in such cases different errors may compensate and the bias would not be detected. As can be seen, the majority of cells show a reasonable approximation (i.e. errors of less than 30%). If large bias were spotted it would be necessary to adjust the model parameters; however, this is not easy as there are no clear-cut rules to do it, and it depends heavily on context.

Table 4.4   Comparison of trips per household (observed/estimated).

| | Number of workers in household | | | |
No. of cars	0	1	2	3 or more
0	0.9/0.9	2.1/2.4	3.4/3.8	5.3/5.6
1	3.2/2.0	3.5/3.4	3.7/4.9	8.5/6.7
2 or more	–	4.1/4.6	4.7/6.0	8.5/7.8

4.2.4   The Problem of Non-Linearity

As we have seen, the linear regression model assumes that each independent variable exerts a linear influence on the dependent variable. It is not easy to detect non-linearity because apparently linear relations may turn out to be non-linear when the presence of other variables is allowed for in the model. Multivariate graphs are useful in this sense; the example of Figure 4.9 presents data for households stratified by car ownership and number of workers. It can be seen that travel behaviour is non-linear with respect to family size.

It is important to mention that there is a class of variables, those of a qualitative nature, which usually shows non-linear behaviour (e.g. type of dwelling, occupation of the head of the household, age, sex). In general there are two methods to incorporate non-linear variables into the model:

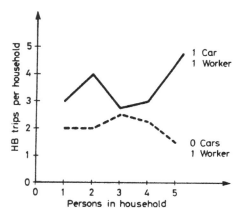

Figure 4.9   An example of non-linearity

1. Transform the variables in order to linearise their effect (e.g. take logarithms, raise to a power). However, selecting the most adequate transformation is not an easy or arbitrary exercise, so care is needed; also, if we are thorough, it can take a lot of time and effort.
2. Use *dummy* variables. In this case the independent variable under consideration is divided into several discrete intervals and each of them is treated separately in the model. In this form it is not necessary to assume that the variable has a linear effect, because each of its portions is considered separately in terms of its effect on travel behaviour. For example, if car ownership was treated in this way, appropriate intervals could be 0, 1 and 2 or more cars per household. As each sampled household can only belong to one of the intervals, the corresponding dummy variable takes a value of 1 in that class and 0 in the others. It is easy to see that only $(n - 1)$ dummy variables are needed to represent n intervals.

Example 4.4 Consider the model of Example 4.3 and assume that variable X_2 is replaced by the following dummies:

Z_1, which takes the value 1 for households with one car and 0 in other cases;

Z_2, which takes the value 1 for households with two or more cars and 0 in other cases.

It is easy to see that non-car-owning households correspond to the case where both Z_1 and Z_2 are 0. The model of the third step in Table 4.3 would now be:

$$Y = 0.84 + 1.41X_1 + 0.75Z_1 + 3.14Z_2 \quad R^2 = 0.387$$
$$\quad\quad (3.6) \quad\quad (8.1) \quad\quad (3.2) \quad\quad (3.5)$$

Even without the better R^2 value, this model would be preferable to the previous one just because the non-linear effect of X_2 (or Z_1 and Z_2) is clearly evident and cannot be ignored. Note that if the coefficients of the dummy variables were for example, 1 and 2, and if the sample never contained more than two cars per household, the effect would be clearly linear. The model is graphically depicted in Figure 4.10.

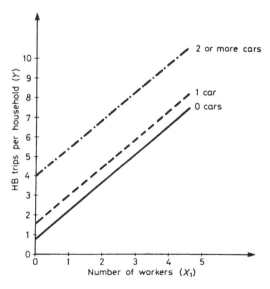

Figure 4.10   Regression model with dummy variables

Looking at Figure 4.10, the following question arises: would it not be preferable to estimate separate regressions for the data on each group, as in that case we would not require each line to have the same slope (i.e. the coefficient of X_1)? The answer is in general *no* unless we had a reasonable amount of data for each class. The fact is that the model with dummies uses all the data, while each separate regression would use only part of it, and this is in general disadvantageous. It is also interesting to mention that the use of dummy variables tends to reduce problems of multicolinearity in the data (see Douglas and Lewis 1971).

4.2.5 Obtaining Zonal Totals

In the case of zonal-based regression models, this is not a problem as the model is estimated precisely at this level. In the case of household-based models, though, an aggregation stage is required. Nevertheless, precisely because the model is linear the aggregation problem is trivially solved by replacing the average zonal values of each independent variable in the model equation and then multiplying it by the number of households in each zone. However, it must be noted that the aggregation stage can be a very complex matter in non-linear models, as we will see in Chapter 9.

Thus, for the third model of Table 4.3 we would have:

$$T_i = H_i(0.91 + 1.44\overline{X}_{1i} + 1.07\overline{X}_{2i})$$

where T_i is the total number of HB trips in zone i, H_i is the total number of households in it and $\overline{X}_{ji}$ is the average value of variable X_j for the zone.

On the other hand, when dummy variables are used, it is also necessary to know the number of households in each class for each zone; for instance, in the model of Example 4.4 we would require:

$$T_i = H_i(0.84 + 1.41\overline{X}_{1i}) + 0.75H_{1i} + 3.14H_{2i}$$

where H_{ji} is the number of households of class j in zone i.

This last expression allows us to appreciate another advantage of using dummy variables over separate regressions. To aggregate the models in that latter case, it would be necessary to estimate the average number of workers per household (X_1) for each car-ownership group in each zone, and this may be complicated for long-term forecasts.

4.2.6 Matching Generations and Attractions

It might be obvious to some readers that the models above do not guarantee, by default, that the total number of trips originating (the *origins O_i*) at all zones will be equal to the total number of trips attracted (the *destinations D_j*) to them, that is the following expression does not necessarily hold:

$$\sum_i O_i = \sum_j D_j \tag{4.13}$$

The problem is that this equation is implicitly required by the next sub-model (i.e. trip distribution) in the structure; it is not possible to have a trip distribution matrix where the total number of trips (T) obtained by summing all rows is different to that obtained when summing all columns (see Chapter 5).

The solution to this difficulty is a pragmatic one which takes advantage of the fact that normally the trip generation models are far 'better' (in every sense of the word) than their trip attraction counterparts. The first usually are fairly sophisticated household-based models with typically good explanatory variables.

The trip attraction models, on the other hand, are at best estimated using zonal data. For this reason, normal practice considers that the total number of trips arising from summing all origins O_i is in fact the correct figure for T; therefore, all destinations D_j are multiplied by a factor f given by:

$$f = T \Big/ \sum_j D_j \qquad (4.14)$$

which obviously ensure that their sum also adds to T.

4.3   Cross-Classification or Category Analysis

4.3.1   The Classical Model

4.3.1.1   Introduction

Although linear regression was the early recommended approach for trip generation, from the late 1960s an alternative method for modelling trip generation appeared and quickly became established as the preferred one in the United Kingdom. The method was known as *category analysis* in the UK (Wootton and Pick 1967) and *cross-classification* in the USA; there it went through a similar development process as the linear regression model, with earliest procedures being at the zonal level and subsequent models based on household information.

The method is based on estimating the response (e.g. the number of trip productions per household for a given purpose) as a function of household attributes. Its basic assumption is that trip generation rates are relatively stable over time for certain household stratifications. The method finds these rates empirically and for this it typically needs large amounts of data; in fact, a critical element is the number of households in each class. Although the method was originally designed to use census data in the UK, a serious problem of the approach remains the need to forecast the number of households in each stratum in the future.

4.3.1.2   Variable Definition and Model Specification

Let $t^p(h)$ be the average number of trips with purpose p (and at a certain time period) made by members of households of type h. Types are defined by the stratification chosen; for example, a cross-classification based on m household sizes and n car ownership classes will yield mn types h. The standard method for computing these cell rates is to allocate households in the calibration data to the individual cell groupings and total, cell by cell, the observed trips $T^p(h)$ by purpose group. The rate $t^p(h)$ is then the total number of trips in cell h, by purpose, divided by the number of households $H(h)$ in it. In mathematical form it is simply as follows:

$$t^p(h) = T^p(h)/H(h) \qquad (4.15)$$

The 'art' of the method lies in choosing the categories such that the standard deviations of the frequency distributions depicted in Figure 4.11 are minimised.

The method has, in principle, the following advantages:

1. Cross-classification groupings are independent of the zone system of the study area.
2. No prior assumptions about the shape of the relationship are required (i.e. they do not even have to be monotonic, let alone linear).
3. Relationships can differ in form from class to class (e.g. the effect of changes in household size for one or two car-owning households may be different).

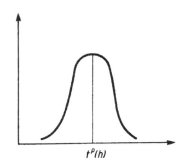

$t^P(h)$

Figure 4.11   Trip-rate distribution for household type

And in common with traditional cross-classification methods it also has several disadvantages:

1. The model does not permit extrapolation beyond its calibration strata, although the lowest or highest class of a variable may be open-ended (e.g. households with two or more cars and five or more residents).
2. There are no statistical goodness-of-fit measures for the model, so only aggregate closeness to the calibration data can be ascertained, but see the discussion in 4.3.2.
3. Unduly large samples are required; otherwise, cell values will vary in reliability because of differences in the numbers of households being available for calibration at each one. For example, in the Monmouthshire Land Use/Transportation Study (see Douglas and Lewis 1971) the estimators for 108 categories (six income levels, three car ownership levels and six household structure levels) shown in Table 4.5 were found, using a sample of 4000 households.

Table 4.5   Household frequency distribution

	No. of categories				
	21	69	9	7	2
No. of households surveyed	0	1–49	50–99	100–199	200+

Accepted wisdom suggests that at least 50 observations per cell are required to estimate the mean reliably; thus, this criterion would be satisfied in only 18 of the 108 cells for a sample of 4000 households. There may be some scope for using stratified sampling to guarantee more evenly distributed sample sizes in each category. This involves, however, additional survey costs.

4. There is no effective way to choose among variables for classification, or to choose best groupings of a given variable; the minimisation of standard deviations hinted at in Figure 4.11 would require an extensive 'trial and error' procedure which may be considered infeasible in practical studies.

4.3.1.3   Model Application at Aggregate Level

Let us denote by q the person type (i.e. with and without a car), by $a_i(h)$ the number of households of type h in zone i, and by $H^q(h)$ the set of households of type h containing persons of type q. With this we can write the trip productions with purpose p by person type q in zone i, O_i^{qp}, as follows:

$$O_i^{qp} = \sum_{h \in H^q(h)} a_i(h)t^P(h) \qquad (4.16)$$

To verify how the model works it is possible to compare these modelled values with observed values from the calibration sample. Inevitable errors are due to the use of averages for the rates t^p (h); one would expect a better stratification (in the sense of minimising the standard deviation in Figure 4.11) to produce smaller errors.

There are various ways of defining household categories. The first application in the UK (Wootton and Pick 1967) employed 108 categories as follows: six income levels, three car ownership levels (0, 1 and 2 or more cars per household) and six household structure groupings, as in Table 4.6.

Table 4.6   Example of household structure grouping

Group	No. employed	Other adults
1	0	1
2	0	2 or more
3	1	1 or less
4	1	2 or more
5	2 or more	1 or less
6	2 or more	2 or more

The problem is clearly how to predict the number of households in each category in the future. The method most commonly used (see Wilson 1974) consists in, firstly, defining and fitting to the calibration data, probability distributions for income (I), car ownership (C) and household structure (S); secondly, using these to build a joint probability function of belonging to household type $h = (I, C, S)$. Thus, if the joint distributions function is denoted by ϕ $(h) = \phi$ (I, C, S), the number of households in zone i belonging to class h, a_i (h), is simply given by:

$$a_i(h) = H_i \cdot \phi(h) \tag{4.17}$$

where H_i is the total number of households in the zone. This household estimation model may be partially tested by running it with the base-year data used in calibration. The total trips estimated with equation (4.16), but with simulated values for a_i (h), can then be checked against the actual observations.

One further disadvantage of the method can be added at this stage:

5. If it is required to increase the number of stratifying variables, it might be necessary to increase the sample enormously. For example, if another variable was added to the *original* application discussed above and this was divided into three levels, the number of categories would increase from 108 to 324 (and recall the discussion on Table 4.5).

4.3.2   Improvements to the Basic Model

4.3.2.1   Equivalence between Category Analysis and Linear Regression

Some of the limitations of the basic model above may be overcome by noting that Category Analysis estimators can be obtained using a linear regression model with dummy variables representing each category (Goodman 1973). This result can be easily shown recalling that if a dummy variable is defined for each category this is equivalent to running separate least squares models with no more variables than an intercept. Then, from equation (4.5), it follows directly that the least square estimator is identical to the average of **Y** for each category.

(continued)

Given this equivalence, the disadvantage of not having statistical goodness-of-fit measures for the model disappears. One can use, for example, the $\overline{R}^2$ measure (4.12) to compare different potential category structures; however, a small shift is needed. To make $\overline{R}^2$ comparable among different models and be constrained between 0 and 1, we need to consider an intercept, but in that case the model would not be identifiable because the category dummies will add up to one (i.e. equal to the intercept). As noted by Guevara and Thomas (2007), this can be solved by setting one of the categories as the base and using dummies for all the others. Then, the model intercept corresponds to the estimated trip rate of the base category and the estimators associated with each other dummy variable will correspond to the difference between the trip rate of the respective category and that one used as a base. The $\overline{R}^2$ calculated in this way and also the F-test may be used to compare different groupings of alternative variables for stratification.

Equally, the analyst may use the t-test (4.8) as a statistical measure of the reliability of the estimates in each category. The analyst may consider valid, for example, only stratifications for which the estimators are different to zero at the 95% confidence level.

Another practical limitation of the basic model is that in some cases the number of observations in the sample is too small or even inexistent, precluding the estimations of trip rates for some categories. However, if the analyst is interested in having an estimate for such categories there is the following alternative; if it is assumed that the impact of an additional variable level in the number of trips is independent of other variables, one can formulate a linear model that depends on each of the variables' levels. For example, if we consider that the number of trips depends on Income and Car ownership, and that those variables are divided into two levels (Low and High income; 0 and 1+ cars), the following linear model could be formulated:

$$Trips_i = \theta_{IL} I_{Low\_i} + \theta_{IH} I_{High\_i} + \theta_{M0} M_{0\_i} + \theta_{M1} M_{1\_i} + e_i$$

where I_{Low} is equal to 1 if the household belongs to the low income category and zero otherwise. Other variables are defined equivalently and e_i corresponds to an error term.

However, it can be noted that this model is not estimable since there is a problem of multicolinearity, as $I_{Low} + I_{High} = M_0 + M_1 = 1$. To achieve estimation the model needs to be normalised, that is, to use some of the categories as a base or reference. This can be achieved, for example, by considering model (4.18), where we also included an intercept to make the $\overline{R}^2$ comparable among different models:

$$Trips_i = \alpha + \alpha_{IH} I_{High\_i} + \alpha_{M1} M_{1\_i} + e_i \qquad (4.18)$$

It follows that even if we do not have, for example, observations for households of low income and high motorisation, their trip rates may be calculated as $\hat{\alpha} + \hat{\alpha}_{M1}$ if we accept the hypothesis of a linear effect of income and car ownership in the number of trips.

On the other hand, even if there are enough observations to estimate the trip rates for all categories, the analyst may be interested in estimating a model such as (4.18); as it involves the estimation of fewer parameters with the same data, it would result in estimators with smaller variance. This hypothesis can be tested, for example, by adding a non linear interaction calculated as the product of I_{Low} times M_1 or, equivalently, a dummy variable for the combined effect of belonging to a high income motorised household. This alternative model can be shown to be equivalent to considering one dummy variable per category.

$$Trips_i = \alpha + \alpha_{IH} I_{High\_i} + \alpha_{M1} M_{1\_i} + \alpha_{MI} I_{High\_i} \cdot M_{1\_i} + e_i'$$

The validity of this linear assumption can be tested by checking the joint significance of the interaction variables through a F-test, as described in 4.2.1.5. In this case $k = 4$ and $r = 1$, because the unrestricted model involves the estimation of four coefficients and only one of them is constrained to zero in the restricted model (4.18). The SSR_R corresponds to model (4.18) and the SSR_U to the model including $I_{Low} \bullet M_1$. If the statistic is smaller than the critical value $F_{r,n-k}$, the null hypothesis is accepted meaning that the linear model (4.18) is acceptable. If the statistic is larger than $F_{r,n-k}$ the model with interactions should be considered. It is worth noting that since in this case the interaction term involves the inclusion of only one additional variable, the t-statistic (4.8) could also be used but the F-test is the tool applied in general.

Another interesting test, beyond the linearity assumption, has to do with the possibility of using an additional variable for classification say, household size. In such case, the following model may be estimated, where S_L would take the value one if the household size is large and zero otherwise, say:

$$Trips_i = \alpha + \alpha_{IH} I_{High\_i} + \alpha_{M1} M_{1\_i} + \alpha_{S3} S_{L\_i} + e_i''$$

To find out whether the inclusion of household size is a good idea, an F-test can be used in general and a t-test in the particular case of needing just one additional variable to do it; the significance level to be used in this latter case deserves some attention, though. The usual procedure is to consider it as small as possible, generally 5%, to reduce Type I errors (*i.e.* rejecting the null hypothesis when it is true). However, as excluding household size may cause endogeneity, because the variable may be correlated with income or car ownership, the cost of excluding it when it should be there (Type II error) may be higher than the cost of including it if it should not be considered (Type I error). Since there is a trade off between Type I and Type II errors, it may be advisable to consider a significance level of 10% or even 20% in this case.

Another possibility is that the analyst may want to explore the validity of the threshold used to define Low and High income strata, say. This can be easily achieved by running a regression considering the alternative thresholds and comparing the $\overline{R}^2$ of both models. In general, the model with the larger $\overline{R}^2$ should be preferred, as it will explain a larger portion of the variance. However, if the model with the larger $\overline{R}^2$ results in unreasonable signs or size of coefficients, or if it affects their significance, the alternative should be chosen instead.

Guevara and Thomas (2007) point out that even after all the potential improvements to Category Analysis described above, the $\overline{R}^2$ of this type of models tend to be very low (i.e. less than 0.2 or 0.3). This is not surprising since the model is indeed extremely simplistic. The consideration of more realistic relationships between explanatory variables and the number of trips may be attained by means of linear regression. The models described in the next subsection represent an improvement in that direction.

4.3.2.2 Regression Analysis for Household Strata

A mixture of cross-classification and regression modelling of trip generation may be the most appropriate approach on certain occasions. For example, in an area where the distribution of income is unequal it may be important to measure the differential impact of policies on different income groups; therefore it may be necessary to model travel demand for each income group separately throughout the entire modelling process. Assume now that in the same area car ownership is increasing fast and, as usual, it is not clear how correlated these two variables are; a useful way out may be to postulate regression models based on variables describing the size and make-up of different households, for a stratification according to the two previous variables.

Example 4.5 Table 4.7 presents the 13 income and car-ownership categories (C_i) defined in ESTRAUS (1989) for the Greater Santiago 1977 origin-destination data. As can be seen, the bulk of the data corresponds to households with no cars and low income. Also note that categories 7 and 10 have rather few data points; this is, unfortunately, a general problem of this approach. Even smaller samples for very low income and high car ownership led to the aggregation of some categories at this range.

Table 4.7   Stratification of the 1977 Santiago sample

| Household income (US$/month) | Household car ownership | | | |
	0	1	2+	Total
< 125	6 564 (C_1)	215 (C_2)		6 779
125–250	4 464 (C_3)	627 (C_4)		5 091
250–500	1 532 (C_5)	716 (C_6)	87 (C_7)	2 334
500–750	305 (C_8)	436 (C_9)	118 (C_{10})	859
> 750	169 (C_{11})	380 (C_{12})	301 (C_{13})	790
Total	12 974	2 373	506	15 853

The independent variables available for analysis (i.e. after leaving out the stratifying variables) included variables of the *stage in the family cycle* variety, which we will discuss in section 4.4. However, after extensive specification searches it was found that the most significant variables were: number of workers (divided into four classes depending on earnings and type of job), number of students and number of residents.

Linear regression models estimated with these variables for each of the 13 categories were judged satisfactory on the basis of correct signs, small intercepts, reasonable significance levels and R^2 values (e.g. between 0.401 for category 4, and 0.682 for category 7; see Hall *et al.* 1987).

Finally, one assumption that may be lifted in this case and that may improve even more the adjustment and the quality of the models, is to accept that some coefficients may be the same across categories. This will involve the joint estimation of the 13 models and necessarily produce and increase in efficiency, that is, a reduction in the variance of the estimators.

4.3.3   The Person-category Approach

4.3.3.1   Introduction

This is an alternative to the household-based models discussed above, which was originally proposed by Supernak (1979). It was argued that this approach offered the following advantages (Supernak *et al.* 1983):

1. A person-level trip generation model is compatible with other components of the classical transport demand modelling system, which is based on trip makers rather than on households.
2. It allows a cross-classification scheme that uses all important variables and yields a manageable number of classes; this in turn allows class representation to be forecast more easily.
3. The sample size required to develop a person-category model can be several times smaller than that required to estimate a household-category model.
4. Demographic changes can be more easily accounted for in a person-category model as, for example, certain key demographic variables (such as age) are virtually impossible to define at household level.
5. Person categories are easier to forecast than household categories as the latter require forecasts about household formation and family size; these tasks are altogether avoided in the case of person categories. In general the bulk of the trips are made by people older than 18 years of age; this population is easier to forecast 15 to 20 years ahead as only migration and survival rates are needed to do so.

The major limitation that a person-category model may have relates precisely to the main reason why household-based models were chosen to replace zonal-based models at the end of the 1960s; this is the difficulty of introducing household interaction effects and household money costs and money budgets into a person-based model. However, Supernak *et al.* (1983) argue that it is not clear how vital these considerations are and how they can be effectively incorporated even in a household-based model; in fact, from our discussion in sections 4.2.3 and 4.3.1 it is clear that this is done in an implicit fashion only.

4.3.3.2   Variable Definition and Model Specification

Let t_j be the trip rate, that is, the number of trips made during a certain time period by (the average) person in category j; t_{jp} is the trip rate by purpose p. T_i is the total number of trips made by the inhabitants of zone i (all categories together). N_i is the number of inhabitants of zone i, and α_{ji} is the percentage of inhabitants of zone i belonging to category j. Therefore the following basic relationship exists:

$$T_i = N_i \sum_j \alpha_{ji} t_j \tag{4.19}$$

As in other methods, trips are divided into home-based (HB) and non-home-based (NHB), and can be further divided by purpose (p) which may apply to both HB and NHB trips.

Model development entails the following stages:

1. Consideration of several variables which are expected to be important for explaining differences in personal mobility. Also, definition of plausible person categories using these variables.
2. Preliminary analysis of trip rates in order to find out which variables have the least explanatory power and can be excluded from the model. This is done by comparing the trip rates of categories which are differentiated by the analysed variable only and testing whether their differences are statistically significant.
3. Detailed analysis of trip characteristics to find variables that define similar categories. Variables which do not provide substantial explanation of the data variance, or variables that duplicate the explanation provided by other better variables (i.e. easier to forecast or more policy responsive) are excluded. The exercise is conducted under the constraint that the number of final categories should not exceed a certain practical maximum (for example, 15 classes).

For this analysis the following measures may be used: the coefficient of correlation (R_{jk}), slope (m_{jk}) and intercept (a_{jk}) of the regression $t_{jp} = a_{jk} + m_{jk}\, t_{kp}$. The categories j and k may be treated as similar if these measures satisfy the following conditions (Supernak *et al.* 1983):

$$R_{jk} > 0.900$$
$$0.75 < m_{jk} < 1.25 \tag{4.20}$$
$$a_{jk} < 0.10$$

These conditions are quite demanding and may be changed.

4.3.3.3   Model Application at the Aggregate Level

Zonal home-based productions are computed in a straightforward manner using equation (4.19), or a more disaggregated version explicitly including trip purpose if desired. However, the estimation of trip attractions in general and NHB trip productions at the zonal level is more involved and requires the development of *ad hoc* methods heavily dependent on the type of information available at each application (see Supernak 1979 for a Polish example).

4.4   Trip Generation and Accessibility

As we mentioned in Chapter 1, the classical specification of the urban transport planning (four-stage) model incorporates an iterative process between trip distribution and assignment which leaves trip generation unaltered. This is true even in the case of more contemporary forms which attempt to solve the complex supply-demand equilibration problem appropriately, as we will discuss in Chapter 11. A major disadvantage of this approach is that changes to the network are assumed to have no effects on trip productions and attractions. For example, this would mean that the extension of an underground line to a location which had no service previously would not generate more trips between that zone and the rest. Although this assumption may hold for compulsory trips, it may not hold in the case of discretionary trips (e.g. consider the case of shopping trips and a new line connecting a low-income zone with the city's central market, which features more competitive prices than the zone's local shops).

To solve this problem, modellers have attempted to incorporate a measure of accessibility (i.e. ease or difficulty of making trips to/from each zone) into trip generation equations; the aim is to replace $O_i^n = f\left(H_i^n\right)$ by $O_i^n = f\left(H_i^n, A_i^n\right)$, where H_i^n are household characteristics and A_i^n is a measure of accessibility by person type.

Typical accessibility measures take the general form:

$$A_i^n = \sum_j f(E_j^n, C_{ij})$$

where E_i^n is a measure of attraction of zone j and C_{ij} the generalised cost of travel between zones i and j. A typical analytical expression used to this end has been:

$$A_i^n = \sum_j E_j^n \exp(-\beta C_{ij})$$

where β is a calibration parameter from the gravity model, as discussed in Chapter 5.

Unfortunately this procedure has seldom produced the expected results in the case of aggregate urban modelling applications because the estimated parameters of the accessibility variable have either been non-significant or of the wrong sign. This issue has remained highly topical for many years and it is clearly related to two interesting and yet unresolved problems: model dynamics and modelling with longitudinal instead of cross-sectional data (Chapter 1). Ortúzar *et al.* (2000b) give an interesting discussion of the problem and offer an example of what can be gained by using stated preference data in this context.

New emphasis was given to elastic trip generation models by work done in the UK on induced traffic in the assessment of trunk road schemes (Department of Transport 1997). This work has led to the study of trip generation methods which are sensitive to changes in accessibility, as it is recognised that the classical methods are not adequate in this sense. Daly (1997) proposes a three-component framework for trip generation:

- The individual in his/her household context formulates an activity pattern for the period to be modelled, say a day. Out-of-home activities are of course the only activities that generate trips.
- The out-of-home activities are organised into 'sojourns', which are defined as stays at a specific location, each of which has a primary purpose (and possibly secondary purposes at the same location too).
- A travel plan is formulated to link the sojourns, in particular deciding which require to be visited by separate home-based tours (two or more trips) and which can be linked with other sojourns by non-home-based trips.

From here it appears reasonable to try and model the number of sojourns generated by a household or person, and to split those sojourns between home-based tours and non-home-based trips. One practical modelling point that follows immediately from this framework is that the dependent variables (i.e.

number of tours and/or trips, or alternatively number of sojourns that can be reached) will be integers: 0, 1, 2, 3, etc. Moreover, the decision between whether to travel or not (i.e. between 0 and 1) can be expected to be taken on a different basis from the decision on whether to make more than one trip (the former decision is whether to take part in an activity *at all*, and the latter is how to organise the time and location *given* that some participation will take place). Another point is that travel by all modes needs to be included to ensure that all out-of-home activities are considered; thus the exclusion of short trips or trips by non-motorised modes will detract from the quality of the model. This is therefore consistent with the contemporary approach to O–D data collection discussed in Chapter 3.

The variables to be included in the model are the same as in the classical methods discussed above, but it is hoped that accessibility can be incorporated. However, there may be a negative cross influence between home-based and non-home-based accessibilities; for example, if home-based trips can be made easily (i.e. in a small town) then fewer non-home-based trips will be needed.

Predictions of number of sojourns must be made for each travel purpose (and note that the variables influencing each type may vary). Logically we should model compulsory purposes first and then the non-mandatory purposes can be modelled, conditional upon the decisions made for the compulsory purposes. Similarly, the choice between meeting a travel need by a home-based tour or a non-home-based trip as a detour on a previously planned tour should be modelled explicitly (Algers *et al.* 1995), but independent models are conceivable in the interest of simplicity. Finally, although trip frequency models may be set up to describe the behaviour of complete households (i.e. considering all the interactions that may be relevant to the number of trips made), the development of person-based models is much simpler in practice, given the data that is usually available.

4.5   The Frequency Choice Logit Model

Daly (1997) discussed several models, concluding that the most adequate was one with a Logit form (see Chapter 7) and which would predict the total number of trips by first calculating the probability that each individual would choose to make a trip. The total travel volume can then be obtained by multiplying the number of individuals of each type by their probabilities of making a trip. The extension needed to deal with individuals that make more than one trip is presented subsequently.

If V is the utility of making a trip (assuming that the utility of not travelling is zero, with no loss of generality), the probability of making a trip is given by:

$$P_1 = \frac{1}{1 + \exp(-V)}$$

where V is usually specified as being a linear function of unknown parameters θ:

$$V = \sum_k \theta_k X_k$$

where X are measured data items such as income, car ownership and household size, and the accessibility that needs to be input in a form consistent with utility-maximising theory (i.e. the theory behind the Logit model). For this, the preferred form is to use a result by Williams (1977) made popular by Ben-Akiva and Lerman (1979), which states that the correct form of accessibility is the 'logsum' of the destination (or mode) choice model; furthermore, and as we discuss in section 7.4, because in this case we have a Nested Logit structure, the parameter multiplying the logsum accessibility variable must lie between 0 and 1. If this condition is not met, the model predictions may violate common sense (Williams and Senior 1977).

(continued)

The Logit model represents the choice of each individual whether or not to make a trip, and this means it is particularly suited to dealing with disaggregate data. But aggregate data can also be used; then the probabilities P represent proportions rather than probabilities. However, to obtain the best chance of finding a significant relationship between accessibility and number of trips use disaggregate data as it preserves the maximum amount of variance. In order to model higher trip frequencies, Daly (1997) proposes the use of a hierarchical structure representing an indefinite number of choices (Figure 4.12).

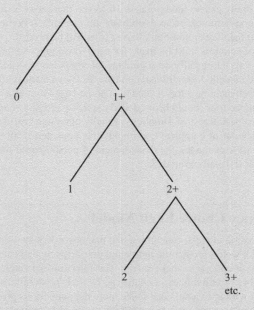

0   1+

1   2+

2   3+
                     etc.

Figure 4.12 'Stop-go' trip generation model

At each hierarchical level, the choice is whether to make further journeys or to stop at the present number (hence the name 'stop-go model'). Because of the possibly strong difference in behaviour between the 0/1+ choice and the remaining choices, it has been found preferable to model the first choice using a separate model. However, because there are often little data on travellers making multiple journeys, it is also necessary to model the remaining choices with a single 'stop-go' model (i.e. which predicts the same probability of stopping at every level of the hierarchy).

It has been found that applying this model system is straightforward. If the probability of making any journeys is p (from the 0/1+ model) and the probability of choosing the 'go' option at each subsequent stage is q (from the stop-go model), then the expected number of journeys is simply:

$$t = p/(1 - q)$$

The method has been applied in several studies in Europe (Daly 1997), obtaining coefficients for the accessibility variable ranging from 0.07 to 0.33 for various trip purposes. A more aggregate version of the model, using linear regression on trips observed at an intercept survey of most roads to the

North of Chile, also gave good results yielding accessibility measures (of the log-sum type) with significant coefficients of the proper sign and magnitude (Iglesias *et al.* 2008).

4.6   Forecasting Variables in Trip Generation Analysis

The choice of variables used to predict (household) trip generation rates has long been an area of concern for transportation planners; these variables typically include household numbers, household size (and/or structure), number of vehicles owned and income. However, interest arose in the early 1980s on research aimed at enriching trip generation models with theories and methods from the behavioural sciences. The major hypothesis behind this work was that the social circumstances in which individuals live should have a considerable bearing on the opportunities and constraints they face in making activity choices; the latter in turn, may lead to differing travel behaviour. For example, it is clear that whether a person lives alone or not should affect the opportunities to coordinate and trade-off activities with others in order to satisfy their travel necessities. Thus, a married couple with young pre-school children will generally find themselves less mobile than a similar couple without children or with older children who require less intensive care. Elderly and retired persons living with younger adults are likely to be more active outside the home than elderly people living alone or with persons roughly their own age.

At the household level the situation is quite similar: households of unrelated individuals, for example, tend to follow a pattern of activities that is less influenced by the presence of other household members (and which normally leads to more frequent trips) than is the case of households of related individuals (obviously with similar size, and other characteristics). This is due to the reduced coordination among different members and also to the fact that their activity patterns typically involve fewer home-centred activities.

One way of introducing these notions into the modelling of trip generation is to develop a set of household types that effectively captures these distinctions and then add this measure to the equations predicting household behaviour. One possible approach considers the age structure of the household and its lifestyle. The approach is consistent with the idea that travel is a *derived demand* and that travel behaviour is part of a larger allocation of time and money to activities in separate locations. For example, the concept of *lifestyle* can be made operational as the allocation of varying amounts of time to different (activity) purposes both within and outside the home, where travel is just part of this time allocation (see Allaman *et al.* 1982). It appears that the time allocation of individuals varies systematically across various segments of the population, such as age, sex, marital status and even race; this may be because different household structures place different demands on individuals.

One set of hypotheses that can be tested empirically is whether the major break points (or stages) in the life (or family) cycle are consistent with major changes of time allocation. For example, the break points may be:

- the appearance of pre-school children;
- the time when the youngest child reaches school age;
- the time when a youth leaves home and either lives alone, with other young adults, or marries;
- the time when all the children of a couple have left home but the couple has not yet retired;
- the time when all members of a household have reached retirement age.

It is usually illuminating to compare households at one stage of this life cycle with households of the immediately preceding stage.

The concepts of lifestyle and stage of family cycle are important from two points of view: first, that of identifying stable groupings (based on age or sex) with different activity schedules and consequently demands for travel; second, that of allowing the tracing of systematic changes which may be based on demographic variations (e.g. changes in age structure, marital or employment status). Numerous demographic trends of significance in terms of travel behaviour have been receiving increasing attention since the early 1980s (see Spielberg *et al.* 1981). One of the most significant for predicting travel behaviour is the changing ratio of households to population, particularly in industrialised nations. Although the rate of population growth has been falling steadily since the 1980s, the rate of household formation has increased in some cases. This is due, among other reasons, to increases in the number of single-parent households and the number of persons who are setting up individual households. Therefore, travel forecasting methodologies which implicitly assume stable ratios of households to population (as is often the case) will be severely affected by this structural shift in the demographic composition of society.

Another trend which has been well discussed is the overall ageing of the population, again particularly in industrialised nations. This is important because age tends to be associated with a decline in mobility and a change in lifestyle. It is interesting to note though, that differences in trip generation by age may reflect in part the so-called *cohort effects*. This means that older people may travel less, simply because they always did so, rather than because of their age. However, this effect may be largest for people over 65 and declining trip generation rates for other age groups probably reflect a true decrease in the propensity of travel.

Finally, another trend worth noting is the increase in the proportion of women joining the labour force. Its significance for transportation planning and forecasting stems from two effects. The first is simply the direct employment effect, where time allocation and consequently travel behaviour are profoundly influenced by the requirements of actually being employed. The second one is more subtle and concerns changes in household roles and their impacts on lifestyle, particularly for couples with children.

To end this section it is interesting to mention that the ideas discussed above led to a proposal for incorporating a household structure variable in trip generation modelling, which was tested with real data (Allaman, *et al.* 1982). The household structure categories proposed were based on the age, sex, marital status and last name of each household member. These variables allowed the determination of the presence or absence of dependents in the household, the number and type of adults present, and the relationship among household members. However, although models using this variable were pronounced a considerable improvement over traditional practice by Allaman *et al.* (1982), further tests with a different data set performed by McDonald and Stopher (1983) led to its rejection. This was not only on the basis of statistical evidence but also on policy sensitivity (i.e. it is difficult to use household structure as a policy variable) and ease of forecasting grounds (i.e. forecasting at zone level, particularly to obtain a distribution of households by household structure category, appears to be very problematic). McDonald and Stopher (1983) argue that in these two senses a variable of the housing type variety should be preferred and it is bound to be easier to use by a local government planning agency.

4.7   Stability and Updating of Trip Generation Parameters

4.7.1   Temporal Stability

Transport models, in general, are developed to assist in the formulation and evaluation of transport plans and projects. Although on many occasions use has been made of descriptive statistics for examining travel trends, most developments have used cross-sectional data to express the amount of travel in terms of explanatory factors; these factors need to be both plausible and easy to forecast for the model to be policy sensitive in the design-year. A key (often implicit) assumption of this approach is that the model parameters will remain constant (or stable) between base and design years.

Several studies have examined this assumption in a trip generation context, finding in general that it cannot be rejected when trips by all modes are considered together (see Kannel and Heathington 1973; Smith and Cleveland 1976), even in the case of the rather crude zonal-based models (although these are not recommended anyway, for reasons similar to those discussed in section 4.2.2; see Downes and Gyenes 1976). However, later analyses reported different results. For example, Hall *et al.* (1987) compared observed trip rates and regression coefficients of models fitted to household data collected for Santiago in 1977 and 1986, and found them significantly different. Copley and Lowe (1981) reported that although trip rates by bus for certain types of household categories seemed reasonably stable over time, car trip rates appeared to be highly correlated with changes in real fuel prices. The latter has the following potential implications:

1. If there is non-zero elasticity of car trip rates to fuel prices, the usual assumption of constant trip rates in a period of rapidly increasing petrol prices could lead to serious over-provision of highway facilities. If, on the other hand, fuel prices were to fall in real terms, the constant trip rates assumption would lead to under-provision (which is precisely what was experienced in the UK and other industrialised countries towards the end of the 1980s).
2. Furthermore, the balance between future investments in public and private transport facilities may be judged incorrectly if based on the assumption of constant trip rates over time.

Clearly then, the correct estimation of the effect of fuel prices on trip rates (and of any other similar *longitudinal* effects) is of fundamental importance for policy analysis. Unfortunately it cannot be tackled with the cross-sectional data sets typically available for transportation studies.

Another factor affecting the stability of trip generation models over time is the evidence available on changes in travel behaviour. We do change our mind and the set of activities we would like to achieve is not fixed. Behavioural change programmes have an effect in reducing the number of trips (often by combining them into more efficient tours on a different day of the week) and in transferring some of them to more environmentally friendly modes. These techniques work because the individual does benefit from saving time spent travelling. There is some evidence that even without these interventions people do change their travel behaviour in response to easier home working and greater awareness of health and environmental issues. The opportunity for changes seems to be most clear when a major intervention into the transport and activity system takes place. This would partially explain, for example, the greater than expected shift to public transport when Congestion Charging was first introduced in London.

4.7.2   Geographic Stability

Temporal stability is often difficult to examine because data (of similar quality) are required for the same area at two different points in time. Thus on many occasions it may be easier to examine geographic stability (or transferability) as data on two different locations might become available (for example, if two institutions located in different areas decide to conduct a joint research project). Geographic transferability should be seen as an important attribute of any travel demand model for the following reasons:

1. It would suggest the existence of certain repeatable regularities in travel behaviour which can be picked up and reflected by the model.
2. It would indicate a higher probability that temporal stability also exists; this, as we saw, is essential for any forecasting model.
3. It may allow reducing substantially the need for costly full-scale transportation surveys on different metropolitan areas (see the discussion on Chapter 9).

It is clear that not all travel characteristics can be transferable between different areas or cities; for example, the average work trip duration is obviously context dependent, that is, it should be a function of area size, shape and the distributions of workplaces and residential zones over space. However, transferability of trip rates should not be seen as unrealistic: trips reflect needs for individuals' participation in various activities outside the home and if trip rates are related to homogeneous groups of people, they can be expected to remain stable and geographically transferable within the same cultural context.

The transferability of trip generation models (typically trip rates on a household-category analysis framework) has been tested relatively rarely, producing normally unsatisfactory results (see Caldwell and Demetski 1980; Daor 1981); the few successful examples have considered only part of the trips, for example trips made by car (see Ashley 1978). On the other hand, Supernak (1979, 1981) reported the successful transferability of the personal-category trip generation model, both for Polish and American conditions. Finally, Rose and Koppelman (1984) examined the transferability of a discrete choice trip generation model, allowing for adjustment of modal constants using local data. One of their conclusions was that context similarity appeared to be an important determinant of model transferability; also, because their results showed considerable variability, they caution that great care must be taken in order to ensure that the transferred model is usable in the new context.

4.7.3 Bayesian Updating of Trip Generation Parameters

Assume we want to estimate a trip generation model but lack funds to collect appropriate survey data; a possible (but inadequate) solution is to use a model estimated for another (hopefully similar) area directly. However, it would be highly desirable to modify it in order to reflect local conditions more accurately.

This can be done by means of Bayesian techniques for updating the original model parameters using information from a small sample in the application context. Bayesian updating considers a *prior* distribution (i.e. that of the original parameters to be updated), new information (i.e. to be obtained from the small sample) and a *posterior* distribution corresponding to the updated model parameters for the new context. Updating techniques are very important in a continuous planning framework; we will see this theme appearing in various parts of this book.

Consider, for example, the problem of updating trip rates by household categories; following Mahmassani and Sinha (1981) we will employ the notation in Table 4.8.

Table 4.8   Bayesian updating notation for trip generation

Variable	Prior information	New information
Mean trip rate	t_1	t_s
No. of observations	n_1	n_s
Trip rate variance	S_1^2	S_s^2

The mean trip rate of a category (or cell), is of course the average of a sample of household trip rates. According to the *Central Limit Theorem*, if the number of observations in a cell is at least 30, the sample distribution of the cell (mean) trip rates may be considered distributed Normal independently of the distribution of the household trip rates. Therefore, the prior distribution of the cell trip rates for the original model is $N(t_1, S_1^2/n_1)$, because t_1 and S_1^2/n_1 are unbiased estimators of its mean and variance. Similarly, the cells for the small sample (new information) may be considered distributed Normal with parameters t_s and S_s^2/n_s.

Bayes' theorem states that if the prior and sample distributions are Normal with known variances σ^2, then the posterior (updated) distribution of the mean trip rates is also Normal with the following parameters:

$$t_2 = \frac{1/\sigma_1^2}{1/\sigma_1^2 + 1/\sigma_s^2}t_1 + \frac{1/\sigma_s^2}{1/\sigma_1^2 + 1/\sigma_s^2}t_s \tag{4.21}$$

$$\sigma_2^2 = \frac{1}{1/\sigma_1^2 + 1/\sigma_s^2} \tag{4.22}$$

which, substituting by the known values S^2 and n, yield:

$$t_2 = \frac{n_1 S_s^2 t_1 + n_s S_1^2 t_s}{n_1 S_s^2 + n_s S_1^2} \tag{4.23}$$

$$\sigma_2^2 = \frac{S_1^2 S_s^2}{n_1 S_s^2 + n_s S_1^2} \tag{4.24}$$

It is important to emphasise that this distribution is not that of the individual trip rates of each household in the corresponding cell, but that of the mean of the trip rates of the cell. In fact the distribution of the individual rates is not known; the only information we have is that they share the same (posterior) mean t_2.

Example 4.6 The mean trip rate, its variance and the number of observations for two household categories, obtained in a study undertaken 10 years ago are shown below:

	Household categories	
Variable (prior data)	1	2
Trips per day	8	5
No. of observations	65	300
Trip rate variance	64	15
Mean trip variance	0.98	0.05

It is felt that these values might be slightly out of date for direct use today, but there are not enough funds to embark on a full-scale survey. A small stratified sample is finally taken, which yields the values shown below:

	Household categories	
Variable (new data)	1	2
Trips per day	12	6
No. of observations	30	30
Trip rate variance	144	36
Mean trip variance	4.80	1.20

The reader can check that by applying equations (4.23) and (4.24) it is possible to estimate the following trip rate values and variances:

(continued)

	Household categories	
	1	2
Posterior		
Trip rate (trips/day)	8.68	5.04
Variance	0.82	0.05

Exercises

4.1 Consider a zone with the following characteristics:

Household type	No.	Income ($/month)	Inhabitants	Trips/day
0 cars	180	4 000	4	6
1 car	80	18 000	4	8
2 or more cars	40	50 000	6	11

Due to a decrease in import duties and a real income increase of 30% it is expected that in five years time 50% of households without a car would acquire one. Estimate how many trips would the zone generate in that case; check whether your method is truly the best available.

4.2 Consider the following trip attraction models estimated using a standard computing package (t-ratios are given in parentheses);

$$Y = \begin{array}{l} 123.2 + 0.89X_1 \\ (5.2) \quad\ (7.3) \end{array} \qquad\qquad R^2 = 0.900$$

$$Y = \begin{array}{l} 40.1 + 0.14X_2 + 0.61X_3 + 0.25X_4 \\ (6.4) \quad (1.9) \quad\ (2.4) \quad\ (1.8) \end{array} \quad R^2 = 0.925$$

$$Y = \begin{array}{l} -1.7 + 2.57X_1 - 1.78X_4 \\ (-0.6) \quad (9.9) \quad (-9.3) \end{array} \qquad R^2 = 0.996$$

where Y are work trips attracted to the zone, X_1 is total employment in the zone, X_2 is industrial employment in the zone, X_3 is commercial employment in the zone and X_4 is service employment.
Choose the most appropriate model, explaining clearly why (i.e. considering all its pros and cons).

4.3 Consider the following two AM peak work trip generation models, estimated by household linear regression:

$$y = \begin{array}{l} 0.50 + 2.0x_1 + 1.5x_2 \\ (2.5) \quad (6.9) \quad (5.6) \end{array} \qquad\qquad R^2 = 0.589$$

$$y = \begin{array}{l} 0.01 + 2.3x_1 + 1.1Z_1 + 4.1Z_2 \\ (0.9) \quad (4.6) \quad (1.9) \quad (3.4) \end{array} \quad R^2 = 0.601$$

where y are household trips to work in the morning peak, x_1 is the number of workers in the household, x_2 is the number of cars in the household, Z_1 is a dummy variable which takes the value of 1 if the household has one car and Z_2 is a dummy which takes the value of 1 if the household has two or more cars.
(a) Choose one of the models explaining clearly the reasoning behind your decision.
(b) Graphically depict both models using appropriate axis.

(c) If a zone has 1000 households (with an average of two workers per household), of which 50% has no cars, 35% has only one car and the rest exactly two cars, estimate the total number of trips generated by the zone, O_i, with both models. Discuss your results.

4.4 The following table presents data collected in the last household O–D survey (made ten years ago) for three particular zones:

Zone	Residents/HH	Workers/HH	Mean Income	Population
I	2.0	1.0	50 000	20 000
II	3.0	2.0	70 000	60 000
III	2.5	2.0	100 000	100 000

Ten years ago two household-based trip generation models were estimated using this data. The first was a linear regression model given by:

$$y = 0.2 + 0.5x_1 + 1.1Z_1 \quad R^2 = 0.78$$

where y are household peak hour trips, x_1 is the number of workers in the household and Z_1 is a dummy variable which takes the value of 1 for high income ($> 70\,000$) households and 0 in other cases.

The second was a category analysis model based on two income strata (low and high income) and two levels of family structure (1 or less and 2 or more workers per household). The estimated trip rates are given in the following table:

Family structure	Income	
	Low	High
1 or less	0.8	1.0
2 or more	1.2	2.3

If the total number of trips generated today during the peak hour by the three zones are given by:

Zone	Peak hour trips
I	8 200
II	24 300
III	92 500

and it is estimated that the zone characteristics (income, number of households and family structure) have remained stable, decide which model is best. Explain your answer.

5

Trip Distribution Modelling

We have seen how trip generation models can be used to estimate the total number of trips emanating from a zone (origins, productions) and those attracted to each zone (destinations, attractions). Productions and attractions provide an idea of the level of trip making in a study area but this is seldom enough for modelling and decision making. What is needed is a better idea of the pattern of trip making, from where to where do trips take place, the modes of transport chosen and, as we shall see in Chapter 10, the routes taken.

The pattern of travel can be represented, at this stage, in at least two different ways. The first one is as a 'trip matrix' or 'trip table'. This stores the trips made from an Origin to a Destination during a particular time period; it is also called an Origin Destination (O-D) matrix and may be disaggregated by person type and purpose or perhaps the activity undertaken at each end of the trip. This representation is needed for all assignment models.

The second way of presenting a trip pattern is to consider the factors that generate and attract trips, i.e. on a Production-Attraction (P-A) basis, with Home generally being treated as the 'producing' end, and Work, Shop etc as the 'attracting' end. By necessity, a P-A matrix will cover a longer time span, (usually a day) than an O-D matrix. Take for example a journey to school and back. On an O-D basis this will generate one trip in the morning from Home to School and one back in the afternoon; on a P-A basis the Home end will generate two school trips and the School end will attract two school trips during the day. Note that the P-A treatment is closer, but not equivalent, to the idea of tours.

Trip patterns obtained through intercept surveys (i.e. roadside interviews or public transport questionnaires) will result in O-D matrices which are probably partial; not all O-D pairs would have been sampled. Even the combination of intercept and home interview surveys will fail to produce matrices where all cells have been sampled. Modelling is required to generate fuller matrices in either P-A or O-D format.

A number of methods have been put forward over the years to distribute trips (from a trip generation model) among destinations; some of the simplest are only suitable for short-term, tactical studies where no major changes in the accessibility provided by the network are envisaged. Others seem to respond better to changes in network cost and are therefore suggested for longer-term strategic studies or for tactical ones involving important changes in relative transport prices; these are often P-A based.

Trip Distribution is often seen as an aggregate problem with an aggregate model for its solution. In fact, most of its treatment in this chapter shares that view. However, the choice of destination can also be treated as a discrete choice (disaggregate) problem, and treated with models at the level of the individual. This is discussed in greater detail in subsequent chapters.

Modelling Transport, Fourth Edition. Juan de Dios Ortúzar and Luis G. Willumsen.
© 2011 John Wiley & Sons, Ltd. Published 2011 by John Wiley & Sons, Ltd.

This chapter starts by detailing additional definitions and notation used; these include the idea of generalised costs of travel. The next section introduces methods which respond only to relative growth rates at origins and destinations; these are suitable for short-term trend extrapolation. Section 5.3 discusses a family of synthetic models, the best known being the gravity model. Approaches to model generation, in particular the entropy-maximising formalism, are presented in section 5.4. An important aspect of the use of synthetic models is their calibration, that is the task of fixing their parameters so that the base-year travel pattern is well represented by the model; this is examined in section 5.5. Section 5.6 presents a variation on the gravity model calibration theme which enables more general forms for the model. Other synthetic models have also been proposed and the most important of them, the intervening-opportunities model, is explored in section 5.7. Finally, the chapter concludes with some practical issues in distribution modelling.

5.1 Definitions and Notation

It is now customary to represent the trip pattern in a study area by means of a trip matrix. This is essentially a two-dimensional array of cells where rows and columns represent each of the z zones in the study area (including external zones), as shown in Table 5.1.

The cells of each row i contain the trips originating in that zone which have as destinations the zones in the corresponding columns. The main diagonal corresponds to intra-zonal trips. Therefore: T_{ij} is the number of trips between origin i and destination j; the total array is $\{T_{ij}\}$ or $\mathbf{T}$; O_i is the total number of trips originating in zone i, and D_j is the total number of trips attracted to zone j. P_i is the number of trips produced or generated in a zone i and Q_j those attracted to zone j.

We shall use lower case letters, t_{ij}, o_i and d_j to indicate observations from a sample or from an earlier study; capital letters will represent our target, or the values we are trying to model for the corresponding modelling period.

The matrices can be further disaggregated, for example, by person type (n) and/or by mode (k). Therefore:

T_{ij}^{kn} are trips from i to j by mode k and person type n;

O_i^{kn} is the total number of trips originating at zone i by mode k and person type n, and so on.

Table 5.1 A general form of a two-dimensional trip matrix

| Origins | Destinations | | | | | $\sum_i T_{ij}$ |
	1	2	3	$\ldots j$	$\ldots z$	
1	T_{11}	T_{12}	T_{13}	$\ldots T_{1j}$	$\ldots T_{1z}$	O_1
2	T_{21}	T_{22}	T_{23}	$\ldots T_{2j}$	$\ldots T_{2z}$	O_2
3	T_{31}	T_{32}	T_{33}	$\ldots T_{3j}$	$\ldots T_{3z}$	O_3
$\vdots$						
i	T_{i1}	T_{i2}	T_{i3}	$\ldots T_{ij}$	$\ldots T_{iz}$	O_i
$\vdots$						
z	T_{z1}	T_{z2}	T_{z3}	$\ldots T_{zj}$	$\ldots T_{zz}$	O_z
$\sum_i T_{ij}$	D_1	D_2	D_3	$\ldots D_j$	$\ldots D_z$	$\sum_{ij} T_{ij} = T$

Summation over sub- or superscripts will be indicated implicitly by omission, e.g.

$$T_{ij}^n = \sum_k T_{ij}^{kn}$$

$$T = \sum_{ij} T_{ij} \quad \text{and} \quad t = \sum_{ij} t_{ij}$$

In some cases it may be of interest to distinguish the proportion of trips using a particular mode and the cost of travelling between two points:

p_{ij}^k is the proportion of trips from i to j by mode k;

c_{ij}^k is the cost of travelling between i and j by mode k.

The sum of the trips in a row should equal the total number of trips emanating from that zone; the sum of the trips in a column should correspond to the number of trips attracted to that zone. These conditions can be written as:

$$\sum_j T_{ij} = O_i \tag{5.1a}$$

$$\sum_i T_{ij} = D_j \tag{5.1b}$$

If reliable information is available to estimate both O_i and D_j then the model must satisfy both conditions; in this case the model is said to be doubly constrained. In some cases there will be information only about one of these constraints, for example to estimate all the O_i's, and therefore the model will be said to be singly constrained. Thus a model can be origin or production constrained if the O_i's, are available, or destination or attraction constrained if the D_j's are at hand.

The cost element may be considered in terms of distance, time or money units. It is often convenient to use a measure combining all the main attributes related to the disutility of a journey and this is normally referred to as the *generalised cost of travel*. This is typically a linear function of the attributes of the journey weighted by coefficients which attempt to represent their relative importance as perceived by the traveller. One possible representation of this for mode k is (omitting superscript k for simplicity):

$$C_{ij} = a_1 t_{ij}^v + a_2 t_{ij}^w + a_3 t_{ij}^t + a_4 t_{ij}^n + a_5 F_{ij} + a_6 \phi_j + \delta \tag{5.2}$$

where

t_{ij}^v is the in-vehicle travel time between i and j;
t_{ij}^w is the walking time to and from stops (stations) or from parking area/lot;
t_{ij}^t is the waiting time at stops (or time spent searching for a parking space);
t_{ij}^n is the interchange time, if any;
F_{ij} is a monetary charge: the fare charged to travel between i and j or the cost of using the car for that journey, including any tolls or congestion charges (note that car operating costs are often not well perceived and that electronic means of payment tend to blur somehow the link between use and payment);
ϕ_j is a terminal (typically parking) cost associated with the journey from i to j;
δ is a *modal penalty*, a parameter representing all other attributes not included in the generalised measure so far, e.g. safety, comfort and convenience;
$a_1 \ldots_6$ are weights attached to each element of cost; they have dimensions appropriate for conversion of all attributes to common units, e.g. money or time.

If the generalised cost is measured in money units ($a_5 = 1$) then a_1 is sometimes interpreted as the *value of time* (or more precisely the *value of in-vehicle time*) as its units are money/time. In that case, a_2 and a_3 would be the values of walking and waiting time respectively, and in many practical studies they have been taken to be two or three times the expected value of a_1.

The generalised cost of travel, as expressed here, represents an interesting compromise between subjective and objective disutility of movement. It is meant to represent the disutility of travel as perceived by the trip maker; in that sense the value of time should be a perceived value rather than an objective, resource-based, value. However, the coefficients $a_1 \ldots _6$ used are often provided externally to the modelling process, sometimes specified by government. This presumes stability and transferability of values for which there is, so far, only limited evidence.

As generalised costs may be measured in money or time units it is relatively easy to convert one into the other. For example, if the generalised cost is measured in time units, a_1 would be 1.0, $a_{2 \ldots 3}$ would probably be between 2.0 and 3.0, and $a_{5 \ldots 6}$ would represent something like the 'duration of money'.

There are some theoretical and practical advantages in measuring generalised cost in time units. Consider, for example, the effect of income levels increasing with time; this would increase the *value of time* and therefore increase generalised costs and apparently make the same destination more expensive. If, on the other hand, generalised costs are measured in time units, increased income levels would appear to reduce the cost of reaching the same destination, and this seems intuitively more acceptable. There are formal reasons in evaluation to prefer expressing generalised cost in time units; the interested reader is referred to the excellent book by Jara-Díaz (2007).

A distribution model tries to estimate the number of trips in each of the matrix cells on the basis of any information available. Different distribution models have been proposed for different sets of problems and conditions. We shall explore, first, models which are mainly useful in updating a trip matrix, or in forecasting a future trip matrix, where information is only available in terms of future trip rates or growth factors. We shall then study more general models, in particular the gravity model family. We shall finally explore the possibility of developing modal-split models from similar principles.

5.2   Growth-Factor Methods

Let us consider first a situation where we have a basic trip matrix **t**, perhaps obtained from a previous study or estimated from recent survey data. We would like to estimate the matrix corresponding to the design year, say 10 years into the future. We may have information about the growth rate to be expected in this 10-year period for the whole study area; alternatively, we may have information on the likely growth in the number of trips originating and/or attracted to each zone. Depending on this information we may be able to use different growth-factor methods in our estimation of future trip patterns.

5.2.1   Uniform Growth Factor

If the only information available is about a general growth rate τ for the whole of the study area, then we can only assume that it will apply to each cell in the matrix:

$$T_{ij} = \tau \cdot t_{ij} \text{ for each pair } i \text{ and } j \tag{5.3}$$

Of course $\tau = T/t$, i.e. the ratio of expanded over previous total number of trips.

Example 5.1 Consider the simple four-by-four base-year trip matrix of Table 5.2. If the growth in traffic in the study area is expected to be of 20% in the next three years, it is a simple matter to multiply all cell values by 1.2 to obtain a new matrix as in Table 5.3.

Table 5.2 Base-year trip matrix

	1	2	3	4	$\sum_j$
1	5	50	100	200	355
2	50	5	100	300	455
3	50	100	5	100	255
4	100	200	250	20	570
$\sum_i$	205	355	455	620	1635

The assumption of uniform growth is generally unrealistic except perhaps for very short time spans of, say, one or two years. In most other cases one would expect differential growth for different parts of the study area.

Table 5.3 Future estimated trip matrix with $\tau = 1.2$

	1	2	3	4	$\sum_j$
1	6	60	120	240	426
2	60	6	120	360	546
3	60	120	6	120	306
4	120	240	300	24	684
$\sum_i$	246	426	546	744	1962

5.2.2 Singly Constrained Growth-Factor Methods

Consider the situation where information is available on the expected growth in trips originating in each zone, for example shopping trips. In this case it would be possible to apply this origin-specific growth factor (τ_i) to the corresponding rows in the trip matrix. The same approach can be followed if the information is available for trips attracted to each zone; in this case the destination-specific growth factors (τ_j) would be applied to the corresponding columns. This can be written as:

$$T_{ij} = \tau_i \cdot t_{ij} \text{ for origin-specific factors} \tag{5.4}$$

$$T_{ij} = \tau_j \cdot t_{ij} \text{ for destination-specific factors} \tag{5.5}$$

Example 5.2 Consider Table 5.4, a revised version of Table 5.2 with growth predicted for origins:

Table 5.4 Origin-constrained growth trip table

	1	2	3	4	$\sum_j$	Target O_i
1	5	50	100	200	355	400
2	50	5	100	300	455	460
3	50	100	5	100	255	400
4	100	200	250	20	570	702
$\sum_i$	205	355	455	620	1635	1962

This problem can be solved immediately by multiplying each row by the ratio of target O_i over the base year total (Σ_j), thus giving the results in Table 5.5.

Table 5.5 Expanded origin-constrained growth trip table

	1	2	3	4	$\sum_j$	Target O_i
1	5.6	56.3	112.7	225.4	400	400
2	50.5	5.1	101.1	303.3	460	460
3	78.4	156.9	7.8	156.9	400	400
4	123.2	246.3	307.9	24.6	702	702
$\sum_i$	257.7	464.6	529.5	701.2	1962	1962

5.2.3   Doubly Constrained Growth Factors

An interesting problem is generated when information is available on the future number of trips origi-nating and terminating in each zone. This implies different growth rates for trips in and out of each zone and consequently having two sets of growth factors for each zone, say τ_i and Γ_j. The application of an 'average' growth factor, say $F_{ij} = 0.5\,(\tau_i + \Gamma_j)$ is only a poor compromise as none of the two targets or trip-end constraints would be satisfied. Historically a number of iterative methods have been proposed to obtain an estimated trip matrix which satisfies both sets of trip-end constraints, or the two sets of growth factors, which is the same thing.

All these methods involve calculating a set of intermediate correction coefficients which are then applied to cell entries in each row or column as appropriate. After applying these corrections to say, each row, the totals for each column are calculated and compared with the target values. If the differences are significant, new correction coefficients are calculated and applied as necessary.

In transport these methods are known by their authors as Fratar in the US and Furness elsewhere. For example Furness (1965) introduced 'balancing factors' A_i and B_j as follows:

$$T_{ij} = t_{ij} \cdot \tau_i \cdot \Gamma_j \cdot A_i \cdot B_j \tag{5.6}$$

or incorporating the growth rates into new variables a_i and b_j:

$$T_{ij} = t_{ij} \cdot a_i \cdot b_j \tag{5.7}$$

with $a_i = \tau_i\,A_i$ and $b_j = \Gamma_j\,B_j$.

The factors a_i and b_j (or A_i and B_j) must be calculated so that the constraints (5.1) are satisfied. This is achieved in an iterative process which in outline is as follows:

1. set all $b_j = 1.0$ and solve for a_i; in this context, 'solve for a_i' means find the correction factors a_i that satisfy the trip generation constraints;
2. with the latest a_i solve for b_j, e.g. satisfy the trip attraction constraints;
3. keeping the b_j's fixed, solve for a_i and repeat steps (2) and (3) until the changes are sufficiently small.

This method produces solutions within 3 to 5% of the target values in a few iterations when certain conditions are met. A tighter degree of convergence may be important from the perspective of model system consistency, see Chapter 11. This method is often called a 'bi-proportional algorithm' because of the nature of the corrections involved. The problem is not restricted to transport; techniques to solve it have also been 'invented', among others, by Kruithof (1937) for telephone traffic and Bacharach (1970) for updating input-output matrices in economics. The best treatment of its mathematical properties seems to be due to Bregman (see Lamond and Stewart 1981).

It will be shown below that this method is a special case of entropy-maximising models of the gravity type if the effect of distance or separation between zones is excluded. But in any case, the Furness method tries to produce the minimum corrections to the base-year matrix **t** necessary to satisfy the future year trip-end constraints.

The most important condition required for the convergence of this method is that the growth rates produce target values T_i and T_j such that

$$\sum_i \tau_i \sum_j t_{ij} = \sum_i \Gamma_j \sum_i t_{ij} = T \tag{5.8}$$

Enforcing this condition may require correcting trip-end estimates produced by the trip generation models.

Example 5.3 Table 5.6 represents a doubly constrained growth factor problem:

Table 5.6   Doubly constrained matrix expansion problem

	1	2	3	4	$\sum_j$	Target O_i
1	5	50	100	200	355	400
2	50	5	100	300	455	460
3	50	100	5	100	255	400
4	100	200	250	20	570	702
$\sum_i$	205	355	455	620	1635	
Target D_j	260	400	500	802		1962

The solution to this problem, after three iterations on rows and columns (three sets of corrections for all rows and three for all columns), is shown in Table 5.7:

Table 5.7   Solution to the doubly constrained matrix expansion problem

	1	2	3	4	$\sum_j$	Target O_i
1	5.25	44.12	98.24	254.25	401.85	400
2	45.30	3.81	84.78	329.11	462.99	460
3	77.04	129.50	7.21	186.58	400.34	400
4	132.41	222.57	309.77	32.07	696.82	702
$\sum_i$	260.00	400.00	500.00	802.00	1962	
Target D_j	260	400	500	802		1962

Note that this estimated matrix is within 1% of meeting the target trip ends, more than enough accuracy for this problem.

5.2.4   Advantages and Limitations of Growth-Factor Methods

Growth-factor methods are simple to understand and make direct use of observed trip matrices and forecasts of trip-end growth. They preserve the observations as much as is consistent with the information available on growth rates. This advantage is also their limitation as they are probably only reasonable for short-term planning horizons or when changes in transport costs are not to be expected.

Growth-factor methods require the same database as synthetic methods, namely an observed (sampled) trip matrix; this is an expensive data item. The methods are heavily dependent on the accuracy of the base-year trip matrix. As we have seen, this is never very high for individual cell entries and therefore the resulting matrices are no more reliable than the sampled or observed ones. Any error in the base-year may well be amplified by the application of successive correction factors. Moreover, if parts of the base-year matrix are unobserved, they will remain so in the forecasts. Therefore, these methods cannot be used to fill in unobserved cells of partially observed trip matrices.

Another, important, limitation is that the methods do not take into account changes in transport costs due to improvements (or new congestion) in the network. Therefore they are of limited use in the analysis of policy options involving new modes, new links, pricing policies and new zones.

5.3   Synthetic or Gravity Models

5.3.1   The Gravity Distribution Model

Distribution models of a different kind have been developed to assist in forecasting future trip patterns when important changes in the network take place. They start from assumptions about group trip making behaviour and the way this is influenced by external factors such as total trip ends and distance travelled. The best known of these models is the gravity model, originally generated from an analogy with Newton's gravitational law. These models estimate trips for each cell in the matrix without directly using the observed trip pattern; therefore they are sometimes called synthetic as opposed to growth-factor models.

Probably the first rigorous use of a gravity model was by Casey (1955), who suggested such an approach to synthesise shopping trips and catchment areas between towns in a region. In its simplest formulation the model has the following functional form:

$$T_{ij} = \frac{\alpha P_i P_j}{d_{ij}^2} \qquad (5.9)$$

where P_i and P_j are the populations of the towns of origin and destination, d_{ij} is the distance between i and j, and α is a proportionality factor (with units trips·distance^2/population^2).

This was soon considered to be too simplistic an analogy with the gravitational law and early improvements included the use of total trip ends (O_i and D_j) instead of total populations, and a parameter n for calibration as the power for d_{ij}. This new parameter was not restricted to being an integer and different studies estimated values between 0.6 and 3.5.

The model was further generalised by assuming that the effect of distance or 'separation' could be modelled better by a decreasing function, to be specified, of the distance or travel cost between the zones. This can be written as:

$$T_{ij} = \alpha O_i D_j f(c_{ij}) \qquad (5.10)$$

where $f(c_{ij})$ is a generalised function of the travel costs with one or more parameters for calibration. This function often receives the name of 'deterrence function' because it represents the disincentive to travel as distance (time) or cost increases. Popular versions for this function are:

$$f(c_{ij}) = \exp(-\beta c_{ij}) \qquad \text{exponential function} \qquad (5.11)$$
$$f(c_{ij}) = c_{ij}^{-n} \qquad \text{power function} \qquad (5.12)$$
$$f(c_{ij}) = c_{ij}^{n} \exp(-\beta c_{ij}) \quad \text{combined function} \qquad (5.13)$$

The general form of these functions for different values of their parameters is shown in Figure 5.1.

Figure 5.1   Different deterrence functions

5.3.2   Singly and Doubly Constrained Models

The need to ensure that the restrictions (5.1) are met requires replacing the single proportionality factor α by two sets of balancing factors A_i and B_j as in the Furness model, yielding:

$$T_{ij} = A_i O_i B_j D_j f(c_{ij}) \tag{5.14}$$

In a similar vein one can again subsume O_i and D_j into these factors and rewrite the model as:

$$T_{ij} = a_i b_j f(c_{ij}) \tag{5.15}$$

The expression in (5.14) or (5.15) is the classical version of the doubly constrained gravity model. Singly constrained versions, either origin or destination constrained, can be produced by making one set of balancing factors A_i or B_j equal to one. For an origin-constrained model, $B_j = 1.0$ for all j, and

$$A_i = 1 \bigg/ \sum_j D_j f(c_{ij}) \tag{5.16}$$

In the case of the doubly constrained model the values of the balancing factors are:

$$A_i = 1 \bigg/ \sum_j B_j D_j f(c_{ij}) \tag{5.17}$$

$$B_j = 1 \bigg/ \sum_i A_i O_i f(c_{ij}) \tag{5.18}$$

The balancing factors are, therefore, interdependent; this means that the calculation of one set requires the values of the other set. This suggests an iterative process analogous to Furness's which works well in practice: given set of values for the deterrence function $f(c_{ij})$, start with all $B_j = 1$, solve for A_i and then use these values to re-estimate the B_j's; repeat until convergence is achieved.

A more general version of the deterrence function accepts empirical values for it and these depend only on the generalised cost of travel. To this end, travel costs are aggregated into a small number (say 10 or 15) of cost ranges or cost bins, indicated by a superscript m. The deterrence function then becomes:

$$f(c_{ij}) = \sum_m F^m \delta_{ij}^m \tag{5.19}$$

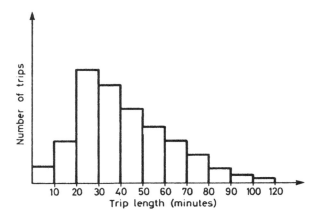

Figure 5.2   Typical trip length distribution in urban areas

where F^m is the mean value for cost bin m, and δ_{ij}^m is equal to 1 if the cost of travelling between i and j falls in the range m, and equal to 0 otherwise.

The formulations (5.11) and (5.12) have one parameter for calibration; formulation (5.13) has two, β and n, and formulation (5.19) has as many parameters as cost bins. These parameters are estimated so that the results from the model reproduce, as closely as possible, the trip length (cost) distribution (TLD) of the observations. A theoretical reason for this requirement is offered below, but meanwhile it is enough to note that the greater the number of parameters, the easier it is to obtain a closer fit with the sampled trip length distribution.

It has been observed, in particular in urban areas, that in the case of motorised trips, the trip length distribution has a shape of the form depicted in Figure 5.2. This shows that there are few short motorised trips, followed by a larger number of medium-length trips; as distance (cost) increases, the number of trips decays again with a few very long trips. The negative exponential and power functions reproduce reasonably well the second part of the curve but not the first. That is one of the reasons behind the combined formulation which is more likely to fit better both parts of the TLD. The greater flexibility of the cost-bin formulation permits an even better fit. However, the approach requires the assumption that the same TLD will be maintained in the future; this is similar but more stringent to requiring β to be the same for the base and the forecasting years.

It is interesting to note that the bulk of the representational and policy relevance advantages of the gravity model lies in the deterrence function; the rest is very much like the Furness method.

5.4   The Entropy-Maximising Approach

5.4.1   Entropy and Model Generation

We shall introduce now the entropy-maximisation approach which has been used in the generation of a wide range of models, including the gravity model, shopping models and location models. The approach has a number of followers and detractors but it is generally acknowledged as one of the important contributions to improved modelling in transport. There are several ways of presenting the approach; we have chosen an intuitive rather than strictly mathematical formulation. For a stricter presentation and references to related and alternative approaches, see Wilson (1974).

Consider a system made up of a large number of distinct elements. A full description of such a system requires the complete specification of its *micro* states, as each is distinct and separable. This would involve, for example, identifying each individual traveller, its origin, destination, mode, time of journey, and so on. However, for many practical purposes it may be sufficient to work on the basis of a more aggregate or *meso* state specification; following our example, a meso state may just specify the *number* of trips between each origin and each destination. In general, there will be numerous and different micro states which produce the same meso state: John Smith and Pedro Pérez, living in the same zone, may exchange destinations generating different micro states but keeping the same meso state.

There is always an even higher level of aggregation, a *macro* state, for example the total number of trips on particular links, or the total trips generated and attracted to each zone. To obtain reliable measures of trip making activity it is often easier to make observations at this higher level of aggregation. In fact, most of our current information about a system is precisely at this level. In a similar way, estimates about the future are usually restricted to macro-state descriptions because of the uncertainties involved in forecasting at more disaggregate levels: for example, it is easier to forecast the population per zone than the number of households in a particular category residing in each zone.

The basis of the method is to accept that, unless we have information to the contrary, all micro states consistent with our information about macro states are equally likely to occur. This is in fact a sensible assumption given our ignorance about meso and micro states. A good way of enforcing consistency with our knowledge about macro states is to express our information as equality constraints in a mathematical programme. As we are interested in the meso-state descriptions of the system, we would like to identify those meso states which are most likely, given our constraints about the macro states.

It is possible to show, see Wilson (1970), that the number of micro states $W \{T_{ij}\}$ associated with the meso state T_{ij} is given by:

$$W\{T_{ij}\} = \frac{T!}{\prod_{ij} T_{ij}!} \qquad (5.20)$$

As it is assumed that all micro states are equally likely, the most probable meso state would be the one that can be generated in a greater number of ways.

Therefore, what is needed is a technique to identify the values of $\{T_{ij}\}$ which maximise W in (5.20). For convenience we seek to maximise a monotonic function of W, namely $\log W$, as both problems have the same maximum. Therefore:

$$\log W = \log \frac{T!}{\prod_{ij} T_{ij}!} = \log T! - \sum_{ij} \log T_{ij}! \qquad (5.21)$$

Stirling's (short) approximation for $\log X! = X \log X - X$, can be used to make it easier to optimise:

$$\log W = \log T! - \sum_{ij} (T_{ij} \log T_{ij} - T_{ij}) \qquad (5.22)$$

Usually the term $\log T!$ is a constant, therefore it can be omitted from the optimisation problem. The rest of the equation is often referred to as the *entropy function*:

$$\log W' = - \sum_{ij} (T_{ij} \log T_{ij} - T_{ij}) \qquad (5.23)$$

Maximising $\log W'$, subject to constraints corresponding to our knowledge about the macro states, enables us to generate models to estimate the most likely meso states, in our case the most likely matrix **T**. The key to this model generation method is, therefore, the identification of suitable micro, meso and macro state descriptions, together with the macro level constraints that must be met by the solution to the optimisation problem.

In some cases, there may be additional information in the form of prior or old values for the meso states, for example an outdated trip matrix **t**. The problem may be recast with this information and the revised objective function becomes:

$$\log W'' = - \sum_{ij} (T_{ij} \log T_{ij}/t_{ij} - T_{ij} + t_{ij}) \tag{5.24}$$

This is an interesting function in which each element in the summation takes the value zero if $T_{ij} = t_{ij}$ and otherwise is a positive value which increases with the difference between **T** and **t**. Therefore $-\log W'''$ is a good measure of the difference between **T** and **t**; it can further be shown that

$$- \log W'' \approx 0.5 \sum_{ij} \frac{(T_{ij} - t_{ij})^2}{t_{ij}} \tag{5.25}$$

where the right-hand side is another good measure of the difference between prior and estimated meso states. Models can be generated minimising $-\log W'''$ subject to constraints reflecting our knowledge about macro states. The resulting model is the one with the meso states closest to the prior meso states, in the sense of equation (5.24) or approximately (5.25), and which satisfies the macro state constraints.

5.4.2   Generation of the Gravity Model

Consider the definition of micro, meso and macro states from the discussion above. The problem becomes the maximisation of $\log W'$ subject to the following two sets of constraints corresponding to the meso states:

$$O_i - \sum_j T_{ij} = 0 \tag{5.26}$$

$$D_j - \sum_i T_{ij} = 0 \tag{5.27}$$

These two sets of constraints reflect our knowledge about trip productions and attractions in the zones of the study area. We are only interested in matrix entries that can be interpreted as trips, therefore we need to introduce the additional constraint that:

$$T_{ij} \geq 0$$

The constrained maximisation problem can be handled forming the Lagrangian:

$$L = \log W' + \sum_i \alpha'_i \left\{ O_i - \sum_i T_{ij} \right\} + \sum_j \alpha''_i \left\{ D_j - \sum_i T_{ij} \right\} \tag{5.28}$$

Taking the first partial derivatives with respect to T_{ij} and equating them to zero we obtain:

$$\frac{\partial L}{\partial T_{ij}} = - \log T_{ij} - \alpha'_i - \alpha''_j = 0 \tag{5.29}$$

therefore

$$T_{ij} = \exp\left(-\alpha'_i - \alpha''_j\right) = \exp\left(-\alpha'_i\right) \exp\left(-\alpha''_j\right)$$

The values of the Lagrange multipliers are easy to find; making a simple change of variables:

$$A_i O_i = \exp\left(-\alpha'_i\right) \quad \text{and} \quad B_j D_j = \exp\left(-\alpha''_j\right)$$

we obtain

$$T_{ij} = A_i O_i B_j D_j \tag{5.30}$$

On the other hand, the use of $-\log W''$ as an objective function generates the model:

$$T_{ij} = A_i O_i B_j O_j t_{ij} \tag{5.31}$$

which is, of course, the basic Furness model. The version resulting in equation (5.30) corresponds to the case when there is no prior information, e.g. all $t_{ij} = 1$. These two models are close to but not yet the gravity model. What is missing is the deterrence function term. Its introduction requires an additional constraint:

$$\sum_{ij} T_{ij} c_{ij} = C$$

where C is the (unknown) total expenditure in travel in the system (in generalised cost units if they are in use). Restating this constraint as

$$C - \sum_{ij} T_{ij} c_{ij} = 0 \tag{5.32}$$

one can maximise $\log W'$ subject to (5.26), (5.27) and (5.32), and using the same constrained optimisation technique it is possible to obtain the Lagrangian:

$$L = \log W' + \sum_i \alpha_i' \left\{ O_i - \sum_j T_{ij} \right\} + \sum_j \alpha_j'' \left\{ D_j - \sum_i T_{ij} \right\} + \beta \left\{ C - \sum_{ij} T_{ij} c_{ij} \right\} \tag{5.33}$$

Again, taking its first partial derivatives with respect to T_{ij} and equating them to zero gives

$$\frac{\partial L}{\partial T_{ij}} = -\log T_{ij} - \alpha_i' - \alpha_j'' - \beta c_{ij} = 0 \tag{5.34}$$

therefore

$$T_{ij} = \exp\left(-\alpha_i' - \alpha_j'' - \beta c_{ij}\right) = \exp\left(-\alpha_i'\right) exp\left(-\alpha_j''\right) exp\left(-\beta c_{ij}\right) \tag{5.35}$$

Making the same change of variables as before one obtains:

$$T_{ij} = A_i O_i B_j D_j \exp\left(-\beta c_{ij}\right) \tag{5.36}$$

which is the classic gravity model. The values for the balancing factors can be derived from the constraints as:

$$A_i = 1 \Big/ \left[\sum_i B_j D_j \exp\left(-\beta c_{ij}\right) \right] \quad \text{and} \quad B_j = 1 \Big/ \left[\sum_i A_i O_i \exp\left(-\beta c_{ij}\right) \right]$$

If one of (5.26a) or (5.26b) is omitted from the constraints a singly constrained gravity model is obtained.

The Lagrange multipliers α_i' and α_j'' are the dual variables of the trip generation and attraction constraints and relate to the variations in entropy for a unit variation in trip generation and attraction. The value of β is related to the satisfaction of condition (5.32). In general C can only be estimated and therefore β is left as a parameter for calibration in order to adjust the model to each specific area. Values of β cannot, therefore, be easily borrowed from one place to another. A useful first estimate for the value of β is one over the average travel cost; in effect, β is precisely measured in inverse of travel cost units.

The use of a different cost constraint, such as (5.37) instead of (5.32),

$$C' - \sum_{ij} T_{ij} \log c_{ij} = 0 \tag{5.37}$$

results in a model of the form

$$T_{ij} = A_i O_i B_j D_j \exp\left(-\beta' \log c_{ij}\right) = A_i O_i B_j D_j c_{ij}^{-\beta'} \tag{5.38}$$

i.e. the gravity model with an inverse power deterrence function!

The reader can verify that the use of constraints (5.32) and (5.37) leads to a gravity model with a combined deterrence function. A further interesting approach is to disaggregate constraint (5.32) into several trip cost groups or bins indicated, as before, by a superscript m:

$$C^m - \sum_{ij} T_{ij} c_{ij} \delta_{ij}^m = 0 \quad \text{for each } m \tag{5.39}$$

The maximisation of (5.23) subject to (5.26), (5.27) and (5.39) leads to:

$$T_{ij} = A_i O_i B_j D_j \sum_m F^m \delta_{ij}^m = a_i b_j \sum_m F^m \delta_{ij}^m \tag{5.40}$$

which is, of course, the gravity model with a cost-bin deterrence function. This model has some attractive properties, which will be discussed in section 5.6.

5.4.3 Properties of the Gravity Model

As can be seen, entropy maximisation is quite a flexible approach for model generation. A whole family of distribution models can be generated by casting the problem in a mathematical programming framework: the maximisation of an entropy function subject to linear constraints representing our level of knowledge about the system. The use of this formalism has many advantages:

1. It provides a more rigorous way of specifying the mathematical properties of the resulting model. For example, it can be shown that the objective function is always convex; it can be shown also that, provided the constraints used, say (5.26) and (5.27) have a feasible solution space, the optimisation problem has a unique solution even if the set of parameters A_i and B_j is not unique (one is redundant).
2. The use of a mathematical programming framework also facilitates the use of a standard tool-kit of solution methods and the analysis of the efficiency of alternative algorithms.
3. The theoretical framework used to generate the model also assists in providing an improved interpretation of the solutions generated by it. We have seen that the gravity model can be generated from analogies with the physical world or from entropy-maximising considerations; the latter are closely related to information theory, to error measures and to maximum likelihood in statistics, and the three provide alternative ways of generating the same mathematical form of the gravity model. Although the functional form is the same, each theoretical framework provides a different interpretation to the problem and the solution found. Each may be more appropriate in specific circumstances. We shall come back to this *equifinality issue* in Chapter 8.
4. The fact that the gravity model can be generated in a number of different ways does not make it 'correct'. The appropriateness of the model depends on the acceptability of the assumptions required for its generation and their interpretation. No model is ever appropriate or correct in itself, it can only be more or less suitable to handle a decision question given our understanding of the problem, of the options or schemes to be tested, the information available or collectable at a justifiable cost, and the time and resources securable for analysis; see the discussion on calibration and validation below.

It is interesting to contrast the classical gravity model as in equation (5.36) with Furness's method as derived above in equation (5.31). We can see that one possible interpretation of the deterrence function is to provide a synthetic set of prior entries for each cell in the trip matrix (i.e. use of $\exp\left(-\beta c_{ij}\right)$ instead of t_{ij}). Both the deterrence function and the prior matrix t_{ij} take the role of providing 'structure' to the resulting trip matrix. This can be seen more clearly if one multiplies and divides the right-hand side of equation (5.31) by T and subsumes this constant in the balancing factors:

$$T_{ij} = T a_i b_j t_{ij} / T = a'_i b'_j p_{ij} \qquad (5.41)$$

where $p_{ij} = t_{ij}/T$, thus giving a better-defined meaning to 'structure' as the proportion of the total trips allocated to each origin-destination pair.

Example 5.4 It is useful to illustrate the gravity model with an example related to the problem of expanding a trip matrix. Consider the cost matrix of Table 5.8 together with the total trip ends as in Table 5.6, and attempt to estimate the parameters a_i and b_j of a gravity model of the type:

$$T_{ij} = a_i b_j \exp(-\beta c_{ij})$$

Table 5.8 A cost matrix and trip-end totals for a gravity model estimation

	Cost matrix (minutes)				
	1	2	3	4	Target O_i
1	3	11	18	22	400
2	12	3	12	19	460
3	15.5	13	5	7	400
4	24	18	8	5	702
Target D_j	260	400	500	802	1962

given the information that the best value of β is 0.10. The first step would be to build a matrix of the values $\exp(-\beta c_{ij})$, as in Table 5.9.

Table 5.9 The matrix $\exp(-\beta c_{ij})$ and sums to prepare for a gravity model run

	$\exp(-\beta c_{ij})$				
	1	2	3	4	$\sum_j$
1	0.74	0.33	0.17	0.11	1.35
2	0.30	0.74	0.30	0.15	1.49
3	0.21	0.27	0.61	0.50	1.59
4	0.09	0.17	0.45	0.61	1.31
$\sum_i$	1.34	1.51	1.52	1.36	5.74

Base	1	2	3	4	$\sum_j$	Target	Ratio
1	253.12	113.73	56.48	37.86	461.19	400	0.87
2	102.91	253.12	102.91	51.10	510.04	460	0.90
3	72.52	93.12	207.23	169.67	542.54	400	0.74
4	31.00	56.48	153.52	207.23	448.23	702	1.57
$\sum_i$	459.54	516.45	520.15	465.87	1962.00		
Target	260	400	500	802			
Ratio	0.57	0.77	0.96	1.72			

With these values we can calculate the resulting total 'trips' (5.74) and then expand each cell in the matrix by the ratio 1962/5.74 = 341.67. This produces a matrix of base trips which now has to be adjusted to match trip-end totals. This process is the same as Furness iterations. The values for a_i and b_j are the product of the corresponding correction factors; these factors will then be multiplied by the basic expansion factor 341.67. The resulting gravity model matrix is given in Table 5.10.

Table 5.10 The resulting gravity model matrix with trip length distribution

	1	2	3	4	$\sum_j$	Target	Ratio	a_i
1	155.73	99.00	64.46	74.17	393.36	400	1.02	1.17
2	57.54	200.22	106.73	90.98	455.56	460	1.01	1.07
3	25.87	47.01	137.16	192.77	402.81	400	0.99	0.68
4	20.86	53.77	191.65	444.08	710.37	702	0.99	1.28
$\sum_i$	260.00	400.00	500.00	802.00	1962.00			
Target	260	400	500	802				
Ratio	1.00	1.00	1.00	1.00				
b_j	179.17	253.50	332.37	570.53				

Ranges (min)							
Cost	**1.0–4.0**	**4.1–8.0**	**8.1–12.0**	**12.1–16.0**	**16.1–20.0**	**20.1–24**	**Sum**
Trips	355.9	965.7	263.3	72.9	209.2	95.0	1962

The reader may wish to verify that the balancing factors a_i and b_j are only unique to a multiplicative constant. It is also possible to calculate, as usual, the standard balancing factors A_i and B_j dividing each corresponding a_i and b_j by the target values O_i and D_j.

5.4.4 Production-Attraction Format

Note that the Gravity model can also be used with a Production-Attraction format. In fact, there are some very good reasons to prefer the P-A format in demand modelling. The P-A approach is closer to dealing with simple tours (from and to home) rather than trips. Travellers would consider, in choosing their destination, the cost of getting there *and* returning home and not just the outward journey. The gravity model is then treated in the same way as for trips although travel costs and the interpretation of the results are, of course, different. In this case, one should use an average of the costs of travelling between the two zones. These costs should correspond to the correct time periods: the generalised cost of the outward and inward journeys. These times will depend on the trip purpose. In an aggregate model, these times can only be an average as some travellers will have earlier and others later, the two legs of the tour. The correct average measure, an inclusive value or logsum, will be discussed at a later chapter. Note further, that using the sum of the two costs (outward and return trips) for an extended gravity model with a new β' is not consistent with the most accepted theories of behaviour.

The resulting P-A matrix will have to be converted into a directional O-D matrix in order to perform the assignment procedure. To achieve this it is essential to have the distribution of the times for outbound and inbound trips, the best source of which will come from a good set of home interviews; during intercept surveys the answers to the question about 'return' trips are fairly unreliable. If we are only interested in the 24-hour case, the two demand matrices are practically the same as it is assumed that each production–attraction trip is made once in each direction during the day. This is of course, an approximation but probably a reasonable one.

However, when a shorter-period OD matrix is required, some trips will be made in the production-to-attraction direction while others only in the opposite one. Two different approaches can be used to overcome this problem. The first is very simplistic and requires to produce a matrix for just a single purpose, typically 'to work', and then assume that these trips follow just one direction of travel, thus producing, for example, the morning journey to work from production to attraction. Survey data must be used to correct for shift work, flexible working hours and trips for other purposes being made during the morning peak; however, the pattern of the morning peak is still dominated by this journey-to-work purpose. The second and better approach is to use survey data directly to determine the proportions of the matrices for each purpose which are deemed appropriate for the part of the day under consideration. For example, a typical morning peak matrix may consist of 70% production-to-attraction movements and only 15% of attraction-to-production movements.

There is a case for handling the mode choice model also in a P-A format. The same argument used for the gravity model applies here. The choice of mode of travel is surely dependent on *all* trips of the tour; at least the P/A format captures the attributes of two of these trips. This argument is even stronger for any advanced 'time of travel choice' model.

5.4.5 Segmentation

The gravity model can be applied with different levels of segmentation. The most obvious one is by journey purpose as different 'generators and attractors' will apply for Journey to Work, to School, Shopping and Other.

It may also be desirable to segment by person type, at least 'car owners' and 'non car owners' as they are likely to have slightly different influences in trip patterns and would certainly perceive costs in different ways. Most non-car owners will perceive public transport costs as the measure of separation. Car owners, on the other hand, will be influenced by a combination of car and public transport costs, their two basic options. In this case, an appropriate average of these should be incorporated in the model, again a logsum as discussed later.

Although this segmentation, car and non-car owners, is possible at the production end it is not quite appropriate for the attraction end, especially in forecasting mode. Therefore, we will have these two segments competing for a set of job (and education) places at the attraction end. This requires a very simple extension of the gravity model equivalent at using an asymmetric matrix of 2NxN.

5.5 Calibration of Gravity Models

5.5.1 Calibration and Validation

Before using a gravity distribution model it is necessary to calibrate it; this just makes sure that its parameters are such that the model comes as close as possible to reproducing the base-year trip pattern. Calibration is, however, a very different process from validation of a model.

In the case of *calibration* one is conditioned by the functional form and the number of parameters of the chosen model. For example, the classical gravity model has the parameters A_i, B_j, and β (that is $Z + Z + 1$ parameters, Z being the number of zones). The parameters A_i and B_j are calibrated during the estimation of the gravity model, as part of the direct effort to satisfy constraints (5.1). Note that at least one of the A_i or B_j is redundant as there is an additional condition $\sum_i O_i = \sum_j D_j = T$, and therefore one of the (5.1) constraints is linearly dependent on the rest. The parameter β, on the other hand, must be calibrated independently, as we do not have complete information about the total expenditure C in the study area. If we had this information, we could have used it directly without having to estimate β by other means. If the combined deterrence function (5.13) is used, we would have an additional parameter and therefore some additional flexibility in calibrating the gravity model.

The *validation* task is different. In this case one wants to make sure the model is appropriate for the decisions likely to be tested with it. It may be that the gravity model is not a sufficiently good representation of reality for the purpose of examining a particular set of decisions. It follows from this that the validation task depends on the nature of the policies and projects to be assessed.

A general strategy for validating a model would then be to check whether it can reproduce a known state of the system with sufficient accuracy. As the future is definitively not known, this task is sometimes attempted by trying to estimate some well- documented state in the past, say a matrix from an earlier study. However, it is seldom the case that such a past state is sufficiently well documented. Therefore, less demanding validation tests incorporating data not used during estimation are often employed, for example: to check whether the number of trips across important screenlines or along main roads is well reproduced.

5.5.2 Calibration Techniques

As we have seen, the parameters A_i and B_j are estimated as part of the Furness (bi-proportional) balancing factor operations. The parameter β must be calibrated to make sure that the trip length distribution (TLD) is reproduced as closely as possible. This is a tall order for a single parameter. We shall see later how to improve on this but meantime, what is needed is a practical technique to estimate the best value for β, say β^*.

A naive approach to this task is simply to 'guess' or to 'borrow' a value for β, run the gravity model and then extract the modelled trip length distribution (MTLD). This should be compared with the observed trip length distribution (OTLD). If they are not sufficiently close, a new guess for β can be used and the process repeated until a satisfactory fit between MTLD and OTLD is achieved; this would then be taken as the best value β^*. Note that a set of home or roadside interviews will produce OTLDs with much greater accuracy than that of individual cell entries in the trip matrix, because the sampling rate for trip lengths is in effect much higher in this case.

The naive approach is not, however, very practical. Running a doubly constrained gravity model is time consuming and the approach provides no guidance on how to choose a better value for β if the current one is not satisfactory. Conventional curve-fitting techniques are unlikely to work well because the gravity model is not just non-linear but also complex analytically; the A_i's and B_j's are also functions of β through the two sets of equations (5.17) and (5.18).

A number of calibration techniques have been proposed and implemented in different software packages. The most important ones were compared by Williams (1976), who found that a technique due to Hyman (1969) was particularly robust and efficient. We shall describe briefly here Hyman's method.

At any stage in the calibration process a trip matrix $\mathbf{T}$ (β), function of the current estimate of β, is available. This matrix also defines a total number of trips $\sum_{ij} T_{ij}(\beta) = T(\beta)$. The method is based on the following requirement for β:

$$c(\beta) = \sum_{ij}[T_{ij}(\beta)c_{ij}]/T(\beta) = c^* = \sum_{ij}(N_{ij}C_{ij})\bigg/\sum_{ij}N_{ij} \tag{5.42}$$

where c^* is the mean cost from the OTLD and N_{ij} is the observed (and expanded) number of trips for each origin destination pair. The method can be described as follows:

1. Start the first iteration making $m = 0$ and an initial estimate of $\beta_0 = 1/c^*$.

2. Using the value of β_0 calculate a trip matrix using the standard gravity model. Obtain the mean modelled trip cost c_0 and estimate a better value for β as follows:

$$\beta_m = \beta_0 c_0 / c^*$$

3. Make $m = m + 1$. Using the latest value for β (i.e. β_{m-1}) calculate a trip matrix using a standard gravity model and obtain the new mean modelled trip cost c_{m-1} and compare it with c^*. If they are sufficiently close, stop and accept β_{m-1} as the best estimate for this parameter; otherwise go to step 4.

4. Obtain a better estimate of β as:

$$\beta_{m+1} = \frac{(c^* - c_{m-1})\beta_m - (c^* - c_m)\beta_{m-1}}{c_m - c_{m-1}}$$

5. Repeat steps 3 and 4 as necessary, i.e. until the last mean modelled cost c_{m-1} is sufficiently close to the observed value c^*.

The recalculations in step 3 are made to approximate closer to the equality in (5.42). A few improvements can be introduced to this method, in particular from the computational point of view. Hyman's approach has been shown to be robust and to offer, in general, advantages over alternative algorithms.

5.6 The Tri-proportional Approach

5.6.1 Bi-proportional Fitting

We have seen in section 5.4.2 how Furness's method can be derived from a mathematical programming framework. This non-linear mathematical program can be solved by a number of algorithms, including Newton's method. However, it is possible to show that the method originally proposed by Furness is indeed a practical and efficient algorithm, in particular for large matrices. The method is often referred to as the bi-proportional algorithm as it involves successive corrections by rows and then columns to satisfy the constraints; the algorithm stops when the corrections are small enough, i.e. when the constraints are met within reasonable tolerances.

The conditions necessary for the existence of a unique solution are that constraints (5.26a) and (5.26b) define a feasible solution space in non-negative T_{ij}'s. This requires $\sum_i O_i = \sum_j D_j$ but this is not a sufficient condition. The model has a multiplicative form and therefore it preserves the zeros present in the prior matrix $\{t_{ij}\}$. The existence of many zero entries in the prior matrix may prevent the satisfaction of one or more constraints. In summary, the product $a_k \, b_j \, c_k$ is unique but not each individual factor; there are two-degrees of indeterminacy (say α and β) that can have arbitrary values without affecting the value of the product:

$$a_i \alpha b_j \beta c_k / \alpha \beta = a_i b_j c_k$$

Example 5.5 Consider the case where a previously empty zone k is expected to see development in the future, thus originating and attracting trips. The cell entries for t_{ik} and t_{kj} would have been zero whilst the future O_k and D_k are non-zero. Therefore in this case there are no possible multiplicative correction factors capable of generating a matrix satisfying the constraints for zone k. It may be possible, however, to replace these empty cell values by 'guesses', i.e. suitable values borrowed from similar zones. Nevertheless, the presence of zeros in the prior matrix may cause subtler but no less difficult problems. If we try to solve the problem in Example 5.1 but with the prior matrix in Table 5.11, it will be found that this problem has no feasible solution in non-negative T_{ij}; there are only 11 unknowns and

7 independent constraints but the position of the zeros is such that there is no feasible solution and the bi-proportional algorithm oscillates without converging.

Table 5.11   A revised version of the doubly constrained growth factor problem in Table 5.6

	1	2	3	4	$\sum_j$	Target O_i
1	5	50	100	200	355	400
2	0	50	0	0	50	460
3	50	100	5	100	255	400
4	100	200	250	20	570	702
$\sum_i$	155	400	355	320	1230	
Target D_j	260	400	500	802	1962	

Readers familiar with linear algebra will be able to describe this problem in terms of the rank of the original and an augmented matrix containing the last column in Table 5.7. Furthermore, the reader may verify that after 10 iterations with this problem the corrected matrix stands as in Table 5.12:

Table 5.12   The matrix from problem in Table 5.11 after 10 Furness iterations

	1	2	3	4	$\sum_j$	Target O_i
1	3.4	0.7	61.0	355.3	420	400
2	0	388.2	0	0	388	460
3	65.5	2.8	5.9	345.7	420	400
4	191.2	8.3	433.1	101.0	734	702
$\sum_i$	260	400	500	802	1962	
Target D_j	260	400	500	802		1962

Several comments can be made at this stage:

1. The matrix after 10 iterations looks quite different from the prior one, thus casting some doubt about the realism, either of the old matrix, its zeros or the new trip-end totals.
2. The main problem seems to be in the second row, where there is a big difference (about 20%) between target and modelled total. There is no way this row can add up to 460 as the only non-zero cell entry has a maximum of 400 trips. The constraints do not generate a feasible solution space.
3. The problem seems ill-conditioned, e.g. a small change in a cell entry can make the problem a feasible one and produce a fairly different trip matrix. For example, the zero in cell $t_{2,4}$ could have arisen because of the sample used; replacing this zero by a 1 produces the matrix in Table 5.13 after the same 10 iterations. This is a much improved match with a fairly different matrix. In fact, it matches the targets with better than 1% accuracy. There is now a feasible solution space.

Real matrices are often sparse and the occurrence of this type of difficulty cannot be discarded as an academic problem. Failure to converge in a few iterations may well indicate that the presence and location of zeros in the prior matrix prevents the existence of a feasible solution with the new trip ends.

Table 5.13   The matrix from problem in Table 5.11 plus a single trip in cell
2, 4 after 10 Furness iterations

	1	2	3	4	$\sum_{j}$	Target O_i
1	4.1	4.5	76.2	315.4	400	400
2	0	339.2	0	119.1	458	460
3	77.3	17.0	7.2	298.5	400	400
4	178.6	39.3	416.6	68.9	703	702
$\sum_{i}$	260	400	500	802	1962	
Target D_j	260	400	500	802		1962

5.6.2   A Tri-proportional Problem

We have already presented the gravity model with a very flexible deterrence function that takes discrete
values constrained by a functional form for each cost bin. This was written in equation (5.40) as:

$$T_{ij} = a_i b_j \sum_m F^m \delta_{ij}^m$$

The main advantages of this model are its flexibility and the ease of calibration. In effect, we can
define any number of cost bins and the deterrence function can take any positive value for them; we
could even represent situations where, for example, there are few short trips, many intermediate trips,
few long trips and again a larger number of long-distance commuting trips.

The calibration of this model requires finding suitable values for the deterrence factor F^m for each
cost bin so that the number of trips undertaken for that distance is as close as possible to the observed
number. This task is, in fact, very similar to the problem of grossing up a matrix to match trip-end totals.
In this case we can start with a unity value for the deterrence factors and then correct these and the
parameters a_i and b_j until the trip ends and the TLD constraints are met. It seems natural to extend the
bi-proportional algorithm to handle this third dimension (cost bins) and utilise a tri-proportional method
to calibrate the model.

The principles behind the technique were proposed by Evans and Kirby (1974). Murchland (1977)
has shown that the application of successive corrections on a two-, three- or multi-dimensional space
conforms to just one of a group of possible algorithms to solve this type of problems; furthermore, the
method is simple to program and does not make excessive demands on computer memory.

Example 5.6 The tri-proportional algorithm can be illustrated with the problem stated in Table 5.8 and
with the trip length distribution (cost-bin) targets of Table 5.14.

Table 5.14   TLD target values for a tri-proportional gravity model calibration

	Ranges					
	1.0–4.0	4.1–8.0	8.1–12.0	12.1–16.0	16.1–20.0	20.1–24+
TLD	365	962	160	150	230	95

The model can then be solved using balancing operations to match trip targets by origin, destination
and cost bin. After five complete iterations, the matrix and modelled trips by cost bin T_k shown in
Table 5.15 are obtained.

Table 5.15 The matrix from problem in Table 5.14 after five iterations, including values for balancing factors a_i, b_j and F^k

	1	2	3	4	$\sum_j$	a_i
1	161.6	102.5	60.8	72.5	397.4	1.27
2	56.5	199.4	101.2	101.0	458.0	1.13
3	18.9	48.7	116.7	217.1	401.4	0.60
4	23.0	49.5	221.3	411.5	705.3	1.14
$\sum_i$	260	400	500	802	1962	
b_j	0.57	0.70	0.87	1.63		

	Ranges					
	1.0–4.0	4.1–8.0	8.1–12.0	12.1–16.0	16.1–20.0	20.1–24+
TLD	365	962	160	150	230	95
T_k	360.9	966.5	159.0	149.8	230.3	95.5
F_k	224.55	220.13	87.54	102.05	54.66	34.90

Of course, in this case the balancing factors are again not unique, at least up to two arbitrary multiplicative constants. Another way of expressing this is to say the balancing factors have two *degrees of indeterminacy*, the two multiplicative constants. It is easy to see that if we multiply each a_i by a factor Γ and each b_j by another factor τ, and then divide each F^k by $\Gamma \tau$, the modelled matrix will remain unchanged.

5.6.3 Partial Matrix Techniques

The tri-proportional calibration method has been used with a full trip length distribution, i.e. one that has an entry from observations in each cell. It would certainly be advantageous if one could calibrate a suitable gravity model without requiring a complete or full trip matrix. This is particularly important as we know that the cost of collecting data to obtain a complete trip matrix is rather high; furthermore, the accuracy of some of the cell entries is not very high and in calibration we actually use aggregations of the data, namely the TLD and the total trip ends O_i and D_j. Having explored the preferred methods for calibration, it should be clear that the possibility of calibrating gravity models with an incomplete or partial matrix does actually exist. For example, we can calibrate a gravity model with exponential cost function just with the total trip ends and a good estimate of the average trip cost, c^*.

The calibration of a gravity model with general deterrence function using the tri-proportional method is even more attractive in this case, as we could use just roadside interviews on cordons and screen-lines to obtain good TLDs and trip ends for some but not all the zones in the study area. There would be no need to use trip generation models except for forecasting purposes.

Example 5.7 The basic idea above can be described with the aid of a 3×3 matrix. Consider first a bi-proportional case where the full matrix-updating problem is to adjust a base-year matrix as follows:

adjusted to   produces

$$
\begin{array}{ccc}
a & b & c \\
d & e & f \\
g & h & i
\end{array}
\qquad
\begin{array}{c}
P \\
Q \\
R
\end{array}
\qquad
\begin{array}{ccc}
A & B & C \\
D & E & F \\
G & H & I
\end{array}
$$

$$
\begin{array}{ccc}
S & T & U
\end{array}
$$

In the case of a partial matrix, for example a survey where entries a and h cannot be observed, we would adjust only to trip ends excluding the corresponding total:

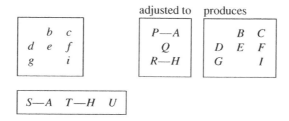

To fill in the missing cells we could use a gravity model; in the case of this example, one without deterrence function:

$$T_{ij} = a_i b_j$$

The estimated values of a_i and b_j (using data from the observed cells) would then be used to fill in these cells.

An extension to the tri-proportional case is almost trivial. Kirby (1979) has shown that there are two basic conditions required for a valid application of this approach:

1. The gravity model must fit both the available data we have and the data that are not available, i.e. the model must be a good model for the two regions of the matrix: the observed and the unobserved.
2. The two regions of the matrix should not be separable, i.e. it should not be possible to split the matrix into two or more independent matrices, typically:

	Internal	External
Internal	× ×	* *
External	* *	× ×

The problem is that each separate area has the two (or three in the tri-proportional case) degrees of indeterminacy and therefore the balancing factors cannot produce unique products, and hence trip estimates. This problem is also referred to as the *non-identifiability* of unique products for unobserved cell entries. As the figure above shows, this is likely to occur when roadside interviews take place only on a cordon to a study area. The provision of interviews on a screen-line will probably eliminate the problem as it would generate observations for the 'internal-internal' matrix.

5.7   Other Synthetic Models

5.7.1   Generalisations of the Gravity Model

The classic gravity model is by far the most commonly used aggregate trip distribution model. It has a number of theoretical advantages and there is no lack of suitable software to calibrate and use it. It can be easily extended further to incorporate more than one person type and it can even be used to model certain types of freight movements. However, the classic gravity model does not exhaust all the theoretical possibilities. We explore here three other approaches which, although they are much less used, offer real alternatives to the classic gravity model. The first one is simply a generalisation of the gravity model itself; the second one is the intervening-opportunities model, and the third one the family of direct demand models discussed in Chapter 6.

A number of authors have suggested extending the classic gravity model to account for not just the deterrent effect of distance but also for the fact that the farther away one is willing to travel the greater the number of opportunities to satisfy your needs.

Fang and Tsao (1995) suggested an entropy distribution model with quadratic costs:

$$T_{ij} = A_i B_j O_i D_j e^{-\beta C_{ij} - \lambda T_{ij} C_{ij}} \tag{5.43}$$

$$A_i = \frac{1}{\sum_j B_j D_j e^{-\beta C_{ij} - \lambda T_{ij} C_{ij}}}, \quad B_j = \frac{1}{\sum_j A_i O_i e^{-\beta C_{ij} - \lambda T_{ij} C_{ij}}} \tag{5.44}$$

They call it a self-deterrent gravity model. The inclusion of a 'congestion term' $\lambda T_{ij} C_{ij}$ in the exponent is the main extension to the classic model. The parameters β and λ are expected to have the same sign; if they have a different sign this would indicate that certain trips have economies of scale: they become more attractive the greater number of people undertaking them. If $\lambda = 0$ we have the classic gravity model.

De Grange *et al.* (2010) generalised this approach and proposed to:

$$\min_{\{T_{ij}\}} \; Z = \sum_{ij} T_{ij} C_{ij} + \frac{1}{\beta} \sum_{ij} T_{ij} (\ln T_{ij} - 1) - \frac{\rho}{\beta} \sum_{ij} T_{ij} \ln S_{ij} + \frac{\lambda}{2\beta} \sum_{ij} C_{ij} T_{ij}^2$$

$$s.t. \tag{5.45}$$

$$\sum_j T_{ij} = O_i \quad (\mu_i)$$

$$\sum_i T_{ij} = D_j \quad (\gamma_j)$$

where

$$S_{ij} = \sum_{\substack{k=1 \\ k \neq i, k \neq j}}^{n} D_k e^{-C_{jk}} \tag{5.46}$$

This term S_{ij} represents the accessibility to destinations as perceived from the origin i. Applying the optimality conditions to (5.45) results in

$$T_{ij} = A_i B_j O_i D_j (S_{ij})^\rho e^{-\beta C_{ij} - \lambda T_{ij} C_{ij}} \tag{5.47}$$

$$A_i = \frac{1}{\sum_j B_j D_j (S_{ij})^\rho e^{-\beta C_{ij} - \lambda T_{ij} C_{ij}}} \tag{5.48}$$

$$B_j = \frac{1}{\sum_i A_i O_i (S_{ij})^\rho e^{-\beta C_{ij} - \lambda T_{ij} C_{ij}}} \tag{5.49}$$

They find the best fit parameters using Maximum Likelihood techniques.

Here if calibration results in $\rho = 0$ we find Fang and Tsao's model. This is a very general model that De Grange *et al.* estimated for Santiago using different levels of aggregation.

5.7.2 Intervening Opportunities Model

The basic idea behind the intervening-opportunities model is that trip making is not explicitly related to distance but to the relative accessibility of opportunities for satisfying the objective of the trip. The original proponent of this approach was Stouffer (1940), who also applied his ideas to migration and the location of services and residences. But it was Schneider (1959) who developed the theory in the way it is presented today.

Consider first a zone of origin i and rank all possible destinations in order of increasing distance from i. Then look at one origin–destination pair (i, j), where j is the mth destination in order of distance from i. There are $m - 1$ alternative destinations actually closer (more accessible) to i. A trip maker would certainly consider those destinations as possible locations to satisfy the need giving rise to the journey: these are the *intervening opportunities* influencing a destination choice. Let α be the probability of a trip maker being satisfied with a single opportunity; the probability of her being attracted by a zone with D opportunities is then αD.

Consider now the probability q_i^m of not being satisfied by any of the opportunities offered by the mth destinations away from i. This is equal to the probability of not being satisfied by the first, nor the second, and so on up to the mth:

$$q_i^m = q_i^{m-1}(1 - \alpha D_i^m)$$ (5.50)

therefore, omitting the subscript i for simplicity we get

$$\frac{q^m - q^{m-1}}{q^m} = -\alpha D^m$$ (5.51)

Now, if we make x_m the cumulative attractions of the intervening opportunities at the mth destination:

$$x_m = \sum_m D^m$$

we can rewrite (5.50) as

$$\frac{q^m - q^{m-1}}{q^{m-1}} = -\alpha \left[x_{m-1} - x_m \right]$$ (5.52)

The limit of this expression for infinitesimally small increments is, of course,

$$\frac{\mathrm{d}q(x)}{q(x)} = -\alpha \, \mathrm{d}x$$ (5.53)

Integrating (5.53) we obtain:

$$\log q(x) = -\alpha x + \text{constant}$$

or

$$q(x) = A_i \exp(-\alpha x)$$ (5.54)

where A_i is a parameter for calibration. This relationship expresses the chance of a trip purpose not being satisfied by any of the m destinations ($m = 1, \ldots, M$) from i as a negative exponential function of the accumulated or intervening opportunities at that distance from the origin. The trips T_{ij}^m from i to a destination j (which happens to be the mth away from i) is then proportional to the probability of not being satisfied by any of the $m - 1$ closer opportunities minus the probability of not being satisfied by any of the opportunities up to the mth destination:

$$T_{ij}^m = O_i[q_i(x_{m-1}) - q_i(x_m)]$$
$$T_{ij}^m = O_i A_i[\exp(-\alpha x_{m-1}) - \exp(-\alpha x_m)] \tag{5.55}$$

It is easy to show that the constant A_i must be equal to

$$A_i = 1/[1 - \exp(-\alpha x_M)] \tag{5.56}$$

to ensure that the trip end constraints are satisfied. The complete model then becomes:

$$T_{ij}^m = O_i \frac{[\exp(-\alpha x_{m-1}) \exp(-\alpha x_m)]}{[1 - \exp(-\alpha x_M)} \tag{5.57}$$

Wilson (1970) has shown that this expression can also be derived from entropy-maximisation considerations.

The intervening-opportunities model is interesting because it starts from different first principles in its derivation: it uses distance as an ordinal variable instead of a continuous cardinal one as in the gravity model. It explicitly considers the opportunities available to satisfy a trip purpose at increased distance from the origin. However, the model is not often used in practice, probably for the following reasons:

- the theoretical basis is less well known and possibly more difficult to understand by practitioners;
- the idea of matrices with destinations ranked by distance from the origin (the nth cell for origin i is not destination n but the nth destination away from i) is more difficult to handle in practice;
- the theoretical and practical advantages of this function over the gravity model are not overwhelming;
- lack of suitable software.

In Chapter 12 we will discuss a more general version of this model that combines gravity and intervening-opportunities features. This is due to Wills (1986) and lets the data decide which combination of the two models fits reality better. However, the computational complexity of this new model is considerable.

5.7.3 Disaggregate Approaches

The whole discussion about distribution models has been cast in terms of zonal based productions/attractions and origins and destinations. We may have increased disaggregation by considering journey purposes and simple person types (with and without a car). Couched in these terms we obtain the number of trips undertaken between each OD pair. It can be argued, as it is in Chapters 7 to 9, that this is too coarse to capture the rich characteristics of travel behaviour; to achieve this capture we need to mode to the level of individuals, or at least, representative individuals.

In this context we do not deal with the number of trips to a particular destination but rather with the probability that a (representative) individual would choose a particular destination to satisfy some basic need. These disaggregate models are probabilistic although they may share apparently similar functional forms.

For example, a disaggregate model that would consider the choice of destination as discussed in Chapter 7, is likely to have a multinomial logit model structure with a form similar to a singly constrained gravity model.

5.8   Practical Considerations

We have discussed a number of frequently used models to associate origins and destinations and estimate the number of trips between them. While doing so, we have omitted a number of practical considerations that must necessarily affect the accuracy attainable from the use of such models. These stem from the inherent limitations of our modelling framework and our inability to include detailed descriptions of reality in the models. We shall discuss these features under the general headings below.

5.8.1   Sparse Matrices

Observed trip matrices are almost always sparse, i.e. they have a large number of empty cells, and it is easy to see why. A study area with 500 zones (250 000 cells) may have some 2.5 million expected total trips during a peak hour. This yields an average of 10 trips per cell; however, some OD pairs are more likely to contain trips than others, in particular from residential to high employment areas, thus leaving numerous cells with a very low number of expected trips. Consider now the method used to observe this trip matrix, perhaps roadside interviews. If the sampling rate is 20% (1 in 5) then the chances of making no observations on a particular OD pair are very high.

This sampled trip matrix will then be expanded, probably using information about the exact sampling ratios in each interview station. The problem generated when expanding empty cells has already been alluded to in section 5.3.4. It may be possible to fill in gaps in the matrix through the use of a partial matrix approach; alternatively, it may be desirable to 'seed' empty cells with a low number and use an alternative matrix expansion method such as that discussed in Chapter 12. It is important to realise, however, that 'observed' trip matrices normally contain a large number of errors and that these will be amplified by the expansion process.

5.8.2   Treatment of External Zones

It may be quite reasonable to postulate the suitability of a synthetic trip distribution model in a study area, in particular for internal-to-internal trips. However, a significant proportion of the trips may have at least one end outside the area. The suitability of a model which depends on trip distance or cost, a variable essentially undefined for external trips, is thus debatable.

Common practice in such cases is to take these trips outside the synthetic modelling process: roadside interviews are undertaken on cordon points at the entrance/ exit to the study area. The resulting matrix of external-external (E − E) and external-internal (E − I) trips is then updated and forecast using growth factor methods, in particular those of Furness. However, a number of trip ends from the trip generation/attraction models correspond to the E − I trips and these must be subtracted from the trip-end totals for inclusion as constraints to the synthetic models.

5.8.3   Intra-zonal Trips

A similar problem occurs with intra-zonal trips. Given the limitations of any zoning system, the cost values given to centroid connectors are a very crude but necessary approximation to those experienced in reality. The idea of an intra-zonal trip cost is then poorly represented by these centroid connector costs. Some commercial software allows the user to add/subtract terminal costs to facilitate better modelling of these trips; the idea is that by manipulating these intra-zonal costs one would make the gravity model fit better. However, this is not very good; it is actually preferable to remove intra-zonal trips from the synthetic modelling process and to forecast those using even simpler approaches. This typically assumes that intra-zonal trips are a fixed proportion of the trip ends calculated by the trip generation models.

Moreover, intra-zonal trips are not normally loaded onto the network as they move from a centroid to itself. This makes it less essential to model them in detail. However, in reality, some of these trips use the modelled network. Nevertheless, this problem is probably of significance only for rather coarse zoning systems.

5.8.4 Journey Purposes

Different models are normally used for different trip purposes and/or person types. Typically, the journey to work will be modelled using a doubly constrained gravity model while almost all other purposes will be modelled using singly constrained models. This is because it is often difficult to estimate trip attractions accurately for shopping, recreational and social trips and therefore proxies for trip attractiveness are used: retail floor space, recreational areas, population.

Some trip purposes may be more sensitive to cost and therefore deserve the use of different values for the deterrence function.

5.8.5 K Factors

The gravity model can provide a reasonable representation of trip patterns provided they can be explained mainly by the size of the generation and attraction power of zones and the deterrence to travel generated by distance (generalised cost). We recognise that most individual decisions on residential location and/or choice of employment incorporate many other factors; therefore, the gravity model could only model destination choice at an aggregate level if the importance of these other factors were much reduced on aggregation. However, there are always aggregate effects that do not conform to a simple gravity model. In some circumstances, there may be pairs of zones which have a special association in terms of trip making; for example, a major manufacturer may be located in one zone and most of its employees in another, perhaps as a result of a housing estate developed by the company. In this case, it is likely that more trips will take place between these two points than predicted by any model failing to consider this association, for example the gravity model. This has led to the introduction of an additional set of parameters K_{ij} to the gravity model as follows:

$$T_{ij} = K_{ij}A_iO_iB_{ij}\exp\left(-\beta c_{ij}\right) \tag{5.58}$$

Some practical studies have used these K factors in an attempt to improve the calibration of the model. This, of course, they do; with the full set of K factors we now have even more flexibility than necessary to reproduce the observed trip matrix; in fact, just the K factors are enough to achieve this; the other parameters are surplus to requirement: K_{ij} factors identical to the observed T_{ij} will do the trick; but then we no longer have a model nor any forecasting ability left.

The best advice that can be given in respect of K factors is: try to avoid them. If a study area has a small number of zone pairs (say, less than 5% of the total) with a special trip making association which is likely to remain in the future, then the use of a few K factors might be justified, sparingly and cautiously. But the use of a model with a full set of K factors cannot be justified. However, K factors are related to incremental forms of the gravity model as we discuss in Chapter 12.

5.8.6 Errors in Modelling

It would appear that many of these practical issues reduce the accuracy of the modelling process. This is, in effect, true and it constitutes a reflection of the contrast between the limitations of the state of the art in transport modelling and the complexities and inherent uncertainties of present and future human behaviour. These practical issues are not restricted to distribution modelling; they are present, in one form or another, in other parts of the modelling process.

Because many of the cells in a trip matrix will have small values, say between 0 and 5 in the sample and perhaps 20 to 30 in the expanded or synthesised matrix, their corresponding errors will be relatively large. A small number of studies have tackled the task of calibrating synthetic models and then comparing the resulting trip matrices with observed ones. An investigation by Sikdar and Hutchinson (1981) used data from 28 study areas in Canada to calibrate and test doubly constrained gravity models. The researchers found that the performance of these models was poor, equivalent to a randomly introduced error in the observations of about 75 to 100%; these results reinforce the call for caution in using the results of such models. This should not be entirely surprising; to model a trip matrix with the use of a few parameters (twice the number of zones for an exponential deterrence function) is a very tall order. This is certainly one of the reasons why few studies nowadays make use of the gravity model in its conventional form. In many cases, however, it is desirable to consider how changes in transport costs would influence trip patterns, in particular for more optional purposes like shopping and recreation. In these cases, the idea of using pivot point or incremental versions of the gravity model becomes more attractive, see Chapter 12.

The treatment of errors in modelling has received attention for some time. There seem to be two methods deserving consideration in this field: statistical and simulation approaches. Statistical methods are very powerful but they are not always easy to develop or to implement. They follow the lines suggested in Chapter 3 when discussing the role of data errors in the overall accuracy of the modelling process. Errors in the data are then traced through to errors in the outputs of the models. The UK Department of Transport provides advice in the *Traffic Appraisal Manual* (Department of Transport 1985) on the sensitivity of distribution models to errors in the input data. To some extent the simplest problem is to follow data errors, in particular those due to sampling, through the process of building matrices. A more demanding problem is to follow these errors when a synthetic distribution model is used. One of the advances of the early 1980s was the development of approximate analytical techniques to estimate the output errors due to sampling variability. For example, the work of Gunn *et al.* (1980) established approximate expressions for the confidence interval for cell estimates for the tri-proportional formulation of the gravity model. The 95% confidence interval for the number of trips in a cell (i, j) is given by the range $\{C_{ij}/T_{ij} \text{ to } T_{ij} C_{ij}\}$, where C_{ij} is a *confidence factor* given by:

$$C_{ij} = \exp\left(2\left[1\bigg/\sum_{ij} n_{ijk} + 1\bigg/\sum_{jk} n_{ijk} + 1\bigg/\sum_{ki} n_{ijk}\right]^{0.5}\right) \qquad (5.59)$$

where n_{ijk} are the number of trips sampled in the observed cells and therefore the summations are over observed cells only. This expression covers only errors due to sampling; data collection and processing errors are likely to increase the range. Moreover, there are other sources of error in the model estimates which are more difficult to quantify; these are mis-specification errors, due to the fact that the model is only a simplified and imperfect representation of reality. Mis-specification errors will, again, increase the range for any confidence interval estimates.

Simulation techniques may play a useful role in cases where analytical expressions for confidence intervals of model output do not exist and are difficult to develop. One can calibrate a model assuming that the data available contain no errors; one would then introduced controlled, but realistic, variability in the data and recalibrate the model. This process could be repeated several times to obtain a range of parameters, each calibrated with a slightly different set of 'survey' data. This process is, of course, quite expensive in time and computer resources and it is therefore attempted mostly for research purposes. However, this type of research can provide useful insights into the stability of model parameters to data errors.

A simpler use of Monte Carlo simulation is in testing the sensitivity of model output to input data in a forecasting mode. One knows that future planning data are bound to contain errors; the use of simulation in this case involves the introduction of reasonable 'noise' into these data and then running the model with each of these future data sets. The results provide an idea of the sensitivity of model output to errors in these planning variables. As no recalibration is involved (the model is assumed to be calibrated with no errors in the base year) the demand on time and resources, although large, is less than in the previous case.

5.8.7 The Stability of Trip Matrices

The stability of trip matrices over time is an issue seldom discussed in transport demand modelling. We know from experience that reality is not entirely repeatable from day to day. We can observe significant day-to-day variations at the level of traffic flows on any link in a network. One would typically expect some 10% variation on flow levels on similar days and on the same day of the week over similar weeks (i.e. excluding seasonal variations). These variations are easily observed, as permanent and semi-permanent automatic traffic counters are easy to install and maintain and are mostly reliable. These variations in traffic flows may result from at least two sources: variations in the trip matrices that originate them and day-to-day changes in route choice. The question arises, therefore, about the extent of day-to-day variations at the level of trip matrix cell values. This information is much more difficult to come by as very rarely repeated data is collected on trip matrices, in the same location, on different days.

Traffic counts are the result of an aggregation of trips into trip matrices and therefore this aggregation process will tend to compensate some of the random variations at the trip matrix level. Leonard and Tough (1979) report on the collection of detailed origin–destination (trip table and traffic count) data on four consecutive days in the centre of Reading, UK. The data was collected to help in the development of a detailed simulation model. Observers recorded car number plates, thus tracking the route vehicles took through the centre of Reading together with their points of entry/exit and parking. Therefore, there were no interview or reporting errors but only a 10% sample was collected over four days (Monday to Thursday) for some 80 links and 40 zones. However, the data was independently analysed by Willumsen (1982) to look at day-to-day variations at link flow and OD matrix level. He used the percentage mean absolute error (%MAE) for both traffic levels and trip matrices:

$$\% \, \text{MAE} = 100\,\% \times \left(\sum_a |V^a - V^b| \Big/ \sum_a V^a \right) \tag{5.60}$$

and

$$\% \, \text{MAE} = 100\,\% \times \left(\sum_{ij} |T_{ij}^a - T_{ij}^b| \Big/ \sum_{ij} T_{ij}^a \right) \tag{5.61}$$

where the indices a and b relate to *observed* flows V and OD trips T_{ij} on different days. Willumsen (1982) found that typical variations were:

	Tuesday		Wednesday		Thursday	
% MAE	Link	Matrix	Link	Matrix	Link	Matrix
Monday	11	76	11	72	12	75
Tuesday			13	68	14	85
Wednesday					12	70

Here we see that the day-to-day variations at flow level are consistent with expectations, whereas those at the trip matrix level are much larger. This is partly because, at trip matrix level, we are dealing with small values and sparse matrices, but even then the evidence suggests that variations at this level can be quite significant.

This is a rather 'inconvenient truth' because it weakens the case for collecting travel data on different days and putting it all together in an 'average (usually working) day'. The representativeness of this average day is seldom questioned and we have few tools to consider it seriously. This limitation applies to all our approaches: aggregate, disaggregate and activity based models.

These results suggest that efforts to obtain a very accurate trip matrix may not be warranted as it will only be a snapshot. The objective for a destination choice model in this context should not be to replicate an observed or underlying trip matrix, but to estimate one that captures the main features of the underlying trip matrices that, when loaded onto the network, produce link flows consistent with observations.

The results also suggest that one should be more careful when testing how the value of a scheme or plan changes with variations in the estimated trip matrix used during assessment. Sensitivity analysis seems a particularly appropriate way to investigate the effects of varying the trip matrix.

Exercises

5.1 A small study area has been divided into four zones and a limited survey has resulted in the following trip matrix:

	1	2	3	4
1	–	60	275	571
2	50	–	410	443
3	123	61	–	47
4	205	265	75	–

Estimates for future total trip ends for each zone are as given below:

Zones	Estimated future origins	Estimated future destinations
1	1200	670
2	1050	730
3	380	950
4	770	995

Use an appropriate growth-factor method to estimate future inter-zonal movements.
Hint: check conditions for convergence of the chosen method first.

5.2 A study area has been divided into three large zones, A and B on one side of a river and C on the other side. It is thought that travel demand between these zones will depend on whether or not the O−D pair is at the same side of the river. A small sample home interview survey has been undertaken with the following results:

Blank entries indicate unobserved cells.

Origin	Destination		
	A	B	C
A	12	10	8
B		5	3
C	4		7

Assume a model of the type $T_{ij} = R_i S_j F_k$ where the parameter F_k can be used to represent the fact that the O−D pair is on the same side of the river or not. Calibrate such a model using a tri-proportional algorithm and fill the empty cells in the matrix above.

5.3 A transport study is being undertaken incorporating four cities A, B, C and D. The travel costs between these cities in generalised time units are given below; note that intra-urban movements are excluded from this study:

Origin	Destination			
	A	B	C	D
A	–	1.23	1.85	2.67
B	1.23	–	2.48	1.21
C	1.85	2.48	–	1.44
D	2.67	1.21	1.44	–

Roadside interviews have been undertaken at several sites and the number of drivers interviewed is shown below together with their respective origins and destinations. Blank entries indicate unobserved cells.

Origin	Destination			
	A	B	C	D
A	–	6		2
B		–	1	4
C	8		–	8
D	6	18	6	–

Assume now that a gravity model of the type $Tij = R_i S_j F_k$ is to be used for this study area with only two cost bins. The first cost bin will cover trips costing between 0 and 1.9 and the second trips costing more than 1.9. Calibrate such a model using a tri-proportional method on this partial matrix. Provide estimates of the parameters R_i, S_j and F_k and of the missing entries in the matrix, excluding intra-urban trips. Are these estimates unique?

6

Modal Split and Direct Demand Models

6.1 Introduction

In this chapter we shall discuss firstly mode choice as an aggregate problem. It is interesting to see how far we can get using similar approaches to those pursued in deriving and using trip distribution models. We will also examine methods to estimate generation, distribution and modal split simultaneously, the so-called *direct demand* models. Finally, we will examine the need for consistency between the parameters and structure of distribution and mode choice models, a topic often disregarded by practitioners at their peril.

The choice of transport mode is probably one of the most important classic model stages in transport planning. This is because of the key role played by public transport in policy making. Almost without exception travelling in public transport modes uses road space more efficiently and produce fewer accidents and emissions than using a private car. Furthermore, underground and other rail-based modes do not require additional road space (although they may require a reserve of some kind) and therefore do not contribute to road congestion. Moreover, if some drivers could be persuaded to use public transport instead of cars the rest of the car users would benefit from improved levels of service. It is unlikely that all car owners wishing to use their cars could be accommodated in urban areas without sacrificing large parts of the fabric to roads and parking space.

The issue of mode choice, therefore, is probably the single most important element in transport planning and policy making. It affects the general efficiency with which we can travel in urban areas, the amount of urban space devoted to transport functions, and whether a range of choices is available to travellers. The issue is equally important in inter-urban transport as again rail modes can provide a more efficient mode of transport (in terms of resources consumed, including space), but there is also a trend to increase travel by road.

It is important then to develop and use models which are sensitive to those attributes of travel that influence individual choices of mode. We will see how far this necessity can be satisfied using aggregate approaches, where alternative policies need to be expressed as modifications to useful if rather inflexible functions like the generalised cost of travel.

Modelling Transport, Fourth Edition. Juan de Dios Ortúzar and Luis G. Willumsen.
© 2011 John Wiley & Sons, Ltd. Published 2011 by John Wiley & Sons, Ltd.

6.2   Factors Influencing the Choice of Mode

The factors influencing mode choice may be classified into three groups:

1. Characteristics of the trip maker. The following features are generally believed to be important:
   * car availability and/or ownership;
   * possession of a driving licence;
   * household structure (young couple, couple with children, retired, singles, etc.);
   * income;
   * decisions made elsewhere, for example the need to use a car at work, take children to school, etc;
   * residential density.
2. Characteristics of the journey. Mode choice is strongly influenced by:
   * the trip purpose; for example, the journey to work is normally easier to undertake by public transport than other journeys because of its regularity and the adjustment possible in the long run;
   * time of the day, when the journey is undertaken; late trips are more difficult to accommodate by public transport;
   * whether the trip is undertaken alone or with others.
3. Characteristics of the transport facility. These can be divided into two categories. Firstly, quantitative factors such as:
   * components of travel time: in-vehicle, waiting and walking times by each mode;
   * components of monetary costs (fares, tolls, fuel and other operating costs);
   * availability and cost of parking;
   * reliability of travel time and regularity of service.
   Secondly, qualitative factors which are less easy (or impossible) to measure in practice, such as:
   * comfort and convenience;
   * safety, protection, security;
   * the demands of the driving task;
   * opportunities to undertake other activities during travel (use the phone, read, etc.).

Note that we have described these in terms of journeys or trips. A richer concept is that of tours with trips as their components. It is clear that the choice of mode is made more on a tour basis (that is considering the requirements of all trips) than on a single trip. If one chooses the car for the first leg of a tour this is likely to remain the choice for the other legs. A good mode choice model would be based at least on simple tours (from home and back) and should include the most important of these factors. It is easy to visualise how the concept of generalised cost can be used to represent several of the quantitative factors included under 3.

Mode choice models can be *aggregate* if they are based on zonal (and inter-zonal) information. We can also have *disaggregate* models if they are based on household and/or individual data (see Chapter 7).

A simplistic but useful way to think about mode choice is as follows. Given that somebody knows where it is going (Destination) this person has many alternative 'routes' to get there; some involve just driving the car whereas others may require to walk to a subway station, take the train, alight at some other station and, say, walk to the final destination (plus many other combinations of modes and routes). This person can then choose the lowest generalised cost option, among all of these, and in doing so the physical route and the combination of modes would be found. If all people think the same we would have an 'all or nothing' route and mode choice model. Alas, life is not so simple for a number of reasons:

* Some people do not have a car available so their choice set will be more limited; this would be the minimum segmentation required.

- As we will see, congestion, both on roads and in public transport, make the choice of more than one route a necessity.
- Generalised costs cannot hope to capture all the relevant elements that determine mode choice; this is particularly relevant in the case of the choice between car and public transport that focus on parameters additional to those relevant for route choice.
- Different people would perceive costs in different ways and would seek to minimise a different 'version' of generalised costs (time versus money minimisers, for example); we must allow, therefore, for a degree of dispersion in choices to consider other factors, not fully visible to the analyst, in mode preferences.
- The modelled costs in a zonal based model are only centroid-to-centroid averages of the costs (time, money) actually perceived by individuals; for example, some may live closer to a rail station and therefore be more inclined to use public transport.

The combined effect of these influences is dispersion in the choices of mode made at each Origin-Destination pair. The nature of this dispersion is influenced by the three groups of factors and conditions mentioned above.

6.3   Trip-end Modal-split Models

In the past, in particular in the USA, personal characteristics were thought to be the most important determinants of mode choice and therefore attempts were made to apply modal-split models immediately after trip generation. In this way the different characteristics of the individuals could be preserved and used to estimate modal split: for example, the different groups after a category analysis model. As at that level there was no indication to where those trips might go the characteristics of the journey and modes were omitted from these models.

This was consistent with a general planning view that as income grew most people would acquire cars and would want to use them. The objective of transport planning was to forecast this growth in demand for car trips so that investment could be planned to satisfy it. This was characterised as the 'predict and provide' approach to transport planning, today considered a blind and dangerous alley. The modal-split models of that time related the choice of mode only to features like income, residential density and car ownership. In some cases the availability of reasonable public transport was included in the form of an accessibility index.

In the short run these models could be very accurate, in particular if public transport was available in a similar way throughout the study area and there was little congestion. However, this type of model is, to a large extent, defeatist in the sense of being insensitive to policy decisions; it appears that there is nothing the decision maker can do to influence the choice of mode. Improving public transport, restricting parking, charging for the use of roads, none of these would have any effect on modal split according to these trip-end models. What was missing was a way to incorporate the aggregate characteristics of alternative modes to make choice more policy sensitive.

6.4   Trip Interchange Heuristics Modal-split Models

Modal-split modelling in Europe was dominated, almost from the beginning, by post-distribution models; that is, models applied after the gravity or other distribution model. This has the advantage of facilitating the inclusion of the characteristics of the journey and that of the alternative modes available to undertake them. However, they make it more difficult to include the characteristics of the trip maker as they may have already been aggregated in the trip matrix (or matrices).

The first models included only one or two characteristics of the journey, typically (in-vehicle) travel time. It was observed that an S-shaped curve seemed to represent this kind of behaviour better, as in Figure 6.1, showing the proportion of trips by mode 1 (T_{ij}^1 / T_{ij}) against the cost or time difference.

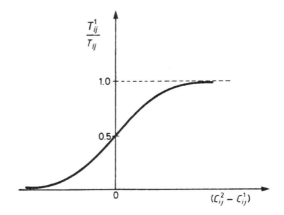

Figure 6.1   Empirical Modal-split curve

These were empirical curves, obtained directly from the data and following a similar approach to the curves used to estimate what proportion of travellers would be diverted to use a (longer but faster) bypass route: hence their name of *diversion curves*. For example, the London Transportation Study (Phase III) used diversion curves for trips to the central area and non- central trips (the former more likely to be made by public transport) and for different trip purposes.

Another approach has been to use, by analogy, a version of Kirchhoff formulation in electricity. The proportion of trip makers between origin i and destination j that chooses model k as a function of the respective generalised costs by model k, C_{ij}^k is given by:

$$P_{ij}^k = \frac{(C_{ij}^k)^{-n}}{\sum_{1}^{k} (C_{ij}^k)^{-n}} \tag{6.1}$$

where n is a parameter to be calibrated or transferred from another location or time (values for n between 4 and 9 have been suggested for both mode and route choice models of this nature). With a judicious choice of n this formulation produces a curve not too dissimilar from the Logit equation. Kirchhoff's model can be derived from entropy maximisation principles assuming that the generalised costs are perceived in a logarithmic fashion as in equation (5.37) and (5.38). The interested reader can also verify that this formulation is consistent with the Box-Cox transformation on the utility function of a Logit model when $\tau = 0$ (see section 8.3).

Model (6.1) is sometimes considered attractive because the choice of mode (or route) depends on the ratio of costs (to a power) and not on their difference. As we will see, one of the issues with most Logit models is that a 5 minute difference in a 30 minute journey has the same effect as a 5 minute difference in a 6 hour trip.

One normal limitation of these models is that they can only be used for trip matrices of travellers with a choice available to them. This often means the matrix of car-available persons, although modal split can also be applied to the choice between different public-transport modes.

The above models have limited theoretical basis and therefore their interpretation and forecasting ability must be in doubt. Further, as these models are aggregate they are unlikely to model in full the constraints and characteristics of the modes available to individual households.

6.5   Synthetic Models

6.5.1   Distribution and Modal-split Models

The entropy-maximising approach can be used to generate models of distribution and mode choice simultaneously. In order to do this we need to cast the entropy-maximising problem in terms of, for example, two modes as follows:

$$\text{Maximise } \log W\{T_{ij}^k\} = -\sum_{ijk} \left(T_{ij}^k \log T_{ij}^k - T_{ij}^k\right) \tag{6.2}$$

subject to

$$\sum_{jk} T_{ij}^k - O_i = 0 \tag{6.3}$$

$$\sum_{ik} T_{ij}^k - D_j = 0 \tag{6.4}$$

$$\sum_{ijk} T_{ij}^k C_{ij}^k - C = 0 \tag{6.5}$$

It is easy to see that this problem leads to the solution:

$$T_{ij}^k = A_i O_i B_j D_j \exp\left(-\beta C_{ij}^k\right) \tag{6.6}$$

$$P_{ij}^1 = \frac{T_{ij}^1}{T_{ij}} = \frac{\exp\left(-\beta C_{ij}^1\right)}{\exp\left(-\beta C_{ij}^1\right) + \exp\left(-\beta C_{ij}^2\right)} \tag{6.7}$$

where P_{ij}^1 is the proportion of trips travelling from i to j via mode 1. The functional form in (6.7) is known as Logit and it is discussed in greater detail in the next chapter. However, it is useful to reflect here on some of its properties:

- it generates an S-shaped curve, similar to some of the empirical diversion curves of Figure 6.1;
- if $C_1 = C_2$, then $P_1 = P_2 = 0.5$;
- if C_2 is much greater than C_1, then P_1 tends to 1.0;
- the model can easily be extended to multiple modes.

$$P_{ij}^1 = \frac{\exp\left(-\beta C_{ij}^1\right)}{\sum_k \exp\left(-\beta C_{ij}^k\right)} \tag{6.8}$$

It is obvious that in this formulation β plays a double role. It acts as the parameter controlling dispersion in mode choice and also in the choice between destinations at different distances from the origin. This is probably asking too much of a single parameter, even if underpinned by a known theoretical basis. Therefore a more practical joint distribution/modal-split model has been used in many studies. This has the form (Wilson 1974):

$$T_{ij}^{kn} = A_i^n O_i^n B_j D_j \exp(-\beta_n K_{ij}^n) \frac{\exp(-\lambda_n C_{ij}^k)}{\sum_k \exp(-\lambda_n C_{ij}^{k'})} \tag{6.9}$$

where K_{ij}^n is the *composite cost* of travelling between i and j as perceived by person type n. In principle this composite cost may be specified in different ways; for example, it could be taken to be the minimum of the two costs or, perhaps better, the weighted average of these:

$$K = \sum_k P^k C^k \quad (i, j \text{ and } n \text{ omitted for simplicity})$$

However, it is interesting to note that some of the formulations used in many early studies are, in fact, inappropriate. The mode choice or mode split component of this is a binary choice model of the form:

$$P_{ij}^1 = \frac{\exp(-\lambda C_{ij}^1)}{\exp(-\lambda C_{ij}^1) + \exp(-\lambda C_{ij}^2)} = \frac{1}{1 + \exp(-\lambda(C_{ij}^2 - C_{ij}^1))}$$

The right hand side of the equation shows clearly that the proportion choosing one mode depends only on the differences in generalised costs. This property suggests, in some cases, segmenting the demand by trip length so that, say, a 5 minute saving is more important in a short trip than in a long one. There are, however, other ways of compensating for this.

The proportion of trips using one mode, say 1, is shown in Figure 6.2 as a function of the scaling parameter λ and the differences in costs between the two modes. The figure also includes the plot of the Kirchhoff model (solid line) with the power value of -8. Note that in this case, it is important that the cost range for each of the modes is 5 to 120 generalised min. Note that the greater the value of λ the closer is the Logit model to an 'all or nothing' allocation of trips to the cheapest mode.

Figure 6.2 Proportion selecting mode 1 for Logit model with different λ and Kirchhoff with power of -8

Example 6.1 Consider the weighted average form above and examine what happens when a new, more expensive mode $(C_2 > C_1)$ is added to an existing unimodal system. In the initial state we would have:

$$K = \sum_k P^k C^k = C^1$$

and in the final state, i.e. after the introduction of mode 2:

$$K^* = P^1 C^1 + P^2 C^2$$

However, by definition $P^1 + P^2 = 1$ and therefore:

$$K^* = (1 - P^2)C^1 + P^2 C^2 = C^1 + P^2(C^2 - C^1)$$
$$K^* = K + P^2(C^2 - C^1)$$

Now, as both P^2 and $(C^2 - C^1)$ are greater than zero, we conclude that $K^* > K$, which is nonsensical as the introduction of a new option, even if it is more expensive, should not increase the composite costs; at worst they should remain the same. The use of the wrong composite costs will lead to mis-specified models.

6.5.2 Distribution and Modal-split Structures

Williams (1977) has shown that the only correct specification, consistent with the prevailing theory of rational choice behaviour (see section 7.2), is:

$$K^n_{ij} = \frac{-1}{\lambda_n} \log \sum_k \exp(-\lambda_n C^k_{ij}) \tag{6.10}$$

where the following restriction must be satisfied:

$$\beta_n \leq \lambda_n \tag{6.11}$$

We will come back to this restriction in Chapter 7. Intuitively, it means that the importance of costs is more critical in the choice of mode than in the choice of destination. If this is not the case, the model structure in (6.9), simultaneous or sequential, would be inappropriate. The composite cost measure (6.10) has the following properties:

- $K \leq \text{Min}_k \{C^k\}$

- $\underset{\lambda \to \infty}{\text{Lim}}\ K = \text{Min}_k \{C_k\}$, that is 'all-or-nothing' mode choice

- $\frac{dK}{dC^k} = P^k$

The first of these properties means that when a new alternative is added, even if it is very unattractive in principle, the composite costs will either reduce (somebody must like it) or at most remain the same. The second property highlights the importance of λ_n as a weight attached in the choice to generalised costs. For a very large λ_n the model will predict an 'all-or-nothing' choice of the least generalised cost alternative.

The model ((6.9)–(6.11)) is frequently found in practice in aggregate applications, in particular in urban areas. One of the problems in practice, however, is that modellers sometimes fail to check whether the restriction (6.11) is satisfied. As the destination and mode choice models may have been calibrated independently, it is quite possible that the restriction is not satisfied. If this is the case, the combined models (gravity and then mode choice) will produce pathological results. This structure, often described as G/D/MS/A (Generation, Distribution, Mode Split and Assignment), would be wrong if $\beta > \lambda$; in that case the structure G/MS/D/A would be probably the correct one.

It has been found in practice that the structure itself may be different for different journey purposes. Typically, the correct structure would be G/D/MS/A for journey to work and G/MS/D/A for other purposes. This would reflect a condition where it is easier to change destination for, say, a shopping trip than to change mode.

Note that in the case of the G/MS/D/A structure one starts by calculating the composite cost by mode n of reaching all destinations from each origin i:

$$K_i^n = \frac{-1}{\beta} \log \left(\sum_j \exp(-\beta C_{ij}^n) \right) \tag{6.12}$$

Then, this composite costs are used to obtain mode splits by origin i.

$$P_i^1 = \frac{1}{1 + \exp(-\lambda(K_i^2 - K_i^1))} \tag{6.13}$$

Separate gravity models are developed using the costs of each mode with mode-specific trip generations but sharing the attraction trip ends. Although this is a generation based mode choice model it takes into account fully the costs of reaching each destination by each mode.

For many applications these aggregate models remain valid and in use. However, for a more refined handling of personal characteristics and preferences we now have disaggregate models which respond better to the key elements in mode choice and make a more efficient use of data collection efforts; these are discussed in Chapters 7 to 9.

6.5.3   Multimodal-split Models

Figure 6.3 depicts possible model structures for choices involving more than two modes. The N-way structure which became very popular in disaggregate modelling work, as we will see in Chapter 7, is the simplest; however, because it assumes that all alternatives have equal 'weight', it can lead to problems when some of the options are more similar than others (i.e. they are correlated), as demonstrated by the famous blue bus-red bus example (Mayberry 1973).

Example 6.2 Consider a city where 50% of travellers choose car (C) and 50% choose bus (B). In terms of model (6.8), which is an N-way structure, this means that $C_C = C_B$. Let us now assume that the manager of the bus company, in a stroke of marketing genius, decides to paint half the buses red (RB) and half of them blue (BB), but manages to maintain the same level of service as before. This means that $C_{RB} = C_{BB}$, and as the car mode has not changed this value is still equal to C_C. It is interesting to note that model (6.8) now predicts:

$$P_C = \frac{\exp(-\beta C_C)}{\exp(-\beta C_C) + \exp(-\beta C_{RB}) + \exp(-\beta C_{BB})} = 0.33$$

when one would expect P_C to remain 0.5, and the buses to share the other half of the market equally between red and blue buses. The example is, of course, exaggerated but serves well to show the problems of the N-way structure in the presence of correlated options (in this case completely correlated). We will come back to this in Chapter 7.

The 'added-mode' structure, depicted in Figure 6.3b, was used by many 'pragmatic' practitioners in the later 1960s and early 1970s; however, it has been shown to give different results depending on which mode is taken as the added one (Langdon 1976). Also, work using Monte Carlo simulation has shown

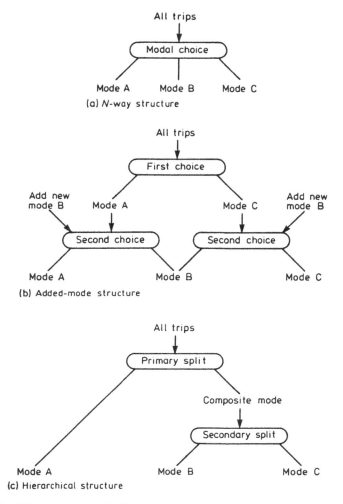

Figure 6.3   Multimodal model structures: (a) *N*-way structure, (b) added-mode structure, (c) hierarchical structure

that the added mode form with better performance in the base year is not necessarily the one to perform best in the future under certain policy changes (Ortúzar 1980a).

The third possibility, depicted in Figure 6.3c, is the hierarchical or nested structure. Here the options which have common elements (i.e. are more similar than others or correlated) are taken together in a primary split (i.e. public transport). After they have been 'separated' from the uncorrelated option, they are subdivided in a secondary split. In fact, this was the standard practice in the 1960s and early 1970s, but with the short-coming that the composite costs for the 'public-transport' mode were normally taken as the minimum of costs of the bus and rail modes for each zone pair and that the secondary split was achieved through a minimum-cost 'all-or-nothing' assignment. This 'pragmatic' procedure essentially implies an infinite value for the dispersion parameter of the submodal-split function, whereas it has normally been found that it has a value of the same order as the dispersion parameter in the primary split, but satisfying (6.11).

Example 6.3 A hierarchical structure model for the red bus/blue bus problem of Example 6.2 would have the following expression:

$$P_C = \frac{1}{1 + \exp\{-\lambda_1(C_B - C_C)\}}; \quad P_B = 1 - P_C$$

$$P_{R/B} = \frac{1}{1 + \exp\{-\lambda_2(C_{BB} - C_{RB})\}}$$

$$P_{B/B} = 1 - P_{R/B}$$

with

$$C_B = \frac{-1}{\lambda_2} \log[\exp(-\lambda_2 C_{RB}) + \exp(-\lambda_2 C_{BB})]$$

where P_C is the probability of choosing car, as before, $(1 - P_C) P_{R/B}$ the probability of selecting red bus and $(1 - P_C) P_{B/B}$ the probability of selecting blue bus; the λs are the primary and secondary split parameters. It is easy to see that, if $C_B = C_C$, this model correctly assigns a probability of 0.5 to the car option and 0.25 to each of the bus modes. However, the value of the composite bus cost C_B is not the same as the cost of the red or blue buses (C_{RB} and C_{BB}). The former depends on the value of λ_2 and for the red bus/blue bus problem it would be:

$$C_B = C_{BB} - \frac{1}{\lambda_2} \log 2$$

Therefore the composite cost of bus will always be cheaper than the cost of the blue or red buses. The dispersion parameter λ_2 allows users to choose options that do not minimise the observed part of the generalised cost function, because of other variables not included in the model.

Consider an O–D pair where the costs of travelling by bus (red or blue) and car are all the same and equal to 50 generalised minutes. Assume also that λ_2 is 0.9. In this case the value of the composite cost C_B is not 50 but 49.23 and the proportion choosing car will depend on the value of λ_1 as shown in the following table.

λ_1	P_C
0.001	0.500
0.005	0.499
0.010	0.498
0.050	0.490
0.100	0.481
0.500	0.405
0.600	0.386
0.700	0.368
0.800	0.351
0.900	0.333

It can be seen that if $\lambda_1 = \lambda_2$ then $P_C = 1/3$, the same result as in a trinomial Logit model; this is expected because the nested structure collapses to the simple Logit model (section 7.4). However, for small values, say $\lambda_1 = 0.1$, the hierarchical or nested structure predicts proportions choosing car (48%) closer to the expected 50%. Is this 2% loss due to travellers with a strong colour preference or those influenced by any change (new paint) in the supply of a service?

6.5.4 Calibration of Binary Logit Models

Consider a model of choice between car and public transport with generalised costs of travel, C_{ij}^k, given by an expression such as (5.2). As discussed in Chapter 5, the weights **a** attached to each element of cost are considered given and calibration only involves finding the 'best-fit' values for the dispersion parameter λ and modal penalty δ (assumed associated with the second mode).

Let us assume that we have C_{ij}^1 and C_{ij}^2 as the 'known' part of the generalised cost for each mode and O–D pair. If we also have information about the proportions choosing each mode for each (i, j) pair, P_{ijk}^*, we can estimate the values of λ and δ using linear regression as follows. The modelled proportions **P** for each (i, j) pair, dropping the (i, j) indices for convenience, are:

$$P_1 = \frac{1}{1 + \exp\{-\lambda(C_2 + \delta - C_1)\}}$$

$$P_2 = 1 - P_1 = \frac{\exp\{-\lambda(C_2 + \delta - C_1)\}}{1 + \exp\{-\lambda(C_2 + \delta - C_1)\}} \tag{6.14}$$

Therefore, taking the ratio of both proportions yields:

$$P_1/(1 - P_1) = 1/\exp\{-\lambda(C_2 + \delta - C_1)\} = \exp\{\lambda(C_2 + \delta - C_1)\}$$

and taking logarithms of both sides and rearranging, we get:

$$\log[P_1/(1 - P_1)] = \lambda(C_2 - C_1) + \lambda\delta \tag{6.15}$$

where we have observed data for **P** and **C**, and therefore the only unknowns are λ and δ (this is well known as the Berkson-Theil transformation). These values could be calibrated by linear regression with the left-hand side of (6.15) acting as the dependent variable and $(C_2 - C_1)$ as the independent one; then λ is the slope of the line and $\lambda\delta$ is the intercept. Note that if we assume the weights **a** in the generalised cost function to be unknown, we can still calibrate the model using (6.15) and multiple linear regression. In this case the calibrated weights would include the dispersion coefficient λ. Other and often better calibration methods are discussed in the next section.

Example 6.4 Data about aggregate mode choice between five zone pairs is presented in the first four columns of Table 6.1; the last two columns of the table give the values needed for the left-hand side of equation (6.15).

This information can be plotted following (6.15) as in Figure 6.4, where it can be deduced that $\lambda \approx 0.72$ and $\delta \approx 3.15$.

Table 6.1 Aggregate binary split data

Zone pair	P_1 (%)	P_2 (%)	C_1	C_2	$\log[P_1/(1 - P_1)]$
1	51.0	49.0	21.0	18.0	0.04
2	57.0	43.0	15.8	13.1	0.29
3	80.0	20.0	15.9	14.7	1.39
4	71.0	29.0	18.2	16.4	0.90
5	63.0	37.0	11.0	8.5	0.53

Figure 6.4   Best-fit line for the data in Table 6.1

6.5.5   Calibration of Hierarchical Modal-split Models

This is usually performed in a heuristic or recursive fashion, starting with the submodal split and proceeding upwards to the primary split. A general discussion on the merits of this approach in comparison with the theoretically better simultaneous estimation is postponed until Chapter 7. Within this general approach there are several possible calibration procedures. It has been shown (see Domencich and McFadden 1975) that maximum likelihood estimates are preferable to least squares estimates, both on theoretical and practical grounds. This is particularly true when working with large data sets. However, when dealing with aggregate data sources it is usually convenient to group the information into suitable classes for analysis (i.e. cost-difference bins). More importantly, the normally available 'factored-up' data are, by definition, raw sampled data which have been manipulated and multiplied by some empirically derived factors. This can cause discrepancies when several data sources with different factors are employed, but the important point at this stage is that the real data set is very small. Hartley and Ortúzar (1980) compared various procedures, and found that maximum likelihood produced not only the most accurate calibration results but also the more efficient ones in terms of computer time.

Let us consider a trinomial problem involving choice between, for example, car, bus and rail. Let us also assume that the last two modes are suspected of being correlated due to their 'public-transport' nature. The heuristic calibration proceeds as follows. First λ_2 is found for the submodal split (bus vs. rail) as explained in Example 6.4 and its value is used to calculate the public-transport composite costs needed for the primary split using an expression such as that for Example 6.3.

For zone pairs where there is a choice of mode (e.g. trips by both modes are possible), trips are classified into cost- difference bins of a certain minimum size. Those trips with no choice of mode are excluded from the calibration. Between cost bins with trips allocated to them, there can be cost bins without any trips; therefore, bins are aggregated into bigger bins until each bin contains some trips. Finally, a weighted representative cost is calculated for each bin.

Then, if N is the total number of bins, n_k the observed number of trips in cost-difference interval k, r_k the observed number of trips by the first mode in the interval, and

$$P_k = 1/[1 + \exp(-Y_k)]$$

the probability of choosing the first mode in interval k, with $Y_k = ax_k + b$, x_k the representative cost of bin k, and a and b parameters to be estimated (i.e. $\lambda = a$ and the modal penalty $\delta = b/a$), the logarithm of the likelihood function (see Chapter 8 for more details) can be written as:

$$L = \text{Constant} + \sum_k [(n_k - r_k)\log(1 - P_k) + r_k \log P_k] \tag{6.16}$$

The maximisation procedure makes use of the first and second derivatives of (6.16) with respect to the parameters, which in this simple case have straightforward analytical expressions:

$$\frac{\partial L}{\partial a} = \sum_k (r_k - n_k P_k)x_k$$

$$\frac{\partial L}{\partial b} = \sum_k (r_k - n_k P_k)$$

$$\frac{\partial^2 L}{\partial a^2} = -\sum_k n_k P_k(1 - P_k)x_k^2$$

$$\frac{\partial^2 L}{\partial b^2} = -\sum_k n_k P_k(1 - P_k)$$

$$\frac{\partial^2 L}{\partial a \partial b} = -\sum_k n_k P_k(1 - P_k)x_k$$

Knowing the values of the derivatives, any search algorithm will find the maximum without difficulty. Maximisation routines require starting values for the parameters, together with an indication of how far they are from the optimum. The efficiency of calibration typically depends strongly upon the accuracy of these estimates. One procedure for generating close first estimates is to find the equi-probability cost (see Bates *et al.* 1978), where the probability of choosing either mode is 0.5.

Before closing this chapter, one must consider an alternative approach offering to consolidate in a single model the features of two or three of the classic sub-models.

6.6 Direct Demand Models

6.6.1 Introduction

The conventional sequential methodology requires the estimation of relatively well-defined sub-models. An alternative approach is to develop directly a model subsuming trip generation, distribution and mode choice. This is, of course very attractive in itself as it avoids some of the pitfalls of the sequential approach. For example, it has been claimed that gravity models suffer from the problem of having to cope with the errors in trip-end totals and those generated by poorly estimated intra-zonal trips. A direct demand model, as it is calibrated simultaneously for the three sub-models, would not suffer from this drawback.

Direct demand models can be of two types: purely direct, which use a single estimated equation to relate travel demand directly to mode, journey and person attributes; and a quasi-direct approach which employs a form of separation between mode split and total (O–D) travel demand. Direct demand models are closely related to general econometric models of demand and have long been inspired by research in that area.

6.6.2   Direct Demand Models

The earliest forms of direct demand models were of the multiplicative kind. The SARC (Kraft 1968) model, for example, estimates demand as a multiplicative function of activity and socioeconomic variables for each zone pair and level-of-service attributes of the modes serving them:

$$T_{ijk} = \phi(P_i P_j)^{\theta_{k1}}(I_i I_j)^{\theta_{k2}} \prod_m [(t_{ij}^m)^{\alpha_{km}^1}(c_{ij}^m)^{\alpha_{km}^2}] \tag{6.17}$$

where P is population, I income, t and c travel time and cost of travel between i and j by mode k, and ϕ, θ and α parameters of the model. This complex expression may be rewritten in simpler form, defining the following composite variables (Manheim 1979):

$$L_{ijm} = (t_{ij}^m)^{\alpha_{km}^1}(c_{ij}^m)^{\alpha_{km}^2}$$
$$Y_{ik} = P_i^{\theta_{k1}} I_i^{\theta_{k2}}$$
$$Z_{jk} = P_j^{\theta_{k1}} I_j^{\theta_{k2}}$$

With these changes of variables (6.17) becomes:

$$T_{ijk} = \phi Y_{ik} Z_{jk} \prod_m L_{ijm}$$

and this transformation eases the interpretation of the model parameters. For example, ϕ is just a scale parameter which depends on the purpose of the trips examined. θ_{k1} and θ_{k2} are elasticities of demand with respect to population and income respectively; we would expect them to be of positive sign. α_{km}^1 and α_{km}^2 are demand elasticities with respect to time and cost of travelling; the direct elasticities (i.e. when k equals m) should be negative and the cross-elasticities of positive sign.

The model is very attractive in principle as it handles generation, distribution and modal split simultaneously, including attributes of competing modes and a wide range of level of service and activity variables. Its main problem is the large number of parameters needed to cash in on these advantages. Alternative forms, containing linear and exponential terms in addition to multiplicative ones, have been suggested by Domencich et al. (1968).

Example 6.5 Consider the following demand function:

$$T_{12} = 10\,000 t_{12}^\alpha c_{12}^\beta q_{12}^\mu$$

where time t is measured in hours, the fare c in dollars and the service frequency q in trips/day. The estimated parameter values are $\alpha = -2$, $\beta = -1$ and $\mu = 0.8$ (note that all the signs are correct according to intuition). The operator wants to increase the fares by 20%; what changes should he make to the level of service in order to keep the same volume of trips if all other things remain equal?

Let us define $L_{12} = L = t^{-2} c^{-1} q^{0.8}$; we know that if L remains constant the total volume T_{12} will not vary (*ceteris paribus*). We also know that the elasticities $E(L, x)$ of the level of service (and hence demand) with respect to each attribute x (time, cost and frequency) are respectively -2, -1 and 0.8.

Now, if only c varies, we have that $L = k/c$, where k is a constant; therefore, a 20% increase in c means a new level of service $L' = k/1.2c$ or $L'/L = 0.833$. That is, a decrease of 16.67% in L. In order to offset this, the operator must introduce changes to the travel time, frequency of service or both. Now, from the definition of elasticity (see Chapter 2) we have that:

$$\Delta L^{(c)} \approx E(L, c)L\Delta c/c \approx -L\Delta c/c$$
$$\Delta L^{(t)} \approx E(L, t)L\Delta t/t \approx -2L\Delta t/t$$
$$\Delta L^{(q)} \approx E(L, q)L\Delta q/q \approx 0.8L\Delta q/q$$

Therefore if we want $\Delta L^{(c)}$ to be equal to $- \Delta L^{(q)}$, we require:

$$-L\Delta c/c \approx -0.8L\Delta q/q$$

that is:

$$\Delta q/q \approx 1.25\Delta c/c \approx 1.25 \times 0.20 = 0.25 \text{ or } 25\%$$

If we are prepared to vary both frequency and travel time, we would require:

$$\Delta L^{(c)} = -(\Delta L^{(q)} + \Delta L^{(t)})$$

that is

$$2\Delta t/t = 0.8\Delta q/q - 0.20$$

a straight line of feasible solutions which is shown in Figure 6.5.

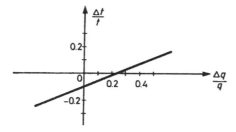

Figure 6.5   Feasible solutions for Example 6.5

Many different variants of direct demand models have been attempted on a heuristic basis. Its use has been mainly in the inter-urban context with very few applications in urban areas. Usually the logarithms of the number of trips and explanatory variables are taken to make the direct demand model log-linear and therefore estimable using generalised linear model software like GLIM.

Direct demand models are certainly an attractive proposition, in particular in areas where the zones are large, for example inter-urban studies. Timberlake (1988) has discussed the use of direct demand models in developing countries and found them better than conventional approaches. For example, in the Karthoum-Wad Medani Corridor in Sudan, the direct demand model gave a better fit than a gravity model because of the unique traffic characteristics exhibited by Karthoum and Port Sudan in comparison with the rest of the country. The direct demand model was able to accommodate these differences better than the gravity model.

6.6.3   An Update on Direct Demand Modelling

Recent versions of the direct demand model brings them closer to the demand component of the classic transport model, albeit still in an interurban context and uses the choice paradigm explained in Chapter 7 to a full extent. Data, coming typically from an intercept origin-destination survey (supplemented by any household data available) are used to estimate a combined frequency-mode-destination choice model where the structure is of Nested Logit form. In particular, a disaggregate version of the combined distribution-modal split model of section 6.5.1 is coupled, through a composite 'accessibility' variable, to the choice of frequency (or trip generation). This has allowed to successfully incorporating a measure of accessibility (i.e. related to the ease or difficulty of travelling from a given zone to the rest of the study area) at the trip generation stage, solving the problem of inelastic demand discussed in section 1.5 (see RAND 2004).

Example 6.6 A direct demand model for the North of Chile macro zone (i.e. 117 zones corresponding to local authorities in a territory of some 1,800 km length housing 67% of the country's population) was estimated using intercept data (Iglesias *et al.*, 2008). The model structure is shown in Figure 6.6; the composite accessibility measure in the trip frequency choice component was the log-sum (see section 7.4) of the destination-mode choice component, and estimation yield correct coefficients (i.e. greater than zero and less than 1.0) for this variable in all user classes. A distinction was made between home-based and non-home based trips as well as between trips of three different class lengths: short trips (less than 150 km), medium (between 150 and 500 km) and long trips (greater than 500 km); the probability of belonging to a length of trip class was modelled as a trinomial Logit with utilities depending on the zone characteristics and its accessibility.

The destination-mode choice component (where eight modes were considered: four types of buses, car, shared taxi, train and airplane) was a Nested Logit model with the following general utility form:

$$V^*_{M_{zj}} = V^{g,a,l}_{M_{zj}} + \log \left(S_{1j} + y_2 S_{2j} + \ldots + y_{117} S_{117j} \right) \tag{6.18}$$

where $V^*_{M_{zj}}$ stands for the utility of travelling by mode M ($M = 1, \ldots, 8$) from origin z to destination j ($j = 1, \ldots, 117$). The first term on the right includes individual characteristics (size of group g, possession of car in the household a, and possession of driving license l) and the level-of-service variables for each destination-mode combination (measured at the zone level). The second term introduces size variables S related to the destination attractiveness as a weighted sum inside a logarithm (Daly 1982a).

Based on (6.18) the representative accessibility for a given zone (Z) and trip length (L) by user class (g, a, l) was defined as shown in (6.19) following Williams (1977). The zones included in the summation over j were exclusively those corresponding to the length of trip considered (i.e. the set J_L) for each given

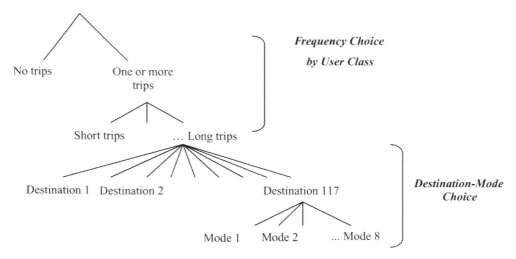

Figure 6.6 Contemporary Direct Demand Model

origin; the structural parameter φ has to be greater than zero and less than or equal to one (the reason for this is discussed in section 7.4):

$$Acc_{Z,L}^{g,a,l} = \frac{1}{\varphi} \cdot \log \left(\sum_{\forall j \in J_L : j \neq Z} e^{\varphi \log \left(\sum_{M=1,8} e^{U_M{}^*_{Zj}} \right)} \right) \qquad (6.19)$$

As the application was made using aggregate data, the accessibility measure for each zone and length of trip was calculated as the weighted sum of representative accessibilities, as follows:

$$Acc_{Z,L} = \sum_{g=1}^{9} \sum_{a=1}^{2} \sum_{l=1}^{2} \left[\left(P_{G=g} \cdot P_{A=a} \cdot P_{Lic=l} \right) \cdot Acc_{Z,L}^{g,a,l} \right] \qquad (6.20)$$

where P_G is the probability of having a given group size (1 to 9 people), which was calculated on the basis of the observed distribution of group sizes by time of year (normal and summer) and type of trip (to work and other); P_A is the probability of having a car in the household and P_{Lic} the probability of having a driving license. The model was applied successfully and details may be consulted in Iglesias *et al.* (2008).

Exercises

6.1 A mode choice survey has been undertaken on a corridor connecting four residential areas A, B, C and D with three employment areas U, V and W. The corridor is served by a good rail link and a reasonable road network. The three employment zones are in a heavily congested area and therefore journeys by rail there are often faster than by car. The information collected during the survey is summarised below:

O–D pair	By car				By rail			Proportion by car
	X_1	X_2	X_3	X_4	X_1	X_2	X_3	
A–U	23	3	120	40	19	10	72	0.82
B–U	20	3	96	40	17	8	64	0.80
C–U	18	3	80	40	14	10	28	0.88
D–U	15	3	68	40	14	12	20	0.95
A–V	26	4	152	60	23	10	104	0.72
B–V	19	4	96	60	18	9	72	0.90
C–V	14	4	60	60	11	9	36	0.76
D–V	12	4	56	60	12	11	28	0.93
A–W	30	5	160	80	25	10	120	0.51
B–W	20	5	100	80	16	8	92	0.56
C–W	15	5	64	80	12	9	36	0.58
D–W	10	5	52	80	8	9	24	0.64

where the costs per trip per passenger are as follows:

X_1 = in-vehicle travel time in minutes (line haul plus feeder mode, if any)
X_2 = excess time (walking plus waiting) in minutes
X_3 = out-of-pocket travel costs (petrol or fares), in pence
X_4 = parking costs associated with a one way trip, in pence.

(a) Calibrate a Logit modal-split model assuming that the value of travel time is 8 pence per minute and that the value of excess time is twice as much.

(b) Estimate the impact on modal split on each O–D pair of an increase in petrol prices which doubles the perceived cost of running a car (X_3).

(c) Estimate the shift in modal split which could be obtained if no fares were charged on the rail system.

6.2 An inter-urban mode choice study is being undertaken for people with a choice between car and rail. The figures below were obtained as a result of a survey on five origin-destination pairs A to E:

| O–D | Elements of cost by each mode | | | | Proportion choosing car |
| | Car | | Rail | | |
	X_1	X_2	X_1	X_2	
A	3.05	9.90	2.50	9.70	0.80
B	4.05	13.10	2.02	14.00	0.51
C	3.25	9.30	2.25	8.60	0.57
D	3.50	11.20	2.75	10.30	0.71
E	2.45	6.10	2.04	4.70	0.63

where X_1 is the travel time (in hours) and X_2 the out-of-pocket cost (in pounds sterling). Assume that the 'value of time' coefficient is 2.00 per hour and calculate the generalised cost of travelling by each mode.

(a) Calibrate a binary Logit modal-split model with these data including the mode specific penalty.

(b) An improved rail service is to be introduced which will reduce travel times by 0.20 of an hour in every journey; by how much could the rail mode increase its fares without losing customers at each O–D pair?

(c) How would you model the introduction of an express coach service between these cities?

6.3 Consider the following trip distribution/modal-split model:

$$V_{ij}^n = A_i O_i B_j D_j \exp(-\beta M_{ij}^n)$$

where

$$M_{ij}^n = -(1/\tau^n) \log \sum_k \exp(-\tau^n C_{ij}^k)$$

and $n = 1$ stands for persons with access to car, $n = 2$, persons without access to car, $k = 1$ stands for car and $k = 2$ for public transport.

If the total number of trips between zones i and j is $V_{ij} = 1000$, compute how many will use car and how many public transport according to the model. The estimated parameter values were found to be: $\tau^1 = 0.10$, $\tau^2 = 0.05$ and $\beta = 0.04$; also, for trips between i and j the modal costs were calculated as: $C_{ij}^1 = 30$ and $C_{ij}^2 = 40$.

6.4 Consider the following modal-split model:

$$P_k = \exp(-\tau C_{ij}^k) \Big/ \sum_m \exp(-\tau C_{ij}^m)$$

with generalised costs given by the following expression:

$$C_{ij}^k = \sum_p \theta_{kp} x_{kp}$$

where θ are parameters weighing the model explanatory variables (time, cost, etc).

(a) Write an expression for the elasticity of P_k with respect to x_{kp}.

(b) Consider now a binary choice situation where the generalised costs have the following concrete expressions:

$$C_{car} = 0.2tt_{car} + 0.1c_{car} + 0.3et_{car}$$

$$C_{bus} = 0.2tt_{bus} + 0.1c_{bus} + 0.3et_{bus} + 0.3$$

where tt is in-vehicle travel time (min), c is travel cost (\$) and et is access time (walking and waiting, min). Assume we know the following average data for the modes:

Mode	Variable		
	tt	*c*	*et*
Car	20	50	0
Bus	30	20	5

Calculate the proportion of people choosing car if $\tau = 0.4$.

6.5 The railway between the towns of A and B spans 800 km through mountainous terrain. The total one-way travel time, t_r, is 20 hrs and currently the fare, c_r, is 600\$/ton. As the service is used at low capacity t_r is a constant, independent of the traffic volume V_r.

There is a lorry service competing with the railway in an approximately parallel route; its average speed is 50 km/hr and it charges a fare of 950\$/ton. There is a project to build a highway in order to replace the present road; it is expected that most of its traffic will continue to be heavy trucks.

The level-of-service function of the new highway has been estimated as:

$$t_t = 7 + 0.08V_t \text{ (hours)}$$

where V_t is the total flow of trucks per hour.

On the other hand the railway has estimated its demand function as follows:

$$(V_r/V_t) = 0.83(t_r/t_t)^{-0.8}(c_r/c_t)^{-1.6}$$

and it is expected that the total volume transported between the two towns, $V_r + V_t$, will remain constant and equal to 200 truck loads/hr in the medium term.

(a) Estimate the current modal split (i.e. volumes transported by rail and lorry).

(b) Estimate modal split if the highway is built.

(c) What would be the modal split if:
   - the railway decreases its fare to 450\$/ton?
   - the lorries were charged a toll of 4\$/ton in order to finance the highway?
   - both changes are simultaneous?

7

Discrete Choice Models

In this chapter we provide a comprehensive introduction to *discrete choice* (i.e. when individuals have to select an option from a finite set of alternatives) modelling methods. We start with some general considerations and move on to explain the theoretical framework, random utility theory, in which these models are cast. This serves us to introduce some basic terminology and to present the individual-modeller 'duality' which is so useful to understanding what the theory postulates. Next we introduce the two most popular discrete choice models: Multinomial and Nested Logit, which taken as a family provides the practitioner with a very powerful modelling tool set. We also discuss other choice models, in particular Mixed Logit which is now recognised as the standard in the field, and also consider the benefits and special problems involved when modelling with panel data and when one wants to incorporate *latent variables*. These are two increasingly important subjects and should shortly become standard practice. Finally, we briefly look at other paradigms which offer an alternative perspective to the classical utility-maximising approach.

The problems of model specification and estimation, both with revealed- and stated-preference data, are considered in sufficient detail for practical analysis in Chapter 8; we provide information about certain issues, such as validation samples, which are seldom found in texts on this subject. The problem of aggregation, from various perspectives, and the important question of model updating and transference (particularly for those interested in a continuous planning approach to transport), are tackled in Chapter 9.

7.1 General Considerations

Aggregate demand transport models, such as those we have discussed in the previous chapters, are either based on observed relations for groups of travellers, or on average relations at the zone level. On the other hand, disaggregate demand models are based on observed choices made by individual travellers or households. It was expected that the use of this framework will enable more realistic models to be developed.

In spite of the pioneering work of researchers such as Warner (1962) or Oi and Shuldiner (1962) who drew attention to apparent serious deficiencies in the conventional methodologies, aggregate models continued to be used, almost unscathed, in the majority of transport projects until the early 1980s. In fact, only then discrete choice models started to be considered as a serious modelling option (see Williams 1981). In general, discrete choice models postulate that:

> *the probability of individuals choosing a given option is a function of their socioeconomic characteristics and the relative attractiveness of the option.*

Modelling Transport, Fourth Edition. Juan de Dios Ortúzar and Luis G. Willumsen.
© 2011 John Wiley & Sons, Ltd. Published 2011 by John Wiley & Sons, Ltd.

To represent the attractiveness of the alternatives the concept of *utility* (which is a convenient theoretical construct defined as what the individual seeks to maximise) is used. Alternatives, *per se*, do not produce utility: this is derived from their characteristics (Lancaster 1966) and those of the individual; for example, the *observable utility* is usually defined as a linear combination of variables, such as:

$$V_{car} = 0.25 - 1.2 \cdot \text{IVT} - 2.5 \cdot \text{ACC} - 0.3 \cdot \text{C/I} + 1.1 \cdot \text{NCAR} \tag{7.1}$$

where each variable represents an attribute of the option or of the traveller. The relative influence of each attribute, in terms of contributing to the overall satisfaction produced by the alternative, is given by its coefficient. For example, a unit change on *access time* (ACC) in (7.1) has approximately twice the impact of a unit change on *in-vehicle travel time* (IVT) and more than seven times the impact of a unit change on the variable *cost/income* (C/I). The variables can also represent characteristics of the individual; for example, we would expect that an individual belonging to a household with a large *number of cars* (NCAR), would be more likely to choose the car option than another belonging to a family with just one vehicle. The *alternative-specific constant*, 0.25 in equation (7.1), is normally interpreted as representing the net influence of all unobserved, or not explicitly included, characteristics of the individual and the option in its utility function. For example, it could include elements such as comfort and convenience which are not easy to measure or observe.

To predict if an alternative will be chosen, according to the model, the value of its utility must be contrasted with those of alternative options and transformed into a probability value between 0 and 1. For this a variety of mathematical transformations exist which are typically characterised for having an S-shaped plot, such as:

Logit $\qquad P_1 = \dfrac{\exp(V_1)}{\exp(V_1) + \exp(V_2)}$

Probit $\qquad P_1 = \displaystyle\int_{-\infty}^{\infty} \int_{-\infty}^{V_1 - V_2 + x_1} \dfrac{\exp\left\{ -\dfrac{1}{2(1-\rho^2)} \left[\left(\dfrac{x_1}{\sigma_1}\right)^2 - \dfrac{2\rho x_1 x_2}{\sigma_1 \sigma_2} + \left(\dfrac{x^2}{\sigma^2}\right)^2 \right] \right\}}{2\pi \sigma_1 \sigma_2 \sqrt{(1-\rho^2)}} \, dx_2 \, dx_1$

where the completely general covariance matrix of the Normal distribution associated with this latter model has the form:

$$\Sigma = \begin{pmatrix} \sigma_1^2 & \rho\sigma_1\sigma_2 \\ \rho\sigma_1\sigma_2 & \sigma_2^2 \end{pmatrix}$$

that is, it allows for correlation (i.e. $\rho \neq 0$) and heteroskedasticity (i.e. different variances and see McCulloch 1985 for a little divertimento) among alternatives.

Discrete choice models cannot be calibrated in general using standard curve-fitting techniques, such as least squares, because their dependent variable P_i is an un-observed probability (between 0 and 1) and the observations are the individual choices (which are either 0 or 1); the only exceptions to this are models for homogeneous groups of individuals, or when the behaviour of every individual is recorded on several occasions, because observed frequencies of choice are also variables between 0 and 1.

Some useful properties of these models were summarised by Spear (1977):

1. Disaggregate demand models (DM) are based on theories of individual behaviour and do not constitute physical analogies of any kind. Therefore, as an attempt is made to explain individual behaviour, an important potential advantage over aggregate models is that it is more likely that DM models are stable (or transferable) in time and space.
2. DM models are estimated using individual data and this has the following implications:
   - DM models may be more efficient than aggregate models in terms of information usage; fewer data points are required as each individual choice is used as an observation. In aggregate modelling one observation is the average of (sometimes) hundreds of individual observations.

- As individual data are used, all the inherent variability in the information can be utilised.
- DM models may be applied, in principle, at any aggregation level; however, although this appears obvious, the aggregation processes are not trivial, as we will discuss in Chapter 9.
- DM models are less likely to suffer from biases due to correlation between aggregate units. A serious problem when using aggregate information is that individual behaviour may be hidden by unidentified characteristics associated with the zones; this is known as *ecological correlation*. The example in Figure 7.1 shows that if a trip generation model was estimated using zonal data, we would obtain that the number of trips decreases with income; however, the opposite would be shown to hold if the data were considered at a household level. This phenomenon, which is of course exaggerated in the figure, might occur for example if the land-use characteristics of zone B are conducive to more trips on foot.

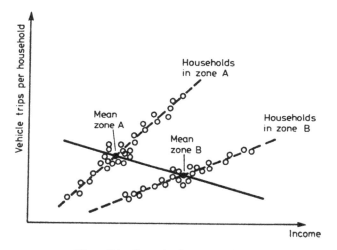

Figure 7.1   Example of ecological fallacy

3. Disaggregate models are probabilistic; furthermore, as they yield the probability of choosing each alternative and do not indicate which one is selected, use must be made of basic probability concepts such as:
   - The expected number of people using a certain travel option equals the sum over each individual of the probabilities of choosing that alternative:

$$N_i = \sum_n P_{in}$$

   - An *independent* set of decisions may be modelled separately considering each one as a conditional choice; then the resulting probabilities can be multiplied to yield joint probabilities for the set, such as in:

$$P(f, d, m, r) = P(f)\, P(d/f)\, P(m/d, f)\, P(r/m, d, f)$$

   with f = frequency; d = destination; m = mode; r = route.
4. The explanatory variables included in the model can have explicitly estimated coefficients. In principle, the utility function allows any number and specification of the explanatory variables, as opposed to the case of the generalised cost function in aggregate models which is generally limited and has several fixed parameters. This has implications such as the following:

- DM models allow a more flexible representation of the policy variables considered relevant for the study.
- The coefficients of the explanatory variables have a direct marginal utility interpretation (i.e. they reflect the relative importance of each attribute).

In the sections that follow and in the next two chapters we will examine in some detail several interesting aspects of discrete choice models, such as their theoretical base, structure, specification, functional form, estimation and aggregation. Notwithstanding, interested readers are advised that there are at least three good books dealing exclusively with this subject (Ben-Akiva and Lerman 1985; Hensher *et al.* 2005; Train 2009).

7.2 Theoretical Framework

The most common theoretical framework or paradigm for generating discrete-choice models is random utility theory (Domencich and McFadden 1975; Williams 1977), which basically postulates that:

1. Individuals belong to a given homogeneous population Q, act rationally and possess perfect information, i.e. they always select that option which maximises their net personal utility (the species has even been identified as '*Homo economicus*') subject to legal, social, physical and/or budgetary (both in time and money terms) constraints.
2. There is a certain set $\mathbf{A} = \{A_1, \ldots, A_j, \ldots, A_N\}$ of available alternatives and a set $\mathbf{X}$ of vectors of measured attributes of the individuals and their alternatives. A given individual q is endowed with a particular set of attributes $\mathbf{x} \in \mathbf{X}$ and in general will face a choice set $\mathbf{A}(q) \in \mathbf{A}$.

   In what follows we will assume that the individual's choice set is predetermined; this implies that the effect of the constraints has already been taken care of and does not affect the process of selection among the available alternatives. Choice-set determination will be considered, together with other important practical issues, in Chapter 8.
3. Each option $A_j \in \mathbf{A}$ has associated a net utility U_{jq} for individual q. The modeller, who is an observer of the system, does not possess complete information about all the elements considered by the individual making a choice; therefore, the modeller assumes that U_{jq} can be represented by two components:
   - a measurable, systematic or representative part V_{jq} which is a function of the measured attributes $\mathbf{x}$; and
   - a random part ε_{jq} which reflects the idiosyncrasies and particular tastes of each individual, together with any measurement or observational errors made by the modeller.
   Thus, the modeller postulates that:

$$U_{jq} = V_{jq} + \varepsilon_{jq} \tag{7.2}$$

which allows two apparent 'irrationalities' to be explained: that two individuals with the same attributes and facing the same choice set may select different options, and that some individuals may not always select what appears to be the best alternative (from the point of view of the attributes considered by the modeller).

For the decomposition (7.2) to be correct we need certain homogeneity in the population under study. In principle we require that *all individuals share the same set of alternatives and face the same constraints* (see Williams and Ortúzar 1982a), and to achieve this we may need to segment the market.

Although we have termed $\mathbf{V}$ *representative* it carries the subscript q because it is a function of the attributes $\mathbf{x}$ and this may vary from individual to individual. On the other hand, without loss of generality it can be assumed that the residuals ε are random variables with mean 0 and a certain probability distribution to be specified. A popular and simple expression for **V** is:

$$V_{jq} = \sum_k \theta_{kj} x_{jkq} \qquad (7.3)$$

where the parameters θ are assumed to be constant for all individuals in the homogeneous set (fixed-coefficients model) but may vary across alternatives. Other possible forms, together with a discussion on how each variable should enter in the utility function, will be presented in Chapter 8.

It is important to emphasise the existence of two points of view in the formulation of the above problem: firstly, that of the individual who calmly weighs all the elements of interest (with no randomness) and selects the most convenient option; secondly, that of the modeller who by observing only some of the above elements needs the residuals ε to explain what otherwise would amount to non-rational behaviour.

4. The individual q selects the maximum-utility alternative, that is, the individual chooses A_j if and only if:

$$U_{jq} \geq U_{iq}, \forall A_i \in \mathbf{A}(q) \qquad (7.4)$$

that is

$$V_{jq} - V_{iq} \geq \varepsilon_{iq} - \varepsilon_{jq} \qquad (7.5)$$

As the analyst ignores the value of $(\varepsilon_{iq} - \varepsilon_{jq})$ it is not possible to determine with certitude if (7.5) holds. Thus the probability of choosing A_j is given by:

$$P_{jq} = \text{Prob}\{\varepsilon_{iq} \leq \varepsilon_{jq} + (V_{jq} - V_{iq}), \forall A_i \in \mathbf{A}(q)\} \qquad (7.6)$$

and as the joint distribution of the residuals ε is not known, it is not possible at this stage to derive an analytical expression for the model. What we do know, however, is that the residuals are random variables with a certain distribution which we can denote by $f(\varepsilon) = f(\varepsilon_1, \ldots, \varepsilon_N)$. Let us note in passing that the distribution of $\mathbf{U}, f(\mathbf{U})$, is the same but with different mean (i.e. **V** rather than 0).

Therefore we can write (7.6) more concisely as:

$$P_{jq} = \int_{R_N} f(\varepsilon)\, d\varepsilon \qquad (7.7)$$

where

$$R_N = \begin{cases} \varepsilon_{iq} \leq \varepsilon_{jq} + (V_{jq} - V_{iq}), & \forall A_i \in \mathbf{A}(q) \\ V_{jq} + \varepsilon_{jq} \geq 0 \end{cases}$$

and different model forms may be generated depending on the distribution of the residuals ε.

An important class of random utility models is that generated by utility functions with independent and identically distributed (IID) residuals. In this case $f(\varepsilon)$ can be decomposed into:

$$f(\varepsilon_1, \ldots, \varepsilon_N) = \prod_n g(\varepsilon_n)$$

where $g(\varepsilon_n)$ is the utility distribution associated with option A_n, and the general expression (7.7) reduces to:

$$P_j = \int_{-\infty}^{\infty} g(\varepsilon_j)\, d(\varepsilon_j) \prod_{i \neq j} \int_{-\infty}^{V_j - V_i + \varepsilon_j} g(\varepsilon_i)\, d\varepsilon_i \qquad (7.8a)$$

where we have extended the range of both integrals to $-\infty$ (a slight inconsistency) in order to solve them.

A two-dimensional geometric interpretation of this model, together with extensions to the more general case of correlation and unequal variances, are presented and discussed by Ortúzar and Williams (1982). Equation (7.8a) can also be expressed as:

$$P_j = \int_{-\infty}^{\infty} g(\varepsilon_j)\mathrm{d}\varepsilon_j \prod_{i \neq j} G(\varepsilon_j + V_j - V_i) \tag{7.8b}$$

with

$$G(x) = \int_{-\infty}^{X} g(x)\,\mathrm{d}x$$

and it is interesting to mention that a large amount of effort has been spent in just trying to find out appropriate forms for g which allow (7.8b) to be solved in closed form.

Note that the IID residuals requisite means that the alternatives should be, in fact, independent. Mixed-mode options, for example car-rail combinations, will usually violate this condition.

7.3 The Multinomial Logit Model (MNL)

This is the simplest and most popular practical discrete choice model (Domencich and McFadden 1975). It can be generated assuming that the random residuals in (7.7) are distributed IID Gumbel (also called Weibull or, more generally, Extreme Value Type I, EV1, as we saw in section 2.5.4.2). With this assumption the choice probabilities are:

$$P_{iq} = \frac{\exp(\beta\, V_{iq})}{\sum_{A_j \varepsilon A(q)} \exp(\beta\, V_{jq})} \tag{7.9}$$

where the utility functions usually have the linear in the parameters form (7.3) and the parameter β (which is normalised to one in practice as it cannot be estimated separately from the θ) is related to the common standard deviation of the EV1 variate by:

$$\beta = \pi/\sigma\sqrt{6} \tag{7.10}$$

In Chapter 9 we will use (7.10) to discuss the problem of bias in forecasts when use is made of data at different levels of aggregation. The fact that β cannot be estimated separately from the parameters θ in V_{iq} is known as *theoretical identification;* all discrete choice models have identification problems, which require to set certain parameters to a given value in order to estimate the model uniquely (see Walker 2002). We will come back to this important issue several times in this chapter.

7.3.1 Specification Searches

To decide which variables $x_k \in \mathbf{x}$ enter the utility function and whether they are of generic type or specific to a particular alternative, a search process is normally employed starting with a theoretically appealing specification (Ortúzar 1982). Then variations are tested at each step to check whether they add explanatory power to the model; we will examine methods for doing this in Chapter 8.

If for all individuals q that have available a given alternative A_j we define one of the values of $\mathbf{x}$ equal to one, the coefficient θ_k corresponding to that variable is interpreted as an alternative specific constant (ASC). Although we may specify an ASC for every option, it is not possible to estimate their N parameters individually due to the way the model works (as shown in Example 7.1). For this reason one alternative is taken as reference (fixing to 0 the value of its parameter without loss of generality) and the remaining $(N - 1)$ values, obtained in the estimation process, are interpreted as relative to that of the reference. This is another theoretical identification issue (Walker 2002; Cherchi and Ortúzar 2008b) associated with the MNL. The rest of the variables **x** may be of one of two kinds:

- generic, if they appear in the utility function of every alternative and their coefficients can be assumed identical i.e. θ_{jk} may be replaced by θ_k;
- specific, if the assumption of equal coefficients θ_k is not sustainable, a typical example occurring when the kth variable only appears in V_j.

It must be noted that the most general case considers specific variables only; the generic ones impose an equality of coefficients condition and this may be statistically tested as we will discuss in Chapter 8.

Example 7.1 Consider the following binary Logit model:

$$P_1 = \exp(V_1)/[\exp(V_1) + \exp(V_2)] = 1/[1 + \exp(V_2 - V_1)]$$

where the observable utilities are postulated as linear functions of two generic variables x_1 and x_2, and two constants (with coefficients θ_3 and θ_4) as follows:

$$V_1 = \theta_1 x_{11} + \theta_2 x_{12} + \theta_3$$
$$V_2 = \theta_1 x_{21} + \theta_2 x_{22} + \theta_4$$

As can be seen from the model expression, the relevant factor is the difference between both utilities:

$$V_2 - V_1 = \theta_1(x_{21} - x_{11}) + \theta_2(x_{22} - x_{12}) + (\theta_4 - \theta_3)$$

and this allows us to deduce the following:

- It is not possible to estimate both θ_3 and θ_4, only their difference; for this reason there is no loss of generality if one is taken as 0 and the other estimated relative to it (this of course applies to any number of alternatives).
- If either x_{1j} or x_{2j} have the same value for both options (as in the case of variables representing individual attributes, such as income, age, sex or number of cars in the household), a generic coefficient cannot be estimated as it would always multiply a zero value. This also applies to level-of-service variables which happen to share a common value for two or more options (for example, public-transport fares in a regulated market). In either case they can only appear in some (but not all) options, or need to enter as specific variables (i.e. with different coefficients for each but one alternative).

The problem posed by individual attributes is further compounded by the fact that it is not always easy or clear to decide in which alternative utility(ies) the variable should appear. Consider the case of a variable such as SEX (i.e. 0 for males, 1 for females) in a mode choice study; if we believe, for example, that males have first call on access to the car for commuting purposes, we would not enter the variable in the utilities of both car driver and car passenger, say. However, we may have no insights on whether to enter it or not in the utilities of other modes such as, for example, bus or metro. The problem is that entering the variable in different ways usually yields different estimation results and choosing the optimum may become a hard combinatorial problem, even for a small number of options and attributes. If we lack insight and there are no theoretical grounds for preferring one form over another, the only way out may be trial and error.

7.3.2 Universal Choice Set Specification

When individuals have different choice sets, it is useful to rewrite the model based on the universal choice set formulation introducing availability variables into the utility function (see Bierlaire *et al.* 2009 for a recent application). Let A_{iq} be 1 if individual q has alternative A_i available and 0 otherwise. For example, if walking is considered not available for distances longer than 3 km, we would have:

$$A_{iq} = \begin{cases} 1 & if \ d_q < 3 \\ 0 & if \ d_q \geq 3 \end{cases}$$

where d_q is the distance to be travelled by individual q. It is possible to write any choice model based on this idea of a universal choice set:

$$P_{iq}\{\mathbf{A}(q)\} = \text{Prob}\left\{U_{iq} \geq U_{jq}, \quad \forall A_j \in \mathbf{A}(q)\right\}$$
$$= \text{Prob}\left\{U_{iq} + \log A_{iq} \geq U_{jq} + \log A_{jq}, \quad \forall A_j \in \mathbf{A}\right\}$$

Thus, when one of the A_{iq} is equal to 1, the additional term does not play any role. But if $A_{iq} = 0$, the inequality is never verified and the probability of choosing the alternative is 0, which makes sense as it is not available. Finally, when $A_{jq} = 0$, for $j \neq i$ the right hand side of the above equation is trivially lower than anything else.

Using this formulation, the Multinomial Logit expression becomes:

$$P_{iq}\{\mathbf{A}(q)\} = \frac{\exp(V_{iq})}{\displaystyle\sum_{A_j \in \mathbf{A}(q)} \exp(V_{jq})} = \frac{\exp(V_{iq} + \log A_{iq})}{\displaystyle\sum_{A_j \in \mathbf{A}} \exp(V_{jq} + \log A_{jq})} = \frac{A_{iq}\exp(V_{iq})}{\displaystyle\sum_{A_j \in \mathbf{A}} A_{jq}\exp(V_{jq})}$$

and this helps to generalise some properties that were originally applicable only to cases where all individuals had the same choice set, as we will see below.

7.3.3   Some Properties of the MNL

The model satisfies the axiom of *independence of irrelevant alternatives* (IIA) which can be stated as:

> *Where any two alternatives have a non-zero probability of being chosen, the ratio of one probability over the other is unaffected by the presence or absence of any additional alternative in the choice set* (Luce and Suppes 1965).

As can be seen, in the MNL case the ratio

$$\frac{P_j}{P_i} = \exp\{\beta(V_j - V_i)\}$$

is indeed a constant independent of the rest of the options. Initially this was considered an advantage of the model, as it allows us to treat quite neatly the *new alternative* problem (i.e. being able to forecast the share of an alternative not present at the calibration stage, if its attributes are known); however, nowadays this property is perceived as a potentially serious disadvantage which makes the model fail in the presence of correlated alternatives (recall the red bus–blue bus problem of Chapter 6). We will come back to this in section 7.4.

If there are too many alternatives, such as in the case of destination choice, it can be shown (McFadden 1978) that unbiased parameters are obtained if the model is estimated with a random sample of the available choice set for each individual (for example, seven destination options per individual). Models without this property may require, even if their estimation process is not complex, a large amount of computing time for more than say 50 options. Unfortunately such a figure is not uncommon in a destination-choice context, if one thinks in zoning systems of normal size, even if the combinatorial problem of forming destination/mode choice options is bypassed.

If the model is estimated with information from a sub-area, or with data from a biased sample, it can be shown (Cosslett 1981) that if the model has a complete set of mode-specific constants, an unbiased model may be obtained just by correcting the constants according to the following expression:

$$K'_i = K_i - \log(q_i/Q_i) \tag{7.11}$$

where q_i is the market share of alternative A_i in the sample and Q_i its market share in the population. All constants must be corrected, including the reference one that is made equal to 0 during estimation.

It is possible to derive fairly simple equations for the direct and cross-elasticities of the model. For example, the direct point elasticity, that is the percentage change in the probability of choosing A_i with respect to a marginal change in a given attribute X_{ikq}, is simply given by:

$$E_{P_{iq}, X_{ikq}} = \theta_{ik} X_{ikq}(1 - P_{iq}) \tag{7.12}$$

while the cross-point elasticity is also simply given by:

$$E_{P_{iq}, X_{jkq}} = -\theta_{jk} X_{jkq} P_{jq} \tag{7.13}$$

that is, the percentage change in the probability of choosing A_i with respect to a marginal change in the value of the kth attribute of alternative A_j, for individual q. Note that as this value is independent from alternative A_i, the cross-elasticities of any option A_i with respect to the attributes X_{jkq} of alternative A_j are equal. This seemingly peculiar result is also due to the IIA property, or more precisely, to the need for IID utility functions in the model generation.

7.4   The Nested Logit Model (NL)

7.4.1   Correlation and Model Structure

In the last section we discussed the MNL model which has a very simple covariance matrix. For example, in the trinomial case it is of the form:

$$\sum = \sigma^2 \begin{pmatrix} 1 & 0 & 0 \\ 0 & 1 & 0 \\ 0 & 0 & 1 \end{pmatrix}$$

This simplicity may give rise to problems in any of the following cases:

- when alternatives are not independent (i.e. there are groups of alternatives more similar than others, such as public-transport modes vs. the private car);
- when the variances of the error terms ε are not equal, i.e. when there is heteroskedasticity (e.g. *between observations*, if some individuals posses a GPS device, and are thus able to measure their times more precisely than others; or *between alternatives*, when certain options have more precise attributes, say waiting times of Metro and bus, see Munizaga *et al.* 2000);
- when there are taste variations among individuals (i.e. if the perception of costs varies with income but we have not measured this variable) in which case we require random coefficient models rather than fixed coefficient models as the MNL;
- when there are multiple responses per individual, as in the case of panel data or stated preference observations; these introduce problems associated with dependency between observations violating one of the assumptions underpinning the MNL model (see Chapter 8).

In these four senses more flexible models such as the Multinomial Probit (MNP) model, which can be derived from a multivariate Normal distribution (rather than IID EV1) or the Mixed Logit (ML) model (we will discuss both in sections 7.5 and 7.6), are completely general because they are endowed with an arbitrary covariance matrix. However, as we will see below, the first is not easy to solve except for cases with up to three alternatives (see Daganzo 1979) and the second requires a more involved estimation method but it is recognized as the current standard in the field.

Notwithstanding, there are certain situations where even if these more powerful models were available, their full generality could be an unnecessary luxury because specific forms for the utility functions suggest themselves. A good example are cases of bi-dimensional choices, such as the combination of destination (D) and mode (M) choice, where alternatives are correlated but taste variations or heteroskedasticity need not be a problem. In these cases the options at each dimension can be denoted as $(D_1, \ldots, D_D)$ and $(M_1, \ldots, M_M)$ with their combination yielding the choice set **A**, whose general element $D_d M_m$ may be a specific destination-mode option to carry out a certain activity.

In this type of context it is interesting to consider functions of the following type (Williams and Ortúzar 1982a):

$$U(d, m) = U_d + U_{dm}$$

where, for example, U_d could correspond to that portion of utility specifically associated with the destination and U_{dm} to the disutility associated with the cost of travelling. If we write the expression above following our previous notation we get:

$$U(d, m) = V(d, m) + \varepsilon(d, m)$$

where

$$V(d, m) = V_d + V_{dm}$$

and

$$\varepsilon(d, m) = \varepsilon_d + \varepsilon_{dm}$$

It can be shown that if the residuals ε are separately IID, under certain conditions the Hierarchical or Nested Logit (NL) model (Williams 1977; Daly and Zachary 1978) is formed:

$$P(d, m) = \frac{\exp\{\beta(V_d + V_d^*)\} \exp(\lambda V_{dm})}{\sum_{d'} \exp\{\beta(V_{d'} + V_{d'}^*)\} \sum_{m'} \exp(\lambda V_{dm'})}$$

with

$$V_d^* = (1/\lambda) \log \sum_{m'} \exp(\lambda V_{dm'})$$

This is precisely the model form used in the destination-mode choice component of contemporary direct demand models as discussed in section 6.6.3. Furthermore, it can easily be shown that if $\beta = \lambda$ (which occurs when $\varepsilon_d = 0$) the NL collapses, as special case, to the single parameter MNL. To understand why this is so, let us first write in full the utility expressions for the first destination in a simple binary mode case:

$$U(1, 1) = V_1 + V_{11} + \varepsilon_1 + \varepsilon_{11}$$
$$U(1, 2) = V_1 + V_{12} + \varepsilon_1 + \varepsilon_{12}$$

As can be seen, the source of correlation is the residual ε_1 which can be found in both $U(1, 1)$ and $U(1, 2)$; therefore when ε_d becomes 0, there is no correlation left and the model is indistinguishable from the MNL.

Finally, it can also be shown that for the model to be internally consistent we require that the following condition holds (Williams 1977):

$$\beta \leq \lambda$$

Models that fail to satisfy this requirement have been shown to produce elasticities of the wrong size and/or sign (Williams and Senior 1977).

7.4.2   Fundamentals of Nested Logit Modelling

In his historical review of the NL model, Ortúzar (2001) mentions several authors whose work predates the model's actual theoretical formulation. Wilson (1970; 1974), Manheim (1973) and Ben-Akiva (1974) all used intuitive versions that – although based on concepts such as marginal probabilities and utility maximisation – did not have a rigorous construction of the functional forms and a clear interpretation of all the model parameters. Domencich and McFadden (1975) generated structured models of Nested Logit form but had an incorrect definition of 'composite utilities'.

Williams (1977) was the first to make an exhaustive analysis of the NL properties, especially composite utilities (or inclusive values), showing that all previous versions had important inconsistencies with micro-economic concepts. He also reformulated the NL and introduced structural conditions associated with its inclusive value parameters, which are necessary for the NL's compatibility with utility maximising theory. With these, he formally derived the NL model as a *descriptive* behavioural model completely coherent with basic micro-economic concepts. Other authors, whose seminal work completed the fundamental theoretical development of the NL, are Daly and Zachary (1978), who worked simultaneously and totally independent from Williams, and McFadden (1978; 1981) who generalised the work of both Williams, and Daly and Zachary. Unfortunately, the latter has given rise to some confusion in terms of estimation and interpretation of results which we discuss below. In what follows we will draw heavily on the definitive study of Carrasco and Ortúzar (2002).

7.4.2.1   The Model of Williams and of Daly-Zachary

As mentioned above, Williams (1977) initially worked with a two-level model in the context of two-dimensional situations, such as destination-mode choice, defining utility functions of the following form:

$$U(i, j) = U_j + U_{i/j} \tag{7.14}$$

where i denotes alternatives at a lower level nest and j the alternative at the upper level that represents that lower level nest. In terms of the representative utility and stochastic terms, (7.14) becomes:

$$U(i, j) = V(i, j) + \varepsilon(i, j)$$

where

$$V(i, j) = V_j + V_{i/j} \quad \text{and} \quad \varepsilon(i, j) = \varepsilon_j + \varepsilon_{i/j}$$

Williams' definition of the stochastic errors may be synthesised as follows:

- The errors ε_j and $\varepsilon_{i/j}$ are independent for all (i, j).
- The errors $\varepsilon_{i/j}$ are identically and independently distributed (IID) EV1 with scale parameter λ.
- The errors ε_j are distributed with variance $\sigma_j{}^2$ and such that the sum of U_j and the maximum of $U_{i/j}$ is distributed EV1 with scale parameter β. It is interesting to mention that such a distribution may not exist (see Carrasco, 2001); also, this derivation is sufficient but many other formulations could lead to the same model.

These assumptions have as a consequence the following relation for the error variances:

$$Var(\varepsilon(i, j)) = Var\left(\varepsilon_j\right) + Var\left(\varepsilon_{i/j}\right) \tag{7.15}$$

which in our case, using (7.10), may be expressed as

$$\frac{\pi^2}{6\beta^2} = \sigma_j^2 + \frac{\pi^2}{6\lambda^2}$$

leading to

$$\frac{\beta}{\lambda} = \left(1 + \frac{6\sigma_j^2\lambda^2}{\pi^2}\right)^{-\frac{1}{2}} \tag{7.16}$$

The above implies the *structural condition* which we had anticipated:

$$\beta \leq \lambda \tag{7.17}$$

Now, if we define the *structural parameter* $\phi = \frac{\beta}{\lambda}$, condition (7.17) becomes:

$$\phi \leq 1 \tag{7.18}$$

and when $\beta = \lambda$ ($\phi = 1$), the NL collapses to the MNL (7.9), as the reader can easily check; but if $\beta > \lambda$ ($\phi > 1$), the hierarchical structure postulated is incompatible with the utility theoretic basis of this formulation.

The above construction may be generalised in two directions:

- Allowing for a different scale parameter λ_j, associated with each nest j, as proposed by Daly and Zachary (1978).
- Allowing for an increase in the number of levels in series and parallel (Williams 1977; Daly and Zachary 1978; Sobel 1980).

A very popular NL specification in practice is one with just two levels of nesting and different scale parameters λ_j in each nest (Figure 7.2), whose functional form is given by:

$$P(i, j) = \frac{\exp\left(\lambda_j V_{i/j}\right)}{\sum\limits_{i' \in j} \exp\left(\lambda_j V_{i'/j}\right)} \cdot \frac{\exp \beta \left\{ \frac{1}{\lambda_j} \log \left(\sum\limits_{i \in j} \exp\left(\lambda_j V_{i/j}\right) \right) \right\}}{\sum\limits_{j'=1}^{m} \exp \beta \left\{ \frac{1}{\lambda_{j'}} \log \left(\sum\limits_{i \in j'} \exp\left(\lambda_{j'} V_{i/j'}\right) \right) \right\}} \tag{7.19}$$

In this case, the structural conditions of the model become:

$$\beta \leq \lambda_j \quad \textit{for all } j \qquad \Leftrightarrow \qquad \phi_j = \frac{\beta}{\lambda_j} \leq 1 \quad \textit{for all } j \tag{7.20}$$

Figure 7.2   A general Nested Logit structure with two levels

The above model (7.19) and conditions (7.20) allow a range of complex choice processes to be modelled, such as location-mode and multi-mode contexts, which allow for different degrees of substitution (responses to policies) within and between the nests.

7.4.2.2   The Formulation of McFadden: The GEV Family

McFadden (1981) generated the NL model as one particular case of the Generalised Extreme Value (GEV) discrete choice model family (which is considered further in section 7.7). The members of this family come from a non-negative function $G(Y_1, Y_2, \ldots, Y_M)$, with $Y_1, Y_2, \ldots, Y_M \geq 0$, which is homogeneous of degree $\mu > 0$, approaches to infinite as any Y_i does and has m cross-partial derivatives which are non-negative for odd m and non-positive for even m. As an aside, note that McFadden originally considered $\mu = 1$, but this was later generalized by Ben-Akiva and Lerman (1985).

If we consider the utility function $U_i = V_i + \varepsilon_i$ for M elemental alternatives, the choice probability may be written as:

$$P_i = \int_{\varepsilon=-\infty}^{\varepsilon=\infty} F_i\left(V_i - V_1 + \varepsilon, \ldots, V_i - V_M + \varepsilon\right) d\varepsilon$$

where F is the cumulative distribution function of the errors $(\varepsilon_1, \ldots, \varepsilon_M)$ and $F_i = \frac{\partial F}{\partial \varepsilon_i}$. Thus, defining the extreme value multivariate distribution:

$$F(\varepsilon_1, \ldots, \varepsilon_M) = \exp\left\{-G\left(e^{-\varepsilon_1}, \ldots, e^{-\varepsilon_M}\right)\right\}$$

P_i, the probability of choosing alternative A_i, is given by:

$$P_i = \frac{e^{V_i} G_i(e^{V_1}, e^{V_2}, \ldots, e^{V_M})}{\mu G(e^{V_1}, e^{V_2}, \ldots, e^{V_M})} \tag{7.21}$$

where G_i is the first derivative of G with respect to $Y_i = \exp(V_i)$. Using the above, McFadden showed that the NL probability function is obtained from the following G function:

$$G\left(e^{V_1}, e^{V_2}, \ldots, e^{V_M}\right) = \left(\sum_{j=1}^{J}\left(\sum_{i \in j} e^{V_{(i,j)}}\right)^{\frac{1}{\mu_j}}\right)^{\mu_j} \tag{7.22}$$

leading to

$$P(i, j) = \frac{\exp\left(\frac{V_{(i,j)}}{\mu_j}\right)}{\sum_{i \in j} \exp\left(\frac{V_{(i,j)}}{\mu_j}\right)} \cdot \frac{\exp \mu_j \ln\left(\sum_{i \in j} \exp\left(\frac{V_{(i,j)}}{\mu_j}\right)\right)}{\sum_{j'=1}^{J} \exp \mu_{j'} \ln\left(\sum_{i \in j'} \exp\left(\frac{V_{(i,j')}}{\mu_{j'}}\right)\right)} \tag{7.23}$$

This probability density function is well-defined (i.e. positive) on the real numbers if the parameter μ_j of the G function (7.22) satisfies the following restriction (McFadden 1981):

$$\mu_j \leq 1 \qquad \forall j$$

Note that this is equivalent to Williams' structural condition (7.20). Furthermore, functional form (7.19) of Williams is equivalent to McFadden's functional form (7.23) if the following relations are established:

$$\beta = 1$$
$$\frac{1}{\lambda_j} = \phi_j = \mu_j \quad \forall j$$

But although the conditions are numerically equivalent, they have different meanings. In Williams' theory, the restriction stems from the definition of the error as the sum of two *separable* terms, one of them EV1 distributed with lower variance than that of the total error, allowing the NL function to satisfy the basic integrability conditions required to be consistent with utility maximisation.

On the other hand, McFadden's condition is directly related to the restriction that the GEV-based probability density function has to be compatible with random utility theory; thus, in his context the definition of the error as the sum of two independent components, and the condition this imposes on their variances, *are not necessary*. This aspect was mentioned – in an indirect way – by Daganzo and Kusnic (1993), who stated that although the conditional probability may be derived with a Logit form, it is not necessary that the conditional error distribution should be EV1.

7.4.3   The NL in Practice

As a modelling tool the NL may be usefully presented in the following fashion (Ortúzar 1980b; Sobel 1980):

1. Its structure is characterised by grouping all subsets of correlated (or more similar) options in hierarchies or nests. Each nest, in turn, is represented by a *composite alternative* which competes with the others available to the individual (the example in Figure 7.2 considers two levels of nesting and four nests).
2. The introduction of information from lower nests in the next higher nests is done by means of the utilities of the composite alternatives; these are, by definition, equal to the expected maximum utility (EMU) of the options belonging to the nest and have the following expression:

$$\text{EMU}_j = \log \sum_k \exp(V_k/\phi_j)$$

Therefore the composite utility of nest j is:

$$V_j = \phi_j \cdot \text{EMU}_j$$

where ϕ_j are *structural* parameters to be estimated.

3. The probability that individual q selects option A_i in nest j may be computed as the product of the marginal probability of choosing the composite alternative N_J (in the higher nest) and the conditional probability of choosing option A_i in the lower nest, given that q selected the composite alternative.
4. If there is only one nest, the internal diagnosis condition (7.17) is expressed in this new notation as:

$$0 < \phi \le 1 \tag{7.24}$$

and let us briefly see why it needs to hold. If $\phi < 0$, an increase in the utility of an alternative in the nest, which should increase the value of EMU, would actually diminish the probability of selecting the nest; if $\phi = 0$, such an increase would not affect the nest's probability of being selected, as EMU would not affect the choice between car and public transport.

On the other hand, if $\phi > 1$ an increase in the utility of an alternative in the nest would tend to increase not only its selection probability but also those of the rest of the options in the nest (but note that the real reason is that β cannot be greater than λ as shown in expression 7.16). Finally, if $\phi = 1$ which is the equivalent to $\beta = \lambda$, the NL model becomes mathematically equivalent to the MNL; in such cases (i.e. when $\phi \approx 1$) it is more efficient to recalibrate the model as an MNL, as the latter has fewer parameters.

But NL models are not limited to just two hierarchical levels; in cases with more nesting levels, such as in Figure 7.3, we need at each branch of the structure:

$$0 < \phi_1 \leq \phi_2 \leq \ldots \leq \phi_s \leq 1 \tag{7.25}$$

where ϕ_1 correspond to the most inclusive parameter and ϕ_s to that of the highest level. Note that there are no relations to be expected between the structural parameters pertaining to different branches.

Figure 7.3   Nested Logit model with several nests

Limitations of the NL
- In common with the MNL it is not a random coefficients model, so it cannot cope with taste variations among individuals without explicit market segmentation. Neither can it treat heteroskedastic options, as the error variances of each alternative are assumed to be the same.
- It can only handle as many interdependencies among options as nests have been specified in the structure; furthermore, alternatives in one nest cannot be correlated with alternatives in another nest (this cross-correlation effect, which might be important to test in a mixed-mode modal choice context, for example, can be handled by more general forms as we will see below).
- The search for the best NL structure may imply the tentative examination of many nesting patterns, as the number of possible structures increases geometrically with the number of options (Sobel 1980). Although *a priori* notions help greatly in this sense (i.e. only theoretically expected nesting patterns should be tried), the modelling exercise might take much longer than with the simple MNL.

7.4.4   Controversies about some Properties of the NL Model

This section discusses some properties of the NL model that were the subject of some controversy in recent literature, in order to illuminate aspects which were confusing and allow a correct use of the model in practice.

7.4.4.1   Specifications which Address the Non Identifiability Problem

As we have mentioned, all discrete choice models are subject to identifiability problems. The NL model (7.19) is not estimable either because it also has an additional degree of freedom; to estimate (7.19), it is necessary to 'fix' one of the scale factors.

Consider, without loss of generality, the two level model (7.19) where β is the parameter at the upper level and λ_j ($j = 1, \ldots, J$) are the corresponding J parameters of the nests. In this case, J structural coefficients may be defined as in (7.20) and the theoretical identifiability problem means

(continued)

that although the ϕ_j parameters can all be estimated, *one* of the $J+1$ scale factors (i.e. the J parameters λ_j plus β) associated with the variance cannot be determined (Carrasco and Ortúzar 2002).

Upper and lower normalisations
According to the above definitions two normalisations can be distinguished: the *upper* one, where β is chosen as the non identifiable parameter and the *lower* one, where *one* of the λ_j parameters (for example that for $j = r, 1 \le r \le J$) is selected.

Consider a typical (linear) representative utility function:

$$\hat{V}_{i/j} = \sum_{k=1}^{K} \hat{\theta}_k x_{(i,j)}^k \tag{7.26}$$

where $x_{(i,j)}^k$ are attributes ($k = 1, \ldots, K$) and $\hat{\theta}_k$ their corresponding *estimated* coefficients. Estimated and population coefficients (if they exist) on the *upper normalisation* are related by:

$$\hat{\theta}_k = \beta\theta_k \quad \forall k \tag{7.27}$$

and in the case of the *lower normalisation* the relation is:

$$\hat{\theta}_k = \lambda_r\theta_k \quad \forall k \tag{7.28}$$

Equations (7.27) and (7.28) allow us to see more clearly that normalisation does not strictly mean to 'define' the parameter as unity (indirectly assuming the value of the variance), but that the 'normalised' parameter multiplies the coefficients of the utility function, 'mixing' with them rather than having a value defined *a priori*.

Now, the specification of the model using the upper normalisation is given by:

$$P(i,j) = \frac{\exp\left(\frac{1}{\phi_j}\hat{V}_{i/j}\right)}{\sum_{i'\in j}\exp\left(\frac{1}{\phi_{j'}}\hat{V}_{i'/j}\right)} \cdot \frac{\exp\phi_j\left(\log\left(\sum_{i\in j}\exp\left(\frac{1}{\phi_j}\hat{V}_{i/j}\right)\right)\right)}{\sum_{j'=1}^{J}\exp\phi_{j'}\left(\log\left(\sum_{i\in j'}\exp\left(\frac{1}{\phi_{j'}}\hat{V}_{i/j'}\right)\right)\right)} \tag{7.29}$$

and using the lower normalisation, it would be:

$$P(i,j) = \frac{\exp\left(\frac{\phi_r}{\phi_j}\hat{V}_{i/j}\right)}{\sum_{i'\in j}\exp\left(\frac{\phi_r}{\phi_j}\hat{V}_{i'/j}\right)} \cdot \frac{\exp\phi_j\left(\log\left(\sum_{i\in j}\exp\left(\frac{\phi_r}{\phi_j}\hat{V}_{i/j}\right)\right)\right)}{\sum_{j'=1}^{m}\exp\phi_{j'}\left(\log\left(\sum_{i\in j'}\exp\left(\frac{\phi_r}{\phi_{j'}}\hat{V}_{i/j'}\right)\right)\right)} \tag{7.30}$$

Equations (7.29) and (7.30) show that, from a practical point of view, the resulting specifications are equivalent to defining the corresponding normalising parameter as one in the NL general functional form (7.19). However, there are two problems here as we discuss in more detail below. First, the option of normalising at the lower level raises the problem of which lower level nest to use. Second, confusion is added when the scales (the parameters λ_j, associated with the EV1 distribution) are unequal as can be seen by comparing equations (7.29) and (7.30). Since an important aspect of modelling is communicating the results to decision makers, this is a non-trivial issue.

Theoretical considerations

Equations (7.31) and (7.32) below describe the relation between both normalisations (Carrasco and Ortúzar 2002):

$$\hat{\phi}_j^{up} = \hat{\phi}_j^{low} \quad \forall j \tag{7.31}$$

$$\hat{\theta}^{up} = \hat{\phi}_r \hat{\theta}^{low} \tag{7.32}$$

where *up* and *low* denote the corresponding upper and lower normalisations. The equations show that *both* specifications are equivalent and therefore compatible with utility maximising principles. Nevertheless, it is interesting to note that depending on the chosen normalisation there will be differences between the estimated values of the coefficients. However, this dissimilarity is not relevant in cases where the main interest is the ratio of coefficients, such as the *value of time* (Gaudry *et al.*, 1989), as the scale factors cancel out and therefore the same result, independent of the normalisation, is obtained. Also the model elasticities are indistinguishable if the normalisations are executed properly (Daly 2001).

However, the dissimilarity may be important if we wish to compare a given NL coefficient such as the marginal utility of income (i.e. the coefficient of the cost variable with a minus sign in the typical wage rate specification, see section 8.3.2) with its MNL counterpart. In this case it is *only possible to compare directly the MNL estimated coefficients $\hat{\theta}$ with the upper normalisation coefficients*; this is because the former are the product of the population coefficients θ and the scale parameter associated with the errors, as we already saw, and equations (7.27)–(7.28) show that only $\hat{\theta}^{up}$ involves the parameter β associated with the *total* variance of the EV1 distributed errors in the NL case. Those of the lower normalisation are the product of θ and the parameter λ_r, which is only related to the variance of the normalised nest.

A final aspect to consider is the possibility of comparing the NL functional forms of Williams-Daly and Zachary and McFadden in this context; the equations above clearly show that only the upper normalisation allows a direct comparison between the coefficients of both specifications. In conclusion, although both normalisations are consistent with the theory there are interesting reasons to prefer the upper normalisation:

 i) The possibility of establishing a direct comparison between NL and MNL coefficients.
 ii) The simplicity of having the *only* parameter related to total variance as reference.
iii) The simpler functional form of the probability in this case.

7.4.4.2 UMNL and NNNL Specifications

Now we discuss the controversy famously raised by Koppelman and Wen (1998a; 1998b) about two forms of the model found in the literature: the UMNL (Utility Maximising Nested Logit) and NNNL (Non Normalised Nested Logit) specifications (Daly 2001; Hensher and Greene 2002; Hunt 2000; Koppelman *et al.* 2001).

Functional forms of the specifications

The UMNL specification is defined as McFadden's model; thus, in the simple case of a two-level tree structure, the NL probability function is given by:

$$P(i, j) = \frac{\exp\left(\frac{\hat{V}_{i'/j}}{\phi_j}\right)}{\sum_{i' \in j} \exp\left(\frac{\hat{V}_{i'/j}}{\phi_j}\right)} \cdot \frac{\exp \phi_j \left(\log \left(\sum_{i' \in j'} \exp\left(\frac{\hat{V}_{i'/j'}}{\phi_{j'}}\right)\right)\right)}{\sum_{j'=1}^{J} \exp \phi_{j'} \left(\log \left(\sum_{i' \in j'} \exp\left(\frac{\hat{V}_{i'/j'}}{\phi_{j'}}\right)\right)\right)} \tag{7.33}$$

(continued)

The NNNL specification is at the root of the pioneering *simultaneous estimation* method designed by Daly (1987). In the NNNL specification the probability function is the same as above but omitting the inverse of the structural parameter in the elemental alternatives, that is:

$$P(i, j) = \frac{\exp\left(\hat{V}_{i/j}\right)}{\sum_{i' \in j} \exp\left(\hat{V}_{i'/j}\right)} \cdot \frac{\exp \phi_j \left(\log \left(\sum_{i' \in j} \exp\left(\hat{V}_{i'/j}\right)\right)\right)}{\sum_{j'=1}^{J} \exp \phi_{j'} \left(\log \left(\sum_{i' \in j'} \exp\left(\hat{V}_{i'/j'}\right)\right)\right)} \qquad (7.34)$$

This specification is not compatible with the fundamental properties of the NL if the correlation is not the same on every nest and/or if the model has generic variables. In this latter case it is easy to see that if a constant is added to the utilities of all options, the choice probabilities do not remain invariant, as is required by the *translational invariance* property.

Contrary to what Hensher and Greene (2002) and Hunt (2000) appeared to claim, in the sense that the identifiability problem was the source of this controversy and that the way to solve it was normalising in a proper way, Carrasco and Ortúzar (2002) showed that the problem was *not* related to the NL identifiability issue; this is easy to check if the functional form (7.34) is compared with those in (7.29) and (7.30), where we can see the differences between both normalised functional forms and the NNNL specification.

Comparing the two specifications
A first element of discussion in the literature was the consistency of both specifications with utility maximising theory. It is clear that the UMNL is compatible with theory because it is a particular case of the GEV family. On the other hand, although the NNNL may be consistent with a general idea of utility maximisation, it is clear that it is *not* compatible with one of the fundamental properties within the frameworks of either Williams or McFadden, if the utility function has generic coefficients.

However, it is interesting to note that there are some particular conditions that allow the NNNL to be equivalent to the UMNL. These are (Carrasco and Ortúzar 2002):

i) When trees have equal structural parameters ϕ on each level; in the framework of Williams-Daly and Zachary this means that all the parameters λ_j associated with the stochastic errors within each nest (and their respective correlation) have the same value.
ii) When the model does not have generic coefficients in a linear specification for the utility function of options at different levels and/or nests. This second condition is less restrictive and may be valid in many real cases.

In addition, it is possible to modify the NNNL specification so that it becomes compatible with utility theory in those cases where it is not equivalent to the UMNL; this involves the inclusion of dummy nodes and links as is standard recommended practice in ALOGIT (Daly 1992), but adds certain restrictions (see Hensher and Greene 2002; Koppelman and Wen 1998a). Also, as the number of structural parameters grows, the more complex the NNNL modified (artificial) tree becomes.

To sum up, the UMNL specification is preferable because it does not need the above changes (which could be complex in more complex tree structures) and its coefficients are directly comparable with those of other models (such as the MNL), having all the advantages of the upper normalisation. One reason that has been argued in favour of the NNNL specification is computational efficiency as it has a simpler likelihood function (Daly 1987), but the main one is that it is used by ALOGIT which is probably the most popular estimation package in practice. Other important practical packages are

LIMDEP (Economic Software Inc. 1995) and Biogeme (Bierlaire 2009) but it is also possible to estimate almost any model using GAUSS (Aptech Systems 1994) or similar packages, although this is less practical for data banks of large size.

7.4.4.3 On the Limits of the Structural Parameters

This section considers the controversy arising from the observation by Börsch-Supan (1990b) who suggested that under special circumstances the structural parameter ϕ could be larger than one, violating the structural condition (7.24). The compatibility of the NL with the basic theoretical conditions is an important issue which has been extensively studied on the literature. Williams and Ortúzar (1982b) presented the necessity of these conditions as a rigorous and unambiguous argument to reject a model, where goodness-of-fit can be a necessary condition, but not enough to validate a model. In fact, an important property of discrete choice models (and the NL belongs to them) is precisely the successful marriage between an explicit theory of behaviour with a micro representation, allowing the constructive use of statistical goodness-of-fit measures for model specification and testing. Thus, an inconsistent model would be a theoretical setback of at least 30 years.

Then, it is important to study if the results of Börsch-Supan (BS) are consistent with theory, especially focusing on its general theoretical consequences and the interpretation in empirical cases. These aspects were not explored by the various authors who cited the BS conditions (for example, Herriges and Kling 1995; 1996; Koppelman and Wen 1998b), but were dealt with conclusively by Carrasco and Ortúzar (2002).

BS proposed extension and further corrections

The consistency conditions with utility maximisation analysed by BS, derive from the work of McFadden (1981). One of these conditions is:

$$\frac{(-1)^{I-1}\partial^{I-1}P_i(V)}{\partial V_1 \ldots \partial V_{i-1}\partial V_{i+1} \ldots \partial V_I} \geq 0 \quad \forall V \in R^I \tag{7.35}$$

where R is the set of real numbers, $V = (V_1, \ldots, V_I)$ is the vector of representative utilities of I alternatives and P_i is the probability of choosing alternative A_i. In passing, note that BS did not consider that the sign alternates; this was corrected by Herriges and Kling (1995).

Equation (7.35) ensures that the probability density function cannot be negative and is equivalent to $0 < \phi \leq 1$ if the condition is valid for all the representative utilities $V \in R^I$. BS argued that the need for condition (7.35) to hold for all R^I is overly restrictive because economic theory (and practical experience) would suggest that only a subset of data points is used for modelling ('relevant subset'). This subset should include the data points used to estimate the model and to examine potential policy changes. As a consequence of this approach it would become feasible that a NL with structural parameters larger than one could be consistent with utility maximising theory. However, it is nearly impossible to find a data set that allows a NL with structural parameters larger than one to be consistent with utility maximisation, as this only happens if the relevant subset is not empty.

Herriges and Kling (1995) not only corrected the omission of signs by BS in (7.35) but presented the necessary conditions for consistency with utility maximisation in two level NL models, as follows:

$$P_j \geq \tau_j \tag{7.36}$$

$$2(\tau_j - P_j)^2 + \tau_j P_j \geq \tau_j \tag{7.37}$$

$$6(P_j - \tau_j)^3 + \tau_j[2(P_j - 1) - \tau_j](1 - P_j) \geq 0 \tag{7.38}$$

(continued)

with

$$\tau_j \equiv \frac{(\phi_j - 1)}{\phi_J}.$$

These conditions result from the differentiation of the NL functional form (7.23) using restriction (7.35) for the first, second and third partial derivatives. Inequality (7.36) must be satisfied for nests with two or more options, (7.37) for nests with three or more alternatives and (7.38) for nests with four options or more. Conditions (7.36)-(7.38) would replace (7.35) when testing consistency with utility maximisation under the framework of McFadden (1981).

Interpretation and applicability of the BS extension
First note that this framework does not prevent in any way the use of $\phi = 1$ as a method to test if a NL collapses to the MNL (which is curiously ignored sometimes). Further, having a structural parameter greater than one implies a greater degree of substitution between nests than within them. However, it may be possible that another tree (which correctly considers a greater degree of substitution within nests) could be postulated. It also implies negative values for the covariance and correlation between nested alternatives. However, both these interpretations seem to be more statistical than behavioural, as Train *et al.* (1987) argued.

Without doubt the most important consequence of allowing the structural parameters to be larger than one is that it denies their use as a test for establishing a hierarchical relationship between the different nesting levels. This has consequences not only on behavioural interpretation terms, but also in terms of the search for the best tree structure (i.e. the information provided by the structural parameter values is quite useful to define upper and lower levels when this is not obvious). Thus, it is important to remark that the BS framework *is not possible* to use if the NL model postulates the separability on choice levels (for example, destination-mode choice), where the variance condition (7.20) of Williams is a *fundamental* property to understand behaviour.

Herriges and Kling (1996) have made the only empirical investigation of the BS extension reported in the literature. To test model consistency with utility maximising, they explored three different procedures, progressively more restrictive. The first two failed, and the third imposed restrictions (7.36) and (7.37) *ex ante*, estimating the coefficient vector from a Bayesian perspective. In practical terms this is equivalent to estimating the NL without prior information, yielding a coefficient vector distributed Normal with mean and covariance matrix taken from the estimation. The procedure generated a large number of Normal distributed values but only the draws satisfying the consistency conditions were retained. So, although by construction all parameters were consistent with theory, some were calculated with a very low percentage of the generated values.

Important objections about the real applicability of the above procedures (and in general about the applicability of the BS extension) were formulated by Carrasco and Ortúzar (2002). Further, the same kind of choice context was later treated successfully using the more flexible Mixed Logit model (Train 1998).

7.4.4.4 Two Further Issues

In this section we will consider two final, relatively minor, controversies about the NL that have also been discussed in recent literature.

Alternative definition of model parameters
Hensher and Greene (2002) and Hunt (2000) proposed an alternative definition of the NL coefficients which may lead to some confusion about certain properties of the model. They defined a scale

parameter for each nest (γ_j) and another at the alternative level (μ_j), associated with each nest, as in Figure 7.4.

alternatives i (μ_j associated with each nest they belong)

Figure 7.4   NL alternative parameter definition

The most important difference between this specification and the Nested Logit of Williams (1977) is the incorporation of parameters at the level of the elemental alternatives instead of the unique parameter associated with the root. Thus, this alternative specification has $2J$ scale parameters instead of the traditional $J + 1$ (with J being the number of nests).

In this alternative vision the choice probability is given by:

$$P_{ij} = \frac{\exp\left(\mu_j V_{i/j}\right)}{\sum_{i' \in j} \exp\left(\mu_j V_{i'/j}\right)} \cdot \frac{\exp \gamma_j \left\{ \dfrac{1}{\mu_j} \log \left(\sum_{i \in j} \exp\left(\mu_j V_{i/j}\right) \right) \right\}}{\sum_{j'=1}^{m} \exp \gamma_{j'} \left\{ \dfrac{1}{\mu_{j'}} \log \left(\sum_{i \in j'} \exp\left(\mu_{j'} V_{i/j'}\right) \right) \right\}} \tag{7.39}$$

If we redefine the variables μ as λ and γ as β above – without considering their theoretical meaning – it is possible to get a very similar specification to that of Williams (7.19), except for the fact that here we get different β_j parameters for each nest

This specification would indirectly allow modelling heteroskedasticity between alternatives, generalising the NL of Williams-Daly and Zachary and McFadden (see Example 4 in section 7.6). However, the specification is inconvenient in practice because it leads to some confusion when solving the identifiability problem. In this case it would be necessary to normalise J parameters; i.e. either define *all* parameters β_j (upper normalisation) or *all* parameters λ_j (lower normalisation) as non estimable. This is different from the normalisations discussed before, where to solve the identifiability problem it was necessary to define *only one* parameter as non identifiable. Another interesting point is that the alternative model's upper normalisation is the same as that of Williams (i.e. it is correct); but the lower normalisation leads to a NNNL specification with all the problems explained before.

In addition, a different definition of the scale parameters also implies some issues on 'partial degenerated' structures (i.e. trees with some nests containing only one alternative). Under the Williams-Daly and Zachary and McFadden theoretical frameworks, if an option A_k is 'degenerated' its corresponding structural parameter ϕ_k is equal to one because λ_k is equal to β (i.e. the nest that contains the alternative 'collapses' to the upper level). This result is *independent from the normalisation*. However, this basic theoretical interpretation is not possible if there is a parameter β_j specific to each nest. If the upper

(continued)

normalisation is used, Hunt (2000) argues that λ_k becomes 'non identifiable', but what actually occurs is that it has collapsed to β, which is non identifiable by definition (Carrasco and Ortúzar 2002). This confusion becomes even worse if the lower normalisation is used because the parameter ϕ_k results non-estimable, being necessary to *define* its value as unity to get consistency with the theory (Hunt 2000); however, in this case the NNNL specification is obtained which suffers the problems mentioned above.

Heteroskedasticity and correlation
Börsch-Supan (1990a) and Hensher and Louviere (1998), report results which suggest that the specification of the NL tree structure could, in some cases, be even more strongly influenced by heteroskedasticity (i.e. different variances between options) than by correlation. On the other hand, Munizaga *et al.* (2000) report surprisingly good behaviour of the NL in the presence of heteroskedasticity between alternatives (but not in the presence of heteroskedasticity between observations), showing a low predictive capacity only for radical policy changes in their Monte Carlo simulation study.

Furthermore, Hensher and Louviere (1998) and Hensher (1999) propose a new method of specifying a NL tree structure (i.e. a way to define which alternative should belong to each nest) based on the scale differences between the options. They use the Heteroskedastic Extreme Value model (Bhat 1995) as a 'search engine' in order to define nestings of alternatives with similar variance. It is interesting to note that the tree specification process in the NL does not have a rigorous procedure (with standard steps) and traditionally it has been based on the idea of grouping alternatives that theoretically (or intuitively) appear to be correlated (see Ortúzar 1982).

Hensher (1999) argues that a statistical rationale for nesting could be related to differential patterns of variance between subsets of alternatives. However, this argument is theoretically suspect because the differential patterns between subsets of options in the NL are based on the different value of *correlation* rather than variances. In fact, in a two level model as (7.19) the scale parameter defining nesting is λ_j (which is *only* related to correlation) and not β (which is associated with both correlation and variance). Therefore, Hensher's tree specification method should be rejected as it is based on a property that the NL does not possess (i.e. heteroskedasticity).

Finally, if there are grounds to believe *a priori* that heteroskedasticity could be an important issue in modelling on a given context, there are more general models that can handle this effect in theory (and with even better results than the NL for simulated data, see Munizaga *et al* 2000), such as Mixed Logit (Train 2009) or Multinomial Probit (Daganzo 1979), which are nowadays less problematic to estimate and a little less problematic to interpret than in the past.

7.5   The Multinomial Probit Model

As we mentioned in section 7.4.1, in the MNP model the stochastic residuals ε of (7.2) are distributed multivariate Normal with mean zero and an arbitrary covariance matrix, i.e. in this case the variances may be different and the error terms may be correlated in any fashion. The problem is, of course, that this generality does not allow us to write the model in a simple closed form as in the MNL (except for the binary case); therefore to solve it numerically we need approximations or, more effectively, simulation.

7.5.1   The Binary Probit Model

In this case we can write the utility expressions (7.2) as:

$$U_1(\theta, \mathbf{Z}) = V_1(\theta, \mathbf{Z}) + \varepsilon_1(\theta, \mathbf{Z})$$
$$U_2(\theta, \mathbf{Z}) = V_2(\theta, \mathbf{Z}) + \varepsilon_2(\theta, \mathbf{Z})$$

where ε $(\theta, \mathbf{Z})$ is distributed bivariate $N(0, \Sigma)$ with

$$\sum = \begin{pmatrix} \sigma_1^2 & \rho\sigma_1\sigma_2 \\ \rho\sigma_1\sigma_2 & \sigma_2^2 \end{pmatrix}$$

where ρ is the correlation coefficient between U_1 and U_2. From (7.6), the probability of choosing option 1 is given by:

$$P_1(\theta, \mathbf{Z}) = \text{Prob}\{\varepsilon_2 - \varepsilon_1 \leq V_1 - V_2\}$$

but as the Normal distribution is closed to addition and subtraction (as the EV1 is closed to maximisation) we have that $\varepsilon_2 - \varepsilon_1$ is distributed univariate $N(0, \sigma_\varepsilon)$, where:

$$\sigma_\varepsilon^2 = \sigma_1^2 + \sigma_2^2 - 2\rho\sigma_1\sigma_2$$

Dividing $(\varepsilon_2 - \varepsilon_1)$ by σ_ε we obtain a standard $N(0, 1)$ variable; therefore we can write the binary Probit choice probability concisely as:

$$P_1(\theta, \mathbf{Z}) = \Phi[(V_1 - V_2)/\sigma_\varepsilon] \tag{7.40}$$

where $\Phi[x]$ is the cumulative standard Normal distribution which has tabulated values. Although this is indeed a simple model, it is completely general for binary choice. Note, however, that equation (7.40) is not directly estimable as the parameters θ in the representative utilities $\mathbf{V}$ cannot be estimated separately from the standard deviation σ_ε. In fact, just as occurred in the MNL and NL models, there is an identifiability problem and one would need to normalise before obtaining an estimate of the model parameters. Bunch (1991) looks at this problem for the general MNP model, and Walker (2002) provides a good discussion about the issue of identifiability in general.

7.5.2 Multinomial Probit and Taste Variations

As we noted in sections 7.3 and 7.4, a potentially important problem of fixed-coefficient random utility models, such as the MNL and NL, is their inability to treat the problem of random taste variations among individuals without explicit market segmentation. In what follows we will first show with an example what is meant by this and then we will proceed to show how the MNP handles the problem.

Example 7.2 Consider a mode choice model with two explanatory variables, cost (c) and time (t) and the following postulated utility function:

$$U = \alpha t + \beta c + \varepsilon$$

Let us suppose, however, that the perception of costs in the population varies with income (I), i.e. poorer individuals are more sensitive to cost changes, such that the true utility function is:

$$U = \alpha t + \phi c / I + \varepsilon$$

It can easily be seen, comparing both expressions, that the model will be correct only if β can be considered as a random variable with exactly the same distribution as ϕ/I in the population; in this case then, the model contains random taste variations.

The problem of random taste variations is normally very serious, as has been clearly illustrated by Horowitz (1981), and may be considered as a special case of one well-known specification error, the omission of a relevant explanatory variable, which we discussed in Chapter 3.

Let us consider again the utility function (7.3) which is linear in the parameters, as discussed in section 7.2. It's most general case considers the parameter set θ to be a random vector distributed across the population; in this case the residuals may be modelled as alternative specific parameters, hence the variables ε in (7.2) may be omitted without loss of generality and the equation can be written more concisely as:

$$U_j = \sum_k \theta_k x_{jk} \qquad (7.41)$$

which is a very general linear specification as it allows for taste variations across the population. If the vector θ is distributed multivariate Normal, the choice model resulting from (7.41) is of MNP form (see Daganzo 1979). Various procedures for estimating this model were discussed by Sheffi *et al.* (1982), Langdon (1984) and others, and we will look at this in Chapter 8.

7.5.3 Comparing Independent Probit and Logit Models

When estimating a MNP model (and it is easy to see it in the binary case) the parameters obtained are:

$$\beta_i^P = \frac{\theta_i}{\sigma_\varepsilon} \quad \text{with} \quad \sigma_\varepsilon^2 = \sigma_1^2 + \sigma_2^2 - 2\rho\sigma_1\sigma_2$$

On the other hand, we know from equation (7.10) that when estimating a MNL the parameters obtained are:

$$\beta_i^L = \lambda\theta_i \quad \text{with} \quad \lambda = \frac{\pi}{\sigma\sqrt{6}}$$

therefore we have that:

$$\beta_i^L = \frac{\theta_i \pi}{\sigma\sqrt{6}}$$

Now, to compare both models we need to estimate a MNP model with a covariance matrix similar to that of the MNL (i.e. an Independent and Identical Probit). In this case $\sigma_\varepsilon^2 = \sigma^2 + \sigma^2$, which implies that $\sigma_\varepsilon = \sigma\sqrt{2}$ and thus $\beta_i^P = \theta_i/\sigma\sqrt{2}$. Therefore, in order to compare both sets of parameters, we should multiply the β_i^P by a factor that makes them equal to $\beta_i^L = \theta_i/\sigma\sqrt{6}$; and this is achieved using the factor:

$$K = \frac{\sigma\pi\sqrt{2}}{\sigma\sqrt{6}} = \frac{\pi}{\sqrt{3}} \qquad (7.42)$$

Therefore if one wants to compare the estimated coefficients of a MNL and an IID Probit model, those belonging to the second structure must be scaled by the factor $\pi/\sqrt{3}$. We have successfully used this method to test the correctness of an experimental code to estimate MNP models (e.g. Munizaga *et al.* 2000).

7.6   The Mixed Logit Model

This appears to be the model for the new millennium. Although its current form originated from the parallel work of two research groups in the 90s (Ben-Akiva and Bolduc 1996; McFadden and Train 2000), the original formulation of the model, as Hedonic or Random Parameters Logit, was made much earlier (Cardell and Reddy 1977; Cardell and Dunbar 1980).

7.6.1   Model Formulation

The Mixed Logit (ML) model can be derived under several behavioural specifications, each providing a particular interpretation. Train (2009) correctly states that the model is *defined* on the basis of the functional form for its choice probabilities. As such, the ML label is applicable to any model the probabilities of which can be expressed as an integral of standard Logit probabilities over a distribution of the parameters, such as:

$$P_{iq} = \int L_{iq}(\theta) f(\theta) d\theta \qquad (7.43)$$

where $L_{iq}(\theta)$ is typically an MNL probability evaluated at a set of parameters θ and their density function, $f(\theta)$, is known as 'mixing distribution'. If $f(\theta)$ is degenerate at fixed parameters $\mathbf{b}$ (i.e. it equals one for $\theta = \mathbf{b}$ and zero for $\theta \neq \mathbf{b}$), the choice probability (7.43) becomes the simple MNL.

If, on the other hand, the mixing distribution is discrete (i.e. if θ takes M values labelled $b_1, \ldots, b_M$ with probabilities s_m that $\theta_m = b_m$), the ML becomes the *latent class* model with applications in psychology and marketing; this is useful when there are distinct segments in the population, each with their own choice behaviour (Train, 2009). In section 7.6.4 we will look at a class of ML models where the mixing distribution lies somewhere between the typical continuous form, below, and the latent class model.

Now, in most ML applications $f(\theta)$ has been taken as continuous with mean $\mathbf{b}$ and covariance matrix $\boldsymbol{\Sigma}$, and modellers have attempted just to estimate these 'population parameters' without taking advantage of one of the most powerful features of the model, that is, estimating the θ that enter in the Logit component (kernel) for each individual; this can be done directly or conditional on the population parameters, $\mathbf{b}$ and $\boldsymbol{\Sigma}$, as we will discuss in Chapter 8.

7.6.2   Model Specifications

7.6.2.1   Basic Formulations

The ML model has two basic forms. The first is the *error components* (EC) version, the utility function of which is characterised by an error term with two elements. One (ε_{qjt}) allows the MNL probability to be obtained (and as such has the usual IID EV1 distribution), while the other has a distribution which can be freely chosen by the modeller, depending on the phenomenon he needs to reproduce. In this case the utility of option j ($j = 1, \ldots, J$) for individual q in choice situation t ($t = 1, \ldots, T$) is given by:

$$\mathbf{U}_{jqt} = \theta_{jt}\mathbf{X}_{jqt} + \boldsymbol{\Omega}_{jqt}\mathbf{Y}_{jqt} + \varepsilon_{jqt} \qquad (7.44)$$

where θ are fixed parameters and $\mathbf{X}$ are observable attributes, $\boldsymbol{\Omega}_{jqt}$ is a vector of random elements with a distribution specified by the modeller, with zero mean and unknown covariance matrix, and $\mathbf{Y}_{jqt}$ is a vector of attributes unknown (in value and nature) to the modeller. Thus, without loss of generality they can be taken as equal to one for all alternatives or for groups of them. Given this, the covariance matrix of the model utilities is:

$$Cov(\mathbf{U}_{jqt}) = Cov(\boldsymbol{\Omega}_{jqt}) + (\pi^2/6\lambda^2) \cdot \mathbf{I}_J$$

where $\mathbf{I}_J$ is a $J \times J$ identity matrix. An adequate choice of $\mathbf{Y}_{jqt}$ allows different error structures such as correlation, cross-correlation, heteroskedasticity, dynamics and even auto-regressive errors to be treated (Hensher and Greene 2003; Train 2009; Walker 2001). Indeed, it has been proven that the ML can approximate any discrete choice model derived from a random utility maximisation model as closely as one pleases (Dalal and Klein 1998; McFadden and Train 2000); this, in fact, led to the demise of the MNP model as a serious candidate in this area. On the other hand, to obtain the simple MNL model $\mathbf{Y}_{jqt}$ has to be zero such that there is no correlation among alternatives.

Example 7.3 To generate a heteroskedastic version of the MNL, one simply needs to specify the following utility function:

$$U_{iq} = \theta\mathbf{X}_{iq} + \sigma_i \Omega_{iq} + \varepsilon_{iq} \quad \text{with} \ \Omega_{iq} \sim IID N(0, 1)$$

and as the errors Ω and ε are independent, it is easy to see that the covariance matrix of the utilities U has the following form (for simplicity we are taking a trinomial case):

$$\Sigma = \begin{bmatrix} \sigma_1^2 + \pi^2/6\lambda^2 & 0 & 0 \\ 0 & \sigma_2^2\pi^2/6\lambda^2 & 0 \\ 0 & 0 & \sigma_3^2 + \pi^2/6\lambda^2 \end{bmatrix}$$

where λ is the scale factor associated with the EV1 errors.

To generate a heteroskedastic version of the NL model, one would need a similarly simple specification; assume a five alternatives case, where the first two are correlated and the last two are also correlated (the third is independent of all others):

$$U_{1q} = \mathbf{X}_{1q}\,\theta + \sigma_1\,\eta_{1q} + \varepsilon_{1q} \qquad U_{2q} = \mathbf{X}_{2q}\,\theta + \sigma_1\,\eta_{1q} + \varepsilon_{2q}$$
$$U_{3q} = \mathbf{X}_{3q}\,\theta + \sigma_2\,\eta_{2q} + \varepsilon_{3q}$$
$$U_{4q} = \mathbf{X}_{4q}\,\theta + \sigma_3\,\eta_{3q} + \varepsilon_{4q} \qquad U_{5q} = \mathbf{X}_{5q}\,\theta + \sigma_3\,\eta_{3q} + \varepsilon_{5q}$$

In this case it is again easy to see that the covariance matrix of the model utilities is given by:

$$\Sigma = \begin{bmatrix} \sigma_1^2 + \pi^2/6\lambda^2 & \sigma_1^2 & 0 & 0 & 0 \\ \sigma_1^2 & \sigma_1^2 + \pi^2/6\lambda^2 & 0 & 0 & 0 \\ 0 & 0 & \sigma_2^2 + \pi^2/6\lambda^2 & 0 & 0 \\ 0 & 0 & 0 & \sigma_3^2 + \pi^2/6\lambda^2 & \sigma_3^2 \\ 0 & 0 & 0 & \sigma_3^2 & \sigma_3^2 + \pi^2/6\lambda^2 \end{bmatrix}$$

and correlation is due to the presence of the common unobservable elements in the utilities of the correlated alternatives; note that replicating the true NL, which is homoskedastic, is more involved (see Munizaga and Alvarez-Daziano 2000).

The second, more classical, version of the ML model considers a *random coefficients* (RC) structure, in which the marginal utility parameters are different for each sampled individual q, but do not vary across choice situations; this last assumption may be relaxed if choice situations are significantly separated along time, as taste parameters could then be presumed to alter (Hess and Rose 2009). So, in this case we have:

$$U_{jqt} = \theta_q \mathbf{X}_{jqt} + \varepsilon_{jqt} \tag{7.45}$$

and the parameters vary over individuals with density $f(\theta)$. This specification yields the choice probabilities (7.43) naturally. Note that the presence of the vector $\mathbf{X}$ in the covariance matrix does not allow the modeller to control for it, but helps to ease an important problem of the model, its identification, which we discuss in section 7.6.3.

The EC and RC specifications are formally equivalent as the coefficients θ_q can be decomposed into their means ($\mathbf{b}$) and deviations, denoted s_q, such that:

$$U_{jqt} = \mathbf{b}\mathbf{X}_{jqt} + s_q \mathbf{X}_{jqt} + \varepsilon_{jqt} \tag{7.46}$$

which has error components defined by $\mathbf{Y}_{jqt} = \mathbf{X}_{jqt}$; conversely, we can also start from the EC specification and get the RC specification. However, though formally equivalent, the manner in which the modeller looks at the phenomenon under study affects the model specification. For example, if the main interest is to represent appropriate substitution patterns through an EC specification, the emphasis will be placed on specifying variables that can induce correlation in a parsimonious fashion, not necessarily considering tastes variations or too many explanatory variables. In fact, Train (2009) wisely states that '... *there is a natural limit on how much one can learn about things that are not seen*', but this is sometimes overlooked by even the most skilful econometricians who focus on the error terms at the expense of correct specification of the observed utility component and ensuring the data are of appropriate quality.

7.6.2.2 More Advanced Formulations

An important issue concerning the apparent dual representation of the model (i.e. EC or RC), as noted recently by many analysts, is that the two versions of the model may give rise to *confounding* effects. As most discrete choice models, the ML is based on the linear-in-parameters-with-additive-disturbances (LPAD) structure, where individuals are assumed to compensate (trade-off) the effects of good and bad attributes even when there are many situations where compensatory rules do not hold (see Cantillo and Ortúzar, 2005). For example, an omitted structure (i.e. any interaction between two variables) will be captured by the error terms, but it may be confused with random heterogeneity for a given attribute in the RC version.

On the other hand, as the model works on the basis of differences between alternatives, it does not matter whether an attribute is included in one alternative or in all others except that one, as long as the relative difference between them does not change. This property, in conjunction with the compensatory rule, may lead to another confounding effect: between correlation and heterogeneity in tastes and response, which can appear in estimated models and produce misleading forecasts, as discussed recently by Cherchi and Ortúzar (2008a). In order to understand the correct underlying structure and to test whether heterogeneity is really present, they recommend estimating alternative specifications and comparing results, looking carefully at the absolute value of the random parameters, and the relative values of the alternative specific constants and correlation coefficients. They also found that a significant specific random parameter may not actually reveal variation in tastes, but correlation among competing alternatives, cautioning that this is especially important if the model is intended as a forecasting tool. These findings complement the observation by Hess *et al.* (2005a) that the assumptions made with regard to error structure can have significant impacts on willingness-to-pay indicators.

The RC and EC specifications can also be combined easily, allowing for the joint modelling of random taste heterogeneity and inter-alternative correlation. This however, as mentioned above, comes at the cost of important issues in identification, and also a heightened cost of estimation and application when using error components for the representation of correlation. While integration over mixture distributions is necessary in the representation of continuous random taste heterogeneity, this is not strictly the case for inter alternative correlation. Indeed, just as, conditional on a given value of the taste coefficients, a typical RC specification allowing for random taste heterogeneity reduces to a MNL model, a model allowing for inter-alternative correlation in addition to random taste heterogeneity can in this case be seen to reduce to a given GEV model (assuming that an appropriate GEV model actually exists). As such, the correlation structure can be represented with the help of a GEV model, while the random taste heterogeneity is accommodated through integration over the assumed distribution of the taste coefficients. The use of the choice probability of a more complicated GEV model instead of the MNL as the integrand in (7.43), leads to a more general type of a GEV mixture model, of which the typical RC specification is simply the most basic form.

In a more general GEV mixture, we would simply replace the MNL choice probability inside the integral by say a NL choice probability. Such model can be estimated using, for example, Biogeme (Bierlaire 2009), and is useful for cases where we need to allow for correlation between, say, train and bus in a car-train-bus mode choice context, while additionally allowing for random variations across respondents in the time and cost sensitivities; in such cases, a NL model could deal with the former, while a RC Mixed Logit could deal with the latter. A general GEV mixture ML can deal with both at the same time, without the need for additional error components.

Applications of this approach include for example Chernew *et al.* (2001), Bhat and Guo (2004) and Hess *et al.* (2005a). In such a GEV mixture model, the number of random terms, and hence the

(continued)

number of dimensions of integration (and thus simulation) is limited to the number of random taste coefficients, whereas, in the EC specification, one additional random term is in principle needed for representing each separate nest.

Finally, it is interesting to mention that problems of a similar nature have been encountered when modelling jointly state dependence (i.e. the state at a given moment depends on the previous state(s) of the system) and preference heterogeneity. Smith (2005) concluded that one should be cautious in interpreting random parameters if researchers are unable to model state dependence. Nevertheless, he also stated that if a more elaborate parameterization of preference heterogeneity is used, excluding state dependence may magnify the apparent preference heterogeneity in the model but not necessarily generate it where none exists. To some extent this could be viewed as the converse of the problem explored by Heckman (1981), where the emphasis was on the emergence of spurious state dependence if heterogeneity was not modelled properly.

7.6.3   Identification Problems

A seminal reference for the 'identification problem' is the work of Walker (2001; 2002), who noted in passing that even the most famous econometricians have been guilty of overlooking this issue in some applications. Nowadays analysts are more cautious and test for this problem in usual practice, but new evidence has appeared showing that it is multifaceted with no easy recipes available to avoid it.

The nature of the problem is that there are infinite sets of restrictions that can be imposed to identify a given set of parameters to be estimated. For example, in the case of the MNL model the problem only relates to the impossibility of estimating the scale parameter β (which has to be normalised), and that one of the alternative specific constants (ASC) needs to be taken as zero (i.e. that of the *reference* alternative). Note that even in this simple case there are 'good practice' rules to follow, i.e. choose as reference the alternative more universally available (Ortúzar 1982).

For more complex models, such as the ML, apart from the above considerations that apply to the vector θ, we also need to examine the identification of the unrestricted parameters of the error distribution. This could be done by studying the *Fisher information matrix* (i.e. the matrix of expected values of the second derivatives of the log-likelihood function), but this requires estimating the model, something which is not always possible. In fact, there are two types of identification problem: the *theoretical identification*, which is inherent to the model specification regardless of the data at hand, and the *empirical identification* that depends on the information used to estimate the model. Although much has been written about the first, the second has recently surfaced as a serious problem deserving more attention.

7.6.3.1   Theoretical Identification

This problem is usually associated with the presence of too many parameters, i.e. the model cannot be estimated simply because of its implicit structure. By looking at the covariance matrix of utility differences, Walker (2001) generalised the work of Bunch (1991) for the MNP and provided an outstanding analysis of the three conditions (order, rank and equality) that must hold for the ML model to be identifiable.

In particular, the *order condition* is a necessary condition and establishes the maximum number of parameters that can be estimated based on the number of alternatives (J) in the choice set. In the EC version of the model this condition states that there are at most $J(J-1)/2 - 1$ parameters estimable in the disturbance; this is equal to the number of unique elements in the covariance matrix of utility differences (as it is symmetric), minus one to set the scale.

The *rank condition* refers to the rank of the covariance matrix of utility differences. This is a sufficient condition and establishes the actual number of parameters that can be estimated as the number of independent equations available to do it. If this condition holds, the previous one necessarily holds but it is trivial to apply and useful to highlight any obvious identification problems. In many cases, these two conditions can be applied by simple visual inspection of the covariance matrix of utility differences; Walker (2002) also notes that when restrictions are needed for the covariance matrix terms it is desirable that these point to the MNL being a special case of the ML (i.e. if only two variances can be estimated, the restriction on the third is that it should equal zero; furthermore, the choice of which variance should be zero is not arbitrary – she recommends choosing that which obtained the lowest value in an estimation run without considering the identifiability problem).

Finally, the *equality condition* (formerly called positive definiteness) is used to verify that the chosen normalization, based on the identification restrictions imposed by the rank and order conditions, is valid in the sense that the resulting unique solution does in fact maximise the log-likelihood. Walker *et al.* (2007) note that this condition is particular to the ML model due to the special structure of its disturbance (the sum of an IID EV1 component and another with a different distribution).

It is important to note that the theoretical identification problem is only crucial for the EC version of the ML model, and does not exist when the RC version is specified for continuous attributes of the competing alternatives. In the RC model the random parameters are associated with some known (by the modeller) attributes and thus there is always some information that allows theoretically identifying extra parameters. But whether the full covariance structure can be estimated or not, will depend on the quality of the information as discussed below.

7.6.3.2   Empirical Identification

This problem, instead, occurs when the model is estimable in principle but the data cannot support it. In theory, the parameters can be empirically identified if the number of observations and draws in the simulated maximum likelihood procedure required to estimate the model (which we will discuss in Chapter 8) are sufficiently large to provide enough information. However, in practice, researchers face datasets with a limited number of observations and must apply a finite number of draws. Therefore, it becomes an empirically important question to check whether a given dataset can support the model at hand.

Ben-Akiva and Bolduc (1996) and Walker (2001), noted that an identification problem can arise when a low number of draws is used, and they and others, such as Hensher and Greene (2003), emphasised the necessity of verifying the stability of parameter estimates as the number of draws increased (thereby assuring that the bias was sufficiently reduced). More recently, Munizaga and Alvarez-Daziano (2005) have confirmed, using simulated data, that small sample sizes can lead to erroneous conclusions about the model's covariance structure (a warning in relation to the sample size required to recover parameters by simulation was given nearly 25 years ago by Williams and Ortúzar, 1982a). Chiou and Walker (2007) demonstrated that a low number of draws in the simulation process can mask identification issues leading to biased estimation results, even when a large number (i.e. 1000) of draws is used.

Finally, Cherchi and Ortúzar (2008b) used simulated data to analyse the extent to which the empirical identification problem depended on the variability of the data, the degree of heterogeneity of the taste parameters, the sample size and the number of choice tasks for each individual. They found that identification problems appeared if a variable had low variability between alternatives;

(continued)

they also found that models are quite unstable in the case of low variability, and – deceptively – very often cannot be estimated unless very few draws (i.e. as low as 30) are used, clearly a problem of identification and a procedure that results in a suspicious model. Contrariwise, if the difference in attributes has a high standard deviation (i.e. four times the mean), the identification problem disappears for any number of draws. Also, the identification problem does not depend on the degree of variability inherent in the random parameters but only in the richness of the associated data. Finally, they found that the capability of the ML to reproduce random heterogeneity increases when more than one choice is available for each individual (as in SP or panel data, except when there are identical repeated observations), and in that case the effect of sample size on empirical identification reduced considerably.

7.7   Other Choice Models and Paradigms

7.7.1   Other Choice Models

As we saw in section 7.4, each alternative in a Nested Logit (NL) model is a member of only one nest. This is a restriction that can be inappropriate as, for example, mixed modes (such as park & ride) could be correlated both to car and to rail.

To tackle this problem various types of GEV models have been formulated with what Train (2009) calls 'overlapping nests', such that a given alternative can belong to more than one nest. For example, Vovsha (1997), Bhat (1998), and Ben-Akiva and Bierlaire (1999) have developed a Cross-Nested Logit (CNL) model, managing to implement an original idea of Williams (1977), the Cross-Correlated Logit model, that was solved numerically by Williams and Ortúzar (1982a) but was not used ever since.

Chu (1989) proposed the Paired Combination Logit (PCL), in which each pair of alternatives constitutes a nest with its own correlation; thus, each alternative is a member of $J-1$ nests. Koppelman and Wen (2000) examined this relatively simple but flexible structure and found that it outperformed both NL and MNL in their application. All these models can be derived as members of the GEV family (McFadden 1981), as shown for the NL in section 7.4.2.2.

As in general, all these models can be approximated by the ML we will leave this topic here and refer readers to Train's excellent book for more details.

7.7.2   Choice by Elimination and Satisfaction

In Chapter 8 we discuss the problem of specification and functional form giving particular emphasis to the linear-in-the-parameters form which has accompanied the vast majority of disaggregate demand (normally of MNL structure) applications. Owing to a growing body of criticism directed at linear-in-the-parameters forms, the early 1980s witnessed an interest in the specification and estimation of non-linear formulations of varying designs. Commentary on the functional characteristics of these forms was intertwined with statements about alternative models of the decision process considered to underpin choice models.

One typical view was that because linear-in-the-parameters forms are associated with a compensatory decision-making process (i.e. a change in one or more of the attributes may be compensated by changes in the others), models cannot be appropriately specified for decision processes characterised by perception of discontinuities which are more plausibly of a non-compensatory nature (i.e. where good aspects of an alternative may not be allowed to compensate for bad aspects which are ranked higher in importance in the selection procedure, simply because that alternative may be eliminated earlier in the search process; see the discussion in Golob and Richardson 1981).

Example 7.4 Let us consider a set of individuals, confronted by a choice, to be endowed with a set of objectives **G** and a set of constraints **B**. A general multi-criterion problem can then be formally stated as:

$$\text{Max}_{(\text{options})} \left\{ F_1(Z_1^1) \ldots F_1(Z_N^1) \right\}$$

$$\vdots$$

$$\text{Max}_{(\text{options})} \left\{ F_k(Z_1^k) \ldots F_k(Z_N^k) \right\} \tag{7.47}$$

$$\vdots$$

$$\text{Max}_{(\text{options})} \left\{ F_K(Z_1^K) \ldots F_K(Z_N^K) \right\}$$

subject to the vector of constraints:

$$\mathbf{f(Z)} \leq \mathbf{B} \tag{7.48}$$

in which $F_k(Z_j^k)$ is the value of the criterion function associated with attribute Z_j^k of option A_j. For example, we might be interested in finding a mode, in a choice set of size N, which minimises travel time and cost, maximises comfort and safety, and so on. These attributes associated with any particular alternative might, in addition, be required to satisfy absolute constraints such as (7.48).

If a single alternative is found which simultaneously satisfies these optimality criteria (i.e. it optimises the K functions in expression 7.47) and whose attributes are feasible in terms of (7.48), then an unambiguous optimal solution is obtained. In general, however, there will be conflicts between objectives (i.e. options superior in some respects and inferior in others).

A number of important questions can be posed before a choice model based on this multi-criterion problem may be constructed:

- What strategies might be adopted to solve the problem?
- Are there differences in the strategies adopted by different individuals in a given population?
- How can these strategies be formally represented?
- How should the aggregation over the population be conducted to produce a model to be estimated with individual data?

The last point is especially important because choice models are derived by aggregating over the actions of individuals within the population, and while any or all of them may indulge in a non-compensatory decision process, it may or may not be appropriate to characterise the sum total of these decisions and the resultant choice model in these terms (see the discussion in Williams and Ortúzar 1982a).

We will just refer here to the first of these issues, namely how an individual confronted by a hypothetical decision context may resolve the multi-criterion problem. There is of course a wide literature dispersed over several fields, which involves the application of decision theory to problems of this kind. We will mention three methods, starting with the best known, simplest and most widely used approach, the trade-off strategy which forms the basis for compensatory decision models.

7.7.2.1 Compensatory Rule

Here the preferred option is selected by optimising a single objective function $O = O(F_1, F_2, \ldots, F_K)$. If the F_k functions are simply the attributes $\mathbf{Z}^k$, or linear transformations of them, O may be written as:

$$O = O \left(\sum_k \theta_k Z_1^k, \ldots, \sum_k \theta_k Z_j^k, \ldots, \sum_k \theta_k Z_N^k \right) \tag{7.49}$$

and the conventional linear trade-off problem is addressed. The parameters θ are determined from either stated or revealed preferences of the individual decision maker. One of the characteristics of this trade-off approach is its symmetric treatment of the objective functions.

7.7.2.2 Non-Compensatory Rules

An alternative general approach is to treat the objective functions (7.47) asymmetrically by either ranking them or converting some or all to constraints by introducing norms or thresholds. That is, we might require that any acceptable alternative has, for example, an associated travel cost not exceeding a particular amount; formally the restriction is imposed that:

$$Z_1^k, \ldots, Z_j^k, \ldots, Z_N^k \leq Z^k \tag{7.50}$$

in which Z^k is a maximum (or minimum when the inequality sign is reversed) satisfactory value for the attribute. The creation of norms or thresholds restricts the range of feasible alternatives which individuals are considered to impose on their decision process.

Choice by Elimination   In this case it is assumed that individuals possess both a ranking of attributes (e.g. cost is more important than waiting time, which in turn is more important than walking time, etc.) and minimum acceptable values or thresholds (7.50) for each. For example, the decision process may solve the multi-criterion problem in the following fashion: first the highest ranked attribute is considered and all alternatives not satisfying the threshold restriction are eliminated (even though they may excel in lesser ranked attributes); the process is repeated until only one option is left, or a group which satisfies all the threshold constraints among which one is selected in a compensatory manner (see Tverski 1972; Cantillo and Ortúzar 2005).

Satisficing Behaviour   There are, however, a great many ways in which the above search strategy may be organised; for example, it might be that a complex cyclic process is used by the individual whereby the thresholds become sequentially modified until a unique alternative is found. Equally, a *satisficing* mechanism might operate in which the individual might be prepared to curtail the search at any point according to a pre-specified rule, in which case some or all of the attributes or alternatives may not be considered. Indeed, when the notion of satisficing (Simon 1957; Eilon 1972) is applied to travel-related decisions involving location, the decision model is closely associated with the acquisition of information in the search process.

As Young and Richardson (1980) remarked, a search may be characterised by an elimination process based on attributes or one based on alternatives. In the former, attributes are selected in turn and options are processed, and maintained or rejected depending on the values of these attributes; in the latter, alternatives are considered in turn and their bundle of attributes examined. At any stage of the process options which do not satisfy norms or other constraints are eliminated. A more detailed consideration of decision strategies is given by Foerster (1979) and Williams and Ortúzar (1982a). Denstadli *et al.* (2011) discuss different decision strategies and go on to characterise the decision process of individuals confronted with different types of choice tasks, by recording their verbalised thoughts while completing them.

7.7.3   Habit and Hysteresis

At the end of the 1970s there was considerable interest in the relevance and role of habit in travel choice behaviour, particularly in cases of relocation (i.e. migration) or other phenomena granting a fresh look at the individual's choice set. Empirical evidence (Blase 1979) suggested that the effect of habit can be of practical significance and the problem should be treated seriously. The interest on this issue has not abated as most commuter trips have a tendency to be repeated over time, thus acquiring a potentially important inertia component (Lanzendorf 2003; Pendyala *et al.* 2001).

The existence of habit, or what might be considered as inertia accompanying the decision process of an individual, is possibly the most insidious of behavioural aspects which represent divergences from the traditional assumptions underpinning choice models, for it appears directly in the response context. In order to examine the effects and implications of habit it is appropriate to return to the assumptions behind the conventional cross-sectional approach.

Figure 7.5a reproduces the S-shaped curve relevant to binary choice. For a given difference in utility $(V_2 - V_1)$ there exists a certain unique probability of choice; under conditions of change $(V_2' - V_1')$, the probability will correspond to that observed for that utility difference in the base year, i.e. the response is determined from the cross- sectional dispersion. An implication of this assumption is that response to a particular policy or change will be exactly reversed if the stimulus is removed; the stimulus–response relation is symmetric with respect to the sign and size of the stimulus.

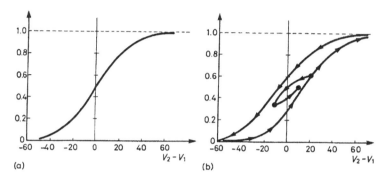

Figure 7.5   Influence of habit in cross-sectional models: (a) Logit response curve, (b) Hysteresis curve for habit effect

If habit exists it will affect those members of the population who are currently associated with an option experiencing a stimulus to the relative advantage of another alternative. This introduces a basic asymmetry into response behaviour and gives rise to the phenomenon of hysteresis (Goodwin 1977), as pictured in Figure 7.5b. In this case the present state of the population identified in terms of the market share of each alternative depends not only on the utility values V_2 and V_1, but on how these variables attained their current value.

Formally, the state of the system P may be expressed as a path integral in the space of utility components **V**; the value of the integral is path independent when habit is absent but path dependent when it is present (see the discussion in Williams and Ortúzar 1982a). These ideas have been taken into an operational model by Cantillo *et al.* (2007), which is a precursor of models for panel data that we will examine in the following section.

7.7.4   Modelling with Panel Data

The long-term planning of transport systems, especially when decisions about substantial changes are involved, requires special demand models. However, most demand models to date have used readily available cross-sectional data, which do not allow for an appropriate consideration of temporal effects as information is considered only for a single point in time. This limitation may be especially restrictive when personal routines are habitual (such as in the cases discussed in the previous section) or when a substantial intervention is planned for a system.

Unfortunately, in transport this is more the norm than the exception as, on one hand, mode choice in a stable context (particularly for non-discretionary trips to work or study) is a process that is more a habit than a plan (Wood *et al.* 2002). On the other hand, changes introduced to transport systems are becoming more common every day. Some famous examples are the electronic road pricing system in Singapore (Menon *et al.* 1993), the congestion charge in London (Banister 2003), Transmilenio in Colombia (Steer Davies Gleave 2000) and, more recently, Transantiago in Chile (Muñoz *et al.* 2009).

In contrast, panel data is an alternative that has significant and relevant advantages (Golob *et al.* 1997; Stopher and Greaves 2004). For example, it is possible to introduce temporal effects, as panels collect information at several successive times retaining the same individuals for the entire series of surveys.

7.7.4.1 Panel Data Models

Although panel data models have been estimated in the past using fairly typical discrete choice functions (notable exceptions are Daganzo and Sheffi 1979; Johnson and Hensher 1982), the presence of repeated observations makes it more appropriate to use a flexible model formulation, accounting for correlation among observations belonging to the same individual. Thus, when more than one observation per individual is available, we need to take into account the sequence of choices, made by the respondent.

Revelt and Train (1998) proposed a ML framework which accommodates inter-respondent heterogeneity but assumes intra-respondent homogeneity in tastes (i.e. it includes the effect of repeated choices by assuming that tastes vary across respondents, but stay constant across observations for the same respondent); this *ML panel probability*, is given by the following product of ML probabilities:

$$P_{jq} = \int_{\theta_q} \prod_{t=1}^{T} \left(\frac{e^{V_{jqt}(\theta_q)}}{\sum_{A_i \in \mathbf{A}^t(q)} e^{V_{iqt}(\theta_q)}} \right) f\left(\theta_q \,|\mathbf{b}, \, \mathbf{\Sigma} \right) d\theta_q \tag{7.51}$$

where V_{iqt} is the observable component of the utility of option A_i for individual q at time t; $\mathbf{A}^t(q)$ is the choice set of individual q at time t; T is the number of periods (*waves*) in the panel, and $f(\cdot)$ is the mixing distribution, with means $\mathbf{b}$ and covariance matrix $\mathbf{\Sigma}$ (i.e. the population parameters) of the coefficients to be estimated in $\mathbf{V}$.

Hess and Rose (2009) relaxed the assumption of intra-respondent homogeneity of tastes, proposing a choice probability with the following form:

$$P_{jq} \int_{\alpha_q} \prod_{t=1}^{T} \left(\int_{\gamma_{q,t}} \frac{e^{V_{jqt}(\theta_q)}}{\sum_{A_i \in \mathbf{A}^t(q)} e^{V_{iqt}^t(\theta_q)}} g(\gamma_{q,t}|\mathbf{\Sigma}_\gamma) d\gamma_{q,t} \right) h(\alpha_q|\mathbf{\Sigma}_\alpha) d\alpha_q \tag{7.52}$$

where θ are now a function of α_q, which varies over respondents with density $h(\alpha_q|\mathbf{\Sigma}_\alpha)$, and $\gamma_{q,t}$, which varies over all choices with density $g(\gamma_{q,t}|\mathbf{\Sigma}_\gamma)$. This model has integrals inside and outside the product over periods; the latter accounts for inter-respondent heterogeneity as in the previous model (Revelt and Train 1998), while the inside integral accounts for intra-respondent heterogeneity. However, this formulation is more demanding in terms of estimation time and currently available packages just allow using a simplified version of it (see Hess and Rose 2009). Fortunately, the need for assuming intra-respondent heterogeneity is not that pressing, as it is reasonable to expect that in the short to medium term respondent tastes will probably stay the same.

On the other hand, several empirical applications have shown that the inclusion of inter-respondent heterogeneity in the random parameters leads to very significant improvements in model fit and a greater ability to retrieve taste heterogeneity. In fact, this is also the most common approach to deal with stated preference data that includes multiple choices for each respondent as we discuss in section 8.7.2.7 (the approach has been implemented in the majority of estimation packages). Although the influence of repeated observations (i.e. inter-respondent heterogeneity in tastes) can be considered directly via the estimation of random parameters, there might be extra correlation across repeated observations besides the effect of the random parameters. Thus, even though random parameters and error components might induce confounding effects, they might also account for slightly different effects. In fact, as long as both effects are significant, the pure error-panel component accounts for correlation in the preference for alternatives, while the random parameters account for correlation in tastes (Yáñez *et al.* 2010b).

7.7.4.2 Efficiency and Repeated Observations

Efficiency, in general, can be measured by the Fisher information matrix $\mathbf{I}$ (see Example 3.15); this is inversely related to sample size, the attribute values associated with the estimated parameters and the probability associated with the chosen alternative (McFadden 1974). Rose and Bliemer (2008) analysed the effect of the number of alternatives, attributes, and attribute levels on the optimal sample size for SC experiments in MNL models, as part of their search for the design with highest asymptotic efficiency of the estimated parameters. They found that only the range of attribute levels could offer an explanation for some problems of convergence encountered in their experiments. Cherchi and Ortúzar (2008b) demonstrated that while efficiency clearly improves with sample size, data variability does not always increase it.

In contrast, the repeated observations in a short-survey panel, for example, will increase the number of observations but might reduce data variability, because observations that are identical do not bring new information about attribute trade-offs. Thus, when using panel data it is important to understand how efficiency is influenced by the repeated observations and up to what point these are actually beneficial. This is also crucial to determine the length of a multi-day panel survey, which is something that has not been explored much up to date. Moreover, as in panel data each individual provides more than one observation, it is necessary to account for correlation among these and this has a different effect depending on how the repeated observations are treated. Cherchi *et al.* (2009) found that the effect of correlation is, to a large extent given by the repeated observations.

In Chapter 8 we will see that when the parameters of a discrete choice model are estimated by maximum likelihood, the expected value of the variance of the *kth* estimated parameter (i.e. the *kth* element of the diagonal of the Fisher information matrix) is given by:

$$E\left[\frac{\partial^2 \ell(\theta)}{\partial \theta_k^2}\right] \cong \sum_{q=1}^{Q} \sum_{A_j \in \mathbf{A}(q)} \left[\frac{\partial^2 \left(g_{jq} \ln P_{jq}(\mathbf{x}_{jq}, \theta)\right)}{\partial \theta_k^2}\right]_{\theta=\hat{\theta}} \tag{7.53}$$

where $\ell(\theta) = \ln \prod_q P_{jq}^{c_{jq}}$ is the log-likelihood function with respect to the parameters θ evaluated at their estimated values, P_{jq} is the probability that individual q chooses alternative A_j among the alternatives belonging to her choice set $\mathbf{A}(q)$, $\mathbf{x}_{jq}$ are the level-of-service and socio-economic attributes, and g_{jq} equals one if A_j is the alternative actually chosen by individual q and zero otherwise.

Equation (7.53) shows that the efficiency of the estimated parameters depends on sample size, the values of the attributes associated with the estimated parameters and the probability of the chosen

(continued)

alternative. As the Logit probability also depends, among other things, on the data variability and on the variance of the error term (through the scale factor), understanding the sensitivity of the efficiency of the estimated parameters is a complex task. Cherchi and Ortúzar (2008b) analysed how the efficiency of the estimated parameters varied for RP and SC data.

Looking at expressions for a single element of the Fisher information matrix, as above, is convenient for a theoretical discussion of the efficiency issue, because they illustrate what elements influence the matrix $\mathbf{I}$. However, in practice it would be better to measure the statistical efficiency of the expected outcomes of models as in the experimental design literature (Rose and Bliemer 2009), by computing the negative inverse of the Fisher information matrix (i.e. the asymptotic covariance matrix, $\mathbf{S}^2$) and then computing the D-error (see section 3.4.2.3); a smaller D-error yields more efficient estimates.

Let us consider, for simplicity, a binary Logit model (i.e. with 'fixed' parameters). The variance of the parameters estimated with panel data is given by:

$$\text{var}(\hat{\theta}) = -\frac{1}{\sum_q \sum_t \Delta x_{jqt}^2 \hat{P}_{jqt}(1 - \hat{P}_{jqt})} \tag{7.54}$$

where Δx_{jqt}^2 is the attribute difference between both alternatives in period t. However, in contrast to the case of, for example SC data, when using information from a short survey panel the attribute values will be identical for the same individual in the period (i.e. five days of the week). Thus, in such cases we will have that $\Delta x_{jqt} = \Delta x_{jq} \forall t$ and the variance of the parameters will simplify to:

$$\text{var}(\hat{\theta}) = -\frac{1}{\sum_q \sum_t \Delta x_{jq}^2 \hat{P}_{jqt}(1 - \hat{P}_{jqt})}$$

These equations show that the variance depends clearly on the number of repeated observations as well as on the data variability and number of observations. However, the efficiency of the parameters increases with the variability of the attributes only for scale factors over 0.5. This, which might seem counterintuitive, is due to the effect that the scale factor has on the variability of the data, because efficiency reduces as data variability diminishes; and is also due to the second order function of the probability, that tends to zero as the probability of the chosen alternative approximates one.

It is important to note that a panel with identical repeated observations for each individual is a special case. In fact, in terms of the above discussion having equal observations repeated a certain number of times increases only marginally the variability of the attributes. In particular, if N is the number of observations and R is the number of times these are repeated for each individual, the variance of the attributes (Δx_{jq}) for N and RN observations is related by the following expression (Yáñez et al. 2010b):

$$\frac{\text{var}(R\Delta x_{jq})}{\text{var}(\Delta x_{jq})} = \frac{(RN - R)}{(RN - 1)} \tag{7.55}$$

Hence, identical repeated observations should not in theory influence the efficiency of the estimated parameters. This result may be confirmed by computing the D-error.

The extension of this result to the ML case (which we need to properly estimate models with panel data) is not difficult. In the ML model, the variance of the mean of the random parameters is more complex, but the structure is basically the same (Cherchi and Ortúzar 2008b). It is still inversely related to the square value of the attributes associated with each parameter (as in the case of the fixed parameters model), to the number of repeated observations, and is also a function of the probabilities (Bliemer and Rose 2010).

Using observed data, Yáñez *et al.* (2010b) show that the inclusion of intra-respondent hetero-geneity requires more observations, which means that the repeated observations can affect the definition of model structure. Therefore, a potential benefit of considering a longer multi-day survey in a short-survey panel context is the highest probability to capture different kinds of heterogeneity among observations.

Complementary, results from simulated data have shown that having repeated observations in a data panel increases the efficiency of the estimated parameters only because this increases the sample dimensions. Therefore, based on the results from real and synthetic data, it is possible to say that there is a trade-off between the higher probability of capturing effects (different types of heteroskedasticity) in a longer multi-day-panel sample, and the risk of a decreased capability of reproducing true phenomena (as this worsens in the presence of repeated observations).

Finally, a suggestion on the definition of the length of a multi-day-panel survey would be to consider not only the number of individuals, but also the level of routine expected. This last factor seems to be especially important in a short-survey panel context, as these panels commonly feature a large proportion of identical observations, which are actually harmful, i.e. they reduce the capability of reproducing the true phenomenon. Thus, even though having more observations per respondent requires smaller sample sizes to establish the statistical significance of the parameter estimates derived from choice data (Rose *et al.*, 2009b), Yáñez *et al.* (2010b) show that this is effectively true if and only if the level of routine is not strong.

7.7.4.3 Dealing with Temporal Effects

One of the temporal effects more often discussed in the literature is *habit*, leading to *inertia* (Goodwin 1977; Blase 1979; Williams and Ortúzar 1982a); Daganzo and Sheffi (1979) proposed a MNP formulation to treat this phenomenon which was later implemented by Johnson and Hensher (1982) for a two-wave panel in Australia. More recently, the discrete choice modelling field has seen significant advances in terms of incorporating inertia, examples of that are: a model including prior behaviour on a time-series context (Swait *et al.* 2004), a model including inertia on a two-wave panel formulation (Cantillo *et al.* 2007), and a *planning-and-action* model considering inertia as an effect of previous plans (Ben-Akiva 2009). All these studies refer to cases where there are no changes in the transport system (i.e. a stable choice environment).

The changing choice environment defined by the *Santiago Panel* (Yáñez *et al.* 2010a), with data before and after the introduction of Transantiago (Muñoz *et al.* 2009), acted like a *shock* to the system and required the introduction of another temporal effect beyond inertia. Assuming that the shock effect could reduce or even overcome the effect of inertia, Yáñez *et al.* (2010d) formulated a model incorporating the effects of three forces involved in the choice process: (1) the relative values of the modal attributes, (2) the inertia effect, and (3) the shock resulting from an abrupt policy intervention.

In their model, inertia is a function of the previous valuation of the options and its effect may vary for each wave and among individuals due to systematic or purely random effects. Furthermore, the effect might be positive or negative; the former representing the 'typical' inertia effect in the absence of changes, the latter indicating the preference for changing that might occur after a significant variation in the system.

On the other hand, after a shock individuals may modify their valuation process, altering their utility function. The shock effect is a function of the difference between the utility of an option evaluated at the current wave w, and its utility evaluated at the preceding wave ($w-1$); hence, the effect is expected to be negative when the alternative worsens (making its utility lower), and positive when

(continued)

it improves. The perception of the shock may also be different for each wave and may vary among individuals due to systematic or random effects. In particular, the shock effect should have the highest value immediately after the introduction of the new policy, and then its magnitude should attenuate.

According to these assumptions, let the utility associated with each option A_j at wave $w = 1$ (i.e. the base situation) be the sum of observable (V_{jq1}) and non-observable components (ζ_{jq1}). Then, the probability of choosing option $A_j \in \mathbf{A}^1(q)$ at wave $w = 1$ will be, as usual:

$$P_q\left(A_j^1\right) = \text{Prob}\left((V_{jq1} + \zeta_{jq1}) - (V_{iq1} + \zeta_{iq1}) \geq 0, \quad \forall A_i^1 \in A^1(q)\right) \tag{7.56}$$

where $\mathbf{A}^1(q)$ is the choice set of individual q in wave $w = 1$. In subsequent waves, the option chosen in the previous wave will be denoted by A_r; temporal effects will be also included to detect how the choices in a given wave (w) are influenced by the choices made in a previous one ($w-1$).

If $\tilde{U}_{jqw}$ denotes the utility that individual q associates to a generic option A_j on wave w ($w = 2, 3$, etc.). This utility will include inertia and shock effects, such that:

$$\tilde{U}_{jqw} = U_{jqw} - I_{jrq}^w + S_{jq}^w \tag{7.57}$$

where I stands for inertia and S for shock, and there are several ways to express them. In particular, Yáñez *et al.* (2010d) proposed the following general expressions:

$$I_{jrq}^w = \left(\theta_{Ij}^w + \delta_{iq} \cdot \sigma_{Ij}^w + \theta_{I\_SE} \cdot SE_I\right) \cdot \left(V_{rq(w-1)} - V_{jq(w-1)}\right) \tag{7.58}$$

$$S_{jq}^w = \left(\theta_{Sj}^w + \delta_{Sq} \cdot \sigma_{Sj}^w + \theta_{S\_SE} \cdot SE_S\right) \cdot \left(V_{jqw} - V_{jq(w-1)}\right) \tag{7.59}$$

where θ_{Ij}^w and θ_{Sj}^w are the population means, and σ_{Ij}^w and σ_{Sj}^w the standard deviations of the inertia and shock parameters respectively, for option A_j on wave w; SE_I and SE_S are socioeconomic variables, with parameters θ_{I_SE} and θ_{S_SE} respectively; these allow for systematic variations of the inertia and shock parameters, δ_{iq}, δ_{Sq} are the standard factors to introduce panel correlation (note that these could be included either as random parameters or error components), and $\mathbf{V}$ are the observable components of the utility function without temporal effects.

Note that if I_{jrq}^w is greater than zero inertia exists; while, if I_{jrq}^w is negative, it implies that the individual has a high disposition to change. Also, note that (7.58) assumes a zero inertia effect on wave w for the option chosen on wave ($w-1$). It means: $\tilde{U}_{rqw} = U_{rqw} + S_{rq}^w$.

In the presence of inertia and shock, the probability to change from A_r (i.e. the option chosen in the previous wave) to A_j (i.e. the 'candidate option') for individual q on wave w, is given by:

$$P_{jqw} = \text{Prob}\left(\tilde{U}_{jqw} - \tilde{U}_{rqw} \geq 0 \quad \text{and} \quad \tilde{U}_{jqw} - \tilde{U}_{iqw} \geq 0, \quad \forall A_i^w \in \mathbf{A}^w(q), \text{ except } r = j\right) \tag{7.60}$$

while, the probability to remain with A_r is given by:

$$P_{rqw} = \text{Prob}\left(\tilde{U}_{rqw} - \tilde{U}_{jqw} \geq 0\right)$$

In this formulation, and as usual in current practice, option attributes and socioeconomic characteristics are associated with parameters that could be either fixed or random; on the other hand, the non-observable component ζ_{jqw} is a random error term that can be formulated as $\zeta_{jqw} = \upsilon_q + \varepsilon_{jqw}$, where υ_q is a random effect specific to the individual and ε_{jqw} is, once more, the typical random error distributed IID EV1.

With all the above, the probability of choosing option A_j on wave w, ($\forall w > 1$) can be written as:

$$
\begin{aligned}
P_{jqw} = \exp &\left(V_{jqw} - \left(\theta_{Ij}^w + \delta_{iq} \cdot \sigma_{Ij}^w + \theta_{I\_SE} \cdot SE_I\right) \cdot \left(V_{rq(w-1)} - V_{jq(w-1)}\right)\right. \\
&\left. + \left(\theta_{Sj}^w + \delta_{Sq} \cdot \sigma_{Sj}^w + \theta_{S\_SE} \cdot SE_S\right) \cdot \left(V_{jqw} - V_{jq(w-1)}\right)\right) \\
&\cdot \left[\sum_i \left(\exp\left(V_{iqw} - \left(\theta_{Ii}^w + \delta_{iq} \cdot \sigma_{Ii}^w + \theta_{I\_SE} \cdot SE_I\right) \cdot \left(V_{rq(w-1)} - V_{iq(w-1)}\right)\right.\right.\right. \\
&\left.\left.\left. + \left(\theta_{Si}^w + \delta_{Sq} \cdot \sigma_{Si}^w + \theta_{S\_SE} \cdot SE_S\right) \cdot \left(V_{iqw} - V_{iq(w-1)}\right)\right)\right)\right]^{-1}
\end{aligned}
\tag{7.61}
$$

where if $j = r$, then $(V_{rq(w-1)} - V_{jq(w-1)}) = 0$ and, as previously discussed, inertia is zero while the shock effect would still be active. Actually, the shock effect S_{jq}^w is null if either the shock parameter is itself null ($\theta_{Sj}^w = 0$) or if the utility of option A_j does not change between consecutive waves, i.e. $V_{jqw} = V_{jq(w-1)}$.

Note that equation (7.61) is a general formulation that can accommodate panel correlation either in the representative utility V_{jqw} (using random parameters), error term (as an error component), or in the inertia and shock effects (again using random parameters). But for empirical estimation it is not possible to consider all these panel correlation forms at the same time. In fact, since the inertia and shock parameters multiply the expressions $\Delta V_I = (V_{rq(w-1)} - V_{jq(w-1)})$ and $\Delta V_S = (V_{jqw} - V_{jq(w-1)})$ respectively, randomness cannot be added in the representative utility and temporal effects at the same time; we will come back to these issues in Chapter 8.

As individual responses present panel correlation, given a sequence of choices A_j^w, one for each wave, the probability that a person follows this sequence is given by:

$$P_q\left(A_j^1 \wedge A_j^2 \wedge \ldots A_j^W\right) = \prod_{w=1}^{W} P_{jqw} \qquad (7.62)$$

and as inertia, shock and panel correlation are actually unknown, the probability of this sequence of choices is of Mixed Logit form; we will look at ways to estimate this model in Chapter 8.

7.7.5 Hybrid Choice Models Incorporating Latent Variables

The inclusion of subjective elements in discrete choice models re-emerged recently as an analysis and discussion topic, after losing some of the importance that made it a subject in the early 80s (see for example Ortúzar and Hutt 1984; McFadden 1986). Thus, *hybrid choice models* have been proposed considering not only tangible attributes of the alternatives (classic explanatory variables) as in traditional choice models, but also more intangible elements associated with users' perceptions and attitudes (including happiness), expressed through latent variables (Morikawa and Sasaki 1998; Ashok *et al.* 2002; Abou-Zeid and Ben-Akiva 2009).

To estimate models with both kinds of variables, two methods have been developed: the sequential approach, on which the latent variables are constructed before entering into the discrete choice model as a further regular variable (Ashok *et al.* 2002; Vredin Johansson *et al.* 2005; Raveau *et al.* 2010) and the simultaneous approach, where both processes are done at once (Bolduc *et al.*, 2008; Raveau *et al.* 2009). It has been argued that the second approach should result in more efficient estimators of the involved parameters (Ben-Akiva *et al.* 2002), but it has been used less often due to its greater complexity and because currently available software does not allow to exploit the full capabilities of the base discrete choice model as we will see below. We will come back to these issues in Chapter 8.

7.7.5.1 Modelling with Latent Variables

Latent variables are factors that, although they influence individual behaviour and perceptions, cannot be quantified in practice (e.g. safety, comfort, reliability). This is because of either their intangibility, as these variables do not have a measurement scale, or their intrinsic subjectivity (i.e. different persons may perceive them differently). Identification of latent variables requires supplementing a standard survey with questions that capture users' perceptions about some aspects of the alternatives (and the

(continued)

choice context). The answers to these questions generate perception indicators that serve to identify the latent variables. Otherwise, these latent variables could not be measured.

To make use of latent variables (Bollen 1989) a MIMIC (Multiple Indicator Multiple Cause) model is estimated, where the latent variables (η_{ilq}) are explained by characteristics s_{iqr} from the users and from the alternatives through *structural equations* such as (7.63); at the same time, the latent variables explain the perception indicators (y_{ipq}) through *measurement equations* as (7.64):

$$\eta_{ilq} = \sum_r \alpha_{ilr} \cdot s_{iqr} + v_{ilq} \tag{7.63}$$

$$y_{ipq} = \sum_l \gamma_{ilp} \cdot \eta_{ilq} + \zeta_{ipq} \tag{7.64}$$

where the index i refers to an alternative, q to an individual, l to a latent variable, r to an explanatory variable and p to an indicator; α_{ilr} and γ_{ilp} are parameters to be estimated, while v_{ilq} and ζ_{ipq} are error terms with mean zero and standard deviation to be estimated. As the η_{ilq} terms are unknown, both equations must be considered jointly in the parameter estimation process.

7.7.5.2 Hybrid Discrete Choice Model

When latent variables η_{ilq} are considered, the systematic or representative utility V_{iq} in equation (7.2) incorporates them together with the objective attributes x_{ikq} (i.e. travel time or fare, as well as socioeconomic characteristics of the individual), leading to a utility function such as:

$$V_{iq} = \sum_k \theta_{ik} \cdot x_{ikq} + \sum_l \beta_{il} \cdot \eta_{ilq} \tag{7.65}$$

where θ_{ik} and β_{il} are parameters to be estimated. However, Since the η_{ilq} variables are unknown the model must be estimated jointly with the MIMIC model's structural (7.63) and measurement (7.64) equations. Finally, to characterise individual decisions binary variables g_{iq}, that take values according to (7.72), have to be defined:

$$g_{iq} = \begin{cases} 1 & \text{if } U_{iq} \geq U_{jq}, \quad \forall j \in \mathbf{A}(q) \\ 0 & \text{in other case} \end{cases} \tag{7.66}$$

where, as usual, $\mathbf{A}(q)$ is the set of available alternatives for individual q.

Note that as the latent variables are on the right-hand side (i.e. as explanatory or independent variables) both in the utility function (7.65) and in the measurement equation (7.64) of the MIMIC model, there will not be endogeneity for simultaneous determination even if the errors (of either equation) are correlated (see Guevara and Ben-Akiva 2006).

In Chapter 8 we will discuss the two methods available to estimate these hybrid models in practice, and comment on some interesting findings.

Exercises

7.1 There is interest to study the behaviour of a group of travellers in relation to two transport options A and B, with travel times t_a and t_b respectively. It has been postulated that each traveller experiments the following net utilities from each option:

$$U_a = \alpha t_a + \beta I$$
$$U_b = \alpha t_b$$

where α and β are known parameters and I is the traveller's personal income.

Although there is no reliable data about the income of each traveller, it is known that the variable I has the following distribution in the population:

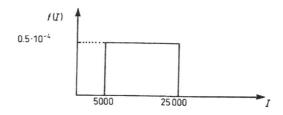

If $\alpha = -0.5$ and $\beta = 2.10^{-4}$, find out the probability function of choosing option A for a given traveller, as a function of the value of $(t_b - t_a)$; sketch the function in appropriate axis.

7.2 Consider a binary Logit model for car and bus, where the following representative utility functions have been estimated with a sample of 750 individuals belonging to a particular sector of an urban area:

$$V_c = 3.5 - 0.25t_c - 0.42e_c - 0.1c_c$$
$$V_b = -0.25t_b - 0.42e_b - 0.1c_b$$

where t is in-vehicle travel time (min), e is access time (min) and c is travel cost ($\$$).

Assume the following average data is known:

Mode	Variable		
	t	e	c
Car	25	5	140
Bus	40	8	50

If you are informed that the number of individuals choosing each option in the sector and in the complete area are respectively as follows:

Option	Number of individuals choosing option i	
	Sample	Population
Car	283	17 100
Bus	467	68 900

(a) Indicate what correction would be necessary to apply to the model and write its final formulation.
(b) Calculate the percent variation in the probability of choosing car if the bus fares go up by 25%.
(c) Find out what would happen if, on the contrary, the car costs increase by 100%.

7.3 Compute the probabilities of choosing car, bus, shared taxi and underground, according to the following Nested Logit model:

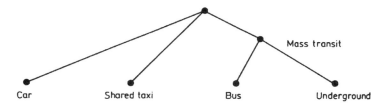

with the following utility functions:
(a) High nest

$$V_c = -0.03t_c - 0.02c_c + 1.25$$
$$V_{st} = -0.03t_{st} - 0.02c_{st} - 0.20$$
$$V_{mt} = 0.60EMU$$

(b) Mass transit nest

$$V_b = -0.04t_b - 0.03c_b + 0.5$$
$$V_u = -0.04t_u - 0.03c_u$$

and for the average variable values presented in the following table:

Mode	Time (t)	Cost/income (c)
Car	4.5	23.0
Shared taxi	5.5	15.0
Bus	7.5	5.5
Underground	5.5	3.6

7.4 The binary Probit model has the following expression:

$$P_1 = \Phi \{(V_1 - V_2)/\sqrt{\sigma_1^2 + \sigma_2^2 - 2\rho\sigma_1\sigma_2}\}$$

Using this result write down the probability of choosing option one in the following binary model:

$$U_i = \theta X_i + \varepsilon_i$$

where the ε are distributed IID standard Normal, for the following cases:
(a) If the value of θ is fixed and equal to 3.
(b) If θ is distributed Normal $N(3, 1)$ and is independent of the ε.

8

Specification and Estimation of Discrete Choice Models

8.1 Introduction

The previous chapter provided an overview of discrete choice modelling and an introduction to different model forms and theoretical frameworks for individual decisions. This chapter is devoted to a discussion of two key issues: how to fully specify a discrete or disaggregate model (DM) and how to estimate such a model once properly specified.

The search for a suitable model specification involves selecting the structure of the model (MNL, NL, ML, etc.), the explanatory variables to consider, the form in which they enter the utility functions (linear, non-linear) and the identification of the individual's choice set (alternatives perceived as available). In broad terms the objectives of a specification search include realism, economy, theoretical consistency and policy sensitivity. In other words, we search for a realistic model, which does not require too many data and computer resources, does not produce counter-intuitive results and is appropriate to the decision context in which it is to be used. Early aggregate models such as those discussed in Chapters 5 and 6 were often critically portrayed as policy insensitive, either because key variables have been completely left out of the model or because important model components have been specified as insensitive to certain policies (e.g. consider the problem of inelastic trip generation). Most of the features of model specification are susceptible to analysis and experimentation (see Leamer 1978) but they are also strongly dependent on study context and data availability.

In this chapter we start by considering how to identify the set of options available to individuals: choice-set determination. This is a key problem as we usually estimate DM by means of the (generally) observed individual choices between alternatives. These should be the alternatives actually considered, consciously or unconsciously, by the individual. The omission of seemingly unimportant options on the grounds of costs may bias results. For example, in the vast majority of aggregate studies only binary choice between car and public transport has been considered with the consequence that the multimodal problem could not be treated seriously; in fact, in many cases the consideration of alternative public-transport options was relegated to the assignment stage employing multipath allocation of trips to sub-modal network links. In the same vein, the inclusion of alternatives which are actually ignored by certain groups (say walking more than 500 metres for high income individuals), could also bias model estimation.

Modelling Transport, Fourth Edition. Juan de Dios Ortúzar and Luis G. Willumsen.
© 2011 John Wiley & Sons, Ltd. Published 2011 by John Wiley & Sons, Ltd.

Section 8.3 then considers the other elements of model specification and in particular functional form and model structure. The criteria of economy, realism, theoretical consistency and decision-making context play a key role in complementing the experience and intuition of the modeller during specification searches. An additional, and often over-riding element, is the availability of specialised software. In fact, one of the reasons behind the immense popularity of the linear-in-the-parameters Multinomial Logit (MNL) model is that it can be easily estimated with normally available software; this was not the case, for many years, for more general structures or functional forms which presented much greater difficulties (Daganzo 1979; Liem and Gaudry 1987).

The increasing availability of good software to select and estimate these models has certainly alleviated this problem. However, one issue to which we will return is that although we may be able to successfully estimate the parameters of widely different models with a given data set, these (and their implied elasticities) will tend to be different and we often lack the means to discriminate between them, at least with cross-sectional data. Another important issue is that of interpretation of results. More complex/richer models are even more dependent on data quality than their simpler counterparts, and the insights they offer on individual behaviour often require experienced analysts to interpret them correctly.

The final specification will then depend heavily on the modeller's experience and theoretical understanding, and context-specific factors such as: time and resources available for the modelling activity, degree of correlation among alternatives, heterogeneity of preferences and required degree of accuracy of the forecasts. It must be borne in mind that using an inadequate model, such as the MNL when the hypotheses needed to generate it do not hold, may lead to serious errors (Williams and Ortúzar 1982a).

Section 8.4 concentrates on the statistical estimation of discrete choice models using data from random and choice-based samples and including methods to validate models and compare different model structures; we also consider here the estimation of hybrid choice models with latent variables. Section 8.5 discusses two methods available to estimate the Multinomial Probit model, and section 8.6 discusses in depth the estimation of the Mixed Logit model, including its application to modelling with panel data. The chapter concludes with considerations relevant to model estimation and forecasting with stated preference data and the joint estimation of RP-SP models.

8.2   Choice-Set Determination

One of the first problems an analyst has to solve, given a typical revealed-preferences cross-sectional data set, is that of deciding which alternatives are available to each individual in the sample. It has been noted that this is one of the most difficult of all the issues to resolve, because it reflects the dilemma the modeller has to tackle in arriving at a suitable trade-off between modelling relevance and modelling complexity; usually however, data availability acts as a yardstick.

8.2.1   Choice-set Size

It is extremely difficult to decide on an individual's choice set unless one asks the respondent directly; therefore the problem is closely connected with the dilemma of whether to use reported or measured data, as discussed in Chapter 3. Although in mode choice modelling the number of alternatives is usually small, rendering the problem less severe, in other cases such as destination choice, the identification of options in the choice set is a crucial matter. This is not simply because the total number of alternatives is usually very high, as we will see below, but because we face the added problem of how to measure/represent the attractiveness of each option. Ways of managing a large choice set include:

1. Taking into account only subsets of the options which are effectively chosen in the sample (i.e. in a sampling framework such as the one used by Ben-Akiva 1977).

2. Using the *brute force* method, which assumes that everybody has all alternatives available and hence lets the model decide that the choice probabilities of unrealistic options are low or zero.

   Both approaches have disadvantages. For example, in case 1 it is possible to miss realistic alternatives which are not chosen owing to the specific sample or sampling technique; in case 2 the inclusion of too many alternatives may affect the discriminatory capacities of the model, in the sense that a model capable of dealing with unrealistic options may not be able to describe adequately the choices among the realistic ones (see Ruijgrok 1979). Other methods to deal with the choice set-size problem are:

3. The aggregation across options, such as in a destination choice model based on zonal data.
4. Assuming continuity across alternatives, such as in the work of Ben-Akiva and Watanatada (1980).

8.2.2   Choice-set Formation

Another problem in this realm is that the decision maker being modelled may well choose from a relatively limited set; in this sense if the analyst models choices which are actually ignored by the individual, some alternatives will be given a positive probability even if they have no chance of being selected in practice. Moreover, consider the case of modelling the behaviour of a group of individuals who vary a great deal in terms of their knowledge of potential destinations (owing perhaps to varying lengths of residence in the area); because of this, model coefficients which attempt to describe the relationship between predicted utilities and observed choices may be influenced as much by variation in choice sets among individuals (which are not fully accounted for in the model) as by variations in actual preferences (which are accounted for). Because changes in the nature of the destinations may affect choice set and preferences to different degrees, this confusion may be likely to play havoc with the use of the model in forecasting or with the possibility of transferring it over time and space.

Ways to handle this problem include:

1. The use of heuristic or deterministic choice-set generation rules which permit the exclusion of certain alternatives (i.e. bus is not available if the nearest stop is more than some distance away) and which may be validated using data from the sample.
2. The collection of choice-set information directly from the sample, simply by asking respondents about their perception of available options (it has been found preferable to ask which options, out of a previously researched list, are not available and why).
3. The use of random choice sets, whereby choice probabilities are considered to be the result of a two-stage process: firstly, a choice-set generating process, in which the probability distribution function over all possible choice sets is defined; and secondly, conditional on a specific choice set, a probability of choice for each alternative is defined (see the discussions by Lerman 1984 and Richardson 1982).

Non-compensatory protocols, such as satisfaction, lexicographic behaviour and elimination by aspects, may often be more appropriate than compensatory behaviour, as we saw in Chapter 7. In fact, many choice processes may be seen as a mixture of compensatory and non-compensatory protocols, and this is especially the case when the number of physically available options is large. In this context, Morikawa (1996) developed a hybrid model that applies compensatory and non-compensatory decision rules with a relatively large number of alternatives in a model where the decision process is divided into a choice-set formation stage and a choice stage. Choice-set formation is modelled by a random constraints model that has a non-compensatory nature among constraints, and the choice stage is described by a Multinomial Logit model. This approach gave good results when applied to destination choice of vacation trips with up to 18 alternatives.

8.3   Specification and Functional Form

The search for the best model specification is also related to functional form. Although it may be argued that the linear function (7.3) is probably adequate in many contexts, there are others such as destination choice where non-linear functions are deemed more appropriate (Foerster 1981; Daly 1982a). The problems in this case are: firstly, that in general there is no guarantee that the parameter-estimation routine will converge to unique values and, secondly, that suitable software is not readily available. Another specification issue related to functional form is how the explanatory variables should enter the utility function, even if this is linear in the parameters.

Three approaches have been proposed in the literature to handle the functional form question:

- The use of stated preference in real or laboratory experiments to determine the most appropriate form of the utility function (Lerman and Louviere 1978); we will briefly come back to this in section 8.7.
- The use of statistical transformations, such as the Box–Cox method, letting the data 'decide' to a certain extent an appropriate form (Gaudry and Wills 1978).
- The constructive use of econometric theory to derive functional form (Train and McFadden 1978; Jara-Díaz and Farah 1987; Jara Díaz 2007); this is perhaps the most attractive proposition as the final functional form can be tied up to evaluation measures of user benefit.

As we will see later, it is important to note that, in general, non-linear forms imply different trade-offs to those normally associated with concepts such as the value of time (Bruzelius 1979); also, it is easy to imagine that model elasticities and explanatory power may vary dramatically with functional form.

8.3.1   Functional Form and Transformations

Linear-in-the-parameters expressions such as (7.3) usually contain a mixture of quantitative and qualitative variables (where the latter are normally specified as dummies, i.e. sex, age, income level), and the problems are how to enter both and where to enter the latter, as we have already discussed. In other words, it would be more appropriate to write (7.3) as:

$$V_{jq} = \sum_k \theta_{kj} f_{kj}(x_{kjq}) \tag{8.1}$$

which is still linear in the parameters, but makes it explicit that the functional form of the **x** variables is somewhat arbitrary. Usual practice consists in entering the variables in raw form (i.e. time rather than 1/time or its logarithm) but this could have some consequence if the model response is sensitive to functional form.

If we do not have theoretical reasons to back up a given form, it appears interesting to let the data indicate which could be an appropriate one. A class of transformations widely used in econometrics has been successfully adapted for use in transport modelling (see Gaudry and Wills 1978; Liem and Gaudry 1987). We will review two examples, the second one being a generalisation of the first:

8.3.1.1   Basic Box–Cox Transformation

The transformation $x^{(\tau)}$ of a positive variable x, given by:

$$x^{(\tau)} = \begin{cases} (x^\tau - 1)/\tau, & \text{if } \tau \neq 0 \\ \log x, & \text{if } \tau = 0 \end{cases} \tag{8.2}$$

is continuous for all possible τ values. With this we can rewrite equation (8.1) as:

$$V_{jq} = \sum_k \theta_{kj} x_{kjq}^{(\tau_k)} \tag{8.3}$$

and it is easy to see that if $\tau_1 = \tau_2 = \ldots = \tau_k = 1$, (8.3) reduces to the typical linear form (7.3); furthermore, if all $\tau_k = 0$, we obtain the widely used log-linear form. Therefore both traditional forms are only special cases of (8.3).

8.3.1.2 Box–Tukey Transformation

The basic transformation (8.2) is only defined for $x > 0$; a more general form, for variables that may take negative or zero values, is given by:

$$(x + \mu)^{(\tau)} = \begin{cases} [(x + \mu)^{\tau} - 1]/^{\tau}, & \text{if } \tau \neq 0 \\ \log(x + \mu), & \text{if } \tau = 0 \end{cases} \tag{8.4}$$

where μ is just a translational constant chosen to ensure that $(x + \mu) > 0$ for all observations.

The values of τ must satisfy certain conditions if the model is to be consistent with microeconomic theory. In particular, it is instructive to derive what restrictions exist in the case of attributes such as travel time (which produce disutility) or the number of cars in the household (which should increase the probability of choosing car), to ensure decreasing marginal utilities as the theory demands. This small challenge is left for the interested reader.

It can be shown that if an MNL is specified with functional form (8.4) and restricting all τ to be equal, its elasticities are given by:

$$E_{P_J,x_{ki}} = (\delta_{ji} - P_j)x_{ki}\theta_k(x_{ki} + \mu)^{\tau-1} \tag{8.5}$$

with δ_{ji} equal to 1 if $j = i$ and 0 otherwise. Although it is obvious from (8.5) that the elasticities depend on the values of τ and μ, it is not clear how large the effect might be as the values of θ also vary.

In Chapter 15 we will discuss the consequences of using Box–Cox models in the derivation of subjective values of time (Gaudry *et al.* 1989).

8.3.2 Theoretical Considerations and Functional Form

Although we have made it clear that in any particular study, data limitations and resource restrictions often play a vital role, it is important to consider the influence of theory in the construction of a demand function. In what follows we will show how the constructive use of economic theory helps to solve the important problem of how to incorporate a key variable, such as income, in a utility function. Throughout we will assume a linear-in-the-parameters form and will not be concerned with model structure, but the analysis may be generalised at a later stage.

The conventional approach to understanding the roles of income, time and cost of travel within the discrete choice framework, is based on the work of Train and McFadden (1978); they established the microeconomic foundations of the theory by considering the case of individuals who choose between leisure (L) and goods consumed (G); the trade-off appears once the link between G and income (I) is formulated: they assume that I depends on the number of hours worked (W). Thus, increasing W allows G to increase, diminishing L. More formally the problem is stated as follows:

$$\text{Max} U(G, L)$$

subject to:

$$G + c_i = wW \atop W + L + t_i = T \Bigg\} \forall A_i \in \mathbf{A} \qquad (8.6)$$

where U is the individual utility function, w is the real wage rate (the amount the individual gets paid per hour), c_i and t_i are the money and time spent per trip respectively, $\mathbf{A}$ is the choice set and T is a reference period; the unknowns are G, L and W.

If U in problem (8.6) is given a fairly general form, such as Cobb–Douglas, finding its maximum with respect to $A_i \in \mathbf{A}$, is equivalent to finding the maximum of $(-c_i/w - t_i)$ among other possibilities. This is the origin of the widely used cost/wage rate variable in discrete-mode choice models, for which cost/income has been used as a proxy in many applications. The possibility of adapting working hours to attain a desired level of income plays a key role in the above derivation; thus, as W is endogenously determined and w is given exogenously, income becomes endogenous. This formulation assumes that the cost of travelling is negligible in relation to income, i.e. that there is no income effect.

However, for many individuals (particularly in emerging countries) both income and working hours are fixed and there may be income effects. In such cases it can be shown that the maximum of U depends on the value of $(-c_i/g - t_i)$ among other possibilities (Jara-Díaz and Farah 1987), where g is an *expenditure rate* defined in general by:

$$g = I/(T - W) \qquad (8.7)$$

The presence of such an income variable, reflecting purchasing power in the utility specification, indicates that the marginal utility of income varies with income, i.e. the model allows for an income effect. Besides, it is interesting to mention that empirical tests have shown that this new specification consistently outperforms the conventional wage-rate specification, even for individuals with no income effect (Jara-Díaz and Ortúzar 1989). More complex theoretical derivations of functional form, even for general joint models of activities (time use) and mode choice can be derived in similar fashion (see Munizaga *et al.* 2006; Jara-Díaz 2007).

8.3.3   Intrinsic Non-linearities: Destination Choice

Let us treat the singly constrained gravity model (5.14)–(5.18) we examined in Chapter 5 in a disaggregate manner by considering each individual trip maker in zone i as making one of the O_i trips originating in that zone. In this case the probability that a person will make the choice of travelling to zone j is simply:

$$P_j = T_{ij}/O_i = D_j f_{ij} / \sum_d D_d f_{id} \qquad (8.8)$$

Now if we define:

$$V_d = \log (D_d f_{id}) = \log D_d + \log f_{id} \qquad (8.9)$$

the model is seen to be exactly equivalent to the Multinomial Logit model (7.9). Thus the conventional origin-constrained gravity model may be represented by the disaggregate MNL without any loss of generality (Daly 1982a); note that (8.9) imposes no restrictions on the specification of the separation function f_{ij}. As we saw in Chapter 5, probably the most common function used in practice is the negative exponential of c_{ij}, the generalised cost of travelling between zones i and j; it is interesting to mention that when this form is substituted in (8.9) we obtain:

$$V_d = \log D_d - \beta c_{id} \qquad (8.10)$$

which is in fact linear in the parameter β. The problem of non-linearity arises due to the presence of D_d which may contain variables of the *size* variety that describe not the quality but the number of elementary choices within k and are typical of cases, such as choice of destination, where aggregation of alternatives is required (Daly 1982a). An example of this type of form was presented in Example 6.6.

8.4 Statistical Estimation

This section considers methods for the estimation of DM together with the goodness-of-fit statistics to be used in this task. Model estimation methods need to be adapted to the sampling framework used to generate the observations. This is necessary to improve estimation efficiency and avoid bias.

8.4.1 Estimation of Models from Random Samples

To estimate the coefficients θ_k in (7.3) the maximum likelihood method that we saw in section 2.5.4 is normally used. This method is based on the idea that although a sample could originate from several populations, a particular sample has a higher probability of having been drawn from a certain population than from others. Therefore the maximum likelihood estimates are the set of parameters which will generate the observed sample most often.

Let us assume a sample of Q individuals for which we observe their choice (0 or 1) and the values of x_{jkq} for each available alternative, such that for example:

individual 1 selects alternative 2

individual 2 selects alternative 3

individual 3 selects alternative 2

individual 4 selects alternative 1, etc.

As the observations are independent the likelihood function is given by the product of the model probabilities that each individual chooses the option they actually selected:

$$L(\theta) = P_{21} P_{32} P_{23} P_{14} \ldots$$

Defining the following dummy variable:

$$g_{jq} = \begin{cases} 1 & \text{if } A_j \text{ was chosen by } q \\ 0 & \text{otherwise} \end{cases} \tag{8.11}$$

the above expression may be written more generally as:

$$L(\theta) = \prod_{q=1}^{Q} \prod_{A_j \in \mathbf{A}(q)} (P_{jq})^{g_{jq}} \tag{8.12}$$

To maximise this function we proceed as usual, differentiating $L(\theta)$ partially with respect to the parameters θ and equating the derivative to 0. As in other cases we normally maximise $l(\theta)$, the natural logarithm of $L(\theta)$, which is more manageable and yields the same optima θ^*.

Therefore, the function we seek to maximise is (Ortúzar 1982):

$$l(\theta) = \log L(\theta) = \sum_{q=1}^{Q} \sum_{A_j \in \mathbf{A}(q)} g_{jq} \log P_{jq} \tag{8.13}$$

When $l(\Theta)$ is maximised, a set of estimated parameters Θ^* is obtained which is asymptotically distributed $N(\Theta, \mathbf{S}^2)$ where:

$$\mathbf{S}^2 = -\left(E\left(\frac{\partial^2 l(\Theta)}{\partial \Theta^2}\right)\right)^{-1} \tag{8.14}$$

Also $LR = -2 \cdot l(\Theta)$ is asymptotically distributed χ^2 with Q degrees of freedom (see Ben-Akiva and Lerman 1985). All this indicates that even though Θ^* may be biased in small samples, the bias is small for large enough samples (normally, samples of 500 to 1000 observations are more than adequate).

Now, although we have an explicit expression for the covariance matrix $\mathbf{S}^2$, determining the parameters Θ^* involves an iterative process. In the case of linear-in-the-parameters MNL models the function is well behaved, so the process converges quickly and always to a unique maximum; this explains why software to estimate this model is so easily available. Unfortunately this is not the case for other discrete choice models the estimation processes of which are more involved; therefore in what follows we will mainly refer to this simpler model.

Substituting the MNL expression (7.9) in (8.13), it can be shown that if the variable set includes an alternative specific constant for option A_j we have:

$$\sum_q g_{jq} = \sum_q P_{jq}$$

and this allows us to deduce that as alternative specific constants tend to capture the effect of variables not considered in the modelling, they ensure that the model always reproduces the aggregate market shares of each alternative. Therefore it is not appropriate to compare, as a goodness-of-fit indicator, the sum of the probabilities of choosing one option with the total number of observations that selected it, because this condition will be satisfied automatically by a MNL model with a full set of constants. As it is also not appropriate to compare the model probabilities with the g_{jq} values (which are either 0 or 1), a goodness-of-fit measure such as R^2 in ordinary least squares, which is based on estimated residuals, cannot be defined.

Example 8.1 Consider a simple binary-choice case with a sample of just three observations (as proposed by Lerman 1984); let us also assume that there is only one attribute x, such that:

$$P_{1q} = 1/\{1 + \exp[\theta(x_{2q} - x_{1q})]\}; \quad P_{2q} = 1 - P_{1q}$$

and also that we observed the following choices and values:

Observation (q)	Choice	x_{1q}	x_{2q}
1	1	5	3
2	1	1	2
3	2	3	4

In this case for any given value of θ, the log-likelihood function for the sample is given by:

$$l(\theta) = \log(P_{11}) + \log(P_{12}) + \log(P_{23})$$

and replacing the values we obtain:

$$l(\theta) = 10\theta - \log(e^{5\theta} + e^{3\theta}) - \log(e^{\theta} + e^{2\theta}) - \log(e^{3\theta} + e^{4\theta})$$

Figure 8.1 shows the results of plotting $l(\theta)$ for different values of θ. The optimum, $\theta^* = 0.756$, allows us to predict the following probabilities:

Observation (q)	P_{1q}	P_{2q}
1	0.82	0.18
2	0.32	0.68
3	0.32	0.68

Therefore, if we adopt a criterion in which individuals are assigned to that option which has maximum utility, this would result in an incorrect prediction for the second observation.

Figure 8.1   Variation of $(l)\,\theta$ with θ

We mentioned that the maximum likelihood parameters θ^* are asymptotically distributed Normal with covariance matrix S^2. In general the well-understood properties of the maximum likelihood method for well-behaved likelihood functions allow, as in multiple regression, a number of statistical tests which are of major importance:

8.4.1.1   The t-test for Significance of any Component θ_k^* of θ^*

Equation (8.14) implies that θ_k^* has an estimated variance s_{kk}^2, where $S^2 = \{s_{kk}^2\}$, which is calculated during estimation. Thus if its mean $\theta_k = 0$,

$$t = \theta_k^*/s_{kk} \tag{8.15}$$

has a standard Normal distribution N(0,1). For this reason it is possible to test whether θ_k^* is significantly different from zero (it is not exactly a t-test as we are taking advantage of a large-sample approximation and t is tested with the Normal distribution). Sufficiently large values of t (typically bigger than 1.96 for

95% confidence levels) lead to the rejection of the null hypothesis $\theta_k = 0$ and hence to accepting that the kth attribute has a significant effect.

The variable selection process followed during the specification searches of discrete choice models normally considers both formal statistical tests, such as the above one, and more informal (but even more important) tests such as examining the sign of the estimated coefficient to judge whether it conforms to *a priori* notions or theory. In this sense it is worth noting that rejection of a variable with a proper sign crucially depends on its importance; for example, let us note that the set of available explanatory variables can be usefully divided into two classes:

- highly relevant or policy variables, which have either a solid theoretical backing and/or which are crucial to model forecasting;
- other explanatory variables, which are either not crucial for policy evaluation (for example gender), or for which there are no theoretical reasons to justify or reject their inclusion.

Table 8.1 depicts the cases that might occur when considering the possible interactions in the above framework, and the solutions recommended by current practice. Consider first the case of rejecting a variable of type Other with correct sign; this may depend on its significance level (i.e. it may only be significant at the 85% level) and usual practice is to leave it out if it is not significant at the 80% level.

Table 8.1   Variable selection cases

		Variable	
		Policy	**Other**
Correct sign	Significant	Include	Include
	Not significant	Include	May reject
Wrong sign	Significant	Big problem	Reject
	Not significant	Problem	Reject

Current practice also recommends including a relevant (i.e. Policy type) variable with a correct sign even if it fails any significance test. The reason is that the estimated coefficient is the best approximation available for its real value; the lack of significance may just be caused by lack of enough data.

Variables of the Other class with a wrong sign are always rejected; however, as variables of the Policy type must be included at almost any cost, current practice dictates in their case model re-estimation, fixing their value to an acceptable one obtained in a study elsewhere. This will be an easy task if the variable is also non-significant, but might be very difficult otherwise as the fixed value will tend to produce important changes in the rest of the model coefficients.

Let us consider the role of socio-economic variables like gender, age, profession and occupation in discrete choice models. The usual way of introducing these variables was as additive constants, to one or more of the utilities of the alternatives (but not to all, unless they have specific coefficients), based on the modeller's experience and common sense, as in:

$$
\begin{aligned}
V_{1q} &= \alpha t_{1q} + \beta c_{1q} + \gamma f_{1q} + \ldots + \sum_l s_{lq} \\
V_{2q} &= \alpha t_{2q} + \beta c_{2q} + \gamma f_{2q} + \ldots
\end{aligned}
\tag{8.16}
$$

where, for example, t is time, c is cost, f is frequency and the dummy variables s_{lq} represent socio-economic characteristics of the individuals q. In this case the socio-economic data serve to improve the explanation of choice but do not provide any bonus in terms of using the model to estimate subjective values or willingness to pay, i.e. the ratio of the parameters of time and cost (Gaudry *et al.* 1989). It is also normally found that very few of these Other type variables provide enough explanation to be kept in the models.

An alternative and much better procedure is to parameterise the coefficients of each attribute in the model using socio-economic variables; in this case (8.16) changes to:

$$V_{iq} = \left(\alpha_0 + \sum_l \alpha_l s_{lq}\right) t_{iq} + \left(\beta_0 + \sum_l \beta_l s_{lq}\right) c_{iq} + \left(\gamma_0 + \sum_l \gamma_l s_{lq}\right) f_{iq} \quad (i = 1, 2) \qquad (8.17)$$

Now, dummy variables s_{lq} refer to the socio-economic characteristic l (i.e. gender) of individual q. This is both a simple and interesting manner of incorporating socio- economic variables, while at the same time helping in computing value functions which vary for each individual. Fowkes and Wardman (1988) proposed this method as a way of segmenting by individual tastes. Equation (8.17) states that given the characteristics of the individual, different coefficients will be obtained for a given attribute; note that the same socio-economic variable can appear in the expression corresponding to each coefficient. And note how this formulation does not imply that tastes are randomly distributed in the population; on the contrary, it assumes that the taste parameters (α, β and χ) depend on the individual characteristics in a deterministic manner; it has been popularised as *systematic taste variations* (Rizzi and Ortúzar 2003). This parameterisation allows for the incorporation of taste heterogeneity in an economical way, using computer programs widely available, rather than having to rely on a more complex function such as the Mixed Logit model.

Example 8.2 Table 8.2 presents two models. The first uses the method explained in equation (8.16) and the second uses the new method of equation (8.17). The sample size was 1631 stated preference observations (Rizzi and Ortúzar 2003) about route choice in the presence of the following attributes: accident risk, toll charge and travel time.

The socio-economic (SE) variables considered were sex (one for males), age (three dummies, with value one if the person's age was in the range considered) and night/day (one if the person travelled by day) in the case of the accident risk variable; and high income (one if the respondent's income was high) in the case of the toll variable. These binary variables were entered in the utility function of the safest route in the first model, but were assumed to interact with the base coefficients of either risk or toll in the case of model 2.

Looking at the results, it is obvious that the more flexible parameterisation of model 2 is superior to the traditional way of incorporating SE variables. Note how the results suggest that women value safety more than men, as do people with progressively higher age; on the other hand, if the trip takes place at night, the value of safety also should increase according to model 2. Finally, it is worth noting that the marginal utility of income (i.e. the toll coefficient with the opposite sign) correctly decreases for high-income individuals.

8.4.1.2   The Likelihood Ratio Test

A number of important model properties may be expressed as linear restrictions on a more general linear in the parameters model. Some important examples are:

- Are attributes generic? As mentioned in section 7.3, there are two main types of explanatory variables, generic and specific; the former have the same weight or meaning in all alternatives, whereas the latter have a different, specific, meaning in each of the choice options and therefore can take on a zero value for certain elements of the choice set.

Table 8.2   Alternative ways of entering socio-economic variables

Variables (*t*-ratios)	Model 1	Model 2
Risk of death	$-2.41E + 05$	$-2.18E + 05$
	(-5.6)	(-3.4)
Sex	-0.4233	$1.29E + 05$
	(-3.3)	(2.9)
Age$_1$ (30–49)	0.4605	$-1.76E + 05$
	(3.5)	(-3.5)
Age$_2$ (50–65)	1.02	$-3.75E + 05$
	(5.8)	(-6.0)
Age$_3$ (> 65)	1.48	$-5.49E+05$
	(2.8)	(-3.0)
Day/night	0.2097	$-8.45E+04$
	(2.5)	(-3.1)
Travel time (h)	-3.318	-3.738
	(-13.9)	(-14.0)
Toll charge (US$)	-0.702	-0.826
	(-9.9)	(-10.8)
High income	–	$4.13E-04$
		(3.4)
ρ^2 (c)	0.1545	0.1703

- Sample homogeneity. It is possible to test whether or not the same model coefficients are appropriate for two subpopulations (say living north and south of a river). For this a general model using different coefficients for the two populations is formulated and equality of coefficients may be tested as a set of linear restrictions.

Example 8.3 Let us assume a model with three alternatives, car, bus and rail, and the following choice influencing variables: travel time (TT) and out-of-pocket cost (OPC). Then a general form of the model would be:

$$V_{car} = \theta_1 \, TT_{car} + \theta_2 \, OPC_{car}$$
$$V_{bus} = \theta_3 \, TT_{bus} + \theta_4 \, OPC_{bus}$$
$$V_{rail} = \theta_5 \, TT_{rail} + \theta_6 \, OPC_{rail}$$

However, it might be hypothesised that costs (but not times, say) should be generic. This can be expressed by writing the hypothesis as two linear equations in the parameters:

$$\theta_2 - \theta_4 = 0$$
$$\theta_2 - \theta_6 = 0$$

In general it is possible to express the possibility of having generic attributes as linear restrictions on a more general model. For extensive use of this type of test, refer to Dehghani and Talvitie (1980). Some programs, for example Biogeme (Bierlaire 2009), present as a standard output a covariance/correlation analysis of pairs of estimated parameters θ_i and θ_j, sorted according to a t-test value constructed as follows:

$$t^* = \frac{\theta_i - \theta_j}{\sqrt{(\sigma_i^2 + \sigma_j^2 + 2\rho\sigma_i\sigma_j)}}$$

where σ are their standard errors and ρ the correlation coefficient between both estimates; if this test is accepted (i.e. when the value of t^* is less than a critical value, say 1.96 for the typical 95% level) the two parameters are not significantly different and are, thus, candidates for being treated as generic (if they refer to the same attribute in two alternatives).

Because of the properties of the maximum likelihood method, it is very easy to test any such hypotheses, expressed as linear restrictions, by means of the well-known likelihood ratio test (LR). To perform the test the estimation program is first run for the more general case to produce estimates θ^* and the log-likelihood at convergence $l^*(\theta)$. It is then run again to attain estimates θ_r^* of θ and the new log-likelihood at maximum $l^*(\theta_r)$ for the restricted case. Then if the restricted model under consideration is a correct specification, the LR statistic,

$$-2\{l^*(\theta_r) - l^*(\theta)\}$$

is asymptotically distributed χ^2 with r degrees of freedom, where r is the number of linear restrictions; rejection of the null hypothesis implies that the restricted model is erroneous. It is important to note that to carry out this test we require one model to be a restricted or nested version of the other. Train (1977) offers examples of use of this test to study questions of non-linearity, non-generic attributes and heterogeneity. Horowitz (1982) has discussed the power and properties of the test in great detail and should be consulted for further reference.

8.4.1.3 The Overall Test of Fit

A special case of likelihood ratio test is to verify whether the estimated model is superior to a model where all the components of θ are equal to zero. This model is known as the equally likely (EL) model and satisfies:

$$P_{jq} = 1/N_q$$

with N_q the choice set size of individual q. The test is not helpful in general because we know that a model with alternative-specific constants (ASC) will reproduce the data better than a purely random function. For this reason a more rigorous test of this class is to verify whether all variables, except the ASC, are 0. This better *reference* or *null* model is the market share (MS) model, where all the explanatory variables are 0 but the model has a full set of ASC; in this case we get:

$$P_{jq} = MS_j$$

where MS_j is the market share of option A_j.

Let us first look at the test for the EL model because it is simpler than that for the MS model. Consider a model with k parameters and with, as usual, a log-likelihood value at convergence of $l^*(\theta)$, and denote by $l^*(0)$ the log-likelihood value of the associated EL model; then under the null hypothesis $\theta = 0$ we have that the LR statistic:

$$-2\{l^*(0) - l^*(\theta)\}$$

is distributed χ^2 with k degrees of freedom; therefore we can choose a significance level (say 95%) and check whether LR is less than or equal to the critical value of $\chi^2 (k, 95\%)$, in which case the null hypothesis would be accepted.

However, we already hinted that the test is weak because as it is always rejected it only means that the parameters θ explain the data better than a model with no significant explanatory power. Actually the best feature of this test is its low cost as $l^*(0)$ does not require a special program run since it is usually computed as the initial log-likelihood value by most search algorithms.

To carry out the test with the market share model we must compute $l^*(C)$ it's log-likelihood value at convergence; if there are $(k - c)$ parameters which are not specific constants, the appropriate value of LR is compared with $\chi^2(k - c, 95\%)$ in this case. In general an extra run of the estimation routine is required to calculate $l^*(C)$ except for models where all individuals face the same choice set, in which case it has the following closed form equation:

$$l^*(C) = \sum_j Q_j \log(Q_j/Q) \tag{8.18}$$

where Q_j is the number of individuals choosing option A_j.

Figure 8.2 shows the notional relation between the values of the log-likelihood function, for the set of parameters that maximise it, $l^*(\theta)$, for the two previous models, $l^*(0)$ and $l^*(C)$ respectively, and for a fully saturated (perfect) model with an obvious value $l(*) = 0$.

Figure 8.2   Notional relation between log-likelihood values

8.4.1.4   The ρ^2 Index

Although it is not possible to build an index such as R^2 in this case, it is always interesting to have an index which varies between 0 (no fit) and 1 (perfect fit) in order to compare alternative models. An index that satisfies some of the above characteristics was initially defined as:

$$\rho^2 = 1 - \frac{l^*(\theta)}{l^*(0)} \tag{8.19}$$

However, although its meaning is clear in the limits (0 and 1) it does not have an intuitive interpretation for intermediate values as in the case of R^2; in fact, values around 0.4 are usually considered excellent fits.

Because a ρ^2 index may in principle be computed relative to any null hypothesis, it is important to choose an appropriate one. For example, it can be shown that the minimum values of ρ^2 in (8.19), in models with specific constants, vary with the proportion of individuals choosing each alternative. Taking a simple binary case, Table 8.3 shows the minimum values of ρ^2 for different proportions choosing option 1 (Tardiff 1976). It can be seen that ρ^2 is only appropriate when both options are chosen in the same proportion.

These values mean, for example, that a model estimated with a 0.9/0.1 sample yielding a ρ^2 value of 0.55, would be undoubtedly much weaker than a model yielding a value of 0.25 from a sample with an equal split. Fortunately, Tardiff (1976) proposed a simple adjustment that allows us to solve this difficulty; it consists of calculating the index with respect to the market share model:

$$\bar{\rho}^2 = 1 - \frac{l^*(\theta)}{l^*(C)} \tag{8.20}$$

This *corrected* ρ^2 lies between 0 and 1, is comparable across different samples and is related to the χ^2 distribution.

Table 8.3 Minimum ρ^2 for various relative frequencies

Sample proportion selecting the first alternative	Minimum value of ρ^2
0.50	0.00
0.60	0.03
0.70	0.12
0.80	0.28
0.90	0.53
0.95	0.71

Ben-Akiva and Lerman (1985) propose another correction to the ρ^2 index; this is usually referred as *adjusted* ρ^2 and it is defined as:

$$\rho_{adj}^2 = 1 - \frac{l^*(\theta) - K}{l^*(0)}$$

which takes into account the number of parameters estimated. However, it is still based on the likelihood of the equally-likely model so it maintains the main problems of the original ρ^2.

8.4.1.5 The Percentage Right or First Preference Recovery (FPR) Measure

This is an aggregate measure that simply computes the proportion of individuals effectively choosing the option with the highest modelled utility. FPR is easy to understand and can readily by compared with the chance recovery (CR) given by the equally likely model:

$$CR = \sum_q (1/N_q)/Q$$

Note that if all individuals have a choice set of equal size N, then $CR = 1/N$. FPR can also be compared with the market share recovery (MSR) predicted by the best null model (Hauser 1978):

$$MSR = \sum_{Aj} (MS_j)^2$$

Disadvantages of the index are exemplified by the fact that although an FPR of 55% may be good in general, it is certainly not so in a binary market; also an FPR of 90% is normally good in the binary case, but not if one of the options has a market share of 95%. Another problem with the index, worth noting in the sense of not being an unambiguous indicator of model reliability, is that too high a value of FPR should lead to model rejection as well as a too low value; to understand this point it is necessary to define the expected value of FPR for a specific model as:

$$ER = \sum_q P_q \tag{8.21}$$

where P_q is the calculated (maximum) probability associated with the best option for individual q. Also, because FPR is an independent binomial random event for individual q, occurring with probability $1/N_q$ in the CR case and P_q in the ER case, their variances are given respectively by:

$$Var\,(CR) = (1/N_q)(1 - 1/N_q) \tag{8.22}$$

and

$$\text{Var(ER)} = P_q(1 - P_q) \tag{8.23}$$

Thus, a computed value of FPR for a given model can be compared with CR and ER; if the three measures are relatively close (given their estimated variances) the model is *reasonable but uninformative*; if FPR and ER are similar and larger than CR, the model is *reasonable and informative*; finally, if FPR and ER are not similar, the model does not explain the variation in the data and should be rejected whether FPR is larger or smaller than ER (see Gunn and Bates 1982).

8.4.1.6 Working with Validation Samples

As we already mentioned in Chapter 5, the performance of any model should be judged against data other than that being used to specify and estimate it and, ideally, taken at another point in time (perhaps after the introduction of a policy in order to assess the model response properties). This is true for any model. We will define a subsample of the data, or preferably, another sample *not used* during estimation, as a *validation sample*.

We will first briefly describe a procedure to estimate the minimum size of such a validation sample (ideally to be subtracted from the total sample available for the study) conditional on allowing us to detect a difference between the performance of two or more models, when there is a true difference between them. The method, which is based on the FPR concept, was devised by Hugh Gunn and first applied by Ortúzar (1983).

		Model 2	
		Not FPR	**FPR**
	Not FPR	n_{11}	n_{12}
Model 1	**FPR**	n_{21}	n_{22}

Consider the 2×2 table layout shown above, where n_{ij} is the number of individuals assigned to cell (i, j). For all individuals in a validation sample, choice probabilities and FPR are calculated for each of two models under investigation and the cells of the table are filled appropriately (e.g. assigning to cell $(1,1)$ if not FPR in both models, and so on). We are interested in the null hypothesis that the probabilities with which individuals fall into cells $(1,2)$ and $(2,1)$ are equal, for in that case the implication on simple FPR is that the two models are equivalent; on this null hypothesis the following statistic M is distributed χ^2 with one degree of freedom (see Foerster 1979):

$$M = \frac{(n_{12} - n_{21})^2}{n_{12} + n_{21}} \tag{8.24}$$

Thus, a test of the equivalence of the two models in terms of FPR is simply given by computing M and comparing the result with $\chi^2 (1, 95\%)$; if M is less than the appropriate critical value of χ^2 (3.84 for the usual 95% confidence level) we cannot reject the null hypothesis and we conclude that the models are equivalent on these terms.

Given this procedure we can select whichever level of confidence seems appropriate for the assertion that the two models under comparison differ in respect of the expected number of FPR. This gives us control over the fraction of times that we will incorrectly assert a difference between similar models. As usual, the aim of choosing a particular sample size is to ensure a corresponding control over the proportion of times we will make the other type of error, namely incorrectly concluding that there is no difference between different models.

Now, to calculate the probability of an error of the second type we need to decide what the minimum difference we should like to detect is; with this we can calculate the sample size needed to reduce the chance of errors of the second kind to an acceptable level for models which differ by exactly this minimum amount, or more.

Example 8.4 Consider the case of two models such that, on average, model 2 produces 10 extra FPR per 100 individuals modelled as compared to model 1. Note that here it does not matter whether this arises as a result of model 1 having 20% FPR and model 2 having 30% FPR, or the first 80% and the second 90%; in other words, both models can be inadequate.

In this simple case n_{21} is zero and M simply becomes n_{12}. If we are ensuring 95% confidence that any difference we establish could not have arisen by chance from equivalent models, we will compare n_{12} with the χ^2 value for one degree of freedom (3.84); for any given sample size n, the probability that r individuals will be assigned to cell (1, 2) is simply the binomial probability $\binom{n}{r} p^r (1 - p)^{(n-r)}$ where p denotes the probability of an individual chosen at random being assigned to cell (1, 2) i.e. the minimum difference we wish to detect.

Given n and taking $p = 0.05$ as usual, we can calculate the probabilities of 0, 1, 2, and 3 individuals being assigned and sum these to give the total probability of accepting the null hypothesis (i.e. committing an error of the second kind). Table 8.4 gives the resulting probabilities for different sample sizes.

Table 8.4 Probability of an error of the second kind for given sample size and models as defined

Sample size	Minimum difference 5% Prob (error II)
50	0.75
100	0.26
150	0.05
200	0.01
250	0.00

It is clear that the required validation sample size needs to be relatively large given that typical estimation data sets have only a few hundred observations. Also recall that Table 8.4 is for the simple case of one model being better than or equal to the other in each observation; the method of course may easily be extended to cases where both the (1, 2) and (2, 1) cells have non-zero probability.

An especially helpful feature of validation samples is that provided their size is adequate the issue of ranking non-nested models (see section 8.4.3) is easily resolved, as likelihood ratio tests can be performed on the sample regardless of any difference in model structure parameters. This is because the condition of one model being a generalisation of the other is only required for tests with the same data used for estimation (Gunn and Bates 1982; Ortúzar 1983).

Example 8.5 Let us assume that we are interested in an option with low market share at present and that we have two model specifications (models A and B) for a six-alternative choice situation. The two models have similar FPR but one always predicts that option badly and the others a bit better, compared with the second model that gives reasonable predictions for all options. In this case we can use a validation sample and estimate, for each individual in it, the choice probabilities for each option by the two models; the alternative actually chosen is, as usual, an observed piece of information. In order to investigate the consistency of the predictions with the data, we can compare them with proportions calculated from the sample.

Table 8.5 presents the values N_{ij} and O_{ij} (where i indicates a probability band and j an option). N_{ij} is the number of observations to which the model assigned a probability in band i to alternative A_j; O_{ij} is the observed number of choices of option A_j to which the model assigned a probability in that band.

Table 8.5 Modelled choices by probability band

Predicted probability band (i)	0–0.1		0.1–0.2		0.2–0.3		...		0.9–1.0	
Alternative (j)	N_{1j}	O_{1j}	N_{2j}	O_{2j}	N_{3j}	O_{3i}	...	...	N_{10j}	O_{10j}
Model A										
1	0	0	8	0	11	0	...	...	0	0
2	40	0	0	0	0	0	...	...	0	0
3	94	0	0	0	0	0	...	...	0	0
...	...	...	...	...	...	...	...	...	...	...
...	...	...	...	...	...	...	...	...	...	...
6	55	6	11	3	58	14	...	...	0	0
Total		6		6		24	...	...		0
Model B										
1	9	0	5	0	0	0			0	0
2	36	0	4	0	0	0			0	0
...	...	...	...	...	...	...	...	...	...	...
...	...	...	...	...	...	...	...	...	...	...
...	...	...	...	...	...	...	...	...	...	...
6	43	3	44	7	18	8			0	0
Total		6		11		13	...	...		15

Table 8.6 builds on the previous table and presents the values E_{ij} and O_{ij}, where E_{ij} is given by:

$$E_{ij} = N_{ij} \times \bar{p}_i$$

which corresponds to the expected value of the number of individuals choosing option A_j with probability in the band i, associated with a mean probability $\bar{p}_i$. For example, in the case highlighted in the table we have that $E_{36} = 58 \times 0.25 = 14.5$, as 0.25 is the mean value of probability band 3 (i.e. between 0.2 and 0.3).

To compare the values in Table 8.6 it is possible to apply a χ^2 test defined as follows (Gunn and Bates 1982):

$$\chi^2_{\text{cell}} = \sum_{ij} \frac{(O_{ij} - E_{ij})^2}{E_{ij}} \text{ with } ij - 1 \text{ degrees of freedom}$$

It is possible in principle to apply the test to each cell in the matrix if $E_{ij} > 5$, as the test is not valid otherwise. For this reason, and given the limited size of validation samples, it may be necessary to aggregate cells but unfortunately there are no clear-cut methods to do it. The reader may check that different aggregation strategies lead to different results.

Table 8.6   Expected proportions by probability band

Predicted probability band (i)	0–0.1		0.1–0.2		0.2–0.3		. . .		0.9–1.0	
Alternative (j)	E_{1j}	O_{1j}	E_{2j}	O_{2j}	E_{3j}	O_{3j}	. . .	. . .	E_{10j}	O_{10j}
Model A										
1	0	0	1.2	0	2.75	0	. . .	. . .	0	0
2	2	0	0	0	0	0	. . .	. . .	0	0
3	4.7	0	0	0	0	0	. . .	. . .	0	0
. . .	. . .	. . .	. . .	. . .	. . .	. . .	. . .	. . .	. . .	. . .
. . .	. . .	. . .	. . .	. . .	. . .	. . .	. . .	. . .	. . .	. . .
6	2.75	6	1.65	3	14.5	14	. . .	. . .	0	0
Total	9.45	6	4.05	6	29	24	. . .	. . .	0	0
Model B										
1	0.45	0	0.75	0	0	0			0	0
2	1.8	0	0.6	0	0	0			0	0
3	4.05	0	1.65	0	0.5	0	. . .	. . .	0	0
. . .	. . .	. . .	. . .	. . .	. . .	. . .	. . .	. . .	. . .	. . .
. . .	. . .	. . .	. . .	. . .	. . .	. . .	. . .	. . .	. . .	. . .
6	2.15	3	6.6	7	4.5	8	. . .	. . .	0	0
Total	9.5	6	11.7	11	7.5	13	. . .	. . .	15.2	15

A less informative case, but one that it is usually possible to carry out, is to compare expected and observed totals for each column in Table 8.6, $E_i = \sum_j E_{ij}$ and $O_i = \sum_j O_{ij}$ respectively, using the index:

$$\chi^2_{\text{FPR}} = \sum_{i=1}^{m} \frac{(O_i - E_i)^2}{E_i} \tag{8.25}$$

where m is the number of columns with $E_i > 5$. In this case the appropriate number of degrees of freedom is $m - 1$, and χ^2_{FPR} may be compared with the critical value $\chi^2_{0,95;m-1}$. If $\chi^2_{\text{FPR}} < \chi^2_{0,95;m-1}$ then the null hypothesis that the model is consistent with the data is accepted. If, according to the previous test, two or more models are acceptable then it is possible to discriminate between them using the direct likelihood ratio test (Gunn and Bates 1982; Ortúzar 1983):

$$\frac{L_A}{L_B} = \frac{\prod_i \bar{p}_i^{O_i} (\text{model A})}{\prod_i \bar{p}_i^{O_i} (\text{model B})} \tag{8.26}$$

If we applied this test to the data of Table 8.6, we would get:

$$\frac{L_A}{L_B} = \frac{(0.05)^6 \times (0.15)^6 \times (0.25)^{24} \times \ldots \times (0.95)^0}{(0.05)^6 \times (0.15)^{11} \times (0.25)^{13} \times \ldots \times (0.95)^{15}} = 0.0455$$

and we would say that the data are approximately 22 times (that is 1/0.0455) more probable under model B that under model A. This means that we would prefer the second model although both yield predictions which are consistent with the data.

8.4.2 Estimation of Models from Choice-based Samples

As mentioned in Chapter 3, estimating a model from a choice-based sample may be of great interest because the data-collection costs are often considerably lower than those for typical random or stratified samples. The problem of finding a tractable estimation procedure possessing desirable statistical properties is not an easy one, and the state of practice has been provided by the excellent papers of Coslett (1981) and Manski and McFadden (1981).

It has been found in general that maximum likelihood estimators specific to choice-based sampling are impractical, except in very restricted circumstances, due to computational intractability. However, if it can be assumed that the analyst knows the fraction of the decision-making population selecting each alternative, then a tractable method can be introduced. The approach modifies the familiar maximum likelihood estimator of random sampling by weighting the contribution of each observation to the log-likelihood by the ratio Q_i/S_i, where the numerator is the fraction of the population selecting option A_i and the denominator the analogous fraction for the choice-based sample.

Manski and Lerman (1977) have shown that the un-weighted random-sample maximum likelihood estimator is generally inconsistent when applied to choice-based samples and in most choice models this inconsistency affects all parameter estimates. However, as we saw in section 7.3.2, for simple MNL models with a full set of alternative-specific constants the inconsistency is fully confined to the estimates of these dummy variables. In this case, the estimates obtained without weighting are more efficient than the estimates obtained with the weighted sample. Therefore, it is good practice to estimate the parameters of the MNL model without weighting the sample, and to correct the constants afterwards. Bierlaire *et al.* (2008) show that this property does not apply to Generalised Extreme Value models (including Nested Logit and Cross-Nested Logit models). They propose a simple estimator for these models that does not require the weighting of the sample nor knowledge of the actual market shares.

8.4.3 Estimation of Hybrid Choice Models with Latent Variables

As mentioned in section 7.7.5, two approaches have been proposed to estimate hybrid choice models; they differ in how the available information is used.

8.4.3.1 Sequential Estimation

In sequential estimation the problem is treated in two stages. First, the MIMIC model (Bollen 1989) discussed in section 7.7.5.1 is solved to obtain parameter estimators for the equations relating the latent variables to the explanatory variables and the perception indicators. Then, using these parameters in equation (7.63), expected values for the latent variables of each individual and alternative are obtained. In turn, these expected latent variable values are added to the set of typical variables of the discrete choice model, as in equation (7.65), and their parameters estimated together with those of the traditional variables in a second stage.

Although this method has the disadvantage of not using all the available information jointly, its application is clear and simple, which is why it is the most used method in practice (Ashok *et al.* 2002; Vredin Johansson *et al.* 2005; Raveau *et al.* 2010). Furthermore, giving currently available software, this method allows estimating more flexible hybrid models than the simultaneous approach (see Yáñez *et al.* 2009). Nevertheless, it is argued that a potentially serious problem of the approach is that it may result in biased estimators for the parameters involved (Bollen 1989); similarly, it has been noted that the method tends to underestimate the parameters' standard deviations, resulting in estimators with a statistical significance higher than their real contribution to the model. This notwithstanding, the problem can be solved by means of a statistical correction to the parameters' variances (Murphy and Topel 1985), but it is not an easy process. It is interesting to mention though, that using both real and simulated data Raveau *et al.* (2010) estimated parameters which were totally consistent in both cases, but were able to estimate substantially more flexible models than in the case of simultaneous estimation due to current software limitations.

8.4.3.2   Simultaneous Estimation

In this approach the joint estimation is done by maximising the likelihood of the probability of replicating the individual choices based on the representative utility proposed by the modeller; that is, Prob $(g_{iq}|V_{iq})$, where g_{iq} is equal to one if individual q chooses option A_i. Now, recall equation (7.65) for the hybrid discrete choice model:

$$V_{iq} = \sum_k \theta_{ik} \cdot x_{ikq} + \sum_l \beta_{il} \cdot \eta_{ilq}$$

where as usual, $\mathbf{x}$ are level-of-service attributes, η are the latent variables, to be estimated jointly with the structural (7.63) and measurement (7.64) equations, and θ and β are parameters to be estimated.

From (7.65), the conditional probability above can be expressed in terms of the variables and parameters of the discrete choice model. However, as the latent variables are not observed it is necessary to integrate over their whole variation range, conditioning them by their explanatory variables. Thus, the choice probability is given by (8.27), where $h(\cdot)$ is the probability density function of the latent variables:

$$\text{Prob}\left(g_{iq} \mid x_{ikq}, s_{iqr}, \theta_{ik}, \beta_{il}, \alpha_{ilr}\right) = \int_{\eta_{ilq}} \text{Prob}\left(g_{iq} \mid x_{ikq}, \eta_{ilq}, \theta_{ik}, \beta_{il}\right) \cdot h\left(\eta_{ilq} \mid s_{iqr}, \alpha_{ilr}\right) \cdot d\eta_{ilq}$$

$$(8.27)$$

and $\mathbf{s}$ and α are the socio-economic variables (and their parameters) explaining the latent variables in structural equation (7.63). However, to estimate the model it is necessary also to introduce the information provided by the perception indicators $\mathbf{y}$ in the measurement equation (7.64), since otherwise the model would not be identifiable. The indicators are not explanatory variables of the model; instead, they are endogenous to the latent variables. This implies that the choice probability used during estimation is given by (8.28), where $f(\cdot)$ is the probability density function of the indicators.

$$\text{Prob}\left(g_{iq}, y_{ipq} \mid x_{ikq}, s_{iqr}, \theta_{ik}, \beta_{il}, \alpha_{ilr}, \gamma_{ipq}\right)$$

$$= \int_{\eta_{ilq}} \text{Prob}\left(g_{iq} \mid x_{ikq}, \eta_{ilq}, \theta_{ik}, \beta_{il}\right) \cdot f\left(y_{ipq} \mid \eta_{ilq}, \gamma_{ipq}\right) \cdot h\left(\eta_{ilq} \mid s_{iqr}, \alpha_{ilr}\right) \cdot d\eta_{ilq} \qquad (8.28)$$

Once the functional form of the discrete choice model is defined, the simulated maximum likelihood method can be used for estimation (Bolduc and Alvarez-Daziano 2009; Bolduc and Giroux 2005); we will examine the method in depth in section 8.5.2, but as we will see there are difficult practical problems in this case due to the particular form of the estimation problem (see Hess and Rose 2009).

Example 8.6   A recent urban mode choice study considered ten transport modes, both pure and combined, for journey to work trips: car-driver, car-passenger, shared taxi, bus, underground, combinations of the previous four with underground and shared taxi/bus. For each available mode,

(continued)

information was precisely measured about walking, waiting and in-vehicle time, trip cost, and number of transfers made. Regarding users' information, socioeconomic variables such as age, education level and income, were obtained. Respondents were also asked to evaluate different characteristics of the modes, generating perception indicators to allow us including latent variables in the model.

Three latent variables were considered: *accessibility/comfort*, *reliability* and *safety*; the effects of each were captured through seven perception indicators based on the evaluation of several aspects of the pure modes: (i) safety regarding accidents, (ii) safety regarding theft, (iii) ease of access, (iv) comfort during the trip, (v) availability of suitable information, (vi) possibility of calculating the travel time prior to the trip, and (vii) possibility of calculating the waiting time prior to the trip.

Four explanatory variables were finally included in the MIMIC model: education level, age, sex, and monthly income. The MIMIC model's structural relations were studied using factor analysis to guarantee their correct specification. Figure 8.3 illustrates the results of this process (Raveau *et al*. 2010).

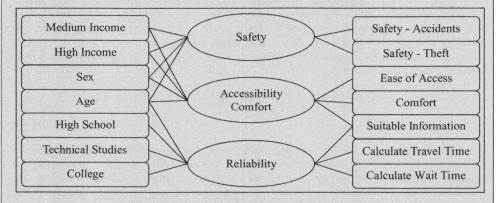

Figure 8.3   Latent variables model relationships

The representative utility function included the number of transfers during the trip as well as the different time variables obtained from the survey; in the case of travel time, systematic taste variations according to the respondent's sex were found (the variable Sex, takes the value one for males). Travel cost was standardized by the individual's wage rate. This was the best specification obtained among several formulations studied. The model also included a complete set of alternative specific constants (ASC).

The hybrid model was estimated both sequentially and simultaneously. Table 8.7 presents the results together with those of an equivalent MNL model without latent variables (the ASC are not reported, interested readers are referred to Raveau *et al*. 2010). The simultaneously estimated parameters where obtained using an experimental hybrid choice model estimation software (Bolduc and Giroux 2005).

The signs of all estimated parameters are consistent with microeconomic theory. When using the simultaneous method, all variables are statistically significant at least at the 90% confidence level, but not all variables are statistically significant in the sequential hybrid model or in the model without latent variables; the waiting time variable is especially problematic.

Table 8.7   Hybrid choice model estimation results

Parameter	Without Latent Variables	Hybrid Model Sequential	Hybrid Model Simultaneous
Cost/wage rate	−0.027 (−8.13)	−0.028 (−6.46)	−0.032 (−7.32)
Travel time	−0.033 (−4.82)	−0.007 (−4.25)	−0.006 (−4.67)
Sex interaction with time	0.030 (2.98)	−0.001 (−2.91)	−0.001 (−3.01)
Waiting time	−0.009 (−0.53)	−0.013 (−0.58)	−0.015 (−1.69)
Walking time	−0.016 (−1.80)	−0.019 (−1.69)	−0.022 (−2.89)
No. of transfers	−1.110 (−8.20)	−1.060 (−7.85)	−1.102 (−8.21)
Accessibility-comfort	–	0.590 (4.23)	0.622 (3.79)
Reliability	–	0.339 (2.91)	0.441 (2.70)
Safety	–	0.582 (2.01)	0.613 (1.87)
Log-likelihood	−105,567.06	−55,578.85	−47,883.43

Both hybrid models show that men are slightly more sensitive to travel time than women, but the model without latent variables shows precisely the opposite effect and the large difference (in magnitude) of the marginal utility of travel time for women is certainly suspect. The ASC obtained for the model without latent variables were more significant than those obtained for the hybrid models. This is an expected result since the model without latent variables has fewer explanatory variables, and so the constants must explain (as far as possible) the missing information according to the individual choice patterns. Finally, it is important to mention the clear superiority (in terms of log-likelihood) of the simultaneous model over the sequential model. In addition, the sequential model is significantly better than the model without latent variables (i.e. a gain of 50,000 in log-likelihood for just three degrees of freedom).

The example above serves to illustrate another view of *endogeneity* in latent variable modelling (recalled the comment made in section 7.7.5.2). The 'true' model in the population has attributes, such as safety and comfort, that are not measurable in practice, are probably relevant in decision making, and are correlated with the observed variables (possibly mainly with cost). This makes the observed variables correlated with the error term, i.e. they are endogenous, and hence their estimates should not be consistent in a model without latent variables.

If these unobserved variables were identically and independently distributed (IID) among modes and attributes (even with different means by mode), there would be no problem as the endogeneity could be resolved using ASC. But if the unobserved variables are not IID among modes and individuals, a solution is precisely to treat them as latent variables. And, as they cannot be observed, they need to be measured indirectly by means of additional questions or 'indicators'. In fact, as the latent variables actually serve to correct the endogeneity problem caused by omitted variables, this could explain the very large increase in fit shown in the example above.

8.4.4   Comparison of Non-nested Models

The likelihood ratio test outlined in section 8.4.1.2 requires testing a model against a parametric generalisation of itself, that is, it requires the model to be *nested*. Models with utility functions having significantly different functional forms, or models based on different behavioural paradigms, cannot be compared by this test.

It is easy to conceive of situations in which it would be useful to test a given model against another which is not a parametric generalisation of itself. The following example, provided by Horowitz (1982), is very illustrative.

Example 8.7 Consider one model with a representative utility function specified as:

$$V = \theta_1 x_1 + \theta_2 x_2$$

and another with a representative utility function given by:

$$W = \theta_3 x_3 x_4$$

and assume we want to test both models to determine which explains the data best. Clearly, there is no value of θ_3 that causes V and W to coincide for all values of θ_1 and θ_2 and the attributes **x**. If both models belong to the same general family, however, it is possible to construct a hybrid function; for example, in our case we could form a model with a measured utility Z containing both V and W as special cases:

$$Z = \theta_1 x_1 + \theta_2 x_2 + \theta_3 x_3 x_4$$

and using a log-likelihood ratio tests both models could be compared against the hybrid; the first one would correspond to the hypothesis $\theta_3 = 0$ and the second to the hypotheses $\theta_1 = \theta_2 = 0$.

Horowitz (1982) discusses several other tests at length, including cases where the competing models do not belong to the same general family. But recall that in the presence of a validation sample the issue may be particularly easily resolved, as discussed by Gunn and Bates (1982).

8.5 Estimating the Multinomial Probit Model

Flexible choice models, such as Multinomial Probit (MNP) and Mixed Logit (ML) do not have a closed form, so their choice probabilities are characterised by a multiple integral that is not easy to solve efficiently.

8.5.1 Numerical Integration

The choice probability for a general random utility model may be expressed as follows:

$$P_i(\mathbf{\theta}, \mathbf{x}) = \int_{u_1=-\infty}^{u_i} \int_{u_2=-\infty}^{u_i} \cdots \int_{-\infty}^{\infty} \cdots \int_{u_J=-\infty}^{u_i} f(\mathbf{u}) \, du_J \ldots du_1 \tag{8.29}$$

where $f(\mathbf{u})$ is the joint distribution function of the option utilities. For example, in the case of the MNP model we have:

$$f(\mathbf{u}) = \mathrm{MVN}(\mathbf{V}, \mathbf{\Sigma}) = \left[(2\pi)^J \, |\mathbf{\Sigma}| \right]^{-1/2} \exp\left\{ -\frac{1}{2}(\mathbf{u} - \mathbf{V})\mathbf{\Sigma}^{-1}(\mathbf{u} - \mathbf{V})^{\mathrm{T}} \right\} \tag{8.30}$$

To integrate numerically, the region of integration must first be divided into a series of elements of differential size. Then the area under the curve is approximated, for each element, as the equivalent mean rectangle (given the element and its height); finally, the value of the integral is the sum of these areas. Although the difficulty of the problem increases geometrically with the dimensionality of the integral, in the majority of cases this dimensionality may be reduced for the MNP because:

(i) If a change of variables is made, expressing all elements of the integral as the difference between the utility of the alternative under consideration and the others, this yields a vector $\hat{\mathbf{u}}$ of just $J - 1$ components (that are also distributed Normal) given by:

$$\hat{u}_k = u_k - u_i$$
$$\cdots \qquad \text{(assume we are evaluating } P_i)$$
$$\hat{u}_{J-1} = u_J - u_i$$

Then the probability of choosing A_i will be:

$$P_i(\hat{\mathbf{u}}, \hat{\boldsymbol{\Sigma}}) = \text{Prob}\{\hat{u}_k < 0, \forall A_k \in \mathbf{A}\}$$

and the integral reduces to:

$$\int_{\hat{u}_1=-\infty}^{0} \cdots \int_{\hat{u}_{J-1}=-\infty}^{0} (2\pi^{J-1}|\hat{\boldsymbol{\Sigma}}|)^{-1/2} \exp\left\{-\frac{1}{2}(\hat{\mathbf{u}} - \hat{\mathbf{V}})\hat{\boldsymbol{\Sigma}}^{-1}(\hat{\mathbf{u}} - \hat{\mathbf{V}})^{\mathrm{T}}\right\}$$

with $\hat{\mathbf{V}}$ and $\hat{\boldsymbol{\Sigma}}$ the vector of means and the covariance matrix of the new variables.

(ii) Make a Choleski decomposition (see section 2.5.4.1), which in practical terms also reduces the integral dimensionality by one, because it allows us to separate the integrals and the first, corresponding to A_i, is equal to one.

Numerical integration is the most accurate method to solve these problems, but it is only feasible at a reasonable cost for problems with a maximum of four alternatives. It may also have problems of (computer) approximation if one or more choice probabilities are close to zero. For these reasons it is generally used only as a standard of comparison for the other methods.

8.5.2 Simulated Maximum Likelihood

8.5.2.1 The Basic Approach

Lerman and Manski (1981) originally proposed the evaluation of the MNP choice probability $P_i(\mathbf{V}, \boldsymbol{\Sigma})$ by generating a number of draws $\mathbf{U}$, from MVN $(\mathbf{V}, \boldsymbol{\Sigma})$, counting a *success* when U_i was the highest value. For a sufficiently large number of draws, the proportion of successes approximates the choice probability (see Figure 8.4). Thus, the method was theoretically simple but unfortunately had several problems in practice:

(i) If the number of successes is equal to zero (an event that could occur in certain circumstances), the log-likelihood tends to infinity and the method collapses. To solve this problem, Lerman and Manski (1981) suggested replacing the ratio of the number of successes over the total number of

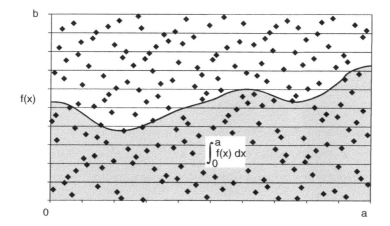

Figure 8.4   Solving an integral through Monte Carlo simulation

draws (i.e. the estimate of the choice probability) by the quantity $(N_i + 1)/(N + J)$ where N_i is the number of successes, N the sample size (number of draws) and J the number of options. However, this introduces bias (as the correct estimator of P_i is obviously N_i/N). The bias is small in large problems but it could be considerable in more practical problems.

(ii) The relative error associated with this simulation method is inversely proportional to the square root of the number of successes. This implies that many draws have to be made and it was computationally too demanding for real-life problems in the past.

However, at the beginning of the 1990s this approach found renewed favour through a series of advances in the simulation of multivariate processes in discrete choice models (Börsch-Supan and Hajivassiliou 1993). There is also an alternative approach (McFadden 1989; Pakes and Pollard 1989) which avoids evaluating the multiple integral by replacing the choice probability in the moments equation by an unbiased simulator. This *simulated moments* method may be considered a precursor of the Mixed Logit or *error components* model, the estimation of which we will review in section 8.6.

8.5.3 Advanced Techniques

Using advanced integration techniques based on Monte Carlo simulation developed by several authors, Börsch-Supan and Hajivassiliou (1993) proposed the GHK simulator. This has the essential property of producing unbiased simulated probabilities that lie strictly between zero and one, and that are also continuous and differentiable functions of the model parameters. Furthermore, the computational effort increases only linearly with the dimensionality of the integral and is independent of the true probabilities. The simulator is based on recursively decreasing the problem dimension, and for this it has to generate repetitions of a truncated uni-dimensional Normal distribution.

For the MNP model, the method started with the model reduced in one dimension after subtracting the utility of the chosen option from the remaining utilities for each observation (i.e. $U_1 - U_c$ where c is the chosen option); as we already mentioned, this transformed utility is also Normal distributed. In mathematical terms the transformation simply consists of pre-multiplying the vector of utilities by a matrix $\mathbf{A}$, which is equal to minus the identity matrix and incorporating a column of ones in the position corresponding to the chosen option. Then the resulting vector can be freed of the row corresponding to the chosen alternative because it only contains zeros.

In this way, the transformed systematic utility is given by $\mathbf{V}^* = \mathbf{AV}$, and the error term distributes Normal with zero mean and covariance matrix given by $\mathbf{M} = \mathbf{A\Sigma A}'$. In turn, $\mathbf{M}$ can be decomposed by applying the Choleski decomposition to produce a lower triangular matrix $\mathbf{L}$ and a superior matrix $\mathbf{L}'$, such that $\mathbf{LL}' = \mathbf{M}$. The GHK simulator was developed to simulate the probability that a Normal random variable lies within limits a and b:

$$\mathbf{u} \sim N\left(\mathbf{\theta} \cdot \mathbf{x}, \sum\right) \quad \text{subject to } a \leq \mathbf{AU} \leq b$$

Instead of simulating for these variables, the process is performed for:

$$\mathbf{e} \sim N(0, \mathbf{I}) \text{ subject to } a^* \equiv a - \mathbf{A\theta x} \leq \mathbf{Le} \leq b^* \equiv b - \mathbf{A\theta x}$$

and, thanks to the triangular structure of $\mathbf{L}$, the restrictions are recursive:

$$e_1 \sim N(0, \mathbf{I}) \quad \text{subject to } a_1^* \leq l_{11}e_1 \leq b_1^* \quad \Leftrightarrow \quad a_1^*/l_{11} \leq e_1 \leq b_1^*/l_{11}$$
$$e_2 \sim N(0, \mathbf{I}) \quad \text{subject to } a_2^* \leq l_{21}e_1 + l_{22}e_2 \leq b_2^* \quad \Leftrightarrow \quad (a_2^* - l_{21}e_1)/l_{22} \leq e_2 \leq (b_2^* - l_{21}e_1)/l_{22}$$
$$\text{etc.}$$

In this form the e_i values can be generated sequentially with a univariate truncated simulator. Finally, the random vector of interest, $\mathbf{u}^*$, can be defined as:

$$\mathbf{u}^* = \Theta\mathbf{x} + \mathbf{A}^{-1}\mathbf{Le}$$

This vector has a covariance matrix given by $\mathbf{A}^{-1}\mathbf{LL'A}^{-1'} = \mathbf{A}^{-1}\mathbf{A}\Sigma\mathbf{A'A}^{-1'} = \Sigma$ and is subject, by construction, to the condition $a \leq \mathbf{Au}^* \leq b$. Börsch-Supan and Hajivassiliou (1993) show that although the generation of draws of $\mathbf{u}^*$ is biased, the contribution of each observation to the likelihood function (i.e. the probability that $\mathbf{Au}$ is between a and b) is simulated correctly by the probability that $a^* \leq \mathbf{Le} \leq b^*$.

To speed the process and reduce the variance of the choice probabilities that are eventually calculated, it is possible to use *antithetic draws* (see the discussion by Train 2009). If we consider that P_{iq} can be approximated as the average of the probabilities (P_{iq}^0 and P_{iq}^1) corresponding to two sets of repetitions of random variables, then it can be seen that:

$$Var(P_{iq}) = Var\left(\frac{P_{iq}^0 + P_{iq}^1}{2}\right) = \frac{1}{4}Var\left(P_{iq}^0\right) + \frac{1}{4}Var\left(P_{iq}^1\right) + \frac{1}{2}Cov\left(P_{iq}^0, P_{iq}^1\right)$$

Thus, if both sets are independent then the covariance is zero, but if they are negatively correlated then the covariance will be less than zero. This suggests the ideal situation of generating a series of random numbers to calculate probabilities and then, as an antithetic, to use the same series but with the opposite sign to generate the new set of probabilities. Not only does this achieve savings in random number generation, but it also computes choice probabilities with a smaller variance.

In the case of the MNP, where we are interested in evaluating the probability that the utility of the chosen option is higher than those of the remaining options in the choice set of each individual, the lower limit a^* is equal to zero and the upper limit b^* is infinity. The likelihood function is, as usual, the product of the probabilities of choosing the chosen option for each individual. Experimental programs to estimate the MNP model using the GHK simulator have been written in GAUSS (Aptech Systems 1994), and have been validated using simulated data (e.g. Munizaga *et al.* 2000).

The optimisation problem to solve in this case is not necessarily convex, so convergence to a unique optimum is not guaranteed. For example, among the routines offered in GAUSS, the Newton-Raphson method was the more robust in convergence terms (although somewhat slow) and the fastest method was the Berndt-Hall-Hall-Hausman algorithm (Berndt *et al.* 1974), although it did not always converge.

A practical issue of interest is that it is highly convenient to start by considering a very simple model, where only the parameters of the representative utility function are estimated (even starting with initial values taken from an MNL), and then re-estimate the model liberating the covariance matrix parameters one by one. This is not a sequential estimation, because at the last iteration the complete model is estimated, but a useful strategy to obtain the best initial point for what is in general a very complex optimisation problem (among other things, the log-likelihood surface is relatively flat and full of local optima).

8.6   Estimating the Mixed Logit Model

In section 7.6 we presented the Mixed Logit (ML) model and made reference to its two specifications, as error components (EC) model (7.47) and as random coefficients (RC) model (7.48). However, we also saw that both were formally equivalent (7.49) and the manner in which the modeller looks at the phenomenon under study will decide which form is more appropriate in any given case. Interestingly, there are also two general methods for estimating the model, the *classical* approach (using simulated maximum likelihood) and the *Bayesian* approach. Also, recall that there are two sets of parameters that in principle can be estimated, the *population* parameters (i.e. the vector of means and the covariance matrix associated with the mixing distribution), and the *individual* parameters which have a distribution over the population conditional on the former.

First we present the classical approach, incorporating the latest developments in the field of estimation via simulated maximum likelihood methods (Bhat 2001; Train 2009), including the framework by which population distribution parameters combined with information from individual choices can lead to consistent estimates of individual marginal utilities (Revelt and Train 2000). Secondly, we present the hierarchical Bayes estimation procedure, which has seen remarkable development over the last decade (Allenby and Rossi 1999; Sawtooth Software 1999; Huber and Train 2001; Andrews *et al* 2002; Sillano and Ortúzar 2005; Godoy and Ortúzar 2008).

8.6.1   Classical Estimation

By classical estimation we refer to the maximum likelihood procedure commonly used to estimate flexible discrete choice models (Train 2009).

8.6.1.1   *Estimation of Population Parameters*

Consider the most general case, of having available a sequence of T choices per individual (i.e. as in stated choice or panel data), denoted by $\mathbf{c}_q = (c_{1q}, \ldots, c_{Tq})$, where $c_{tq} = i$ if $U_{iqt} > U_{jqt} \; \forall \; A_j \neq A_i$. In a typical ML model, the conditional probability of observing an individual q stating a sequence $\mathbf{c}_q$ of choices, given *fixed* values for the model parameters $\bar{\theta}_q$, is given by a product of Logit functions:

$$\Lambda(\mathbf{c}_q|\theta_q) = \prod_{t=1}^{T} \frac{\exp(\lambda \cdot \bar{\theta}_q \cdot \mathbf{x}_{iqt})}{\sum_{j=1}^{J} \exp(\lambda \cdot \bar{\theta}_q \cdot \mathbf{x}_{jqt})} \tag{8.31}$$

where λ is the MNL's scale factor that has to be normalised as usual.

Now since θ_q is unknown, the unconditional probability of choice is given by the integration of (8.31) weighted by the density distribution of θ_q over the population:

$$\mathbf{P}_q(\mathbf{c}_q) = \int \Lambda(\mathbf{c}_q|\theta_q) f(\theta_q|\mathbf{b}, \Sigma) \, d\theta_q \tag{8.32}$$

where $f(\cdot)$ is the multivariate distribution of θ_q over the sampled population. If covariance terms are not specified, Σ is a diagonal matrix. Note that the majority of applications use diagonal matrices as results seem not to be affected strongly by this assumption (Sillano and Ortúzar 2005).

The log-likelihood function in $\mathbf{b}$ and Σ is:

$$l(\mathbf{b}, \Sigma) = \sum_{q=1}^{Q} \log \mathbf{P}_q(\mathbf{c}_q) \tag{8.33}$$

but as the probabilities $\mathbf{P}_q$ do not have a closed form they are approximated through simulation ($\mathbf{SP}_q$), where draws are taken from the mixing distribution $f(\cdot)$ weighted by the Logit probability, and then averaged up (McFadden and Train 2000):

$$\mathbf{SP}_q\left(\mathbf{c}_{qt}|f(\bullet|\mathbf{b}, \Sigma)\right) = \frac{1}{R} \sum_r \left(\prod_t \frac{\exp(\theta_q^r \cdot x_{iqt})}{\sum_{A_j \in \mathbf{A}(q)} \exp(\theta_q^r \cdot x_{jqt})} \right) \tag{8.34}$$

The issue of how many draws R and how should they be generated to improve the efficiency of the simulation is discussed below (Bhat 2003; Hess *et al.* 2006). The simulated log-likelihood function is given by:

$$sl(\mathbf{b}, \mathbf{\Sigma}) = \sum_{q=1}^{Q} \log \mathbf{SP}_q(\mathbf{c}_q) \tag{8.35}$$

Under regularity conditions this estimator is consistent and asymptotically Normal; furthermore, when the number of repetitions grows more rapidly than the square root of the number of observations, the estimator is asymptotically equivalent to the maximum likelihood estimator (Hajivassiliou and Ruud 1994). Other useful properties of the estimator are being twice differentiable (which helps in the numerical search of the optimum) and being strictly positive, so the log-likelihood function is always defined. Note that the same procedure would be followed if the ML had another Logit kernel, say a NL function or any more general and appropriate GEV model, instead of the MNL.

Different forms of 'smart' drawing techniques (i.e. Halton or other low discrepancy sequences, antithetic draws, quasi-random sampling, etc.) can be used to reduce the simulation variance and to improve the efficiency of the estimation (Hajivasiliou and Ruud 1994, Bhat 2003; Hensher and Greene 2003); we will refer briefly to this issue in section 8.6.4. Train (1998) presents a good example of the use of this model and offers an experimental estimation code, written in GAUSS, which can be downloaded from his web page. Another piece of free software available for estimating the ML is Biogeme (Bierlaire 2009), which offers many capabilities and allows the estimation of several other discrete choice models. Finally, new releases of the leading packages ALOGIT and LIMDEP include modules to estimate ML models, and these are several times faster than the more experimental codes available, definitely making the model a practical proposition.

8.6.1.2   *Estimating Individual Parameters*

Numerical procedures are used to find the maximum likelihood estimators for $\mathbf{b}$ and $\mathbf{\Sigma}$ above. These parameters define a frequency distribution for the θ_q over the population. To obtain actual point estimates for each θ_q a second procedure, described originally by Revelt and Train (2000), is required as follows.

The conditional density $h(\theta_q|\mathbf{c}_q, \mathbf{b}, \mathbf{\Sigma})$ of any θ_q given a sequence of T_q choices $\mathbf{c}_q$ and the population parameters $\mathbf{b}$ and $\mathbf{\Sigma}$, may be expressed by Bayes' rule as:

$$h(\theta_q|\mathbf{c}_q, \mathbf{b}, \mathbf{\Sigma}) = \frac{\mathbf{P}_q(\mathbf{c}_q|\theta_q) f(\theta_q|\mathbf{b}, \mathbf{\Sigma})}{P_n(\mathbf{c}_q|\mathbf{b}, \mathbf{\Sigma})} \tag{8.36}$$

The conditional expectations of θ_q result from integrating over its domain. This integral can be approximated by simulation, averaging weighted draws θ_q^r from the population density function $f(\theta_q|\mathbf{b}, \mathbf{\Sigma})$. The simulated expectations $\mathbf{SE}$ of the individual parameters are then given by:

$$\mathbf{SE}(\theta_q|\mathbf{c}_q, \mathbf{b}, \mathbf{\Sigma}) = \frac{\sum_{r=1}^{R} \theta_n^r \mathbf{P}_q(\mathbf{c}_q|\theta_n^r)}{\sum_{r=1}^{R} \mathbf{P}_q(\mathbf{c}_q|\theta_n^r)} \tag{8.37}$$

Revelt and Train (2000) also proposed, but did not apply, an alternative simulation method to condition individual level choices. Consider the expression for $h(\theta_q|\mathbf{c}_q, \mathbf{b}, \mathbf{\Sigma})$ in (8.36). The

(continued)

denominator is a constant value since it does not involve θ_q, so a proportionality relation can be established as:

$$h(\theta_q \mid \mathbf{c}_q, \mathbf{b}, \boldsymbol{\Sigma}) \propto \mathbf{P}_q(\mathbf{c}_q \mid \theta_q) f(\theta_q \mid \mathbf{b}, \boldsymbol{\Sigma}) \qquad (8.38)$$

Draws from the posterior $h(\theta_q \mid \mathbf{c}_q, \mathbf{b}, \boldsymbol{\Sigma})$ can then be obtained using the Metropolis-Hastings algorithm (Chib and Greenberg 1995), with successive iterations improving the fit of the θ_q to the observed individual choices. During this process the prior $f(\theta_q \mid \mathbf{b}, \boldsymbol{\Sigma})$, i.e. the parameter distribution obtained by maximum likelihood, remains fixed; it provides information about the population distribution of θ_q. After a sufficient number of *burn-out* iterations to ensure that a steady state has been reached (typically a few thousands, Kass *et al.* 1998; Godoy and Ortúzar 2008), only one every *m* of the sampled values generated is stored to avoid potential correlation among them; *m* is a result of the analysis of convergence (Raftery and Lewis 1992).

From these values a sampling distribution for $h(\theta_q \mid \mathbf{c}_q, \mathbf{b}, \boldsymbol{\Sigma})$ can be built, and inferences about the mean and standard deviation values can be obtained (Godoy and Ortúzar 2008). Sillano and Ortúzar (2005) favoured this latter procedure for implementation purposes and used WinBUGS (Spiegelhalter *et al.* 2001), a software package that can also be freely downloaded from the web.

Thus, the outcome of the estimation process is two sets of parameters: $\mathbf{b}$ and $\boldsymbol{\Sigma}$, the population parameters obtained by simulated maximum likelihood and θ_q, the individual parameters for $q = 1, \ldots, Q$, estimated via conditioning the observed individual choices on the estimated population parameters. It is surprising that the large majority of applications of ML models stop short of reaching the full capability of the model, by not going to this second stage.

8.6.2   Bayesian Estimation

Use of the Bayesian statistic paradigm for estimating the ML model gained much interest at the beginning of the century (Train 2001; Huber and Train 2001; Sawtooth Software 1999; Sillano and Ortúzar 2005) but has surprisingly lost appeal in recent years, together with the possibility of estimating individual rather than just population parameters. In fact, the ability to estimate individual part-worths appeared initially as its main appeal, but the estimation procedure has subsequently shown further advantages (Godoy and Ortúzar 2008). The Bayesian approach considers the parameters as stochastic variables so applying Bayes' rule of conditional probability, a posterior distribution for θ_q conditional on observed data and prior beliefs about these parameters can be estimated; let us denote this distribution by $\pi(\mathbf{b}, \boldsymbol{\Sigma} \mid \mathbf{c}_q)$.

Now, let $\psi(\mathbf{b}, \boldsymbol{\Sigma})$ represent the analyst's prior knowledge about the distribution of $\mathbf{b}$ and $\boldsymbol{\Sigma}$; typically a Normal distribution is used for the means $\mathbf{b}$ and an Inverted Wishart distribution for the variances in $\boldsymbol{\Sigma}$ (Allenby 1997). Then, consider a likelihood function for the observed sequence of choices conditional on fixed values of $\mathbf{b}$ and $\boldsymbol{\Sigma}$. By Bayes' rule, the posterior distribution for θ_q, $\mathbf{b}$ and $\boldsymbol{\Sigma}$ must be proportional to:

$$\prod_{q=1}^{Q} \Lambda(\mathbf{c}_q \mid \theta_q) f(\theta_q \mid \mathbf{b}, \boldsymbol{\Sigma}) \psi(\mathbf{b}, \boldsymbol{\Sigma}) \qquad (8.39)$$

Although it is possible to draw directly from $\pi(\mathbf{b}, \boldsymbol{\Sigma} \mid \mathbf{c}_q)$ with the Metropolis-Hastings (MH) algorithm, this would be computationally very slow. Indeed, it would be necessary to calculate (8.39) at every iteration of the MH algorithm but the choice probability inside is an integral without a closed form resolution and must be approximated through simulation; thus, an iteration of the MH algorithm would require simulation for each individual q. That could be time consuming and affect the properties of the resulting estimator.

Drawing from $\pi(\mathbf{b}, \boldsymbol{\Sigma}|\mathbf{c}_q)$ becomes fast and simple if each $\boldsymbol{\theta}_q$ is considered to be a parameter along with $\mathbf{b}$ and $\boldsymbol{\Sigma}$, and Gibbs sampling is used for the three sets of parameters for each individual. The posterior distribution in this case is:

$$\pi(\mathbf{b}, \boldsymbol{\Sigma}|\mathbf{c}_q) \propto \prod_{q=1}^{Q} \Lambda(\mathbf{c}_q|\boldsymbol{\theta}_q) f(\boldsymbol{\theta}_q|\mathbf{b}, \boldsymbol{\Sigma}) \psi(\mathbf{b}, \boldsymbol{\Sigma}) \tag{8.40}$$

The sequential procedure simulates each set of parameters from the following conditional posterior distributions:

- The conditional posterior for $\mathbf{b}$ is $\pi(\mathbf{b}|\boldsymbol{\Sigma}, \boldsymbol{\theta}_q \ \forall q)$ and this distributes $N(\bar{\boldsymbol{\theta}}, \boldsymbol{\Sigma}|Q)$ where $\bar{\boldsymbol{\theta}} = \sum_q \boldsymbol{\theta}_q/Q$.
- The conditional posterior for $\boldsymbol{\Sigma}$ is $\pi(\boldsymbol{\Sigma}|\mathbf{b}, \boldsymbol{\theta}_q \ \forall q)$ and this distributes inverted Wishart $IW\left(K + Q, \frac{K \cdot J + Q \cdot \bar{S}}{K+Q}\right)$ where $\bar{S} = \sum_q (\boldsymbol{\theta}_q - \mathbf{b})(\boldsymbol{\theta}_q - \mathbf{b})^T/Q$.
- The conditional posterior for $\boldsymbol{\theta}_q$ is given by $\pi(\mathbf{b}, \boldsymbol{\Sigma}|\mathbf{c}_q) \propto \prod_{q=1}^{Q} \Lambda(\mathbf{c}_q|\boldsymbol{\theta}_q) f(\boldsymbol{\theta}_q|\mathbf{b}, \boldsymbol{\Sigma})$.

Then, rth iteration of the Gibbs sampler consists on the following steps: (1) Draw $\mathbf{b}^r$ from $N(\bar{\boldsymbol{\theta}}^{r-1}, \boldsymbol{\Sigma}|Q)$; (2) Draw $\boldsymbol{\Sigma}^r$ from $IW\left(K + Q, \frac{K \cdot J + Q \cdot \bar{S}^{r-1}}{K+Q}\right)$; (3) For each individual q draw $\boldsymbol{\theta}_q^r$ using one iteration of the MH algorithm, starting from the draw at the previous iteration and using the Normal density $f(\boldsymbol{\theta}_q|\mathbf{b}, \boldsymbol{\Sigma})$. These three steps are repeated many times. The resulting values converge to draws from the joint posterior of $\mathbf{b}$, $\boldsymbol{\Sigma}$ and $\boldsymbol{\theta}_q \ \forall q$. Once the converged draws from the posterior are obtained, the mean and standard deviation of the draws can be calculated to obtain estimates and standard errors of the parameters.

Train (2001) discusses how the posterior means from the Bayesian estimation can be analysed from a classical perspective. This is thanks to the Bernstein-von Mises theorem which states that, asymptotically, the posterior distribution of a Bayesian estimator converges to a Normal distribution which is the same as the asymptotic distribution of the maximum likelihood estimator (e.g. the standard deviation of the posterior distribution of the Bayesian estimator can be taken as the classical standard error of a maximum likelihood estimator). This means that classical statistical analysis (for example the construction of t-statistics to analyse the significance of an estimated parameter) can be performed on Bayesian estimators without compromising the interpretation of the results.

Bayesian estimation has certain advantages over the classical approach:

- No numerical maximisation routines are necessary; rather, draws from the posterior distribution are taken until convergence is achieved.
- As the number of attributes considered in the utility expression grows, the number of elements in the covariance matrix $\boldsymbol{\Sigma}$ rises exponentially increasing computation time in the classical approach. However, the Bayesian method can handle a full covariance matrix almost as easily as a restricted one, with computation time rising just as the number of parameters.
- Identification issues are related with the lack of orthogonality in the effects of the random variables and not with the number of independent equations representing these. This means that an identification problem may rise when the effect of a certain variable in the structural utility formulation is confused with the effect of another variable, but not because of insufficient sample points.

(continued)

> The Bayesian estimation procedure is available as an experimental code on Ken Train's website, and can be also implemented in WinBUGS. This package incorporates Gibbs sampling protocols and the Metropolis-Hastings sampling algorithm but lacks a convergence analysis that has to be performed separately (Godoy and Ortúzar 2008 provide useful advice on how to do this properly).

Example 8.8 Stated preference data on residential location choice considered the following attributes: travel time to work (by the parents), travel time to study (by the children), rent or mortgage of the flat, and a variable related to atmospheric pollution in the zone (days of alert, see Ortúzar and Rodríguez 2002). Seventy-five families were asked to express their location preferences for a flat of otherwise exactly the same characteristics, finally yielding 648 usable responses (i.e. some observations were discarded in the data cleaning process). MNL and ML models were estimated with this data, and Table 8.8 shows the results for the classical estimation of population parameters in the ML (Sillano and Ortúzar 2005).

Table 8.8   Model results for location choice analysis

Attributes		Parameters (t-test)	
		MNL	**ML1**
Travel time to work	Mean	−0.00417 (−10.6)	−0.009924 (−7.9)
	Std. Dev.		0.005734 (4.5)
Travel time to study	Mean	−0.00250 (−7.8)	−0.005769 (−8.2)
	Std. Dev.		0.002656 (2.7)
Days of alert (environment)	Mean	−0.27370 (−11.0)	−0.478625 (−6.8)
	Std. Dev.		0.405665 (4.7)
Rent/Mortgage	Mean	−0.02641 (−12.5)	−0.057396 (−7.0)
	Std. Dev.		0.047482 (6.2)
Inertia	Mean	0.89690 (5.9)	1.053245 (5.5)
Log-likelihood		−849.6	−747.0

In model ML1, the nine choices from each household were considered, correctly, as repeated choice observations and it was assumed that the parameters distributed IID Normal; the Inertia variable (a dummy which took the value one if the household ranked their current location first) received a non significant standard deviation, and for that reason the model was re-estimated with Inertia as a fixed parameter for all individuals.

A number of issues are important from this table. First, although the MNL model would be judged adequate by any seasoned analyst, there is a notable increase in log-likelihood (more than a hundred points) for the addition of only four parameters in ML1; a large part of this increase is due to the proper consideration of repeated observations in ML1 (further evidence to this fact has already been discussed above).

Second, note the substantial increase in parameter values from the MNL model to ML1; this (expected) result is due to the 'lurking' scale factor λ corresponding to the IID EV1 error in both models, and it is illustrative to discuss it. The MNL parameters do not vary among individuals (when it is clear from ML1 that this should be the case); so as the MNL EV1 error is picking up this, its variance is high and, correspondingly, the MNL scale factor is small. Conversely, the ML EV1 error only has to pick up other, remaining, sources of error so its variance is low, and its scale factor large.

Third, the estimated standard deviations are not only significant but relatively large in magnitude (in comparison to the mean parameter estimates). So, the portion of the population for which the model would assign an incorrect parameter sign can be estimated as the cumulative mass function of the frequency distribution of the parameter evaluated at zero (i.e. for supposedly negative parameters, the area under the frequency curve between zero and positive infinity). In this case, ML1 would account for 4% of the population having positive *Time-to-work* parameters, 1% of the population having positive *Time-to-study* parameters, 12% of the population having positive *Days-of-alert* parameters, and 11% of the population having positive *Rent* parameters. This problem may be overcome in various ways, for example, using a log-normal distribution (effectively constraining the parameters to be negative), but this has a series of undesirable properties as we will discuss in more depth below.

Sillano and Ortúzar (2005) went on to estimate individual parameters (see for example Figure 8.5, for *Time-to-work* and *Days-of-alert*) finding that the above expected proportions were overestimated (e.g. none in the first case and only three out of 75 households in the second); furthermore, the individual parameters for the offending households were not significantly different from zero; hence they could be considered as null values for those households, and the sign assumptions could be maintained.

A final issue relating to Example 8.8 is that while the distribution in Figure 8.5a looks acceptably like a Normal distribution (given the small sample size), that in Figure 8.5b certainly does not. This means that a certain amount of error must be expected when analysing a discrete set of values using a continuous distribution.

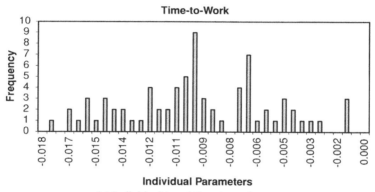

(a) Individual travel time to work parameters

(b) Individual of days of alert parameters

Figure 8.5   Distribution of individual parameters in the population for ML1

8.6.3   Choice of a Mixing Distribution

The most popular distribution in ML applications has been the Normal, but some analysts have claimed that the fact it is unbounded imposes unacceptable conditions on the signs of the estimated parameters. This a debatable issue, as precisely because it is unbounded it may help reveal 'outliers' or observations which are plainly wrong (in some sense, for example badly coded, or not consistent with the compensatory choice paradigm, see Sælensminde 1999). Furthermore, as we just saw in the example above, even if a proportion of individuals would appear to receive a wrong sign given the estimated population parameters, when we move to the stage of estimating individual parameters the number of cases in this condition might turn to be (i) very low and (ii) their parameters are not significantly different from zero.

An extensive literature exists on this subject but the question is still open. Recent advances are the work of Train and Sonnier (2005) on bounded distributions of correlated part-worths; that of Dong and Koppelman (2004) and Hess *et al.* (2007), on continuous vs. discrete mixing distributions, and Fosgerau and Bierlaire (2007) on semi-nonparametric (SNP) tests. Fosgearu and Hess (2010) make a useful comparison of different approaches.

8.6.3.1   Alternative Mixing Distributions

Mixing distributions can be split into two main groups (Hess *et al.* 2005b): those with fixed bounds, such as the Lognormal, Gamma and Rayleigh, and those with bounds that are estimated during model fitting, such as the Uniform, Triangular and, more recently, the Johnson S_β, which has many interesting properties (Train and Sonnier 2005). In ML model estimation, it is important to choose the correct distribution to reproduce the heterogeneity underlying population preferences.

In real cases almost all attributes have associated a parameter which is logically bounded, either because it can be only positive or negative, or because it cannot be unboundedly large. Train and Sonnier (2005) formulate ML models with partworths that are transformations of normally distributed (latent) terms, where the transformation induces bounds. The Johnson S_β distribution has several advantages in this sense; its density can be shaped like a Lognormal with an upper bound and thinner tails below the bound, but it is more flexible as it can also be shaped like a plateau with a fairly flat area between drop-offs on each side, and can even be bi-modal.

When a lower bound other than zero is specified, the distribution is useful for attributes that some people like and others dislike but for which there is a limit on how much the person values having or avoiding it. Even more interesting, the bounds of the Johnson S_β distribution can be estimated as parameters, rather than specified by the modeller. However, this last property requires a more complex model estimation process and identification becomes an issue (as the difference between upper and lower bounds is closely related to the variance of the latent Normal term). For these reasons and in some cases also depending if it is practical or theoretical work, some analysts prefer to use simpler and more robust forms such as the Triangular or Rayleigh distributions (Hensher 2006).

Finally, a number of applications have also looked at incorporating deterministic heterogeneity components into the distribution of the random terms, either in the mean or the standard deviation, hence allowing the modeller to relate the variation of random coefficients to individual-specific observed attributes i.e. akin to what we called systematic taste variations in equation (8.17). As an example, in a standard framework we would possibly use $\theta \sim N(\mathbf{b}, \Sigma)$, but here we would additionally specify $\mathbf{b} \sim f(\mathbf{s})$ and $\Sigma \sim g(\mathbf{s})$, making the parameters of the distribution a function of socio-demographic variables $\mathbf{s}$. This can be useful either in a random coefficients as well as error components context; a recent example of such an approach is given by Greene *et al.* (2006).

8.6.3.2 Discrete Mixtures and Latent Class Modelling

Dong and Koppelman (2004) represented heterogeneity with a discrete distribution with a finite number of supports; in this case $f(\theta)$ is replaced by a mass-point distribution with weight at mass point m given by π_m. Then, by replacing the integration in (8.32) with a sum over a finite number of mass points M, the ML model can be expressed as:

$$\mathbf{P}_{iq} = \sum_{m=1}^{M} \frac{\exp(\theta_i^m \mathbf{x}_{iq})}{\displaystyle\sum_{A_j \in \mathbf{A}(q)} \exp(\theta_j^m \mathbf{x}_{jq})} \cdot \pi_m \tag{8.41}$$

so, it is a weighted average of Logit probabilities computed at each possible value of θ (the weights are the probabilities of θ to be at each value θ^m), and can be estimated by maximum likelihood *without* simulation. Using simulated data they found that model (8.41) was inferior to the conventional ML whether the true distribution was continuous or discrete; furthermore, they found that the model estimates could be misleading if the true distribution was, in fact, continuous; however, their model allowed the identification of heterogeneity which was not discovered by the continuous version of the ML.

Hess *et al.* (2007) generalised this approach by letting the MNL probability be any more general GEV function; they divided the set θ into two parts, one, $\tilde{\theta}$, containing deterministic parameters and another, $\hat{\theta}$, with K random parameters with a discrete distribution; within this set, $\hat{\theta}_k$ has m_k mass points, $\hat{\theta}_k^n, n = 1, \ldots, m_k$, each of them associated with a probability π_k^n, where the following conditions must be imposed:

$$0 \le \pi_k^n \le 1, \quad k = 1, \ldots, K; \, n = 1, \ldots, m_k \quad \text{and} \quad \sum_{n=1}^{m_k} \pi_k^n = 1, \quad k = 1, \ldots, K \tag{8.42}$$

They discuss several extensions that offer more modelling flexibility, but note that some may lead to parameter over-specification, impairing estimation. They also note that the non-concavity of the log-likelihood function in this case does not allow the identification of a global maximum, even for discrete mixtures of the simple MNL model; thus, they advise the performance of several estimations from various starting points and recommend, as good practice, the use of staring values different from 0 or 1 for the π_k^n parameters.

If the class allocations are linked to socio-demographic variables, we obtain a latent class (LC) model (see for example Hess *et al.* 2009). In an LC model the heterogeneity in tastes across respondents is accommodated by making use of separate classes with different values for the vector of taste coefficients θ. Specifically, in an LC model with M classes, we would have M instances of the vector θ, say θ^1 to θ^M, with a possibility of some of the elements in θ staying constant across some of the classes.

An LC model uses a probabilistic class allocation model, where respondent q belongs to class m with probability $\pi_{q,m}$ and where $0 \le \pi_{q,m} \le 1$, for all m and $\Sigma_m \pi_{q,m} = 1$. LC models are generally specified with an underlying MNL model, but can easily be adapted for more general underlying structures such as Nested Logit (NL) or Cross-Nested Logit (CNL).

Let $P_{iq}(\theta_m)$ give the probability of respondent q choosing alternative A_i conditional on her falling into class m. The unconditional (on m) choice probability for alternative A_i and respondent q is then given by:

$$\mathbf{P}_q(A_i | \theta_1, \ldots, \theta_M) = \sum_{m=1}^{M} \pi_{q,m} P_{iq}(\theta_m) \tag{8.43}$$

(continued)

i.e. the weighted sum of choice probabilities across the M classes, with the class allocation probabilities being used as weights. Unlike with the ML model, no simulation is required in the estimation of LC models.

This specification can easily be extended to a situation with multiple choices per respondent, where, when making the same assumption of intra-respondent homogeneity as in the Revelt and Train (1998) work for continuous ML, we obtain:

$$\mathbf{P}_q(A_i|\theta_1, \ldots, \theta_M) = \sum_{m=1}^{M} \pi_{q,m} \left(\prod_{t=1}^{T_q} P_{iqt}(\theta_m) \right) \qquad (8.44)$$

In the most basic version of an LC model, the class allocation probabilities are constant across respondents such that $\pi_{q,m} = \pi_m$ for all q. The resulting model then corresponds to a discrete mixture analogue to the ML model, as discussed above.

The real flexibility, however, arises when the class allocation probabilities are not constant across respondents but a class allocation model is used to link these probabilities to characteristics of the respondents. Typically, these characteristics would take the form of socio-demographic variables such as income, age and employment status. With $\mathbf{s}_q$ giving the concerned vector of characteristics for respondent q, and with the class allocation model taking on a MNL form, the probability of respondent q falling into class m would be given by:

$$\pi_{q,m} = \frac{\exp\left(\delta_m + g(\beta_m, \mathbf{s}_q)\right)}{\sum_{l=1}^{M} \exp\left(\delta_l + g(\beta_l, \mathbf{s}_q)\right)} \qquad (8.45)$$

where δ_m is a class-specific constant, β_m a vector of parameters to be estimated and $g(\cdot)$ gives the functional form of the *utility* function for the class allocation model.

Here a major difference arises between class allocation models and choice models. In a choice model, the attributes vary across alternatives while the estimated coefficients (with a few exceptions) stay constant across alternatives. In a class allocation model, the attributes normally stay constant across classes while the parameters vary across classes. This allows the model to allocate respondents to different classes depending on their socio-demographic characteristics. For example, a situation where high income and low income respondents are allocated differently to two classes could be represented with a positive income coefficient for the first class and a negative income coefficient for the second class. Finally, we can mention that it is possible to combine latent class and ML structures, leading to latent class models with some continuous elements, as for example done by Walker and Li (2007).

To end this part, we note the work of Fosgerau and Bierlaire (2007) who propose the use of semi-nonparametric (SNP) techniques to test if a given mixing distribution is appropriate. The SNP models offer the advantage, over conventional ML, that the structure does not need to be specified *a priori*. In particular, they introduce parametric assumptions like the specification of some relationship to be a linear combination of independent variables while perhaps the errors remain nonparametric. SNP models are not based on local approximations; instead, they use series to approximate functions such as densities. The number of SNP terms must be chosen in advance; increasing this number makes the model more general but increases the demand on the data. Fosgearu and Bierlaire (2007) found that two or three terms give a large degree of flexibility, which may be sufficient for most purposes, while one SNP term is not always sufficient to reject a false null hypothesis.

8.6.4   Random and Quasi Random Numbers

The multidimensional integral (8.31) has to be solved via simulated maximum likelihood, and this can be very time consuming in real large-scale model estimation. As a consequence, several methods have been devised to help in this task including the use of cheaper (in time) quasi Monte Carlo approaches, based on the generation of *low discrepancy* or *quasi-random* sequences (see Niederreiter 1992) as they allow more accurate integration approximations than classical Monte Carlo samplings (Train 2009).

Figure 8.6a shows the uneven coverage of the space of integration by the typical pseudo-random numbers generated automatically by computers (300 points in two dimensions). Figure 8.6b shows the much better coverage of Halton numbers, one of the early sequences used by researchers, in this case.

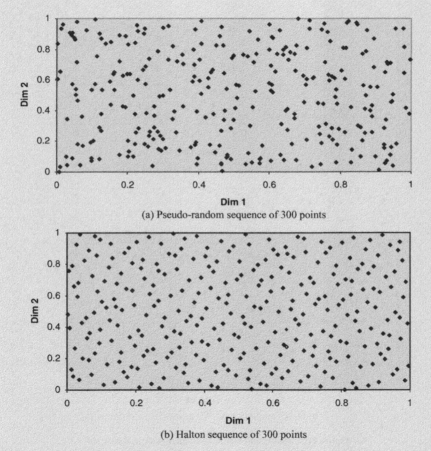

(a) Pseudo-random sequence of 300 points

(b) Halton sequence of 300 points

Figure 8.6   Pseudo-random and Halton coverage in two dimensions

(continued)

In fact, it has been reported that only 125 Halton numbers can provide the equivalent coverage of 1000 pseudo-random numbers (see Train 2009). Now, although Halton sequences ruled for a while, it was soon shown that their coverage of the integration domain rapidly deteriorated for higher integration dimensions (Silva 2002); for example, Figure 8.7a shows the Halton sequence pattern for an example with several dimensions and this should be compared with the Sobol sequence pattern for the same number of dimensions (Figure 8.7b).

(a) Halton sequence of 300 points in 15 dimensions

(b) Sobol sequence of 300 points in 15 dimensions

Figure 8.7   Halton and Sobol coverage in many dimensions

This fostered a search for new sequences, including work on scrambled and shuffled Halton sequences (Bhat 2003), Sobol sequences (shown to be superior to the former by Silva 2002) and, more recently, Modified Latin Hypercube sampling (Hess *et al.*, 2006), which appears to be the preferred method today.

A related but different approach has been taken by Bastin *et al.* (2006) who capitalise on the desirable aspects of pure Monte Carlo techniques while significantly improving their efficiency. They proposed a new algorithm for stochastic programming based on 'trust-region' techniques (a well-known method in nonlinear non concave optimisation, which proved reliable and efficient for both constrained and unconstrained problems). They also allowed for an adaptive variation of the number of draws used in successive iterations, yielding an algorithm with comparable execution time to existing methods for similar results.

Numerical experimentation suggests that the choice of optimisation framework is of crucial importance; also, that the strategy of using a variable number of draws in the estimation of choice probabilities gives significant gains in optimisation time (compared with the usual approach of using fixed draws), and additional information on the closeness between the Monte Carlo approximation and the true function, while not suffering of non-uniform coverage in high integration dimensions. However, the field is still young in this sense and many research directions remain wide open.

8.6.5   Estimation of Panel Data Models

As we saw in Chapter 7, panel data offer major advantages over cross-sectional data; in particular, having repeated observations from the same individual generally allows for more accurate measurement of changes in individual mobility. Furthermore, as we commented in section 7.7.4.1, the inclusion of intra-respondent heterogeneity, which is only possible if there are multiple observations per individual, leads to significant improvements in model fit.

Given the potential of panel data structures the challenge is to make the most of such potential capturing as many effects as possible. In this sense, the most general formulation for panel data model estimation is that proposed by Hess and Rose (2009) in equation (7.55); this considers not only inter-respondent heterogeneity as in the more classical specification (7.54), but also intra-respondent heterogeneity of tastes. Hess and Train (2010) consider various alternatives to estimate models under this complex formulation, and note that even with state-of-the art computers and optimization techniques the full generality afforded by the formulation may lead to very long estimation times.

Surprisingly, even this general panel formulation (that considers two dimensions of heterogeneity) accommodates heterogeneity only via the estimation of random parameters. Thus, random parameters θ_q account also for correlation in tastes. However, as we commented in section 7.7.4.1, there might be extra correlation across multiple observations besides the effect of the random parameters. For example, the inclusion of pure panel component errors may also account for correlation in the preferences for alternatives, as proven by recent work by Yáñez *et al.* (2010b). They analysed the impact of panel sample size and repeated observations on both the model capability to reproduce the true phenomenon and the probability to capture different kinds of heterogeneity among observations. They found that their best model accommodated inter-respondent heterogeneity through random parameters and intra-respondent heterogeneity through pure-error components.

For practical purposes, another important issue regarding panel correlation has to do with model estimation using available software. The usual way to incorporate panel correlation under the pure error-components approach consists of adding an error component to $(J - 1)$ of the available alternatives; otherwise, for identifiability reasons the model cannot be estimated (Walker 2001). However, this methodology may lead to biased results as it requires choosing, arbitrarily, a single

(continued)

reference alternative for the error-component (i.e. one not having a pure panel error-component) in all cases. The reason is that this is equivalent to assuming that this reference alternative has the same alternative specific constant (ASC) for all observations, while the remaining ones have different ASC values among observations. Moreover, even using the best recommended normalisation (i.e. selecting as *error-component reference alternative* that option obtaining the minimum variance in a model run without considering identifiability, see section 7.6.3.1), this approach leads to a heteroskesdastic Nested Logit model, as it correlates the $(J - 1)$ alternatives including error components.

One way to avoid this problem is to modify this traditional estimation method by randomly selecting the error-component reference alternative for each individual (or each observation in the case of allowing for intra-respondent heterogeneity).

For this we need first to choose randomly and exogenously (i.e. before model estimation) an error-component reference alternative for each individual (or for each observation in the case of needing to accommodate intra-respondent heterogeneity). Then, we need to create J binary variables that take the value one only for the error-component reference alternative in either case of respondent heterogeneity. Finally, the pure error-component term is included in the utility function of each alternative multiplied by the corresponding binary variable.

Example 8.9 Consider the following utility function:

$$U_{iqd} = \alpha_i + \sum_j X_{iqk}^d \cdot \theta_{iqk} + \zeta_{iqd}$$

where the error component has the form $\zeta_{iqd} = \upsilon_{qd} + \varepsilon_{iqd}$. Here ε_{iqd} is a random term distributed IID EV1, as usual, and υ_{qd} is a random effect which may be specific to the individual (i.e. just υ_q), in which case we assume panel correlation as inter-respondent heterogeneity. But we could also make it variable among observations (υ_{qd}), in which case we would assume intra-respondent heterogeneity.

As the θ vector has means θ_{ik} and standard deviations σ_{iqk} the utility function can be rewritten as:

$$U_{iqd} = (\alpha_i + \upsilon_{qd}) + \sum_k (\theta_{ik} + \sigma_{iqk}) X_{iqk}^d + \varepsilon_{iqd}$$

where X_{iqk}^d is the kth level-of-service attribute of option A_i for individual q on day d. This equation shows that both random coefficients and error components are separable. Indeed, the random coefficients allow tastes to vary across respondents in the sample, but stay constant across observations for the same respondent (Revelt and Train 1998). On the other hand, the 'pure' error components (which also capture heterogeneity) affect the values of the alternative specific constants (ASC). Thus, the error component υ_{qd} has the power to increase/decrease the relative weight of the ASC in relation to the explanatory variables in the utility function.

Now, confounding effects are implicit in the ML structure as we saw in section 7.6.2 and should not strictly depend on whether they do or do not account for random tastes. On the contrary, Cherchi and Ortúzar (2008b) have shown that decomposing randomness in as many components as possible helps to reveal the confounding effects.

8.7   Modelling with Stated-Preference Data

In Chapter 3 we discussed the experimental design and the data collection process of stated choice data in some detail; we made scant reference to traditional conjoint analysis (rank and rating data) and left contingent valuation for Chapter 15. In section 8.3 we noted that stated preference experiments could be instrumental in helping to decide the most appropriate functional form to model a given choice situation.

In this section we will first briefly review how this can be done and then we will proceed to discuss what changes are introduced to discrete choice modelling estimation by the use of stated-preference data.

8.7.1 Identifying Functional Form

The travel-demand model estimation literature is heavily oriented towards the problem of estimating a set of model parameters given a functional specification; only occasionally are alternative model structures tested. The favoured functional forms are those which can be deduced from (economic) first principles and also satisfy the condition of being easily estimable; for this reason the vast majority of studies has considered linear (in the parameters) utility functions. A notable exception to this rule is the increasing use of transformations to search for functional form but, as we saw in section 8.3, in these cases the computational problem of model estimation is greatly increased; in fact, estimation methods have only been developed for the simpler MNL model in this case.

In contrast, the literature in the area of psychological measurement procedures that use laboratory or interview data, has been deeply concerned with questions of functional form for a long time (see Louviere 1988a). In these studies subjects are asked to make judgements about hypothetical alternatives; for example, in a mode choice context they may be asked to select the preferred alternative from a hypothetical set, or to rank the options, or to associate a level of utility to each of them.

Because an individual can be asked to make a fairly large number of judgements in a single interview, the experiment designer can explore, for example, the effects on response of changes to one variable while keeping all the others constant. This allows a much more detailed assessment of functional form, since the analyst can almost trace the shape of response with respect to each variable. A very interesting finding of such studies is that for any particular decision, functional forms tend to be fairly stable across the population even though the values of their parameters can vary widely (see Meyer *et al.* 1978).

Let us assume that travel behaviour is influenced by a set of independent factors which may be quantitative or qualitative in nature. Following Lerman and Louviere (1978), let us denote the set of G quantitative factors for option A_i by $\mathbf{D}_i = \{D_{ig}\}$ and the set of H qualitative factors by $\mathbf{E}_i = \{E_{jh}\}$. The total number of factors is $K = G + H$, and the entire attribute vector $\mathbf{X}_i = \{X_{ki}\}$ is simply $\mathbf{D}_i$ and $\mathbf{E}_i$.

Let us also assume that each factor has associated with it a certain value (which may be obtained by some or other measurement process) and that the utility of this quantity as perceived by the individual is $u_{ki} = f_{ki}(X_{ki})$, where f is a perception function.

Consider now an experimental context where we observe the response to a combination of $(D_{1i}, \ldots, D_{Gi}; E_{1i}, \ldots, E_{Hi})$ on a psychological measurement scale. If we assume that this response measure is connected to the utility U_i of option A_i by some algebraic combination rule, we can write:

$$U_i = p_i(u_{1i}, \ldots, u_{Ki}) \tag{8.46}$$

Finally if we postulate that the vector of responses $\mathbf{U} = \{U_i\}$ is connected to non-experimental (i.e. observed) behaviour B by another algebraic function, we can write:

$$B = w(\mathbf{U}) \tag{8.47}$$

and by substituting, we get:

$$B = w\{p[f(\mathbf{D}, \mathbf{E})]\} \tag{8.48}$$

As this is too general a formulation for modelling purposes, in practical applications one must make explicit assumptions about the functions f, p and w, and deduce their consequences.

Now, for the purposes of developing an appropriate functional form, the critical component of this approach is the specification of equation (8.46). Alternative forms, such as multiplicative or linear cases, may be tested and selected by means of analysis of variance; however, in order to successfully apply it two conditions must be satisfied: first, the pattern of statistical significance of the utility responses to various combinations of the independent variables must be of a specific nature in order to permit diagnosis or testing of model form; second, corresponding graphical evidence must support the diagnosis or test.

Example 8.10 Consider a residential location model where individuals are assumed to trade off the total cost of travel (including travel times) with house price, independently of one another, i.e. it is assumed that they combine the effects of the two variables linearly. This hypothesis may be tested directly by an analysis of variance. Suppressing the option index i for simplicity, we can write:

$$U_{mn} = U_m^1 + U_n^2 + \varepsilon_{mn}$$

where U_l^k are utility values assigned to the lth level of the kth attribute in a factorial design, U_{mn} stands for the overall utility assigned by individuals to combinations of levels of both attributes, and ε_{mn} is a random term with zero mean.

A test for independence of the two effects corresponds to a test of the significance of the interaction effect $U_m^1 U_n^2$. As Lerman and Louviere (1978) point out, in an analysis of variance this is a global test for any and all interactions between both variables; thus if the interaction effect is not significant, the hypothesis of linear form cannot be rejected. If the interaction is significant, on the other hand, it implies that a simple linear combination is not appropriate.

This test should be accompanied by a graphical plot of the interaction. If the linear hypothesis (no interaction) is correct, the data should plot as a series of parallel lines when plotted against either utility value. It can be shown that this is true regardless of the form assumed for the marginal relationships (8.46); it can also be shown that this is true for any multi-linear utility model and for any forms less restrictive than simple addition or multiplication.

8.7.2 Stated Preference Data and Discrete Choice Modelling

There are two particular features of SP data that lend the approach to different analysis methods, *vis à vis* other sources of disaggregate data: first, the fact that each respondent may contribute with more than one observation and, second, the different forms in which preferences can be expressed. In Chapter 3 we mentioned in passing that traditional conjoint analysis considers two types of responses: ratings and rankings, but that the field has been clearly dominated by stated choice (SC) data. In the first type of response, the subject is asked to rate each option using a number between 1 and 5 or 10. The result of this exercise may be interpreted as the strength of the individual preference for each alternative. Therefore, normal algebraic operations can be carried out on them, for example extracting a ratio or subtracting one from another. However, this is now believed to be a weak element in SP work as there is no evidence to support the assertion that individual preferences can be elicited and translated into cardinal scales of this kind.

Simpler, and more reliable, tasks are to ask individuals to rank alternatives in order of preference or, much simpler, to make several choices between hypothetical alternatives. In the case of *ranking* experiments the individual is asked to rank a set of N alternatives in order of preference. If r_i denotes the alternative ranked in the ith position, the response implies that:

$$U(r_1) \geq U(r_2) \geq \ldots \geq U(r_N) \tag{8.49}$$

In the case of stated *choice* exercises the individual is only asked to choose his preferred option from the alternatives (two or more) in the choice set; therefore in this case the response corresponds with the usual discrete choice RP approach, except for the fact that both alternatives and choices are hypothetical. Note, however, that his type of exercise can be extended and enriched by allowing respondents to express their degree of confidence in the stated choice. To this end, the respondent is offered a semantic scale, the most typical having five points (*1: Definitively prefer first option; 2: Probably prefer first option; 3: Indifferent; 4: Probably prefer second option; 5: Definitively prefer second option*). This exercise is sometimes also called *rating* in the transport literature although it is actually a generalization of a choice experiment (see for example Ortúzar and Garrido 1994a; b). This generalisation offers advantages and disadvantages: on the one hand it permits a richer range of modelling techniques to be applied to the data; on the other hand, it may weaken the specificity of the choice and that of the response, increasing the difference between experiment and behaviour.

Taking advantage of the special features of SP data there are four broad groups of techniques for analysis:

(i) Naive or graphical methods.
(ii) Least square fitting, including linear regression.
(iii) Non-metric scaling.
(iv) Logit and Probit analysis.

These methods can be used to provide different levels of analysis of SP experiments. In general, all seek to establish the weights attached to each attribute in an (indirect) utility function estimated for each alternative. These weights are sometimes referred to as 'preference weights', 'part utilities' 'part-worths' or simply 'coefficients' associated with each attribute. Once these have been estimated they can be used for various purposes:

(a) To determine the relative importance of the attributes included in the experiment.
(b) An extension of this is the estimate of the rate at which one attribute is traded-off with another (a typical example is the estimation of 'values-of-time' when both time and cost attributes have been included in the experiment); it is also possible to estimate the value of more qualitative attributes like reliability, security levels, and so on; we will come back to this in section 15.4.
(c) To specify utility functions for forecasting models, including questions of model structure.

The nature of SP data and the objective of the analysis will be determining factors in the choice of model estimation techniques.

8.7.2.1   Naive Methods

The naive or graphical methods utilise a simple approach based on the fact that in many designs each level of each attribute appears the same number of times. Therefore, some indication of the relative utility of that attribute-level pair can be obtained by computing the mean average rank, rating or choice score for each option in which it was included and comparing that with similar mean averages for other levels and attributes. In effect, just plotting these means on a graph often gives useful indications about the relative importance of the various attributes included in the experiment. This model does not make use of any statistical theory and therefore fails to give us an indication of the statistical significance of the results.

Example 8.11 Consider an SP exercise comparing three alternative modes of transport, a traditional diesel bus (DB), a modern mini bus (MB) and an electric light rail vehicle (LRT). The attributes included in the SP experiments are in-vehicle travel time, the headway, the fare, and, of course, the vehicle type. The following table shows the different levels to be tested for each attribute:

	Level 1	Level 2	Level 3
Travel Time (min)	25	15	35
Fare (£)	1.30	1.00	1.50
Headway (min)	5	10	20
Vehicle type	DB	LRT	MB

A fractional factorial design is used, and respondents are asked to rate, or score, the alternatives (10 is the highest or best service). The results are as follows:

Travel time	Fare	Headway	Vehicle type	Score
25	1.30	5	DB	8
25	1.00	10	MB	9
25	1.50	20	LRT	4
15	1.30	10	LRT	10
15	1.00	20	DB	7
15	1.50	5	MB	8
35	1.30	20	MB	4
35	1.00	5	LRT	4
35	1.50	10	DB	1

It is now possible to calculate a 'naive' value for each attribute by calculating the average score for that level and attribute and comparing it with the difference in values. For instance, in the case of travel time the following table can be constructed:

Travel Time Level	Value (min)	Difference in values	Average rating	Difference in rating	Rating per minute
1	25	–	21/3	–	–
2	15	−10 (2 − 1)	25/3	4/3 (2−1)	−4/30
3	35	20 (3 − 2)	9/3	−16/3 (3 − 2)	−16/60

and in the case of fares:

Fare Level	Value (£)	Difference in value	Average rating	Differences in rating	Rating per £
1	1.3	–	22/3	–	–
2	1.00	−0.3 (2 − 1)	20/3	−2/3	2.22
3	1.50	0.5 (3 − 2)	13/3	−7/3	−14.3

From this we can estimate the subjective value of time (SVT) as follows: SVT is equal to $(-5/20)/(-14/3) = 0.054$, that is the ratio of ratings per minute over ratings per £. The reader can calculate the values of headway and vehicle type in the same way. Two interesting reflections can follow this very simple example: the values of time or other attributes do depend on the 'difference' being considered, for instance moving from 15 to 25 minutes does not produce the same SVT as moving from 25 to 35 minutes. The second comment is that we have estimated the values of these coefficients using the scores produced by a single respondent; that is, because each interview generates several observations in many cases we can estimate individual rather than sample based models.

The naive method is seldom used in practice, except as a quick way of estimating indicators like the value of time to provide an initial, 'in the field' validation of an experiment. However, this example has served to illustrate some of the ideas behind SP data analysis.

8.7.2.2 Discrete Choice Modelling with Rating Data

The objective of the rating data analyst is to find a quantitative relation between the set of attributes and the response expressed in the semantic scale. For this they need first to associate a numerical value R_m to each sentence m ($m = 1, \ldots, M$) of the scale and postulate a linear model such as:

$$\theta_0 + \theta_1 X_1 + \theta_2 X_2 + \ldots + \theta_K X_k = r_j \tag{8.50}$$

where θ_0 is a constant, X_k is typically the difference between the kth attributes of two competing options in the situation considered; θ_k is the coefficient of X_k and r_j represents a transformation of the response of individual j (i.e. it defines a unique correspondence between the semantic scale and the numerical scale R_m). Thus, when the questionnaire is completed the analyst obtains the chosen values of the dependent variable R_m and knowing the attribute values X_k they can perform a *multiple regression analysis* to estimate the values of θ_k.

Ordinary least squares or weighted and generalised least squares have been used to this end. One of the advantages of using these techniques is the ability to obtain goodness-of-fit indicators and measures of the significance of the model parameters. The main problem with this approach is that there are innumerable numerical scales that could be associated with the response scale. It may occur therefore, that the results of the analysis (estimated coefficients, their ratios and model goodness of fit) will depend on the definition of R_m; this hints at the importance of choosing the scale correctly. This issue will be discussed in greater detail when considering the analysis of extended choice data.

8.7.2.3 Discrete Choice Modelling with Rank Data

Rank data is arguably simpler and more reliable than rating data. Individuals are expected to be able to say that they prefer A to C and C to B with greater confidence and consistency than they can have in assigning scores to each alternative. There are several ways of exploiting rank data.

Monotonic Analysis of Variance or MONANOVA (Kruskal 1965) has been used for many years as a method for non-metric scaling. MONANOVA is a decomposition technique specifically developed to analyse rank order data. The method estimates part utilities iteratively thus estimating 'utility values' corresponding to each alternative. The first of these part utility estimates is generated using the naive method just discussed. These utilities permit the modelling of a ranking of alternatives; a 'stress' measure is used to indicate how much the modelled ranking differs from the ranking actually elicited from each individual. MONANOVA then seeks to improve the estimates of the 'part utilities' in order to reduce the stress (or badness-of-fit) indicator. MONANOVA, as in the naive method, is also capable of generating one model for each individual. Despite its uses, the approach lacks a robust statistical grounding and fails to provide global goodness-of-fit and measures of significance indicators; it also restricts the type of utility function that can be specified and it is less well suited to the development of forecasting models.

A more interesting form of analysing rank data is to convert them into implicit choices. In the case above the rank ACB would be converted into the choices A better than C, C better than B and A also better than B. The data thus transformed can now be analysed using Logit or Probit discrete choice modelling software. For the MNL model this can be done using the following theorem (Luce and Suppes 1965).

$$\text{Prob}(r_1, r_2, r_3, \ldots) = \text{Prob}(r_1/\mathbf{C})\text{Prob}(r_2, r_3, \ldots)$$

where $\text{Prob}(r_1, r_2, r_3, \ldots)$ is the probability of observing that the ranking indicates that r_1 is preferred to r_2 and so on, and $\text{Prob}(r_1/\mathbf{C})$ is the probability of r_1 being chosen from the choice set $\mathbf{C} = \{r_1, r_2, r_3, \ldots\}$.

If the theorem is applied recursively, an expression for the probability of the ranking in terms of N - 1 probabilities of choice is obtained:

$$\text{Prob}(r_1, r_2, r_3, \ldots) = \text{Prob}(r_1/\mathbf{C})\,\text{Prob}(r_2/\mathbf{C} - \{r_1\}) \ldots$$

where, for instance, $\mathbf{C} - \{r_1\}$ indicates the choice set excluding alternative r_1. Using this theory, Chapman and Staelin (1982) proposed that the content of a ranking of choices (8.49) can be exploded into $N - 1$ statistically independent choices as:

$$(U_1 \geq U_n, \; n = 1, 2, \ldots, N)(U_2 \geq U_n, \; n = 2, 3, \ldots, N) \ldots (U_{N-1} \geq U_N) \tag{8.51}$$

and these data can be estimated simply by a MNL routine. However, care must be taken with the following potential problems.

1. As the ranking considers hypothetical options it is likely that the information will contain some noise. This may be particularly serious in the case of less attractive alternatives which are often treated with less care by respondents and bunched together at the bottom of the ranking. This type of behaviour is not consistent with the independence of irrelevant alternatives axiom of the Logit model, so its occurrence must be statistically tested.
2. The rankings must be constructed in decreasing order of preference (i.e. from the best to the worse alternative) by each respondent; failure to do this might generate noisy data which can invalidate the modelling results.

As ranking a set of N options is a difficult task, i.e. it requires $\frac{1}{2}(N^2 + N) - 1$ comparisons, respondents are typically asked to divide the set (normally 9 to 12 options). First into three subsets (i.e. the better, medium and worse options), then to rank the options in each subset, and finally to exchange, say, the last of the first set with the first of the second, if appropriate. This algorithm has been found to ease considerably respondent burden in practice (Galilea and Ortúzar 2005; Ortúzar and Rodríguez 2002).

Problems with this approach have been reported by Ben-Akiva et al. (1992). They found that the response data from different depths of the ranking (i.e. not exploding the full rank) were not equally reliable in the sense of producing statistically significantly different utility estimates. However, this may depend on how carefully designed and conducted the SP experiment is, as Ortúzar and Palma (1992) found that models for the full depth of the ranking consistently produced better results.

To treat this problem in a less ad hoc manner, Bradley and Daly (1994) proposed separating the data into $N - 1$ different groups (n), each corresponding to a level of depth in the ranking (i.e. the first contains the individual preferences when all alternatives are available, and so on). Once the groups are identified, a joint estimation is performed considering different scale factors for each one (i.e. consistent with different variances for the error terms of their utilities). For this, one group has to be defined as reference and the scale factors (μ_n) associated with the rest of the groups represent the ratio between the variance of the error term corresponding to the reference group and that associated with the group under consideration (see the discussion in section 8.7.2.7). Thus, if the error variance associated with group n is the same as that corresponding to the reference group, the scale factor of group n will be equal to one.

Bradley and Daly (1994) arbitrarily defined group one as reference and reached the following important conclusions:

- The magnitude of the scale factors diminished with ranking depth (i.e. the error variance was higher in the case of the less preferred options).
- A likelihood ratio test confirmed that the model with scale factors was superior to the simple Logit model.
- The t-ratios of the explanatory variables fell to about one-third of their values in the simple Logit model (see the discussion in section 8.7.2.6).
- The subjective values of the various attributes in their experimental design changed by as much as 50% when scale factors were considered.

Ortúzar and Rodríguez (2002) tested this approach, finding that results changed significantly in their case depending on which level of ranking depth was selected as reference; however, in all cases the model with scale factors was statistically superior to the simple MNL model. They considered a group-based ranking experiment designed to study the willingness to pay for reductions in atmospheric pollution in a residential location context. The attributes were travel time to work, travel time to study, number of days of environmental alert in the area, and value of the house rent.

Two important findings were that if the fourth depth level (rather than the first, say) was chosen as reference, not only the t-ratios changed (the attribute values and log-likelihood at convergence remained constant) but also the number of significantly different scale factors. In fact, they finally reached the conclusion that the preferred modelling technique was one with only two scale factors: if the first three options are taken as reference, there was one large scale factor for the second set of four options, and a smaller one (i.e. closer to one) for the last three options. This is consistent with the way in which the options were ranked by the individuals and suggests that households were clearer about extreme options rather than middle-of-the-road options.

8.7.2.4   Modelling with Stated Choice Data

In this case we are able to use the whole range of analysis tools available for RP discrete choice modelling; for example, this includes Nested Logit because we are not restricted to only two options nor do we require the IIA property to hold (as in rank orderings) in order to exploit the data fully and also Mixed Logit, which is now the preferred option. We will come back to this issue in more depth below.

An interesting difference between RP and SC data is that the latter, by design, lacks some sources of error. In particular, there is no measurement error since all attribute values are *presented* to respondents (although there may be some perception problems). However, we have already discussed other features of SC surveys that weaken the behavioural value of the data: lack of realism in the decision context and artificiality of the alternatives.

Apart from specification error, which clearly does still apply, there is another potentially serious source of error related to the response itself. Although practical results are generally encouraging, in terms of suggesting that most respondents do understand what is expected of them, there is no guarantee that they are able to complete an SC experiment with complete accuracy. In fact, a good review by Bates (1988a) discusses the following types of potential error applying to all types of SP data:

- Respondent fatigue, which obviously increases with the complexity of the experimental design (see the discussion in Chapter 3).
- Policy response bias, which might occur if the respondent is interested in affecting the outcome of the analysis.
- Self-selectivity bias, when respondents either inadvertently or on purpose, cast their existing behaviour in a better light.

The outcome of all this is that we may have measurement error in the dependent variable, i.e. instead of getting a true estimate of the utility U, we are obtaining some pseudo utility $\ddot{U}$ which can be linked to our general formulation (7.2) by:

$$U_i = V_i + \epsilon_i = \ddot{U}_i + \tau_i \tag{8.52}$$

Assuming homoskedastic τ_i (although it is quite possible that their variance varies across experiments either due to fatigue or learning), the estimation of the parameters of $\mathbf{V}$ presents no problems as (8.52) can be rewritten as:

$$\ddot{U}_i = V_i + (\varepsilon_i - \tau_i) \tag{8.53}$$

and the normal estimation methodology may be employed. The problem comes in forecasting, because in that case we are interested in making estimates of $\mathbf{U}$, and what we would get from applying this model are estimates of $\ddot{\mathbf{U}}$ provided the same distribution of errors apply in the design year. In other words:

... we are making estimates of relative preferences *as expressed in a Stated Preference experiment* rather than of what would occur in the market (Bates, 1988a).

The only way to get round this problem is to apportion the error between ε_i and τ_i, using both SC and RP data to estimate the models, and this is somewhat similar to the problem of using aggregate data in model estimation which we discuss in Chapter 9. Bates (1988a) notes that an understanding of the magnitude of τ_i is of crucial importance to the use of SC in forecasting. Only if it is insignificant in relation to ε_i, could the estimated model be used directly to give forecasts. This calls for special care in the design of the SC experiments to reduce respondent fatigue, enhance realism, prevent policy-response bias and minimise self-selectivity bias. However the problem remains normally serious and so current practice recommends mixed estimation with RP data whenever possible (see Bradley and Daly 1997).

8.7.2.5 *Model Estimation with Generalised Choice Data*

In the case of generalised or extended choice surveys the respondent is allowed to express degrees of confidence in her choices. If conventional Logit modelling is used two models can be estimated, one including only the 'definitely choose' responses and another including also the 'probably choose' responses and the results compared for goodness-of-fit and parameter significance. But note that in either case we would lose the responses marked 'indifferent' and if the choice tasks have been designed to make respondents really think, there might be many in this class and such data loss would be unfortunate.

Alternatively, one can research more closely what is the best transformation of the semantic scale into a numerical one, in the sense of producing the best possible models. Several practitioners have used the following symmetric scale: $R_1 = 2.197$, $R_2 = 0.847$, $R_3 = 0.000$, $R_4 = -0.847$, $R_5 = -2.197$, which corresponds to the Berkson-Theil transformation of the following choice probabilities: 0.1, 0.3, 0.5, 0.7, 0.9 (see for example the review in Bates and Roberts 1983) and became almost standard practice among transport practitioners in the 1990s. However, this is not necessarily the most 'appropriate' scale for any given study and it is important to investigate if scale selection may have a significant effect on the results of the analysis.

Example 8.12 A group of staff and students participated in a generalised SC experiment comparing two options in the following context: a morning trip from home to the university (about 10 km away), involving choice between bus and light rail (an option which does not exist today). For simplicity the experimental design considered only four attributes:

- Travel Cost (varying at three levels).
- Travel Time (varying at two levels).
- Walking Distance (varying at three levels).
- Waiting Time, estimated as half of the public transport headway (varying at two levels).

Thus we had a $3^2\ 2^2$ factorial design and since we were looking for main effects only, we just required nine options in the simple orthogonal case. The following table shows the attribute differences (instead of their absolute values) between the two options; the design (in terms of the options offered) was based on combinations of such differences. This implicitly assumes the resulting model will be generic (e.g. same coefficient for in-vehicle time for each mode) helping to reduce the size of the design.

	Attribute Level Difference		
Bus attribute minus LRT Attribute	**Low**	**Medium**	**High**
Travel cost (Ch$)	−10	60	80
In-vehicle time (min)	15	25	na
Walking distance (blocks)	−7	−3	0
Headway (min)	−3	2	na

Consider now the four probability scales defined in the following table:

	Scale 1	Scale 2	Scale 3	Scale 4
R_1	0.100	0.010	0.300	0.200
R_2	0.300	0.400	0.450	0.400
R_3	0.500	0.500	0.500	0.500
R_4	0.700	0.600	0.850	0.880
R_5	0.900	0.990	0.950	0.970

The next table presents SVT (i.e. coefficient ratios of the parameters of time and cost) derived from models estimated after applying the Berkson–Theil transformation to the four probability scales:

Value of Time	Scale 1	Scale 2	Scale 3	Scale 4
In-vehicle travel	4.01	1.73	3.98	4.11
Waiting	20.68	18.67	23.89	23.24
Walking	23.68	21.63	24.91	24.74
R^2	0.48	0.44	0.46	0.45

As can be seen, scale selection does indeed influence the modelling results. The SVT values do not only differ but belong to models with different goodness of fit to the data. Furthermore, the differences do not seem to depend on whether the scale is symmetrical or not; that is, although one could expect a symmetric scale (like scales 1 and 2) to produce more reasonable results, the fitted models and estimated SVT values reject this notion (Ortúzar and Garrido 1994a).

One way of avoiding the problem described above would be to consider an approach not requiring the analyst to specify the numerical scale *a priori* in order to estimate the model. McKelvey and Zavoina (1975) developed an approach with this feature, called 'Ordinal Probit' which can be easily used but requires specialised software.

Another possibility would be to estimate the response scale during the model fitting process by effectively considering each value of the scale as an additional variable. In this case a coordinate search method may be used, starting with the typical symmetric scale 1 in Example 8.12. The procedure consists simply of changing in turn each point of the scale (say R_i) by a small amount and estimating a linear regression model with the new values. The search continues until R^2 is maximised and the value of R_i is fixed. The procedure is repeated for each point of the scale (save for R_3 which is always kept as 0.5) in an iterative routine until a best fit is found in each case (that with the highest R^2). This process is repeated again to check for differences. Ortúzar and Garrido (1994a) found that the search never involved more than two iterations before convergence (for four different samples), but they could not prove, mathematically, that a global optimal solution is guaranteed. Indeed, the method was used later by Bianchi *et al.* (1998) who found that the method did not converge for their pricing study data.

Example 8.13 The following table shows the original symmetric scale and the scales found after performing the above 'optimal scale linear regression approach' on two samples for the rating experiment of Example 8.12.

Scale	Initial	Students	Staff
R_1	0.1	0.284	0.228
R_2	0.3	0.286	0.278
R_3	0.5	0.500	0.500
R_4	0.7	0.714	0.722
R_5	0.9	0.900	0.842

The results suggest the possibility of testing whether the original number of points in the semantic scale is appropriate. If only one value was used for the first two points of the scale in the optimal scale models (which appear strikingly close) it would be interesting to see what consequences this apparent loss of information brings about. On the plus side a four-point scale would have one parameter fewer to be estimated. The next table shows the optimal values of the new scale obtained when R_1 and R_2 are replaced by a single point R_1'. In these scales, as in the previous ones, the probability value of R_3 was fixed to 0.5 as it corresponds to the point of indifference between both modes.

Scale	Students	Staff
R_1'	0.277	0.121
R_3	0.500	0.500
R_4	0.716	0.776
R_5	0.899	0.922

As can be seen, the scale values in both samples are further apart than in the previous table which suggests that no other point fusion would be necessary. Also, all values appear reasonable in relative terms, i.e. they correspond to increasing probability values from R_1' to R_5.

8.7.2.6   Interactions in SC Modelling

Next we consider a potential although seldom achieved advantage of the SC approach: the possibility of estimating models with non-linear utility functions. The reason for not doing this in practice has been typically one of convenience. SC experiments allowing the incorporation of interactions (and not just main effects) were more complex to design and analyse, and required data that was more difficult to collect.

In discrete choice modelling many potential forms of the utility function can be transformed (e.g. even as a last resort using series approximations) into additive linear forms of the type:

$$V = \theta_1 X_1 + \theta_2 X_2 + \theta_3 X_1^2 + \theta_4 X_1 X_2 + \theta_5 X_1 X_2^3 + \theta_6 X_1 X_2 X_3$$

where X_i are attributes and θ_i are coefficients to be estimated. This function contains linear terms ($\theta_1 X_1$ and $\theta_2 X_2$), non-linear terms ($\theta_3 X_1^2$), interactions with linear effects ($\theta_4 X_1 X_2$ and $\theta_6 X_1 X_2 X_3$) and general interactions ($\theta_5 X_1 X_2^3$). The main effects can be defined as the response to passing to the next level of the variable when the rest of the attributes remain constant (all other things being equal); it is normally postulated that these are the main determinants of changes in choice. In fact, according to Louviere (1988b):

- The main effects explain 80% or more of the data variance.
- Two-term interactions rarely explain more than 2% or 3% of the variance.
- Three-term interactions explain even smaller proportions of the data variance, normally of the order of 0.5% to 1% and rarely over 2% or 3%.
- Higher-order effects explain a minuscule proportion of the data variance.

For these reasons, only main effects are normally considered in practice. On the other hand, there seems to be a consensus that interactions between more than two variables as well as interactions incorporating non-linear effects should be insignificant. Therefore, only two-term interactions are in a kind of limbo and require more attention. Note that if interactions are actually insignificant, a model incorporating only main effects will allow us to obtain precise measurements of individual preferences. However, if the interactions are significant and are not included in the utility specification, their effects will be erroneously attributed to the simple variables. This notwithstanding, as we shall see below, it may happen that when certain interactions are included, their effects dominate that of certain individual variables to the extent that the latter may be left out of the regression (i.e. the variable may end up with a non-significant coefficient or with a counterintuitive sign).

The cost of allowing for interactions in the experimental design is that it becomes more complex (i.e. it requires respondents to evaluate a higher number of hypothetical situations). A good solution in such cases is to use block designs as we saw in section 3.4.2.3. The assumption is that consistent models will be obtained when the total number of responses is considered. To ensure compatible answers the size of each subsample should guarantee that its socio-economic characteristics are representative.

Example 8.14   A generalised SC experiment using a five-point semantic scale was designed to study choice between car-alone and car-pool for campus students (Ortúzar *et al.* 2000c). After extensive piloting, the following attributes were selected:

- *Daily travel time*: this was always higher for car-pool as the student providing the car on the day needed to collect the members of the group in the morning and take them back home in the afternoon.
- *Weekly travel cost*: associated with fuel consumption and estimated on the basis of information about travel distance and type of car (in some cases this value included a parking charge); this was always smaller for car-pool as drivers did not need to use their cars every day of the week in this case.

- *Waiting time*: associated with sharing the trip with a group in the case of car-pool; waiting occurs because the proposed car-pool system implies the complete group arriving at and leaving the campus at the same time, and not all exit hours coincide; note that this time may be used in other activities, because both its duration and day of occurrence are known in advance, given the fixed university schedules.

The attribute levels were defined on the basis of differences between travelling by car and by car-pool. Two levels were used in the case of travel time (i.e. 10 and 20 min more than in the case of car-pool); four levels in the case of travel cost (i.e. three-quarters and half the cost of the car in the case of car-pool, and 25% and 40% more than that cost if a parking charge was included), and three levels for waiting time. In this last case the levels were determined based on the possibility that the group members would not coincide in their lectures. So, waiting times of zero, 30 min (i.e. one member needed to do a small errand) and 90 min (i.e. the extent of a complete lecture module) were considered.

With this, 16 hypothetical situations are needed to estimate main effects only and 24 if two-term interactions are included in a simple orthogonal design. Given these numbers, block designs should be used in both cases (see Caussade *et al.* 2005). In fact, we tested using 16 options directly but found that this confused or bored respondents, leading to too many inconsistencies, confirming the findings of Carson *et al.* (1994).

To model we first looked at the expected signs of the interaction terms (given the special characteristics of the competing options), concluding that their most appropriate definition was as follows:

- T^*C represents interaction between the ratios of travel time and cost by both modes; positive coefficient:

$$T^*C = \frac{\text{Travel time}^{car}\,\text{Cost}^{cp}}{\text{Travel time}^{cp}\,\text{Cost}^{car}}$$

- W^*T represents interaction between the car-pool waiting time and the ratio of travel time by both modes: negative coefficient:

$$W^*T = \text{Waiting time}^{cp}\,\frac{\text{Travel time}^{cp}}{\text{Travel time}^{car}}$$

- W^*C represents interaction between the car-pool waiting time and the ratio of travel cost by both modes: negative coefficient:

$$W^*C = \text{Waiting time}^{cp}\,\frac{\text{Cost}^{cp}}{\text{Cost}^{car}}$$

Table 8.9 shows the results of two Ordinal Probit specifications, the 'best model' (estimated) and the 'preferred model'. The *Inertia* dummy takes the value of one if the respondent was a current car-pool user and g is the expenditure rate, i.e. the ratio between income and free time; see for example Jara-Díaz and Ortúzar (1989). As can be seen only the variables *Sex* (dummy which takes the value of one for males) and *Waiting time* are not significant at the 95% level in the first model; however, if the latter is removed (because its effect is considered by the strong interaction terms) the model improves.

In order to verify the relative importance of the interactions in the utility function, the product of the average value of each normalised variable and its coefficient was calculated. This revealed that the interactions were undoubtedly important, especially T^*C. This procedure was confirmed by calculating the elasticity of the probability of choosing car for various changes in the attribute values (Ortúzar *et al.* 2000c).

Table 8.9   Ordinal Probit model considering interactions

Attributes (*t*-ratios)	Best model	Preferred model
Car-specific constant	1.65418	1.68808
	(9.89)	(10.17)
Travel time (min)	−0.00311	−0.00343
	(−4.57)	(−5.30)
Waiting time (min)	−0.00363	–
	(−1.53)	
Cost/g (min)	−0.06729	−0.06930
	(−7.54)	(−7.83)
Sex	0.11021	0.11372
	(1.92)	(1.98)
Car-pool inertia	−0.40907	−0.41113
	(−6.57)	(−6.61)
T*C	0.70067	0.73763
	(9.67)	(10.73)
W*T	−0.00629	−0.01038
	(−2.11)	(−8.04)
W*C	−0.00124	−0.00157
	(−3.23)	(−4.92)
R^2	0.543	0.541
Sample size	1640	1640

8.7.2.7   The Problem of Repeated Observations

One of the most important attractions of the SC approach is the generation of multiple observations by each individual. However, almost every application in the last millennium considered the responses by a given individual not only independent of those given by the rest of the sample members, but also independent of each other. Although this problem received a little more attention at the end of the 1990s, it is only in recent years that it has been handled correctly using Mixed Logit models.

In the 90s it was generally assumed that these observations were independent, leading to the concept of pseudo-individuals. Clearly, this hypothesis cannot be valid and for many years it was hoped (and believed) that the problem was bounded to obtaining upward biased values of the *t*-ratios associated with the estimated parameters. In this way the solution consisted of proposing correction factors for the resulting *t*-ratios.

By the end of the 1990s more interesting approaches have been proposed and partially tested. For example, Cirillo *et al.* (2000) proposed the use of re-sampling techniques, such as bootstrap and jackknife (Shao and Tu 1995), finding that the jackknife-estimated parameters did not vary much with respect to those estimated assuming independence and that the *t*-ratios diminished, as expected (the bootstrap results were similar but had more noise, particularly for low-re-sampling strategies). Ortúzar *et al.* (2000c) also tested these methods (with all their variations in re-sampling) for four different samples, finding that the parameter values remained practically identical to those estimated with the traditional approach in all cases. However, the standard errors varied inconsistently (i.e. they correctly increased in three cases but decreased in the other). They extensively checked the samples for either outliers or peculiarities in the originally estimated values and found nothing special. Thus, they were forced to conclude that the applicability of these techniques in solving the problem of repeated observations must be put under further scrutiny.

Outwerslot and Rietveld (1996), and independently Abdel-Aty *et al.* (1997), suggested decomposing the total error ε in a random utility model into two mutually exclusive parts: an individual-specific effect that distributes independently among individuals, and an observation-specific effect which distributes independent among individuals and observations (i.e. very much in line with the error components specification of the Mixed Logit model). Inevitably the standard approach led to a multiple integral which was hard to evaluate. To avoid this problem, Outwerslot and Rietveld (1996) used a minimum distance method proposed by Chamberlain (1984), which considers dividing the sample into T randomly selected independent subsamples containing only one observation per person (T is the number of repeated observations per individual). The coefficients of the models estimated for each subsample were then used in a rather complex algorithm to obtain the final model parameters and their variances.

Contrary to expectations, Outwerslot and Rietveld (1996) found that the parameters of their Probit model were different to those of the classic method (although less than 27%) but the *t*-ratios remained practically invariant. Ortúzar *et al.* (2000c) also tested this method, finding that most parameter values decreased, and in some cases considerably, but sometimes they also increased. With respect to the *t*-ratios they found that in general they decreased as expected, but not always and particularly in the case of the specific constants.

Yen *et al.* (1998) developed a method to treat this problem using a generalised dynamic version of the Ordinal Probit model, which allows one to incorporate a measure of the correlation between the responses of a given individual. As comparative issues were not their main concern, Yen *et al.* (1998) did not report whether there were differences between their estimations and those obtained with a standard application of Ordinal Probit.

Current practice accepts that estimation can be handled without problems by a Mixed Logit model, such as (8.32). We will look at the way to do it in the richer case involving joint estimation with RP data in the next section.

8.7.3   Model Estimation with Mixed SC and RP Data

Consider the MNL model (7.9) and the inverse relation that its scale parameter β has with the single standard deviation σ of the Gumbel residuals ε. This relation explains why it is not correct to postulate the same error distribution for estimation and forecasting as mentioned above; the near and extreme right hand side expressions in (8.52) should yield different values for β. This produces 'scale' differences on the parameters and if such equality is improperly assumed we might finish by estimating pseudo utilities instead of 'true' utilities. To avoid this problem it is necessary to adjust the SC data to actual behaviour, exploiting the advantages of the RP data in this sense, and estimating the parameters θ jointly.

In econometrics the estimation of models with different data sources is called 'mixed estimation'. Often these data are divided into two sets: *primary* and *secondary* data. The primary data provide direct information about the main modelling parameters. The secondary data provide additional (indirect) information about the parameters. For example, in discrete choice modelling the primary data could be information coming from a survey at the disaggregate level, and the secondary one could be data coming from an aggregate survey. In our case RP data constitute the primary set, since these data capture the actual behaviour of the individuals, and SC data constitute the secondary set.

8.7.3.1   *Estimation without Considering Correlation among Repeated Observations*

Although we know that this is not correct nowadays, it is still informative to learn how this important task was first undertaken. Ben-Akiva and Morikawa (1990) developed a framework which postulates that the difference between the errors in the RP and SC domains may be represented as a function of the variances of the errors ε and η associated with each data set respectively. This can be written as follows:

$$\sigma_\varepsilon^2 = \mu^2 \sigma_\eta^2 \tag{8.54}$$

where μ is an unknown *scale coefficient*. This leads to the following utility functions for a certain alternative A_i:

$$\begin{aligned} U_i^{RP} &= \boldsymbol{\theta}\mathbf{x}_i^{RP} + \boldsymbol{\alpha}\mathbf{y}_i^{RP} + \varepsilon_i \\ \mu U_i^{SC} &= \mu\left(\boldsymbol{\theta}\mathbf{x}_i^{SC} + \boldsymbol{\phi}\mathbf{z}_i^{SC} + \eta_i\right) \end{aligned} \tag{8.55}$$

where $\boldsymbol{\alpha}$, $\boldsymbol{\phi}$ and $\boldsymbol{\theta}$ are sets of parameters to be estimated; $\mathbf{x}^{RP}$ and $\mathbf{x}^{SC}$ are attributes (of both alternatives and individuals) at the RP and SC levels respectively. $\mathbf{y}^{RP}$ and $\mathbf{z}^{SC}$ are attributes which only belong to the RP or SC sets respectively (notice that vector $\mathbf{x}$ is common to both types of data).

The consideration of the utility functions (8.55) allows homogenising the type of error, as multiplying the SC utility by μ makes the associated stochastic error (η_i) to have the same variance as the corresponding RP error (from 8.54). Thus, assuming that both stochastic errors have IID EV1 distributions with zero mean but with a different variance, the choice probabilities at each domain would be given by (Morikawa *et al.* 1992):

$$\begin{aligned} P_i^{RP} &= \frac{\exp\left(\boldsymbol{\theta}\mathbf{x}_i^{RP} + \boldsymbol{\alpha}\mathbf{y}_i^{RP}\right)}{\sum_j \exp\left(\boldsymbol{\theta}\mathbf{x}_j^{RP} + \boldsymbol{\alpha}\mathbf{y}_j^{RP}\right)} \\ P_i^{SC} &= \frac{\exp\mu\left(\boldsymbol{\theta}\mathbf{x}_i^{SC} + \boldsymbol{\phi}\mathbf{z}_i^{SC}\right)}{\sum_j \exp\mu\left(\boldsymbol{\theta}\mathbf{x}_j^{SC} + \boldsymbol{\phi}\mathbf{z}_j^{SC}\right)} \end{aligned} \tag{8.56}$$

From these exressions it is possible to postulate a joint likelihood function which should be maximised to yield the parameter estimates. The reader might have noted that equations (8.56) have incorporated some assumptions:

(i) the scale parameter of the RP model has been normalised;
(ii) the sale parameter of the SC model should be identical to μ.

In fact, the real assumptions are different but when the joint model is estimated arrive at the same result. Yáñez *et al.* (2010c) provide a good discussion on this issue in their analysis of mixed RP-SC models in forecasting.

Choosing the Attributes with the Same Parameter in Both Domains Deciding which attributes should belong to set $\mathbf{x}$ is not straightforward. In principle, though, the only candidates are those attributes that being measured in practice (i.e. travel time, cost, waiting time) also appear in the SC tasks. If all 'common' attributes are taken as members of $\mathbf{x}$ we speak of the *full data enrichment approach*; if only some common attributes end up belonging to $\mathbf{x}$ we have the *partial data enrichment approach*. To decide this matter, Louviere *et al.* (2000) recommend the following procedure:

- First, estimate (separately) the two models associated with equation (8.55), under the assumption that the errors distribute IID EV1 (obviously without including the unknown scale factor μ in the second case); this will yield two sets of parameters, $\boldsymbol{\theta}^{RP}$ and $\boldsymbol{\theta}^{SC}$, for all the attributes which are common to both domains.
- As we know, these two sets of parameter estimates cannot be equal, in principle, as they contain the unknown scale parameters β associated with the MNL model in each domain; however, the idea is to find out if they are different over and above this scale problem, in which case they should not be joined under set $\mathbf{x}$.

- From equation (8.54), and recalling the inverse relation between the scale parameter of the MNL and the variance of its IID EV1 error (7.10), the reader can readily deduce that if both parameters were equal (apart from scale), their relationship should be: $\theta^{SC} = \mu\,\theta^{RP}$.
- Based on this relation, if we plot the estimated parameters in the two domains we should expect them falling inside the elliptical region shown in Figure 8.8; and values outside it (as those shown in stars) would not be available for set **x**.

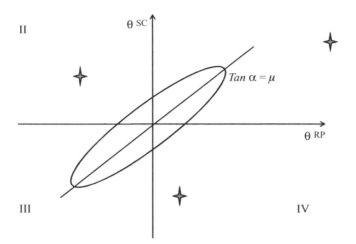

Figure 8.8   Plotting the parameters from both domains

- However, note that even if some parameters fall outside the range the attributes involved could still be considered part of **x**; this would be the case if one of the estimates is not significantly different from zero, as in that case fixing its value to be equal to its counterpart in the other domain would bring no problems. In fact, as we will see below, this can be tested using a LR test (as we saw in section 8.4.1.2).

The joint likelihood function incorporating the two models in equation (8.56) simultaneously is a highly non-linear function, because μ is multiplying not only the attributes but also the SC parameters. Two approaches were devised during the 1990s to solve this problem, the simultaneous estimation method (Bradley and Daly 1997) and various forms of sequential estimation method (Ben-Akiva and Morikawa 1990; Swait et al. 1994). We will only mention the former here as it was the most popular in practice until very recently.

The simultaneous estimation method consists of constructing an artificial tree with twice as many alternatives as there are in reality. Half of these are labelled RP alternatives, the other half are SC alternatives. The utility functions are $\mathbf{U}^{RP}$ and $\mathbf{U}^{SC}$ (as in 8.55). As indicated in Figure 8.9, the RP alternatives are placed just below the root of the tree; however, the SC alternatives are each placed in a single-alternative nest; we will see now why this is so important. Observe that in this case, for an RP observation the SC alternatives are unavailable and the choice is modelled as in a standard MNL or NL model. For an SC observation, the RP alternatives are also unavailable and the choice is modelled by a Nested (tree) Logit structure. For this reason the method came to be known as the 'nested logit trick' (Louviere et al. 2000).

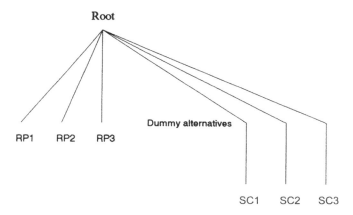

Figure 8.9   Artificial tree structure for joint RP and SC estimation

For the SC observations, the mean utility of each of the dummy composite alternatives is computed as usual (see Daly 1987):

$$V^{\text{COMP}} = \mu \log \sum e^{V^{\text{SC}}} \tag{8.57}$$

where the sum is taken over all of the alternatives in the nest corresponding to the composite alternative (i.e. in this case only one) and

$$V^{\text{SC}} = U^{\text{SC}} - \eta = \boldsymbol{\theta} \cdot \mathbf{x}^{\text{SC}} + \boldsymbol{\phi} \cdot \mathbf{z}^{\text{SC}} \tag{8.58}$$

is simply the measured part of the SC utility. Then, because each nest contains only one alternative in this specification,

$$V^{\text{COMP}} = \mu \, V^{\text{SC}} = \mu \, \boldsymbol{\theta} \cdot \mathbf{x}^{\text{SC}} + \mu \, \boldsymbol{\phi} \cdot \mathbf{z}^{\text{SC}} \tag{8.59}$$

which is exactly the form required as long as the value of μ is constrained to be the same for each of the dummy alternatives. Since the dummy composite alternatives are placed just below the root of the tree, as are the RP alternatives, a standard estimation procedure will ensure that μ is estimated to obtain uniform variance at this level. It is important to note that this artificial construction does not require the usual consistency assumptions for NL models (i.e. that μ should not exceed one), because the individuals are not modelled as choosing from the whole choice set. However, as noted before, the value of μ may be taken as providing an indication of which data set is more accurate.

Partial or Fuller Data Enrichment   To test whether a given common attribute to both data sets can form part of set $\mathbf{x}$, it is possible to use a likelihood ratio test. Let $l^*(\boldsymbol{\theta}^{\text{RP}}, \boldsymbol{\alpha})$ be the log-likelihood at convergence for the model with RP data only, $l^*(\boldsymbol{\theta}^{\text{SC}}, \boldsymbol{\phi})$ the same for the model with only SC data, and $l^*(\boldsymbol{\theta}, \boldsymbol{\alpha}, \boldsymbol{\phi}, \mu)$ the log-likelihood at convergence of the joint RP/SC model. If the k common parameters are equal, then:

$$LR = -2 \left\{ l^* \left(\boldsymbol{\theta}^{\text{RP}}, \boldsymbol{\alpha} \right) + l^* \left(\boldsymbol{\theta}^{\text{SC}}, \boldsymbol{\phi} \right) - l^* (\boldsymbol{\theta}, \boldsymbol{\alpha}, \boldsymbol{\phi}, \mu) \right\}$$

distributes χ^2 with k degrees of freedom. If LR is greater than the critical value of χ_k^2 for the required confidence level, the test is rejected and one (or more) attribute should be taken out of the set $\mathbf{x}$ and be specified with a different parameter in both domains for the joint estimation, and the test is repeated.

8.7.3.2   Joint Estimation Considering Correlation between Repeated Observations

The Mixed Logit (ML) model offers much in terms of the appropriate mixing of revealed and stated preference data. The traditional approach above, employing an artificial NL structure (e.g. 'the nested logit trick'), suffered from at least two important deficiencies; first, the stated choices of the same individual were considered independent and second, the stated choices of individuals who also responded to the RP survey (generally only part of the sample is in this category), were unrelated to their RP choices.

Bhat and Castelar (2002) were probably the first to formulate and apply a unified Mixed Logit framework for joint RP/SC model estimation that could accommodate a flexible competition pattern across alternatives, scale differences in the RP and SC choice contexts, heterogeneity across individuals, state dependence of the stated choices on the revealed choices, and heterogeneity across individuals in the state dependence effects. Their likelihood function has two levels of integration because they postulate an EC formulation that generates inter-alternative correlation operating at the choice level, and also random coefficients that accommodate taste variations across individuals and operates at the individual level. Using real data, they found – among other things – that heterogeneity and state dependence effects were tempered when included simultaneously, indicating confounding of true and spurious state dependence. They also found that the better specified model significantly outperformed more restrictive structures.

Train and Wilson (2008) improved on the above by postulating a ML model that explicitly incorporates the fact that SC experiments are usually constructed on the basis of RP choices. Thus, they address an important issue that could be a source of inconsistency in estimation. For example, Bhat and Castelar (2002) included a state dependence variable in the form of a dummy for the choice in the RP setting that enters the SC model; however, they did not account for the fact that this variable is correlated with unobserved factors insofar as any unobserved factors from the RP setting carry over into the SC setting. This is equivalent to entering a lagged dependent variable in time series data and estimating by ordinary least squares; i.e. it is fine only if the unobserved factors are not correlated over time. Train and Wilson (2008) develop an appropriate method to use when the lagged dependent variable is included and unobserved factors are correlated over time. Thus, it is a discrete-choice-model analogue of the method of estimating regression with lagged dependent variables and serially correlated errors. Following this analogy, note that allowing for random coefficients and different variance of the error term does not change the fact that entering a lagged dependent variable (or variables created from it) is inconsistent when errors are serially correlated.

8.7.3.3   Forecasting with Joint RP-SC Models

A key issue in forecasting with joint RP-SC models is how to treat the alternative specific constants (ASC). Cherchi and Ortúzar (2006) provide an in-depth discussion of the problem for the following three cases:

- when the RP and SC alternatives are exactly the same;
- when the SC data include new alternatives (i.e. not present in the base year), and
- when the SC design implies substantial changes, such that alternatives sharing the same label (e.g. normal train and a substantially improved fast train service) could represent new options.

They concluded that the first case is trivial as the ASC corresponding to the RP domain should be used without rescaling but, of course, adjusted to match the market shares of the base year. In the second case, if the analyst truly believes that the SC data reproduce correctly the market shares of the population in forecasting, then the ASC (both for the existing and for the new alternatives) should be adjusted to match the market shares in the SC data. Conversely, if the market shares to match are unknown then the analyst must rely on estimation results, i.e. as long as the usual theoretical restrictions of the model are satisfied, it might be useful to draw further considerations on the ASC specification from the model that provides the best statistical fit. Finally, if the SC design implies substantial changes, such that alternatives sharing the same label could represent new options, and there is uncertainty as to the extent they are actually different, then best fit and analyst's judgement, seem to be the only guide.

On the other hand, regardless of the way the ASC are specified (i.e. generic or specific), depending on the results for each specific context the application of a mixed RP/SC model in forecasting implies some limitations on the scenarios to be tested. In particular:

- If ASC which are specific to RP and SC are estimated, forecasts can only be made for scenarios involving structural characteristics not inferior to those described in the SC design, and in that case rescaled SC-ASC should be used.
- If specific ASC for both domains are estimated and a scenario not involving structural changes is considered, the RP-ASC should be used.
- Finally, if constrained generic ASC are estimated for both domains, scenarios involving structural changes (for those alternatives with constrained RP/SP ASC) should not be tested, unless we obtain ASC with a fairly close value from estimation.

Cherchi and Ortúzar (2011) considered the problem of forecasting with a joint RP-SC Mixed Logit model allowing for random taste heterogeneity. They note that although a basic assumption when pooling RP and SC data is that they share the same underlying behaviour, often the partial preference homogeneity approach (i.e. the parameters are not constrained to be equal in both data sets) gives better results because some attributes can only be measured/estimated properly in one set, or because differences in the nature of attributes produce different, and highly significant, estimated parameters in both sets.

Note that the differences between RP and SC results might be implicit in the need for using SC data in the first place; indeed, they may represent exactly what we look for when using SC data. Consider the case when we want to forecast the effects of structural changes (i.e. departs from the current real market) and utilities are not linear in the attributes. The effect of the partial enrichment approach will be more evident when one attempts to consider the various components of individual heterogeneity, because to estimate complex behaviour we need datasets that are both fairly rich and fairly large; unfortunately, this is often *not* the case for RP data.

However, the partial preference homogeneity approach implies problems in forecasting, as the model used for prediction is not the same as the estimated one; thus, it is crucial to carefully check if the estimated model parameters fulfil the microeconomic conditions on the marginal utilities for any scenario to be tested (see the discussion in Chechi and Ortúzar 2010).

Yáñez *et al.* (2010c) extended the analysis of partial preference homogeneity to the correlation structure among alternatives, i.e. how to deal with the problem of finding different correlation structures revealed in the RP and SC data sets. They also discuss the problem of the normalisation in the joint RP/SC model (i.e. defining an appropriate scale) and its effect in estimation and forecasting. They consider several cases from the most simple, when alternatives are independent in both the RP and SC datasets, to the most complicated one when the two datasets present different inter-alternative correlation structures. They show that from a theoretical point of view both the lower and upper normalisations of the Nested Logit model (see section 7.4.4) are equivalent in this case, but their practical convenience is limited to the simple case of independent alternatives in both datasets;

(continued)

furthermore, they confirm the results of Carrasco and Ortúzar (2002) that the upper normalisation is more intuitive. Moreover, although any inter-alternative correlation structure between the alternatives in the RP and SC domains can in principle be estimated, they recommend using a generic one for the existing alternatives in both data sets, if possible; otherwise, the model might not be consistent in prediction. Furthermore, assuming different structural parameters in the RP and SC data sets means that the unobserved components of the utilities (that make some alternatives to be perceived as more similar than others), are not the same in both cases. This can be justified when the systematic utilities are specified differently, but it should not occur when alternatives have the same specification.

Finally, in terms of model use in forecasting Yáñez *et al.* (2010c) provide the following recommendations:

- If the joint RP-SC model structure assumes the same structural parameters for both data environments, their estimated values can be used directly in forecasting.
- If the joint RP-SC model structure assumes the same structural parameters for both environments except for alternatives present only in the SC case (i.e. usually new alternatives), the whole correlated (or uncorrelated) structure of the SC-alternatives needs to be moved into the RP domain to make forecasts, but the structural parameter does not need to be scaled because it was already estimated scaled by the RP scale parameter, and it is associated with the EMU term.
- If the joint RP-SC model structure allows for different structural parameters in both environments (the most general and most complicated case), the general advice is to use the structure estimated with each data set (RP or SC); however, to be consistent, the structural parameters should be associated with utilities measured in the same environment. This means that if we have more faith in the SC data, we should move both the SC structural parameters and the utilities associated with the alternatives in the nest to the RP environment.

Example 8.15 Consider the introduction of a new high speed rail (HSR) interurban line to compete with car, bus and airplane. Further, given the competitiveness among different services, the following groups of alternatives were identified: three bus alternatives (conventional bus, executive bus, sleeper bus), three plane alternatives (to represent different pairs of airports available at the two main cities affected by the new service), and two HSR alternatives (conventional and executive).

A RP-SC survey was conducted with the final aim of forecasting the demand for the new mode. After estimating separate models for each data set, different inter-alternative correlation structures for the RP and SC data were found. In particular, the SC data presented a clear and strong correlation between the two HSR services and between the three bus alternatives; while the inter-alternative correlation among plane alternatives in both cases and between the RP bus alternatives was not significant.

Following the discussion about model consistency for prediction purposes above (Yáñez *et al.* 2010c), the correlation structure in both environments should be constrained to establish a unique and consistent structure. This offers three possibilities: (i) all alternatives are considered independent (model 'MNL'); (ii) all alternatives are considered independent except the two new ones in the SC environment (model 'MNL-NL Rail'); (iii) the bus alternatives are considered correlated with the same structural parameter in both data sets, and the two HSR alternatives are correlated in the SC case (model 'NL').

To evaluate the effect of establishing different inter-alternative correlation patterns on demand predictions, we can calculate the variation in aggregate market shares for various simple policies: Table 8.10 shows that all models predict a decrease in the market shares of the existing modes following a reduction in the HSR fares. However, there are important differences in the magnitude of the changes (the policy impacts are evidently greater for the MNL model). Indeed, for a 50% reduction in HSR fares, if we erroneously assume the MNL the estimated percent change in the aggregate HSR share (ΔP_j) is 50% larger than if both HSR alternatives are assumed to be correlated. However, the differences between the results for the two nested models are not large.

Table 8.10 Forecast Effects of Including Inter-alternative Correlation

Model	Attribute	HSR Fare			Airplane Fare			Bus Fare		
	% change	−50	−25	−10	−50	−25	−10	−50	−25	−10
MNL	Car	−0.460	−0.230	−0.090	−0.390	−0.180	−0.060	−0.077	−0.039	−0.016
	Airplane	−0.410	−0.210	−0.080	1.050	0.460	0.170	−0.048	−0.024	−0.009
	Bus	−0.490	−0.230	−0.080	−0.400	−0.160	−0.050	0.157	0.078	0.031
	HSR	0.630	0.310	0.120	−0.360	−0.170	−0.060	−0.057	−0.028	−0.011
MNL-NL Rail	Car	−0.103	−0.051	−0.020	−0.139	−0.066	−0.026	−0.007	−0.003	−0.001
	Airplane	−0.081	−0.040	−0.016	0.200	0.093	0.036	−0.002	−0.001	−0.001
	Bus	−0.214	−0.108	−0.042	−0.215	−0.088	−0.031	0.048	0.024	0.010
	HSR	0.139	0.070	0.027	−0.138	−0.066	−0.026	−0.005	−0.003	−0.001
NL	Car	−0.100	−0.050	−0.020	−0.140	−0.060	−0.030	−0.007	−0.003	−0.001
	Airplane	−0.080	−0.040	−0.010	0.190	0.090	0.030	−0.003	−0.001	−0.001
	Bus	−0.200	−0.100	−0.040	−0.210	−0.090	−0.030	0.050	0.025	0.010
	HSR	0.130	0.070	0.030	−0.140	−0.070	−0.030	−0.006	−0.003	−0.001

Based on this example, we could say that the models that simply follow the correlation structure detected for the RP data, without considering what the SC data might reveal in this sense, may over estimate the potential market shares of new alternatives. Thus, and as a conclusion, SC data may not only help to improve the specification of representative utility in estimation, but also to define the most appropriate correlation structure of a forecasting model.

Exercises

8.1 Consider the following mode choice model:

$$V_1 = \theta_1 t_1 + \theta_3 c_1 + \theta_4 Nc + \theta_7$$
$$V_2 = \theta_1 t_2 + \theta_2 e_2 + \theta_5 c_2 + \theta_8$$
$$V_3 = \theta_1 t_3 + \theta_2 e_3 + \theta_6 c_3$$

where t_k is in-vehicle travel time, e_k is access time, c_k is cost divided by income and Nc is the number of cars in the household.

(a) Indicate which variables are generic, which are specific and what is the real meaning of θ_7 and θ_8.

(b) Discuss the implications of having obtained the following values during model estimation:

$$\theta_1 = -0.115 \qquad \theta_2 = -0.207 \qquad \theta_3 = -0.301$$
$$\theta_4 = 1.730 \qquad \theta_5 = 0.476 \qquad \theta_6 = -0.301$$
$$\theta_7 = -1.250 \qquad \theta_8 = 2.513$$

8.2 During specification searches you obtained the set of mode choice models for car (1), bus (2) and underground (3), shown in the table below; the units of time and cost/income are minutes, sex is a dummy variable which takes the value of 1 for males and 0 for females; EMU is the expected maximum utility of the transit nest (bus-underground).

Variable (option entered)	Coefficient (t-ratio)			
	MNL-1	**MNL-2**	**HL-1**	**HL-2**
Car time (1)	−0.112	–	−0.114	–
	(−6.10)		(−6.00)	
Transit time (2,3)	0.006	–	−0.001	–
	(1.25)		(−0.94)	
Travel time (1–3)	–	−0.071	–	−0.083
		(−3.34)		(−3.60)
Cost/income (1–3)	−0.031	−0.040	−0.035	−0.033
	(−2.56)	(−3.52)	(−2.83)	(−3.10)
No. of cars (1)	1.671	1.823	1.764	1.965
	(4.21)	(4.80)	(4.12)	(5.14)
Sex (2,3)	−0.752	−0.776	−0.739	−0.701
	(−1.87)	(−1.98)	(−2.01)	(−1.83)
EMU	–	–	0.875	0.800
			(5.12)	(13.4)
ρ^2	0.412	0.284	0.376	0.315

(a) Indicate which model you prefer explaining very clearly why.
(b) The sample you used for estimation comprised 1000 individuals having all alternatives available. If 250 choose car, 600 choose bus and the rest underground, compute $l^*(0)$, the log-likelihood value for the equally likely model, and $l^*(C)$, the log-likelihood for the constants only model.

8.3 You were asked to estimate a Multinomial Logit (MNL) model and an Independent Probit (IP) model with the same data set; imagine (as it is not possible to estimate σ in practice) that you obtained the values shown in the following table:

Parameters	MNL	IP
θ_1	1.285	1.698
θ_2	−0.026	−0.034
θ_3	−0.123	−0.162
σ^2	Not applicable	2.870

Indicate whether these results appear to be consistent; if your answer is affirmative explain which the cause of the differences is. If your answer is negative, explain why.

8.4 While conducting an SP survey you asked three individuals to rank the three options whose attributes are given below:

Option	Travel time (min)	Fare ($)
1. High speed train	30	10
2. Express train	40	8
3. Luxury Coach	60	5

After completing the survey you obtained the following results:

Individual	Ranking
1	1, 2, 3
2	2, 3, 1
3	2, 1, 3

You are interested in estimating a MNL model with linear in the parameters utility function given by:

$$V_i = \theta_1 t_i + \theta_2 c_i$$

If you are told that $\theta_1 = -0.03$, find a maximum likelihood estimate for θ_2. Discuss your results.

9

Model Aggregation and Transferability

9.1 Introduction

The planning and evaluation of transport improvements requires models both to deliver forecasts and to examine their sensitivity with respect to changes in the values of key variables under the control of the analyst. The forecasts themselves normally need to be aggregate, i.e. to represent the behaviour of an entire population or market segment.

In many practical studies the models used have been of the classical aggregate four-stage form despite many (and often justified) criticisms about their inflexibility, inaccuracy and cost. One important reason for this persistence, apart from their familiarity (e.g. they have been considered accepted practice for many years) is that they offer a tool for the complete modelling process, from data collection through to the provision of forecasts of flows on links. This has not often been the case with disaggregate model approaches, perhaps because the data necessary to make aggregate forecasts with them is not readily available (see the discussion by Daly and Ortúzar 1990).

In an econometric interpretation of demand models, the aggregation over *unobservable* factors (either attributes or personal characteristics) results in a probabilistic decision model and the aggregation over the *distribution* of observables results in the conventional aggregate or macro relations (Williams and Ortúzar 1982b). Cast in these terms, the difficulty of the aggregation problem depends on how the components of the system are described within the frame of reference employed by the modeller; it is this framework which will determine the degree of variability to be accounted for in a *causal* relation. To give an example, if the framework used by the analyst is that provided by the entropy-maximising approach we saw in Chapter 5, the explanation of the statistical dispersion in a given data set will be very different to that provided by another modeller using a random utility approach, even if they both finish with identical model functions; this *equifinality issue* is discussed by Williams (1981).

In the case of disaggregate random utility models the aggregation problem is how to obtain from data at the level of the individual, aggregate measures such as market shares of different modes, flows on links, and so on. This can be achieved in one of two ways, by having the process of aggregating individual data either before or after model estimation, as shown in Figure 9.1.

In the first case we have variations of the classical aggregate approach, which can be easily criticised for being inefficient in the use of the data, not accounting for their full variability and for risking statistical distortion such as the ecological fallacy discussed in section 7.1. The second approach answers most of

Modelling Transport, Fourth Edition. Juan de Dios Ortúzar and Luis G. Willumsen.
© 2011 John Wiley & Sons, Ltd. Published 2011 by John Wiley & Sons, Ltd.

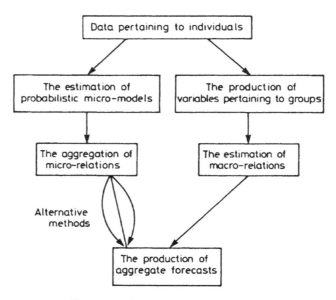

Figure 9.1   Alternative aggregation strategies

the above criticisms; the question that remains is how exactly to perform the aggregation operation over the micro relations.

Daly and Ortúzar (1990) have studied the problem of aggregation of exogenous data in some depth. They concluded that in the case of models representing the behaviour of more than one individual (as is the case with the classical aggregate model) some degree of aggregation of the exogenous data is inevitable and the issue becomes one of to what extent greater accuracy (i.e. smaller zones) is desirable. However, when the model represents the behaviour of a single individual it is conceivable that exogenous data can be obtained and used separately for each traveller; therefore the issue is whether it is preferable on cost or other grounds to use less accurate data; their findings support the notion that the cost/accuracy trade-off is heavily dependent on context. For example, it is clear that for mode choice modelling and short-term forecasting the use of highly disaggregate data is desirable; however, the plot thickens considerably for other choice contexts and longer-term forecasting. The next two sections will consider aggregation bias and forecasting methods in greater detail.

9.2   Aggregation Bias and Forecasting

Let us consider, for simplicity, the Multinomial Logit (MNL) model (7.9) we derived in section 7.3 and the inverse relation (7.10) that its parameter β has with the single standard deviation σ of the residuals ε. If we also consider the typical linear form (7.3) for the measurable utilities **V** it is easy to see that it is not possible to estimate β separately from the parameters θ; in fact the calibration process will yield estimates

$$\hat{\theta} = \beta\, \theta \qquad\qquad (9.1)$$

which correspond to the marginal utilities θ deflated by σ.

We are interested in examining the effect of the manner in which the attributes **x** (or at least some of them) are calculated, measured or codified, on the estimated demand functions. As usual we will assume that the MNL model (7.9) is well specified (i.e. there are no taste variations or correlation problems).

Let us assume now that we replace one of the attributes, for example x_1, by an aggregate estimate z_1, where:

$$x_{1i} = z_{1i} + \tau_1 \tag{9.2}$$

where the τ_i are distributed $(0, \sigma_\tau)$; then replacing (7.3) and (9.2) in (7.2) we get (note that we had dropped the individual index q for simplicity):

$$U_i = \theta_1 z_{1i} + \sum_k \theta_k x_{ki} + \delta_i \tag{9.3}$$

where the error term $\delta_i = \theta_1 \tau_1 + \varepsilon_i$ has variance $(\theta_1^2 \sigma_\tau^2 + \sigma^2)$. Thus, if we re-estimated the model in this case the coefficient estimates would not be

$$\hat{\theta}_k = \frac{\pi \sqrt{6} \cdot \theta_k}{\sigma} \tag{9.4}$$

as before, where but

$$\hat{\hat{\theta}}_k = \frac{\pi \sqrt{6} \cdot \theta_k}{\sqrt{(\theta_1^2 \sigma_\tau^2 + \sigma^2)}} \tag{9.5}$$

that is to say, $\hat{\hat{\theta}}_k \le \hat{\theta}_k, \forall k$. This is normally known as *aggregation bias* and has led to the recommendation that use of average zonal variables for estimating disaggregate demand models should be avoided whenever possible (see for example Horowitz 1981). The previous analysis may be extended to examine the consequences of this bias in forecasting, as in the following example taken from Gunn (1985a).

Example 9.1 Consider a choice situation modelled by an MNL model such as (7.9) and assume that attribute x_{1j} is doubled, *ceteris paribus*, for each option. It is clear that neither θ nor σ are affected by this; so if the model was re-estimated with a new data bank containing a consistent choice set, we would obtain exactly the same values $\hat{\theta}_k$ from the original context again, and so these would predict satisfactorily in the new context.

Consider now what would happen if after doubling x_{1j}, each of these values was replaced by its aggregate estimate z_{1j} (for example, the zonal average). In this case we would obtain equation (9.3) again, but the variance of δ_i would now be $(\theta_1^2 4\sigma_\tau^2 + \sigma^2)$; in other words, if the model was re-estimated with the new data it would yield coefficients with expected values given by

$$\hat{\theta}'_k = \frac{\pi \sqrt{6} \cdot \theta_k}{\sqrt{(\theta_1^2 4\sigma_\tau^2 + \sigma^2)}} \tag{9.6}$$

that is $\hat{\hat{\theta}}_k > \hat{\theta}'_k$ and the $\hat{\hat{\theta}}$ would produce greater than normal predictions in these conditions. Alternatively, attribute reduction policies would imply under-predictions of the model calibrated with aggregate data (see Ortúzar and Ivelic 1987).

9.3   Confidence Intervals for Predictions

As we saw in Chapter 8, the maximum likelihood estimated parameters $\hat{\theta}$ of a discrete choice model are asymptotically distributed $N(\theta, S^2)$, where θ are the population parameters and S^2 their covariance matrix given by (8.14), that is:

$$S^2 = -\left(E\left(\frac{\partial^2 l(\theta)}{\partial \theta^2} \right) \right)^{-1}$$

(continued)

i.e. the negative inverse of the Fisher information matrix $\mathbf{I}$. From this knowledge, it is straightforward to compute confidence intervals for the estimated parameters, on the basis of the well-known property that quadratic forms distribute χ^2 with degrees of freedom equal to the number of variables in the vector of interest (see 2.5.4.1).

Applying this to our estimated parameters we obtain the following quadratic form:

$$QF(\hat{\theta}, \theta) = (\hat{\theta} - \theta) \cdot \mathbf{I} \cdot (\hat{\theta} - \theta)^{\mathsf{T}}$$

that distributes asymptotically χ^2 with K degrees of freedom (K is the number of estimated parameters).

Therefore, if we can apply the asymptotic assumption a confidence region at the 95% level for the set of estimated parameters is given by the values of θ that satisfy (9.7):

$$QF(\hat{\theta}, \theta) \leq \chi^2_{K,95\%} \tag{9.7}$$

However, converting the above region into a confidence region for the estimated probabilities is not easy, as the relation between parameters and probabilities is not linear.

It is interesting to mention that the immense majority of discrete choice model applications have failed to produce confidence intervals for the estimated probabilities, although two methods for doing it have been available for many years (Horowitz 1980):

- Approximate the choice probabilities by a first order Taylor series expansion; in practice this is equivalent to assume that the relation between probabilities and parameters is linear (which is, of course, untrue). This is a fairly usual approach in mathematical statistics because it is easy to implement and inexpensive in computational terms.
- Solve a non-linear programming problem; although this is a bit more complex and expensive is subject to less errors than the previous method.

9.3.1 Linear Approximation

If $\hat{\mathbf{P}}$ is the estimated and $\mathbf{P}$ is the true value of the choice probabilities, the Taylor series approximation is given by:

$$\hat{\mathbf{P}} = \mathbf{P} + (\hat{\theta} - \theta)\frac{\partial \mathbf{P}}{\partial \theta} + \Delta \tag{9.8}$$

where the expected value of $\hat{\mathbf{P}}$ is equal to $\mathbf{P}$ and Δ is a residual term.

Now as the parameters θ distribute asymptotically Normal, $\hat{\mathbf{P}}$ also distributes asymptotically Normal as (9.8) is a linear transformation. Thus $\hat{\mathbf{P}} \sim N(\mathbf{P}, \mathbf{W})$, where:

$$\mathbf{W} = \left(\frac{\partial \mathbf{P}}{\partial \theta}\right) \mathbf{S}^2 \left(\frac{\partial \mathbf{P}}{\partial \theta}\right)^{\mathsf{T}}$$

and the numerical value of $\mathbf{S}^2$ can be estimated substituting $\hat{\theta}$ by θ in the derivatives. This approach is actually called the *Delta Method* in statistics (see Greene 2003) and it has been used in practice for some time.

Given the above, if $Z_{\alpha/2}$ denotes the percentile $(1 - \alpha/2)$ of the standard Normal distribution, then if the asymptotic assumption can be assumed to hold a confidence region of $100(1 - \alpha)$ for $\mathbf{P}$ would be given by:

$$\hat{\mathbf{P}} - Z_{\alpha/2}|\mathbf{W}|^{1/2} \leq \mathbf{P} \leq \hat{\mathbf{P}} + Z_{\alpha/2}|\mathbf{W}|^{1/2} \tag{9.9}$$

The main problem of this quick and simple method is that it can lead to erroneous results in the case of non-linear functions and when the asymptotic assumptions do not hold (Horowitz, 1980).

Example 9.2 Consider the following simple single-parameter Logit model:

$$P_1(x) = \frac{1}{1 + \exp(\theta x)}$$

where x is an independent variable and $P_1(x)$ the probability of choosing the first option. Assume that the maximum likelihood estimate of θ was $\hat{\theta} = 3$ and that its sample variance was equal to 1. Then, if $x = 0.01$ (i.e. a case where non-linearity is not of great concern), the reader can check that equation (9.9) yields the following confidence interval at the 95% level:

$$0.4876 \leq \mathbf{P} \leq 0.4974$$

However, if $x = 1$ (i.e. a situation where non-linearity should bite) we would get:

$$-0.041 \leq \mathbf{P} \leq 0.136$$

an interval which is clearly erroneous as it allows for negative values of P.

Daly and de Jong (2006) have given a better interpretation of the method, based on the idea that at the maximum likelihood values of the parameters, the measure is no less exact than the original estimates; this notion can be applied to estimate the confidence intervals of various important measures, such as user benefits, and it is easy to use to estimate the errors of predicted market shares in the MNL model. However, Daly and de Jong caution that when the model is no longer MNL and/or the sample being expanded is large, with complicated calculations for the weights attached to each observation in the aggregation procedure, the amount of calculations involved can be prohibitive and a sampling approach can be necessary, as discussed by de Jong *et al.*, 2007).

9.3.2 Non Linear Programming

The simplest way to formulate this method is as follows. Let $P_i(\theta)$ be the probability of choosing alternative $A_i \in \mathbf{A}$, where the total number of alternatives in the choice set is J; the decision variables take fixed values $\mathbf{x}$ and the parameters are, as usual, θ. Consider that $b_i(\alpha)$ and $B_i(\alpha)$ are the results of the following non-linear problems:

$$b_i(\alpha) = Min \, P_i(\theta), \quad i = 1, \ldots, J$$
$$B_i(\alpha) = Max \, P_i(\theta), \quad i = 1, \ldots, J$$

$$\text{subject to } H(\hat{\theta}, \theta) \leq \chi^2_{K,(1-\alpha)}$$

where the maximisation and minimisations operations are done for different values of θ. In this case, the following inequalities define a rectangular confidence region for P_i at the $100(1 - \alpha)$ level:

$$b_i(\alpha) \leq P_i \leq B_i(\alpha) \quad i = 1, \ldots, J \tag{9.10}$$

This method tends to produce larger confidence regions than the previous one and it is also harder to implement; but it has the advantage of yielding always reasonable (in the sense of not inconsistent) confidence intervals.

9.4   Aggregation Methods

While a disaggregate model allows us to estimate individual choice probabilities, we are normally more interested in the prediction of aggregate travel behaviour. If the choice model was linear the aggregation process would be trivial, amounting only to replacing the average of the explanatory variables for the group in the disaggregate model equation; see for example the aggregation of household-based trip generation models in Chapter 4. However, if the model is non-linear, this method, called *naive aggregation*, will generally produce bias as shown in Figure 9.2. The correct aggregate probability for a group of two individuals A and B is $(P_A + P_B)/2$; the naive method yields a probability P_C given by: $P[(V_A + V_B)/2]$. As can be seen, if the model was linear both values would coincide.

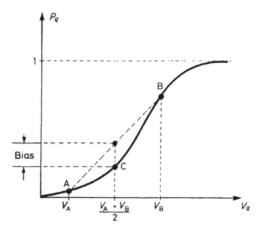

Figure 9.2   Bias of the naive aggregation method

Discrete choice models such as those we have discussed can be represented in general by:

$$P_{jq} = f_j(\mathbf{x}_q)$$

where P_{jq} is the probability that individual q selects option A_j, $\mathbf{x}_q$ is the set of variables influencing their decision, and f_j is the choice function for A_j (for example, the MNL).

For a population of Q individuals the aggregate proportion choosing A_j, according to the model, is the expected value (or enumeration) of the probabilities of each individual in the population:

$$P_{jq} = \frac{1}{Q} \sum_q f_j(\mathbf{x}_q) \qquad (9.11)$$

Unfortunately this method would require an impossibly large data set. However, if we accept that the sample used to estimate the model is a good representative of the population, we can use a modified version of (9.11) and refer to *sample enumeration* as in (9.12):

$$MS_j = \sum_{q=1}^{Q_S} w_q\, f_j(\mathbf{x}_q) \qquad (9.12)$$

where MS_j is the predicted market share of alternative A_j in the population, Q_S is the sample size and w_q the expansion factor corresponding to observation q in the sample.

This is a good practical method for moderate size choice sets and is excellent for mode choice models in short-term predictions. However, the method is not so useful in the long term because it does not allow us to address overall contexts which are very different to that of the base year (it assumes that the distribution of the attributes will not differ from that of the sample in the future); it is also unable to produce aggregate zone-to-zone flows necessary for the estimation of demand at the link level.

Example 9.3 Consider the model of Example 9.2, with the same estimated parameter (i.e. $\hat{\theta} = 3$) but with a variance equal to 4.0; assume also that the sample was composed of five individuals with the following values of x:

Individual	x
1	0.89
2	0.75
3	−0.25
4	0.80
5	−0.40

In this case, and assuming that the expansion factors were all equal to 100, the reader can check that applying (9.12) the market share of the first alternative would be approximately equal to 169 (i.e. out of 500). To estimate the precision of this estimate we could apply the Delta Method mentioned in the previous section. For the case of market shares estimated from a simple MNL model, Daly and de Jong (1996) show that the appropriate expression is:

$$Var(MS_j) = \mathbf{MS}'_j \cdot \mathbf{S}^2 \cdot \mathbf{MS}'^{\mathrm{T}}_j \tag{9.13}$$

where $\mathbf{MS}'_j$ is the vector of first derivatives of the estimated market share with respect to the parameters; thus, its *kth* element is given by:

$$ms'_{jk} = \sum_q w_q \cdot \partial P_{jq}/\partial \theta_k$$

In the case of our example, the equations above simplify substantially as we only have one parameter. Thus, $\mathbf{S}^2$ is equal to the scalar 4.0 and considering the expression of our binary Logit model, we have that for alternative A_1 we would get:

$$ms'_1 = 100^* \sum_q \partial P_{1q}/\partial \theta = 100^* \sum_q P_{1q}(1 - P_{1q})x_q$$

The reader can easily check that $ms'_1 = 5.398$ in this case, and so the estimated variance for the market shares of the first alternative equals 58.27. Thus, a 95% confidence interval for the market share would be approximately equal to:

$$169 - 1.96 \cdot \sqrt{58.27} \leq MS_1 \leq 169 + 1.96 \cdot \sqrt{58.27} \rightarrow 154 \leq MS_1 \leq 184$$

The reader can also check that if the same calculation was done for the second alternative, the confidence interval would be $316 \leq MS_1 \leq 346$, and so in this case there would be a perfect match (which, in general, may not be the case).

To cope with the problem of having a sample that is only good for the relatively short term, the *artificial sample* enumeration approach may be used (see Daly and Gunn 1986; Daly 1998). An artificial sample is one in which personal characteristics (believed to be representative of the population of the study area) of members of existing households are matched with characteristics of

(continued)

---

a number of locations also believed to be typical of the area. Thus, the marginal distributions of both personal and location characteristics are by construction typical of the study area; the approximation made is that the joint distribution of these characteristics can be represented by the product of the two marginal distributions.

Given suitable networks, zoning systems and planning data, the marginal distributions of locations can be those of actual locations in the study area (details of their accessibility to available destinations are needed); if the locations are distributed over the whole of the study area, we can be reasonably confident of overall representativeness in large samples (see Gunn 1985b).

For personal characteristics the following steps are needed to achieve realism:

1. Actual households' members are drawn at random from a large nationally representative data set (e.g. a census).
2. For each zone of the study area, different expansion factors are found for each of these households such that the expanded sample corresponds as closely as possible to known or forecast aggregate totals (i.e. of variables such as numbers of workers, numbers of individuals by sex and age grouping, etc.).
3. The expansion factors, or more commonly the number of households in each group, are chosen such that the overall distribution of households in terms of a given stratification (say size, number of workers and age of head of household) is not too different from the overall national average (note that when classifying data in this form there are several impossible strata, e.g. households of size 1 with more than 1 worker). Daly (1998) compares the two most used methods to do this, iterative proportional fitting (what we called Furness method in section 5.2.3), which has been extensively applied in practice (see the discussion by Beckman *et al.* 1995) and quadratic optimisation, as in (9.14), which was judged an improvement over the previous one. Daly (1998) also provides details about the steps to follow in the construction and use of the prototypical example and several successful examples about its application:

$$S(N_i) = \sum_k W_k \left[\sum_i (X_{ik} N_i - Y_k)^2 \right] + \sum_i (N_i - R_i)^2 \qquad (9.14)$$

Here N_i is the required number of households in stratum i; W_k is a weight chosen to increase or decrease the importance of the fit to the kth variable (e.g. number of workers, number of males between 18 and 65, etc.); X_{ik} is the average value of variable k for household stratum i; Y_k is the average (observed) value of variable k for (each zone of) the study area; and R_i the number of households in stratum i in the base-year sample. Various other constraints can be put on the process, as discussed by Gunn (1985b).

The artificial sample replicates the population of each zone of the study area; thus aggregate forecasts can simply be obtained by applying the enumeration method to these data. The interested reader can find more on the creation of synthetic samples, although not exactly for the same purposes as discussed here, in the work of Guo and Bhat (2007) and Ye *et al.* (2009). We also discuss the generation of synthetic samples in a little more detail in section 14.5.

Another practical method is known as the *classification approach*, which consists in approximating (9.11) by a finite number of relatively homogeneous classes, as in:

$$P_{jq} = \sum_c f_j(\mathbf{X}_c) Q_c / Q \qquad (9.15)$$

where $\mathbf{X}_c$ is the mean of the variable set vector for subgroup c and Q_c/Q the proportion of individuals in the subgroup.

The accuracy of the method depends on the number of classes c and their selection criteria (in the limit it equals the naive method, when the number of subgroups $c = 1$, and the enumeration method, when $c = Q$). Interesting but not often practical methods to define the classes have been reported (McFadden and Reid 1975) but the approach is recommended for cases where sample enumeration is not appropriate (Koppelman 1976).

An obvious method to define classes is to use as market segmenting variables those that present the greatest variance or those which limit in some way the available choice set of each individual. Thus in the mode choice case good variables are the number of cars per household and family income.

9.5 Model Updating or Transference

9.5.1 Introduction

During the 1980s a substantial body of literature emerged with empirical evidence about the stability (or, in most cases, lack of it) of parameters of disaggregate travel demand models, across space, cultures and time (see for example Gunn *et al.* 1985; Koppelman and Wilmott 1982; Koppelman *et al.* 1985a, b). The reasons were simple: firstly, evidence of stable values of estimated parameters could provide a direct indication of model validity; secondly, a model that is not stable over time is likely to produce inaccurate predictions; finally, and not less importantly, transferable models should allow for more cost-effective analyses of transport plans and policies.

Because it is unrealistic to expect an operational model in the social sciences to be perfectly specified, it is quite obvious that any estimated model is in principle context dependent. For this reason, it is not very useful to look for perfect model stability and to consider model transferability in terms of equality of parameter values in different contexts (although many studies initially took this view; see Galbraith and Hensher 1982; Ortúzar 1986).

A more appropriate view considers model transfer as a practical approach to the problem of estimating a model for a study area with little resources or a small available sample. In this sense the model-transfer approach is based on the idea that estimated parameters from a previous study may provide useful information for estimating the same model in a new area, even when their true parameter values are not expected to remain the same. Now, as transferred models cannot be expected to be perfectly applicable in a new context, updating procedures to modify their parameters are needed so that they represent behaviour in the application context more accurately. Depending on the information available in the new environment, different updating procedures may be applied (see Ben-Akiva and Bolduc 1987).

9.5.2 Methods to Evaluate Model Transferability

If we define transferability as the usefulness of a transferred model, information or theory in a new context, we can attempt to measure it by comparing the model parameters and, more interestingly, its performance in the two contexts. For this we will assume that we have estimated the parameters independently in the two contexts; we will also assume that we would like to measure the errors involved in using the first model in the second context. The following tests and measures were used in such analyses in many practical studies (Galbraith and Hensher 1982; Koppelman and Wilmott 1982; Ortúzar *et al.* 1986).

9.5.2.1 Test of Model Parameter for Equality

To evaluate the absolute difference between coefficients of a given model estimated in two different contexts, the t^*-statistics have been used; if (9.16) holds, the null hypothesis that this difference is zero cannot be rejected at the 95% level:

$$t^* = \frac{\theta_i - \theta_j}{\sqrt{(\theta_i/t_i)^2 + (\theta_j/t_j)^2}} < 1.96 \qquad (9.16)$$

where θ_k denotes coefficients, t_k their t-ratios, i stands for the original context and j for the new context; note that this is the same test we saw in Example 8.3, but here there is no possible correlation among the parameters as they belong to two different contexts. Galbraith and Hensher (1982) recommended the application of this test only to parameters with low standard error (high t-ratio); otherwise, the t^*-statistic may reject the alternative hypothesis (i.e. the parameters are different) even if they exhibit substantial differences. However, note that this statistics suffers from the scale problem, as it cannot be possible to assume that the variances of the error components in both contexts will be the same; thus, one cannot be sure if differences are real or just a scaling problem. We will consider more appropriate methods below.

9.5.2.2 Disaggregate Transferability Measures

These are based on the ability of a transferred model to describe individual observed choices in the new context and rely on measures of log-likelihood as those that were depicted in Figure 8.2. In addition we need to define $l_j^*(\theta_i)$ as the log of the likelihood that the observed data in the application context j were generated by the transferred model estimated in context i; note that we need to denote the measures previously used in Chapter 8 as $l_j^*(\theta_j)$, $l_j^*(C)$ and $l_j^*(0)$ respectively. Figure 9.3 shows the expected relation among these values.

Figure 9.3   Expected relation between log-likelihood values

A natural measure of the transferability of a model estimated in context i for the application in context j, is the difference in log-likelihood (i.e. likelihood ratio) between this model and one originally estimated in context j: $-\{l_j^*(\theta_i) - l_j^*(\theta_j)\}$. This measure has been used to build two specific indices of transferability:

1. Transferability test statistics (TTS), defined by Atherton and Ben-Akiva (1976) as twice the difference in log-likelihood identified above:

$$\text{TTS} = -2\{l_j^*(\theta_i) - l_j^*(\theta_j)\} \qquad (9.17)$$

   This statistic is distributed χ^2 with degrees of freedom equal to the number of model parameters, under the assumption that the parameter vector of the transferred model is fixed. The test is not symmetric; therefore it is both possible and reasonable to accept transferability in one direction, between a pair of contexts, but reject it in the other direction.
2. Transfer index (TI), which describes the degree to which the log-likelihood of the transferred model exceeds a null or reference model (such as the market shares model), relative to the improvement provided by a model developed in the new context. It was defined by Koppelman and Wilmott (1982) as:

$$\text{TI}_j(\theta_i) = \frac{l_j^*(\theta_i) - l_j^*(C)}{l_j^*(\theta_j) - l_j^*(C)} \qquad (9.18)$$

TI has an upper bound of one (which is obtained when the transferred model is as accurate as the local one), but does not have a lower bound; negative values imply only that the transferred model is worse than the local reference model.

The two measures defined above are interrelated by their dependence on the difference in log-likelihood between transferred and local models. However, they offer different perspectives on model transferability: TI provides a relative measure and TTS a statistical test measure (Koppelman and Wilmott 1982).

9.5.3   Updating with Disaggregate Data

The most general presentation of the MNL model (7.9) with linear utility functions **V** given by (7.3), considers not only the explicit inclusion of relation (7.10) – as we saw in section 9.2 – but also the explicit inclusion of a set of location parameters w_i as in:

$$P_{iq} = \frac{\exp\left[(w_i + \theta \mathbf{X}_{iq})/\sigma\right]}{\sum_j \exp\left[(w_j + \theta \mathbf{X}_{jq})/\sigma\right]} \tag{9.19}$$

where the location parameters represent the mode of the distribution of errors for each alternative, the scale parameter σ is the standard deviation of the distribution of the error term, and the parameters θ the attribute weightings employed by the individual in evaluating alternatives (note that strictly speaking, we are missing the constant $\pi/\sqrt{6}$ in 9.19).

In his analysis of model mis-specification, Tardiff (1979) shows that the omission of explanatory variables should have the following effects:

- shift the mean of the error distribution, represented in the model by w_i, and increase its variance reflected by σ;
- bias the estimates of the parameters associated with the included variables.

When comparing models which are incompletely specified, in different contexts, it is expected that the differences in the mean values of the error distribution will be relatively large, the differences in the error standard deviation will be smaller, and the differences in the parameter estimates the smallest. Thus, efforts to improve model transfer to a specific application environment should emphasise adjustment of constants first, parameter scale second and relative values of the parameter last; this has been confirmed by several practical studies using both aggregate and disaggregate data (Gur 1982; Dehghani and Talvitie 1983; Koppelman *et al.* 1985b; Gunn and Pol 1986).

The parameters in equation (9.19) are of course not uniquely identifiable and therefore cannot all be estimated; as we have seen, in the case of the alternative specific constants one is arbitrarily (and with no loss of generality) set to 0. Also, it is not possible to estimate σ but only the ratios $\mathbf{w}/\sigma$ and θ/σ; defining these ratios by $\mu = \mathbf{w}/\sigma$ and $\phi = \theta/\sigma$, we obtain the more familiar version of the MNL model as:

$$P_{iq} = \frac{\exp\left(\mu_i + \phi \mathbf{X}_{iq}\right)}{\sum_{Aj \in A(q)} \exp\left(\mu_j + \phi \mathbf{X}_{jq}\right)} \tag{9.20}$$

where one of the μ_i must be constrained to zero.

(continued)

9.5.3.1 Updating the Constants

Parameter estimates for a choice model are obtained by maximising a log-likelihood expression such as (8.13), where embedded in the probability function P_{iq} are expressions for the representative utility of each option formulated as:

$$V_{iq} = \mu_i + \phi \mathbf{X}_{iq} \qquad (9.21)$$

Let us denote as Φ_T a set of parameters estimated in one context to be transferred to a new application context; in this case the transferred portion of the utility function can be defined as (Koppelman *et al.* 1985b):

$$Z_{iq}^A = \phi_T \mathbf{X}_{iq}^A \qquad (9.22)$$

where $\mathbf{X}_{iq}^A$ is a vector of attributes of alternative A_i for individual q in the application context (A). The updating of alternative specific constants is accomplished by modifying the utility function in equation (9.21) for the application context to:

$$V_{iq}^A = \mu_i^A + Z_{iq}^A \qquad (9.23)$$

where V_{iq}^A is the representative utility of option A_i in the application context and μ_i^A its updated alternative specific constant. To estimate the updated value of the constants it is necessary to maximise the log-likelihood function:

$$l(\mu^A) = \sum_q \sum_{A_j \in \mathbf{A}(q)} g_{jq} \log P_{jq}(\mathbf{Z}_q^A, \mu^A) \qquad (9.24)$$

whereas before, g_{jq} is defined by:

$$g_{jq} = \begin{cases} 1 & \text{if } A_j \text{ was chosen by } q \\ 0 & \text{otherwise} \end{cases}$$

9.5.3.2 Updating of Constants and Scale

The methodology just outlined can be trivially extended to adjust the scale of the transferred parameters as well as the constants. The coefficient of Z_{iq}^A in equation (9.23) was restricted to one in the preceding approach; to update the parameter scale, that restriction is relaxed yielding the following representative utility (Koppelman *et al.* 1985b):

$$V_{iq}^A = \mu_i^A + \lambda^A Z_{iq}^A \qquad (9.25)$$

where λ^A is the scaling parameter for the application context relative to the estimation, or original, context. In this case the log-likelihood function to be maximised is as (9.24) but including the extra parameter λ^A. Note that this adjusts the scale of the explanatory variables but does not affect their relative importance. Practical applications of this method have been reported by Gunn *et al.* (1985) and a discussion of further refinements to this problem can be found in Ben-Akiva and Bolduc (1987).

9.5.4 Updating with Aggregate Data

Consider the same problem as before with the exception that no disaggregate data are available in the application context; however, assume we possess data on observed market shares P_{jq}^*, and also average values for the explanatory variables $\bar{X}_{jz}$, for certain groups $\mathbf{Z}$ (say residents of a given zone) in both contexts.

Consider a naive aggregation in the original context, where the measured utility of option A_j for a given group z is given by:

$$\tilde{V}_{jz} = \mu_j + \phi \, \bar{\mathbf{X}}_{jz} \qquad (9.26)$$

Updating both alternative constants and scale in this case, requires first to compute non-constant utility for the application context as:

$$\bar{\mathbf{Z}}_{jz}^A = \phi_T \, \bar{\mathbf{X}}_{jz}^A \qquad (9.27)$$

then postulate an expression for the representative utility of group z in the application context as:

$$\tilde{V}_{jz}^A = \mu_j^A + \tau^A \, \bar{\mathbf{Z}}_{jz}^A \qquad (9.28)$$

where μ^A and τ^A are chosen so as to maximise the following log-likelihood function (Koppelman et al. 1985a):

$$l(\mu^A, \tau^A) = \sum_z W_z \sum_j P_{jz}^* \log P_{jz}(\bar{\mathbf{Z}}_{jz}^A, \mu^A, \tau^A) \qquad (9.29)$$

with W_z a weight, usually the number of observations, which indicates the relative importance of the group in the data set. Other (more suspect) methods to update the constants only have been proposed by Dehghani and Talvitie (1983) and Gur (1982).

The aggregation issue in the presentation above is not trivial as it is well known that the naive method may introduce severe bias. In this sense it is interesting to mention that the methodology just discussed is wholly consistent with the aggregation approach implicit in most aggregate transport studies (recall Figure 9.1 and the discussion in Chapter 5). There, disaggregate model parameters have been traditionally used as fixed coefficients of generalised cost functions, and later *scale* and *bias* parameters have been fitted using aggregate data (Williams and Ortúzar 1982b).

It is also of interest to note that a more elaborate version of this approach has also been used in practice; for example, in the Greater Santiago Strategic Transport Study (ESTRAUS 1989) disaggregate mode choice parameters were firstly estimated with a mixture of data for 1983 to 1986 (Ortúzar and Ivelic 1988); these were used to build generalised cost functions the scale and bias parameters of which were then calibrated using 1977 network and survey data (the only O-D and network data available at the time). Finally, the resulting aggregate distribution and modal-split models were validated using volume counts and other aggregate information for 1986.

An interesting alternative, if available, is the use of purposely designed synthetic samples in an enumeration approach such as we discussed in section 9.4 (Gunn et al. 1982). An important advantage of this method is that no major adjustments need to be made to the disaggregate models if the artificial sample provides unbiased information to the model system.

Exercises

9.1 A group of 800 heads of household with different income levels and located in various parts of an urban area, are confronted with choice between two transport services A and B, for travelling to the central business district. The first, which is more oriented to the population segment with higher income, has a cost C_a and the second a cost C_b.

It has been estimated that the utilities of each alternative are given by the following linear functions:

$$U_a = -0.30C_a + 3.23I$$
$$U_b = -0.30C_b$$

where I is family income (1000$/week).

Estimate the number of households that would choose service A using the following information:

Family income (100$/week)	Number of households	C_a ($)	C_b ($)
Between 1 and 2	450	150	120
Between 2 and 3	250	175	145
Between 3 and 4	100	160	130

9.2 Consider the urban corridor depicted in the figure

which has the following characteristics:
- Underground and highway run parallel to each other.
- There are underground stations at each zone.
- The households in the corridor have different income levels, different car ownership and different access to the underground, as shown in Table 9.1.

We are interested in the trips between zone 1 and the centre of town. We are informed that a binary Logit model has been estimated yielding the following representative utilities:

$$V_c = -2.0 + 9 \times 10^{-5}I + 2.84CO - 0.03t_c - 0.68e_c/d - 50.0c_c/I$$
$$V_u = -0.03t_u - 0.68e_u/d - 50.0c_u/I$$

where t is in-vehicle travel time (min), e is access time (min), c is cost ($), d is distance (km), I is income ($/month) and CO is the number of cars divided by the number of licences in the household.

Underground trips are divided according to access into U (DA), underground with direct access (i.e. on foot), and U (CA), underground with car access. The levels of service by individuals travelling between zone 1 and the centre are summarised in Table 9.2.

Table 9.1   Distribution of households with trips between zone 1 and the centre

CO	Access	Income 5000	10000	15000	Total
	U (DA)	0	0	350	350
1.0	U (CA)	0	50	150	200
	Total	0	50	500	550
	U (DA)	150	100	0	250
0.5	U (CA)	200	0	0	200
	Total	350	100	0	450
	U (DA)	150	100	350	600
Total	U (CA)	200	50	150	400
	Total	350	150	500	1000

Table 9.2 Levels of service

	t_c	e_c	c_c	t_u	e_u	c_u	d
U (DA)	11.3	5	122.5	14	8	50	14.5
U (CA)	14.2	5	131.3	22	15	75	16.3

Find out, using an appropriate method, the aggregate probability (i.e. for the whole population) of choosing underground.

9.3 Consider a binary Logit model for car and bus with the following representative utility functions:

$$V_c = 1.35 - 0.03t_c - 0.15c_c$$
$$V_b = -0.03t_b - 0.15c_b$$

where t is total travel time (min) and c is travel cost divided by income (min). Assume the data in Table 9.3 is known about individuals from zone A travelling to work at zone C:

(a) Find out the aggregate proportion choosing car by the naive aggregation method and by the sample enumeration method. Compute the naive aggregation error in this case.

(b) Find now the aggregate proportion using car by the classification method (using income as stratification variable). Plot your results and those of the naive aggregation method; discuss your graph.

(c) Compare all your results and discuss them critically.

Table 9.3 Individual data

Individual	Chosen option	Income level	t_c (min)	t_b (min)	c_c (min)	c_b (min)
1	Car	High	47.5	83.2	14.8	7.0
2	Car	High	30.2	45.0	10.4	5.0
3	Car	High	22.2	30.4	12.6	4.0
4	Bus	High	45.0	50.6	8.2	5.0
5	Bus	Low	15.3	20.5	50.0	17.0
6	Car	Low	34.8	50.2	55.0	35.0
7	Bus	Low	65.5	100.5	200.3	53.5
8	Bus	Low	12.0	14.0	44.6	17.0

9.4 You are interested in transferring the model of Exercise 9.3 to a new context, where you have taken a small sample of five individuals whose characteristics are presented in the following table:

Individual	Chosen Option	t_c (min)	t_b (min)	c_c (min)	c_b (min)
1	Car	37.5	70.2	16.8	10.0
2	Car	20.2	30.0	16.4	8.0
3	Car	12.0	15.4	18.6	7.0
4	Bus	35.0	35.6	14.2	8.0
5	Bus	5.3	6.5	56.0	20.0

Assuming there are no mode specific constants, estimate the value of τ, the transfer scale parameter, using the data above. Discuss your result.

10

Assignment

10.1 Basic Concepts

10.1.1 Introduction

The last six chapters have dealt in detail with the key models currently in use to represent the demand for travel in a study area. This chapter will deal with the assignment of vehicles and people to road and public transport networks following a rather intuitive approach in order to introduce some of the basic relevant ideas. The next chapter will adopt a more formal approach concentrating on equilibrium both at the network and system levels. The network system, and in the case of public transport the characteristics of the services offered such as frequency and capacity, represent the main elements of the supply side in transport. These are more or less fixed in the short run. Over a longer period, transport authorities and operators will change fares, frequencies and vehicle types; road network managers will improve existing (and build new) roads, constrain parking, and introduce tolls and congestion charges. Although these are real representations of supply changes to increased demand, we do not have good models to forecast this type of longer term changes in supply. Our network models fall short of that: they are only cost-models: how transport costs will change with different levels of demand. The task of specifying a better longer term supply system falls to decision makers, planners and analysts.

In conventional economic thinking the actual exchanges of goods and services take place as a result of combining their demand with their supply. The equilibrium point resulting from this combination defines the price at which the goods will be exchanged and their respective flows (quantities exchanged) in the market. The equilibrium point is found when the marginal cost of producing and selling the goods equals the marginal revenue obtained from selling them. Economic theory admits that this equilibrium may never actually happen in practice as the system of prices and production levels is under permanent adjustment to cope with changes in purchasing power, tastes, technology and production techniques. However, the concept of equilibrium is still valuable in understanding the movement of the economy and to forecast its future states.

It is useful to consider the transport system within that context. The (short term) supply side, or more correctly the cost model, is made up of a transport network $S(L, C)$ represented by links L (and their associated nodes) and their costs C. The costs are a function of a number of attributes associated with the links, e.g. *distance, free-flow speed, capacity* and a *speed–flow relationship*, and, in the case of public transport, on *route and service attributes* like fares, frequencies and running times. The demand side is made up of an indication of the number of trips by O–D pair and mode that would be made for a given level of service, i.e. that assumed in their estimation. One of the main elements defining levels of

Modelling Transport, Fourth Edition. Juan de Dios Ortúzar and Luis G. Willumsen.
© 2011 John Wiley & Sons, Ltd. Published 2011 by John Wiley & Sons, Ltd.

service is, in this context, travel time, but often monetary costs (fares, fuel) and features like comfort for the public may be relevant too. If the actual level of service offered by the transport network turns out to be lower than estimated, then a reduction in the demand and perhaps a shift to other destinations, modes and/or times of day would be expected. The speed–flow (or generalised cost–flow) relationship is important as it relates the use of the network to the level of service it can offer.

The public-transport network must be defined in similar terms to the private network. However, it should contain additional specification of the services offered in terms of their routes, capacities, frequency, fares and ideally, though seldom in practice, their quality, reliability and regularity.

In the case of a transport system one can see equilibrium taking place at several levels. The simplest one is equilibrium in the road network where travellers from a fixed trip matrix seek routes to minimise their travel costs (times). This results in their trying alternative routes, exploring new ones and perhaps settling into a relatively stable pattern after much trial and error. This allocation of trips to routes yields a pattern of path and link flows which could be said to be in equilibrium when travellers can no longer find better routes to their destinations: they are already travelling on the best routes available. This is the *road network* equilibrium. A similar, but perhaps less dramatic, phenomenon takes place in public-transport networks where passengers may seek routes (i.e. combinations of services) to reduce their generalised journey costs as affected by overcrowding, waiting and walking times, and in-vehicle times.

There are, however, other (higher) levels of interaction. As car congestion increases, buses operating on the same roads will have their journey times increased as well. This may induce some public-transport users (and bus operators) to change their routes to avoid these delays. These choices interact with those of car drivers as the new arrangements may provide additional capacity in some links and therefore new equilibrium points. These are *multimode network* equilibrium problems and are discussed in Chapter 11.

At an even higher level, the resulting flow pattern may affect choices of mode, destination and time of day for travel. Each of these shifts in demand will induce in turn changes in the corresponding equilibrium points. In modelling terms, the new flow pattern produces levels of service for routes and modes which may or may not be consistent with those assumed in estimating the (presumed) fixed trip matrix. This requires re-estimating the matrix and therefore feeding back the new levels of service into the estimation process to obtain a new one. The process may need to be repeated in a systematic way until the trip matrices (and therefore trip time, destination and mode) are obtained with values for travel costs which are consistent with the flows estimated for each network. This higher level we shall call *system equilibrium* as opposed to *network equilibrium*.

The rest of the chapter is organised as follows. We consider first the problem of assigning a fixed trip matrix to a road network. In order to treat this problem we consider typical characteristics of speed– or cost–flow curves. The assignment problem is split into a route choice model and the loading of the trip matrix onto the identified routes. Different conditions require different loading methods. Stochastic methods allow for variability in drivers' perception of route costs; these methods are discussed in section 10.4. The most interesting deterministic assignment methods try to include consistently the effect of congestion on route choice. This chapter considers only pragmatic methods under the general title of congested assignment in section 10.5; we leave a more rigorous treatment of equilibrium assignment for Chapter 11. Section 10.6 considers the problems and approaches required to model public-transport assignment.

10.1.2   Definitions and Notation

Some further notation will be introduced as required but the basic elements used in this chapter are:

T_{ijr} is the number of trips between i and j via route r,

V_a is the flow on link a in vehicles per hour (vph), or passenger car units (pcu) per hour, where typically a bus is equivalent to between 2 and 3 pcu and trucks between 3 and 4 pcu,

$C(V_a)$ is the cost-flow relationship for link a,

$c(V_a)$ is the actual cost for a particular level of flow V_a; the cost when $V_a = 0$ is referred to as *free-flow* cost,

c_{ijr} is the cost of travelling from i to j via route r,

$$\delta_{ijr}^a = \begin{cases} 1 & \text{if link } a \text{ is on path (or route) } r \text{ from } i \text{ to } j \\ 0 & \text{otherwise} \end{cases}$$

A superscript n will be used to indicate a particular iteration in iterative methods. A superscript * will be used to indicate an optimum value, e.g. c_{ij}^* is the minimum cost of travelling between i and j.

In many cases it is important to recognise that there are different road users and they may display different behaviour on the same link. Therefore, we further introduce an additional index (usually u) for *user class*. It is possible to have different user classes for each vehicle type (car, bus, truck) and for different types of drivers as a function of their journey purpose, willingness to pay (income) tolls and parking, and other personal characteristics relevant to the study.

10.1.3   Speed–Flow and Cost–Flow Curves

A familiar relationship in traffic engineering is that relating the speed on a link to its flow. This concept was originally developed for long links in motorways, tunnels or trunk roads. A speed–flow relationship is usually presented as in Figure 10.1; as flow increases, speed tends to decrease after an initial period of little change; when flow approaches *capacity* the rate of reduction in speed increases. Maximum flow is obtained at capacity and when attempts are made to force traffic volumes beyond this value an unstable region with low flows and low speeds is reached.

For practical reasons, in traffic assignment this type of relationship is handled in terms of travel time per unit distance versus flow, or more generally, as a cost–flow relationship, as also shown in Figure 10.1. Traffic assignment methods taking into account congestion effects need a set of suitable functions relating link attributes (capacity, free flow speed) and flow on the network with the resulting speeds or costs. This can be written in general terms as:

$$C_a = C_a(\{\mathbf{V}\}) \tag{10.1}$$

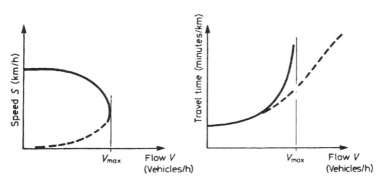

Figure 10.1   Typical speed–flow and cost–flow relationship for a long link

that is the cost on a link a is a function of all the flows $\mathbf{V}$ in the network, i.e. not just the flow on the link itself. This general formulation is relevant in urban areas where there is a good deal of interaction between flows on different links and their corresponding delays, for example at priority junctions or roundabouts. However, this can be simplified if one considers long links, that is, links where most of the travel time takes place on the link rather than at the end junctions. In this case the function is said to be *separable* and we can write:

$$C_a = C_a(V_a) \tag{10.2}$$

that is, the cost on the link depends just on its flow and the link characteristics. This assumption simplifies the estimation of these functions and the development and use of suitable trip assignment techniques. It must be recognised, however, that it becomes much less realistic as one works with denser and more congested urban areas.

A number of general functional forms have been proposed to embody the general relationship in equation (10.2). The fact that our main concern in this section is traffic assignment permits us to concentrate on a smaller set of these functions, in particular those with good mathematical properties. The following are desirable properties from the point of view of traffic assignment:

* Realism; the modelled travel times should be realistic enough.
* The function should be non-decreasing and monotone; increasing flow should not reduce travel time. This is not only reasonable but also desirable, as we shall see below.
* The function should be continuous and differentiable.
* The function should allow the existence of an overload region, i.e. it should not generate infinite travel time, even when flow is equal or greater than capacity. This may happen as part of an iterative process when more traffic is assigned to a link than its capacity; a high positive value for travel time should be produced but infinity will generate overflow in computer programs, an undesirable occurrence. More-over, short-term overload can certainly happen in practice without generating anything approaching infinite delay! The dotted line in the cost–flow curve in Figure 10.1 simulates this.
* For practical reasons the cost–flow relationship should be easy to transfer from one context to another; the use of engineering parameters like free-flow speed, capacity, and number of junctions per kilometre is therefore desirable.

One would expect the cost–flow relationship to be an increasing function with flow, except perhaps at very low flow levels when travel times may remain constant despite small increases in traffic volume. The total operating cost on a link will then be given by $V_a C_a(V_a)$; it is interesting to consider the corresponding marginal cost, that is, the contribution to total cost made by the marginal addition of one vehicle to the stream:

$$C_{ma} = \frac{\partial \left[V_a C_a(V_a) \right]}{\partial V_a} = C_a(V_a) + V_a \frac{\partial C_a(V_a)}{\partial V_a} \tag{10.3}$$

On the right-hand side we have two terms, the first one corresponding to the average cost on the link and the second to the contribution to delay to other traffic made by the marginal vehicle. This is an external effect and corresponds to the additional costs incurred by other users of the link when a new car is added to it. As the cost–flow curve is an increasing one this contribution is always greater than zero. It is also clear that in economic terms the average and marginal costs will only be the same in the flat part of the cost–flow curve, if any.

A number of authors have suggested functional forms for cost–flow relationships. These usually rely on the assumption that one is trying to model steady-state conditions and some kind of average behaviour. Branston (1976) has produced a good review of the practical problems encountered when trying to calibrate these cost–flow functions:

- There are problems with the length of the observation period in particular in congested areas and where an upstream junction acts as bottleneck; the exact location of flow and delay measuring areas plays a critical role in determining the quality of the results obtained.
- The assumption that delays depend only on flow on the link itself is unrealistic in most dense urban networks and this is particularly critical in trying to estimate cost–flow functions.

Branston (1976) also reviews cost–flow curves proposed by other authors. Some of the most used are the following:

1. Smock (1962) for the Detroit Study:

$$t = t_0 \exp(V/Q_s) \tag{10.4}$$

where t is travel time per unit distance (min/km), t_0 is travel time per unit distance under free flow conditions, and Q_s is the steady-state capacity of the link.

2. Overgaard (1967) generalised (10.4) as follows:

$$t = t_0 \alpha^{\beta(V/Q)} \tag{10.5}$$

where Q is the capacity of the link, and α and β are parameters for calibration.

3. The Bureau of Public Roads (1964) in the USA proposed what is probably the most commonly used function of this type:

$$t = t_0 \left[1 + \alpha (V/Q)^\beta \right] \tag{10.6}$$

4. The Department of Transport in the UK has produced a large number of cost–flow curves for a variety of link types in urban, sub-urban and inter-urban roads. Some have a general form which considers first the speed–flow $s(V)$ curve:

$$s(V) = \begin{cases} S_0 & V < F_1 & (10.7a) \\ S_0 - \dfrac{S_0 - S_1}{F_2 - F_1}(V - F_1) & F_1 \le V \le F_2 & (10.7b) \\ S_1/[1 + (S_1/8d)(V/F_2 - 1)] & V > F_2 & (10.7c) \end{cases}$$

where
  S_0 is the free flow speed,
  S_1 is the speed at capacity flow F_2 (or Q),
  F_1 is the maximum flow at which free-flow conditions prevail, and
  d is the distance or length of the link.
Then the time-flow $T(V)$ relationship becomes:

$$T(V) = \begin{cases} d/S_0 & V < F_1 & (10.8a) \\ d/S(V) = \dfrac{d}{S_0 + SS_{01}F_1 - SS_{01}V} & F_1 \le V \le F_2 & (10.8b) \\ d/S_1 + (V/F_2 - 1)/8 & V > F_2 & (10.8c) \end{cases}$$

with SS_{01} given by:

$$SS_{01} = \frac{S_0 - S_1}{F_1 - F_2} \tag{10.9}$$

Typical values for these coefficients (Department of Transport 1985) are given in Table 10.1. In some cases a cut-off point in speed reductions is assumed; for example the speed may be assumed to remain at F_2 for $V > F_2$.

Table 10.1 Typical speed–flow curve coefficients in the UK

Type	S_0 km/h	S_1 km/h	F_1 pcu/h/lane	F_2 pcu/h/lane
Single 2 lane, rural	63	55	400	1400
Dual 2 lane, rural	79	70	1600	2400
Single 2 lane, urban, outer area	45	25	500	1000

5. Akçelik function. All the functions mentioned above tend to underestimate delays at junctions as they concentrate on the links characteristics. Moreover, they also tend to underestimate delays when demand is close or above the capacity of the link. They are less appropriate in urban conditions where junctions play a more important role in determining travel times than the speed mid-link. Akçelik (1991) has suggested a better curve, based on earlier work by Davidson, which tackles these problems much better. When considering conditions close or above saturation the length of the modelling period matters considerably as it influences the length of the 'overflow' curve which in turn drives delay. Akçelik's function applies to v/c ratios above and below 1:

$$t = t_0 + \{0.25T\} \left[(x - 1) + \sqrt{(x - 1)^2 + \frac{8J_A}{Q_j T} x} \right] \tag{10.10}$$

where:
   T is the flow modelling period (typically one hour),
   Q_j is the capacity at the junction; if the saturation flow is Q_s then $Q_j = Q_s g/cy$,
   g is the length of the green period at the junction and cy is the cycle length in the same units,
   x is the degree of saturation $= V/Q_j$,
   J_A is a delay parameter.

In principle there is no upper limit on the value of x that could be input above since this equation is designed to approximate the delays due to queuing when demand exceeds capacity. The equation explicitly takes into account the delays caused by queuing and can be applied to any facility type. The assumptions are that there is no queue at the start of the analysis period, and there is no peaking of demand within the analysis period (T).

The delay parameter J_A is a function of the number of delay-causing elements in the section of road and the variability of the demand. Akçelik suggests lower values of J_A for freeways and coordinated signal systems. Higher values would apply to secondary roads and isolated intersections.

The value of J_A can be computed if the difference in the rate of travel (hours per km) between capacity and free flow conditions on the facility is known. Substituting $x = 1$ in the above equation and solving for J_A yields:

$$J_A = \frac{2Q_j}{T} (t_c - t_0)^2 \tag{10.11}$$

where t_c is the rate of travel at capacity (hours per km).

All the above speed or cost–flow curves produce information about travel time on a link. However, it is recognised that most users might wish to minimise a combination of link attributes including time and distance. Conventional practice recommends the use of a simplified version of the generalised cost concept, namely a linear weighted combination of time and distance:

$$C_a = \alpha(\text{travel time})_a + \beta(\text{link distance})_a \tag{10.12}$$

This cost could be measured in generalised time or generalised money units. It is also possible to include an out-of-pocket expenditure element, for example a toll to be applied on a given link.

The calibration of cost–flow relationships is time consuming and requires a good deal of high-quality data: observations of travel times on links under different flow levels. For this reason, this is rarely attempted and many countries have developed their own functions. See also the limitations of link-based cost–flow functions in urban areas as discussed in section 11.3.

Suh *et al.* (1990) have put forward an innovative approach to estimate cost–flow curves based on traffic counts; they use a bi-level optimisation method that, in essence, seeks to establish the parameters for the cost–flow curves minimising a measure of difference between assigned and observed flows. The value of this approach is limited by the errors in the assignment process as discussed, again, in section 11.3: e.g. errors in the network, trip matrix, in the assumption of perfect information and that all users perceive link costs in the same way. Thus the estimated cost–flow curves incorporate these errors and are, therefore, difficult to transfer to other areas or even schemes.

10.2   Traffic Assignment Methods

10.2.1   Introduction

During the classic traffic assignment stage a set of rules or principles is used to load a fixed trip matrix onto the network and thus produce a set of links flows. This is not, however, the only relevant output from the assignment stage; this has several objectives which are useful to consider in detail. Not all of them receive the same emphasis in all situations nor can all be achieved with the same level of accuracy. The main objectives are:

1. Primary:
   * to obtain good *aggregate* network measures, e.g. total motorway flows, total revenue by bus service;
   * to estimate zone-to-zone travel costs (times) for a given level of demand;
   * to obtain *reasonable* link flows and to identify heavily congested links.
2. Secondary:
   * to estimate the routes used between each O–D pair;
   * to analyse which O–D pairs use a particular link or route;
   * to obtain turning movements for the design of future junctions.

In general terms we shall attain the primary objectives more accurately than the secondary ones. Even within objectives we are likely to be more accurate with those earlier in the list. This is essentially because our models are more likely to estimate aggregate than disaggregate values correctly.

The basic inputs required for assignment models are:

* A trip matrix expressing estimated demand. This will normally be a peak-hour matrix in urban congested areas, and perhaps other matrices for other peak and off-peak periods. A 24-hour matrix is sometimes used for assignment of uncongested networks. The conversion of 24-hour matrices into single hours is seldom satisfactory in terms of congestion, as these matrices are symmetric and single-hour trips seldom are. The matrices themselves may be available in terms of person trips; therefore, they should be converted into vehicle trips as capacity and speed–flow relationships are described in these terms.
* A network, namely links and their properties, including speed–flow curves.
* Principles or route selection rules thought to be relevant to the problem in question.

The traffic assignment methods involve a set of rules on how to identify desirable routes (fastest, lowest generalised cost) to connect origin to destination and then a systematic way of allocating O-D trips to these routes so that certain features of reality are achieved. In the next sections we will discuss these methods from a practical viewpoint identifying its strengths and weaknesses. In Chapter 11 we will adopt a more rigorous approach setting up the assignment task as an optimisation problem and discussing solution algorithms in a more systematic way.

10.2.2   Route Choice

The basic premise in assignment is the assumption of a rational traveller, i.e. one choosing the route which offers the least perceived (and anticipated) individual costs. A number of factors are thought to influence the choice of route when driving between two points; these include journey time, distance, monetary cost (fuel and others), congestion and queues, type of manoeuvres required, type of road (motorway, trunk road, secondary road), scenery, signposting, road works, reliability of travel time and habit. The production of a *generalised* cost expression incorporating all these elements is a difficult task. Furthermore, it is not practical to try to model all of them in a traffic assignment model, and therefore approximations are inevitable.

The most common approximation is to consider only two factors in route choice: *time* and *monetary cost*; further, monetary cost is often deemed proportional to travel distance. The majority of traffic assignment programs allow the user to allocate weights to travel time and distance in order to represent drivers' perceptions of these two factors. The weighted sum of these two values then becomes a generalised cost used to estimate route choice. There is evidence to suggest that, at least for urban car traffic, time is the dominant factor in route choice. Outram and Thompson (1978) compared drivers' stated objectives with their actual performance in route choice. They found that the proportion of drivers being successful in achieving their objectives was relatively low. They also found that the combination of time and distance gave the best explanation of route choice. However, even if we allow the combination of time and distance in a generalised cost function, we can only explain something of the order of 60 to 80% of the routes actually observed in practice. As the marginal contribution of other factors in untangling route choice is very small, the unexplained part must be attributed to factors like differences in perception, imperfect information on route costs or simply errors.

The fact that different drivers often choose different routes when travelling between the same two points may be ascribed to three different types of reasons:

1. Differences in individual perceptions of what constitutes the 'best route'; some may wish to minimise time, others fuel consumption and many a combination of both and this introduces a variety in route choices.
2. The level of knowledge of alternative routes varies and this introduces apparent irrationality (from the point of view of the observer) in the choices.
3. Congestion effects affecting shorter routes first and making their generalised costs comparable to initially less attractive routes.

We normally handle the first issue through multiple user classes, the second through 'stochastic effects' and the third one via congested assignment and equilibrium.

Example 10.1   Consider an idealised town with a low-capacity through route (1000 vehicles per hour) and a high-capacity bypass, as in Figure 10.2. The bypass is a longer but faster route with a capacity of 3000 vph. Assume that during the morning peak 3500 drivers approach the town and that everyone would like to use the shortest route, i.e. via the town centre. It is clear that it would not be possible

Figure 10.2 Town served by a bypass and a town centre route

for all of them to do so as the route would become too congested even before its ultimate capacity is reached. Many would opt then for second choice to avoid long queues and delays. Presumably drivers would experiment with the two routes until they find a more or less stable arrangement when none can improve their time by switching to the other route. This is a typical case of Wardrop's equilibrium, which is discussed in greater detail below. Diversion across routes in this case is due to *capacity restraint*.

However, not all 3500 drivers will think alike; some would always prefer the bypass because of its uninterrupted flow conditions or its scenery, where as others would value other features of the town-centre route. These differences in objectives can be modelled using multiple user classes. The differences in perceptions and knowledge would also lead to a spread of routes and such effect is customarily referred to as the *stochastic* element in route choice.

Particular types of models are more suited to representing one or more of the above influences. A possible classification of traffic assignment methods is given in Table 10.2. The details and characteristics of each method are discussed below.

Each assignment method has several steps which must be treated in turn. Their basic functions are:

- To identify a set of routes which might be considered attractive to drivers; these routes are stored in a particular data structure called a *tree* and therefore this task is often called the *tree-building stage*.
- To assign suitable proportions of the trip matrix to these routes or trees; this results in flows on the links in the network.
- To search for convergence; many techniques follow an iterative pattern of successive approximations to an ideal solution, e.g. Wardrop's equilibrium; convergence to this solution must be monitored to decide when to stop the iterative process.

Table 10.2 Classification scheme for traffic assignment

		Stochastic effects included?	
		No	Yes
Single user class	No capacity restraint	All-or-nothing	Pure stochastic: Dial's, Burrell's
	With capacity restraint	Wardrop's equilibrium	Stochastic user equilibrium SUE
Multiple user classes	No capacity restraint	All-or-nothing with multiple user classes	Multiple user classes stochastic: Dial's, Burrell's
	With capacity restraint	Wardrop's equilibrium with multiple user classes	Stochastic user equilibrium with multiple user classes

10.2.3 Tree Building

Tree building is an important stage in any assignment method for two related reasons. First, it is performed many times in most algorithms, at least once per iteration. Second, a good tree-building algorithm can save a great deal of computer time and costs. By a good algorithm we mean an efficient one which is also well programmed in a suitable language. Van Vliet (1978) has produced a good discussion of the most widely used algorithms for tree building and this section is based on his paper.

There are two basic algorithms in general use for finding the shortest (cheapest) paths in road networks, one due to Moore (1957) and one due to Dijkstra (1959). The two will be discussed using a more convenient node-oriented notation: the length (cost) of a link between A and B in the network is denoted by $d_{A,B}$. The path or route is defined by a series of connected nodes, A-C-D-H, etc., whilst the length of the path is the arithmetic sum of the corresponding link lengths in the path. Let d_A denote the minimum distance from the origin of the tree S to the node or centroid A; P_A is the *predecessor* or *backnode* of A so that the link (P_A, A) is part of the shortest path from S to A.

The procedure for building a minimum path tree from S to all other nodes may be described as follows:

Initialisation Set all $d_A = \infty$ (a suitable large number depending on computer and compiler) except d_S which is set equal to 0; set up a *loose-end table* L to contain nodes already reached by the algorithm but not fully explored as predecessors for further nodes. They are the tip of the tree as branches grow to reach all nodes. Initialise all entries L_i in L to zero, and all P_A to a suitable default value.

Procedure Starting with the origin S as the 'current' node $= A$:

1. Examine each link (A, B) from the current node A in turn and, if $d_A + d_{A,B} < d_B$ then set a new value for $d_B = d_A + d_{A,B}$, make $P_B = A$ and add B to L;
2. Remove A from L, if the loose-end table is empty, stop; otherwise,
3. Select another node from the loose-end table and return to step 1 with it as the current node.

Three comments should be made at this stage. First, routes are in general not allowed to use centroids; therefore in step 1, B would not be added to L if it was a centroid. Second, the essential difference between Moore's and Dijkstra's algorithms lies in the procedure for selecting a node from L. Moore selects the top entry, that is the oldest entry in the table; Dijkstra selects the node nearest to the origin, i.e. the node L_i such that d_{L_i} is a minimum. This requires some additional calculations (including sorting of nodes) but ensures that each link is examined once and only once. It is well known that Dijkstra's algorithm is superior to Moore's, in particular for larger networks; it is however, more difficult to program. Finally, trees are often stored in the computer in one of two forms: as a set of ordered *backnodes* in which A is the backnode of B if link (A, B) forms part of the tree; or as a set of *backlinks* with a similar definition.

Van Vliet (1977) also identified a lesser known algorithm which performs very well even in large networks: D'Esopo's algorithm, as described and tested by Pape (1974). D'Esopo's uses a 'two-ended' loose-end table so that node B is entered at one or other end depending on its 'status'. If B had not been previously reached by the tree then it is entered at the bottom of L; if it is currently on the table no entry is made; but, if it has already been entered to L, examined and removed from the table then it is entered at the top. A simple array can be used to record the status with three potential values ($+1$, 0 or -1) representing each case for each node. As shown by Van Vliet (1977), D'Esopo's algorithm can reduce CPU times by 50% relative to Moore's. Furthermore, its performance is very close and often better compared with that of the best implementations of Dijkstra's; it has the added advantage of being much simpler to program.

Trees have two important additional uses in transport planning. They are often employed to extract cost information in a network. For example, the total travel time between two zones can be obtained by following the sequence of links in the tree connecting them and accumulating their travel times. This operation is often referred to as 'skimming' a tree. Trees built for, say, travel time can be skimmed for other attributes, for example generalised cost, distance, number of nodes, etc. Trees can also be used to produce information on which O–D pairs are likely to use a particular link. This facility, often called a 'selected link analysis', permits the identification of who is likely to be affected by a network change. Moreover, it can also be used to cordon a trip matrix for a smaller study area; in this case the selected links are used to identify entry and exit points to the small study area and the trees to combine the original zones into single external ones for the new sub-area.

10.3   All-or-nothing Assignment

The simplest route choice and assignment method is 'all-or-nothing' assignment. This method assumes that there are no congestion effects, that all drivers consider the same attributes for route choice and that they perceive and weigh them in the same way. The absence of congestion effects means that link costs are fixed; the assumption that all drivers perceive the same costs means that every driver from i to j must choose the same route. Therefore, all drivers are assigned to one route between i and j and no driver is assigned to other, less attractive, routes. These assumptions are probably reasonable in sparse and uncongested networks where there are few alternative routes and they are very different in cost.

The assignment algorithm itself is the procedure that *loads* the matrix $\mathbf{T}$ to the shortest path trees and produces the flows $V_{A,B}$ on links (between nodes A and B). All load algorithms start with an initialisation stage, in this case making all $V_{A,B} = 0$ and then apply one of two basic variations: pair-by-pair methods and once-through approaches.

Pair-by-pair   This is probably the simplest but not necessarily the most efficient method. In this case we start from an origin and take each destination in turn. First, we initialise all $V_{A,B} = 0$. Then for each pair (i, j):

1. Set B to the destination j;
2. If (A, B) is the backlink of B then increment $V_{A,B}$ by T_{ij}, i.e. make $V_{A,B} = V_{A,B} + T_{ij}$;
3. Set B to A;
4. If $A = i$ terminate (i.e. process the next (i, j) pair), otherwise return to step 2.

Once-through   This is sometimes called a 'cascade' method as it loads accumulated flow from nodes to links following the minimum cost trees from an origin i. Let V_A be the cumulative flow at node A:

1. Set all $V_A = 0$ except for the destinations j for which $V_j = T_{ij}$.
2. Set B equal to the most distant node from i.
3. Increment V_A by V_B where A is the backnode of B, i.e. make $V_A = V_A + V_B$.
4. Increment $V_{A,B}$ by V_B, i.e. make $V_{A,B} = V_{A,B} + V_B$.
5. Set B equal to the next most distant node; if $B = i$ then the origin has been reached, begin processing the next origin, otherwise proceed with step 3.

In this form V_B represents the total number of trips from i passing through node B en route to destinations further away from i. By selecting nodes in reverse order of distance, each node is processed once only. This algorithm requires the trees to be stored in terms of backnodes ordered by distance from the origin.

Example 10.2 Consider the simple network in Figure 10.3 and its associated trip matrix: A-C = 400, A-D = 200, B-C = 300 and B-D = 100. Section (a) shows the travel costs (times) on each link; section (b) the corresponding trees based on these costs together with the contributions to the total flow after assignment; these are shown in section (c).

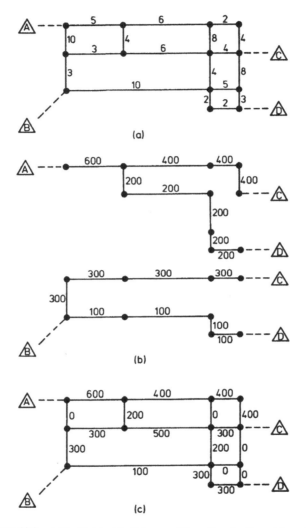

Figure 10.3 A simple network, its trees and flows from loading a trip matrix

All-or-nothing assignment is generally of limited interest to the planner; it may be used to represent some sort of 'desire line', i.e. what drivers would like to do in the absence of congestion. However, its main usefulness is as a basic building block for other types of assignment techniques, e.g. equilibrium and stochastic methods.

10.4   Stochastic Methods

Stochastic methods of traffic assignment emphasise the variability in drivers' perceptions of costs and the composite measure they seek to minimise (distance, travel time, generalised costs). Stochastic methods need to consider second-best routes (in terms of engineering or modelled costs); this generates additional problems as the number of alternative second-best routes between each O–D pair may be extremely large. Several methods have been proposed to incorporate these aspects but only two have relatively widespread acceptance: *simulation-based* and *proportion-based* methods. The first uses ideas from stochastic (Monte Carlo) simulation to introduce variability in perceived costs. The proportion-based methods, on the other hand, allocate flows to alternative routes from proportions calculated using logit-like expressions.

10.4.1   Simulation-Based Methods

A number of techniques use Monte Carlo simulation to represent the variability in drivers' perceptions of link costs; in particular, the method developed by Burrell (1968) has been widely used for many years. These techniques usually rely on the following assumptions:

- For each link in a network one should distinguish objective or engineering costs as measured/estimated by an observer (modeller) and subjective costs as perceived by each driver. It is further assumed that there is a distribution of perceived costs for each link with the engineering costs as the mean, as shown in Figure 10.4.

Figure 10.4   Distribution of perceived costs on a link

The various implementations of these ideas differ in their assumptions about the shape of these distributions: while Burrell's assumes a uniform distribution, other models hypothesise a Normal distribution. In either case one also needs to assume or calibrate a standard deviation or range for the distribution of perceived costs.

- The distributions of perceived costs are assumed to be independent.
- Drivers are assumed to choose the route that minimises their perceived route costs, which are obtained as the sum of the individual link costs.

A general description of these algorithms would be as follows. First, select a distribution (and spread parameter, σ) for the perceived costs on each link; then, split the population travelling along each O–D pair into N segments, each assumed to perceive the same costs.

1. Make $n = 0$.
2. Make $n = n + 1$.
3. For each $i - j$ pair:
   - Compute perceived costs for each link by sampling from the corresponding distributions of costs by means of random numbers.
   - Build the minimum perceived cost path from i to j and assign T_{ij}/N trips to it accumulating the resulting flows on the network.
4. If $n = N$ stop, otherwise go to step 2.

In practice many short-cuts are taken to reduce computation times, for example:

- Generate new sets of random costs per origin and not per O–D pair.
- Use N equal to just 3 or 5 and generate one set of random costs for each matrix and not for each O–D pair or origin.
- Use small values for N, even 1 in some circumstances.

This type of approach uses simulation in order to reduce the number of second-best routes to be considered. If a wider range of routes is thought necessary, one can increase the value of N and/or the spread parameter in the distribution of link costs. Burrell's approach has the advantage of generating cheap routes more often than more expensive ones: if a route is expensive it is much less likely to appear as the cheapest as a result of the stochastic variations in link costs. Although the uniform distribution is efficient in computer time, it is not very realistic. A better function, but more expensive in terms of CPU time, is the Normal distribution with variance proportional to the mean engineering costs.

As in all Monte Carlo methods, the final results are dependent on the series of random numbers used in the simulation. Increasing the value of N reduces this problem. There are, however, more serious difficulties with this approach:

- The link perceived costs are not independent, as drivers usually have preferences, for example, for motorway links or to avoid priority junctions or minor roads. The assumption of independence in perceived costs may lead to unrealistic switching between parallel routes connected by minor roads.
- No explicit allowance is made for congestion effects.

In compensation, these methods often produce a reasonable spread of trips, are relatively simple to program and do not require the choice or estimation of speed–flow relationships (which may turn out to be a problem in some cases).

10.4.2 Proportional Stochastic Methods

Virtually all these methods are based on a loading algorithm which splits trips arriving at a node between all possible exit nodes, as opposed to the all-or-nothing method which assigns all trips to a single exit node. Very often the implementation of these methods reverses the problem so that the division of trips at a node is actually based upon where the trips are coming from rather than where they are going to. Consider node B in Figure 10.5; there are a number of possible entry points denoted by A_1, A_2, A_3, A_4 and A_5 for trips from I to J.

The 'splitting factors' f_i are defined by:

$$f_i = 0 \qquad \text{if } d_{A_i} \geq d_B$$
$$0 < f_i \leq 1 \quad \text{if } d_{A_i} < d_{B_i}$$

Figure 10.5   A node (B) and links feeding trips into it

where d_{Ai} represents the minimum cost of travel from the origin i to node A_i. The first condition requires that f_i should be zero if an entry node A_i is further from the origin than B, therefore ensuring that trips are allocated to routes which take them efficiently away from the origin. The trips T_B that pass through B are divided according to the equation:

$$F(A_i, B) = \frac{T_B f_i}{\sum_i f_i} \tag{10.13}$$

The assignment procedure is now equivalent to the cascade method for all-or-nothing assignment. Implementations of these ideas differ mainly in the way in which they define the splitting function f_i. The single-path method due to Dial (1971) requires that:

$$f_i = \exp(-\Omega \delta d_i) \tag{10.14}$$

where δd_i is the extra cost incurred in travelling from the origin to node B via node A_i rather than via the minimum cost route. In this way, if A_i is in the minimum-cost route, δd_i is equal to zero and $f_i = 1$. Nodes that lie on more expensive routes have $\delta d_i > 0$ and their f_i values are less than 1. In this way shorter routes are favoured over more expensive ones.

Dial originally described a double-pass algorithm which effectively uses a logit-type formulation to split trips from i to j among alternative routes r:

$$T_{ijr} = \frac{T_{ij} \exp(-\Omega C_{ijr})}{\sum_r \exp(-\Omega C_{ijr})} \tag{10.15}$$

The parameter Ω can be used to control the spread of trips among routes.

The algorithm involves a forward and a backward pass:

1. The forward pass: take each node A in ascending order of d_A and define a weight for each exit link (A, B) such that:

   $w_{(A, B)} = W_A \exp(-\Omega \delta d_{(A, B)})$ if $d_A < d_B$ or zero otherwise; W_A is the accumulated weight at A defined as:

   $$W_A = \sum_{A'} w_{(A', A)} \text{ and } W_I = 1 \quad [A' \text{ is a predecessor of A}]$$

2. The backward pass: identical to the single-pass algorithm with the exception that the weights $w_{(A, B)}$ are used to work out the split of trips rather than the splitting factors f_i.

Example 10.3 A practical problem with Dial's assignment is that it assumes that all routes are equally likely candidates and for this reason it is biased against trunk routes as opposed to secondary links. Consider the problem of a town served by a bypass and a town-centre route with three small variations as illustrated in Figure 10.6. Assume also that there are 4000 trips from A to B and that all routes have approximately the same cost.

Figure 10.6 Town served by a bypass and three city-centre routes

   In this case Dial's algorithm would split the 4000 trips as follows: 1000 via the bypass and 1000 via each of the town-centre routes. However, most users would regard this problem as one with only two alternatives: bypass or town centre. Recall the discussion about the independence of irrelevant alternatives property of the logit model in Chapter 5. Dial's runs into trouble when it considers every possible route even if some permutations or combinations of links may differ just in a few percentage points of their total cost. In behavioural terms Dial ignores the correlation between similar routes. In practice, Dial tends to allocate more traffic to dense sections of the network with short links, compared with sparser parts of the network with relatively longer links. In fact, coding strategies for networks can affect the allocation of flows.

10.4.3   Emerging Approaches

Research is still active seeking to integrate stochastic assignment methods closer with developments in discrete choice. The problem is generally split into three components: (a) how to identify a feasible, efficient and distinct set of routes that would be considered by drivers when making their choices, (b) how to estimate the parameters of route choice models, and (c) how to integrate more efficiently the choice mechanism into an equilibrium assignment framework.

   An excellent paper by Prato (2009) reviews current approaches to accomplish these three tasks. The methods discussed remain mostly in the research realm but take advantage of the advances in discrete choice discussed in Chapters 7 to 9. One of the key problems is the difficulty associated with collecting good data, in particular for revealed preferences/choices. The provision of GPS units in an experimental setting may help to alleviate this constraint.

   Let us assume that trip makers limit their choices among a certain number (K) of minimum cost paths avoiding extremely costly alternatives. For both estimation and prediction purposes, we need to caution against the possibility of generating routes that are either too circuitous or very similar, as both types would be unattractive to drivers or not really perceived as different. A number of techniques have been developed to avoid these problems and generate acyclic and heterogeneous paths, but they still have to face two further problems. First, all drivers travelling between the same O-D pair will share the same generated choice set (and one would expect differences among them resulting from personal constraints and preferences). Second, the measures of route attractiveness are basically subjective relying on the experience of the researcher that controls their inclusion.

   An alternative choice set generation approach is based on doing repeated shortest path searches in the network using random extraction of link generalised costs and individual preferences from probability distributions. The various methods here (all of a heuristic nature) produce solutions that are stochastic and where O-D pairs are processed simultaneously (Prato 2009). Moreover, stochastic path generation is generally a case of importance sampling because the selection probability of a route depends on the properties of the route itself, such as length or travel time.

The original version of this approach is Burrell's method which we discussed earlier. Generalisations include the use of different probability distribution functions to sample from and different sequences for that sampling. A more interesting enhancement, the double stochastic approach, allows travellers to perceive path costs with error recognising that different drivers have different preferences. Accordingly, the generation function has a random term for the generalised cost function and a random term for the traveller taste heterogeneity. Bovy and Fiorenzo-Catalano (2007) propose a trip utility function as the basis for a doubly stochastic generation function. Relevant routes are created through optimal path searches in the network by stochastically varying network attributes and attribute preferences. Variation in link impedances reflects differences in the knowledge and perception of link attributes among travellers. Variation in the parameter values reflects differences among travellers in their utility function.

Constrained enumeration methods rely on the behavioural assumption that drivers choose routes according to rules other than the minimum cost path. Prato and Bekhor (2006) propose a branch and bound algorithm where the branching rule requires the definition of thresholds. A directional threshold excludes from consideration links that take the driver significantly away from the destination and closer to the origin of the trip. A temporal threshold rejects paths that travellers would consider unrealistic because of excessive travel time. Other thresholds discard routes that include large detours or have overlapping paths that travellers would not consider as separate alternatives.

The application of an assignment technique to these identified routes presents another problem. The sets of alternative routes generated with the described path generation techniques are usually quite large since all relevant routes are possibly included and some irrelevant routes are probably created. Intuitively, the number of alternatives in the choice set plays a role in the estimation of discrete choice models within the route choice context.

Accordingly, route choice models should exhibit robustness in utility parameter estimates with respect to choice set size. For estimation purposes, this model requirement would allow the definition of choice sets with a reasonable number of attractive alternatives in order to obtain reliable model estimates. Dense urban networks with many (say a 100) alternatives show a high degree of similarity among alternative routes. For this reason, most of the literature focuses on the correlation between alternatives, which alters choice probabilities of overlapping routes. This problem has been discussed at length in Chapters 7 and 8; we only present here some of the relevant implications for assignment.

Cascetta *et al.* (1996) propose a modification of the MNL model, in which a commonality factor measures the degree of similarity of each route with other routes in the choice set C. The expression of the probability P_k of choosing route k within the choice set C reflects the simple logit structure of the model:

$$P_k = \frac{\exp(V_k + \beta_{CF}CF_k)}{\sum_{l \in C} \exp(V_l + \beta_{CF}CF_l)} \tag{10.16}$$

where V_k and V_l are the utility functions of routes k and l, respectively, CF_k and CF_l are the commonality factors, and β_{CF} is a parameter to be estimated.

Ben-Akiva and Bierlaire (1999) propose the Path-Size Logit (PSL) model for an application of discrete choice theory for aggregate alternatives, already used in other transport contexts such as destination choice. In the PSL model, the expression of the probability of choosing route k within the alternative paths is:

$$P_k = \frac{\exp(V_k + \beta_{PS}PS_k)}{\sum_{l \in C} \exp(V_l + \beta_{PS}PS_l)} \tag{10.17}$$

Despite its similarity with (10.16) the interpretation is different and there are different expressions proposed for the path size. The CF factor reduces the attractiveness of a path because it shares elements of others, where the *PS* index identifies what proportion of a path is unique.

Generalised Extreme Value (GEV) models allow similarities within the stochastic part of the utility function and relate the network topology to the specific coefficients that characterise their tree structure. A small number of models using this approach have been suggested to handle probabilistic assignment; for example a Cross Nested Logit (Prashker and Bekhor 2000) or a Generalised Nested Logit (Bekhor and Prashker 2001). However, they have significant computational demands and use complex nested structures making them more difficult to implement.

One of the most attractive treatments of this problem uses the Mixed Logit (ML) model discussed in Chapter 7. Here the unobserved factors can be decomposed into a part that contains correlation and heteroscedasticity, and another part that is IID extreme value. The most straightforward derivation assumes that the probability for an individual n of choosing route k has the same form of the standard MNL, but it is conditional on the distribution of the coefficients β_n where $f(\beta)$ is the mixing distribution of β over the population. The unconditional probability is computed by simulation:

$$P_{nk} = \frac{1}{D} \sum_{d=1}^{D} \frac{\exp(\beta_d' X n_k)}{\sum_{l \in C_d} \exp(\beta_d' X_{nl})} \tag{10.18}$$

where β_d indicates a draw d from the distribution of β and D is the number of draws.

Another adaptation to route choice of the ML model assumes that the covariance of path utilities is proportional to the length by which paths overlap (Bekhor *et al.* 2002). Extending from the derivation of the ML model with factor analytic approach, the probability of choosing route k given a vector δ of standard Normal variables is given by:

$$P_k = \Lambda(k \mid \delta) = \frac{\exp(\mu(\beta X_k + F_k T \delta))}{\sum_{l \in C_n} \exp(\mu(\beta X_l + F_l T \delta))} \tag{10.19}$$

where $\beta(1 \times B)$ is the column vector of parameters, X_k is the k-*th* row of the matrix of explanatory variables $X(J \times B)$, F_k is the k-*th* row of the factor loadings matrix $F_{(J \times M)}$ (J paths and M network elements), $T_{(M \times M)}$ is a diagonal matrix of covariance parameters σ_m, $\delta_{(M \times 1)}$ is a vector of standard Normal variables. Bekhor *et al.* (2002) assume that the link-specific factors are IID Normal and that the variance is proportional to the link length; the F matrix corresponds to the link-path incidence matrix and the T matrix corresponds to the link-factor variance matrix. Accordingly, the covariance parameter σ shared by each link is estimated. Other variations on this theme have been proposed but they are computationally demanding and there are difficulties in obtaining significant estimates for the parameters.

In summary, currently emerging route choice models offer advantages and disadvantages. From a computational perspective MNL modifications, such as PSL (10.17), are not challenging, but GEV models are more demanding because of the estimation of structural coefficients within complex model structures. ML models introduce additional complexity because of the need to simulate choice probabilities and the absence of an equivalent mathematical formulation of the Stochastic User Equilibrium problem. From a behavioural perspective GEV and ML models depend on theoretical formulations of the correlation structure among alternative routes. GEV models seem preferable because of the superior theoretical foundation with respect to the MNL-modifications and relatively lighter computational demands compared to ML models.

10.5   Congested Assignment

10.5.1   Wardrop's equilibrium

If one ignores stochastic effects and concentrates on capacity restraint as a generator of a spread of trips on a network, one should consider a different set of models. For a start, capacity restraint models have to make use of functions relating flow to the cost (time) of travel on a link. These models usually attempt, with different degrees of success, to approximate to the equilibrium conditions as formally enunciated by Wardrop (1952) as a 'criterion':

> *The journey times on all routes actually used are equal, and less than those which would be experienced by a single vehicle on any unused route.*

This was later on expressed more formally as:

> *Under equilibrium conditions traffic arranges itself in congested networks in such a way that no individual trip maker can reduce his path costs by switching routes.*

If all trip makers perceive costs in the same way and seek the same objective (single user class, no stochastic effects):

> *Under equilibrium conditions traffic arranges itself in congested networks such that all used routes between an O–D pair have equal and minimum costs while all unused routes have greater or equal costs.*

This is usually referred to as Wardrop's first principle, or simply Wardrop's equilibrium. It is easy to see that if these conditions did not hold, at least some drivers would be able to reduce their costs by switching to other routes.

Example 10.4 Consider again the case of a bypass and a single town-centre route of Figure 10.2. Assume now that the absolute capacity restriction for each route is replaced with two corresponding time–flow relationships as illustrated in Figure 10.7.

Figure 10.7   Time–flow relationships for Figure 10.2

The flows on the two routes will satisfy Wardrop's equilibrium when the corresponding 'costs' are identical. In this case it is relatively simple to write two equations for travel time versus flow and equate them to find the equilibrium solution, for example:

$$t_b = 15 + 0.005 V_b \qquad (10.20a)$$

$$t_t = 10 + 0.02 V_t \qquad (10.20b)$$

where t_b and t_t are travel 'costs' (time in minutes) via the bypass and the town-centre routes respectively, and V_b and V_t are their corresponding flows.

By equating t_b to t_t it is possible to find, in this simple case, the direct solution to Wardrop's equilibrium as a function of the total flow $V_b + V_t = V$:

$$15 + 0.005 V_b = 10 + 0.02 (V - V_b)$$

that is:

$$V_b = 0.8 \, V - 200 \qquad (10.21)$$

Expression (10.21) has meaning only for non-negative flows, i.e. for V greater than or equal to $200/0.8 = 250$. For $V < 250$, $Ct < C_b$, $V_b = 0$ and $V_t = V$, i.e. all traffic chooses the town-centre route. For situations where $V > 250$ the two routes will be used; for example, the reader can verify that for $V = 2000$ the equilibrium flows are $V_b = 1400$ and $V_t = 600$ and the 'costs' by each route are 22 min.

The same idea would apply to flows on networks where the costs of travel by each of the routes used between two points are the same under Wardrop's equilibrium. The problem is, of course, that in anything but the simplest cases it is not possible to solve the equilibrium flows algebraically; rather an algorithmic solution method is required.

Several techniques have been proposed as reasonable approximations to Wardrop's equilibrium: some of them are simple heuristic approaches and the most interesting ones follow a more rigorous mathematical programming framework. In order to compare these algorithms against each other the following properties are of interest:

- Is the solution stable?
- Does it converge to the correct solution (Wardrop's equilibrium)?
- Is it efficient in terms of computational requirements?

The indicator δ, defined in the following equation, is often used to measure how close a solution is to Wardrop's equilibrium:

$$\delta = \frac{\sum\limits_{ijr} T_{ijr}(C_{ijr} - C_{ij}^*)}{\sum\limits_{ij} T_{ij} C_{ij}^*} \qquad (10.22)$$

where $C_{ijr} - C_{ij}^*$ is the excess cost of travel over a particular route relative to the minimum cost of travel for that (i, j) pair. These costs are calculated after the last iteration has been performed and total flows obtained for each link. Therefore δ is a measure of the total cost of excess travel via less than optimal routes, with denominator introduced so that the measure is recorded in relative rather than absolute terms.

Wardrop (1952) proposed an alternative way of assigning traffic onto a network and this is usually referred to as his second principle:

Under social equilibrium conditions traffic should be arranged in congested networks in such a way that the average (or total) travel cost is minimised.

This is a *design* principle, in contrast with his first principle which endeavours to model the behaviour of individual drivers trying to minimise their own trip costs. The second principle is oriented towards transport planners and engineers trying to manage traffic to minimise travel costs and therefore achieve an optimum *social equilibrium*. In general the flows resulting from the two principles are not the same but one can only expect, in practice, traffic to arrange it following an approximation to Wardrop's first principle, i.e. *selfish* or *users' equilibrium*. It is interesting to note that the basic objective of congestion charging and road pricing is to get closer to Wardrop's second principle. Indeed, most methodologies for pricing congestion start by assessing what tolls should be charged on each link to achieve this equilibrium.

10.5.2 Hard and Soft Speed-Change Methods

Some of the first heuristic methods still maintained the idea of assigning all trips per O–D pair to a single route (all-or-nothing assignment), but acknowledged the fact that speeds, and therefore travel times, responded to flow levels. The simplest of these methods involves just recalculating link travel times after an all-or-nothing assignment so that they are consistent with the current flow levels. A new all-or-nothing assignment is then performed with the new costs and trees. It is easy to see that in general this is a poor approach as the chosen routes will oscillate and the flow pattern will, in general, never converge. In the case of the town-centre bypass problem of Example 10.4 with, say, $V > 250$, the flows would oscillate between all via the town centre in one iteration and all via the bypass in the next one. This phenomenon will be repeated in larger networks although in some cases it may be more difficult to identify.

In an attempt to dampen such route and flow oscillations it has been proposed to use an average speed of two or more all-or-nothing assignments to perform the next iteration. This if often called a *soft* speed change as opposed to the *hard* speed change of the original method. However, this may only provide an apparent improvement as the main weakness of these two approaches is that they still assign all traffic to a single route for each O–D pair, therefore contradicting Wardrop's principle. Taking again the case of Example 10.4, it can easily be seen that the soft speed-change method will still load all traffic alternatively via one route and then the other in the next iteration. Both methods produce unstable solutions, are inherently non-convergent and the use of soft speed changes will only attempt to disguise this fact in larger networks.

10.5.3 Incremental Assignment

This is a more interesting and realistic approach. In this case the modeller divides the total trip matrix **T** into a number of fractional matrices by applying a set of proportional factors p_n such that $\Sigma_n\, p_n = 1$. The fractional matrices are then loaded, incrementally, onto successive trees, each calculated using link costs from the last accumulated flows. Typical values for p_n are: 0.4, 0.3, 0.2 and 0.1. The algorithm can be written as follows:

1. Select an initial set of current link costs, usually free-flow travel times. Initialise all flows $V_a = 0$; select a set of fractions p_n of the trip matrix **T** such that $\Sigma_n\, p_n = 1$; make $n = 0$.
2. Build the set of minimum cost trees (one for each origin) using the current costs; make $n = n + 1$.
3. Load $\mathbf{T}_n = p_n\, \mathbf{T}$ all-or-nothing to these trees, obtaining a set of auxiliary flows F_a; accumulate flows on each link:

$$V_a^n = V_a^{n-1} + F_a$$

4. Calculate a new set of current link costs based on the flows V_a^n; if not all fractions of T have been assigned proceed to step 2; otherwise stop.

This algorithm does not necessarily converge to Wardrop's equilibrium solution even if the number of fractions p is large and the size of the increments (p_n **T**) is small. Incremental loading techniques suffer from the limitation that once a flow has been assigned to a link it is not removed and loaded onto another one; therefore, if one of the initial iterations assigns too much flow on a link for Wardrop's equilibrium conditions to be met (for example, because the link is short but has very low capacity), then the algorithm will not converge to the correct solution.

However, incremental loading has two advantages:

- it is very easy to program;
- its results may be interpreted as the build-up of congestion for the peak period.

Example 10.5 Consider again the problem of the two routes, town centre and bypass, of Example 10.4. We split the demand of 2000 trips into four increments of 0.4, 0.3, 0.2 and 0.1 of this demand, i.e. 800, 600, 400 and 200 trips. At each increment we calculate the new travel costs using equations (10.16). The following table summarises the results of this algorithm:

N	Increment	Flow town	Cost town	Flow bypass	Cost bypass
0	0	0	10	0	15
1	800	800	26	0	15
2	600	800	26	600	18
3	400	800	26	1000	20
4	200	800	26	1200	21

It can be seen that the algorithm does not converge, in this case, to the correct equilibrium solution. This is because once the wrong flow (800) has been loaded onto the town-centre route, this method cannot reduce it; therefore the flow and cost via the town centre remain overestimated. As a matter of interest, the value of the δ indicator for the solution above is:

$$\delta = [800\,(26 - 21) + 1200\,(21 - 21)]\,/(2000 \times 21) = 0.095$$

The reader can verify that using smaller increments would produce closer solutions to true equilibrium. Note that if one starts with an increment of 0.3 times the total demand, the solution is true equilibrium; however, this is just a chance occurrence in this case.

10.5.4   Method of Successive Averages

Iterative algorithms were developed, at least partially, to overcome the problem of allocating too much traffic to low-capacity links. In an iterative assignment algorithm the 'current' flow on a link is calculated as a linear combination of the current flow on the previous iteration and an auxiliary flow resulting from an all-or-nothing assignment in the present iteration. The algorithm can be described by the following steps:

1. Select a suitable initial set of current link costs, usually free-flow travel times. Initialise all flows $V_a = 0$; make $n = 0$.
2. Build the set of minimum cost trees with the current costs; make $n = n + 1$.
3. Load the whole of the matrix **T** all-or-nothing to these trees obtaining a set of auxiliary flows F_a.

4. Calculate the current flows as:

$$V_a^n = (1 - \phi)V_a^{n-1} + \phi F_a$$
$$\text{with } 0 \le \phi \le 1 \tag{10.23}$$

5. Calculate a new set of current link costs based on the flows V_a^n. If the flows (or current link costs) have not changed significantly in two consecutive iterations, stop; otherwise proceed to step 2. Alternatively, the indicator δ in (10.22) could be used to decide whether to stop or not. Another, less good but quite common, criterion for stopping is simply to fix the maximum number of iterations; δ should be calculated in this case as well to know how close the solution is to Wardrop's equilibrium.

Iterative assignment algorithms differ in the method used to give a value to ϕ. A simple rule is to make it constant, for example $\phi = 0.5$. A much better approach due to Smock (1962), is to make $\phi = 1/n$. The reader may verify that equal weight is given to each auxiliary flow F_a in this case; for this reason, the algorithm is also known as the method of successive averages (MSA). It has been shown (see, for example, Sheffi 1985) that making $\phi = 1/n$ produces a solution convergent to Wardrop's equilibrium, albeit not a very efficient one. As we shall see in Chapter 11, the Frank–Wolfe algorithm estimates optimal values for ϕ in order to guarantee and speed up convergence.

Example 10.6 Consider the same bypass versus town-centre problem of Example 10.5 and use $\phi = 1/n$. The following table summarises the steps in the MSA algorithm.

Iteration		ϕ	Flow town	Cost town	Flow bypass	Cost bypass
1	F		2000		0	
	V^n	1	2000	50	0	15
2	F		0		2000	
	V^n	1/2	1000	30	1000	20
3	F		0		2000	
	V^n	1/3	667	23.3	1333	21.7
4	F		0		2000	
	V^n	1/4	500	20	1500	22.5
5	F		2000		0	
	V^n	1/5	800	26	1200	21
6	F		0		2000	
	V^n	1/6	667	23.3	1333	21.7
7	F		0		2000	
	V^n	1/7	572	21.4	1428	22.1
8	F		2000		0	
	V^n	1/8	750	25	1250	21.25
9	F		0		2000	
	V^n	1/9	667	23.3	1333	21.7
10	F		0		2000	
	V^n	0.1	600	22	1400	22

It can be seen that it takes a number of iterations to approximate to the right solution. Of course, the value of δ after iteration 10 is zero in this case. However, the reader will note that the algorithm was close to the correct equilibrium solutions in iterations 3, 6 and 9 but only reached it in iteration 10. This is due to the rigid nature of the rule to calculate ϕ. For more realistic networks the number of iterations needed to reach satisfactory convergence may be very high.

Another lesson from this simple example is that fixing the maximum number of iterations is not a good approach from the point of view of evaluation. Link and total costs can vary considerably in successive iterations and this may affect the feasibility of a scheme.

10.5.5 Braess's Paradox

The basic ideas about Wardrop's first and second principles are often illustrated using Braess's Paradox; although strictly speaking not a paradox it is nearly as famous as the 'blue bus/red bus' conundrum. The paradox was first proposed by Dietrich Braess in 1968 but it is mostly known through its translation into English in Braess *et al.* (2005). It demonstrates that under certain conditions adding capacity to a road network when drivers seek to minimise their own costs can actually make everybody worse off.

Consider the simple network depicted in Figure 10.8.

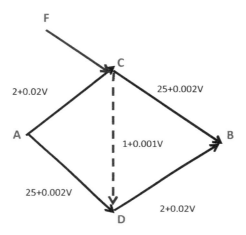

Figure 10.8   A simple network to illustrate Braess's Paradox

The linear relationship associated with each link represents the travel time-flow formulation in minutes. Solid arrows indicate existing links and the dotted arrow a planned link. Assume first that there are 1000 cars wishing to travel between A and B and none from F. The logical route choice under these conditions is for 500 cars to use the ACB route and the other 500 the ADB route. Both costs are the same: $2 + 10 + 25 + 1 = 38$ min. Consider now what happens when a new, high capacity link, is built between C and D. Under these conditions all drivers would choose to start on the AC path as under the most loaded conditions it would cost $2 + 20 + 1 + 1 = 24$ min to reach D when it takes at least 25 min if the AD route is used. At C, and for the same reasons, every rational driver would take the CD route as it would take at most $1 + 1 + 2 + 20 = 24$ min to reach B, one minute less than the most optimistic conditions for the CB route.

The total cost from A to B for each driver would then be $2 + 20 + 1 + 1 + 2 + 20 = 46$ min. In effect, 8 min longer than before the link CD was built. If all drivers could agree not to use the CD link they would all be better off. However, if starting from the original position (500 on each route) one driver chooses to use link CD he would be better off as from C it would only take $1 + 0.001 + 2 + 10.02 = 13.021$ to reach B, much less than the $25 + 1$ that the CB route offers.

So, the ACDB path represents a selfish equilibrium condition (Wardrop's First Principle) but this is such that everybody is worse off than before the new link was built. If, by any chance, there are 1000 vehicles travelling from F to D the travel time on that link would be at least 2 min. Now the choices open to our original drivers are a bit less clear. Starting from A and assuming the worst conditions (all drivers choose the ACD route) the cost of reaching D is practically the same via the new link or the old AD route (25 min). This would suggest that the original equilibrium could be restored and everybody benefit from 38 min journeys. However, if one driver at C chooses the new link he would benefit again with $2 + 0.001 + 2 + 10.02 = 24.021$ min to reach B instead of the expected $25 + 1$ for the direct route. This suggest that either drivers will find each day quite different conditions on their routes due to uncoordinated experimentation, with some days resulting in very poor choices, or that the new stable conditions will revert to all drivers using the centre route and spending now 47 min each day to reach B, a new equilibrium condition. Introducing a toll on link CD could make things better. What is the minimum toll that will produce the best selfish and social (Wardrop's Second Principle) equilibrium conditions? Of course, to avoid charging drivers from F unnecessarily, the toll should be imposed only to vehicles taking the ACD turning at the top.

If the conditions that lead to Braess's paradox happen in practice as well as in textbooks, it would be interesting to identify the perverse links (perhaps built because it was feasible rather than desirable) and to either toll them or close them to vehicular traffic. Steinberg and Zangwill (1983) developed necessary and sufficient conditions for Braess' paradox to occur when a new route or link is added. They concluded that these conditions were not unusual and that they were likely to occur in practice. Youn et al. (2008) studied the cities of New York, Boston and London, established routes where these conditions were likely to be present and pointed out roads that could be closed to traffic to reduce travel times.

10.6   Public-Transport Assignment

10.6.1   Introduction

In this section the problems associated with route choice and assignment for passengers using public-transport networks will be discussed. These problems are, in many ways, more difficult than those encountered by private-transport assignment; computer requirements tend to be heavier and even the best methods require important simplifying assumptions. Recent years have seen significant improvements in transit assignment techniques leading to better public-transport service provision and operational efficiency.

We shall discuss first the issues that make public-transport assignment different from private vehicle route choice; then, we will outline some of the approaches that have been implemented to tackle them in practice.

10.6.2   Issues in Public-Transport Assignment

10.6.2.1   Supply

The network of public-transport services is different from that of private cars. It includes, as links, sections of the bus or rail services running between two stops or stations. The concept of link capacity is now related to the capacity of each unit (bus, train) and its corresponding frequency. The travel time has an in-vehicle component as well as components for waiting at stops and walking to and from them. Many of the public-transport sections will use road links, e.g. most buses and some light rail-transit (LRT) services with street running. There will be other public-transport sections or services which will use completely different links, e.g. busways, segregated rail track, etc. The nature of these links generally produces a more complex network, an example of which is given in Figure 10.9.

Figure 10.9   An example of a public-transport network

10.6.2.2   Passengers

In public-transport route choice we are dealing with the movement of passengers and not of vehicles. Passengers can walk to a stop, interchange between two services and even drive part of the way to board a public-transport service later. This calls for the need to provide and specify walk and transfer links between different services, different public-transport modes (bus, rail) and between public-and private-transport facilities (e.g. 'Park & Ride').

10.6.2.3   Monetary Costs

In private car networks it is usually assumed that the monetary cost is directly associated to fuel consumption, which in turn is directly proportional to travel distance. These are both approximations but they are usually accepted as drivers do not perceive these costs in such a direct way as a passenger buying a ticket when starting a bus journey. Modern payment systems based on smart cards or mobile phones allow more complex fare structures and these have been introduced in many public-transport operations: fares variable with distance, flat fares (independent of distance travelled), zonal fares (for one or more specific geographic zones), combination and transfer tickets (valid for two or more services), time limit fares (e.g. valid for any number of boardings in an hour), daily, weekly and other season tickets for a fixed service or covering one or more zones and modes. This wide range of fares places difficult requirements on route choice and assignment models, as monetary costs do not depend directly on distance but in general on the location of the origin and destination, and on the route chosen.

10.6.2.4 The Definition of Generalised Costs

In the case of public-transport assignment the generalised cost of travelling may be defined as follows:

$$C_{ij} = a_1 t_{ij}^v + a_2 t_{ij}^w + a_3 t_{ij}^t + a_4 t_{ij}^n + a_1 \delta^n + a_5 F_{ij} \tag{10.24}$$

where

t_{ij}^v is in-vehicle travel time between i and j,
t_{ij}^w is walking time to and from stops (stations),
t_{ij}^t is waiting time at stops,
t_{ij}^n is interchange time,
δ^n is an intrinsic 'penalty' or resistance to interchange, measured in time units (typically around 5 generalised min),
F_{ij} is fare charged to travel between i and j, and
a_1 to a_5 are coefficients associated to the elements of generalised cost above.

Usually either a_1 or a_5 is equal to 1.0 in order to measure generalised costs in time or monetary units respectively. Again, it is usual to find that a_2, a_3 and a_4 are taken to be two to three times the value of a_1 as passengers dislike a minute spent walking or waiting more than if spent travelling in-vehicle.

In modelling terms, the software should be able to handle these variables and produce good estimates of each of the component times (in-vehicle, walking, waiting, transfer) if they are not provided externally. In-vehicle travel time depends on the speed attainable and the number and duration of stops en route; walking time, which depends on proximity to the best stop, is in some cases approximated by an average value for a whole zone; interchange time depends on station/stop configuration and separation; waiting time depends essentially on the frequency of the service and its reliability. A general formulation for waiting time is:

$$t^w = \frac{(h^2 + \sigma^2)}{2h} \tag{10.25}$$

where h is the expected headway of the service and σ its standard deviation (the less regular a service, the greater the expected waiting time). This formulation assumes that passengers arrive at random at the stop and that no passenger fails to board the next bus because of lack of space in it. This 'bus congestion' problem is difficult to solve but algorithms incapable of handling it will tend to produce unrealistic loadings in terms of actual service capacity, see De Cea and Fernández (1989). If the service is perfectly regular, i.e. $\sigma = 0$, then the expected waiting time is half of the headway. It is known, however, that if the frequency of the service is low, passengers will try to arrive just a few minutes before the next departure, thus setting an upper limit to the expected waiting time of perhaps 5 to 10 minutes; how close to the timetabled departure are passengers aiming to come will depend, of course, on the reliability of the service.

10.6.2.5   The Common Lines Problem

This is probably one of the most difficult and typical problems of public-transport assignment. The problem arises when for at least some O–D pairs there are sections in a path which have more than one parallel service offered and passengers can choose the one suiting them better. This choice is often not trivial for passengers ('I wish I had known that an express service was going to come three minutes after the slow one I have taken!'), nor simple from a modelling point of view. We are used to the idea that a driver chooses a single path from a choice set of all possible paths. In the case of public-transport passengers, they may choose a *set of paths* and let the vehicle that arrives first determine which of the paths they will actually use. The choice is therefore more complex and calls for a more detailed treatment.

A full review of the most suitable algorithms for public-transport assignment is outside the scope of this book. Instead, we shall discuss the main approaches to modelling route choice first and then assignment; not surprisingly, these different approaches result from the treatment they give to some of the issues above, in particular to the parallel or common lines problem, and to the choice of all-or-nothing, stochastic or capacity restraint-assignment methods.

10.6.2.6   Frequency or Schedule Based Route Choice

When the frequency of a public transport service is reasonably high, say every 10 min for an urban context and 15 or 20 min for the inter-urban case, travellers will, in general, not use or memorise a timetable (if it exists) but just turn up at the stop for a short waiting time. In these cases, it may be quite appropriate to use the frequency of the services as sufficient descriptor to estimate waiting times. However, this approach would be less appropriate for larger headways where trip makers are more likely to plan their access to arrive just a few minutes before the bus/train is due, according to their schedule (timetable). This can be taken into account by capping the waiting time to a maximum of, say, 10 min depending on context. However, this also fails to take full account of two situations. The first one is the provision in practice of irregular frequencies, for example a timetabled service at 5, 15, 20, 35, 45 and 50 minutes past the hour. The second one is the opportunity to provide well coordinated services even under low frequency schedules; for example, timing a half-hourly bus service to a rail station to arrive there 5 min before the train departs for a main destination.

10.6.3   Modelling Public-Transport Route Choice

It is worthwhile defining some terms such as route, line and section in a bit more detail before embarking on a discussion of the route choice problem in the presence of common lines.

A *public-transport (or transit) line*, or simply a *line*, is a fleet of vehicles that run between two points (terminals) on a network. They generally have the same characteristics of size, capacity, speed, etc. Vehicles stop at each node in their path to allow passengers to alight and board. Therefore, each transit line is defined by the vehicle characteristics, the sequence of nodes it serves and its frequency (or timetable).

A *line section* is any portion of a public-transport line between two, not necessarily consecutive, nodes.

A *public-transport route* is any path a user can follow on the transit network in order to travel between two nodes. The portion of a route between two consecutive transfer nodes is called a *route section*, and each route section has associated a set of *attractive* or *common lines*.

Consider now the simple case of an origin A and destination J connected by three transit services: lines 1, 2 and 3 as in Figure 10.10; they follow different routes and offer travel times of 20, 17 and 18 minutes to reach the desired destination. The frequency of each line is six services/hour; this means an expected waiting time of 5 min assuming perfectly regular services and random arrival of travellers. A traveller will then face three alternative segments in his journey (either from origin to destination or as

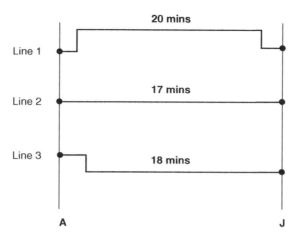

Figure 10.10   A basic section of public transport services showing travel times

a stage of a trip with one or more transfers): Line 1 with an expected travel time of 25 min (20 plus 5), Lines 2 and 3 with 22 and 23 min respectively.

A naive 'all-or-nothing' route choice will assign all travellers to Line 2 to minimise travel time. On the other hand, a more realistic approach would be to allocate the probability of boarding proportional to its frequency given that travellers are faced with actually 18 useful services per hour. Now the average waiting time is 3 min and 20 seconds (18/60) and the average travel time is 18 min and 20 seconds. The total expected travel time is 21 min 40 seconds. This approach produces a smaller expected travel time than the naive one if the travel times are similar (as they are when the lines follow the same sequence of nodes) but a large difference in travel times will result in a larger expected value: check with travel times of 17, 20 and 30 min.

Note that one can also build the network recognising that waiting (and walking) times are valued as about twice in-vehicle times (IVT) producing slightly different results above in terms of generalised times. In a longer route over a transit network we would also add transfer penalties and additional waiting times for some routes; moreover, in many cases additional fares would be charged with each transfer and this can also be added to the computation of generalised cost per link. Boarding penalties are often used to represent these effects.

Transit assignment methods can then be divided into:

• Naive all-or-nothing approaches that would only be acceptable for sparse and long distance travel networks.
• Multi-path approaches, for example the allocation of trips to paths proportional to the perceived service frequencies as outlined above.
• Equilibrium assignment methods with or without a stochastic element in them; these focus on congestion effects on public transport systems.

The all-or-nothing approach, despite its simplicity, may be very useful in refining a transit network, often a more subtle task than debugging a road network, a task that benefits from many good databases.

There are many versions of multipath approaches implemented in current software, some better than others at handling the issues discussed above. An interesting approach is one that allows travellers to adopt, as they do in many cases, a flexible strategy to reach their final destination. A *strategy* is a set of rules that allows the traveller to reach his destination.

Example 10.7 Consider the public-transport network of Figure 10.11; a simple strategy could be:

Figure 10.11   A simple public-transport network with transfers

- Take line 2 to stop H; transfer to line 3 and then exit at stop J;
    A more complex one may take the form:
- Wait up to 3 min for a line 5 vehicle or up to 4 min for a line 2 vehicle; otherwise take line 1; if line 5 is taken and you see a line 4 vehicle at stop F then board it and alight at J; if no line 4 vehicle at F continue to J; if line 2 vehicle was taken then transfer at H to line 4 if about to depart, otherwise wait for line 3 to reach J; etc.

In general terms a good flexible strategy will produce shorter expected travel times than the choice of the single path that minimises travel time; the choice of this single minimum path has been for many years the conventional approach to the problem.

In contrast, a more realistic flexible strategy allows the passenger to take advantage of the variability of waiting times and the opportunistic choice of a good, but low-frequency, service. This is well illustrated in Spiess and Florian (1989).

One can then define, for each node, the set of attractive lines that would be part of a good strategy to reach a given destination j. Given a strategy, an actual trip is then carried out according to a mechanism like:

1. Set i to origin node;
2. Board the first arriving vehicle from the set of attractive lines at i;
3. Alight at a predetermined node;
4. If not yet at destination, set i to the current node and return to step 2; otherwise the trip is completed.

Note that although this mechanism has a well-defined destination node, the origin is not part of the strategy. A strategy is the set of rules that enables travellers to reach their destination starting from any node in the network. This treatment is helped by the following additional notation:

S_{jk} = set of line sections connecting directly nodes j and k;
L_j^+ = set of outgoing (ingoing if $-$ instead of $+$ is used) line sections from node j;
v_s = flow on line section s;

t_s = in-vehicle travel time on line section s;
f_s = frequency associated to line section s;
g_j = number of trips going to destination node j;
V_{jk} = total flow on route section jk.

We can now identify the set of attractive routes emanating from node j using the dummy variable X_s which takes the value 1 if the line section s, belonging to the set of sections from j to k, is attractive, and zero otherwise. Then, for a given pair of nodes jk the associated values X_s ($s \in S_{jk}$) define the optimum or attractive set of lines towards k.

The total waiting time for users travelling from j to k can be written as:

$$w_{jk} = \frac{V_{jk}}{\sum\limits_{s \in S_j k} f_s X_s} \tag{10.26}$$

The problem of finding an optimum strategy for travelling from all origins to a destination can now be written as:

$$\text{Minimise} \sum_s v_s t_s + \sum_{jk} w_{jk} \tag{10.27}$$

subject to:

$$\sum_{s \in L_j^+} v_s + g_j = \sum_{s \in L_j^-} V_s \tag{10.28}$$

$$v_s = \frac{X_s f_s V_{jk}}{\sum\limits_{s \in S_i j} f_s X_s} = X_s f_s w_{jk} \tag{10.29}$$

The first term of the objective function (10.27) represents the in-vehicle travel time while the second is the total waiting time. This objective function is linear in the variables v_s and w_{jk} and the main problem seems to be generated by the non-linear constraints (10.29). Spiess (1983) has shown that these constraints can be relaxed as follows:

$$v_s \leq f_s w_{jk} \tag{10.30}$$

We can further introduce constraints (10.29) into the objective function:

$$\text{Minimise} \sum_{jk} \frac{V_{jk} \left\{ \sum\limits_s t_s X_s f_s + 1 \right\}}{\sum\limits_{s \in S_{ij}} f_s X_s} \tag{10.31}$$

subject to (10.28). This is a (0,1) hyperbolic programming problem.

Two different approaches can be followed here. The one proposed by Spiess and Florian (1989) is based on the linear programming version of this problem, whilst that proposed by De Cea and Fernández (1989) uses the hyperbolic programming (non-linear) formulation. If there are no congestion or capacity problems, the tasks above can be simplified as the set of optimal strategies will not depend on the actual flows. The Florian–Spiess algorithm has been implemented in EMME/2 (Babin *et al.* 1982) and the De Cea–Fernández algorithm in ESTRAUS (De Cea *et al.* 2005). Some tests show that the De Cea–Fernández approach is about 2.5 times faster than the Florian–Spiess method and nearly 50 times faster than the best conventional approach. This improvement in performance, which is crucial to model realistic size problems, is achieved at the cost of additional memory requirements, not a significant requirement today.

10.6.4 Assignment of Transit Trips

Once the best set of line segments to join origin and destination have been identified, one needs to consider the assignment of trips to them. Most programs seek to obtain a reasonable and realistic spread of trips among feasible routes. Conventional approaches, not dealing with the common lines problem explicitly, adopted a number of measures to generate this wider spread of trips. For example: to distinguish explicitly the different access points (bus stops, stations) for each zone and to build trees from each of them (and not just from the centroids) to all destinations. In this way several alternative routes are identified, one via each different access point. Passengers can then be assigned to these routes using a multinomial logit function of the costs of joining origin and destination via each path.

Spiess and Florian (1989) perform the assignment stage following the identified optimal strategies. This is achieved by assigning to each link the proportion of the volume accumulated to the upstream node that corresponds to the frequency served by the link. De Cea and Fernández (1989) follow a similar approach but in two stages:

1. First, once the set of common lines for all (i, j) pairs have been identified a new network is built on the basis of *nodes* and *route sections*. Note that route sections contain only the lines that minimise the total expected travel time for the section; they have an associated travel time (t_r) and a frequency (f_r) corresponding to the sum of the attractive frequencies (those in the common lines). With these two elements it is possible to obtain a composite cost of travelling along this route section and therefore an efficient private-transport tree-building algorithm can be used to find the best paths. Loading onto these trees results in a set of *route section flows* v_r.
2. Second, we can decompose the route section flows into their *line section* components:

$$v_s = \frac{f_s v_r}{f_r} \tag{10.32}$$

The treatment so far has not discussed the problems associated with special fare systems. If the fare system is proportional to the distance travelled, this is not a major problem as it is normally possible to convert it to time units and add them to the travel time on each link. However, this type of fare structure is hardly common. A flat fare system could also be accommodated but the treatment of more complex schemes (from a modelling point of view) may pose additional problems for algorithm design.

In most practical cases it will not be possible to model the whole complexity of fare systems and some approximate shortcuts will have to be taken in accordance with the most common type of ticket used. For example, in the case of a zonal fare system assignment may be performed on the basis of time alone and the fare cost added at the end. This may still ignore the importance of special pass holders but is probably good enough for places like London.

Finally, we must stress that public-transport assignment suffers, in general, from similar weaknesses to those identified for road networks. Furthermore, it is fair to say that congested assignment is less well developed for transit networks. There are two effects in play here: first, the limited capacity of the units (buses, trains) may prevent some travellers from implementing their optimal strategies, thus increasing their travel times; second, there is interaction between public transport and private cars sharing the same road network–increased traffic on one mode will affect travel times on the other as well. We will consider some approaches to deal with these issues in the next chapter.

10.7   Limitations of the Classic Methods

In previous sections we have described the most important classic methods for traffic assignment. Before considering more detailed and to some extent advanced methods, it is worthwhile reviewing what are seen as the main limitations of these approaches. These deficiencies may come from different sources.

10.7.1   Limitations in the Node-link Model of the Road Network

These include the fact that not all real road links are considered in the network (incomplete networks), 'end effects' due to the aggregation of trip ends into zones represented by single centroids, banned and penalised turning movements not specified in the network, and the fact that intra-zonal trips are not fully treated.

The main problem with incomplete networks arises in heavily congested areas where some of the medium- and long-distance trips will use minor roads as 'rat runs'; a new road scheme may relieve congestion and attract some of these rat-run trips, which will seem to be 'generated journeys' when they are not. Even when great care is taken in connecting the network to zone centroids, end effects are inevitable. These will make estimated link volumes less reliable in the vicinity of centroid connectors, probably overestimating the flows.

It is possible to expand simple nodes to represent all turning movements at a junction and then penalise or remove those links representing banned manoeuvres. An example of a fully expanded junction is given in Figure 10.12; any particularly difficult manoeuvre, e.g. an opposed turn, can then be penalised by associating a longer delay to it. Good software provides efficient ways of automatically expanding junction representations and banning or penalising movements; alternatively, this must be done by hand in the network-building stage itself. In either case, it is likely that some turning movements will not be properly treated.

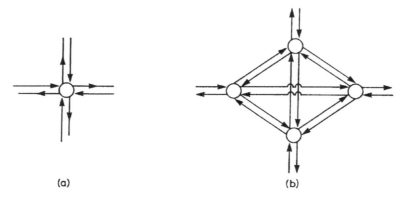

(a)   (b)

Figure 10.12   Representation of a junction as a simple node (a) and expanded showing all turning movements (b)

The treatment of intra-zonal movements is also a source of problems: some of them could make use of main links in the road network but they will not appear in the network model. It is difficult to devise a good method to account for them in assignment.

All of these problems are more difficult to handle when the zones are large and the network representation sparse. As usual, greater resolution in network and zonal definition will increase realism but at the cost of data collection, processing and interpretation.

10.7.2   Errors in Defining Average Perceived Costs

We do not have enough evidence about how these are likely to change with time, journey purpose, length of journey, income, predictability and the environment. Moreover, when we wish to forecast components of cost, for example fuel consumption, we rely on simplifying assumptions which may give rise to additional errors.

10.7.3   Not all Trip Makers Perceive Costs in the Same Way

Our stochastic methods are an approximation to this phenomenon but even they must limit the number of randomisations for reasons of economy. Another possibility is to consider several different user classes, each with its own set of perceived costs.

It is possible to express the deterministic equilibrium assignment problem with multiple user classes, each with its own set of parameters defining perceived link costs; see for example the work of Leurent (1998). Convergence to a unique solution is achieved under analogous conditions to those required for the single-user problem. Moreover, the problem and the solution can also be extended to elastic demand modelling (combined mode choice and assignment, for example). Most modern software packages offer this type of facility.

The modelling of multiple user classes (each with different willingness to pay for a better service) is often quite critical in demand studies for private sector facilities and services like high-speed rail links or toll roads. In the context of toll roads, some users may have high willingness to pay for services because their costs are covered by their employers; others may be very price-sensitive because of personal income or cash constraints. These different user classes can be well represented in these cases, although good stated preference/revealed preference studies will be required to determine the correct parameters for each model.

10.7.4   The Assumption of Perfect Information about Costs in all Parts of the Network

Although this is common to all models it is essentially overoptimistic, at least until the widespread use of road transport informatics makes more realistic modelling a possibility. Drivers have only partial information about traffic conditions on the same route last time they used it and on problems in other parts of the network depending on their own experience, disposition to explore new routes and the use of traffic information services. Moreover, there is evidence that many drivers are heavily influenced by road signs in their choice of route and that sometimes signed routes are not the cheapest (Wootton *et al.* 1981). Current methods ignore these effects. The future influence of variable message signs and more advanced route guidance technology over part of the vehicle fleet, is likely to place new requirements for traffic assignment methods (see several articles in this field in Papageorgiou, 1991).

10.7.5   Day-to-day Variations in Demand

These probably prevent true equilibrium ever being reached in practice. In that sense Wardrop's equilibrium represents 'average' behaviour if all travellers think alike and have perfect information. Its solution, however, has enough desirable properties of stability and interpretation to warrant its use in practice; however, it is still only an approximation to the traffic conditions on any one day.

In the same vein, there are time variations in demand and flow within each day. This makes 24-hour models very poor in terms of traffic assignment, and therefore travel times and costs. The use of peak and off-peak periods for modelling and assignment is essential in congested urban areas but even then we know that the build-up of congestion produces important changes in travel time in very short time frames. Moreover, a 10-minute delay in departure for the same journey may produce a much greater delay on arrival at the destination because of increased congestion in the network. The costs on links change dynamically in response to traffic: some drivers understand this well and plan their journeys accordingly; others lack the necessary experience. In reality, the route choice problem has strong time-dependent elements but practical dynamic assignment techniques are as yet in their first steps.

10.7.6 Imperfect Estimation of Changes in Travel Time with Changes in the Estimated Flow on Links

This is partly due to the nature of the cost–flow relationships used. As stated in section 10.1.3, it is normally assumed that the travel time on a link depends only on the flow on the link itself. At least in urban areas, the delay on a link depends in general on flow on other links too, for example at a priority junction, thus creating interaction effects. This assumption will be discussed later as it requires better delay models than those assumed in conventional cost–flow relationships.

10.7.7 The Dynamic Nature of Traffic

Most classic assignment methods assume the existence of a trip matrix that is valid over a modelling period, say one hour in the peak. Traffic is then assigned to the network under the assumption of steady state conditions over that period. In practice, however, traffic behaviour is dynamic and 'steady state' is only a useful simplification. Consider, for example, a road that provide access to a town centre and that most drivers will like to reach it around 9:00 AM. Figure 10.13 represents an idealised diagram of traffic along this road starting from a place 60 minutes away from the town centre.

Figure 10.13   Simplified traffic volumes on a link

As can be seen, traffic at each time is different from the average conditions assumed in any classic assignment model. In a real network, with more entry and exit points, real traffic is more like a series of 'surges' or 'waves' that interact at junctions and at bottlenecks generating a different set of optimal routes depending on the time of the day and how far ahead the user is able to estimate delays on alternative routes.

Figure 10.13 is somewhat of an oversimplification for illustration purposes. In reality the waves will be fatter and as more traffic joins the main road to a town centre the volumes will increase faster than suggested in the figure. The same phenomenon takes place in public transport systems. The most congested section in an underground will be closer to the most desirable destination. One of the consequences of this is that great care must be taken when allocating trips to a particular time interval (say 8 to 9 AM peak): a different result will be achieved in assignment if trips are allocated according to the time they start, the time they arrive to a destination or an average of the two. When public transport congestion is an issue it is advisable to allocate trips according to the time of arrival to their destination as this is where the most severe congestion usually takes place.

Moreover, a different set of conditions will be generated if there is a bottleneck limiting capacity to some 4000 cars/hour at a distance of 30 min from the town centre. In this case the time-profile of traffic will look more like that in Figure 10.14. In this case, not all traffic will be able to get through the bottleneck in one go; queues will build up that will be cleared once demand falls below the 4000 cars/hour limit.

Figure 10.14   Simplified Traffic volumes on a link with a capacity bottleneck

The assumption, prevalent in most classic assignment models, that the whole matrix will clear the network in the time interval modelled is likely to underestimate delays, even with very good volume-delay formulations. Ideally, a good traffic assignment model should be able to handle these dynamic queues and pass on demand to the next time interval with a better estimation of total delay.

10.7.8   Input Errors

The accuracy of an assignment model depends also on the accuracy of other elements in the transport model, in particular that of the trip matrix to be loaded. This matrix will inevitably contain many errors and discrepancies, whether this is a synthetic one obtained from a gravity model or a carefully observed one using extensive surveys. Errors in the conversions from passengers to vehicle trip matrices also limit the accuracy of traffic assignment. This conversion is usually assumed to be a uniform (and constant over time) occupancy rate for each type of vehicle and perhaps journey purpose. Simple observations will show that this is only an average with significant variations over regions.

To some extent most of these difficulties can be overcome, at least partially, with appropriate tools but at a cost in data collection, analysis and running time; however, in some cases it may be more difficult to interpret results. Moreover, sometimes these improvements may not provide the reassurance that we have finally reached true equilibrium conditions so that results do not depend on some arbitrary decision on the number of iterations or a similar measure. We will discuss a more rigorous approach in the next chapter.

10.8   Practical Considerations

The assignment sub-model is critical in the implementation of the whole transport modelling package. However, in contrast with the other three sub-models there is no standard calibration procedure to make sure the assignment stage reproduces observations as closely as possible. The most likely candidate for external validation of the model is the use of traffic or cordon counts. The following procedure seems applicable to all kind of assignment packages, including public-transport and equilibrium methods as discussed in the next chapter.

Goodness of fit for assignment   Assignment is critical in that is relatively simple to cast doubts about the quality of a model because it does not reproduce a particular observation, perhaps flows on a link well known to the decision maker. There are a number of ways to present the quality of an assignment run for a particular time period. Most of them are based on comparing modelled with observed flows, either at link level or on one or more screen-lines. It is good practice to start by plotting observed versus modelled link flows and fitting the best straight line to them (Figure 10.15).

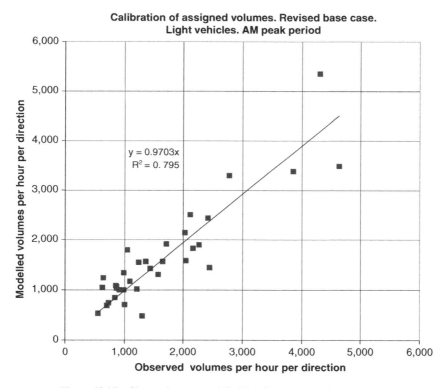

Figure 10.15   Observed versus modelled link flows and best fit straight line

One would also show the corresponding R^2 (the closer to 1 the better) and the slope and intercept. The closer the slope to 1 the better (here it is good at 0.97) and the closer the intercept to zero the better. The cloud of points and the parameters above will help identify any bias in the results.

Transport authorities in different countries adopt different indicators and thresholds to judge the overall fitness of an assignment model. These often take the form of measures of differences between observed and modelled flows; for example the Root Mean Squared Error (RMSE) in absolute or percentage terms. A difficult issue is always how to account for variations in flows in a network when some of them are very large (say on a motorway) and some offer lower flows, for example on local links. The GEH 'statistic' gets its name from Geoffrey E. Havers (who proposed it in the 1970s while working as a transport planner in London); it has been suggested to overcome this difficulty. Although its mathematical form is similar to a chi-squared test, is not a true statistical test. Rather, it is an empirical formula that has proven useful for a variety of traffic analysis purposes.

The GEH measure is defined as:

$$GEH = \sqrt{\frac{(O_i - E_i)^2}{0.5 \cdot (O_i + E_i)}}$$ (10.33)

where O_i are observed values and E_i modelled or estimated values for one variable i.

This may be seen as the square root of the product of the absolute difference (O-E) and the relative difference (O−E)/0.5 (O+E). The reason for using this statistic is the inability of both the absolute difference and the relative difference to cope with a wide range of flows. For example, an absolute difference of 100 pcu/hour may be considered a big difference if the flows are of the order of 100 pcu/hour but completely unimportant for flows of the order of several thousand vehicles an hour. Equally, a 10% error in 100 pcu/hour may not be important whereas a 10% error in, say, 6000 pcu/hour might mean the difference between building an extra lane or not.

Generally speaking the GEH statistic will be less sensitive to these problems as a modeller would probably feel that an error of 20 in 100 would be roughly as bad as an error of 90 in 2000, and both would have a GEH of around 2.

Note that this indicator is not a-dimensional. This means that the recommendation below applies only to hourly traffic flows. If peak period (often 3 hours) or daily flows are used we will exaggerate the acceptability of the results. Equally, the pass criteria below should not be used for other purposes like, for example, total screen-line or cordon flows, for the same reason.

For traffic modelling work in the 'baseline' scenario, a GEH of less than 5.0 is considered a good match between the modelled and observed hourly volumes (flows of longer or shorter durations should be converted to hourly equivalents to use these thresholds). Guidance on what is required for a good model validation varies among countries. In general terms between 60% and 85% of the volumes in a traffic model should have a GEH less than 5.0. GEH in the range of 5.0 to 10.0 may warrant investigation. If the GEH is greater than 10.0, there is a high probability that there is a problem with the travel demand model, the data or both. In the case of screen-lines GEH values greater than 4.0 would indicate poor fit.

However, if the range of flows one is interested in is, say below 500 (an hour/day or whatever), these thresholds would be too generous and a more demanding one should be sought. Figure 10.16 illustrates how the GEH value changes for different variations in flows (5, 10, 20 and 30%) and at different flow levels (50 to 4000 vehicles/hour).

Another indicator that must be checked is the model ability to reproduce the travel times observed during the travel time surveys. The best way of presenting these are to plot observed and modelled cumulative times along the routes travelled during the survey.

Check and Double-check the Network   This is the most important source of error in traffic assignment. There are numerous potential errors in coding a network: the omission of links and nodes previously thought irrelevant, miscoding of distances, use of wrong directions, missing turning-movement penalties, specification of incorrect capacities and time-flow curves, etc. Good software packages will flag many of these errors on input; the use of graphic displays of the network and even better, graphic editing of

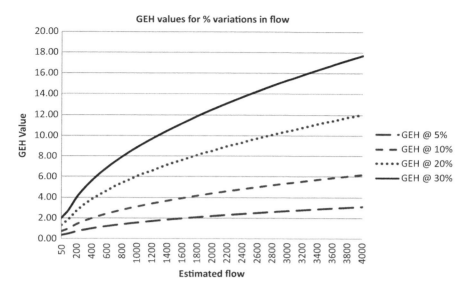

Figure 10.16 GEH indicator values for different flow levels and percentage variations

networks is very important. It is easy to underestimate the time taken to input and check a network for a particular study. Any facility likely to speed up and increase the accuracy of the process is worth many professional days.

An additional method for checking a network consists in loading a unit trip matrix (i.e. with a single trip per cell) and then checking modelled flows. This will facilitate the identification of unused links (perhaps because they were coded with too slow speeds, or too long distances) and also heavily used ones; these serve as pointers for coding errors. The printing, or even better plotting, of minimum path trees is also a useful aid for network checking. Odd shortest routes and unreachable nodes will also help to identify sources of problems.

Improve the connection of centroids to the network if some routes look too strange. Keep in mind, however, that under congested conditions other routes will become attractive and be used. In the case of detailed (microsimulation) assignment models there will be additional sources of problems as more local data are needed. The same applies to public-transport assignment where the connection to bus stops or stations is critical for good route choice representations; the same is true of interchange facilities, frequencies and speeds. The basic rule is: before going to the next step in model fitting make sure all the observable (measurable) data are correctly represented in the network. Check connectivity first, then link attributes and then detailed data like saturation flows, signal timings, and so on.

Fit the Generalised Cost Function   Assign weights to time, distance and any other variables included in it (link status, scenic quality, etc.). Use the GEH measure to assess goodness of fit. This can be applied to cordon counts or to groups of traffic counts on parts of the network thought to be most critical, say primary and secondary roads. The value of the statistic for the whole network also provides an indication of overall fit.

Usually a good starting point is to assume that time alone explains route choice: use this assumption, run a complete assignment and then calculate the statistics above. Then begin increasing the weight attached to distance (or other factors) and recalculate the statistics so that the choice of parameters that produces the best fit can be made. One must resist the temptation of improving the fit at one step by trivial alteration of link speeds or turning penalties at this stage, as this reduces the value of the model for forecasting purposes. True errors discovered at this stage must, of course, be corrected; the model should then be re-run for other generalised cost coefficients as well.

Note that the statistic proposed above gives greater weight to a given absolute difference at low flow levels that at high ones. If this is undesirable, collect it for different flow ranges. The percentage of over and under-estimations of flows can give some indication of bias which, if present, should be investigated more thoroughly. Note too, that if the link capacities were well identified and coded and there is considerable congestion, then equilibrium assignment will tend to produce a good fit with observed flows, even to the extent of masking a few errors in other sub-models.

There may be evidence suggesting that different weights should be applied to different user classes, for example, that heavy lorries are more sensitive to distance and gradient than cars. In that case, the classes should be assigned separately onto the network using their best coefficients in each case.

In the case of public-transport assignment the relative weights of walking, waiting and in-vehicle time are part of this calibration process. Interchange penalties play a similar role and provide an additional element for making the model more realistic. Passenger counts at interchanges and stops should be considered separately for the calibration of these weights. An approach similar to that of Suh *et al.* (1990) may well prove advantageous in fitting generalised cost functions once all other errors have been reduced to a minimum. Alvarez (1995) has studied analytical optimisation methods to achieve the best fit with good results.

Fine-tune the Assignment Model   This involves finding the best dispersion parameters for stochastic assignment models. Particular care should be exercised at this stage, as depending on the implementation these parameters may have different interpretation and even dimensions. The documentation of the programs should be examined in detail to guide us in this task.

Detailed urban assignment models like those described in the next chapter offer additional opportunities for fine-tuning. These make them powerful but may also inadvertently hide more fundamental errors in coding. Examples of this type are the fine-tuning of gap acceptance parameters at some junctions, the representation of opposed turning movements at traffic signals, and so on. Particular care should be taken to make sure these modifications correspond to actual traffic engineering conditions on the ground and not to fudge factors included simply to improve the fit of the model.

It must be recognised that no assignment model will ever reproduce the observations exactly. There will be always variability in the traffic counts themselves, errors in the trip matrices used and a proportion of the actual route choice behaviour which will remain unexplained. What matters, however, is that the resulting costs are as accurate as possible and that the model rests on a sound basis to compare alternative tactical or strategic schemes as required.

Exercises

10.1 The road network represented in Figure 10.17 links two residential areas A and B with two major shopping centres L and M. Travel times between nodes are depicted in minutes and all links are two-way. Assume first that the costs on these links do not depend on traffic levels.

   (a) Use a systematic procedure to find the quickest routes between origins A and B and destinations L and M; calculate the corresponding travel times.

   (b) During a Saturday morning peak hour the numbers of vehicle movements from A and B to L and M are as follows:

$$A - L = 600 \qquad A - M = 400$$
$$B - L = 300 \qquad B - M = 400$$

   Estimate the traffic flow on each link during this period.

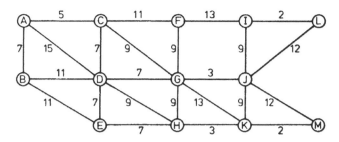

Figure 10.17   Simple network for Exercise 10.1

(c) Consider now that travel time on each link increases by 0.02 of a minute for each vehicle/hour of flow. Use an incremental loading technique to obtain a capacity-restrained set of flows. Calculate final travel times for each O–D pair.

(d) Use an iterative loading procedure to obtain flows and costs under the conditions (c) above.

10.2 A study area contains two residential zones A and B and three workplace zones J, K and L. The zones are connected by a road network as shown in Figure 10.18, which also depicts travel costs in either direction; these are independent of the traffic flows.

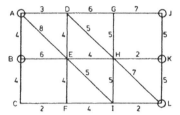

Figure 10.18   Simple network for Exercise 10.2

(a) Use a systematic procedure to find the cheapest routes from nodes A and B to destinations J, K and L and obtain the matrix of travel costs **C**.

(b) The total number of trips originating and terminating in each zone during the morning peak are given by:

Origin	Trips	Destination	Trips
A	1000	J	700
B	2000	K	1000
		L	1300

Run an origin-constrained gravity model in which the deterrence function is proportional to $\exp(-0.1\ C_{ij})$ and obtain a trip matrix. Use this matrix to calculate flows on all the links of the network.

(c) Run a doubly constrained gravity model with the same type of deterrence function and obtain a new trip matrix and link flows. Compare your results of (b) and (c).

10.3 Consider the simple network in Figure 10.19 where there are 100 vehicles per hour travelling from A to X and 500 from B to X. The travel time versus flow relationships are depicted in the figure in minutes and the flow q in vehicles per hour.

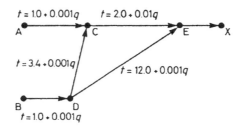

Figure 10.19   Simple network for Exercise 10.3

(a) Use an incremental loading technique with fractions 40, 30, 20 and 10% of the total demand to obtain an approximation to equilibrium assignment.

(b) Use an iterative loading procedure to achieve the same objective. How many iterations do you need to achieve a good degree of convergence.

11

Equilibrium and Dynamic Assignment

11.1 Introduction

In Chapter 10 we introduced assignment techniques for both private vehicles and public transport. We identified three main reasons for the spread of routes between each O–D pair that can be observed in practice. The first one is the different objectives of drivers: time or cost minimisers for example. The second was imperfect perceptions of drivers about travel and link costs. The third reason resides in congestion effects, and we used Wardrop's principles as a general framework to discuss this issue. Wardrop's first principle states that under congested conditions drivers will choose routes until no one can reduce their costs by switching to another path; if all drivers perceive costs in the same way, this produces equilibrium conditions where all the routes used between two points have the same and minimum cost and all those not used have equal or greater cost.

Congested assignment techniques as discussed in the previous chapter try to approximate to this type of equilibrium. We saw that these heuristic methods often failed to achieve true Wardrop's equilibrium; therefore the problem deserves a better treatment. In section 11.2 we will cast equilibrium assignment in a more rigorous mathematical programming framework. This section is restricted to problems where the delay on a link depends only on flows on the link itself; however, extensions to stochastic user equilibrium and to social equilibrium will also be discussed there. Section 11.3 extends the treatment of equilibrium to mode choice and distribution modelling; the objective here is to make sure that the travel times implied in the costs used to run these models are consistent with those generated during assignment. The naive iteration or feedback of the last three sub-models is known not to lead naturally to equilibrium conditions as it is somewhat akin to hard speed-change methods for congested assignment. Improved methods and practical considerations are included in this section. Section 11.4 extends equilibrium assignment to problems where the delay on a link depends on the flow on the link itself and on other flows. This more general formulation is more appropriate to urban areas where the delay at, say a roundabout approach depends on circulating flows in the junction too. Section 11.5 considers the most appropriate way of handling some of the dynamic aspects of traffic assignment including micro-simulation techniques. Finally section 11.6 looks into the issue of departure time modelling and extends previous formulations to cover this important behavioural response.

11.2   Equilibrium

In this section methods specifically designed to achieve traffic assignment solutions satisfying Wardrop's first principle are discussed. We shall follow a combination of intuitive and analytical arguments but we shall not pursue the latter beyond what is necessary to understand and use equilibrium assignment techniques; readers interested in the more theoretical aspects of equilibrium assignment are directed to the excellent book by Sheffi (1985) or the more recent text by Bell and Iida (1997).

In what follows we seek first to establish a more formal formulation of the assignment problem, often using mathematical programming, and then we explore its properties and the solution methods that can be used to solve it; this often involves some kind of iterative method and the issue of degree of convergence to the right solution is therefore important. Finally, we look at some practical issues and extensions to the problems we have considered.

11.2.1   A Mathematical Programming Approach

Consider first some of the properties of Wardrop's selfish equilibrium, in particular that all routes used (for an O–D pair) should have the same (minimum) travel cost, and that all unused routes should have greater (or at most equal) costs. This can be written as:

$$c_{ijr} \begin{cases} = c_{ij}^* & T_{ijr}^* > 0 \\ \geq c_{ij}^* & T_{ijr}^* = 0 \end{cases}$$

where $\{T_{ijr}^*\}$ is a set of path flows which satisfies Wardrop's first principle and all the costs have been calculated after the $\{T_{ijr}^*\}$ have been loaded. In this case the flows on links result from:

$$V_a = \sum_{ijr} T_{ijr}\, \delta_{ijr}^a \tag{11.1}$$

where δ_{ijr}^a is 1 if path r between i and j uses link a and zero otherwise. The cost along a path can be calculated as:

$$C_{ijr} = \sum_{a} \delta_{ijr}^a\, c_a(V_a^*) \tag{11.2}$$

Although Wardrop presented his principles in 1952 it was not until four years later that Beckman *et al.* (1956) proposed a rigorous framework to express them as a mathematical program; it took several more years before suitable algorithms for practical implementations were proposed and tested.

The mathematical programming approach expresses the problem of generating a Wardrop assignment as one of minimising an objective function subject to constraints representing properties of the flows. The problem can be written as:

$$\text{Minimise } Z\{T_{ijr}\} = \sum_{a} \int_o^{V_a} C_a(\upsilon)\mathrm{d}\upsilon \tag{11.3}$$

subject to

$$\sum_{r} T_{ijr} = T_{ij} \tag{11.4}$$

and

$$T_{ijr} \geq 0 \tag{11.5}$$

The objective function corresponds to the sum of the areas under the cost–flow curves for all links in the network. Why this is a sensible objective to minimise in order to obtain Wardrop's equilibrium, is something we will attempt to show below; but first we must consider the general properties of this mathematical programme.

The two constraints (11.4) and (11.5) have been introduced to make sure we work only on the space of solutions of interest, i.e. non-negative path flows T_{ijr} making up the trip matrix. The role of the second constraint (non-negative trips) is important but not essential as this level of discussion of the problem. The interested reader is referred to Sheffi's book or to some of the classic papers on the topic like Fernández and Friesz (1983) and Florian and Spiess (1982).

It can be shown that the objective function Z is convex as its first and second derivatives are non-negative:

$$\frac{\partial Z}{\partial T_{ijr}} = \frac{\partial}{\partial T_{ijr}} \sum_a \int_0^{V_a} C_a(\upsilon)\,\mathrm{d}\upsilon$$

$$= \sum_a \frac{\mathrm{d}}{\mathrm{d}V_a}\left(\int_0^{V_a} C_a(\upsilon)\,\mathrm{d}\upsilon\right)\frac{\partial V_a}{\partial T_{ijr}}$$

but from (11.1)

$$\frac{\partial V_a}{\partial T_{ijr}} = \delta_{ijr}^a$$

Now, as V_a only depends on T_{ijr} if the path goes through that link,

$$\frac{\mathrm{d}}{\mathrm{d}V_a} \int_0^{V_a} C_a(\upsilon)\,\mathrm{d}\upsilon = C_a(V_a)$$

therefore,

$$\frac{\partial Z}{\partial T_{ijr}} = \sum_a C_a(V_a)\,\delta_{ijr}^a = c_{ijr} \tag{11.6}$$

and the second derivative of Z with respect to the path flows is:

$$\frac{\partial^2 Z}{\partial T_{ijr}^2} = \frac{\partial}{\partial T_{ijr}} \sum_a C_a(V_a)\,\delta_{ijr}^a$$

$$= \sum_a \frac{\mathrm{d}C_a(V_a)}{\mathrm{d}V_a}\frac{\partial V_a}{\partial T_{ijr}}\delta_{ijr}^a$$

$$= \sum_a \frac{\mathrm{d}C_a(V_a)}{\mathrm{d}V_a}\delta_{ijr}^a\,\delta_{ijr}^a \tag{11.7}$$

This expression is greater than or equal to zero only if the derivative of the cost–flow relationship is positive or zero. This is a general requirement for convergence of Wardrop's equilibrium to a unique solution. The meaning of this condition is that the cost–flow curve should not have sections where costs decrease when flows increase.

As the problem identified in (11.3)–(11.5) is a constrained optimisation problem, its solution may be found using a Lagrangian method. The Lagrangian can be written as:

$$L(\{T_{ijr}, \phi_{ij}\}) = Z(\{T_{ijr}\}) + \sum_{ij} \phi_{ij}[T_{ij} - \sum_r T_{ijr}] \tag{11.8}$$

where the ϕ_{ij} are the Lagrange multipliers corresponding to constraints (11.4).

Taking the first derivative of (11.8) with respect to ϕ_{ij} one obtains, of course, the corresponding constraints. Taking the derivative with respect to T_{ijr} and equating it to zero (for optimisation), one has:

$$\frac{\partial L}{\partial T_{ijr}} = \frac{\partial Z}{\partial T_{ijr}} - \phi_{ij} = c_{ijr} - \phi_{ij}$$

Here we have two possibilities with respect to the value of T_{ijr}^* at the optimum. If $T_{ijr}^* = 0$ then

$$\frac{\partial L}{\partial T_{ijr}} \geq 0 \quad \text{as the function is convex}$$

If $T_{ijr}^* \geq 0$ then

$$\frac{\partial L}{\partial T_{ijr}} = 0$$

This can be translated into the following conditions at the optimum:

$$\phi_{ij}^* \leq c_{ijr} \text{ for all } ijr \text{ where } T_{ijr}^* = 0$$
$$\phi_{ij}^* = c_{ijr} \text{ for all } ijr \text{ where } T_{ijr}^* > 0$$

In other words, the ϕ_{ij}^* must be equal to the costs along the routes with positive T_{ijr} and must be less than (or equal) to the costs along the other routes (i.e. where $T_{ijr} = 0$). Therefore, ϕ_{ij}^* is equal to the minimum cost of travelling from i to j: $\phi_{ij}^* = c_{ij}^*$.

In this way, the set of T_{ijr}^* which minimises (11.7) has the following properties:

$$c_{ijr} \geq c_{ij}^* \text{ for all } T_{ijr}^* = 0$$
$$c_{ijr} = c_{ij}^* \text{ for all } T_{ijr}^* > 0$$

Therefore, the solution satisfies Wardrop's first principle.

Example 11.1 Consider again the town-centre/bypass problem of Example 10.4. Figure 11.1 shows the cost–flow relationships and the shaded area is the objective function that we want to minimise. Of course one way to minimise this area is to have no flow $V_b = V_t = 0$, but this solution is not only trivial but of little interest. What we want is the solution that satisfies the total demand (2000 vehicles), and this is shown in Figure 11.2, where the two cost–flow functions are now displayed with the X-axis running in opposite directions and separated by the total flow that must be split between the two routes.

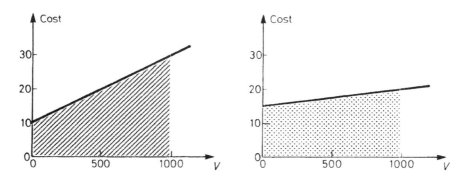

Figure 11.1   Two cost–flow relationships for bypass-town centre problem

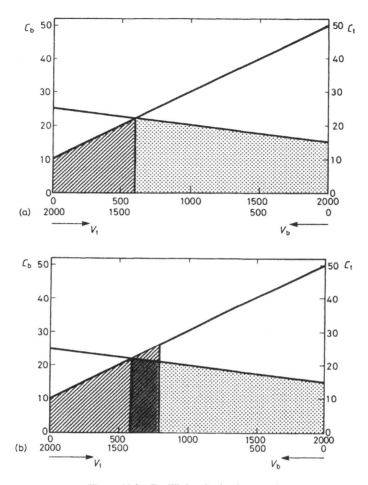

Figure 11.2   Equilibrium in simple network

It can easily be seen in Figure 11.2a that the sum of areas under the cost–flow curves is minimised for $C_b = C_t$; any departure from this point will simply add a new section to the area, as illustrated in Figure 11.2b. As can be seen, the equilibrium solution involves a flow via the town centre of 600 vehicles and 1400 via the bypass. It is worth noting that the cost via each route is 22 minutes and the total expenditure in the network is then 44 000 vehicle-minutes.

In this treatment of equilibrium assignment we have omitted a number of issues; for example, that of uniqueness of the solution. It can be shown that only the link costs c_a^*, inter-zonal costs c_{ij}^* and link flows V_a^* are unique in the optimum. The path flows T_{ijr}^*, however, are in general not unique at all. What this means is that there may be several combinations of paths and trips using them which result in the same link flows and costs; as all used routes (for an O–D pair) have the same minimum cost, the total inter-zonal costs are the same. This can be easily seen if one thinks of several external zones of origin feeding trips into junction A and then exiting to different destinations at junction B in Figure 10.2; although these trips can be distributed in many ways between town-centre and bypass routes under equilibrium conditions, the link flows and costs will remain the same.

11.2.2   Social Equilibrium

Most of what has been discussed so far applies to Wardrop's first principle or *user equilibrium* (UE) problems. Wardrop's second principle specifies that drivers should be persuaded to choose routes in such a way that total (or average) costs are minimised. This is the *social optimum* solution and is a prescription for design rather than a model of driver's behaviour.

It is easy to see that Wardrop's second principle can be embodied in a mathematical programme of the form:

$$\text{Minimise } S\{T_{ijr}\} = \sum_a V_a c_a(\upsilon) \tag{11.9}$$

subject to (11.4) and (11.5).

This objective function can also be expressed in the following form:

$$\text{Minimise } S\{T_{ijr}\} = \sum_a \int_o^{V_a} Cm_a(\upsilon)\,d\upsilon \tag{11.10}$$

where Cm_a is the *marginal cost* of travelling along link a.

This problem can be solved with a simple adaptation to most solution algorithms for the selfish user equilibrium problem. In the case of Frank-Wolfe, the adaptation consists of replacing the objective function used in the estimation of the parameter ϕ in step 4 by (11.10). It is easy to see that the solution to this problem makes all the marginal costs of all the routes used between two points to be equal and minimum.

The solutions to the two problems do not coincide; in other words, the user equilibrium solution generates higher total costs than the social equilibrium solution. The difference lies in the external effects due to congestion. Users perceive only their own personal costs and do not discern the additional delay incurred by other drivers due to extra vehicle on the road. One can envisage electronic road pricing as a possible method to make drivers perceive marginal rather than average costs.

Example 11.2 We take again our town-centre/bypass problem but now seek the flow pattern that minimises total expenditure (or what is equivalent in the case of a fixed trip matrix like this one, minimise average travel costs). The total expenditures are:

$$E_b = V_b(15 + 0.005\,V_b) \text{ via the bypass, and}$$
$$E_t = V_t(10 + 0.02 V_t) \quad \text{via the town centre}$$

The respective marginal costs are

$$\frac{\partial E_b}{\partial V_b} = 15 + 0.01 V_b$$

$$\frac{\partial E_t}{\partial V_t} = 10 + 0.04 V_t$$

Equating the two and taking advantage of the fact that $V_b + V_t = 2000$, one can solve and find that for social equilibrium conditions:

	Town centre	Bypass	Total
Flow	500	1 500	2 000
Marginal cost	30	30	
Average cost	20	22.5	
Expenditure	10 000	33 750	43 750

Note that the total network expenditure is now 250 vehicle-minutes less than the user equilibrium solution found in Example 11.1. Of course, one cannot expect drivers to choose the bypass in these numbers as at least some could reduce their travel costs by choosing the town-centre route. In order to achieve this social optimum one would need to increase user costs by 2.5 minutes via the town-centre, for example by charging the equivalent as a town-centre toll. This would represent simply a transfer from private to social consumption resulting in a saving in the use of resources (time, fuel).

11.2.3 Solution Methods

We have described a mathematical programme and shown its relevance in solving the traffic assignment equilibrium problem. The mathematical programme is non-linear and it can be solved by a number of methods. Although understanding the theory of equilibrium assignment requires some mathematical background, the actual application of the principles and solution algorithms is much less demanding.

A key consideration when looking into solution algorithms is how quickly and well they converge to the correct solution of Wardrop's equilibrium. It is important to select a good convergence criterion to ensure that the solution reached is stable and suitable for project or strategy evaluation. Without this guarantee, small and localised changes in some links may be reflected all over the network and an arbitrary stop in the iterations may result in unreliable results.

Rose *et al.* (1988) researched a variety of convergence criteria and looked into their usefulness to ascertain proximity to the correct solution. They recommend the Relative Gap (RG) as the most reliable measure of convergence:

$$RG = \frac{\sum_a V_a^* c_a - \sum_a V_a^{AON} c_a}{\sum_a V_a^* c_a} \qquad (11.11)$$

where c_a is the cost (time) at the current flow on link a; V_a^{AON} is the all-or-nothing flow on link a and V_a^* is the current flow on link a.

The relative gap is an estimate of the distance between the current solution and the optimal equilibrium solution. This is because the all-or-nothing solution can be seen as a lower bound for the traffic assignment problem. At true equilibrium the relative gap would be zero. As true equilibrium may be too onerous to achieve a number of tests have been proposed to determine how close is 'close enough'. This would depend on the relative size of the user benefits that are being estimated. The general guideline is to make sure that user benefits, in terms of percentage time savings, are at least 10 times the relative gap (in %). Boyce *et al.* (2004) investigated this issue in some practical cases and recommended that the relative gap should be at most 0.1% (0.0001) for satisfactory convergence. This is an exacting requirement, probably too demanding for early stages in the model development process. However, it is good and solid advice for the final stages of model calibration and, in particular, for strategic project evaluation.

Patriksson (1994) developed a good systematic way of looking at the many different algorithms that can be used to solve the mathematical programme for User Equilibrium (11.3–11.5). Those found in practice (i.e. implemented in commercial software) can be grouped into:

a) a linear approximation (Frank-Wolfe);
b) route or path based assignment;
c) origin based assignment.

The most commonly used algorithm is due to Frank and Wolfe. This algorithm can be seen as an improvement on the standard iterative method discussed in section 10.5.4.

11.2.3.1   The Frank–Wolfe Algorithm

This is presented in both conventional and pseudo code format:

1. Select a suitable initial set of current link costs, usually free-flow travel times C_a (0). Initialise all flows $V_a^\circ = 0$; make $n = 0$.
 2. Build the set of minimum cost trees with the current costs; make $n = n + 1$.
 3. Load the whole of the matrix $\mathbf{T}$ of these trees all- or-nothing, obtaining a set of auxiliary flows F_a.
 4. Calculate the current flows as:
 	
 $V_a^n = (1 - \phi)V_a^{n-1} + \phi F_a$
 	
 choosing ϕ such that the value of the objective function Z is minimised.
 5. Calculate a new set of current link costs based on the flows V_a^n; use a good convergence indicator (say RelGap<0.0001) to decide whether to stop or to proceed to step 2.
	Initialisation
	 for every link in the network
	    Let $c_a = C_a$ (0) for all a
	    Let $V_a^n = 0$ for all a and $n = 0$;
	 end for
	
	 Main loop
	 for n = 1 to number of iterations
	    build minimum path trees with
	    V_a^{n-1} flow costs
	    load $\mathbf{T}$ AON and obtain flows F_a
	    estimate ϕ to minimise Z
	    make $V_a^n = (1 - \phi)V_a^{n-1} + \phi F_a$
	    update $c_a = C_a(V_a^n)$
	    if RelGap<0.0001 stop
	 end for

The main improvement over the iterative method is in step 4, where ϕ is calculated using the mathematical programming formulation instead of a fixed rule. In essence, Frank-Wolfe solves a linearised sub-problem to get a good descent direction and finds a new solution using linear search. This is enough to guarantee reasonable convergence to Wardrop's equilibrium.

The Frank–Wolfe algorithm can be visualised as a descent approach to the problem of minimising the objective function. The problem is similar to the establishment of the rules to be followed to find the lowest point of an enclosed valley in thick fog (or more realistically perhaps, to find the peak of a mountain in thick fog, but then one has to use *up* instead of *down* in the rules below). A suitable set of rules for the valley problem would be:

1. Choose what looks like a good downhill direction; in thick fog this will depend essentially on local topography.
2. Walk in that direction until you start to go uphill again.
3. Stop at that point and choose another good downhill direction and proceed to step 2, unless you have found a point with no downhill directions, i.e. the bottom of the valley.

This is essentially what the Frank–Wolfe algorithm does, albeit in a space with many more dimensions. At each step in the iterations we have a current feasible solution (a location in the valley) and the algorithm uses the latest all-or-nothing assignment to provide a descent direction. The use of the latest all-or-nothing assignment to this end can be seen as a local approximation to minimising the objective function Z. Given that the current feasible solution is specified by the path flows $\{T_{ijr}\}$, Frank–Wolfe seeks a second attractive feasible direction $\{W_{ijr}\}$ using a linear (Taylor series expansion) approximation to Z:

$$Z'(\{W_{ijr}\}) = Z(\{T_{ijr}\}) + \sum_{ijr} \frac{\partial Z}{\partial T_{ijr}}(W_{ijr} - T_{ijr})$$

$$= Z(\{T_{ijr}\}) + \sum_{ijr} C_{ijr}W_{ijr} - \sum_{ijr} C_{ijr}T_{ijr} \qquad (11.12)$$

Here the only term which is not fixed by the feasible solution $\{T_{ijr}\}$ is $C_{ijr} W_{ijr}$; so if we wish to minimise a local approximation to Z we must choose routes W_{ijr} such that the corresponding multipliers C_{ijr} are minimised. A way to do this is to choose routes which are currently and locally minimum cost, i.e. all-or-nothing assignment on trees from current costs.

In general terms the Frank–Wolfe algorithm tends to converge rapidly over early iterations but less so as it starts to approach the optimum. Related to this is the problem that link flows tend to oscillate during iterations making it more difficult to achieve the necessary precision in the final solution. It has the advantage of requiring little computer memory as only link variables need to be stored. Modern computers offer plenty of memory so the original advantage is less of a constraint. The slow convergence of Frank-Wolfe is a well-known problem and a number of improvements have been suggested to speed up convergence; see for example the work of Weintraub *et al.* (1985) and Arezki (1986). Alternative solutions methods, as those discussed below, offer better convergence properties, especially for large and congested networks. It is interesting to note that better solutions are often helped by the adoption of a new framework to cast the problem in.

11.2.3.2   Route Based Assignment

There are at least two important algorithms that work on the path-flow (rather than link flow) space. We will present here that due to Jayakrishnan *et al.* (1994) as a 'gradient projection' algorithm. This algorithm uses a transformed objective function which includes the flow conservation constraints into the objective.

The formulation of the algorithm is based on the traffic demand constraints:

$$\sum_r T_{ijr} = T_{ij}$$

The shortest path flows can be expressed as:

$$T_{ij\bar{r}} = T_{ij} - \sum_{r \notin \bar{r}} T_{ijr} \tag{11.13}$$

Now, the optimisation problem can be re-stated as:

$$\min \ \bar{Z}(T_{ij\bar{r}})$$
$$\text{subject to } T_{ijr} \geq 0 \quad \forall \, T_{ijr} \in \bar{T}_{ijr}$$

where $\bar{Z}$ is the new objective function and $\bar{T}_{ijr}$ is the set of non-shortest path flows.

At any (non-optimal) stage in the algorithm a better solution can be found by moving in the negative gradient direction. This is calculated with respect to the flows on the non-shortest paths and a move-size is found using second derivatives with respect to these path flows. For a fuller description of the algorithm see Jayakrishnan *et al.* (1994). Larsson and Patriksson (1992) have developed a related algorithm they call Disaggregate Simplicial Decomposition.

11.2.3.3   Origin Based Assignment

Origin-based Assignment (OBA) represents, in fact, a family of solution methods (Bar-Gera 2002). The basic idea is to define the solution variables in an intermediate way between links and routes. The main solution variables in this algorithm are *origin-based approach proportions*, α_{ia} for every origin i and every link a, such that for every origin i and node p the sum of origin-based approach proportions over all links ending at node p is equal to one. Using origin-based approach proportions, *route proportions* are determined by the product of approach proportions of all the links along the route, that is

$$\gamma_{ijr} = \prod_{a \subseteq r} \alpha_{ia}$$

Route flows are determined by the product of OD flow and route proportion, that is

$$h_{ijr} = T_{ij}\gamma_{ijr}$$

It can be shown (Bar-Gera 2002) that if link a goes from node p to node q, and if the total flow from origin i to node q is g_{iq} then the total flow from origin i that arrives at node q through link a is $f_{ia} = \alpha_{ia}g_{iq}$.

This representation of the solution allows an efficient storage of route flows. A key element in this solution method is that for every origin an *a-cyclic* restricting sub-network is chosen, A_i, such that for origin i approach proportions of links that are not included in Ai are restricted to zero.

The following outline of the algorithm is based on Boyce (2007). Start with trees of minimum cost routes as restricting sub-networks, leading to an all-or-nothing assignment. Next, consider all origins in a sequential order. For each origin the restricting sub-network is updated, and the origin-based approach proportions are adjusted within the given restricting sub-network.

To update a restricting sub-network, unused links are removed; the maximum cost from the origin to node q (v_q) within the restricting sub-network, is calculated for all nodes and all links pq where $v_p < v_q$ are added to the restricting sub-network. Once a new restricting sub-network is found, several computationally intensive steps are needed including reorganisation of the data structure.

As the restricting sub-networks tend to stabilise quickly, it is useful to update origin-based approach proportions while keeping the restricting sub-networks fixed. This is done by introducing *inner iterations*. To update origin-based approach proportions within a given restricting sub-network, a search direction based on shifting flow from high cost alternatives to low cost alternatives is used. In addition to current costs, estimates of cost derivatives are used to improve the search direction in a quasi-Newton fashion.

The full algorithm can be displayed (Boyce 2007) in pseudo code as:

Initialisation:
    for every origin i
        Let A_i be a tree of minimum cost routes under free flow conditions from i
        Let α_{ia} equal 1 for all links in A_i and 0 otherwise. (all-or-nothing assignment)
    end for

Main loop:
    for n = 1 to number of main iterations
        for every origin i
            update restricting subnetwork A_i
            update origin-based approach proportions α_{ia}
        end for
        for m = 1 to number of inner iterations
            for every origin i
                update origin-based approach proportions α_{ia}
            end for
        end for
    end for

Update restricting sub-network for origin i:
  remove unused links from A_i
  for every node p compute the maximum cost v_p from i to p
  for every link $a = [p,q]$
    if $\mathbf{V}p > \mathbf{V}q$ add link a to A_i
  find new topological order for new A_i
  update data structures

Update origin-based approach proportions for origin i:
  compute average costs and Hessian approximations
    for step size 1, 1/2, 1/4, 1/8...
      compute flow shifts and scale by step size
      project and aggregate flow shifts
      if convergence criteria is met then stop
    end for
  apply flow shifts
  update total link flows and link costs

Dial (1999), and Bar Gera and Luzon (2007) have developed variations on this approach that represent improvements on the original algorithm.

The following table compares some of the features of the three general approaches:

	Link based	**Path based**	**Origin based**
Decision space	Link flows	Path flows	Origin based approach proportions and link flows
Memory requirements	Minimum	Greatest	Medium
Speed of convergence	Fast early, slower close to optimum	Fast	Fast

Either the Path- or the Origin-based approaches are to be preferred over the traditional Frank-Wolfe in most cases, depending on what is available in a particular software package.

11.2.4 Stochastic Equilibrium Assignment

We have discussed pure stochastic and pure user-optimised equilibrium traffic assignment models. In the first case a spread of routes between two points is produced because of variability in the perceived routes costs, and in the second because of capacity-restraint effects. One would expect that in reality both types of effects should play a role in route choice. Models which try to include both effects are called stochastic user equilibrium (SUE) models and they seek an equilibrium condition where:

Each user chooses the route with the minimum 'perceived' travel cost; in other words, under SUE no user has a route with lower 'perceived' costs and therefore all stay with their current routes.

The difference between stochastic and Wardrop's user equilibrium is that in SUE models each driver is meant to define 'travel costs' individually instead of using a single definition of costs applicable to all drivers.

In theory, models incorporating stochastic and equilibrium properties look particularly attractive; there are, however, operational and practical difficulties for applying them. From a practical point of view, the most important of these difficulties lies on the convergence properties of these algorithms. To examine this problem, let us define convergence here in the following way: an assignment algorithm is said to be convergent if:

- one starts with a particular set of link costs C_a, for example free-flow costs in the first iteration but calculated costs as a function of flows in subsequent ones; and
- one assigns a matrix using specific rules, say Dial's, and produces new link flows $\{V_a\}$, and then one finds that:

$$C_a = C_a(V_a)$$

In other words, the costs resulting from the new flows are practically the same as those used to find routes and assign traffic. If an algorithm is not convergent the solution (flows and costs) will depend on when the iterative process was stopped, i.e. an arbitrary decision. For example, the next planner dealing with exactly the same problem but specifying a different number of iterations would find different costs; this is obviously not a desirable property for the assessment of transport projects.

It can be shown that under specific circumstances it is possible to formulate convergent SUE algorithms (Sheffi 1985). In fact, a practical algorithm to perform SUE assignment is just an extension of the iterative loading methods (MSA algorithm) described in section 10.5.4. Such an algorithm can be described as follows:

1. Set current costs $C_a = C_a (0)$, i.e. free-flow travel costs, initialise $V_a = 0$ for all a, make $n = 0$.
2. Make $n = n + 1$; build a set of minimum cost trees with the current costs.
3. Assign the trip matrix to the network using the current trees and a suitable stochastic method, e.g. Burrell's; obtain a set of auxiliary flows F_a.
4. Calculate current flows as:

$$V_a^n = (1 - \phi)V_a^{n-1} + \phi F_a$$

   with $\phi = 1/n$.
5. Calculate a new set of current link costs based on the flows V_a^n; if the flows (or current link costs) have not changed significantly in two consecutive iterations, stop; otherwise proceed to step 2.

This algorithm will always tend to produce small changes in flows and costs as ϕ is small for large n. However, it is important to prove that it converges to the right SUE solution.

Sheffi (1985) has shown that this algorithm converges to a SUE solution in the long run, that is, for a large number of iterations, perhaps 50 or more. The convergence of this algorithm is not monotonic because the search direction is only a descent direction on average. The speed of convergence depends on the level of network congestion and on the dispersion parameter.

The convergence of the MSA algorithm for SUE problems is rather slow for congested networks. Sheffi (1985) has also shown that for very congested networks UE provides a good approximation to SUE and is faster in convergence. This suggests that the use of SUE would only be advantageous in low to medium congested assignment problems.

11.2.5   Congested Public Transport Assignment

In the previous chapter we looked at public transport assignment with fixed costs. This means the link costs do not depend on the number of passengers on the bus or train, and they do not depend on the carrying capacity of each unit. It is a reasonable approach in all those cases where the goal of the planning process is to provide enough capacity for all public transport passengers on the routes of their choice. And it has the advantage of facilitating the solution to the public transport assignment problem.

There are, however, situations where it is not feasible to provide enough public transport capacity to preclude congestion, or when that capacity is not present in the base year. In these cases the route choice of the public transport passenger is likely to be influenced by the congestion onboard the vehicles; some travellers will switch from congested to less congested routes, even if the less congested routes are not as attractive in terms of travel time or cost.

We therefore turn our attention to the assignment problem where link travel times are no longer constant. The dependency of link costs on passenger flows may take different forms, but from a solution viewpoint, the simplest and most convenient are continuous non-decreasing functions of the corresponding link flows. This dependence of the link cost on the public transport volume may represent an actual slowing down of the vehicle due to the number of passengers, or it may be interpreted as a generalised cost which includes a 'discomfort' term that increases as the vehicles get crowded.

In this context, the transit assignment problem is no longer separable by destination node, since the link costs depend on the total flow of passengers. The total transit volumes are the sum of the volumes bound for each of the destinations. As the expected cost of any given strategy is no longer fixed, but depends on the total volumes, only strategies with minimal expected cost will be used by the travellers (Wardrop 1952).

Spiess (1983) has shown how the problem above can be formalised and solved by applying the successive linear approximation method (Frank and Wolfe 1956). An important advantage of this method is that only total volumes need to be computed and stored, since the destination-dependent volumes are dealt with implicitly. This approach is easy to implement in packages like EMME/2.

A variant of the macro outlined above is being used at London Transport for modelling crowding in the London Underground. Instead of using one of the default congestion functions based on nominal capacity, the macro has been modified to include the actual profile of train density and passenger load during peak periods (Abraham and Kavanagh 1992).

However, there are conditions where it is not reasonable to assume that link costs depend only on passenger levels on that link. For example, the delay at a stop may depend significantly on the number of passengers already on the public transport unit (bus or train/metro) as some people may not be able to board the first vehicle that comes along. In this case, delays or generalised costs on a link will depend significantly on flows on other links as well; the situation is not entirely dissimilar to junction delays.

In these cases the modelling of congestion should be done using asymmetric generalised cost functions. Here the perceived waiting time for a service (line) for a boarding passenger depends on the number of passengers already on board, or the dwell time of a line at a node depends on the number of boarding and alighting passengers. Although such phenomena do occur in reality, their inclusion into assignment models leads to models without the guarantee of a unique solution.

De Cea and Fernández (2000) developed a multimodal/multiple user class's equilibrium model that incorporates asymmetric generalised cost functions for public assignment (but symmetric functions for road assignment). The model combines destination, mode choice and assignment in an equilibrium framework. Destination and mode choice are treated in a hierarchical logit formulation (destination at the top) and the problem combined with equilibrium assignment with capacity constraints. The problem is formulated as a variational inequality and a diagonalisation algorithm is used for its solution. It is recognised that there is no guarantee of convergence to a unique solution. However, the authors state that they have achieved convergence in all their applications of the model (De Cea *et al.* 2005).

11.3   Transport System Equilibrium

11.3.1   Equilibrium and Feedback

The type of equilibrium problems we have discussed so far concerns just a single mode in a network. Wardrop's first principle models this type of behaviour and a suitable algorithm permits the identification of the routes and flows that will generate consistent costs for all users. As stated before, a similar principle applies to congestion or capacity problems in public transport networks.

The problem becomes more complex when one considers interactions between two or more modes. These may take the following forms:

- Congestion generated by cars will affect bus travel times in certain routes and therefore change assignment strategies for public transport users; congestion generated by buses (and street-running LRT systems) and bus stop operations will affect capacities and speed for cars, and therefore their route choices;
- Interaction due to park-and-ride and kiss-and-ride operations for buses and for segregated track systems. The attractiveness of these mixed-mode operations will depend on road congestion, service frequency and fares (mode and parking) and all of these are, in general, mutually related.

Pragmatic approaches to this problem are usually of the hard or soft speed-change nature discussed in section 10.5: assume bus times and flows fixed and known, assign cars to the network to equilibrium, assign passengers to the transit network, obtain new speeds and travel times and fix them, re-assign, obtain new speeds, re-assign, etc. Of course, if one is not prepared to change the bus frequencies in accordance to demand, the problem may converge soon at this level.

In the case of mixed-mode users the problem is more difficult because they may decide to change their park-and-ride station as a result of congestion in the road network and therefore change the same levels of congestion when they do so. Even if mixed-mode movements are few at present, not including them in the equilibrium procedure may cause severe problems for design-year forecasts in heavily congested networks.

In all the cases above we have kept the assumption of a fixed trip matrix (inelastic demand) for each mode. However, what we have seen in earlier chapters must lead us to treat the assumption of a fixed matrix with caution, at least when we are considering major changes to the transport network or longer timescales. Indeed, the whole point of distribution, mode and time of travel choice models is that demand is elastic, in particular to travel and route generalised costs. This leads us to consider the influence of congestion and delay on mode and destination choice at least: the issue of *System Equilibrium* or at least *Model System Consistency*.

Looking at the whole modelling system in forecasting mode, the generalised costs of travel will be affected by congestion and future interventions like new links and modes. Any assumption about travel costs must be revised after assignment and the system of models should be run again to obtain demand consistent with future costs.

What we have now is a nested set of models and we need to make sure that the travel costs used by all of them are consistent. A naive (in the sense of simple, not pejorative) iterative strategy would be '*run all the models first, obtain new travel times, feedback the new travel times to the models above and repeat until convergence*'. This naive feedback strategy is similar to either hard or soft-speed adjustment methods for assignment discussed briefly in Chapter 10. Similarly, it has all the makings of a *non-convergent* approach. Oscillations are likely to be a feature of this type of technique unless special conditions are met, or we pay considerably more attention to the development and use of better algorithms.

Example 11.3 We consider again the town-centre/bypass problem of Example 10.4 but we now add a rail service that links A to B in 12 minutes ten times an hour. For simplicity let's assume that car occupancy is just 1 person per car and that there are no fares and no fuel or access costs; time is the only cost element. The cost of using rail is then 18 generalised minutes (12 + 2 times 3). The total demand V_T is still 2000 passengers an hour. The choice between rail and car is estimated using a logit model with only one parameter λ. In this case it is very easy (and fast) to reach equilibrium on the road side using the fact that

$$V_b = 0.8\, V_C - 200$$

and

$$t_b = 15 + 0.005 V_b$$

with the total car demand Vc is the total demand minus rail demand: $V_C = V_T - V_R$.

The quality of convergence to equilibrium can be measured by the proportion of total demand that is displaced from one mode to the other each iteration. Convergence is reached when this displaced demand is zero. One would expect that the speed of convergence would depend on the parameter λ as higher values (giving greater weight to cost differences) will make the logit result getting closer to all-or-nothing mode choice. Figure 11.3 shows the number of iterations needed to reach particular levels of convergence as a function of λ for this simple example. As can be seen, low values of λ enable reasonable convergence for some 15-20 iterations. The reader can verify that 10 iterations are enough if λ is 0.05, for example. For larger values of λ, convergence requires solving the whole model 100 times or more. Indeed, in this case for λ greater than 0.34 convergence is never achieved making it unsound to compare any two schemes after an arbitrary number of iterations. For instance, for $\lambda = 0.34$ the number of trips by rail after 100 iterations oscillate between 650 and 950 each time. Of course, one should not

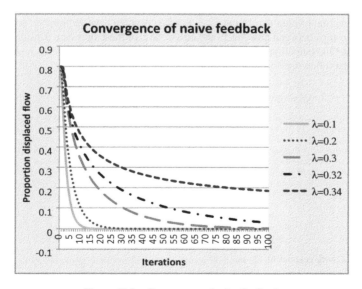

Figure 11.3   Convergence of naive feedback

generalise from this simple example to real networks and problems. However, it shows at least that great care must be placed in organising the interaction between different sub-models.

This is an important issue that, in our experience, is sometimes ignored or handled incorrectly in practice. Moreover, in the USA at least, running the models with 'feedback' is required by Congressional and judicial mandates. Therefore, we attempt here to approach the issue from two complementary perspectives. We try to explain the key components of the problems in both intuitive and a more formal mathematical framework.

First, let us mention that the problem is not even a recent one. It was researched by a British applied mathematician John Murchland and Suzzane Evans (Evans 1976), a graduate student at University College London, in the seventies. She successfully solved the combined distribution and assignment problem in her PhD thesis but after a couple of published papers changed her field of inquire. They introduced the terminology of 'combined models' that has been applied ever since.

One early review of the state of the art is Fernández and Friesz (1983). More recent developments have focussed on developing improved and practical algorithms and recommendations. The next sections are inspired by the work of Professor David Boyce whose efforts to convey the importance of the problem and the rigour required to tackle it correctly are exemplary.

11.3.2   Formulation of the Combined Model System

A useful way of tackling this problem is to frame it as a mathematical programme. We have already done this separately for Distribution, Mode Choice and Equilibrium Assignment; framing a combined mathematical programme seems a natural next step. For convenience, we start first with the combined Mode Choice and Assignment problem.

A reasonable start is to collapse as many sub-models as possible into one, in particular if one can include assignment in the same process. What may be important, however, is not to compromise too much the realism of the modelling process for the sake of expedience in equilibrium, particularly in short-term tactical decision making.

Consider first the problem in general terms where a typical demand curve may be inverted to give travel costs as a function of number of trips $C_{ij} = g_{ij}(T_{ij})$. We then have a combined problem described in terms of relationships between flow levels and costs; some of these flows are trips on real links a and others are flows on O-D pairs (hyperlinks) with the "cost function" above.

Now consider the following objective function:

$$\text{Minimise } Z = \sum_a \int_0^{V_a} c_a(\upsilon)\,\mathrm{d}\upsilon - \sum_{ij} \int_0^{T_{ij}} g_{ij}(t)\,\mathrm{d}t \tag{11.14}$$

subject to

$$T_{ijr} \geq 0$$

$$T_{ij} = \sum_r T_{ijr} \tag{11.15}$$

$$V_a = \sum_{ij} T_{ijr}\delta^a_{ijr} \tag{11.16}$$

The derivative of Z with respect to T_{ijr} is:

$$\frac{\partial Z}{\partial T_{ijr}} = c_{ijr} - g_{ij}$$

We can now consider the behaviour of Z at T_{ijr}^* directly:

$$\text{If } T_{ijr}^* = 0 \quad \text{then } \frac{\partial Z}{\partial T_{ijr}} \geq 0 \quad \text{and} \quad c_{ijr} \geq g_{ij} \tag{11.17a}$$

$$\text{If } T_{ijr}^* > 0 \quad \text{then } \frac{\partial Z}{\partial T_{ijr}} = 0 \quad \text{and} \quad c_{ijr} = g_{ij} \tag{11.17b}$$

Therefore, if a particular path is used, then the path cost specifies a value for the demand curve, so we must have:

$$g_{ij}(T_{ij}) = c_{ij}^*$$

A couple of issues are worth noting here. First, for assignment one usually deals with vehicular flows whereas in distribution and mode choice the main variables are trips. There is a need to account for vehicle occupancy in combined models although this factor has been omitted for simplicity. The inverted demand function could be of a very general form provided the problem remains a convex one. However, in many practical problems it may not be possible to develop suitable closed analytical forms.

Consider a slightly more general case where we add a transit mode b to the system; assume first that the travel times on this transit mode are independent from road speed. The function can be written as follows:

$$\text{Minimise } Z = \eta \sum_a \int_0^{V_a} c_a(\upsilon) \, d\upsilon + \sum_{ij} c_{ij}^b T_{ij}^b \tag{11.18}$$

Subject to

$$\sum_{ijr} \eta T_{ij}^c \delta_{ijr}^a = V_a \quad \text{a flow conservation constraint} \tag{11.19}$$

here η is vehicle occupancy, b indicates the public transport mode and c the car; k is the index for mode, either b or c.

We now add a constraint stating that total O-D flows are split between the two modes and add also a dispersion constraint to ensure that this split is not all-or-nothing, as we cannot account for all factors explaining mode choice.

$$T_{ij} = T_{ij}^c + T_{ij}^b \quad \text{for all } i, j \tag{11.20}$$

and

$$\sum_{ijk} T_{ij}^k \log T_{ij}^k = -S_0 \tag{11.21}$$

$$T_{ij}^k \geq 0 \tag{11.22}$$

Equation (11.21) is a dispersion constraint that allows some flows to use the higher cost mode; the associated parameter (multiplier) would be estimated at calibration to reflect observed dispersion. In this combined problem we have relaxed slightly the user equilibrium conditions to allow trips to choose a different mode (hyper-route) whilst retaining the logit formulation for mode choice.

If we further add an origin-destination choice element to the problem, to relax the assumption that T_{ij} is fixed we have:

$$\sum_{jk} T_{ij}^k = O_i \quad \text{and} \tag{11.23a}$$

$$\sum_{in} T_{ij}^k = D_j \tag{11.23b}$$

The optimality conditions for this model are the same as for User Equilibrium plus:

$$T_{ij}^k = \frac{T_{ij}\exp(-\lambda C_{ij}^{k*})}{\sum_n \exp(-\lambda C_{ij}^{k*})} \tag{11.24}$$

and

$$T_{ij}^k = A_i O_i B_j D_j \exp(-\lambda C_{ij}^{k*}) \tag{11.25}$$

where C_{ij}^{k*} is the user equilibrium cost of travelling from i to j by mode k.

This solution has the same dispersion coefficient λ for mode and destination choice. A second dispersion constraint (on ij) can be added to convert it to a problem where these dispersion constraints are allowed to be different.

The constraint (11.21) can be integrated into the objective function using the Lagrange multiplier λ and retain only the linear constraints:

$$\text{Minimise } Z = \eta \sum_a \int_0^{V_a} c_a(\upsilon)\,d\upsilon + \sum_{ij} c_{ij}^b T_{ij}^b + \frac{1}{\lambda} \sum_{ijm} T_{ij}^m \log T_{ij}^m \tag{11.26}$$

subject to constraints (11.19 to 11.23b).

These are all linear constraints and the objective function is the sum of convex functions. The solution algorithm proposed by Evans (1976) can be generalised as follows (Boyce 2007):

1. Initialisation – Make iteration counter $n = 0$; compute an initial solution for (T_{ijk}^0), (V_a^0). This normally involves using free-flow costs to estimate gravity and mode choice models and assign trips to the networks.
2. Update link costs. $C_a = C_a(V_a^n)$; increment n by 1.
3. Compute new shortest routes from each origin i to destination j and obtain new (C_{ijc}^n), that is car costs.
4. Solve the OD and mode choice model and obtain the sub-problem flows (e_{ijk}^n), sometimes called auxiliary flows, in this case at the Origin-Destination-Mode level.
5. Perform all-or-nothing (AON) assignment of car flows (e_{ijc}^n) to the shortest path from i to j obtaining car flows (g_a^n).
6. Compute the Relative Gap and test for convergence.
7. Perform a line search to determine the optimal step length (weight) λ^n.

$$\text{Minimise } Z = \eta \sum_a \int_0^{V_a^n} c_a(\upsilon)\,d\upsilon + \sum_{ij} c_{ij}^b(T_{ij}^b)^n + \frac{1}{\lambda} \sum_{ijm} (T_{ijm})^n \log(T_{ijm})^n \tag{11.27}$$

where $V_a^n = (1 - \lambda^n)V_a^{n-1} + \lambda^n g_a^n$ and $T_{ijm}^n = (1 - \lambda^n)T_{ijm}^{n-1} + \lambda^n e_{ijm}^n$

8. Update the OD mode and link flows

$$T_{ijm}^n = (1 - \lambda^n)T_{ijm}^{n-1} + \lambda^n e_{ijm}^n$$

$$V_a^n = (1 - \lambda^n)V_a^{n-1} + \lambda^n g_a^n$$

9. Retest the updated value of the objective function for convergence; if not achieved go back to Step 2.

This approach has a critical difference with the naive feedback treatment of the problem. Evans solution is to average flows (on links and trips by mode and O-D pair) rather than just feed-back costs, averaged or otherwise.

Similar formulations have been produced by, among others, Gartner (1980) and Sheffi (1985). This type of approach has been extended further by De Cea *et al.* (2008) who presented very general combined models with hierarchical demand choices using a multi-objective entropy maximisation approach. This is an important and valuable generalisation as it allows for a general hierarchy of choice models in combination with what they correctly call demand-performance models. The choices may include: destination, mode, time of travel, modal transfer point, etc. The main characteristics of their approach are:

i) Demand choices are assumed to have a hierarchical structure where entropy must be maximised to produce the most likely arrangement subject to the corresponding constraint at each level of the nested tree.
ii) A combined demand model incorporating these choices can be incorporated as a multi-objective programming problem; they put the destination choice model at the top of this tree but other arrangements are possible.
iii) All users by class and mode behave according to Wardrop's first principle; the link flow-cost functions are separable and convex.
iv) The combined performance-demand equilibrium models are also formulated as multi-objective programming problems.
v) A convex optimisation problem cannot be formulated if the network cost functions are asymmetric but the problem may be specified as a variational inequality.

With these conditions the set of choice models turns out to be a hierarchical logit model where the scaling parameters must comply with the requirements for their relative values from bottom to the top of the hierarchy.

11.3.3 Solving General Combined Models

The considerations above are particularly useful when it is possible to formulate an appropriate mathematical programming problem with the necessary conditions of convexity and separability or symmetry of the performance-demand functions. The solution algorithm will depend, in general, of the specific formulation in each case. One of the attractions of the naive feedback approach is that it is general enough and does not require assumptions about the characteristics of the model formulation. As we have seen, however, there is no guarantee of convergence to a unique solution that would make it possible to compare strategies or projects.

It is generally accepted that the weights in the generalised Evan's solution method could be replaced by pre-determined weights, for example the rules of the Method of Successive Averages (MSA). The sequence of sub-problem weights or step-sizes λ^n applied in the MSA are 1, 1/2, 1/3, ..., 1/n. The use of this sequence is known to converge to equilibrium albeit at a fairly slow rate.

An alternative approach is to use relatively arbitrary constant weights instead of the MSA sequence. Perhaps surprisingly, the constant weight (CW) approach has been found in practice to converge faster to equilibrium than MSA, see Bar-Gera and Boyce (2006). A more general version of this algorithm could be presented in an intuitive form as follows:

Step 1. Input data, the road and public transport networks, trip end constraints O_i, D_j.
Step 2. Compute an initial solution for iteration $n = 1$ using free-flow costs or another suitable starting point.
Initialise travel costs C_{ijm}^n.
Solve the demand model $e_{ijm}^n = T_{ijm}^n$ (a provisional solution for O-D-mode).
Assign T_{ijc}^n to road network; where the sub-index c indicates cars.

Step 3. Compute a solution for $n = n + 1$ the next iteration.
        Calculate costs on used route-mode combinations C_{ijm}^n.
        Solve the demand model e_{ijm}^n (auxiliary demand volumes).

Step 4. Average trip matrices T_{ijm}^{n-1} and e_{ijm}^n

$$\text{For MSA } T_{ijm}^n = \left(\frac{n-1}{n}\right)T_{ijm}^{n-1} + \left(\frac{1}{n}\right)e_{ijm}^n$$

$$\text{For CW } T_{ijm}^n = wT_{ijm}^{n-1} + (1-w)e_{ijm}^n$$

Step 5. Assign T_{ijc}^n to the road network to the desired degree of convergence and produce V_a^n

Step 6. Check for convergence of e_{ijm}^n to T_{ijm}^{n-1}

Total Misplaced Flow $\qquad\qquad TMF = \sum_{ijm} \left| T_{ijm}^n - e_{ijm}^n \right| \leq E,$ $\qquad\qquad$ (11.28)

or

Root Squared Error $\qquad\qquad RSE = \sqrt{\sum_{ijm} (T_{ijm}^{n-1} - e_{ijm}^n)^2} \leq E$ $\qquad\qquad$ (11.29)

if converged, stop, otherwise continue to Step 3.

This is a general formulation that can be applied with the MSA or CW methodology. Bar-Gera and Boyce (2006) and Boyce *et al.* (2008) report that on real networks the use of constant weights performs better than MSA in terms of speed and consistency of convergence. They recommend the adoption of CWs with w in the range of 0.2 to 0.5. In particular, the same weight $w = 0.25$ performs well for three cases with very different congestion levels. Naive feedback always performed poorly and therefore should be always avoided.

A general (provisional) rule seems to be to use the CW method above testing a few weights w around the 0.25 to find what works best for a particular network and matrices.

Equilibrium in transport systems and markets is not an end in itself. There are good reasons to suspect that equilibrium does not happen in practice, not even at the simplest network level. Real systems are in a permanent state of change, with travellers experimenting new routes, modes and destinations. Families change residences, jobs, shopping and social patterns and lifestyles. However, the state of the art in dynamic modelling of these phenomena is still many years behind that of equilibrium modelling.

The main reason to use models is to provide advice on transport decisions and this requires comparing alternative ways of intervening in the transportation system. Consistency in the use of models to estimate the performance of these interventions is then of capital importance as we wish to compare 'like with like'. Casting the transport modelling effort into a general equilibrium framework seems a prerequisite for ensuring this consistency. It is not, of course, a sufficient condition: there will be cases where partial modelling of the system will be enough to discriminate a good scheme from one that is not so good. However, the state of the art of equilibrium modelling is such that one seldom has to sacrifice too much realism to achieve it.

Computer memory and speed constraints are mostly a thing of the past. Most modern software now provides all the facilities required to seek equilibrium solutions involving route, mode, destination and time-of-departure choice. There seems to be little reason not to use these facilities, at the very least for the final runs used to compare two alternative strategies or schemes.

11.3.4 Monitoring Convergence

The combined problem solution methods discussed above rely on a valid estimate of the degree of convergence to an equilibrium solution. In general terms, two convergence criteria are needed, one for the trip matrix (destination, mode and time of day) and one for the link flow arrays. These are sometimes combined under the banner of 'relative gap' (RG).

Feasible solutions to convex optimisation problems have a lower bound associated to them; this is defined in terms of auxiliary demand matrices e_{ijm}^n and flows V_a^{n-1} and g_a^n. The RG is obtained from the lowest objective function (LOF) value that is always a result from the current iteration, and the best (or highest) lower bound (BLB) which may be the results of an earlier iteration:

$$RG = \frac{(LOF - BLB)}{BLB}$$

These measures are useful for global convergence but are not that easy to interpret intuitively and this makes it difficult to establish a desired level for the tolerance E in (11.28). When dealing with O-D (plus mode and time of travel) volumes, it seems natural to compare the current solution with that resulting from the generalised costs of travel under current conditions. The preferred measure is the total misplaced flows as defined in (11.28) and measured in trips per unit time (hour). This lends itself to an easier intuitive interpretation.

For example, consider a scheme involving the introduction of a new metro station that is expected to attract/generate 2,000 trips per hour during the peak. It will be desirable to know that the solution found to the combined problem is misplacing less than, say 200 trips. Depending on the problem and model, it may be acceptable to have 2,000 misplaced trips over a larger area but figures above, say, 5,000 trips are likely to cast some doubt about comparisons between alternatives.

Assignment accuracy can be ascertained using the distribution of *excess costs* among all used routes:

$$EC_{ijr} = C_{ijr} - C_{ij}^* \tag{11.30}$$

where C_{ijr} is the current cost from origin i to destination j via route r, and C_{ij}^* is the minimum cost between those O-D pairs.

A good measure is the *average excess cost* (AEC)

$$AEC^n = \frac{\sum\limits_{ijr} T_{ijr} EC_{ijr}}{\sum\limits_{ij} T_{ij}}$$

which is equivalent to a normalised gap for the fixed demand problem.

11.4   Traffic Dynamics

11.4.1   The Dynamic Nature of Traffic

In this section we focus a bit more deeply into the nature of vehicular traffic and how this is modelled on conventional and more detailed assignment models. There are three common assumptions used in assignment models that have proved helpful in devising more rigorous mathematical programming formulations and determining the conditions for a unique equilibrium solution:

- The traveller has full knowledge on the generalised costs of travelling on every link and route in the network (perfect information assumption).
- Delays on links can be described using a function of flows on that link alone (separability assumption).
- The demand and flows during a modelled period do not change over time (steady state assumption).

In congested real world networks, none of these assumptions is very realistic. Even with the best GPS guidance, knowledge about travel costs on any network require perfect foresight about the future costs when the traveller actually reaches more distant parts of the network. Stochastic assignment goes some way to address this issue but the introduction of time dependent delays makes it more difficult to handle.

Delays at a junction approach depend not only on own link flows but also, and in some cases chiefly, on flows on other approaches. Priority junctions, roundabouts and even traffic signals, display a degree of interaction among entry links and flows thus leading to delay functions that are non-separable. Junction interaction delay models seek to overcome this limitation at some cost in tractability.

The fact that demand varies over time and that peak flows propagate during the peak period is well known and was already discussed in Chapter 10. One can try to handle this by modelling a short time period, say just the peak hour, where demand can be considered to be more or less uniform. But even then, real capacity constraints in the network create dynamic conditions that standard speed-flow curves cannot handle correctly. Conventional flow-delay curves, as those discussed in section 10.1.3, allow flows to exceed capacity and normal equilibrium assignment assumes that all the demand in a time period reaches its final destination. Reality is different; real capacity constraints generate dynamic queues at bottlenecks that prevent all traffic reaching their destinations during the modelled period. Moreover, these queues remain and grow until demand declines below capacity when they start to clear.

Example 11.4 Take a 5 kms long road corridor that has a junction every kilometre. The capacity of each junction is 2,000 vehicles/hr and the free flow speed is 60 km/hr ($t_0 = 1$ min/km). Assume that a BPR function with $\alpha = 4$ and $\beta = 4$ is a valid representation of delay on these five links. The flow-travel time relationship for each 1 km link is depicted in Figure 11.4. Consider now that a greater capacity road feeds onto this corridor 2,200 vehicles during the peak hour. Using a conventional traffic model based on BPR curves the time at each link (junction) will be 6.86 minutes. As the BPR curves accept flows above capacity, the total time spent on those 5 kilometres will be 34.30 minutes.

However, in reality queues will form at the first junction so that only 2,000 vehicles/hr filter through to the other four junctions during the peak hour. The delays for these 2,000 vehicles will be only 5 minutes per link. If we accept the BPR curve as accurate the total delay will now be 6.86 minutes for the first link

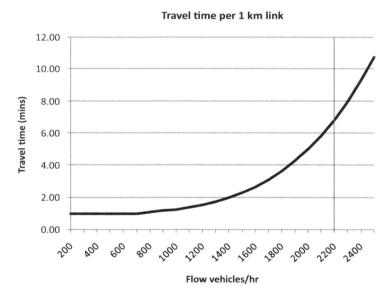

Figure 11.4   BPR curve for 1 km links in Example 11.4

plus 4 times 5 minutes, i.e. 26.86 minutes. In fact 10% of the vehicles will not clear the first junction in the first hour. These would be the last 200 vehicles arriving at that junction but all previous vehicles will also suffer additional delay. This is because the arrival rate is greater than the capacity and the queue will build up to reach the value of 200 vehicles at the end of the hour. These 200 queueing vehicles will take up storage space and, if they need 10 metres each on two lanes, will block the upstream junction causing additional delays. This deterministic analysis assumes regular arrivals and departures; randomness in arrivals will add to these delays.

This simple example illustrates some of the difficulties of modelling traffic assignment accurately using conventional speed-flow curves that allow over-capacity flows. Improved assignment methods will have to consider the physical characteristics of traffic and handle issues like real capacity constraints, the storage capacity of links to handle queues, and queues remaining at the end of a modelling period and spilling over the next time slice. Another problem that is gaining in importance is the role of reliability in the estimation of travel times. This is, of course, central to time-critical journeys like those to catch a flight or attend an important meeting. How best to model this feature is also important in improving assignment models.

11.4.2   Travel Time Reliability

The variability and unreliability of travel time in congested urban areas has become a significant issue for many types of trips. As traffic increases in heavily loaded networks, the travel time required to perform a particular journey becomes particularly difficult to estimate. Under these circumstances, users may have to make large time allowances to avoid missing a plane or a business meeting; for other activities they may just accept the penalty associated with unpredictable delays. It has been argued that one of the key benefits of pricing for road space is to increase the reliability of journey times and therefore produce significant resource savings.

It is possible to use stated preference/revealed preference surveys and data to develop appropriate generalised cost functions incorporating these effects. In this way one could develop a subjective value of unreliability in travel time. This requires a measure of such reliability, for example the expected standard deviation of travel time σ_t or the expected coefficient of variation of travel time. Note the emphasis on expected or subjective measures of travel time variability.

An equally important requirement is to develop models that link travel time variability to congestion and supply conditions (incident management facilities, redundancy in the network, etc.). This is a less well-researched area for a number of reasons. First data collection is often very expensive in this field as one would require repeated journeys (same departure time, same origin and destination) over a large number of days to pick up systematic and random variations; this has to be repeated for several times of the day and several origin–destination pairs. Second, the supply models have to be reasonably simple to be of use in large strategic models, or sensitive to key policy instruments (new traffic control measures, variable message signs/route guidance) for detailed tactical modelling.

Willumsen and Hounsell (1998) report a general study for use in strategic models in the context of road pricing. They used extensive observations in a congested network (London) and extended their value using simulation runs for over 2,000 O–D pairs and over a large number of 'days'. As independent variables they selected actual journey time (JT), free flow travel time (FFTT) and a congestion index, defined as CI = JT/FFTT.

A number of models were calibrated to estimate the standard deviation of travel time under different congested conditions. The authors recommend the following model as offering a good compromise between simplicity and realism:

$$\sigma_t = 0.9\,\text{FFTT}^{0.87}(\text{CI} - 1)$$

In practical terms this model offers a simple form for relating the standard deviation of travel time to network conditions and is relatively insensitive to trip length, therefore offering promise of adaptation to environments different from London. One advantage of this treatment is that journey time variability can be estimated after assignment and then incorporated into other choice models (time of day, mode, and destination choice). Complex interactions between congestion, travel time variability and route choice are then avoided.

11.4.3   Junction Interaction Methods

Classic assignment methods often use the simplification that delays on a link will depend only on flow on the link itself; this is useful to set a straightforward traffic assignment problem convergent to a unique solution. However, this assumption may not be realistic enough for congested urban areas. If one considers the route choice and assignment problems in greater detail, one should search for better delay models and a better treatment of dynamic problems. In addition there is a need to consider the interaction between traffic control and route choice, and to treat different vehicle classes separately. We shall discuss these issues in turn.

11.4.3.1   Improved Delay Models

So far we have considered traffic as a continuous variable operating under steady-state conditions. In reality, traffic is made up of discrete entities (vehicles) which in urban areas form queues at junctions and bottlenecks. If a particular assignment model puts more traffic on a junction than its capacity, it is very likely that the flows downstream will be overestimated; this happens because the junction will actually put an effective upper limit, not recognised by the model, and the modelled flows downstream will be greater than the actual flows. Therefore, potential routes using these links may well be ignored by the model. Double counting of delay and missing of potential routes are a perverse effect of this simplistic treatment of traffic delay.

Two types of improvement are needed here: first, to consider the physical nature of queues at junctions and their effects in limiting traffic downstream; second, the need to model the time-dependent nature of queues at junctions as demand builds up and decays before, during and after the peak period. The second problem can be treated using time-dependent queueing models as proposed by Kimber and Hollis (1979). These approaches model the way in which queues and delay change over time, as traffic demand evolves, and even allow for the presence of queues at the start of a time period of interest.

The first problem requires a physical model of queues and this can be undertaken through a simple conversion of vehicles queued into queue length or, in more detailed models, through the simulation of the actual queues. A critical issue is the ability of these models to represent the situation where a queue begins to block back an upstream junction and the additional delays this generates to other streams.

11.4.3.2   The Interaction between Traffic Control and Delay

This is difficult to treat in detail. Most large urban areas are under area traffic control (ATC) systems, that is, computer control of the traffic signals to reduce delay and, in some cases, create attractive 'green waves'. It is known that such systems are designed to cope well with existing traffic patterns and that travel time savings of between 10 and 20% can be achieved in comparison with non-coordinated systems. The problem is that the traffic flow patterns (flows on links) depend on the set of best routes available and that these depend, in turn, on signal timings at each junction. However, any model attempting to combine assignment and traffic control may run into a number of problems; see for example Allsop and Charlesworth (1977) and M.J. Smith (1979a, 1981).

One possible solution is to run an assignment problem with fixed signal settings, obtain a future set of link flows and then run a program like TRANSYT (Robertson 1969) to optimise the setting for these new flows. The process should be repeated with the new settings, obtaining in turn new flows, with the hope that these iterations will converge to a stable and self-consistent solution. The problem is that the solution depends considerably on the starting point; if a corridor is heavily used in the first iteration, TRANSYT will produce signal timings to reduce delay there, thus encouraging more trips to prefer it in the next iteration. This also tends to favour all-or-nothing type of solutions to the traffic control/assignment problem.

11.4.4 Dynamic Traffic Assignment (DTA)

11.4.4.1 General Requirements

There are some basic requirements for a truly dynamic traffic assignment model. These have been identified by, among others, Heydecker and Addison (2005) as:

- Positivity: we are only really interested in non-negative flows on links, paths, trip matrices and costs.
- Conservation: the model must satisfy flow conservation requirements.
- FIFO: in real traffic the FIFO (First In, First Out) behaviour generally prevails and this must be maintained in the model if proper delays are to be estimated.
- Minimum travel time: flows do not propagate instantaneously.
- Finite clearing time: there are no queues left at the end of the modelling period; infinite delays do not occur (as a standard queueing model might suggest).
- Capacity: there is such a thing as strict capacity constraint in the sense that actual flows cannot exceed it even for a short period of time.
- Causality: delays now are affected by what other vehicles do or have done in the past, not in the future.

These requirements lead to correct flow propagation and the consequent interrelationship between travel time and link outflow. Finite clearing time ensures that no travellers remain on the network indefinitely and that it returns to free-flow conditions after the study period. The causality requirement ensures that response follows stimulus.

Wardrop's user equilibrium principle of route choice can be extended to the dynamic problem as follows:

> *Under equilibrium conditions in networks where congestion varies over time traffic arranges itself so that at each instant the costs incurred by drivers on those routes that are used are equal and no greater than those on any unused route.*

If travellers choose not only route but also departure time, Wardrop's equilibrium expression can be further extended:

> *Under equilibrium conditions in networks where congestion varies over time and travellers can choose their time of travel, traffic arranges itself so that the total cost associated with travel on those route that are used by travellers at the time when they are used, are equal and no greater than those on any route at a time when it is not used.*

It is possible to present the dynamic user equilibrium problem, with or without choice of time of travel, in a closed form. However, practical methods for its solution often rely on modelling discrete time-slices or time-intervals. Therefore, a key element in the development of numerical methods for the solution of dynamic assignments is the transition from the continuous time formulation of the equilibrium conditions to a discrete time formulation for its solution.

The numerical solution method will typically assign a calculated flow at time s, $T_{ij}(s)$, to a path p throughout a time increment $[s, s+\Delta s)$. It is important to use the flows and costs of that time interval to model equilibrium conditions. If the previous costs are used for assignment, the result will not represent the new traffic conditions.

This approach can be applied to a mathematical programming formulation discussed by Han and Heydecker (2006). In this case, the objective $Z(s)$ that is minimised is calculated at each incremental time interval using the flow that is assigned throughout that increment together with the costs $c(s+\Delta s)$ at the end of it. Although somewhat outside the scope of this book, we must mention that the variational inequality formulation developed initially by M.J. Smith (1979b) and Dafermos (1980) provides a practical approach to calculation of dynamic traffic assignments within the present framework. This was introduced for dynamic traffic assignment by Friesz *et al.* (1993) and has been adopted by others since then.

11.4.4.2 Micro and Meso-Simulation

Finding a Dynamic User Equilibrium (DUE) solution for a set of time-varying link and route volumes and travel times that satisfy the Wardrop's equilibrium for a given network and time-varying O-D demand pattern is non-trivial. Each traveller's best route choice depends on congestion levels throughout the trip, and these in turn depend on the route choices and progress through the network of other trip makers who leave at different times. This interdependence means that solutions are found through an iterative process, starting from some initial set of route choices, and gradually improving them. A practical goal of many current DTA models is to find something close to equilibrium within a reasonable amount of time.

Several different approaches have been tried over the years to deliver practical ways of solving these problems. As computer power has increased new and better software has led to more interesting and persuasive solutions. The approaches could be classified under the labels meso and micro-simulation. Meso-simulations models came first. One approach was to use route choice simulation via packets of vehicles released sequentially during a time period, as treating them one by one was not possible at the time. This was the approach followed by CONTRAM (Leonard and Gower 1982); the cost of using each path is calculated from cost-flow and queueing formulae, and the path costs are then updated. This process is iterated until a degree of convergence is achieved.

Another approach was to use platoon dispersion formulations, as those successfully used in TRANSYT (Robertson 1969), to represent the movement of vehicles and their interaction at different types of junctions. This is the approach used in SATURN (Hall *et al.* 1980) by dividing the period of interest into shorter time intervals, typically 10 or 15 minutes long. Each time interval is then treated as a steady-state assignment problem. This captures some of the effects of the build-up of congestion but still assumes that all vehicles in the same time interval are faced with the same set of costs.

Moreover, SATURN cleverly combines a platoon-dispersion simulation module with a good equilibrium assignment module. The simulation module is based on the use of cyclic flow profiles to represent the movement of platoons of vehicles over a network taking good account of the interaction of different flows at roundabouts, signal-controlled and priority junctions. It needs information about the volume on each movement (represented by a link) on the network to estimate capacity, queues and delays. Therefore, an assignment model is required to load a trip matrix onto the network and obtain an estimate of these flows. This is achieved through a separate assignment model which can perform Wardrop's selfish and stochastic user equilibrium assignments. The link between the two is through link volumes (from assignment to simulation) and through speed–flow relationships (from simulation to assignment), as depicted in Figure 11.5.

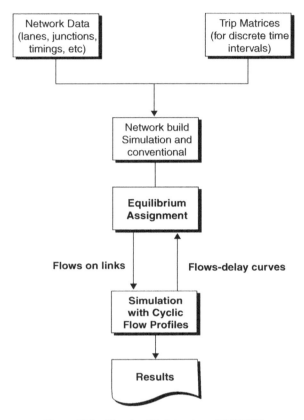

Figure 11.5   The simplified structure of SATURN

The simulation model is used, therefore, to generate suitable cost–flow relationships for the assignment problem. The cost–flow relationships are produced for each link in terms of the flow on the link itself, and take the form of a polynomial:

$$C(V_a) = a_0 + a_1 V_a^n$$

However, these relationships are calculated from the current simulation model so that they take into account the interaction and constraints generated by the flows on the other links in the network. In fact, several iterations of the simulation–assignment cycle must be performed before the whole process converges to a self consistent set of flows and costs.

Improved computer power has meant that it is now possible to simulate the movements of vehicles individually, thus generating a group of micro-simulation models. These models are based on a combination of traffic engineering relationships, car following, lane choice, gap acceptance/merging models including the treatment of pedestrians, motorcycles and trams. Micro-simulation models are very powerful and most of them include a visualisation module that produces good animations of traffic on the network. Micro-simulation models offer a large number of parameters for calibration, including some that relate to the driving culture of a city or country, for example parameters for 'aggression', 'anticipation' and the variability of gap acceptance with queue length.

The visualisation/animation modules are very useful in two main areas. First of all, they provide a good environment to verify the reasonableness of the modelled traffic behaviour and therefore assist model calibration. Second, and this has been most valued, they are very persuasive tools to demonstrate problems and solutions to decision makers. Herein lays a risk. Sometimes a poorly calibrated model may produce very persuasive animations thus supporting solutions that may not be the most appropriate. Some of the best known micro-simulation models include AIMSUN (http://www.aimsun.com), VISSIM (http://www.ptv.de) and PARAMICS (http://www.paramics.com). In general terms these packages have a more detailed simulation of traffic dynamics and delays but as yet a less rigorous treatment of equilibration.

11.4.4.3 Equilibrium and Simulation

There seems to be a degree of conflict between a very detailed and accurate treatment of the dynamics of traffic delay and equilibrium. In a congested and well connected network, like those existing in urban areas, the cost on a link depends not just on the flow on that link but on all other flows in the network (albeit especially on those joining the same junction). The flow delays functions are, therefore, non-separable in the sense that they cannot be written as a function of the flow on the link alone, so we get:

$$C_a = C_a(V_1, V_2, \ldots, V_a, \ldots, V_n)$$

The strict condition for the convergence of this type of scheme requires that the delay on a link depends mainly on the flow on the link itself and more weakly on flows on the other links (Sheffi 1985). In practice, however, this condition is not satisfied as delays at, for example, priority junctions and roundabouts depend primarily on the flows on the links having priority (circulating and main-road flows respectively).

For example, SATURN attempts to *diagonalise* the flow-delay formulations after simulation. If we fix all flows but that on link a and we vary V_a between, say zero and the capacity at a, then we can 'calibrate' a cost–flow relationship that, in this iteration, depends only on V_a. We can then perform a conventional Wardrop equilibrium assignment using, for example, the Frank–Wolfe algorithm, obtain a new set of flows on all links and run the simulation program again.

Example 11.5 Consider the simple network in Figure 11.6 corresponding to two routes from an origin to a destination merging into one. The total flow is 100 vehicles from A to Z.

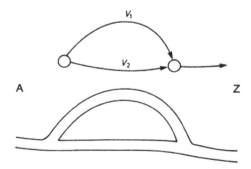

Figure 11.6 A simple network with a merge or give-way junction

Consider first the case in which both streams perform a merge operation; therefore delays on each depend also on flow on the other link. Assume now that the cost–flow relationships are as follows:

$$C_1(V_1, V_2) = 8 + 0.3\,V_1 + 0.2\,V_2$$
$$C_2(V_2, V_1) = 13 + 0.4\,V_2 + 0.2\,V_1$$

This can be solved to find a single equilibrium point at $V_1 = 83.5$ and $V_2 = 16.5$ with a minimum cost of 36.35. However, it is illustrative to show a range for values for V_1 and the corresponding link and total expenditure:

V_1	C_1	C_2	Expenditure
0	28.0	53.0	5300
10	29.0	51.0	4880
20	30.0	49.0	4520
30	31.0	47.0	4220
40	32.0	45.0	3980
50	33.0	43.0	3800
60	34.0	41.0	3680
70	35.0	39.0	3620
80	36.0	37.0	3620
83	36.3	36.4	3632
84	36.4	36.2	3636
90	37.0	35.0	3680
100	38.0	33.0	3800

As can be seen, the solution is a unique, stable equilibrium point. If some flow switches to link 2 then that link has increased delay and therefore drivers will come back to the original route. The same is true if more traffic switches to link 1. The fact that the total expenditure is minimal at another point, approximately $V_1 = 75$, is another example of the difference between social and selfish user equilibrium.

Consider now a slightly different problem with the same type of network. Now the junction is of a give-way type for link 1; link 2 has right of way and therefore its travel time does not depend on flow on link 1. The new relationships are now:

$$C_1(V_1, V_2) = 8 + 0.1V_1 + 0.2V_2$$
$$C_2(V_2, V_1) = 20 + 0.05V_2$$

The same type of table can be used to illustrate possible solutions to this assignment problem as shown below. In this case, the solution $V_1 = 60$ and $V_2 = 40$ is not stable. A switch to link 1 will decrease costs on that link faster than on link 2, therefore precipitating the solution $V_1 = 100$ and $V_2 = 0$. However, a switch in the other direction, that is to link 2, has the opposite effect, increases costs on link 2 slower than on link 1 therefore leading to another solution: $V_1 = 0$ and $V_2 = 100$. These two extreme solutions are stable albeit not with equal costs by each route; however, these two are UE solutions as the costs of the paths not used are greater than the costs on the paths used. Any departure from these extreme points will result in new cost pulling the solution back to the starting point. Note that the equations chosen are simple but not unreasonable. Observe too, that the equation for the non-priority flow shows that delay depends mainly on flow on the priority link, therefore violating the requirement for a unique solution.

V_1	C_1	C_2	Expenditure
0	28	25.0	2500
10	27	24.5	2475
20	26	24.0	2440
30	25	23.5	2395
40	24	23.0	2340
50	23	22.5	2275
60	22	22.0	2200
70	21	21.5	2115
80	20	21.0	2020
90	19	20.5	1915
100	18	20.0	1800

The fact that the solution $V_1 = 100$ is preferable because of lower overall expenditure is only relevant in terms of network design. For example, we may wish to direct drivers to choose link 1 and ignore link 2. Without this advice drivers may find either of the two extremes or even one on a particular occasion and the other the following day. Reality may be non-convergent to a stable equilibrium solution; good assignment models may fail to converge simply because they represent well this feature of reality.

SATURN and models like it therefore, can only be said to provide a reasonable practical approximation to the ideal of Wardrop's equilibrium in congested urban areas. They normally offer practical indicators to estimate how close to a possible equilibrium the iterative process has been able to reach at any one stage. Meso and micro-simulation models do represent, however, the state of the art in detailed traffic assignment for the design of traffic management and other schemes in urban areas.

11.5 Departure Time Choice and Assignment

11.5.1 Introduction

Peak spreading is a phenomenon widely observed in most large urban areas. As congestion increases, drivers start choosing different departure times to avoid the worst delays and therefore the duration of the peak is increased. In very large and congested urban areas it is not uncommon to observe extended peaks (morning and evening) and an interpeak period with quite high levels of flow and delay.

The change of departure time has been recognised in many cases as the second most likely response to changes in travel conditions, the first one being the change of route. This is mostly due to efforts to avoid the worst of congestion but it will increasingly reflect also more enlightened pricing structures for toll roads and road user charging as well as public transport fare systems.

Traditional approaches to modelling this phenomenon have been very simple. It is always possible to adopt pragmatic assumptions about the duration of the peak in the future and how expected demand is going to be spread over this period. This requires only simple factoring of demand for future peak periods in order to generate reasonable levels of congestion and delay. However, these pragmatic approaches lend themselves to arbitrary decisions that will affect the evaluation of schemes and policies, and ignore the fact that travelling at a less desirable time increases the disutility of travel.

We outline the key issues in modelling time of day choice. Because of its close relationship with assignment and delay, this theme integrates both assignment and choice modelling. This section first considers current thinking behind time of travel choice and then looks at the associated supply models. Finally, a simple combined departure time choice and assignment model is presented together with its current limitations and pointers for improvement.

11.5.2 Macro and Micro Departure Time Choice

It is useful to distinguish between macro and micro time of travel choice. Macro time choice involves the selection of travel between broad time periods (say 2–3 hours), for example the decision to go shopping at an off-peak period instead than at 05:00 PM. Micro departure time choice is related to the phenomenon of peak-spreading. As congestion increases in a city, some travellers will choose to depart a bit earlier or a bit later than originally desired in order to avoid the worse of congestion.

In principle, macro departure time choice can be modelled as a logit choice between travelling at different periods. Each period will offer some advantages in terms of desirability and disadvantages in terms of travel time and costs (parking and/or congestion charging). However, if the demand models use the typical division of time into two (say 3 hours) peak periods and an inter-peak the freedom of most trips to transfer between them will be severely constrained: few work trips, for example, could move outside the three-hour peak periods entirely, and such a mechanism might be applied predominantly for discretionary trips as opposed to the journey to work or education.

To model macro choices, it is necessary to know what proportion of each type of trip takes place in each period. This information is best collected from household survey data that contains complete tours. At a macro level, trips must be allocated to a discrete time period even those which start and finish in different periods. An incremental logit model (see Chapter 12) can then be used to modify the total number of trips of each type in each time period according to the changes in the mean generalised costs in each period.

In these cases, it will be important to apply different sensitivity parameters to different trip purposes, since travellers to work/education and business, for example, are less likely to reschedule their activities than shoppers.

11.5.3 Underlying Principles of Micro Departure Time Choice

A basic concept in micro departure time choice is that travellers have a *preferred time* of travel and any shift away from it incurs disutility, known as *schedule disutility*. The preferred time of travel may be defined as the preferred departure time or preferred arrival time, the second one being more important for certain activities (e.g. work with a fixed starting time, business meetings, theatre). The schedule disutility can be added to the travel time disutility to express a combined utility function for travel with variable departure time.

The work of Small (1982) inspires most applications of these ideas. If we focus on arrival time, Small's function takes the following form:

$$U(\tau) = -\alpha \cdot C(\tau) - \beta \cdot \text{SDE}(\tau) - \gamma \cdot \text{SDL}(\tau) - \delta \cdot d_L(\tau)$$

where τ is the arrival time and C is the travel duration, expressed as a function of the arrival time, since traffic conditions vary by time of day. SDE and SDL are called the *early schedule delay* and the *late schedule delay*. SDE and SDL express the difference between the chosen time of arrival and the preferred arrival time (PAT), in the case of early and late arrival respectively. Therefore, SDE and SDL can be defined as:

$$\text{SDE} = \max\,(\text{PAT} - \tau, 0)$$
$$\text{SDL} = \max\,(\tau - \text{PAT}, 0)$$

The parameters α, β, γ and δ are positive. α, β, and γ measure the disutility associated with a unit of increase in C, SDE and SDL respectively; δ is a fixed penalty for late arrival, and d_L is a dummy variable for late arrival (equal 1 if $\tau > \text{PAT}$ and 0 otherwise). The fixed penalty is often omitted from the function and subsumed within the utility parameter for late schedule delay γ.

The utility function defined above can be regarded as the sum of a travel duration term $(-\alpha\, C\,(\tau))$ and a term expressing the variation in utility associated with the arrival time per se $(-\beta\,\text{SDE} -\gamma\,\text{SDL} -\delta\, d_L)$, referred to as the schedule utility term. This term is maximised when travellers arrive at their preferred arrival time $(\tau = \text{PAT}$, making the schedule utility equal to zero). Therefore, when travel duration is constant and no trade-off is possible between travel duration and schedule utility, the distribution of actual arrival times is identical to the distribution of PATs. However, when travel duration varies by time of day, travellers will shift from their preferred arrival time if the schedule disutility is outweighed by the gain from reduced travel time, resulting in a distribution of actual arrival times wider than the distribution of PATs.

The parameters for these combined utility functions can be estimated by stated preference/revealed preference techniques; see for example Small (1982) and Bates (1996). An example is shown in Figure 11.7, where the y-axis is the schedule utility in travel duration units (hours) and the x-axis is the arrival time either earlier or later than PAT (hours).

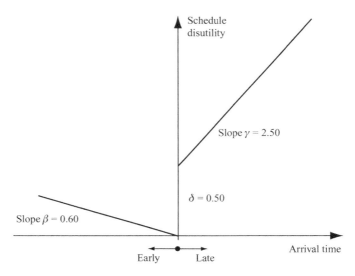

Figure 11.7   Idealised schedule (dis)utility (equivalent minutes of travel time) based on small's basic model

The coefficients in the figure reflect an idealised disutility function. The asymmetry of the function is something observed in many stated preference/revealed preference studies: a 5 min delay in arrival is generally perceived as worse than arriving 5 min too early. In this idealised case, an arrival 30 min earlier than PAT would be justified if the individual could achieve a travel time saving of more than 0.30 h or 18 min. An arrival 30 min later than PAT would incur a fixed penalty equivalent to 30 min of travel time plus an additional penalty of 1.25 h or 90 min. Therefore, this 30 min late arrival would only be justified if the individual could save more than two hours of travel time, an unlikely event. Of course, different (groups of) individuals will have different values for α, β, γ and δ and also different PATs.

The basic formulation where the penalty term is omitted has been extended by Hyman (1997) to include an *indifference band* around the PAT, during which arrivals incur no schedule delay. Hendrickson and Plank (1984) proposed a quadratic form of the utility functions, and Polak *et al.* (1991) proposed a piecewise linear model. Addison and Heydecker (1999) have also examined three classes of smooth functions, namely the sheared hyperbola, the superhyperbola and a simple non-convex function, the Witch of Agnesi. Despite these efforts, most practical applications have relied on linear functions, with or without a fixed disutility for late arrival δ.

11.5.4   Simple Supply/Demand Equilibrium Models

The utility functions associated with travel demand by time of day can be combined with supply characteristics to provide an equilibrium time-dependent demand profile emerging from the interaction of travellers' preferences and choices. Earlier models considered simple network types, consisting of one origin-destination pair connected by a single link with a bottleneck in between (Figure 11.8). Vickrey (1969) examined equilibrium with a fixed number of identical commuters travelling through a single link, where flow is uncongested (travel time is constant and equal to zero for simplicity), except at a bottleneck with fixed capacity that causes delay directly proportional to the length of the queue. Applying the basic principle of equilibrium, namely that no commuter can increase their overall utility by altering their departure time, Arnott *et al.* (1993, 1994) extended Vickrey's model, calculating the resulting departure profile of the commuters.

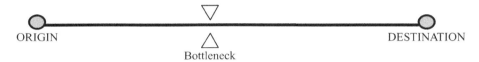

ORIGIN   Bottleneck   DESTINATION

Figure 11.8   Simple network type

Figure 11.9 illustrates the Arnott *et al.* (1994) equilibrium departure profile for homogeneous commuters. This is fully defined by closed-form expressions for the departure rates (q_1, q_2), the arrival times of the first and last arrival (τ_1, τ_2) and the switching time τ_p at which the departure rate changes from q_1 to q_2 and the maximum queue occurs.

Several authors have extended Vickrey's model to account for *heterogeneity* among travellers in their PATs and/or in the parameters associated with travel duration and schedule delay. Hendrickson and

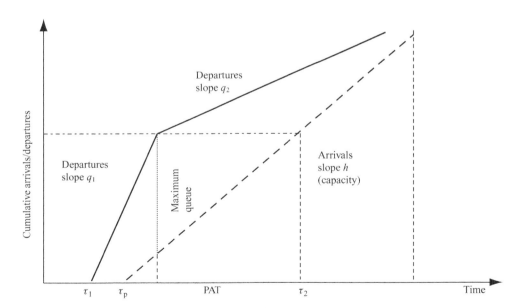

Figure 11.9   Equilibrium departure profile for homogeneous commuters

Kocur (1981) considered Vickrey's problem when travellers have a distribution of PATs. Arnott *et al.* (1994) approached the issue of heterogeneity by segmenting the population into homogeneous subgroups according to the values of their PATs and utility parameters.

11.5.5 Time of Travel Choice and Equilibrium Assignment

Practical applications of these principles require casting the problem in the context of variable demand equilibrium assignment modelling, since it requires both the flows and the level of demand for every time period (slice) to be determined. Given the progress discussed earlier in combining equilibrium assignment formulations with logit choice models, it appears natural to use a similar approach to include time of travel choice as well. Willumsen *et al.* (1993) assumed that, for each O–D pair, $C(\tau)$ was variable in the peak (and calculated through equilibrium assignment) but constant in time periods outside it; this, and the use of linear-in-the-parameters Small-like utility functions, enabled them to cast the problem in a simple combined logit choice and equilibrium assignment formulation. The time of travel choice is then made discrete: travel 'now' or travel during an 'earlier time slice' or during a 'later time slice'. Similar approaches have been put forward by Hendrickson and Plank (1984) and Cascetta *et al.* (1992).

These approaches, although superior to ignoring time of travel choice altogether, have a number of limitations:

- They ignore any interaction *between* time periods. Trips displaced from the peak to other time periods will increase travel times in them and therefore new calculations for $C(\tau)$ will be required. As it is generally not possible to estimate a function for $C(\tau)$ the full treatment for time of day choice is very difficult and therefore, this approach is an approximation of the dynamics between different time periods.
- Time of departure should be a continuous variable and its 'discretisation' into time slices is a coarse approximation. The use of relatively small time intervals (say 15 min instead of peak/off-peak hours) is an improvement but still an approximation.
- Linear-in-the-parameter logit formulations may be particularly flawed in the case of time of day choice. This is because the alternatives are almost certainly correlated in this case, as travel times on one time slice depend on travel times on other time slices.

The problem of interaction between time slices with an improved departure time model has been tackled in a practical manner by HCG *et al.* (2000) in the form of HADES (Heterogeneous Arrival and Departure times based on Equilibrium Scheduling theory). HADES is a departure time model for heterogeneous travellers, which interfaces with external commercial assignment software. Taking into account network travel times and travellers' PATs and utility parameters, HADES produces a time-dependent O–D matrix. The solution to the equilibrium problem is approached by iterating between the demand component (HADES departure time model) and the supply component (external assignment model) as illustrated in Figure 11.10.

The use of logit choice models in departure time choice has been criticised by Bates (1996), among others. This is partly due to the assumption of independence of the random components of the utilities of different alternatives in the multinomial logit model. Additionally, this model cannot accommodate heteroscedasticity problems likely to arise if error terms are proportional to a power of the schedule delay increasing with greater late or early shifts from the PATs.

The assumption of independent random utility components is restrictive and unrealistic in this case as adjacent intervals are very likely to be correlated, because the unobserved attributes (random component of the utility) affect the desirability of the alternatives in a similar way. Small (1987) also stresses the

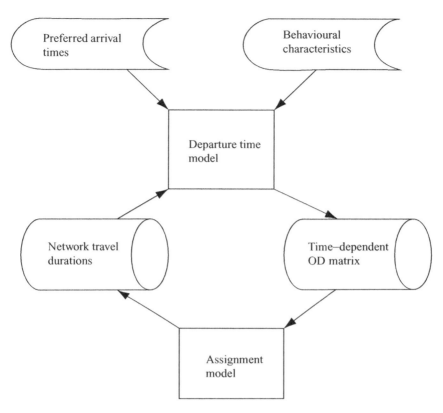

Figure 11.10   Operation of HADES in conjunction with an assignment model

fact that correlation usually arises when the dependent variable is only a discrete representation of an underlying continuous variable; this is the case for the time variable.

However, logit choice models have been extended and improved in a variety of ways to accommodate various patterns of stochastic correlation among alternatives, as discussed in Chapters 7 and 8, relaxing the assumption of the independence of random components across alternatives.

Several researchers have pointed out that a richer representation of time of travel choice behaviour could be achieved within the context of all the activities undertaken by trip makers in each tour. This is certainly correct, but as indicated by Mahmassani (2000), it is not a simple problem and we must perhaps wait until improved passive data collection methods become common practice. Novel measuring techniques using global positioning systems (GPS) and the now ubiquitous mobile phones and personal digital assistants (PDAs) are likely to revolutionise data collection in this field.

11.5.6   Conclusion

The existing literature indicates a diversity of adopted approaches in modelling departure time choice, as well as lack of consensus. There is still much work to be carried out, both theoretically and in practice, to bring this important area of travel behaviour into the mainstream of transport modelling.

Dynamic user equilibrium approaches seem to offer possibilities for incorporating departure time choice, possibly within a stochastic context and a robust network performance sub-model predicting travel times on a continuous basis. However, such an explicit treatment of time requires a detailed description of flow through the network as well as robust solution algorithms; this makes it both analytically and computationally demanding. Its practical implementation in this form must await further developments in these fields.

A better avenue seems to be to try to overcome the limitations of the models proposed in software like HADES, both in terms of their internal consistency and of the time of travel choice model (Polak 1999). However, the clear importance of this behavioural response to congestion (and differential pricing) makes it important to explore practical and better ways of incorporating these effects into most transport models for congested urban areas.

Exercises

11.1 A 12-kilometre expressway connects two urban areas. The supply function for each of the three lanes per direction of the link may be approximated by

$$t = 20 + q/200$$

where t is the travel time in minutes and q the flow per lane in passenger car units (PCU) per hour. The road is normally used by cars and express (non-stop) buses only; the corresponding vehicle travel times are t_c and t_b. The bus service has a peak-hour frequency of one bus per minute. The demand function for car travel has been estimated to be:

$$V_c = 3480 - 60t_c$$

where V_c is the total car flow per hour and direction. In a similar way, the demand function for bus trips is thought to be:

$$V_b = 4200 - 75t_b$$

where V_b is the number of passengers per hour and direction. You may assume that both t_c and t_b can be calculated from the above supply functions and that a bus is equivalent to 2 PCUs.
   (a) What is the initial equilibrium state? If a bus has 60 seats, what is their load factor (occupancy divided by capacity)?
   (b) One of the lanes is now taken for exclusive use by buses. What is the new equilibrium state and the new load factor for buses?
   (c) Discuss the assumptions implicit in the demand functions used above.
11.2 Two cities 60 kilometres apart are connected by a two-way road over which cars operate throughout the day. The peak-hour demand for travel by car between the two cities is thought to be well described by the following function:

$$q = 6000 - 1500t$$

where q is the demand in vehicles per hour and t the travel time in hours. The travel times versus flow relationship for the road is:

$$t = 0.90 \exp(0.0003q)$$

   (a) Estimate how many vehicular and person trips per hour are made under equilibrium conditions if each car carries 1.5 passengers on average.
   (b) A frequent (but slow) rail service is now implemented between the cities, where each train has a nominal capacity of 300 passengers. During the peak hour the rail company is prepared to run

a train every 10 minutes with an estimated travel time of 90 minutes. If passengers are assumed to use the fastest mode available, is this a sensible level of service?

11.3 Consider the network and conditions described in Exercise 10.3.

    (a) Express the objective function of the mathematical programme corresponding to Wardrop's selfish equilibrium in terms of the flows and travel time-flow relationships in the figure.

    (b) Calculate the equilibrium flows on each link and the travel time for each group of travellers. Calculate the value of the objective function above under equilibrium conditions and the total expenditure in travel time in the system.

    (c) Local traffic engineers have decided to install speed restrictions on link C–D so that the new travel time versus flow function is:

$$t = 5.2 + 0.001q$$

Calculate the new equilibrium conditions in terms of flows and travel times and show that under these conditions the total expenditure in travel time in the system is less than in (b).

11.4 The network in Figure 11.11 is loaded during the peak hour with 100 vehicles travelling from A to D. The equations in the network show the travel time on each link in minutes as a function of the flow q on the link in vehicles per hour. All links are unidirectional as shown.

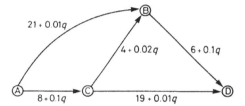

Figure 11.11   Simple network for Exercise 11.4

    (a) Identify the minimum-cost routes used, their flows and their corresponding equilibrium costs. What is the total expenditure in travel time in the network?

    (b) Assume that link CB is pedestrianised and therefore unavailable to vehicular traffic. Identify the new equilibrium flows, costs and total expenditure in travel time in the network.

    (c) Discuss your results.

12

Simplified Transport Demand Models

12.1 Introduction

For many years the main emphasis in transport modelling has been to enrich their behavioural content and improve data-collection methods as a means to enhance their accuracy, realism and reduce costs. A parallel line of research has sought to improve transport modelling by emphasising the use of readily available data and the communicability of simpler model features and results. This stream of research has had an impact in practice as it offers not only reduced costs but also simplified data-collection and processing requirements. The interest in simplified modelling techniques has spanned more than 30 years (see for example the compilation in Ortúzar 1992). As consultants and local authority modellers are often asked to study transport proposals in very short time spans, the development of better and sounder simplified methods will always be welcome.

The idea of using simpler and quick response models is not new. The practice of not using any formal model for transport project assessment is much more prevalent than what official documents and technical literature would lead one to believe. Of course, the idea of not using any formal model simply means that decision makers are using their own, mental models, to make decisions. These may be quite powerful and certainly more sensitive to political and social variables than any formal mathematical effort.

Mental models are formed and refined through observation, analogies, discussions, experimentation and mistakes. Mental models are indeed essential to make use of formal ones, interpret their results and add considerations normally outside their scope. For this end, the limited numerical processing ability of mental models is not a major limitation. However, mental models have two major weaknesses: (i) sometimes they fail completely, for example, to consider the explosive implications of exponential growth or the interconnections between seemingly unrelated decisions on taxation and mode choice; (ii) they cannot normally be 'opened up' to discuss them and qualify the recommendations resulting from their use. They are, therefore, more difficult to transfer to other users.

There is a whole range of modelling approaches in between the extremes of using no formal models at all and employing the most advanced and complex simulation techniques. One of the ways of looking at these is to consider the manner in which different approaches represent space, and hence distance, the key element in transport. Some models ignore space completely. These are usually of the kind concentrating on the financial implications of subsidies, taxation, and so on. They may be simple elasticity models, sometimes used to discuss fare increases or changes to petrol prices and car taxes. In other cases they may include more complex interactions, for example, between road, petrol and car taxes, and car ownership and use.

Some authors have advocated the use of structural modelling techniques; see for example the interesting work of Roberts (1975) in respect of fuel consumption. In this case a directed graph is often used to connect elements in the transport system, for example, the number of cars, fuel tax, improved fuel consumption, pollution emissions and costs. Weights could be attached to these linkages to represent the relative strength of each relationship.

If weights are replaced with formal equations, calibrated from actual observations, one ends up with a non-spatial interaction model. Khan and Willumsen (1986) developed a model of this kind to enhance the study of car ownership in less developed countries; the philosophy behind their model was that in developing countries car ownership should not just be forecast but examined together with its implications for resource allocation to roads and fuel consumption. The model included, in addition to the variables above, functions representing fuel consumption and the need for additional expenditure on road maintenance and new construction. Some of these, in particular construction and the importation of new cars, have severe implications for the balance of payment in these countries and should be explored before deciding on a policy relaxing restrictions to car ownership and use, see chapter 15.

A better representation of space can be obtained with idealised models of the type first proposed by Smeed (1968) and also used by Wardrop (1968) to study, among other policy issues, the limits of car commuting in urban areas. As more people use cars for the journey to work, more space needs to be devoted to roads and parking until radical changes are needed to the nature of the urban area. These models have seldom been used for decision making but have served to illustrate important policy issues.

The next stage in space modelling involves simplifications to more conventional modelling approaches as addressed in this book. Sketch planning models have been developed specifically to provide quick response and limited data-collection requirements; they are discussed in section 12.2. Increasing the degree of realism, we then discuss the idea of using simplified incremental split models in section 12.3. Section 12.4 covers an important group of models which make use of readily available data, in particular traffic counts. The special characteristics of transport systems in corridors enable another type of simplification, as discussed in section 12.5. Finally, the interpretation of model output and the use of models would also benefit from special training techniques; gaming simulation has been put forward as assisting in this area and it is discussed in the last section of this chapter.

12.2   Sketch Planning Methods

Sketch planning models have been put forward as tools for long-range planning by many authors, as reported in OECD (1974) and Sosslau *et al.* (1978). They are models with a greater level of detail than the idealised network approaches mentioned in the previous section but much simpler than conventional computer suites. This feature facilitates the analysis of broad transport and land-use strategies at a coarse level of resolution, without requiring large amounts of data or the rigid assumptions of ideal space models. Their practical implementation ranges from scaled-down conventional aggregate modelling suites of programs to *ad hoc* approaches developed from some simple ideas and assumptions.

Most sketch planning methods rely considerably on the transfer of parameters and relationships from one area or country to another. Only certain aspects of the models are made location dependent, usually network characteristics, population, income levels, and so on. Perhaps at one extreme of sketch planning models are those relying heavily on assumed regularities in human behaviour in the transport field. A typical example of this was the UMOT (Unified Mechanism of Travel) model proposed by Zahavi (1979). This model was based on the assumption that the following relationships were transferable over time and space (regions, countries):

- the average daily travel time per traveller, i.e. an assumption of constant travel time budgets;
- the average daily travel expenditure (money) as a function of income and car ownership, i.e. a money budget relationship;
- the average number of travellers per household as a function of household size and car ownership;
- the unit cost of owning and running a car;
- the speed-flow relationship by road type;
- the threshold of daily travel distance that justifies owning a car.

These relationships were developed by Zahavi following an extensive compilation of data bases from all over the world. UMOT only required as location-specific input the following:

- the number of households and their sizes in the study area;
- the income distribution of households;
- the unit cost of travel by mode;
- the length of the road network in the study area.

An interesting feature of UMOT was that it produced the following results as output:

- car ownership per household by income group;
- aggregate modal choice for the whole city;
- average travel times and speeds;
- other performance indicators like total expenditure and travel times.

UMOT gained some support as a tool for testing broad policy options, for example on fiscal policy (taxation), on fuel and car ownership, pricing policy for public transport and even broad infrastructure investment programmes. However, the model was tested by Downes and Emmerson (1983) and Willumsen and Radovanać (1988), among others, who found that, in general, it did not represent situations in other countries well, not even at a very high level of aggregation. In fact, the transferability of relationships and budgets was not found to be consistent enough to warrant the use of UMOT, even after improvements to the models were implemented by the authors.

Sketch planning techniques seem to offer advantages in terms of simplicity, fast response and low data requirements. However, very often they rely too heavily on the transfer of relationships and parameters from one context to another. This detracts from the analysis unless it is performed only as an initial coarse sketch to select possible solutions for more detailed consideration.

12.3 Incremental Demand Models

A number of approaches have been put forward to perform quick demand analysis of the impact of changes in fares, levels of service (LOS), or other attributes of a particular mode. The best known methods fall under the heading of incremental elasticity analysis and pivot-point modelling. In both cases, the aim is to estimate small changes in demand as a result of (small) changes in one (seldom more) of the LOS attributes, at a given point in time.

12.3.1 Incremental Elasticity Analysis

Consider an initial situation where the level of demand for a mode is T_0, its level of service S_0 (probably a vector including attributes like travel time, fare, waiting time, etc.). The elasticity of demand with respect to LOS (at a given level of demand and LOS) is given by:

$$E_s = \frac{S_0}{T_0} \frac{\partial T}{\partial S} \approx \frac{S_0}{T_0} \frac{T - T_0}{S - S_0} = \frac{S_0}{T_0} \frac{\Delta T}{\Delta S} \tag{12.1}$$

There is an initial distinction between *arc* and *point* elasticities. The right hand side of (12.1) is an expression for arc elasticity. As S approximates S_0 the elasticity will approach the exact *point* value $\frac{\partial T}{\partial S}$. In general, point elasticites are more often estimated from demand models and arc elasticities from time series data. This definition leads to:

$$T - T_0 = \frac{E_s T_0 (S - S_0)}{S_0} \tag{12.2}$$

The left-hand side of this equation is the estimated change in demand for the mode due to a relative change in the level of service of size $(S - S_0)/S_0$. This type of calculation is often used during fare or frequency reviews for public-transport services.

This is, of course, an approximation which assumes that we have calculated E_s *beforehand* (perhaps from time series data), that this elasticity is constant (or that the demand function is linear–not very likely) and that everything else remains the same. This result is a reasonable approximation for small changes in the LOS variables.

Example 12.1 The fare/demand elasticity of public transport is often taken to be -0.30. If a public-transport system carries 200 000 passengers in the peak period at an average fare of 80 pence/trip:

- Estimate the fall in the demand if the average fare increases by 2.5%.
- Find out how sensitive is the result to the elasticity value.

In this case $T_0 = 200\,000$; $E_s = -0.30$, and $(S - S_0)/S_0 = 0.025$, so using (12.2) we get:

$$T - T_0 = -0.30 \times 200\,000 \times 0.025 = -1500 \text{ passengers}$$

If $E_s = -0.2$, the expected reduction in patronage would be 1000 passengers; if it is -0.4, it would then be 2000 passengers.

It is also possible to define a cross-elasticity, that is the change in demand of one alternative (mode, destination, route) when the LOS of another alternative changes; say the change in demand for inter-city rail when air travel fares increase.

We define cross-elasticities of demand for mode i with respect to attributes in mode j as:

$$E_{ij} = \frac{S_j}{T_i} \frac{\partial T_i}{\partial S_j} \approx \frac{S_j}{T_i} \frac{\Delta T_i}{\Delta S_j}$$

Elasticities for a few types of demand functions with respect to changes in one attribute S of the LOS are given below:

Type	Functional Form	Elasticity
Linear	$T = \alpha + \beta S$	$E = \dfrac{\beta S}{T} = \dfrac{1}{1 + \alpha/\beta S}$
Product	$T = \alpha S^\beta$	$E = \beta$
Exponential	$T = \alpha \exp(\beta S)$	$E = \beta S$
Share	$p_i = \dfrac{T_i}{\sum_j T_j}$	$E_{S_i}(p_i) = 1 - p_i$ $E_{S_j}(p_i) = -p_j$

There are plenty of compilations of elasticities from around the world. One of the most useful is that from the Victoria Transport Policy Institute (www.vtpi.org) compiled by Todd Litman (http://www.vtpi.org/tdm/tdm11.htm). One would expect all elasticities with respect of components of generalised cost of travel to be negative (an increase in cost results in a reduction in demand).

Note that point elasticities, estimated from an analytical function, are symmetric: the absolute value of a positive change is the same as that of a negative change. However, we know from experience that this is not the case. The impact of a 10% increase in fares is greater than that of a 10% reduction; people place greater value to a loss than to a gain of the same magnitude, an issue that we also revisit in another chapter. Economists also distinguish between short and long-term elasticities based on the fact that it may be difficult to adapt instantly to some changes in costs. For example, a moderate change in fuel costs may have little impact on travel in the short term as people will continue to travel to work. However, in the long term people will change jobs and/or place of residence and, in considering these choices, they will also take into account travel costs and the availability of public transport, something they could not do in the short run. We would expect, therefore, that long term elasticities will be larger than short term ones: demand should be more elastic in the long term.

It should also be noticed that if the change in costs is large, say a doubling of fuel prices, the additional expenditure incurred by travellers will affect consumption in other goods and services as incomes and budgets are fixed in the short run. This is an 'income effect' and is the change in consumption resulting from changes in one or more prices.

Finally, one can also estimate elasticities of travel demand to changes in attributes of the traveller (for example income levels) or the region (for example GDP). We would expect these to be positive and most likely declining with per capita income levels. Evidence suggests, for example, that transport demand elasticities to GDP are greater than one in emerging countries but less than one in post industrial ones. This is important as it will help to de-couple economic development and traffic growth.

12.3.2 Incremental or Pivot-point Modelling

This method has been developed to estimate future travel demand on the basis of knowledge of the current levels of demand and changes in the LOS variables for each alternative. In this case we require knowing the demand function but not the specific values of the levels of service variables which are not to change; for example, that of parking charges in different parts of a city. The only data needed are the current market shares of each mode and the proposed changes in the LOS variables; then, an incremental form of the demand model is used to 'pivot' around the current situation.

The incremental form of the Multinomial Logit (MNL) mode choice model was first given by Kumar (1980):

$$p'_k = \frac{p^0_k \exp(V_k - V^0_k)}{\sum_j p^0_j \exp(V_j - V^0_j)} = \frac{p^0_k \exp(\Delta V_k)}{\sum_j p^0_j \exp(\Delta V_j)} \tag{12.3}$$

where p'_k is the new proportion of trips using mode k; p^0_k is the original proportion of trips by mode k; and $(\Delta V_k = V_k - V^0_k)$ is the change in the utility of using mode k, in our case generated by changes to the LOS attributes of mode k.

It is also possible to develop incremental forms for the Nested Logit model (Bates *et al.* 1987; Martínez 1987). In this case we will have a change in utility at the lower nest as: $\Delta V_i = \beta(V_i - V^0_i)$ and for choices above the lower nest the change in utility is the composite change over the alternatives at the level below:

$$\Delta V^* = \ln \sum_i p^0_i \exp(\Delta V_i)$$

Example 12.2 Consider a transport system with three modes: car, bus and rail with proportions 0.4, 0.45 and 0.15. Assume that the utility function has the following linear form:

$$V_k = -0.10t_k - 0.20w_k - 0.05C_k/I + \delta_k$$

where t_k stands for in-vehicle travel time, w_k for waiting time and C_k/I, cost divided by income; δ_k is a modal penalty.

Assume also that we are only interested in changes in frequency that would reduce expected waiting time by rail from 10 minutes to 7.5 minutes and increase that of bus from 3 to 4 minutes; therefore we would have for rail:

$$V_r - V_r^0 = -0.2(7.5 - 10) = 0.5$$

and for bus:

$$V_b - V_b^0 = -0.2(4 - 3) = -0.2$$

The change in modal share would then be:

$$p_r' = \{0.15 \exp(0.5)\}/\{0.15 \exp(0.5) + 0.45 \exp(-0.2) + 0.4\}$$

the reader can verify that this produces:

$$p_r' = 0.24 \quad \text{and} \quad p_b' = 0.36$$

In the same vein, the singly constrained incremental gravity model can be written as:

$$T_{ij} = \frac{G_i T_{ij}^0 a_j \exp(-\beta \Delta GC_{ij})}{\sum_l T_{lj}^0 a_j \exp(-\beta \Delta GC_{lj})} \tag{12.4}$$

where G_i is the total trips generated at zone i, ΔGC_{ij} the difference in generalised cost between the base and design years, and a_j growth factors reflecting changes in the destinations j.

Incremental forms for most travel choice models are not, in general, difficult to develop or implement. For example, Abraham *et al.* (1992) report on an incremental model for the whole of London handling both mode and doubly constrained gravity models for different person types and modes. This was implemented in EMME/2 taking advantage of its macro facilities. Other software has similar modules to implement incremental mode, distribution and other Logit choice models (see Willumsen *et al.* 1993).

Incremental or pivot-point model formulations are helpful as we only need to account for changes in the generalised costs or utility functions, not their complete values. Therefore, if we are not introducing new modes modal penalties can be ignored as they cancel out in ΔGC or ΔV. An additional advantage is that the model preserves the current (or base) matrices, therefore retaining any special associations detected in the data but never completely accounted for in a model; this is particularly valuable when dealing with destination choice where the gravity model has never performed sufficiently well. The incremental gravity model is expected to represent changes in the trip pattern resulting from changes in travel costs and generations and attractions.

The way pivot point or incremental models have been described is in accordance with the underlying principles of logit and gravity model development. A similar, but less rigorous, idea is to focus on changes in demand as a result of changes in certain attributes but using absolute models incrementally instead of pivot point models. The main motivation behind this approach lies in the difficulties in calibrating a distribution model that fits observations sufficiently well. It is common practice in many countries, like the UK, to spend considerable resources in collecting origin-destination (O-D) data and developing one or more robust O-D matrices (by trip purpose and time of day). It is very difficult indeed to adopt any type of distribution or destination choice model that would not distort these matrices significantly. It is highly desirable, in these cases, to use the rich information in the 'observed trip matrix' $\left[T_{ij}^0\right]$ fully and attempt to model only *changes* in trip patterns as a function of cost and trip end future states.

In this case, modellers would use absolute models but apply them incrementally. To this end an absolute (usually gravity) model is estimated for the base year $\left[GM_{ij}^0\right]$ and then used for a future year $\left[GM_{ij}^1\right]$. One approach would be to estimate the future matrix as:

$$T_{ij}^1 = \frac{T_{ij}^0}{GM_{ij}^0} GM_{ij}^1 \text{ for all } ij$$

Note that this approach is equivalent to adopting a full set of k factors in a gravity model. The problem with this is that those cells in the base year matrix T^0 that are zero will remain zero in the future; this would be unrealistic for zones that are fairly empty in the base year but are expected to have increased activity in future years. An alternative approach that avoids this problem is to employ an additive form:

$$T_{ij}^1 = T_{ij}^0 + (GM_{ij}^1 - GM_{ij}^0)$$

This has the potential danger that some cells may turn out to have negative values that should be rounded up to zero. The essential feature of these two approaches is capturing any significant difference between the base year output from a calibrated model and the observations, and to pass on these differences to future forecasts.

12.4 Model Estimation from Traffic Counts

12.4.1 Introduction

Conventional methods for collecting origin-destination information from, for example, home or roadside interviews tend to be costly, labour intensive and time disruptive to the trip makers. The problem is even more acute in developing countries, where rapid changes in land use and population shorten the 'shelf-life' of data. The need for developing low-cost methods to estimate the present and future O–D matrices is apparent.

Traffic counts can be seen as the result of combining a trip matrix and a route choice pattern. As such, they provide direct information about the sum of all O–D pairs which use the counted links. Traffic counts are very attractive as a data source because they are non-disruptive to travellers, they are generally available, they are relatively inexpensive to collect, and their automatic collection is well advanced. The idea of estimating trip matrices or demand models from traffic counts deserves serious consideration and the last decades have seen the development of a number of approaches attempting just that.

Consider a study area which is divided into N zones inter-connected by a road network which consists of a series of links and nodes. The trip matrix for this study area consists of N^2 cells, or $(N^2 - N)$ cells if intra-zonal trips can be disregarded. The most important stage for the estimation of a transport demand model from traffic counts is to identify the paths followed by the trips from each origin to each destination. The variable p_{ij}^a is used to define the proportion of trips from zone i to zone j travelling through link a. Thus, the flow (V_a) in a particular link a is the summation of the contributions of all trips between zones to that link. Mathematically, it can be expressed as follows:

$$V_a = \sum_{ij} T_{ij} p_{ij}^a, \quad 0 \le p_{ij}^a \le 1 \tag{12.5}$$

The variable p_{ij}^a can be obtained using various trip assignment techniques ranging from a simple all-or-nothing to a more complicated equilibrium assignment. Given all the p_{ij}^a and all the observed traffic counts ($\hat{V}_a$), there will be N^2 unknown T_{ij} values to be estimated from a set of L simultaneous linear equations (12.5), where L is the total number of traffic counts.

In principle, N^2 independent and consistent traffic counts are required in order to determine uniquely the trip matrix **T**. In practice, the number of observed traffic counts is much less than the number of unknown T_{ij} values. Therefore it is impossible to determine a unique solution to the matrix estimation problem. In general, there will be more than one trip matrix which, when loaded onto the network, will reproduce (satisfy) the traffic counts. Two basic approaches have been proposed to resolve this problem: structured and unstructured methods. In the structured case, the modeller restricts the feasible space for the estimated matrix by imposing a particular structure which is usually provided by an existing travel demand model, for example a gravity or direct-demand model. The unstructured approach relies on general principles, like maximum likelihood or entropy maximisation, to provide the minimum of additional information required to estimate the matrix. These two general approaches will be discussed below, but first we must consider the relationship between route choice and matrix estimation.

12.4.2   Route Choice and Matrix Estimation

Robillard (1975) classified assignment methods for trip matrix estimation from counts under two main groups: *proportional* and *non-proportional* assignment. Proportional assignment methods make the proportion of drivers choosing each route independent from flow levels. The most common example is all-or-nothing assignment and in this case p_{ij}^a is defined as:

$$p_{jq}^a = \begin{cases} 1 & \text{if trips from origin } i \text{ to destination } j \text{ use link } a \\ 0 & \text{otherwise} \end{cases}$$

Pure stochastic assignment methods such as Burrell's and Dial's also fall into this group but in these cases p_{ij}^a can also take intermediate values between 0 and 1.

Non-proportional assignment techniques take explicit account of congestion effects and therefore the proportion of travellers using each link does depend on link flows. Equilibrium and stochastic user equilibrium assignment methods are members of this group.

Non-proportional assignment techniques are thought to be more realistic for congested conditions. However, the advantage of proportional assignment methods is that they permit the separation of the route choice and matrix estimation problem; the proportion of trips using each link p_{ij}^a can be assumed to be independent of the trip matrix to be estimated. In contrast, non-proportional route choice requires the joint or iterative estimation of route choice and trip matrices so that both are consistent. In what follows, we shall assume that proportional assignment methods are a reasonable approximation to route choice; we shall discuss later the extensions needed to cover non-proportional methods.

12.4.3   Transport Model Estimation from Traffic Counts

The calibration of a gravity model was one of the first methods put forward for estimating trip matrices from traffic counts. The basic idea is to postulate a particular form of gravity model and examine what happens when it is assigned onto the network. For example, in the case of inter-urban travel the trip matrix could be:

$$T_{ij} = \frac{\alpha P_i P_j}{d_{ij}^2}$$

where P_j is the population of urban area j, d_{ij} is the distance between both areas and α is a constant for calibration, in this case the only one. If a matrix of this kind is assigned on the network we get:

$$V_a = \sum_{ij} \frac{p_{ij}^a \alpha P_i P_j}{(d_{ij})^2} = \alpha \sum_{ij} \frac{p_{ij}^a P_i P_j}{(d_{ij})^2} \tag{12.6}$$

Note that on the right-hand side of this equation the only unknown is α: the other variables are provided by external data or a good route choice model. One can generalise this model slightly and include other trip generation/attraction factors like employment, industrial production, shopping floor space, and so on. If we denote the gravity part of this model by:

$$G_{ij} = \frac{O_i D_j}{d_{ij}^2}$$

and allow several journey purposes k (or commodities if one is dealing with freight movements), one can write:

$$V_a = \sum_k \sum_{ij} p_{ij}^a \alpha_k O_i^k D_j^k / (d_{ij})^2 = \sum_k \alpha_k \sum_{ij} p_{ij}^a G_{ij}^k \qquad (12.7)$$

Here the α_k are parameters for calibration but the rest of the data are, once more, assumed to be available. It is relatively simple to see that the α_k may be estimated using least squares techniques. In this case we postulate that $V_a' = V_a + \varepsilon_a$, where ε_a is an error term. A change of variable:

$$X_k = \sum_{ij} p_{ij}^a G_{ij}^k$$

permits writing:

$$V_a' = \alpha_0 + \sum_k \alpha_k X_k \qquad (12.8)$$

where α_0 is the intercept and may be deemed to depict the part of the flow not represented by the gravity model, for example local or intra-zonal traffic. This type of approach was followed by the first researchers in this area, Low (1972) for urban areas and Holm *et al.* (1976) for planning inter-urban networks in Denmark.

Equation (12.7) has at least one obvious deficiency. If a particular O_i and a particular D_j are each doubled, then the number of trips between these zones would quadruple when it would be more likely that it should double also. To improve on this the following more conventional model can be used:

$$T_{ij} = \sum_k \left[\alpha_k O_i^k D_j^k A_i^k B_j^k f_{ij}^k \right] \qquad (12.9)$$

where α_k is a scaling parameter which enable us to use different units for T_{ij} and O_i^k, D_j^k. A_i^k and B_j^k are the balancing factors expressed as:

$$A_i^k = \left[\sum_j (B_j^k D_j^k f_{ij}^k) \right]^{-1}$$

$$B_j^k = \left[\sum_i (A_i^k O_i^k f_{ij}^k) \right]^{-1}$$

and f_{ij}^k is a deterrence function, for example $\exp(-\beta_k C_{ij})$.

Estimating this more conventional model from traffic counts represents a greater effort as the parameters for calibration are now A_i^k, B_j^k, β_k and α_k. This calls for alternative calibration methods, for example non-linear regression as used by Högberg (1976) or Robillard (1975).

Tamin and Willumsen (1989) generalised this approach following suggestions from Wills (1986) to combine in a single model features of the gravity and the intervening opportunities (OP) model. Wills proposed a flexible gravity-opportunity (GO) model for trip distribution in which standard forms of the gravity and opportunity models are obtained as special cases. The choice between gravity or opportunity approaches is decided empirically by allowing the estimation of parameters which control the global functional form of the trip distribution mechanism.

We can define a transformation δ^i_{dj} such that δ^i_{dj} equals 1 if destination j is the dth position in ascending order of distance away from i, and zero otherwise, then the ordered (opportunities) trip matrix can be obtained by the following transformation:

$$Z_{id} = \sum_j [\delta^i_{dj} \, T_{ij}] \tag{12.10}$$

While the ordering transformation δ^i_{dj} produces an ordered trip matrix, its inverse $(\delta^i_{dj})^{-1}$ allows the observed trip matrix to be recovered by

$$T_{ij} = \sum_d [(\delta^i_{dj})^{-1} Z_{id}] \tag{12.11}$$

It should be noted that this class of transformation is applicable to any variable based on the O–D matrix, notably the cost matrix and the proportionality factor, in addition to the trip matrix. We can also define a direct Box–Cox transformation such as (8.2) on a variable y as:

$$y^\tau = \begin{cases} (y^\tau - 1)/\tau & \tau \neq 0 \\ \log y & \tau = 0 \end{cases}$$

and an inverse Box–Cox transformation as

$$y^{(1/\tau)} = \begin{cases} (y\tau + 1)^{1/\tau} & \tau \neq 0 \\ \exp y & \tau = 0 \end{cases}$$

These transformations may be combined into a new function which we introduce as a convex combination in μ,

$$y^{(\tau,\mu)} = \mu y^{(\tau)} + (1 - \mu)y^{(1/\tau)}, \quad 0 \leq \mu \leq 1 \tag{12.12}$$

The proposed model can finally be written then as:

$$T_{ij} = \sum_k [\alpha_k O^k_i D^k_j A^k_i B^k_j f^k_{ij}] \tag{12.13}$$

where:

$$f^k_{ij} = \sum_d [(\delta^i_{dj})^{-1} F^k_{id}] \tag{12.14}$$

$$F^k_{id} = \left(\sum_p^d U^k_{ip} \right)^{(\tau,\mu)} - \left(\sum_p^{d-1} U^k_{ip} \right)^{(\tau,\mu)} \tag{12.15}$$

$$U^k_{ip} = \exp[(1 - \tau)\gamma_m \log D^i_{pk} - \beta_m C_{ip}] \tag{12.16}$$

and

$$D^i_{dk} = \sum_j [\delta^i_{dj} D^k_j] \tag{12.17}$$

From this general form several special cases may be derived by setting τ and μ to particular values. Three extreme cases generating specific models are easily identified: the gravity (GR), the pure logarithmic-opportunity (LO) and the pure exponential-opportunity (EO) models.

Three estimation methods were implemented by Tamin and Willumsen (1989) to calibrate the general form from traffic counts, namely: non-linear least squares (NLLS), weighted non-linear least squares (WNLLS) and maximum likelihood (ML). The general model was tested for both freight transport in

Bali, Indonesia (Tamin and Willumsen 1992) and passenger traffic in Ripon, UK (Tamin and Willumsen 1989). In the case of road haulage, even if the traffic counts were not classified by lorry type it was possible to discriminate up to nine different commodity types, one of them empty trucks. In this case proxy data to the O_i^k and D_j^k are required, for example production levels of certain commodities. The parameter α_k then plays the double role of converting these proxies first to tonnes and then to lorries.

The main conclusions from this research were:

- The GO and OP models are more time consuming than the GR model since they require more complicated algebra and procedures which take longer to solve.
- Good fit at the traffic count level produced a general good fit at the trip matrix level as well.
- Although Burrell's stochastic assignment was also used to estimate the p_{ij}^a, it gave no better fit to the traffic counts than all-or-nothing assignment.
- Although the GO was the best model in terms of matching the observed traffic counts, it cannot be guaranteed that it will also produce the best-fit to an independently observed trip matrix. In fact, it was found that the model which gives the best fit at the trip matrix level is the GR gravity model with the NLLS method and Burrell assignment.

Holm *et al.* (1976) have extended the gravity model approach to include some features of equilibrium assignment. They make use of an iterative loading with $\phi = 1/n$ (see section 10.5.4) to obtain the proportion of trips using each link. However, this is only a heuristic approximation as under strict equilibrium conditions the proportions are not, in general, unique.

Of course other, perhaps direct-demand, models could also be used in this type of estimation method. One interesting advantage of this approach is that once a demand model is calibrated it may be used for forecasting purposes too, provided future values for parameters like O_i and D_j are available or estimable.

12.4.4 Matrix Estimation from Traffic Counts

Entropy-maximising and information-minimising techniques have been used as model-building tools in urban, regional and transport planning for many years, particularly after the work of Wilson (1970). For example, we discussed the derivation of the conventional gravity model from an entropy-maximising formalism in Chapter 5. In this context, the entropy-maximising formalism provides a naive, least-biased, trip matrix which is consistent with the information available represented as constraints to a maximisation (of an entropy function) problem. In the case of the gravity model the constraints represent trip-end and total cost information.

This idea was used by Willumsen (1978) to derive a model to estimate trip matrices from traffic counts. The problem can be written as:

$$\text{Maximise } S(T_{ij}) = -\sum_{ij} (T_{ij} \log T_{ij} - T_{ij}) \tag{12.18}$$

subject to:

$$\hat{V}_a - \sum_{ij} T_{ij} p_{ij}^a = 0 \tag{12.19}$$

for each counted link a, and:

$$T_{ij} \geq 0$$

Constraints (12.19) replace the trip-end and cost constraints of the gravity model derivation. The use of Lagrangian methods permits the formal solution to this problem to be found as:

$$T_{ij} = \exp \sum_a (-\tau_a p_{ij}^a) = \prod_a X_a^{p_{ij}^a} \tag{12.20}$$

where τ_a are the Lagrange multipliers corresponding the constraints (traffic counts) and,

$$X_a = \exp(-\tau_a)$$

The availability of an old matrix, or simply a matrix estimated (or cordoned off) from another study could be accommodated to some advantage. Let **t** be this prior matrix, sometimes called a 'reference trip matrix'; the new objective function becomes:

$$\text{Maximise } S_1(T_{ij}/t_{ij}) = -\sum_{ij}(T_{ij}\log T_{ij}/t_{ij} - T_{ij} + t_{ij}) \tag{12.21}$$

subject to the same constraints (12.19) and non-negativity. This objective function is, of course, convex and the term t_{ij}, being a constant, is only there for convenience; it can actually be eliminated from the derivation of the model.

Using the same methodology and change of variables, the formal solution can be seen to be:

$$T_{ij} = t_{ij}\exp\sum_a(-\tau_a p_{ij}^a) = t_{ij}\prod_a X_a^{p_{ij}^a} \tag{12.22}$$

Example 12.3 Consider the simple network depicted in Figure 12.1. This network has two origins (1 and 2) and two destinations (3 and 4). The flows on all links are also shown in this figure.

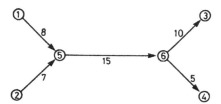

Figure 12.1   Simple network with traffic counts

It can be seen that there are only six (integer) trip matrices that can reproduce the observed flows as shown below.

Matrix	First		Second		Third		Fourth		Fifth		Sixth	
i \ j	3	4	3	4	3	4	3	4	3	4	3	4
1	8	0	7	1	6	2	5	3	4	4	3	5
2	2	5	3	4	4	3	5	2	6	1	7	0
$S(T_{ij})$	−11.07		−7.46		−5.98		−5.78		−6.84		−9.96	
$S_1(T_{ij}/t_{ij})$	−5.79		−3.69		−3.70		−5.07		−7.22		−12.20	

The entropy-maximising formalism seeks to identify the most probable trip matrix consistent with the information available, in this case five traffic counts. Incidentally, the reader can verify that only three of these counts are independent (see section 12.4.5); therefore the problem is, indeed, underspecified.

The values of the objective function $S(T_{ij})$ are also shown in this table. According to this, the most probable trip matrix would be the fourth, $\{5, 3, 5, 2\}$, as it has maximum entropy value. If a prior matrix is available then a second objective function (12.21) should be used. Assume the prior matrix $\{3, 2, 1, 3\}$ is available; the new values from the entropy function are also depicted above. The most probable trip matrix in these circumstances is the second one, $\{7, 1, 3, 4\}$. Of course, in more practical problems we cannot hope to calculate directly the entropy values of all possible matrices. Note, for instance, that reducing the number of counts increases the number of feasible trip matrices. More importantly, flows of the order of hundreds or thousands increase the number of possible (integer) trip matrices enormously. What is needed is an effective solution method not requiring matrix identification.

There are several possible methods to solve model (12.22). The most widely used one is the multi-proportional approach. This is, in essence, an extension of the bi-proportional and tri-proportional methods discussed in Chapter 5. In this case, instead of balancing the trip matrix trying to match trip-end totals (and cost-bin totals in the tri-proportional case), we undertake successive corrections to the prior trip matrix in order to reproduce the observed traffic counts. There is one correction factor X_a for each traffic count and its calculation involves the iterative estimation of these factors until the observed link flows are replicated to within an acceptable tolerance.

If no prior matrix is available, **t** can be taken as unity; in effect, an entropy-maximising formalism may be considered to generate as the most likely trip matrix, one that has the same number of trips in each cell, unless being prevented from achieving this by the constraints. In other words maximising entropy is equivalent to minimising the difference between a uniform target and the estimated matrix.

The detailed analysis of this maximum entropy matrix estimation (ME2) model and that of a related approach, based on information-minimising principles, is given by Van Zuylen and Willumsen (1980). Both models are practically equivalent and share most of their properties. The ME2 model will always reproduce the observations V'_a to within a given tolerance provided the constraints define a feasible space, i.e. equations (12.19) must have at least one solution in non-negative T_{ij}. An additional condition for the prior matrix **t** is discussed below.

It can be shown that minimising the negative of the objective function (12.21) is approximately equivalent to minimising:

$$S_2(T_{ij}/t_{ij}) = \frac{0.5(T_{ij} - t_{ij})^2}{T_{ij}} \tag{12.23}$$

This is an error-like measure of the difference between the values of t_{ij} and T_{ij}. In effect, the negative of $S_1(T_{ij}/t_{ij})$ is also a natural measure of the difference between these cell values: it is zero when $t_{ij} = T_{ij}$ and increasingly positive as the difference increases. In this sense, the estimated matrix is that closest to the prior matrix which when loaded onto the network can reproduce the traffic counts.

The model can accommodate other sources of data provided they can be incorporated as linear constraints. An example of this type may be information about the trip length distribution (TLD) thought to be realistic for the study area. This type of information can be translated into constraints equivalent to those of cost bins, as discussed in Chapter 5; for example:

$$\frac{1}{T} \sum_{ij} T_{ij} \delta_{ij}^k = P_k \tag{12.24}$$

where T is the total number of trips, P_k is the proportion of trips in cost (length) range (bin) k, δ_{ij}^k is 1 if trips between i and j have cost in range k, and zero otherwise.

Public-transport systems with a zonal or other variable fare system permit the introduction of constraints of this type to help estimate the corresponding trip matrices using passenger counts and ticket sales data (see de Cea and Cruz 1986).

Moreover, the mathematical program can also be written with a combination of equality and inequality constraints, thus enhancing the value of this type of approach. For example, the planner may know that the capacity of a link is Q_a but not have a traffic count for it; or that no more than D'_j vehicles can go to a particular destination because of parking capacity there. This type of information can be incorporated as inequality constraints, for example:

$$\sum_{ij} T_{ij} p_{ij}^a \le Q_a \quad \text{for some links } a \tag{12.25}$$

$$\sum_{i} T_{ij} \le D'_j \quad \text{for some destinations } j \tag{12.26}$$

The solution to this program is still a multiplicative model; Lamond and Stewart (1981) have shown how the multi-proportional algorithm can be extended to handle inequality constraints; therefore the same solution method may be used for this expanded model.

One of the features of the (extended) ME2 model is its multiplicative nature. This means that if a cell in the prior matrix is zero it will remain zero in the solution as well. This may be a source of problems if the cell in the prior matrix was zero by chance (i.e. because of the sampling rate adopted in the study) instead of representing an O–D pair with no trips at all. One pragmatic solution to this problem, for very sparse prior matrices, is to 'seed' the empty cells with a small value, for example 0.5 trips. The constraints, through the multi-proportional or other solution algorithm, will then ensure that some of these trips 'grow' to one or more full trips while others regain a zero value.

Example 12.4 Consider the same network as in Example 12.3 but assume now that we only have two traffic counts, on links 5–6 and 2–5 (15 and 7). Table 12.1 shows the multi-proportional algorithm as applied to this problem. The table shows first the full solution for the case of uniform (no) prior matrix, Case A.

Table 12.1 Multiproportional solution for two traffic counts

		Traffic count	Modelled flow	Ratio	\multicolumn{4}{c}{Trips per O–D pair}
					1–3	1–4	2–3	2–4
A	Prior Matrix				1.00	1.00	1.00	1.00
	Iteration	15	4.00	3.750	3.75	3.75	3.75	3.75
	1	7	7.50	0.933			3.50	3.50
	Iteration	15	14.50	1.034	3.88	3.88	3.62	3.62
	2	7	7.24	0.967			3.50	3.50
	Iteration	15	14.76	1.016	3.94	3.94	3.56	3.56
	3	7	7.11	0.984			3.50	3.50
	Iteration	15	14.89	1.008	3.97	3.97	3.53	3.53
	4	7	7.05	0.992			3.50	3.50
	Iteration	15	14.95	1.004	3.99	3.99	3.51	3.51
	5	7	7.03	0.996			3.50	3.50
B	Prior matrix				3.00	2.00	1.00	3.00
	Iteration	15	15.03	0.998	4.81	3.21	1.75	5.24
	5	7	6.98	1.002			1.75	5.25
C	Prior matrix				3.00	2.00	0.00	3.00
	Iteration	15	15.06	0.996	4.82	3.21	0.00	6.97
	6	7	6.97	1.004			0.00	7.00
D	Prior matrix				3.00	2.00	0.50	3.00
	Iteration	15	15.04	0.998	4.81	3.21	1.00	5.99
	6	7	6.98	1.002			1.00	6.00

As can be seen, it takes only five iterations to reach convergence within 5% tolerance. The solution {3.99, 3.99, 3.5, 3.5} does not coincide with the maximum-entropy solution in Example 12.3 because the number of traffic counts is not the same. Case B shows the problem with the prior matrix {3, 2, 1, 3}; again, it takes only five iterations to reach satisfactory convergence. The solution {4.81, 3.21, 1.75, 5.25} is indeed different, thus showing how the information contained in an outdated trip matrix can be used to advantage in matrix estimation; there is something of value in past information worth making use of.

Case C illustrates what happens when there is a zero entry in the trip matrix. There is still a solution but the zero is preserved in it. Finally, Case D shows the effect of 'seeding' the zero in the prior matrix with 0.5. The solution this time, {4.81, 3.21, 1.0, 6.0} affects only trips from the origin previously containing the zero.

Consider now the effect of increasing the number of counts to three by including link 6–3. The corresponding results are depicted in Table 12.2.

First, note that the number of iterations required has now increased. This seems to depend not so much on the actual number of counts used but on how close to removing all flexibility in the matrix these are. In this case three out of four degrees of freedom are removed by these counts. The solution in case A, {5.33, 2.68, 4.67, 2.35}, is the one that maximises $S(T_{ij})$ and if rounded to integers coincides with the solution in Example 12.3.

The solution for case B, {6.55, 1.51, 3.45, 3.58}, has the same properties in respect of $S_1(T_{ij})$. Case C is interesting as it shows that in this opportunity with the inclusion of a zero in the prior the algorithm fails to converge, even after 20 iterations. The reader may verify that forcing cell 2–3 to zero makes the problem unfeasible: there are seven trips out of node 2 but only five are permitted to reach their destination. Case D illustrates the effect of seeding the empty cell with 0.5 trips; the algorithm now converges to a reasonable solution.

Table 12.2   Multiproportional solution for three traffic counts

					Trips per O–D pair			
		Traffic count	**Modelled flow**	**Ratio**	**1–3**	**1–4**	**2–3**	**2–4**
A	Prior matrix				1.00	1.00	1.00	1.00
	Iteration	15	4.00	3.750	3.75	3.75	3.75	3.75
	1	7	7.50	0.933			3.50	3.50
		10	7.25	1.379	5.17		4.83	
	Iteration	15	15.05	0.997	5.32	2.68	4.65	2.35
	10	7	7.00	1.000			4.65	2.35
		10	9.97	1.003	5.33		4.67	
B	Prior matrix				3.00	2.00	1.00	3.00
	Iteration	15	15.11	0.992	6.51	1.51	3.41	3.56
	14	7	6.97	1.004			3.42	3.58
		10	9.94	1.006	6.55		3.45	
C	Prior matrix				3.00	2.00	0.00	3.00
	Iteration	15	17.15	0.875	8.75	0.13	0.00	6.12
	20	7	6.12	1.143			0.00	7.00
		10	8.75	1.143	10.00		0.00	
D	Prior matrix				3.00	2.00	0.50	3.00
	Iteration	15	15.10	0.994	6.98	1.05	2.96	4.01
	19	7	6.97	1.004			2.97	4.03
		10	9.95	1.005		7.01	2.99	

12.4.5   Traffic Counts and Matrix Estimation

One can ask at this stage whether any set of counts is suitable for trip matrix estimation. For example, is it possible that certain combinations of counts make it impossible to estimate a matrix which satisfies them? These problems will be discussed under the headings of independence and inconsistency of traffic counts.

12.4.5.1   Independence

Not all traffic counts contain the same amount of 'information'. For example, in Figure 12.2 traffic link c is made up of the sum of traffic on links a and b. Counting traffic on link c is then redundant and only two counts there can be said to be independent.

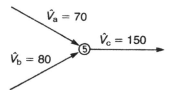

Figure 12.2   Dependent counts

Wherever a flow continuity equation of the type 'flows into' a node equals 'flows out of' the node can be written, its counts will be linearly dependent. In this case it will always be possible to describe one link flow as a linear combination of the rest. Note that a centroid connector attached to node 5 will remove the dependency in Figure 12.2.

12.4.5.2   Inconsistency

Counting errors and the fact that often traffic counts are obtained on different occasions (hours, days or weeks) are likely to lead to inconsistencies in the flows. In other words, the expected flow continuity relationships will not be met. If the count V_c in Figure 12.2 were to be 160 instead of 150, the corresponding equations would be inconsistent and no trip matrix could possibly reproduce these flows. One way of reducing this problem is to allow an error term in the equations or to remove the inconsistencies beforehand.

It is possible to identify two sources for inconsistencies in the link flows. The first one is simply the fact that errors in the counts may lead to situations in which the 'total flow into' a node does not equal the 'total flow out of' the same node, thus not meeting link flow *continuity* conditions. The second source is a mismatch between the assumed traffic assignment model and observed flows. For example, an assignment model may allocate no trips on a link having an observed (perhaps small) flow. In these conditions there will be no trip matrix capable of reproducing the observed link flows using that route choice model.

Example 12.5 It is useful to distinguish between these two types of inconsistency, first at *flow level* and then at *path flow level*. Assume we have observations on the flow of four links (identified by the pair of nodes delimiting them) and we would like to find non-negative trip matrices satisfying these and a route choice model as depicted in Figure 12.3.

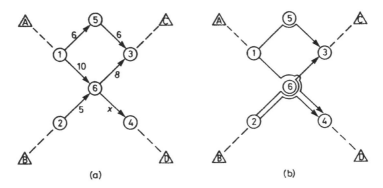

Figure 12.3   An example of path flow inconsistencies with counts: (a) network and flows, (b) assumed route choices

Consider first the case where the count x has been found to be 8, thus making the total flow into node 6 equal to 15, and the flow out of this node equal to 16. These counts are then inconsistent, perhaps because they were taken on different days or simply because of counting errors. We can remove this inconsistency by arbitrarily increasing the flows on links (1, 6) or (2, 6) by one, or by reducing the flows on links (6, 3) or (6, 4) by one. We can be more systematic and make the least adjustments necessary to preserve flow continuity conditions. For example, if what we want to minimise is the sum of the squares of the increments/reductions, then the optimum change is 0.25 on each link.

An alternative approach is to seek a maximum-likelihood solution to this problem, as put forward in Van Zuylen and Willumsen (1980). This assumes that link flows are Poisson distributed and that the observations available are samples on this distribution. Maximum likelihood is then used to generate a model for producing improved and consistent estimates of the flows. On the other hand, model calibration from traffic counts, as discussed in the previous section, makes an explicit allowance for errors in the observed link flows. These methods are not limited, therefore, by independence and consistency problems.

Consider now the case when the count x is 7. It can be seen that the link flow continuity conditions are now met. However, the assumed assignment depicted in Figure 12.3b is incompatible with the flows shown in Figure 12.3a. No feasible trip matrix can reproduce the count of 8 at link (6, 3) because the only path using it, B–C, is limited to a maximum of 5 by link (2, 6).

The set of linear equations corresponding to this example is given by:

$$\begin{array}{llll}
\text{link}\,(1, 5) & T_{\text{AC}} & = 6 & (12.27) \\
\text{link}\,(5, 3) & T_{\text{AC}} & = 6 & (12.28) \\
\text{link}\,(1, 6) & T_{\text{AD}} & = 10 & (12.29) \\
\text{link}\,(2, 6) & T_{\text{BC}} + T_{\text{BD}} & = 5 & (12.30) \\
\text{link}\,(6, 3) & T_{\text{BC}} & = 8 & (12.31) \\
\text{link}\,(6, 4) & T_{\text{AD}} + T_{\text{BD}} & = 7 & (12.32)
\end{array}$$

Clearly equations (12.30) and (12.31) are incompatible with the non-negativity of T_{BC}. The same applies to equations (12.29) and (12.32), making it impossible to solve this set of equations. In simple problems like this, inconsistencies can be ascertained by inspection but in more complex networks they can only be identified by means of row and column operations on the linear equations. For large systems these operations are likely to be expensive in terms of computer requirements.

In this simplistic example it is not difficult to see that the problem originates in the assumed single route between A and C. If two paths were allowed, one via node 5 and the other via node 6, the inconsistency could be removed. Furthermore, the value of the resulting variable p_{AC}^6 cannot be arbitrarily chosen; in effect, a feasible solution requires

$$0.2 \leq p_{AC}^6 \leq 0.5$$

The fact that the *path flow* continuity conditions are not met seems to reflect errors in assignment, whereas the *link flow* discontinuities are a reflection of errors in the traffic counts alone. It seems reasonable then to develop a technique for removing the link flow inconsistencies in the counts in order to ensure that the link flow continuity conditions are met. On the other hand, a reasonable approach to deal with the lack of consistency at the path flow level seems to be the adoption of a better route choice model. In general terms, consistency at the link flow level is a necessary but not sufficient condition for consistency at path flow level. Consistency at path flow level is, however, a sufficient condition for link flow consistency.

The interested reader may verify that there are only seven different (integer) trip matrices which can satisfy the observed flows in the example above.

12.4.6   Limitations of ME2

ME2, probably because of its simplicity, relative efficiency and ease of programming, has been widely implemented and used, particularly in the UK. The model has, however, some known limitations and it is worth exploring them before discussing opportunities to improve it.

One of the limitations arises when traffic has grown (or declined) markedly between the prior (or old) trip matrix and the present. The model estimates the matrix closest to the prior which, when loaded on the network, reproduces the traffic counts but this may lead to distortions. In these cases it is probably better to consider the structure of the prior matrix, say through the proportion of total trips which appear in each cell, and not the absolute number of trips in each O–D pair. One would then try to find a matrix with the closest structure to that of the prior matrix which reproduces the traffic counts when loaded onto the network. This can be approximated by means of a general growth factor first, for example:

$$\tau = \frac{\displaystyle\sum_a \hat{V}_a}{\displaystyle\sum_a \sum_{ij} t_{ij} p_{ij}^a} \tag{12.33}$$

which is then applied to the prior matrix before using the ME2 model. In this way the structure of the prior matrix is preserved as much as possible. The estimation of τ above is only an approximation; for a more rigorous approach see Bell (1983).

A second limitation of ME2 is the fact that it considers the traffic counts as error-free observations on non-stochastic variables. In effect, the model gives complete credence to the traffic counts and uses the prior matrix only to compensate for the fact that they do not contain sufficient information for estimation purposes. However, this may not be very appropriate in practice. For a start, one must acknowledge that traffic counts are certainly not error free. Apart from counting errors there is the problem of time variations (hourly, seasonal, etc.). Traffic counts obtained on different days or at different times can hardly be considered to be observations on a non-stochastic variable.

Willumsen (1984) has suggested an approach to compensate for this second difficulty. It starts from the idea that functions of the type $\{X \log X/Y - X + Y\}$ can be seen as useful measures of the difference between X and Y. He then constructs a composite objective function to satisfy the following:

$$\text{Minimise } S_3 = \sum_{ij} (T_{ij} \log T_{ij}/t_{ij} - T_j + t_{ij}) + \sum_a \phi_a(V_a \log V_a/v_a - V_a + v_a) \tag{12.34}$$

where

V_a is now the 'true' value of the traffic count at a.
υ_a is the value of one observation of the flow made at a.
ϕ_a is a weighting factor which depends on the confidence attached to the observation υ_a.

The use of the Lagrangian method now leads to the solution:

$$T_{ij} = t_{ij} \prod_a X_a^{p_{ij}^a} \tag{12.22}$$

$$V_a = \upsilon_a X_a^{1/\phi_a} \tag{12.35}$$

Again this model can be solved using the multi-proportional algorithm but in this case we also need to correct the observations to obtain a better estimation of the true value of the link flows. Note that if ϕ_a is very large, i.e. we assign a high weight to the counts as we believe them to be very accurate, V_a tends to υ_a; in the limit with $\phi_a = \infty$ we revert to the original model as $V_a = \upsilon_a$. Note that the smaller the value of ϕ_a, the greater the credence given to the prior matrix $\mathbf{t}$.

One would expect that the weights ϕ_a depend on the variability of the observations. Brenninger-Gothe *et al.* (1989) have discussed this model in detail. They have shown that a very natural value for the weights ϕ_a is the variance (or standard deviation) associated with the observations. If these are not available they can be estimated using some assumption about the distribution of the error terms. These authors have further extended the model to consider weights attached to both the prior matrix (μ_{ij}) and the traffic counts (ϕ_a); thus the new objective function becomes:

$$\text{Minimise } S_3 = \sum_j \mu_{ij}(T_{ij} \log T_{ij}/t_{ij} - T_{ij} + t_{ij}) + \sum_a \phi_a(V_a \log V_a/\upsilon_a - V_a + \upsilon_a) \tag{12.36}$$

The main limitations of ME2 can therefore be reduced using reasonably simple methods. However, other authors have proposed alternative approaches to solve the matrix estimation problem, some of which start from a different basic framework.

12.4.7 Improved Matrix Estimation Models

Bell (1983) has formulated a model which tries to preserve the structure of the prior matrix, in the sense described in the previous section, adding a new constraint and thus modifying the mathematical programme as follows:

Minimise $-S_2$ subject to

$$\hat{V}_a - \sum_{ij} T_{ij} p_{ij}^a = 0 \text{ for each counted link } a \tag{12.19}$$

$$\tau = \sum_{ij} T_{ij} \Big/ \sum_{ij} t_{ij} \tag{12.37}$$

and

$$T_{ij} \geq 0$$

In addition to this, Bell suggests the use of a Newton–Raphson method to solve this model with an iterative estimation for τ. Alternatively, one may assume an initial value for τ, solve the standard model using a multi-proportional method and then check if it is consistent with equation (12.37). The cycle should be repeated until the value of τ converges.

The use of a Newton–Raphson algorithm has advantages in terms of computer time and is also useful in tracing the effect of errors in the traffic counts through to the estimated trip matrix (Bell 1983); this type of sensitivity analysis is an alternative to the treatment of errors in the traffic counts suggested above. However, the Newton–Raphson method requires more memory and is therefore restricted to small and medium-size networks.

A variant to the standard objective function (S_1) is either to linearise it using Taylor's expansion or to construct a generalised least squares formulation. In both cases we still try to minimise the difference between prior and estimated matrices subject to the same constraint (12.19). Bell (1984) suggested the Taylor series expansion solution whereas McNeil and Hendrickson (1985) and Cascetta (1984) have put forward versions involving generalised least squares approaches. One problem is that under certain circumstances these models may produce negative entries in the estimated trip matrix, in particular where the prior matrix originally had small values. This is not an uncommon occurrence and therefore this feature is undesirable.

Maher (1983) proposed the use of a Bayesian approach to the trip matrix estimation problem which results in functional forms equivalent to the generalised least squares method. A prior estimate of the trip matrix is updated in the light of a set of traffic counts; both are assumed to be multivariate Normal distributed variables with known covariance.

Spiess (1987) proposed a maximum likelihood model to solve the problem. He considered a specific formulation where for each O–D pair t_{ij} is obtained by observing an independent Poisson process with mean $\Omega_{ij}T_{ij}$. This corresponds to the problem of taking a sample of an existing trip matrix with a sampling rate of $\Omega_{ij} < 1$. The probability of observing t_{ij} is:

$$\text{Prob}[\text{Poisson}(\Omega_{ij}T_{ij}) = t_{ij}] = (\Omega_{ij}T_{ij})^{t_{ij}} \exp(-\Omega_{ij}T_{ij})/t_{ij}! \tag{12.38}$$

The joint probability of observing the sample matrix $\{t_{ij}\}$ is therefore:

$$\text{Prob}[\{t_{ij}\}] = \prod_{ij} \text{Prob}[t_{ij}] = \prod_{ij} (\Omega_{ij}T_{ij})^{t_{ij}} \exp(-\Omega_{ij}T_{ij})/t_{ij}! \tag{12.39}$$

Applying the maximum likelihood estimation technique to this problem requires finding the matrix $\{T_{ij}^*\}$ which satisfies the constraints and yields the maximum probability (12.39) of observing $\{t_{ij}\}$. By taking logarithm of equation (12.39) and adopting the usual convention that $0 \log 0 = 0$, we can formulate the maximum likelihood model as:

$$\text{Max} \sum_{ij} (t_{ij} \log(\Omega_{ij}T_{ij}) - \Omega_{ij}T_{ij} - \log t_{ij}!) \tag{12.40}$$

subject to the usual non-negativity constraints and to equation (12.19). Separating the logarithm into the sum and discarding constant terms one can rewrite (12.40) as:

$$\text{Min} \sum_{ij} (\Omega_{ij}T_{ij} - t_{ij} \log T_{ij}) \tag{12.41}$$

This objective function is convex in T_{ij}; provided the set of constraints is consistent and the flows feasible, then the existence of an optimal solution is assured. The solution may be obtained by any standard solution method for convex programming problems. However, Spiess (1987) has developed an algorithm that exploits some of the specific properties of this problem.

For further comments on this problem and possibilities for extensions see Cascetta and Nguyen (1988) and Willumsen (1991).

12.4.8   Treatment of Non-proportional Assignment

The ME2 model discussed in the preceding sections is based on the assumption that it is possible to obtain the route choice proportions $\{p_{ij}^a\}$ independently from the matrix estimation process. Wherever congestion plays an important role in route choice this assumption becomes questionable as the route choice proportions and the trip matrix become interdependent. Because of its theoretical and practical advantages, equilibrium assignment is the natural framework for extending the ME2 model for the congested network case.

The main problem in incorporating Wardrop's equilibrium into trip matrix estimation is that now the route choice proportions and the trip matrix to be estimated are interdependent. One way of tackling this problem is to adopt an iterative approach: assume a set of route choice proportions $\{p_{ij}^a\}$, estimate a matrix $\mathbf{T}$, load it onto the network and obtain a new set of route choice proportions; repeat the process until route choice proportions and estimated matrices are mutually consistent.

This general scheme can be implemented in different ways. For example, in SATURN (Hall *et al.* 1980) the route choice proportions are estimated using the value ϕ in the Frank–Wolfe algorithm (the optimum linear combination of accumulated and auxiliary flows; see section 11.2.3). It is recognised that in general the path flows under equilibrium conditions are not unique. However, this method assumes them to be unique.

An alternative approach requires restating the original problem in terms of a three-dimensional matrix (origin, destination and route) as follows:

$$\text{Maximise } S_4 = - \sum_{ijr} T_{ijr}(\log T_{ijr}/t_{ijr} - 1) \tag{12.42}$$

subject to

$$\sum_{ijr} T_{ijr}\, \delta_{ijr}^a - \hat{V}_a = 0 \tag{12.43}$$

and

$$T_{ijr} \geq 0$$

where the index r indicates the route or path chosen; δ_{ijr}^a is 1 if route r between i and j uses link a, and zero otherwise.

It is always possible, of course, to reconstruct the O–D matrix $\{T_{ij}\}$ by aggregating the path flow matrices $\{T_{ijr}\}$. Again the solution to this new program is:

$$T_{ijr} = t_{ijr} \prod_a X_a^{\delta_{ijr}^a} \tag{12.44}$$

and

$$T_{ij} = \sum_r T_{ijr} \tag{12.45}$$

The prior path flows may be calculated from the prior trip matrix as $t_{ijr} = t_{ij}/R_{ij}$, where R_{ij} is the number of paths between i and j. In this case, the path flows can take any value as they are not assumed unique. The Frank–Wolfe algorithm for equilibrium assignment is used to identify attractive paths (those selected at each all-or-nothing step) but not to define the strict proportions of the trip matrix using them. This is only a heuristic scheme and a suitable algorithm for its solution is as follows:

1. Assign, using equilibrium assignment methods, a base-year matrix $\{t_{ij}\}$ to the network and save the corresponding routes (trees). Set the cycle counter n to 1.
2. Estimate a trip matrix $\{T_{ij}\}^n$ for iteration n, using independent routes $\{\delta_{ijr}^a\}$ and observed flows $\{\hat{V}_a\}$.
3. Assign $\{T_{ij}\}^n$ to equilibrium, saving the routes (trees) used in the process.
4. Increment n by 1 and return to step 2 unless the changes in routes $\{\delta_{ijr}^a\}$ or estimated matrices have been sufficiently small.

For a test of this approach and a comparison with proportional assignment techniques in the case of a comprehensive data set for Reading in the UK, see Willumsen (1982).

A more general approach has been put forward by Fisk (1988) and Oh (1989), where maximum-entropy matrix estimation and user equilibrium assignment are combined as a single mathematical program.

12.4.9 Quality of Matrix Estimation Results

Matrix adjustment from traffic counts is a powerful group of techniques that provides significant help in developing useful and robust trip matrices. However, in order to use the approach in a sound and reliable manner a number of points require careful attention. One particular aspect to bear in mind is the fact that matrix estimation techniques may try to force an adjusted trip matrix to reproduce traffic counts even if there are significant errors in the network, the assignment method or the counts themselves. The following recommendations reflect our views on pitfalls to avoid when using this type of technique:

- Make sure the network is fully debugged and that all relevant turning movements are well represented.
- Use an assignment method appropriate to the context; this usually means equilibrium assignment.
- Ensure that any prior matrix is reasonable and do not over-rely on one that is not.
- Set aside some 10–15% of the traffic counts for validation of the adjusted trip matrix.
- Ensure all traffic counts are adjusted using seasonal and daily factors to a common representative day and that only relevant vehicle types are included (i.e. do not use passenger car units (pcus) when car trips are needed).
- If possible, assign a level of confidence to each count and allow greater tolerance to those that are less reliable.
- Bear in mind that some bottlenecks may restrict actual traffic on the network to levels below demand (metering effect); it may be better to ignore counts affected by this constraint.
- Apply matrix estimation techniques in small increments and obtain network and matrix statistics at the end of each run: compare number of trips and travel speeds and trust only matrices that do not change these indicators by more than 10%; monitor in particular the trip length distribution before and after matrix estimation as significant changes probably indicate the trip matrix is being distorted by the procedure.
- Use only the validation counts above to report fitness for purpose.
- Never accept a post-matrix estimation trip table without thorough checks on its validity; these methods are powerful and generally easy to use but may distort a perfectly good prior matrix too much and render the results of any scheme test unreliable.

12.4.10   Estimation of Trip Matrix and Mode Choice

The idea of extending this type of approach to matrix and mode choice estimation is attractive. Let us consider a singly constrained destination/mode choice model of the following Logit form:

$$T_{ij} = O_i \frac{S_j \sum_k \exp\left(\sum_p \theta_p X_{ijk}^p\right)}{\sum_d S_d \sum_k \exp\left(\sum_p \theta_p X_{idk}^p\right)} \tag{12.46}$$

where the mode choice component of the model is given by:

$$P_{ij}^k = \frac{\sum_p \exp\left(\sum_p \theta_p X_{ijk}^p\right)}{\sum_m \exp\left(\sum_p \theta_p X_{ijm}^p\right)} \tag{12.47}$$

T_{ij} are trips between zones i and j, O_i is the total number of trips originating at zone i, S_j is a measure of the attractiveness of zone j, P_{ij}^k is the proportion of trips using mode k between zones i and j, X_{ijk}^p is the pth explanatory variable for mode k (for example, in-vehicle travel time) and θ are model parameters.

Although the derivations we will present below are for the simpler MNL case, they can easily be extended to consider the simultaneous estimation of more general Nested Logit forms (Ortúzar and Willumsen 1991).

12.4.10.1 Simple Unimodal Case

Let us consider first a single mode case with just one scale parameter μ, multiplying a 'generalised cost' variable X_{ij}, to be estimated. In this simple case (12.46) reduces to:

$$T_{ij} = O_i \frac{S_j \exp(\mu X_{ij})}{\sum_d S_d \exp(\mu X_{id})} \tag{12.48}$$

Now, assume we possess observations on a set of link flows $\hat{V}_a$, and also that we know, from an assignment model, the proportions P_{ij}^a for all links with observed flows. In such a case we can postulate that equation (12.19) holds and to estimate the value of μ we can, for example, seek to minimise the following normalised non-linear (generalised) least squares function:

$$S = \sum_a \left[\left(\hat{V}_a - \sum_{ij} T_{ij} p_{ij}^a \right) \Big/ \hat{V}_a^2 \right]^2 \tag{12.49}$$

In order to find the minimum we usually require first and second derivatives of S with respect to μ. These are provided by Ortúzar and Willumsen (1991); unfortunately, even in this simple case the derivatives look rather intractable so a unique solution to the problem may be difficult to establish.

12.4.10.2 Updating with Aggregate Modal Shares

Let us consider the transference of model (12.46)–(12.47) with parameters θ estimated in another context; we ignore the original mode-specific constants as they ensure reproduction of the aggregate market shares in that context. Define a transfer utility function as:

$$V_{ijk} = \mu \left(\sum_p \theta_p X_{ijk}^p \right) + M_k \tag{12.50}$$

where X_{ijk}^p are zonal values for the level-of-service and socioeconomic variables in the new context, μ is a scale parameter as before and $\mathbf{M}$ a set of $(K-1)$ mode-specific constants to be estimated; K is the total number of modes.

In this case it is possible to find maximum likelihood estimators for μ and $\mathbf{M}$ but it is possible to guarantee a unique optimum only for fixed μ, i.e. when only the constants are updated.

(continued)

12.4.10.3   Updating with Traffic Counts

The main problems arise in this case if we are interested in mixed-mode combinations but only have counts for the 'pure' modes. For example, consider the case of choice between car, bus, underground and combinations of the latter with the first two. It is obvious that even if we have separate counts for each pure mode, these include observations corresponding to the mixed-mode movements. If we settle for a mode aggregation and are interested in estimating the scale parameter μ and a set of constants for the pure modes, the problem can be solved using a generalised least squares formulation similar to (12.49), as shown by Ortúzar and Willumsen (1991).

12.4.10.4   Updating with Combined Information

Assume we wish to update μ and $\mathbf{M}$ of (12.50) and have available observed aggregate shares P_k and sets of observed passenger counts $\hat{V}$ for each competing mode. The problem can be formulated either as a maximum likelihood or generalised least squares one.

In the first case we will get different functions to maximise and hence different first-order conditions and optima, depending on the assumptions made about the distribution of count errors. The favourite assumptions have been multinomial, independent Poisson and independent Normal (see Tamin and Willumsen 1992). As it can be assumed that data on counts are independent of data on aggregate shares, the log-likelihood function takes the form of a sum of two expressions. If it is assumed that the counts have no error, a final case of interest results which requires maximising a much simpler function subject to (12.19). Expressions for each of these cases are given by Ortúzar and Willumsen (1991); there is no guarantee, however, that either of them leads to a unique optimum.

The generalised least squares formulation has two advantages: the first is that no distributional assumptions are needed on the data set; the second is the possibility of incorporating explicitly differences in the accuracy of each data item prior to estimation. A need for normalising, which is also a feature of this approach, is very evident here given the different order of magnitude of the differences between observed and modelled values for both types of data. For example, the maximum difference in the case of aggregate shares is just 1, while differences in count data may easily run to figures in the hundreds or thousands.

The range of methodologies available in principle to solve this important problem is difficult to evaluate without recourse to experimentation; by the end of 2010 such an exercise had not been reported.

12.5   Marginal and Corridor Models

12.5.1   Introduction

We have seen how conventional modelling approaches often require large amounts of resources (especially computing time and technical expertise), sometimes have a slow response rate, may not be sensitive to some of the policy options needing analysis and may be based on weak theoretical frameworks (see for example, Supernak 1983). In previous chapters we have discussed how to avoid most of these common pitfalls; in this section we wish to explore some shortcuts which can be taken to speed up the response time of modelling exercises.

Having considered some of the simplified approaches in the preceding sections one must recognise that they would seldom satisfy, on their own, all the requirements of a large scale project or major policy change. The use of trip matrix estimation techniques from traffic counts may be acceptable for situations where a fixed-matrix assumption is reasonable, for example, the design of traffic management schemes. However, the adoption of model estimation from traffic counts methods is still weak in terms of modal choice, an important element in most project assessments. Sketch planning methods offer quick response but at a high risk in terms of coarseness of the analysis. It is interesting to explore whether these approaches can be combined to utilise their strengths and avoid their weak points.

The basic idea is to adopt an approach which would use simpler models to provide a planning background and would selectively apply 'state-of-the-art' models to the most relevant decision elements of the problem. Technical journals devote little space to report systematically on the many shortcuts planners and consultants by necessity adopt in practice (Leamer 1978). Some conferences offer better illustrations of these; see for example Ashley *et al.* (1985) and Clancy *et al.* (1985).

The first element in the development of practical simplified approaches is to recognise that there is always some implicit or explicit planning context providing local experience and data. How to utilise these two effectively should always be the first step in this task. The production of sound advice to decision makers under severe time constraints should deal with questions like the following:

- how best to simplify or select models that will appropriately represent the impacts of the project to be analysed;
- how to make adequate use of existing data and local experience;
- how to take advantage of some of the special characteristics of the problem in hand; and
- how to deal with the inevitable biases introduced through the pragmatic answers adopted to the questions above.

12.5.2   Corridor Models

A typical opportunity for simplifying modelling tasks without compromising realism too much is provided in corridor studies. Corridors are strong, basically linear, transport facilities sometimes combining high-capacity and limited-access arterial roads with rail rapid transit or bus-way provisions. The linear nature of the facilities may help to simplify the modelling task; it may be sufficient to model the linear corridor and consider only the points of entry and exit to it as origins and destinations. There may be a major destination at one end of the corridor (the central business district for example) or they may be distributed throughout its length.

In any case, assignment problems will be minimal or non-existent and the modelling effort will be able to concentrate on issues such as mode and, in some cases, destination choice. The basic information needed will be the current flow levels by mode and section of the corridor, data on level of service variables for each mode and section, and the relevant characteristics of travellers. A good deal of these data is obtainable through choice-based interviewing either in-vehicle (train) or at the main destinations (workplace).

The extreme simplification of the network structure generates considerable savings in data collection and coding. The transfer of discrete choice models from other contexts may be undertaken using the techniques discussed in section 9.5. If necessary, trip generation transfer may be performed using the methods discussed in section 4.7; however, in most cases a fixed multimode trip matrix is assumed for these studies. If the study is to cover several years in the future, it may be necessary to use a matrix updating technique based on growth factors, as discussed in Chapter 5.

Corridor modelling with severe capacity constraints requires some care. Bottleneck effects in the corridor should be treated specifically and sometimes micro-assignment models may be applied to them. Direct demand models also appear as suitable choices for this type of problem.

12.5.3 Marginal Demand Models

Faced with problems which cannot be tackled through a full-scale transportation study due to limited resources and time availability, one would like to concentrate efforts on that part of the transport demand most likely to be affected by the project or policy in question. If the project is not corridor based this requires a little more care and attention. However, often the special characteristics of the problem may be utilised to simplify the task in hand. A generalised approach to this problem was proposed by De Cea *et al.* (1986). This approach is outlined below and shown in Figure 12.4.

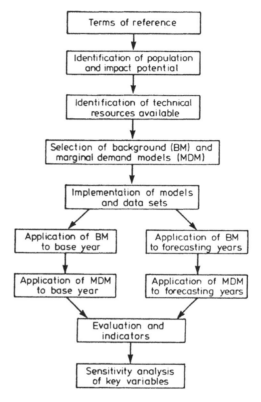

Figure 12.4   Steps in project evaluation using a marginal demand estimation approach

1. Definition of the problem. The terms of reference of the study, if available, should facilitate an identification of the main elements of a problem, be it a particular investment project or the consideration of a new policy option. Terms of reference, however, do not exempt the analyst from identifying the wider implications of the alternatives to be considered.
2. Identification of the relevant population and the impact potential of the project. At this stage, one seeks to identify the most likely impacts of a project or policy option and the sections of the population most likely to be affected. In principle anything is likely to affect everything else, but one should try to identify first-order effects and those most likely to perceive (gain or lose) the costs and benefits of the project.

3. Identification of the technical resources available to analyse the main impacts of the project on the relevant population. The existence of data, perhaps not up to date, other studies and models, and in particular local expertise, can play a key role in providing sound and quick advice to decision makers. Updating data sets and adjusting existing models should normally require fewer resources than starting from a clean slate. Local knowledge may be crucial at this stage.

4. Selection of *background* and *marginal demand* models. A key element of this approach is the use of a coarser background model to estimate the general level of demand, and a finer marginal demand model to identify the specific impacts of the project on that general demand. The choice of background and marginal demand models depends on the nature of the problem and on the technical resources available. The marginal demand model is applied to the relevant population only and should, of course, be able to discriminate the impacts of the project and/or policy options on that population. In selecting these models the feasibility of their implementation and use within the time and resources framework of the study is paramount. The simplifying assumptions adopted at this stage should be properly documented.

5. Implementation of the models and data sets. Background and marginal demand models should then be mounted on a computer together with the data sets to be used and updated as part of the study. In many cases it will be necessary to write short programs to convert data sets to suitable formats and to perform the required tests and report production.

6. Application of the background and marginal demand models to the base year and their validation. This may require some additional data collection, ideally on a small scale.

7. Application of the background and marginal demand models to forecast relevant future years. This will require first forecasting the values of the planning variables for those years and then applying the models with and without the project or under different policy options.

8. Evaluation. Model runs in the previous two steps should provide the indicators required for an evaluation of the options open to decision makers. Attention should be paid to frame this evaluation in terms of good local practice and to produce the indicators which decision makers consider most meaningful.

9. Sensitivity analysis. The simplifying assumptions adopted in previous stages and the uncertainty about the future make it necessary to test how sensitive the advice produced is to changes in the inputs and weights adopted in the study. Budget and time constraints will usually limit the amount of sensitivity tests that can be performed. It is often possible, however, to elicit preferences from decision makers on what they consider to be the most important elements to be examined in these tests. These may take the form of questions like:

> Would the project still be feasible if . . . oil prices double or the discount rate is increased by 2%?

These preferences could then be used to select sensitivity tests complementing those required by the simplifying assumptions adopted above. This is a pragmatic methodology whose virtues and limitations can only be assessed in practice.

De Cea *et al.* (1986) followed this approach to study a possible extension to the Santiago (Chile) underground network. In outline their approach involved:

- The identification of the population of interest as that in zones with walk access to the Metro before or after the potential extension, including mixed-mode journeys.
- The use of trip matrix estimation techniques based on traffic counts to provide background trip matrices for both cars and public transport; use was made of an extensive set of traffic counts supplemented by *ad hoc* surveys at bus stops and Metro stations.

- The transfer of a corridor-based disaggregate mode choice model to the study area through the recalibration of the mode-specific constants. The availability of the corridor model and suitable income data made this possible.
- Economic, financial and environmental evaluation of the project complemented by sensitivity analysis of key parameters.

This complete study was undertaken in four months. The cost-benefit analysis predicted a reasonable return on investment. The extension of the Metro has now been implemented and the results apparently confirm the accuracy of the study.

12.6   Gaming Simulation

Mathematical models do not solve any real-life transport problems: it is the interpretation of mathematical solutions which is useful to make decisions concerning transport problems. Simplified models may help in reducing the effort required to find a mathematical answer and in facilitating the subsequent interpretation of this solution in relation to the real problem. We use conceptual or mental models to understand, interpret and act in our professional life. Mental models are, in effect, a prerequisite for the development and application of mathematical ones run on a computer.

Despite their significance and because of their character, it is difficult to examine mental models and this often leads to quite unmanageable communication problems. Better and richer mental models in the minds of planners and decision makers are probably as important as the use of rigorous and sound behavioural models in the computer, if transport planning is to be improved. Given the key role played by mental models in the use and application of mathematical ones, it seems sensible to investigate techniques for improving the first in order to get better solutions through the second.

But how are mental models acquired, revised, rejected and enhanced? The main factors seem to be formal and informal education, discussions and, above all, practical experience. One of the main problems facing planning education and training is how to provide realistic experience. This is particularly acute in the transport field where the most important consequences of a policy measure or infrastructure project may follow only after considerable time. Besides, it is surprisingly easy to become too involved in the details of particular techniques and lose sight of the wider process where they must fit.

The need for methods of developing a general comprehension of a system rather than detailed information about its parts has been recognised in several fields, particularly in management and business training. Several educational techniques have been developed to this end: case studies, role playing and different types of exercises. Gaming simulation is a particularly attractive technique in this field. It was originally developed for military purposes in the form of war games but since computers became widely available it has spread successfully into management science, politics, sociology, and regional and transport planning.

Educational games are sequential decision-making exercises structured around an artificial environment acting as surrogate for the real world. This artificial environment may be just a set of instructions and graphical material or may involve an elaborate simulation exercise using computer programs, physical models and animated displays. As in real life, games usually have a competitive dimension. This feature can be incorporated in at least two forms: by dividing the players into teams with partially conflicting objectives (e.g. car owners, environmental protection officers, local residents, etc.) or, by facing each player with a computer model of a complex system plus a common set of initial conditions and final objective. Key indicators can then be used to assess the performance of each player in achieving these objectives. The first approach stresses the need for negotiation and compromise whilst the second emphasises efficiency in pursuing objectives. Both methods enhance understanding of complex systems and support the development of learning skills. In both cases the success of players depends on their ability to learn from the outcome of their own decisions that of others and from the effect of unexpected events like a strike or fuel price increases. The final objective of any gaming-simulation exercise is augmenting the ability to learn through the enrichment of the conceptual model every player has of a system. For a good background on gaming-simulation design and experience the reader is directed to Greenblat and Duke (1975) or Taylor (1971), and in the transport field to Ortúzar and Willumsen (1978).

A number of gaming simulations have been developed specifically for the transport field. Some of these cover problems like negotiating the alignment for a new road or planning new public-transport services. Probably the most widely used game in the urban transport management field is GUTS (Willumsen and Ortúzar 1985). The original objectives for this computer-based game were:

- The game should treat the transport sector of an urban area as a system, i.e. it should highlight the interrelations between modes, traffic management and investment decisions, and financial constraints; therefore, the computer program contains relationships conveying these interactions.
- The game should be realistic but manageable; the most common types of investment and traffic management decisions should be included and key financial and resource constraints be simulated.
- The model should allow for a range of alternative and even conflicting objectives to be pursued, and consequently the program should produce not a single but multiple performance indicators; at the same time, the information available to players should not be too different from that commonly available to decision makers.
- The game should stress the importance of continually monitoring the performance of a transport system.
- The model should allow the representation of different types of urban areas in terms of residence, employment, car ownership, income distributions and growth rates, public-transport patronage and related indicators.

GUTS is available as an interactive program for computers with modest memory requirements. The model is based on a simplified, urban area with circular symmetry. Two modes of transport, car and buses, operate freely and in competition; the user can make decisions on public-transport fares and levels of service, the introduction of bus lanes, supplementary licence schemes, parking provision and charges, as well as major investment projects. The program checks these decisions and runs the model to represent one year of operation of the transport system. At the end of the run indicators on flow levels, speeds, modal split, travel time and expenditure by person type are produced, and the financial performance of the bus company is reported. Changes in accessibility levels and the impact of new investment are also simulated, as are unexpected events inducing changes to the cost structure of the transport modes operating in the city. The symmetry condition imposed on the city simplifies the model with advantages in terms of speeding up the learning curve of the user and enhancing running time in the computer.

Games like GUTS can enhance transport planning in a number of ways. First, in their normal training-tool mode, they can be used to educate new recruits to a team and to develop a common language throughout an office. Second, a model of this type may be seen as a simple 'sketch planning' tool valuable in discussing broad policy options and particular conceptions of decision makers. GUTS, and similar programs, are no substitute for full-scale models but may help bridge the gap between broad strategies and specific modelling studies. A third use of tools of this kind is in demonstrating the advantages and limitations of mathematical models. The extremes of total rejection of transport models or their blind acceptance are still present in some political and planning quarters. The evident simplicity of a gaming-simulation exercise combined with its capacity to represent interactions between modes and decisions and decision makers, provide a good example of what the formal modelling approach can offer.

The use and subsequent critique of the game by politicians and planners would help them to understand each other's activities and interests better.

Exercises

12.1 The network in Figure 12.5 represents a small area with two origins A and B and two destinations Y and Z.

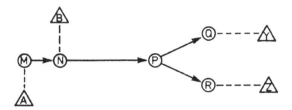

Figure 12.5   Simple network for Exercise 12.1

Traffic counts have been made of the car flows using the network with the following results:

Link	Flow
M–N	400
N–P	700
P–Q	500

(a) Use an entropy-maximising model to estimate a trip matrix from the information above. Assume a suitable prior matrix for this problem if necessary. A 3% error in the modelled flow is considered acceptable for this question.
(b) Repeat the calculations above but assuming the prior matrix is given by:

	Y	Z
A	100	50
B	80	200

12.2 The network in Figure 12.6 represents links connecting two origins A and B to two destinations C and D in a developing country. The populations of the two origins are 10 000 and 20 000 inhabitants respectively and the markets held at C and D are equally attractive in terms of size and prices. The link distances (in km) are indicated in the figure.

Figure 12.6   Simple network for Exercise 12.2

Person counts have been obtained for three links as follows:

Link	Persons/day
A – X	3400
X – Y	11 900
Y – D	4100

Calibrate a model of the type

$$T_{ij} = \frac{b P_i D_j d_{ij}^{-n}}{\sum_j D_j d_{ij}^{-n}}$$

where P_i is the population of zone i, D_j is the attractiveness index for the market in zone j, and d_{ij} is the travel distance between i and j. Try at least two values for the power n, including $n = 2$ and $n = 2.5$.

12.3 Three villages, A, B and C, are connected by a navigable river in an underdeveloped country. Village A has a population of 1000 inhabitants; village B is 30 km downstream of A and has a population of 2000; village C is 10 km down-stream of B and has a population of 300 inhabitants. The value of the goods exchanged in each village per day is 500, 600 and 600 pesos respectively.

Two observers have spent some time making directional counts of passengers travelling in boats along the river with the following results:

River section	Passengers per half day
A – B	45
B – A	60
B – C	360
C – B	560

(a) Calibrate a gravity model of the form suggested in Exercise 12.2, where D_j is replaced by the population of village j. Use $n = 2.0$.
(b) Calibrate a similar model but replace D_j by the value of the goods exchanged in each village per day.
(c) Which model do you think is best? Why?

12.4 The elasticity of the demand for buses to the fare is typically acknowledged to be in the region of -0.3. The average trip maker between zone A and the centre of town (CBD) currently faces a bus fare of $2 per trip; the bus share of all trips between A and the CBD is 60%, other trips use either car or underground.

If the total number of trips between both zones is 2,000 estimate the loss in patronage of the buses if the fare is raised to $3 per trip, all other things being equal, using the incremental logit method. Compare your result with the more crude elasticity calculation; discuss your findings (*Hint*: estimate the parameter θ_c from the data given the simple expression for the logit direct elasticity).

13

Freight Demand Models

13.1 Importance

Most of this book has concentrated on demand modelling for passengers, with a strong emphasis on urban problems. However, freight movements, and in particular road haulage, are an important source of congestion and other traffic problems. The noise and nuisance generated by heavy lorries, the problems created by on-street loading and unloading of goods vehicles to serve shops and premises, and the usual complaint about lorries taking up a good deal of the capacity of inter-urban roads are only some of the problems associated with this type of traffic.

Unfortunately, in urban areas the policy options available to influence road haulage are very limited. They are mainly controls on loading/unloading, on the size of vehicles allowed in certain areas (lorry routeing), special lorry charges, the provision of major freight interchanges, the encouragement of rear access to premises and improved layouts at new developments.

Freight demand modelling may play a particularly important role in developing countries where the efforts to increase exports and to gain access to underdeveloped areas are even more urgent. Facilitating the movement of goods in these cases is likely to have a major impact on economic development. Moreover, the competition between road and rail in some of these countries is a key issue in resource allocation for investment and maintenance.

In the case of inter-urban and international movements there is greater scope for policies to influence freight mode choice and to regulate competition between rail and road. Improved allocation of road user charges and targeting subsidies to key rail or road services, are also an important policy option. The design of these tools may require more refined modelling efforts than those used in urban studies.

One might expect the choices made for freight movements should follow economic rationality alone: minimise a combination travel times and costs appropriate to the value of the goods being transported. In this case the "value of the goods" is not only how much they cost but what are the implications of their delayed or early arrival in terms of storage costs and downstream manufacturing/sales delayed. However, an observation of real flows finds many examples where this economic rationality seems to be difficult to interpret or it is much more subtle and complex than we would expect: moving bottled water all across the world, for example.

One can envisage the complexity of freight movements as the result of four layers of decision and activities. The first layer deals with decisions on productions, destinations, type of product, volume and trade relationships: who produces what, in what quantities and for what intermediate or final consumer. The second layer deals with logistics: decisions on the use and location of inventories and supply chain

Modelling Transport, Fourth Edition. Juan de Dios Ortúzar and Luis G. Willumsen.
© 2011 John Wiley & Sons, Ltd. Published 2011 by John Wiley & Sons, Ltd.

management, for example Just-in-Time contracts and lean manufacturing. The third layer of activities deals with the choice of transport modes, vehicles and multi-modal facilities to deliver the goods according to the previous decisions. Finally, the fourth layer specifies the (multi-modal) transport route followed to reach each particular destination. The actual route chosen may be relatively simple in the case of road haulage but it will be more complex when routeing containers or Intermodal Terminal Units (ITUs) on a rail network, unless there are enough to make up a full trainload.

Attention to each of these layers will depend on the scope and geography of each study. Urban studies will probably focus mostly on the fourth layer and consider the upper three more or less given and identifiable through relatively conventional data collection surveys. Future conditions may imply changes in both route availability and the location of origins and destinations of movements thus focusing on the first and fourth layers.

Regional and international studies will tend to cover the four layers with perhaps a simplified approach at modelling changes in logistic decision making. Trade flows, mode and route choice are likely to be significant focus of attention; data collection and processing will look deeper into these issues.

Given these facts, it appears surprising that much less research has been undertaken on modelling this type of movement than the effort allocated to passenger demand. Why would this be the case? We believe there are several reasons for this:

- There are many aspects of freight demand that make it more difficult to model than passenger movements; some of these are discussed below.
- For some time urban congestion has been highest in the political agenda of most industrialised countries and in this field passenger movements play a more important role than freight.
- The movement of freight involves more actors than the movement of passengers; we have the industrial *firm* or firms sending and receiving the goods, the *shippers* organising the consignment and modes, the *carrier(s)* undertaking the movement and several others running transhipment, storage and custom facilities. In some cases two or more of these may coincide, for example in own-account operations, but there is always scope for conflicting objectives which are difficult to model in detail in practice.
- Recent trends in freight research have emphasised the role it plays in the overall production process, inventory control and management of stocks. These trends are a departure from more traditional passenger modelling techniques and share little in common (see Regan and Garrido 2002).

This chapter summarises approaches to freight demand modelling. It starts with a discussion of the main difficulties associated with modelling freight movements. It then presents what is probably the most traditional approach to the problem, which is to adapt the conventional four-stage aggregate demand model to the case of commodities. Extensions of the disaggregate approach to freight demand are also outlined. The section closes with some practical considerations for the implementation of these ideas. The interested reader is directed to the classic book by Harker (1987) for further details.

13.2   Factors Affecting Goods Movements

As in the case of passenger demand, it is useful to consider first the factors that one would expect to influence freight movements. The following is not an exhaustive list but covers the most important ones.

- Location factors; freight is always a derived demand and usually part of an industrial process. Therefore, the location of sources for raw materials and other inputs to a production process as well as the location of intermediate and final markets for their products, will determine the levels of freight movements involved as well as their origins and destinations.
- The range of products needed and produced is very high, much greater than even the most exaggerated or detailed segmentation of travel demand by person types and journey purposes. A given demand for

bolts cannot be satisfied by providing cashew nuts. There will be very many commodity matrices in any study of freight demand.

- Physical factors. The characteristics and nature of raw materials and end products influences the way in which they can be transported: in bulk, packaged in light vans, in very secure vehicles if the products are of high value, in refrigerated containers if they are perishable. There is a greater variety, therefore, of vehicle types to match commodity classes than in the case of passenger transport.
- Operational factors. The size of the firm, its policy for distribution channels, its geographical dispersion and so on, strongly influence the possible use of different modes and shipping strategies.
- Geographical factors. The location and density of population may influence the distribution of end products.
- Dynamic factors. Seasonal variations in demand and changes in consumers' tastes play an important role in changing goods' movement patterns.
- Pricing factors. As opposed to the case of passenger demand, prices are not, in general, published material because they are much more flexible and subject to negotiations and bargaining power.

13.3   Pricing Freight Services

It is usually quite difficult for the analyst to obtain reliable data about freight charges. For example, in Europe both transport firms and users try to keep them confidential so as to strengthen their position when it comes to renegotiate them. The factors affecting charges or cost imputations, and therefore mode choice, are thought to be:

- The length of the supply contracts. A better price can be obtained if the shipper guarantees demand for one or more years rather than just for one single shipment. The existence of price adjustment clauses helps to extend the lengths of contracts.
- The extent of volume discounts. Following from the above, a contract guaranteeing steady high-volume shipments is likely to benefit from a lower price.
- The importance of terminal facilities. The availability of a rail terminal nearby, or even at the firm, would certainly reduce the cost of shipping by rail; its absence would increase the likelihood of using road transport all the way, without even considering rail or water transport.
- The use of own-account operations, especially road haulage. Some firms prefer this type of operation for reasons other than transport (image, reliability, integration). These firms will tend to extend the use of own-account operation for marginal products rather than consider a completely new mode.
- Some modes are more suited to transport particular commodities. For example, *pipelines* are ideal for bulk liquids and some suspensions and *merry-go-round* (non-stop) trains are very suited for movements from coal-mines to power stations. This closer fit of supply characteristics to demand would certainly influence the charges made for those products.
- Hierarchical transport systems. For example, in the case of petroleum products, use of large tankers to refineries, then small tankers and pipelines to major terminals, rail to other terminals, and lorries to petrol stations and final users. These structures are difficult to modify in the short run as they have evolved over a long period and are well established; thus, their pricing mechanisms may be very difficult to change.

13.4   Data Collection for Freight Studies

As we have seen, the business of moving freight is more complex than that of passengers (see Figure 13.1). Data collection must, therefore, be planned taking into account the key features of goods transport

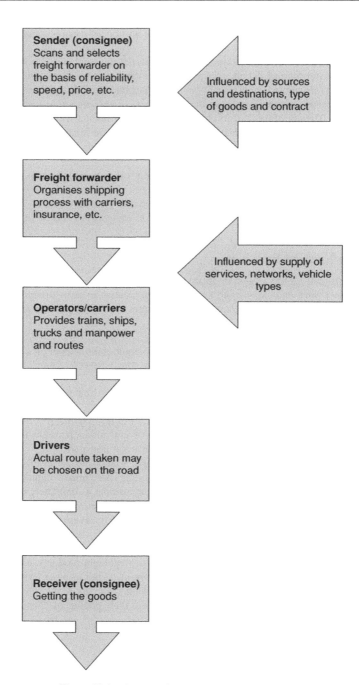

Figure 13.1   Actors and processes in freight movements

to be captured in a particular region or city. What follows is a simplified version of the key participants influencing choices in the movement of goods (Friedrich *et al.* 2003):

- The sender (shipper, consignor) who requires delivery of its goods to a particular destination and puts these goods-units in the care of others (freight forwarder, carrier) to be delivered to a consignee. The sender will decide on a freight forwarder based on reliability, speed of delivery, price and other factors.
- The freight forwarder who organises the shipping process. This firm will provide and schedule uni- or inter-modal transport chains for transporting the goods. To deliver these services, the firm may subcontract carriers or provide an own carrier service.
- The carrier (or carriers) is responsible for the transporting of goods. The carrier will provide the ships, trains and/or vehicles, sometimes in combination, required for the transport operation along a section of the transport chain. The vehicle-units operate on a network connecting origin, hubs and destination. The carrier may specify a route to be followed by vehicle-units.
- The driver guides the vehicle/transport unit along a predefined route. In the case of road transport the driver may decide during the trip to change the route between two points of the journey.
- The consignee is entitled to take delivery of the goods.

Additional actors appear at different stages in this process, for example insurance companies, quality assurance inspectors, customs officials and facilities and intermediate storage units. Some of these services are sometimes provided by shippers or carriers. These complexities are some of the reasons why transport modelling for freight is so drawn-out; it is often difficult to identify exactly who takes actual decisions on mode-combinations or routes and therefore difficult to collect data and develop choice models, aggregate or otherwise.

Whenever goods are transferred from one of these participants to another, a small amount of data is generated and stored in some way. As more of this data is collected and stored electronically it should be easier, in principle, to access and process it. Commercial confidentiality, however, continues to make this task very difficult.

For most urban transport studies it may be enough to collect information at roadside interview sites on the type of vehicle used, the goods transported and their origin and destination. The delivery of goods in urban areas it is often made more complex by the use of distribution/collection tours with multiple stops. This information is difficult to collect at the roadside because of time constraints. This is why it is customary to collect additional information from carriers and from major generators and attractors of goods movements.

In summary, for urban studies the main sources of information would be:

- asking drivers on the road;
- identifying specific carriers (couriers, delivery companies, refuse collectors) and interviewing them about tours and schedules;
- sometimes a mail-back survey may offer a moderate degree of success.

One must bear in mind that in rapidly growing cities construction work provides a significant source and destination of freight movements, including empty vehicles. This is problematic for modelling purposes as future construction activity is almost certainly going to take place in different, and difficult to predict, locations than on the base year.

For regional and international freight studies the movements of interest are somewhat simpler. However, the logistic and multi-modal aspects of decision making is taking a more important role in defining modes, routes and timings and these may be more difficult to model both on the base year and in the future. Additional information is available from waybills and other instruments that accompany consignments and this and may be accessible under favourable circumstances; alas, this is seldom the case.

13.5   Aggregate Freight Demand Modelling

The great majority of freight demand models applied in practice have been of the aggregate kind (see for example Van Es 1982; Friesz *et al.* 1983; Harker 1985). These applications follow the classic four-stage model with some adaptations specific to freight. A typical example of this approach is the work of Kim and Hinkle (1982), who used the American Urban Transport Planning Suite (UTPS) with some adaptations to model state-wide freight movements. In outline this approach involves:

- Estimation of freight generations and attractions by zone.
- Distribution of generated volumes to satisfy 'trip-end' generation and attraction constraints. The usual methods for this task are linear programming or use of a gravity model.
- Assignment of origin-destination movements to modes and routes.

We shall look at these and other factors in some detail below.

13.5.1   Freight Generations and Attractions

The techniques used to obtain total trip ends depend on the scope of the study, the level of aggregation originally envisaged and on the type of products considered:

- Direct survey of demand and supply may be undertaken for major flows for some homogeneous products: sugar, petroleum products, iron ore, coal, cement, fertilisers, grains, etc. These may be forecast using industry or sector studies. This approach is usable for inter-urban movements but is not recommended for urban problems.
- The use of macroeconomic models, for example of the input-output nature, based on regional rather than national data.
- Growth-factor methods, such as those discussed in Chapter 4, are often used in forecasting future trip ends.
- Zonal multiple linear regression is often used to obtain more aggregate measures of freight generations and attractions, in particular in urban areas.
- Demand may be associated with warehouse capacity or with total shopping area at each zone (urban studies) rather than with industrial development.

13.5.2   Distribution Models

Many urban studies simply apply growth-factor methods to observed goods movement matrices, as discussed in Chapter 5. However, many inter-urban freight transport studies have used synthetic aggregate models, even of the direct-demand type. The two aggregate techniques most used in this area are briefly discussed here: a gravity model and a linear programming approach.

In the case of the gravity model it is relatively simple to re-interpret its functional form as:

$$T_{ij}^k = A_i^k B_j^k O_i^k D_j^k \exp\left(-\beta^k C_{ij}^k\right) \tag{13.1}$$

where k is a commodity type index; T_{ij}^k are tonnes of product k moved from i to j; A_i^k, B_j^k are balancing factors with their usual interpretation; O_i^k, D_j^k are supply and demand for product k at zone i (or j); β^k are calibration parameters, one per product k; and C_{ij}^k are generalised transport costs per tonne of product k between zones i and j.

The idea of using a generalised cost function formulation for freight demand is apparently due to Kresge and Roberts (1971). This can be interpreted as follows (omitting superscript k for simplicity):

$$C_{ij} = f_{ij} + b_1 s_{ij} + b_2 \sigma s_{ij} + b_3 w_{ij} + b_4 p_{ij} \qquad (13.2)$$

where f_{ij} is the out-of-pocket charge for using a service from i to j; s_{ij} is door-to-door travel time between i and j; σs_{ij} is the variability of travel time s; w_{ij} is the waiting time or delay from request for service to actual delivery – it may be a long time for maritime transport, for example, and p_{ij} is the probability of loss or damage to goods in transit.

All of these depend on the mode used and to some extent on the commodity being transported. The constants b_n are, in general, proportional to the value of the goods. For example, in the case of the probability of loss the cost is at least the goods value, but probably more, due to penalties for delays in delivery. In the case of delay, variability of delay and transit times, the values of b_n are at least proportional to those of the goods, essentially through increased inventory costs. Modern industrial production techniques, such as those emphasising 'just-in-time' deliveries, try to minimise these elements together with stocking costs. The minimum for b_1 to b_3 is the cost of the interest rate applied to the value of the goods during the time period considered.

In general terms, it is important to consider the relative contribution of transport (generalised) costs to the final cost of a commodity. For example, in the case of wheat, coal, cement and bricks, transport costs are a main element in their final price; however, in the case of convenience foods, consumer goods, chocolates or electronics, transport costs have a low (direct) contribution to price.

A second approach to distribution modelling is *linear programming* (LP). This usually takes the form of a minimisation program: minimise total haulage costs (in money terms, very rarely in terms of generalised costs), subject to supply and demand constraints.

$$\text{Minimise } Z = \sum_{ij} T_{ij} C_{ij} \qquad (13.3)$$

subject to:

$$\sum_i T_{ij} = D_j \qquad (13.4)$$

$$\sum_j T_{ij} = O_i \qquad (13.5)$$

This is the well-known Hitchcock's transportation problem which can be solved efficiently in a very simple way. More advanced formulations may involve non-linear costs and perhaps more elaborate constraints involving a time element and minimum shipment sizes.

This minimisation problem makes some sense from the point of view of a large firm trying to satisfy its customers at a minimum cost. Alternatively, if an industry has several plants with different productions, capacities and costs, the objective function may be to maximise profits or to minimise total cost at the market place. From the point of view of modelling, the LP approach has a better chance of being realistic when:

- the industry is concentrated in a few firms;
- there are low value goods and relatively high transport costs;
- there are few demand points (zones), perhaps a monopsony (a single buyer).

However, it must be recognised that although LP may be a good model for the behaviour of a single client or industrial firm, it cannot hope to represent aggregate behaviour for various commodities. The LP solution will tend to be too sparse, with particular destinations being served only by certain origins.

On the other hand, the gravity model is quite flexible. By changing the value of β it is possible to vary the relative importance of cost compared with supply and demand constraints.

The formal relationship between LP and gravity models has been explored by Evans (1973). She has shown that in the limit, $\beta = 0$ in (13.1) will produce a matrix of movements where transport costs play no role (in fact this is Furness's solution to the growth-factor problem); whereas a very large value for β will generate a solution closer to an LP model, i.e. where transport costs are dominant (in the limit $\beta = \infty$ will reproduce the LP solution). Therefore, it is possible to use the gravity model formulation to represent the whole range of client behaviour for destination choice, from that almost indifferent to transport costs (electronics?) to the behaviour expected in the case of low-cost, high-bulk commodities like cement, sand, and so on, where transport costs are paramount.

13.5.3   Mode Choice

This is essentially a shipper's decision as to which carrier should be used to deliver the goods to their destination. When modelled at this very aggregate level, modal choice is often treated using a Multinomial Logit (MNL) formulation based on generalised costs, as described above. This may turn out to be very approximate because the information can only capture those elements of mode choice incorporated in the generalised costs concept above.

These shippers' decisions are, of course, dependent on the rates charged by carriers, which in turn depend on the volumes they move between each O–D pair. As the size of many consignments is significant in terms of the impact on carriers' rates, there are interactions inside mode choice which go beyond that encountered between passengers and public-transport operators. This problem is often ignored at high levels of aggregation.

In the case of urban freight movements the problem of mode choice is trivial; the coverage provided by non-road modes is extremely limited.

13.5.4   Assignment

In the case of road haulage this is now a carrier's decision sometimes modulated by the driver of each vehicle: the choice of the best route to take the goods from origin to destination. To some extent this is the least difficult of the problems. The use of capacity restraint is probably relevant to most urban situations. In the case of inter-urban movements, on the other hand, it may be sufficient to use a stochastic assignment model. However, it may be argued that different types of vehicles must be modelled in different ways; for example, light vans may be much less sensitive to the hilliness of routes than heavy lorries; also, vehicles carrying perishable goods might give greater priority to minimising time than those carrying, say, bulk coal. The use of multi-class assignment methods may then be warranted to cope with this variety of cost concepts.

Investigations into road haulage route choice have sometimes revealed somewhat unexpected influences on route choice. For example, some newly built toll roads sometimes lack rest, food and refuelling facilities thus making them unattractive routes for long distance drivers. Lorry drivers often prefer to drive at night to avoid the worst of congestion but they are sometimes limited in their choice by deliveries on very narrow time windows.

In the case of rail, trains are sometimes scheduled according to a semi-variable timetable (Roll-on Roll-off trains, mail). In these cases, the algorithms from timetable based public transport assignment can also be applied to rail freight assignment. More often, freight trains operate in response to demand. In this case, a timetable does not exist, not even a line network with headways or frequencies. What is then required is a train formation algorithm to build the train journeys and their implied timetable. This may be an inappropriate level of detail in a regional study and in that case a short or multi path search algorithm may be appropriate.

The use of a shortest path algorithm is likely to require incremental path searches, where the links already used in the previous steps are penalised, in order to prefer routes using other links. This would be important when it is necessary to distinguish different train types (slow local trains, faster direct trains) and contain realistic penalties for shunting operations at transfer locations. It is important to remember that marshalling yards and flat crossings impose capacity constraints to route choice and assignment models.

Intermodal terminals have gained significant market share in the last two decades. They are usually conceived as interconnected by rail corridors although shipping and road haulage also provide services. Intermodal Terminal Units are transferred from one mode to another using gantry cranes and front lifters. Intermodal terminals are often perceived as a set of platforms served by equipment and serving a user catchment area via road and rail networks.

Intermodal assignment requires a multimodal network model where many routes may be used for a specific pair of origins and destinations. A multi-modal route tree concatenates uni-modal route legs into intermodal routes. A route leg describes the part of a journey between two transfer points which does not require a transfer between vehicles. An intermodal freight assignment based on a route tree would consist of the following steps (Friedrich *et al.* 2003):

- Generation of direct route legs between all origins and destinations using a uni-modal search.
- Generation of route legs between transfer points using a uni-modal search.
- Construction of route tree.
- Calculation of generalised costs for all routes including transfer costs.
- Distribution of demand onto routes.

13.5.5 Equilibrium

As in the case of passenger demand, the problem of system or market equilibrium pervades the whole modelling exercise but the techniques to achieve it are still under development. One of the early formulations of this problem is due to Friesz *et al.* (1983) who developed a freight network equilibrium model (FNEM). This model considers explicitly the decisions of both shippers and carriers for an inter-modal freight network with non-linear costs and delay functions that vary with commodity volumes.

FNEM treats shippers and carriers sequentially; shippers are assumed to be user optimisers trying to minimise the delivered price of the commodities they send, and therefore Wardrop's first principle is used to replicate their behaviour. This sub-model is an elastic transport demand model expressed as a mathematical programming problem solvable by the usual extension to the Frank–Wolfe algorithm, as discussed in Chapter 11. The assignment to carriers is performed through the use of a 'perceived' network including only the O–D pairs, transhipment nodes, and associated links considered by shippers in their decisions.

The carrier sub-model uses a full description of the actual transportation networks. Carriers are assumed to be operating-cost minimisers and are modelled using Wardrop's second principle. The flow patterns of individual carriers are aggregated to obtain global network flows.

A similar approach was formulated by Moavenzadeh *et al.* (1983) for planning intercity transport demand in Egypt. In this case the approach is based on the simultaneous transportation equilibrium model (STEM) (Safwat and Magnanti 1988).

At a higher level of analysis, it may well be that the macroeconomic models used to generate the total demand and supply levels, and in some cases the matrix of movements, use transport costs which are inconsistent with those generated by other parts of the model. Consequently, when such models are employed sequentially with a detailed freight network model, the two may well fail to converge to stable solutions.

Harker (1985) formulated a model called the generalised spatial price equilibrium model (GSPEM) which ties together the concepts of spatial process and shipper-carrier equilibrium to simultaneously predict:

- the production and consumption of goods;
- the shippers' routeing of freight traffic; and
- the freight rates.

A variant of the Frank–Wolfe algorithm was developed to solve a particular implementation of this problem and it was applied to a large-scale problem (with approximately 3560 nodes and 14 600 arcs) concerning the US coal economy.

Example 13.1 Three types of aggregate models were estimated by Tamin and Willumsen (1992) for the island of Bali, Indonesia: a gravity (GR), an intervening opportunities (OP) and a combined gravity-opportunities (GO) model. All these models were estimated with five different types of commodities but using traffic counts alone. The resulting freight matrices were then compared with those observed in a major survey of the island. It was found that although the GO model performed slightly better than the pure gravity model, the gain in accuracy did not compensate the greater computational effort. The GR model calibrated in this way was capable of discriminating between the five groups of commodities obtaining a different β value for each. This model was far superior to the simple application of the Furness growth factor method. For more details see Tamin and Willumsen (1992).

13.6 Disaggregate Approaches

Since discrete choice models were developed and applied to model passenger demand, the idea of extending them to cover freight movements also gained currency; see for example Gray (1982) and Van Es (1982). In the case of freight, the demand for transport is seen as that for a number of individual consignments, each with its own characteristics, for which the individual shipper has to take a number of transport-related decisions. Every decision is seen as a choice made from a discrete set of alternatives. There is a number of related choices to be made in each case, e.g. to transport x tonnes at time t of commodity k by transport mode m from origin i to destination j. The carrier would then have to choose the route to perform this task.

The general flexibility of discrete choice modelling permits the construction of very general utility functions for these types of choices. They can include, for example:

- the characteristics of the transport services, such as tariffs, times, reliability, damage and loss, minimum consignment, and so on;
- the attributes of the goods to be transported, such as type of product, volume/weight ratio, value/weight ratio, if the good is perishable, inventory system and ownership;
- the characteristics of the market, such as its relative prices, firm size, availability of loading/unloading facilities, general infrastructure facilities;
- the attributes of the shipping firm, such as its production level, sale prices, plant location, available infrastructure facilities, storage policy, and so on.

This type of approach has found limited application on a national scale. The main reasons for this are the more limited understanding of all the elements involved in developing these utility functions and the very demanding data-collection efforts required to estimate this type of model.

However, its application to particular sub-markets or commodities may provide very valuable insights for policy formulation. For example, Ortúzar (1989) was able to use stated-preference data to examine the question of offering a new service (refrigerated containers) for international maritime cargo. This type of approach has also been used by Fowkes and Tweddle (2000). Efforts in this direction are likely to prove fruitful from both research and practical viewpoints.

13.7 Some Practical Issues

Despite efforts in recent years, freight demand modelling is still less advanced than passenger demand modelling approaches. The leading edge of research and development seems to have been passenger demand forecasting, with freight following its footsteps trying to adapt models to its particular needs.

The problems of data collection may be compounded in the case of freight. For example, data collection for disaggregate approaches suffers from confidentiality and reliability problems. Even collecting data for aggregate modelling represents a much greater effort than that for passenger movements: great dispersion of firms, important daily and seasonal variations, and so on.

Opportunities for extensive roadside interviews are very limited, except at points where long delays are inevitable (waiting for a ferry, for example). In some cases, such as international travel, it may be advantageous to collate data from customs or a collection of waybills.

Because simplified models use low-cost and regularly collected data (traffic counts), it may be possible to run them often enough to update forecasts and provide corrective measures for plans, i.e. they offer opportunities for implementing a continuous planning approach.

In the case of urban freight modelling very simple approaches are normally followed. They are usually based on models of vehicle movements disregarding the commodities shifted, the type of locations served and the underlying economic activities that originate this demand. It is often considered sufficient to obtain a commercial-vehicle matrix using roadside interviews (at cordon and screen-line points) and then to gross it up to the planning horizon by means of growth-factor methods.

Some software packages offer some specialised facilities to solve relatively simple logistic problems like the planning of tours and routes. Others offer more sophisticated modules to optimise the formation of trains, routeing via hubs, using multi-modal networks taking advantage of intermediate storage facilities at different costs.

14

Activity Based Models

14.1 Introduction

Travel has always been seen as 'derived demand'. We rarely travel just for the sake of travelling. We do it in order to satisfy a particular need or requirement at a different location. We can perceive life as a sequence of *activities* undertaken at different *locations,* over a period of a day or even week. To perform these activities we need to make *trips*; these, in turn, are linked by the sequence of activities over time.

The conventional trip-based approach, exemplified in the four stage model, has produced some sound transport systems analyses, with travel demand and network performance procedures determining flows that tend toward equilibrium with input from land use and transport supply. These models can be entirely trip based, or more likely today, based on the estimation of *Productions* and *Attractions* and simple tours to be modelled as such until the assignment stage which is entirely trip based. The use of Productions and Attractions can be seen as a simplified way of handling the link between Travel and Activities (the reason why we travel between two points).

Mitchell and Rapkin (1954) established quite early the link between *travel* and *activities* and called for a comprehensive framework and inquiries into travel behaviour. For a number of reasons these ideas were not taken forward at the time, at least partially because it was difficult to operationalise them for practical planning purposes.

Many authors have contributed to the basic thinking on 'activity analysis'. Among them, one must mention the contribution from Hägerstrand (1970) and Jones (1979). Hägerstrand proposed a time–geographic approach that delineated systems of constraints on activity participation in time and space. The first comprehensive study of activities and travel behaviour was led by Peter Jones at the Transport Studies Unit at Oxford, where the approach was defined and empirically tested, and where initial attempts to model complex travel behaviour were first completed.

Activities take place in space and to reach the desired location people must travel. Looking at trips independently misses some of the behavioural richness of linking activities in different locations and with different time windows or constraints. Some activities can be re-scheduled in time (postponing a trip to the gym) but only within constraints (gym opening hours). Others, like work or school attendance, are more difficult if not impossible to shift. Moreover, some activities may be re-scheduled and re-assigned to different individuals in the household and then to a different day of a week; for example, undertaking a main shopping trip for groceries. It is clear that, at least in principle, getting a better understanding of how people organise activities and the tours that are associated with them, must provide a more solid

Modelling Transport, Fourth Edition. Juan de Dios Ortúzar and Luis G. Willumsen.
© 2011 John Wiley & Sons, Ltd. Published 2011 by John Wiley & Sons, Ltd.

basis for travel demand modelling. This chapter explores how much of that understanding of activities and their schedules can actually be incorporated in operational models and what approaches can be followed to achieve this.

In what follows, we address first the issue of *tours* in greater depth looking also at the activities they link. We then look at activities and how we can model the complex interactions within a household that help schedule them and therefore generate trips. The next section identifies the econometric structures than can be used to represent these scheduling processes. A key element in *Activity Based Modelling* (ABM) is to synthesise detailed populations for both the base and future years. We then discuss approaches to model the daily schedule of activities of members of that population and the downstream tours and trips that result. Finally, we discuss some general points about the approach.

14.2 Activities, Tours and Trips

It is useful to define the key terms in this discussion before advancing further. In this chapter we will consider the following concepts:

- An *activity* is a continuous interaction with the physical environment, a service or person, within the same socio-spatial environment, which is relevant to the sample/observation unit. It includes any pure idle times before or during the activity (e.g. waiting at a doctor's surgery).
- A *stage* is a continuous movement using one mode of transport, more precisely one vehicle. It includes any pure waiting (idle) times immediately before or during that movement (e.g. waiting for a bus, searching for a parking space and making parking manoeuvres).
- A *trip* is a continuous sequence of stages between two activities (a trip can have only one stage, for example a car trip, or more as in a multi-mode trip).
- A *tour* is a sequence of trips starting and ending at the same location; a *trip chain* is the same as a tour but it may not end at the same location.
- A *trip purpose* is defined by the most important activity undertaken at one of the ends of the trip.

Tours may be classified by length and by their 'most relevant' activity, for example: Home Based Tours, Business Based Tours, etc. Consider, for example, an urban area where a classic Production Attraction model is using the following trip purposes:

- HBW (Home Based Work) that includes the journey from work back home;
- HBEd (Home Based Education), including the journey back home;
- HBO (Home Based Other), including shopping, leisure, etc.;
- NHB (Non Home-Based Business);
- NHO (Non Home-Based Other).

Figure 14.1 illustrates the concepts and distinctions between Activities, Trips, Tours, and Purposes. The diagram identifies four typical individuals (A to D) that can undertake six different (aggregated) activities at Home, Work (Factory and Office), Education (School), Shopping and Leisure (Restaurant).

In this diagram, Person A undertakes one tour visiting School, Factory, Shopping mall and then back home. Person A may have taken a child (Person E, not shown) to school and then proceeded to work. In a classic model this tour would appear as four trips, one HBEd (or Escort), one NHB, and two NHOs. Depending on how the data was collected and processed the first two might have been condensed into one HBW trip.

Figure 14.1   Daily activities, tours, trips and purposes

Person B has undertaken also one tour with three trips, HBW, NHO and HBO. A classic model would have captured this sequence a bit better in practice. Person C undertakes two tours. The first is a simple one to work and back and the second is also a simple tour with two HBO trips. These two tours would be perfectly picked up in classical models as two HBW and two HBO trips.

The longest tour is made by Person D who goes to work in an office, visits a factory, goes shopping and finally attends an evening course before returning home. Despite the complexity of the tour, a good classic model would have picked up these trips but not their interrelationships.

The choice of mode is, of course, also related to tour structure and length. If car driver is chosen for the first trip in a tour it is very likely that this will remain the choice for the other legs. A possible exception is for short tours from work (not shown above) where public transport could be chosen for convenience, speed and to avoid parking problems. Similarly, if public transport is chosen for the first leg of a tour, this is likely to remain the choice for the rest of the outing, including taxis.

The description above is appropriate to compare tours and trips but does not give enough information about activities. This is explored further in Figure 14.2 constructed on the basis of the data for Person C.

Person C starts from Leisure at home (although sleep could be considered essential maintenance at home, 8 hours minimum plus breakfast) and then travels to Work; this activity has a strict constraint as starting time but is more flexible on leaving the place of work. C returns home for some Maintenance (rest) and then goes out for a meal at a restaurant that, in this case, has a somewhat flexible start and end as an activity (no strict reservation needed). The role of different time-constraints for the activities just illustrated is made more complex as C is unlikely to want to eat in a restaurant alone. He may wish to coordinate with the spouse for this meal, starting either at home or from a different location.

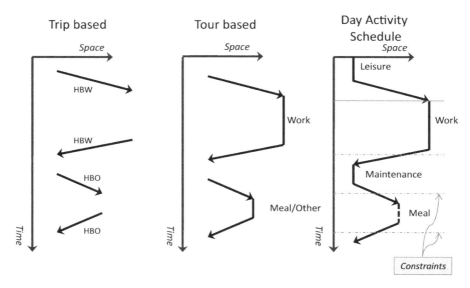

Figure 14.2   Trips, tours and activities of Person C

   In order to capture the behavioural richness of activities we must look at the household and the mutual interaction between trips, tours and activities. We should also consider the following essential aspects of activities and behaviour:

• Travel is derived from the need to change locations between successive activities.
• Individual activities are components of more significant personal projects (shopping for paint contributes to a plan to redecorate your home), reflect longer-term commitments (work, religious attendance) or satisfy some basic physiological or psychological demand such as sleeping or enjoying the company of friends.
• Scheduling activities involves the choice of time, duration, location and access/egress mode for the preferred activities.
• Some activities are 'mandatory' (work, education attendance) and offer limited flexibility in terms of location and duration; others are required to 'maintain' other activities (eating, sleeping at home, shopping, personal business away from home); finally, some activities although discretionary are still essential for a fulfilling life: social, recreational, entertainment.
• Individuals have monetary and time constraints (money and time budgets).
• Individuals schedule their activities in co-ordination with other members of the household or of their social network in order to maximise satisfaction, taking into account short and long term aspirations.
• Individuals are constrained in their scheduling of activities by the resources available to them, in particular vehicles or the availability of good public transport services.
• Individuals are further constrained by the need to be available to others at particular times and locations, either face-to-face (presentation to client) or by phone or videoconference.
• Longer term commitments to other household members, residential locations and work/educational places provide additional constraints to individual choices.

   The task of converting these issues into a workable and reliable activity scheduling process that can be formalised using closed form formulations or more general computer codes is a demanding task. This was identified a long time ago (see for example, Jones *et al.* 1983) but the advent of cheap and widely available computer power has made possible a number of alternative ways of tackling this difficulty.

Before we look into that problem we must consider how to model the individuals that will participate in the choices of activities and tours.

14.3   Tours, Individuals and Representative Individuals

In this section we consider what should be the unit of application of the ABM: households, individuals or 'representative individuals'. To address this question we look first at tours and their complexity.

If the majority of trips in a metropolitan area are of the Person C and D type, a simpler treatment for tours may be sufficient (albeit with some loss of interactions). If the type of tours represented by Persons A and B is significant, say over 15%, then it would be important to address this issue in practice.

A recent study of travel in the Auckland (NZ) region established that overall, 70% of tours have outward and return trips of the same purpose (and corresponding to that of the tour); 57% of tours comprise only two (out and back) legs. For HB Work and Business multi-leg tours, the average is around two extra legs per tour while for the HB Other purposes it is 1.3 extra legs and for Education it is just one extra leg (these extra legs being NHB trips).

Inevitably, longer and more complex tours require additional research, time and resources. It is difficult to make a prior decision about how many different types of tours one should attempt to model. Longer tours are less frequently found but may be more important in the future if policies to manage congestion are implemented: a four-leg tour satisfying three different activities contributes less to congestion than three 'there-and-back' tours doing the same. Depending on the approach, not all possible tours would be included in a model, only the most important and frequent; for example, if 95% of the tours have four trips or less this should probably be the maximum length to model.

The next task is to identify the individuals who would be modelled to undertake these representative tours. In a fully disaggregated approach, these individuals will result from an expansion of a random sample representing the whole universe of travellers. Each of these individuals will have a specific set of characteristics: gender, income, type of work, type of family, car availability, etc. In a disaggregate approach *Sample Enumeration* techniques, as discussed in Chapter 9, will be used to model individuals' choices and to apply these results to the entire population of the study area. This is a major task for the base year (when typically a Household Survey is available) but it becomes an even more demanding task for future forecasting years where the population has to be synthesised. Population synthesis can also be used to replace individual addresses for Home or Work with better spread addresses and even at a more detailed level of resolution than traffic analysis zones (TAZ). The task of fully specifying these present and future individuals is termed 'population synthesis' and is discussed later in this chapter.

An intermediate approach is to identify a number of 'homogeneous behavioural groups' or 'representative individuals', say some 20 segments of the travelling universe. Each group will have a set of characteristics pertaining to travel behaviour but they will still be represented as based on the centroid of a given zone and travelling to other zone centroids. Figure 14.3 illustrates idealised synthetic populations in a zone of a study area.

Both individuals and behavioural groups will require a population synthesis procedure to generate their number and characteristics in each zone (and sub-zone) of the study area in the future on the basis of known land uses in each area. If the key modelling focus is tours and trips, then homogenous behavioural groups may provide sufficient disaggregation. If the interest is on the activities and processes that generate those tours, it is difficult to envisage the use of homogenous behavioural groups as capturing the complexity of these interactions. In fact, one would need to model not just individuals but all the members of a household who take part in these decisions.

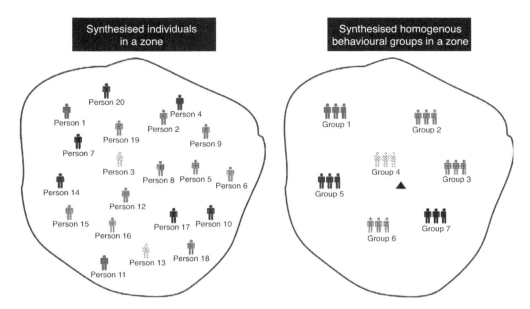

Figure 14.3   Individuals and homogenous behavioural groups in a zone

14.4   The ABM System

It is important to recognise from the outset that the ABM is only part, albeit a key one, of the complete modelling system. For a start, ABM covers only residents in the study area. A good deal of the model system is still aggregated, zonal based, and produces the traditional outputs that are needed for the appraisal of projects and policies. However, because of its finer treatment of activities, long term and short term decision making, tours and mode choice, ABM is able to address, at least in principle, a wider range of policy instruments and behavioural responses. Figure 14.4, adapted from Bowman and Bradley (2008), shows the ABM core and the additional components of a complete transport modelling system:

The ABM core contains the population and long- and medium-term choice simulator, and the person-day simulator. The first one models the long term choices, like normal place of work, car ownership and season ticket commitments. Medium-term issues involve household tasks allocated to individuals (escort children to school, convenience shopping, etc.) The person-day simulator searches for the most appropriate set of activities and tours required to satisfy these tasks. The main output is a list household and person day-tours (including destination, time of travel and mode choices) that, in turn, result in a trip list that goes into a trip aggregator where all trips are consolidated. To these resident trips one needs to add:

- External trips from outside the study area.
- Special generator trips, in particular from an airport or some other large or long distance travel station(s).
- Commercial vehicle trips, including courier deliveries and rubbish collection.
- 'Noise trips', that is trips that are seldom modelled but do exist in the network: empty taxis cruising for passengers, drivers looking for a parking space, emergency services/police vehicles, people out for a 'drive', lost drivers, etc.

The combination of all these trips (except normally the 'noise' group) is used to build time-dependent trip matrices which are assigned to the network to equilibrium; this results in network performance

Figure 14.4   A modelling system with an ABM core

indicators (travel times, etc) which are fed back into the ABM core and trip consolidators modules. Accessibility information is also fed back onto the urban system to influence land use and population attributes.

We will look now at the components of the ABM core itself.

14.5   Population Synthesis

The task of population synthesis is not exclusive to activity based models. Indeed, aggregate and disaggregate models have a ready population with most of the relevant characteristics available for the base year when Household Travel Survey (HTS) data is collected. However, this population needs to be synthesised for future years based on the few properties that are actually forecast by planners, such as: number of people per zone, perhaps income, and with some luck car ownership. Other attributes, like distribution of household sizes, age distribution, school and university attendance, multiple vehicle ownership and so on, need to be estimated, more often than not at the level of the representative households in each zone. This is where the task of population synthesis comes in.

ABM that forecast the activities and travel of urban populations require this task to be undertaken at a higher degree of disaggregation; the most important developments in population synthesisers have been attained seeking this more recent goal. The first task is to create a synthetic population and then simulate the behaviour of the households and persons in that population.

Population synthesis involves generating an artificial population by expanding the disaggregate sample data to mirror known aggregate distributions of household and person variables of interest (recall the discussion in Chapter 9). The process normally starts by creating a base year synthetic population from census and HTS data and then use aggregate demographic and land use forecasts to create a synthetic population for each future year. The synthesis procedure involves two main steps. First a demographic distribution of households is estimated for each transport zone or small census area (zone), and then a matching sample of households is drawn from a set of household records for which nearly complete census information is available.

The demographic distribution is defined discretely by the cartesian product of several categorical control variables (dimensions), with each multidimensional category (or cell) defined as a unique value combination of the one-dimensional control variables. The number of households in each cell is estimated through an iterative proportional fitting procedure analogous to the Furness method. The proportional fitting procedure starts with an initial joint distribution available for (aggregate) census geographical units. It then cycles iteratively through a set of control totals, one total for each category of each control variable.

Example 14.1 Consider a sample data as shown in Table 14.1 below. There are three household sizes and only two income levels. We know from, say, census data that there are 55 households (HH) with low income and 35 HH with high income in that zone, and that there is a total of 20, 40 and 30 HH of each size. Our sample is shown in the 3×2 (say from a HTS) rectangle in the middle.

Table 14.1   Sample data and marginal distributions

		Income			
		Low	High	TOTAL	HH Size Marginals
HH Size	Adjustment				
1		3.00	1.00	4.00	20
2		2.00	4.00	6.00	40
3+		4.00	2.00	6.00	30
Total		9.00	7.00		
Income Marginals		55	35		

The application of a bi-proportional adjustment in this case will lead to solve this population synthesis problem such that after three row and column iterations we get the figures in Table 14.2:

Table 14.2   Adjusted synthesised data after three iterations

		Income			
		Low	High	TOTAL	HH Size Marginals
HH Size	Adjustment				
1	0.997	16.19	3.81	20.00	20
2	1.003	16.59	23.41	40.00	40
3+	0.998	22.17	7.83	30.00	30
Total		54.95	35.05		
Income Marginals		55	35		

This approach can be extended to cover other dimensions like car ownership, number of students at households, number of persons of different type and so on. The adjustments will then be multi-proportional. As we saw in the case of matrix adjustments in Chapter 5, a requirement for this procedure to work is to have consistent control of marginal totals. In this case, the iterative procedure will converge so that all control totals are satisfied and the correlation structure of the initial joint distribution is preserved. Control totals are taken from Census tables for the base year. For the forecast years they will come from demographic and land use forecasts, which may be less detailed. Note that for some model applications, the number of households in each cell may need to be rounded to an integer number.

It is also useful to note that the problem of zero cells or zero marginals that affected trip matrix expansion or matrix estimation, may apply also to the population synthesisers. Similar corrections would need to be applied. In estimating a year distribution, all population synthesisers control for household income, household size and number of workers. Additional household characteristics used as controls in some cases include age, gender of householder, presence of children, and family vs. non-family households.

For an ABM perspective, the process of population synthesis needs entering into a second phase. In this case we need to identify person attributes from within each household; in this case again we will be interested in retaining the person attribute marginal totals for each zone. It is known that the derivation of person attributes can severely affect the accuracy of the subsequent modelling.

This second phase typically includes three or four steps. The first one is to convert into integer the non-integer values for households in zones resulting from the first phase. Second, a Monte Carlo procedure is typically employed to draw the correct number of households of each type from the HTS or an available census sample. Note that as some of the desired data may not be available in the census, or it may not be accessible to the modeller, it is often inevitable that one would sample from the HTS and any activity diary dataset available. Third, the useful household and person variables are extracted from the drawn households and retained for use by the model system. The fourth step is optional. Many implementations of ABM have sought to use a finer level of geography than that offered by conventional TAZ. This optional fourth procedure is used to assign each household to a more precise location (sub-zone) within its geographic unit. The final output from these processes is a synthetic population in which each synthesised household and its members have many clearly defined characteristics of interest for use in the model system, and together they match the estimated demographic distribution within each zone.

14.6   Monte Carlo and Probabilistic Processes

Most ABM use a Monte Carlo process to represent individuals (and vehicles) and their behaviour in a transport system. The name comes from the use of random numbers (as in a roulette) to sample from a population with a known distribution of attribute or characteristics. Pseudo random numbers between 0 and 1 can be generated very easily, for example in Excel using the RAND() function, and these values can be used to sample from any distribution. To create a particular individual one may sample from a 0/1 distribution for Sex, from a Log-Normal distribution for Income, from special distributions for Age, Family Size, Employment, etc, including sampling from a set of possible locations for sub-zone. This is repeated for each individual and then samples are taken for tour length and characteristics, including time of trip making.

Monte Carlo simulations are, therefore, very powerful in that it is possible to represent almost any population, both present and future, and include all characteristics believed to be relevant in order to identify activities and desirable tours. This flexibility comes at two prices. First, that it is often difficult to have full confidence that the resulting model is rigorously 'calibrated' and representative of an external reality that may be different from the ideas of the modeller. Second, the significant computer power and time required to obtain reliable results; this limitation has been more or less removed by the developments in computing (see also the discussion on random and quasi-random number in section 8.6.4). As random numbers are used to represent individual characteristics and travel behaviour, it is not enough to simulate one day (or one hour in a traffic micro-simulation project). It is necessary to repeat the process several times, with different initial random numbers, to gain confidence in the stability of the results.

Monte Carlo simulations can be used to model individual choices within a well structured family of hierarchical logit models for the choice of, for example, activity patterns, tour length, tour characteristics, choice of destination, time of day and mode.

14.7 Structuring Activities and Tours

ABM is in fact an integrated system or combination of several, mostly sequential, econometric sub-models. In principle, the models would be structured to cover:

- Long-term commitments of the household and its members, including the amount and location of work, residential location relative to work, education and friends, preferred types and locations for shopping and leisure, etc.
- Medium-term schedules for each individual in the household reflecting the tasks allocated to them and their specific activity demands, including projects like getting a degree or learning a new skill.
- Daily personal schedule of activities formulated by the individual, although some activities might change during the day in response to changing conditions; this flexibility has been helped by the use of mobile phones.

In practice, most applications start at the Person Day-Activity model and from this activity pattern tours are selected and disaggregated into key components. It is possible to structure these components as a set of nested discrete choice models, frequently of the logit type, as proposed in the key paper by Bowman and Ben-Akiva (1999) that has influenced a number of ABM efforts, in particular TRANSIM. An idealised example of this nested structure is presented in Figure 14.5:

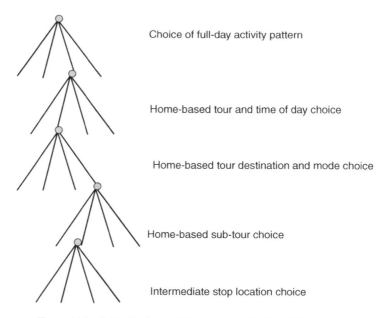

Figure 14.5   Example of nested choice structure for the activity schedule

The lower tier models are conditional on decisions at the higher tiers. As one would expect, the conditional model probabilities flow down the hierarchy whilst logsums, or expected utility values, pass the information from low tier choices up to the top tier ones. Going down one has to calculate probabilities for all alternatives in each nest representing a large number of combinations, as each branch above has many sub-branches below.

Moreover, using logsums in this way means that we need to calculate the utilities of every combination of alternatives going up the tiers before calculating probabilities from the top to the low tier nests. As these alternatives include all destinations, times of travel, mode, tour and sub-tour types, intermediate stops and so on, this is computationally very intensive.

The advantage of using a random utility framework and logit formulations is their solid theoretical background, sound model estimation techniques and user's familiarity with the strengths and limitations of the approach.

Inevitably, the top tier choices are quite critical and, at the same time, complex. The choice of a person full-day activity pattern may involve selecting among many (i.e. 50 or more) pre-identified activity patterns with their respective utility functions estimated from the HTS and travel diary surveys. One of these possible activity patterns is, of course, stay at home; the others require travel in different combinations. It is customary to distinguish between primary activities and tours (work, education) and secondary ones having more flexibility in location, timing and mode.

Taking as an example the seminal ABM for Portland (Oregon, USA) we can appreciate better the scope and complexity of the task (Bowman and Ben Akiva, 1999). The Portland Day Activity Pattern contains 114 alternatives differing in terms of the activities involved and their sequence. The choice set covers Primary Activities and tours, and Secondary tours associated (extensions) to the primary ones:

Primary Activities:

- subsistence (work or education) on tour;
- subsistence (work or education) at home;
- maintenance (shopping, personal business, etc.) on tour;
- maintenance at home;
- discretionary (social, recreational, leisure, etc.) on tour;
- discretionary at home.

If the primary activity is on a tour, the daily activity pattern model also estimates the trip chain type. There are eight possible subsistence tours, four for maintenance and four for discretionary tours. The trip chain type is defined by the number and sequence of stops in the tour (i.e. a simple tour, one or more activities on the way to the primary destination, one or more activities on the way home, and intermediate activities in both directions). For the subsistence tours there is also a work-based sub-tour addition to each of the four tour types above.

Secondary Tours
At the same time as the primary activity and tour type, the daily activity pattern model estimates the number and purposes of the secondary tours with the following alternatives:

- no secondary tours;
- one secondary tour for work or maintenance;
- two or more secondary tours for work or maintenance;
- one secondary tour for discretionary activities;
- two or more tours for discretionary activities;
- two or more tours, at least one for work/maintenance and one for discretionary activities.

As not all tour types apply to all the primary activity types, there are 19 possible combinations of primary tour types; these, times six secondary tour types make up the 114 alternatives.

Besides these choices, one has to consider the timings, duration, modes and destinations/stops for these tours and trips making the set of nested models rather demanding in terms of computational power. Therefore, the methods for solving these large scale nested models become quite critical.

14.8   Solving ABM

In the case of the aggregate classic approach the model is applied using zonal enumeration. For each travel zone, the number of trips by each mode to each destination zone at each time of the day is worked out. This is generally the result of several sub-models: trip generation, proportion of trips going to each destination, modal shares and so on. Model probabilities/shares are used to distribute demand across all feasible alternatives. We may call this approach 'zonal enumeration'.

In the case of ABM we need to use a different approach, either sample enumeration or Monte Carlo microsimulations. Sample enumeration also follows an approach of multiplying conditional probabilities. In this case, however, instead of applying the models separately for each travel zone, we apply them for each household and/or person in a representative sample. Thus, sample enumeration tends to work on a less aggregate scale than zonal enumeration but that is not necessarily the case. Zonal enumeration can be applied also with many different segments of the population in each zone, so that we are essentially working with an expanded sample of representative household/person types.

Sample enumeration, however, enables the retention of more complete information about each person and household in the sample – not just those characteristics that are used to define market segments. As a result, sample enumeration allows a wider range of variables to be included in the models that are applied. Moreover, if models are estimated at the level of the person or household, then sample enumeration applies them at the same level, avoiding aggregation bias.

Bradley *et al.* (1999) looked into the use of sample enumeration and Monte Carlo microsimulation as two alternative methods for solving ABMs using the Portland case as a test bed. Despite using a number of shortcuts for the sample enumeration approach they concluded that it was faster and more practical to use Monte Carlo simulation.

In the case of stochastic Monte Carlo microsimulation one still needs to use the same choice trees and analytical structure but just solves the hierarchical model in a different way. The logsum linkages are also used to calculate utilities up the tiers for each individual up to the full activity/tour pattern. Instead of calculating probabilities for all combinations of alternatives down the tree, samples of activity lists from the survey data are taken and replace the information with choice data from the models. For example, in the case of the Portland model (Bradley *et al.* 1999) the process involves the following steps:

- Draw a random sample of a single full-day activity/tour pattern from the top model probabilities.
- If the primary activity of the day is out-of-home, draw the times of day for the primary tour from the tour time of day model probabilities.
- Use these synthetic choices to sample a corresponding day-long sequence chain of observed activities from the HTS.
- For each tour in the pattern, sample from the destination and mode choice model probabilities to replace the observed destinations and modes in the activity list.
- For any intermediate stops in any tours, apply the intermediate stop location models stochastically to assign locations to those activities.

It should be noted that the only details retained from the HTS activity records are the more precise timings and sequencing of activities, since the time of day models usually deal only with a discrete number of different time periods (e.g. five in the Portland model). All the other observed choices are replaced by synthesised ones sampled from the models.

The use of Monte Carlo microsimulations seems to have become the preferred method for solving this type of ABM. Of course, some questions remain:

- As we are using random numbers, do the starting point and their sequence influence results?
- Do we need to run the model several times with different sequences of random numbers to ensure we obtain reliable and repeatable results?

Bradley *et al.* (1999) investigated these issues offering the following conclusions and recommendations:

1. Always run the model simulating the full population of interest.
2. The differences in results when using a different random number sequence at an aggregate level are minor (1 or 2%).
3. When looking at more focused results, one must bear in mind that if the values are small, for example the number of trips made by a population segment between one group of zones and another group, then the percentage variations are likely to be large.
4. These variations are healthy reminders that all models inevitable contain errors; notwithstanding, stochastic sampling errors are likely to be small compared to other sources of error present in any model (measurement, specification, input forecast errors, for example).

14.9   Refining Activity or Tour Based Models

The description above has focused on the Person Day Activity model only; however, as discussed before, an ABM also includes components for long and medium term decisions. This section provides some additional information on the handling of these and other issues in a complete ABM system; this will have modules for:

- Population synthesis for the geographic allocation of households.
- Longer term decisions: mostly car ownership but in some models also the choice of place of work and education.
- Person/household-daily scheduling, including the choices of activity patterns that span the whole day for a household or person.
- Tour-level choices as discussed in the previous section.
- Trip level choices: intermediate stop locations, mode and timing.
- Consolidation and assignment of trips to their respective networks.

Moreover, the ABM should also provide interrelationships between many of these decisions as, for example, the choice of car as a mode by one person in a household will affect the choices of the other members.

Choice of Usual Place of Work and Education
It is recognised that these are long term decisions that are not adjusted on a day to day basis. The choice of a place to live is implicitly modelled in the population synthesiser so it is included at the top level. The choice of usual place of work, and school or university education, is better modelled at the top of the hierarchy as well and not as part of the person day-schedule; most models today include it at this level. Note that some workers (construction, salespersons) may not have a 'usual' place of work. It is

important to ask in the HTS exercise about the usual place of work even if it was not accessed on that survey day.

Car Ownership
This is usually modelled also at the top level using a disaggregate model based on household and person types as discussed in Chapter 15.

In and Out of Home Activities
The ABM approach recognises that some activities (work, study, maintenance, discretionary) may be, at least partially, undertaken at home and we may wish to identify and include them. Most models, however, focus mainly on out-of-home activities and recognise only the probability that a person will not make any external tours in one day.

The number of out-of-home activities is relatively large, at least seven are usually considered: work, school, escort, shopping, meals, personal business and social-recreational. Additional distinctions are also possible.

Person Day-Patterns Linked Across Household Members
Originally, ABM treated linkages across members of the household implicitly through person type and household composition variables. The use of microsimulation makes it easier to treat these linkages more explicitly. For example, it is possible to simulate the children of a household first and then the adults conditional to what the children do, in particular their educational activities. This will result in escort activities being correctly allocated to adults and children.

Joint activities, such as going together for a meal out, should also be modelled consistently as they are likely to have significant impact on mode choice, for example. This will require a module for joint activity generation and participation; this additional effort must be traded-off against the greater accuracy achievable for trip choices.

Activities Allocated Explicitly Among Members of the Household
In principle, certain activities are undertaken on behalf of the household rather than individually; for example, shopping and escort trips. Modelling how these activities are allocated to different members of the household and at different days of the week has been limited. This task is further hampered by the limitations of survey methods currently in use; these are less useful to identify which activities are more likely to be allocated to different members at different times. With a few exceptions, most ABM do not have a module to allocate these activities to members and assume that they depend only on general household and person characteristics. It is argued that who actually undertakes them is less important than the fact that they are carried out and at certain times and destinations.

Number of Zones Used
In most cases the zones used for developing ABM are similar in size to those of a trip or tour based aggregate model. Ultimately, the car and public transport assignment modules are exactly the same. However, the use of microsimulation facilitates the implementation of finer geographical resolutions. Several models use a finer sub-zone system below that of TAZ. For example, the Portland model uses 20,000 'block faces' and the one in Sacramento 700,000 'parcels' (Bowman 2009). This fine level of disaggregation is useful to define more accurate destination choice alternatives, and estimate mode choice using detailed access to public transport information and level of service data. Note that in this way intra-zonal trips practically disappear from the model.

Time Periods and Time Constraints
Most ABM applications contain tour time-of-day models that reflect some sensitivity of time of travel choice to network conditions. However, there is usually only a limited number (3 to 5) of assignment periods, thus blurring some of the time sensitivity. There is a tendency to use more precise time windows in order to schedule each tour and trip consistently during the day (Bowman 2009). This requires

keeping track of the available time window after blocking off the time required by each activity and associated trips.

This tendency is converging towards half-hour periods, the main constraint being the ability of people to report times accurately. There seems to be a tendency to report times rounded to 10 or 15 minutes intervals. At this level of detail there would be good reasons to move to dynamic assignment.

Network Equilibrium

Given the level of detail of the microsimulation approach for solving ABM it could be argued that there is no role for network equilibrium, as it does not happen in reality and trying to achieve it would distort results. Nevertheless, as indicated earlier, the reason to seek equilibrium solutions is to obtain modelling results that enable the consistent comparison of alternatives. In the case of ABM we compound the problem of iterative processing with the use of Monte Carlo simulations based on computer generated random numbers.

Vovsha *et al.* (2007) have investigated this issue. They looked into a number of alternative methods for ensuring, or at least approaching, convergence of the whole model system. They concluded that the application of the Method of Successive Averages (MSA) to trip consolidated matrices and link flows led to reasonable results after some 8-9 global (feedback) iterations. Further research is necessary in this field.

14.10   Extending Random Utility Approaches

Despite their emphasis on activities the majority of the ABM are essentially microsimulation tour based models using a random utility choice-modelling framework; this has limitations. Current ABM offers only a little in the way real activity scheduling, of negotiations within the household on task allocations and even less in terms of postponing tours to a later day of the week.

There are some experimental models that attempt to go further into treating these issues with greater realism. The most promising approaches depart from the econometric methods solved by sample enumeration or microsimulation. Econometric methods are ultimately based on the idea that individuals seek to optimise their utilities choosing the best among available alternatives. In practice, human behaviour actually recognises the costs of information acquisition, information representation, information processing, and decision making. The new methods seek to represent behaviour and negotiations within this framework and are grouped under the name of Computational Process Models (CPM).

CPM are also microsimulations due to their disaggregate nature, the sequential decision process and the use of heuristics. However, the heuristics employed by CPM involve 'if-then' rules rather than utility-maximizing decision criteria. Models in this line of research are SCHEDULER (Golledge *et al.* 1994), AMOS (Kitamura and Fujii 1998), ALBATROSS (Arentze and Timmermans 2004) and PlanomatX (Feil *et al.* 2009).

Although these models have seen a number of applications, because of their nature they will remain experimental for a while. There are significant differences in the way these models handle the search for improved activity schedules and these rely on assumptions about behaviour and the nature of intra-household negotiations that are difficult to transfer from one context to another. The area of behavioural science is progressing very fast and we are likely to see better models implemented first in a research environment before general adoption for transport decision making and policy development.

15

Key Parameters, Planning Variables and Value Functions

This chapter covers three important aspects of transport modelling. The first is the forecasting of planning variables. These are variables like future population, employment, school places, shopping areas and income distribution, which are needed to make predictions with transport planning models. Sometimes these variables are provided externally to the study; in others they must be estimated as part of the planning exercise. In either case, they play a key role in determining the forecasting ability of the models discussed in this book.

General approaches to obtain these planning variables for aggregate models are discussed in Section 15.1. These are key inputs to a more disaggregate approach to synthesise populations as discussed in Chapter 14. A particular approach to develop these estimates is the use of Land Use Transport Interaction (LUTI) models that aim to capture the mutual influence between changes in accessibility and land use allocation; these are outlined briefly in section 15.2.

One of the most important planning variables is car ownership and this is the subject of section 15.3. Both time-series and econometric models to forecast car ownership are discussed, together with some more recent approaches.

Finally, we refer to value functions used in social project evaluation. First, many issues surrounding the concept, estimation and application of the *value of time* are presented in section 15.4. Then, section 15.5 discusses the concept and methods used to value external effects of transport, such as accidents and pollution. The book would not have been complete without this discussion.

15.1 Forecasting Planning Variables

15.1.1 Introduction

As discussed in Chapter 1, modellers always distinguish between endogenous variables, i.e. those to be forecast as part of the modelling exercise like flows, and exogenous or independent variables. The latter are required to run the models but are supposed to originate externally to the models themselves. Typical examples in the transport field are population, employment, car ownership and income. Values for these variables should be provided for the base year and for each of the years for which forecasts are needed from the transport model.

Modelling Transport, Fourth Edition. Juan de Dios Ortúzar and Luis G. Willumsen.
© 2011 John Wiley & Sons, Ltd. Published 2011 by John Wiley & Sons, Ltd.

The level of detail and disaggregation required for these variables depends on the type of model being used. In general terms an aggregate demand model makes fewer requirements than a disaggregate one in this sense. For example, at the trip generation level an aggregate, zone-based, linear regression model may only require population, car ownership and average income by zone; a cross-classification or category analysis model, on the other hand, will need the number of households in each of the categories used, typically 108 per zone, as we saw in Chapter 4, when the model is stratified by income (6 levels), household structure (6 levels) and car ownership (3 levels).

The importance of these variables in influencing the accuracy of the whole modelling exercise is very high, as established by Mackinder and Evans (1981) in a study of 44 British urban transport studies. It was found that all the models overestimated key indicators of performance but that the most important element in explaining this overestimation was errors in the values used for the planning variables. Model specification errors played a much lesser role in the overall inaccuracies. It appears that the planning variables were often wrong because they followed official global forecasts which were also wrong in the first place.

There are some very good reasons why forecasting planning variables is so difficult. The values of many of them in the future depend on complex interactions with other actors and influences that are very difficult to predict. This is certainly the case with the allocation of population and employment to geographical areas; these future allocations are influenced by interactions among factors such as:

- Population, income and car ownership.
- Levels of employment and their type.
- Land Use Master Plans, zoning and building regulations that affect what can be done, where and at what densities.
- Parking standards (minimum or maximum) for new developments.
- Land parcels available for development (green and brown fields) and their cost.
- The actions of developers regarding new and second hand properties, and the evolution of their 'land banks'.
- Local politicians and decision makers adapting plans and regulations to changing conditions.
- Changing views about what are considered desirable lifestyles and work practices.
- International and local trends on how best to tailor retail and services to a changing population.

The question then arises: how can we reduce as much as possible the errors in these planning variables? This is a difficult problem with no simple or single answer. A full discussion of the techniques available for forecasting these variables is outside the scope of this book; for practical methods the reader may consult England *et al.* (1985). However, we will discuss some of the ideas behind these techniques to appraise their strengths and weaknesses.

15.1.2 Use of Official Forecasts

The apparently simplest option in dealing with planning variables is to use official forecasts. In the UK, for example, there are estimates, at the District Council (and London Borough) level, of:

- population, households, employed residents and employment;
- number of households owning 0, 1 and 2 or more cars;
- private-vehicle trip ends by journey purposes.

The Department of Transport also produces forecasts, from time to time, of future demand expressed as expected vehicle kilometres for different types of vehicles. Other official institutions will provide other types of forecasts for planning variables, at least at a highly aggregate level.

Of course, these forecasts are seldom at a sufficient level of disaggregation to be directly usable in a detailed modelling exercise; however, they do reduce the amount of work needed to generate the required values for the planning variables at zone level. Some of the techniques to achieve this are discussed in the next section.

To some extent the problem with using official forecasts is that they sometimes reflect the expected effect of economic and regional policies whose success may actually depend on other uncontrollable factors like international trade and cooperation. Mackinder and Evans (1981) found that errors in forecasting these global indicators were at the root of the problem of mistakes for the planning variables at the local level.

We shall come back to this problem again. How can we accurately forecast transport activity if there are significant errors in some of the key inputs used in our transport models?

15.1.3   Forecasting Population and Employment

Whenever forecasts of these planning variables are not provided for cities or districts, the planning team will need to develop methods for their estimation. There are several methods that can be used to this end, some more appropriate than others for each particular application.

15.1.3.1   Trend Extrapolation

The direct extrapolation of current trends is the simplest but least satisfactory procedure, even if it is only applied at the level of the whole study area. Trend extrapolation does not take into account decisions already made about the availability of land for future development; it does not value new regional development policies nor does it consider the expected growth in employment in the study area. In addition to this, it does not provide any information about the age structure of the population, an important element in, for example, trip generation modelling.

15.1.3.2   Cohort Survival

A more detailed technique considers deaths, births and immigration, in and out of a study area, to forecast future population:

$$P_{t_1} = P_{t_0} + B_{t_0 t_1} - D_{t_0 t_1} + NI_{t_0 t_1} \tag{15.1}$$

where P_{t_1} is population at time t_1; P_{t_0} is population at time t_0; $B_{t_0 t_1}$ are surviving births in the period t_0 to t_1; $D_{t_0 t_1}$ are deaths in the same period, and $NI_{t_0 t_1}$ is the net migration in the same period.

Used in this very aggregate fashion, equation (15.1) ignores the age structure of the population and could under or over-estimate, for example, the corresponding fertility rates. For this reason the method is usually applied to subgroups of the population, or *cohorts*, and the method becomes a *cohort survival* approach. This involves the following stages:

1. The population is separated into cohorts; males are separated from females and each sex group divided into age strata (usually of five years) to give a population structure for the base year.
2. Fertility rates are then applied to females of child-bearing age.
3. The new-borns are added up and 'sexed' in known proportions.
4. The female and male babies make up the first cohort at the next round of calculations.
5. Survival rates are applied to females and males in all cohorts, starting from the youngest generation; survivors are then 'aged', that is moved forward to the next cohort.
6. The process is repeated, re-starting from stage 2 until the forecasting period has been reached.

If migration of population is to be treated in the forecasts, additional information regarding the sex and age structure of migrants is required. It is easy to see how the method may be adapted to include that new input.

The information demanded by this technique includes the initial number, age/sex structure of the population, and its associated survival, fertility and migration rates. The main source of uncertainty lies in the prediction of the rates, in particular fertility and migration rates.

15.1.3.3   Transitional Probabilities

An interesting alternative approach to cohort survival methods is to follow *family cycles* and use *transitional probabilities* reflecting the chances of moving from one stage in the cycle to another, for example, from married couple with no children to married couple with one child under school age, and from there to married couple with two children, and so on. A whole matrix of transitional probabilities is then built and processed to obtain the population in households at different stages in the family cycle in the forecast years. This approach certainly offers the potential of providing a very detailed account of population growth, very much at the level required for trip generation modelling. However, the uncertainty in the estimation and stability of the transitional probabilities is likely to be greater than that associated with fertility and migration rates in cohort survival methods.

Both cohort survival and transitional probability approaches can be usefully adapted to a continuous planning framework, where periodically collected data about fertility, migration and survival rates, and/or probabilities of changing family cycle status, permit the updating of previous estimates of population in the future and hence the changing of trip generation rates, and so on.

When forecasting employment change we are faced with similar problems. General trends in employment depend on economic policy, international trade and regional incentives. At a more local level aspects like the availability of land and qualified labour force in the study area, play an important role as well as the type of economic activity prevailing there. Moreover, the type and levels of employment also play a key role in determining the levels of income available to the households in the study area, which in turn influence car ownership and trip making behaviour.

15.1.3.4   Economic Base

A useful distinction in employment forecasting is that of *basic* and *non-basic* activities. Non-basic activities are those which are created in response to local demands whereas basic activities are those which require an external stimulus of some kind. Basic activities produce goods or services which are exported to other areas and regions. Non-basic activities produce goods and services to attend the needs of the local population. It is believed that the growth of basic activities creates additional non-basic ones (shops, banks, services, and so on) to satisfy the needs of additional population. The basic activities of a region constitute its *economic base* and strengthening it would result in economic, employment and population growth.

15.1.3.5   Input-Output Analysis

Finally, in forecasting the growth of a particular activity one should also follow the concomitant growth it generates in other industries providing inputs to it. Some of these will be based outside the study area whilst others may be located inside it. The use of an input-output matrix is the traditional method of following these linkages at national or regional levels. Such a matrix depicts how much input from other sectors of the economy is needed to increase output from one particular activity. The availability of such matrices at local level is questionable; the lowest level of disaggregation seems to be a regional one.

15.1.4 The Spatial Location of Population and Employment

Having estimated population and employment (in different subgroups) for the study area, it becomes necessary to allocate them to specific zones in order to apply our transport models. This work is usually carried out in conjunction with local planning authorities who have established plans for future development and re-allocation of land uses to zones in the study area. The use of age or life-cycle specific forecasts is helpful in this process as different types of housing development are more likely to attract different types of families.

The location of employment depends on its nature; for example, industrial development, commercial services, consumer services, and so on. Major changes in the location of economic activities should probably be discussed with those involved in carrying them out. Industrial development may require special sites, good availability of water services and access to major roads and railway/port terminals. In the absence of restrictive planning controls, office employment tends to be located close to good communication facilities and as close as possible to other office developments.

These two examples show that in the final analysis the location of population and employment is not independent of the transport system. Changes in accessibility are likely to affect the potential for development of different parts of a study area. This can be taken into account in the discussions with planning authorities, or more formally, in a more comprehensive model, as outlined in the next section.

In summary, the allocation of population and employment to zones usually requires a combination of formal models and discussions with planning authorities. The practical ways in which these tasks are carried out owes much to heuristic approaches and context-dependent choices. It seems difficult to eliminate current uncertainties about national, regional and local forecasts for these planning variables and this has important implications for the whole planning process.

The issue of disaggregating these allocations at an individual or household level has already been discussed in Chapter 14.

15.2 Land-Use Transport Interaction Modelling

There is an almost universal recognition that transport, in particular via accessibility, and land use are interrelated. One attractive approach to forecasting population and employment and allocating them to zones is, therefore, to internalise these exogenous planning variables in an integrated model of land use and transport. This has been an active area of research since the early 1960s; see for example McLoughlin (1969), Wilson *et al.* (1977), Foot (1981), de la Barra (1989), Echeñique *et al* (1990) and Simmonds (2001). After an initial period of optimistic claims about the success of such models, researchers became more modest in their aspirations (see Mackett 1985).

The importance of the interaction between transport and land use is twofold. First, if transport strategies significantly change accessibility this will change demand for land and generate new development in some areas; these will in turn affect the pattern of trips (trip matrices) and therefore have an impact on the performance of the transport system. These interactions are illustrated in Figure 15.1. Second, changes in the attractiveness of some areas will affect the price of land there; this can be interpreted as the capitalisation of user benefits into land prices and implies a transfer of benefits to land owners. This capitalisation issue raises the question of who gains and who loses as a result of a transport scheme and how a local authority can recover from land owners some of the increase in land prices.

The complexity of the relationships involved and their still fluid theoretical underpinnings have led to a situation where models and software are almost inseparable; an indication, perhaps, of the lack of consensus on what constitutes a good approach. It is impossible to do justice to this specialist modelling area in a book like this; the reader is directed to reviews like those of Wegener (2004) and Hunt *et al.* (2005).

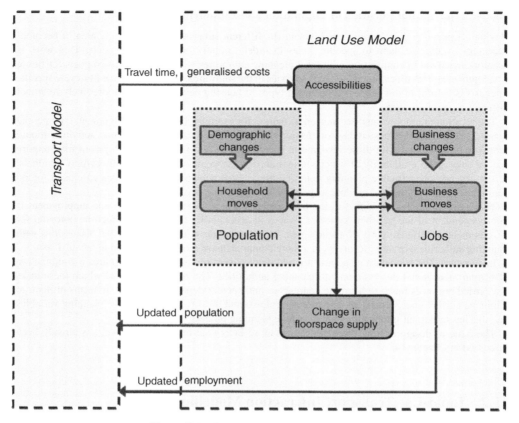

Figure 15.1   Transport and Land Use interactions

Although the basic design structure is similar for most land use models, there are at least four fundamental design features handled differently in them:

- Behavioural or structure-explaining approach.
- Bid-rent or discrete choice approach.
- Aggregate or microsimulation approach; and
- Emphasis in equilibrium or change dynamics.

Behavioural approaches treat relevant behaviour explicitly, for example birth, marriage, job change or relocation. Structure-explaining approaches attempt to model the outcome directly, for example distribution of jobs, without dealing with the processes that lead to that particular distribution. In practice, many models are somewhere between these two approaches.

The bid-rent theory assumes that every actor on the land use market is making bids for a piece of land, and the bidder with the highest offer gets it. Because of transport costs, everybody is willing to bid more for a location with good accessibility and whoever values this more highly, often businesses, gets the most accessible land, usually city centres. In the discrete choice approach households, firms, and developers make choices among a finite set of alternatives for locations, jobs and land, for example.

It has been suggested (NCHRP 2010) that bid-rent approaches work best in markets that are highly competitive and transparent and discrete-choice approaches in markets that react with some time lag and in which users make decisions with imperfect information.

Aggregate models aggregate actors into certain groups (for example households by household type or firms by industry type) and these are assumed to have homogenous preferences. As we have seen in Activity Based Modelling, microsimulation approaches offer advantages in terms of model development and in treating interactions explicitly.

Finally, some modelling approaches are underpinned by general equilibrium considerations whereas others emphasise the fact that change, and the rate at which it happens, is an inherent feature of land use, business and transport markets.

We try to identify here some of the most distinct theoretical components that support this type of model.

15.2.1 The Lowry Model

Many practical applications in the past have followed the lines put forward by Lowry (1965) in the 1960s. His model considers the spatial characteristics of an urban area in terms of three broad sectors of activity: employment in basic industries, employment in population-serving industries, and the household or population sector.

The Lowry model starts by allocating exogenously specified basic employment to zones and then the spatial distribution of households and non-basic employment are assigned using endogenous relationships. In addition, there are constraints on the maximum number of households for each zone (according to local regulations) and on the service employment thresholds for any zone; different types of service employment are assumed to have different minimum thresholds for their viability in any one zone.

The basic equations of the Lowry model can be written as:

$$\mathbf{P} = \mathbf{EA} \tag{15.2}$$

$$\mathbf{E}^S = \mathbf{PB} \tag{15.3}$$

$$\mathbf{E} = \mathbf{E}^b + \mathbf{E}^S \tag{15.4}$$

where $\mathbf{P}$ is a vector of population in each zone i; $\mathbf{E}$ is a row vector for total employment in each zone i, $\mathbf{E}^b$ and $\mathbf{E}^S$ are row vectors for basic and non-basic (service) employment in each zone i; $\mathbf{A}$ and $\mathbf{B}$ are zone-to-zone matrices of workplace-to-household and household-to-service-centre accessibilities.

The accessibility variables have two components, one corresponding to the participation rate in each zone (households per employee for $\mathbf{A}$ and service employment per household for $\mathbf{B}$) and a second corresponding to proper accessibility indices. These are normally calculated as:

$$A'_{ij} = \frac{E_j \exp(-\beta C_{ij})}{\sum_{ij} E_j \exp(-\beta C_{ij})} \tag{15.5}$$

$$B'_{ij} = E_j^S \exp(-\alpha C_{ij}) \sum_{ij} E_j^S \exp(-\alpha C_{ij}) \tag{15.6}$$

which are accessibility indices derived directly from the gravity model; see Chapter 5.

Lowry (1965) proposed a sequential solution to this problem including the constraints and thresholds mentioned above. More recent research efforts have emphasised the simultaneous solution of the same model and its extensions. Most of the latter have to do with additional disaggregation into different person and household types and their treatment over space. For example, certain types of person may be more willing (or capable) to pay for increased accessibility than others, thus influencing land prices and the type of development to be undertaken in different zones.

The integrated land-use and transport model has been implemented in a number of computer suites. In order to keep the problem tractable, some compromise in the level of detail of the transport part of the model is necessary; the hope is that what is lost in richness in the representation of the transport sector is more than compensated for by gains in the forecasting of employment, population and household development in the study area. An important problem of these models, however, is that they may suffer greatly from convergence problems in their extremely complex equilibration mechanisms. For a comparison of different implementations and extensions to this approach the reader should consult Webster *et al.* (1988).

15.2.2   The Bid-Choice Model

A more contemporary approach has been put forward by Martínez (1992) and implemented in a sophisticated package called MUSSA (www.mussa.cl) with several applications in Chile and the USA. The approach follows two modelling streams; the first one, originally proposed by Alonso (1964), is a *bid-auction* model where land is assigned to the highest bidder. The proportion $P_{h/i}$ of customers type h making a successful bid for a given location i depends on whether h's willingness-to-pay WP_{hi} is the highest among the bidders $g \in \mathbf{H}$. The assumption that WP_{hi} is a function of real estate and neighbourhood attributes of the location and socio-economic characteristics of the bidder plus an IID EVI distributed error term, leads to a MNL expression:

$$P_{h/i} = \frac{H_h \exp(\mu\, \mathrm{WP}_{hi})}{\sum_g H_g \exp(\mu\, \mathrm{WP}_{gi})} \tag{15.7}$$

where μ is the usual scaling parameter of the distribution of error terms. The expected market price p_i is equal to the expected maximum bid from potential buyers, given by

$$p_i = (1/\mu)\log\left\{\sum_g H_g \exp(\mu\, \mathrm{WP}_{gi})\right\} \tag{15.8}$$

The second modelling stream is a maximum consumer surplus model or *choice* model, equivalent to Anas (1982)'s maximum utility model. The consumer surplus CS_{hi} of individual h from choosing location i is given by the difference between its willingness-to-pay and the price of the location:

$$\mathrm{CS}_{hi} = \mathrm{WP}_{hi} - p_i$$

Under some simplifying assumptions the proportion $P_{h/i}$ of consumers h choosing location i is given by:

$$P_{h/i} = \frac{S_i \exp[\mu(\mathrm{WP}_{hi} - p_i)]}{\sum_j S_j \exp[\mu(\mathrm{WP}_{hj} - p_i)]} \tag{15.9}$$

Martínez (1991) then proves that the distribution of households and firms obtained from the *bid-auction* model in equations (15.7) and (15.8) is identical to that obtained from the *choice* version in equation (15.9). His *bid-choice* model is then summarised in these equations. These in turn can be simplified further when used at an aggregate level.

The transport system is represented by suitable accessibility (to destinations) and attractiveness (with respect to origins) indices which are included as location attributes in the willingness-to-pay function. The next task is to specify the WP functions; this must be done more or less on a case by case basis as the best function will depend on the availability of data and consumers' behaviour. Real estate developers are assumed to maximise profit, calculated as the price (p_i) minus the development costs (c_i), with profit assumed IID EVI with scale parameter λ, such that the proportion of development allocated to a given zone is a MNL model:

$$P_i = \frac{\exp \lambda(p_i - c_i)}{\sum_i \exp \lambda(p_i - c_i)} \tag{15.10}$$

MUSSA calculates the random bidding and supply market equilibrium (Martínez and Henríquez 2007), where the total number of consumers (households and firms) equals the total number of location units (for residential and non residential use). Each consumer is allocated at one location; consumer behaviour is affected by other consumers' choices (i.e. social externalities and agglomeration economies), and both suppliers and consumers are constrained by external regulation or zoning schemes using a constrained multinomial logit model (Martínez et al. 2009).

The whole model system has been integrated as the Land module in the transport package Cube (www.citilabs.com) which is used in several countries.

15.2.3 Systems Dynamics Approach

Models based on the two formulations above seek, at least to an extent, to achieve equilibrium conditions. The Systems Dynamics (SD) approach places more emphasis on the rate of change and the processes of positive and negative feedback that sometimes result in erroneous insights. The models are based on the pioneering work of Forrester (1969) updated thanks to the availability of low cost software capable of implementing the approach on a wider scale and with refined resolution. We follow here the ideas of Swanson (2003) whose Urban Dynamics Model (UDM) has seen many practical applications, in particular relating transport investment to urban regeneration.

In common with previous models, SD focuses on *accessibility* as a key driver of the attractiveness of a location to business and residents. A good location provides people access to the activities outside the home (including work) they wish to take part in; it provides businesses access to customers, a workforce and markets. A SD model of land use transport interactions will focus on a few features that make a location attractive adding markets for jobs, transport, building residences and business premises.

From the point of view of residents, a location will be more attractive if it provides good access to suitable employment and offers adequate housing. From the point of view of business the main features would be access to a suitable workforce, the availability of adequate premises, and access to markets and suppliers. The combined effect is illustrated in Figure 15.2, which shows the feedback between households and businesses: as the number of households rises, the accessible workforce also rises, increasing the attractiveness of a location for businesses. This will tend to attract more businesses and increase the number of accessible jobs, making areas with good accessibility more attractive to live in. This is an example of positive feedback, as increases in population lead to more business activity that in turn attracts additional households. Of course, other constraints would start to apply, as the supply of premises runs out, accessible land becomes fully utilised and/or congestion becomes severe.

Houses are seen as infrastructure 'stock'; they are built and remain in place for many years, occupying land and providing accommodation either to owners or renters. Houses are built by developers who consider how attractive a location is and whether it will lead to a reasonable financial return; this in turn will depend on their assessment of current and future demand for residences. A similar process applies to the construction or refurbishment of business premises whether they are built by developers or by companies seeking places to expand or to relocate.

Construction will tend to rise as an area becomes more attractive, but there will be delays in the response by developers. Builders need planning permission, land must be prepared, and houses and business premises take time to build. There will be lags, counted in years, between attractive conditions arising and new premises and houses becoming available in the marketplace.

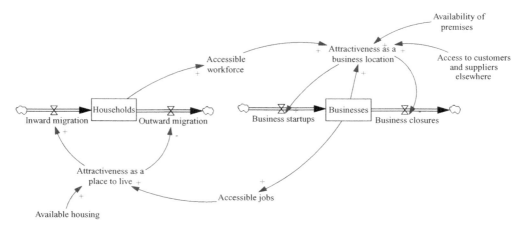

Figure 15.2   Access to business and residences

Such processes of change, their causes and effects, are what SD models are designed to address. Using modern software they are able to distinguish between different person, household and job types, different businesses and their land requirements and different modes of transport. The transport component of the Urban Dynamic Model can handle most of the responses in the classic aggregate transport model including hierarchical mode choice and congested assignment, and it generates a variety of responses in the model: Figure 15.3 illustrates some of the sequences of cause and effect the full UDM recognises. The

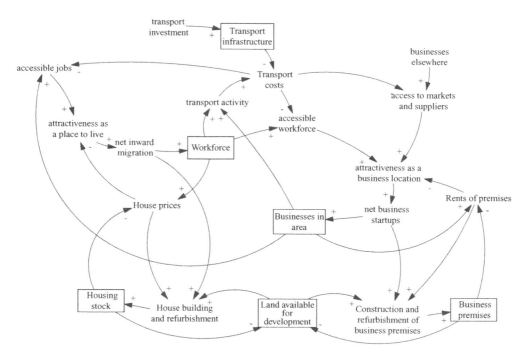

Figure 15.3   Relationships in a Dynamic Urban Model

focus on dynamic change provides some valuable insights into policy development and implementation and a useful tool to track the evolution of markets, businesses and residential location.

15.2.4 Urban Simulation

The advent of powerful computers and low-cost software has led to the development of many modelling approaches based on microsimulation for dynamic assignment and activity based modelling. In the case of land-use transport interactions, the responses of interest with pre-specified probability distributions are analogous to those described above. What microsimulation can offer is the power to incorporate a number of dimensions of both individuals and their choice processes which would otherwise require an excessive amount of disaggregation in model-based accounts. Microsimulation models are relatively easy to understand and implement, and permit tracking of individuals, households, business and parcels of land. The latter feature is, to an extent, misleading as the models will always leave out many variables that are necessary (but even then, never sufficient) to explain individual behaviour.

The MASTER model, developed in the UK by Mackett (1990) and UrbanSim, developed at the University of Washington by a team led by Paul Waddell (Waddell 2002; Waddell *et al.* 2003) are two examples of this type of approach. MASTER, for example, considers households one at a time allowing first for demographic processes including aging, giving birth, dying, divorce, and marriage. Marriage and divorce lead to the creation of new households with divorcees becoming one class of 'forced movers.' Voluntary movers include newly married couples, singles leaving the parental home, and wholly-moving households influenced by changes in life cycle. Both public and private housing markets are recognised, and dwelling occupancies are tracked from one period to the next. Choice of residence zone is based on a weighted function of generalised travel to work costs for the head of household.

UrbanSim simulates households, employees, developers and real estate prices with a similar degree of refinement. Location decisions are based on multinomial logit models. To select a location, a uniform distribution is used to randomly sample a set of nine alternatives in addition to the site with the highest utility. The final location is selected from these ten alternatives. Land values are updated by hedonic regression that estimates how much the individual characteristics of the land contribute to its value.

These are examples of powerful and flexible models. They respond to an urgent need to look closer into the issues of transport and land-use interaction, recovery of surpluses and distribution of benefits, in addition to changes in trip patterns. The widespread availability of general-purpose model estimation software has made possible the development of these models and their increasing application to practical problems.

It has been argued that this type of model is likely to work better where there are fewer constraints to the land market and type of development permitted by local authorities. This is probably the case in many developing countries, as reported by Chadwick (1987). However, as we have seen above, forecasting of planning variables is far from being accurate and its internalisation into an integrated land-use and transport model is unlikely to make it more reliable or robust. Our degree of understanding in this area is probably even more limited than in the transport sector alone. This problem highlights again the advantages of a continuous planning approach where regular updating of forecasts and plans reduces the risk of inaccurate predictions.

15.3 Car-Ownership Forecasting

15.3.1 Background

Although the total number of passenger cars active on the road in industrialised countries almost doubled between 1970 and 1986 (see for example de Jong 1989), the rate of growth was dramatically higher in developing countries and has continued increasing well into the new century. For example, the fall in import duties for small cars of less than 850 cc in Chile (from 120 to only 10%) in 1977, meant that average car ownership in Santiago went up by more than 100% in only five years (see Fernández *et al.* 1983); more recently, a comparison of 1991 and 2001 data for Santiago revealed that car ownership has continued growing at a rate higher than 3% per year (DICTUC 2003). Even if the annual mileage per vehicle had remained constant in this period, it must be noted that the total increase in passenger-car km represents a high cost to society in terms of accidents, fuel, pollution, increased traffic congestion and additional road construction and maintenance costs.

One problem faced by planners of vastly different nations is that forecasts of the number of cars and/or vehicle kilometres for, say, the year 2040, imply that these adverse effects may assume catastrophic proportions. In fact, by the end of the 1980s there were already cities like Athens, Los Angeles, Mexico, Sao Paulo, Seoul and Tokyo which had become notorious for their congestion and pollution problems.

Models to predict changes in car ownership, an essential input to transport planning, have been under development since the early 1940s. It can be said in general that these efforts have been made with the following three different purposes in mind:

- Market research studies for vehicle manufacturers and petrol companies which are of limited interest to transport modellers, as they are more concerned with vehicle attributes like size, engine capacity, and so on.
- Government-sponsored studies seeking to determine the need for new infrastructure (basically high-ways) at a national level; until the end of the 1970s simple time-series models were used for this task.
- Local studies, which are usually part of strategic transport studies, and which have made use of more advanced econometric methods with either cross-sectional and/or longitudinal data.

We will not attempt to cover all aspects of the car-ownership forecasting problem here, as whole books and theses have been devoted to the subject (see for example Mogridge 1983; Train 1986; de Jong 1989). Here we briefly discuss the two following basic methods:

- Time-series extrapolations using aggregate data at a national or regional level (basically the seminal work of John Tanner at the British Transport and Road Research Laboratory).
- Econometric methods using disaggregate data at the household level, as it has been argued that the decision to acquire a car cannot be modelled correctly at the strictly individual level or at the zone level (see for example Bates *et al.* 1978).

Modern methods sometimes incorporate features of both approaches and extend estimates to car usage as well. Critical reviews of these and other methods have been given by Button *et al.* (1982) and de Jong (1989).

15.3.2 Time-series Extrapolations

It seems clear that car ownership rates (e.g. cars/head of population) should not increase indefinitely in time (i.e. in general people with a driving licence are not going to indulge in several cars each); for this reason the increment curves which are usually put forward to model this phenomenon are S-shaped. If the number of cars/person in the USA and in the UK are plotted against time, one can find approximately the shapes depicted in Figure 15.4.

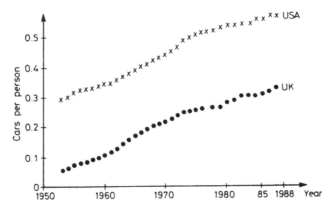

Figure 15.4   Shape of car-ownership increase

One curve which has proved popular in this field is the logistic, pioneered by Tanner (1978). The following three parameters are needed to adjust it:

- C_0, the car-ownership rate in the base year (cars/person);
- g_0, the rate of increase of the car-ownership rate in the base year given by $\frac{1}{C}\frac{dc}{dt}$ evaluated at $t = 0$; and
- S, the saturation level of car ownership.

In logistic curves we have that:

$$\frac{dC}{dt} = aC_t(S - C_t) \qquad (15.11)$$

where a is a constant. Solving this differential equation yields:

$$C_t = \frac{S}{1 + b\exp(-aSt)} \qquad (15.12)$$

where b is an integration constant. To find the values of a and b we can resort to the boundary conditions at $t = 0$; from (15.11) and (15.12) we obtain respectively:

$$g_0 = a(S - C_0) \text{ and } C_0 = \frac{S}{1 + b}$$

and replacing these values in (15.12) we finally get:

$$C_t = \frac{S}{1 + [(S - C_0)/C_0]\exp[-g_0 St/(S - C_0)]} \qquad (15.13)$$

Therefore, knowledge of C_0 and g_0 for one year taken as a base allow us to extrapolate C_t for any future year if S is known; however, S is not known but must be estimated. Tanner's method consists of fitting the following regression line (Figure 15.5):

$$g = \alpha + \beta C_t$$

Saturation corresponds by definition to that instant when the rate of change in the number of cars per capita (g) is zero; in this case we get $S = -\alpha/\beta$, and as we would expect α to be positive and β less than zero, we can deduce that $S > 0$.

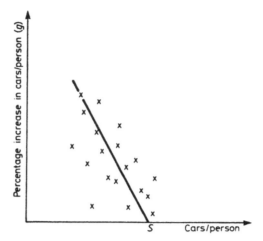

Figure 15.5   Determining the saturation level

Unfortunately constructing the graph of Figure 15.5 with data for the USA and the UK yields Figure 15.6; this implies that the method could work in the latter case, but it is much more doubtful in the former. For this and other reasons, the method was heavily criticised by Button *et al.* (1982).

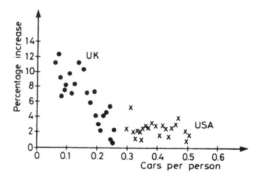

Figure 15.6   Saturation rates for USA and the UK

With the above data, Tanner (1974) estimated S as 0.45 for Great Britain. Table 15.1 compares predictions for 1975 made at different years with the observed figure of 0.25 cars/head in that year. As can be seen, the method is not very reliable.

In summary, the main objections to the logistic extrapolation method are as follows:

1. The model is not sensitive to policy variables. It is impossible to study the effects on car ownership of changes in car prices, road tax and import duties, fuel costs, and so on. Neither does it consider the influence of economic variables; therefore if the correlation among these variables changes in time, perverse results may be obtained (i.e. consider the effect in car-ownership increase brought about by the petrol crisis in 1973, or the aforementioned effect of the reduction of import duties in Chile in 1977).

Table 15.1 Errors in prediction using extrapolation

| Base year | Cars per person | | Predicted growth |
	In base year	Predicted for 1975	Actual growth
1960	0.11	0.28	1.14
1964	0.16	0.32	1.57
1966	0.18	0.31	1.67
1968	0.20	0.30	1.84
1969	0.21	0.28	1.66
1971	0.22	0.27	1.62
1972	0.23	0.26	1.48

2. S is assumed to be a constant; however, this may not be true in practice as attitudes tend to change with time.
3. The model does not yield information about different types of cars or, more importantly for planning purposes, the proportion of people belonging to households with 0, 1 and 2 or more cars.

15.3.3 Econometric Methods

These methods attempt to explain consumer behaviour directly rather than looking at general trends, and normally employ cross-sectional data. We will consider only two methods out of several which have been proposed; for a more comprehensive review see de Jong (1989).

15.3.3.1 The Method of Quarmby and Bates (1970)

This method uses just two independent variables, income and residential density, although it recognises the existence of several other factors of interest, such as household size and vehicle price. The basic relations of the model are:

$$\frac{P_0}{1 - P_0} = \alpha_0 I^{-b_0} D^{c_0} \tag{15.14}$$

$$\frac{P_2}{P_1} = a_1 \exp(b_1 I) D^{-c_1} \tag{15.15}$$

$$P_0 + P_1 + P_2 = 1 \tag{15.16}$$

where I is annual family income (thousands of \$), D is the number of residents per acre and P_i is the probability of owning 0, 1 and 2 or more cars; a_i, b_i and c_i are parameters to be estimated.

Substituting P_1 from (15.16) into (15.15) and taking logarithms we get:

$$\log\{P_2/(1 - P_0 - P_2)\} = \log(a_1) + b_1 I - c_1 \log(D) \tag{15.17}$$

Now, because D is a discrete variable for any given segment it may be considered a constant and (15.17) reduces to:

$$\log\{P_2/(1 - P_0 - P_2)\} = b_1 I + \text{constant}$$

It is instructive to consider that as income (I) increases, so does the left-hand side term of equation (15.17); therefore one can deduce that $(1 - P_0 - P_2)$ tends to zero or, what comes out to be the same, P_2 tends to $(1 - P_0)$. However, as P_0 is nearly zero for high incomes, that would mean that P_2 would tend to 1 and this is obviously incorrect as one would expect a lower limit for it. This upper bound, or saturation level (S) of P_2, may be incorporated to the model by adjusting (15.17), yielding:

$$\log\{P_2/[S(1 - P_0) - P_2]\} = \log(a_1) + b_1 I - c_1 \log(D) \qquad (15.18)$$

where S must be determined empirically; now, as this may be difficult in practice, the usual procedure involves trying different values in a sensitivity analysis. The types of curves obtained by this method are illustrated in Figure 15.7.

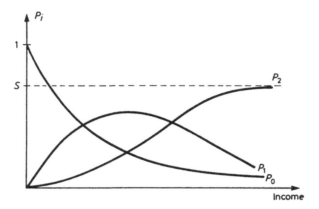

Figure 15.7 Car ownership versus income

Example 15.1 Consider the data in Table 15.2 and assume a value of $S = 0.78$; the problem is to estimate the parameters of the Quarmby and Bates's model for a fixed residential density value.

Table 15.2 Car ownership proportions by income

Income	P_0	P_1	P_2
1	0.61	0.34	0.05
2	0.35	0.47	0.18
3	0.22	0.44	0.34
4	0.16	0.37	0.47
5	0.10	0.30	0.60
6	0.08	0.24	0.68

If we take the logarithm of (15.14) for fixed D (i.e. c_0 is of no interest) we get:

$$\log\{P_0/(1 - P_0)\} = \log(a_0) - b_0 \log(I)$$

and fitting a regression line to the data we obtain $a_0 = 1.74$ and $b_0 = 1.60$. On the other hand, if we replace the value of S in equation (15.18) for constant D, we get:

$$\log\{P_2/[0.78 (1 - P_0) - P_2]\} = \log(a_1) + b_1 I$$

and fitting another regression line to the data we finally obtain $a_1 = 0.10$ and $b_1 = 0.84$.

15.3.3.2   The Regional Highway Transport Model (RHTM) Method

This method (Bates *et al.* 1978) combines the best features of the previous two approaches. First, it is necessary to define the following variables:

- $P(1+)$ = percentage of households with one or more cars, with a saturation level of S $(1+)$;
- $P(2+)$ = percentage of households with two or more cars, with a saturation level of $S(2+)$.

Therefore the previous method's values can be derived as:

$$P_0 = 1 - P(1+)$$
$$P_1 = P(1+) - P(2+)$$
$$P_2 = P(2+)$$

but it must be noted that the saturation levels are different from those of Tanner. The model takes the following form:

$$P_t(1+) = \frac{S(1+)}{1 + \exp\{-a_1(I_t/p_t)^{-b_1}\}} \tag{15.19}$$

$$P_t(2+) = \frac{S(2+)}{1 + \exp\{-a_2 - b_2(I_t/p_t)\}} \tag{15.20}$$

where (I_t/p_t) is annual family income (£/week) deflated by a car price index. The model was calibrated using British data for the period 1969–75, yielding the following parameter values:

$$a_1 = -7.76 \quad b_1 = 2.26 \quad S(1+) = 0.95$$
$$a_2 = -3.76 \quad b_2 = 0.04 \quad S(2+) = 0.60$$

To forecast it is necessary to assume a certain distribution of income (for example, one of the Gamma type); also, to convert the modelled results to cars/person (C_p) it is necessary to use census data. For example, Bates *et al.* (1978) postulated the following conversion rule:

$$C_p = P(1+) + 2.17P(2+)$$

To obtain cars/household we finally require information about the future average number of persons per household.

15.3.3.3   Joint Models of Car and Motorcycle Ownership and Use

Some authors have argued that car ownership should not be considered in isolation of other processes like motorcycle ownership, mode choice or at least car usage as the latter is more critical than ownership. Train (1980) has developed a structured Logit Model of car ownership and mode choice. The work of de Jong has always emphasised the need to model jointly car ownership and use (kilometrage) using different approaches, for example indirect utility (de Jong, 1990) and discrete choice (de Jong 1996).

In a different context, Khan and Willumsen (1986) argued that in developing countries growth in car ownership (and use) commits future resources to additional investment in roads and road maintenance. They insisted that car ownership should be considered as a policy variable rather than an exogenous factor; in order to support these ideas, they developed policy-sensitive models of car ownership and use, and calibrated those using data from different countries and time periods. They studied a number of functional forms, one of the most useful models being:

$$\log C_{1000} = -361 + 70.5 \log \text{GNPH} - 0.373 \log \text{PURTAX} - 2.58 \log \text{OWNTAX}$$
$$-0.682 \log \text{IMPDUTY} - 29.4 \log \text{FUELPR} - 2.04 \log \text{POPDEN}$$
$$R^2 = 0.86$$

where C_{1000} is the number of cars per 1000 inhabitants, GNPH is the gross national product per capita, PURTAX is the purchase tax associated with cars, OWNTAX is the associated ownership tax (road licence), IMPDUTY is the important duty for cars, FUELPR is the price per litre of fuel and POPDEN is the population density.

A second model was developed to estimate annual mileage per car, KM/C. One such model was:

$$\log \text{KM}/\text{C} = 5.76 - 0.434 \log \text{GNHP} - 0.368 \log \text{FUELPR} - 0.67 \log \text{ROADPOP}$$

where ROADPOP is the paved road length per head of population.

Finally, Khan and Willumsen (1986) developed an 'analysis' model where the total number of cars, car-km, fuel consumption, tax revenues, and road maintenance and investment costs are calculated for one or more years in the future. Alternative policies regarding taxation, import duties and road construction can then be compared in terms of their implied costs to the country. The general structure of these models is shown in Figure 15.8.

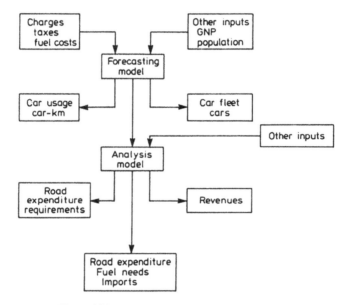

Figure 15.8   Khan and Willumsen's 'analysis' model

It is surprising how little work has been carried out in the modelling of motorcycle ownership. Motorcycles are a much maligned mode of transport despite their importance in many countries, including some in Europe. They have a poor safety record but offer a low cost and efficient mode of transport with smaller requirements for road and parking space than cars. The use of four-stroke engines makes them less polluting than cars and in their modern incarnation as electric two-wheelers they deserve considerably more attention.

One can assume a degree of substitution between motorcycle and car ownership but this cannot be complete. Motorcycles are present in all countries, even in those where car ownership has reached saturation levels. This suggests that the best way to model motorcycle ownership must be jointly with cars. A simple model would assume that the saturation level of motorcycle ownership must be related to the actual level of car ownership: the higher current car ownership the lower the saturation level of motorcycle ownership. One of the authors of this text has developed such a model, out of necessity, to forecast future metro patronage in an Indian city; as the running costs of motorcycles was very similar to the proposed fare of the metro, understanding how motorcycle ownership was likely to evolve was critical to mode choice. Consider, for instance, the context where the saturation level for car ownership is 500 cars/1000 inhabitants, the maximum for motorcycles is 350/1000 and this saturation level declines at half the car ownership level; this will produce a final saturation level for motorcycles of 100 motorcycles per 1000 inhabitants as shown in Figure 15.9.

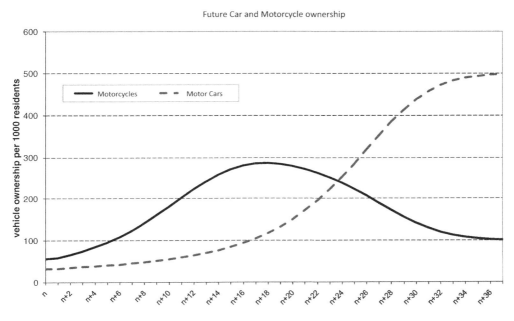

Figure 15.9   Car and motorcycle ownership in an emerging country

Of course, it is possible to use econometric models like those discussed above and link the two levels of ownership through Income (and vehicle prices) but this would require additional data and research.

15.3.4   International Comparisons

Energy use in the transport sector grows faster than in any other sector of the global economy. Of that growth, an increasing proportion originates in emerging countries. This is a reflection of the low levels of car ownership in these countries and the near saturation levels achieved in nations like the United States. It is therefore important to understand better how increases in wealth affect car ownership and use, and how these in turn will affect energy consumption and (until hydrogen becomes commonplace fuel) emissions and greenhouse gases.

Dargay and Gately (1999) have produced comprehensive studies into the effect of income level on car ownership, including international comparisons as part of this process. They used income and car ownership data for the period 1960 to 1992 from 26 countries ranging from the United States to India and China (data was not available for all these years in all countries). Then they searched for suitable functional forms to model car ownership as a function of income level. After experimenting with a number of functional forms, they chose a Gompertz model. The Gompertz equation for long-run vehicle ownership V^* as a function of per capita income I can be written as:

$$V^* = \gamma \exp(\alpha e^{\beta I}) \tag{15.21}$$

where α and β are negative values. The parameter γ defines the saturation level since for $\beta > 0$:

$$\lim_{I \to \infty} V^* = \gamma$$

The parameter α specifies the value of the function at $I = 0$, that is:

$$V^*_{I=0} = \gamma e^{\alpha}$$

Since the saturation level γ cannot be equal to 0, the value of the Gompertz function approaches 0 as α increases negatively.

The Gompertz function has a long-run elasticity that can be calculated by appropriate differentiation:

$$\eta^{LR} = \alpha \beta I e^{\beta I} \tag{15.22}$$

The income level that produces the maximum elasticity is obtained by setting the derivative of the elasticity to 0:

$$I_{\text{ME}} = -1/\beta \tag{15.23}$$

And the maximum elasticity is defined by:

$$\eta^M = -\alpha e^{-1} = -0.3678\alpha \tag{15.24}$$

Dargay and Gately (1999) recognised that vehicle ownership cannot change instantly; there are lags and inertia effects that must be taken into account. They postulated a simple partial adjustment mechanism to account for these lags:

$$V_t = V_{t-1} + \theta(V_t^* - V_{t-1})$$

where θ is the speed of adjustment ($0 < \theta < 1$) and V_t is vehicle ownership at time t. That converts into:

$$V_t = \gamma \theta \exp(\alpha e^{\beta I_t}) + (1 - \theta)V_{t-1} \tag{15.25}$$

For a number of theoretical and practical reasons the authors restrict the values of α, θ and γ to be the same for all countries and allow β to be country-specific. The model then becomes:

$$V_{jt} = \gamma \theta \exp(\alpha e^{\beta_j I_{jt}}) + (1 - \theta)V_{jt-1} \tag{15.26}$$

where the subscript j represents a given country.

Using their data sets they found a common saturation level $\gamma = 0.85$ vehicles per person (and 0.65 cars per person) and a value of $\alpha = -5.9$. They also found the value of $\theta = 0.09$, indicating that 9% of the total response to income takes place within one year. The values of β range from -0.3 to -0.2 in different countries.

From the model, one can estimate the maximum income elasticity to be about 2.4 for cars; this is attained at per capital income levels of about \$5000 (US dollars at their 1985 value) for countries with $\beta = -0.02$.

Given the range of countries in their database, the models developed by Dargay and Gately (1999) are quite useful for application in different countries where there are limited time series available for car ownership forecasting.

15.4 The Value of Travel Time

15.4.1 Introduction

The question 'has time a value?' is answered in the affirmative by most people. A more serious problem is 'what value?' and under what circumstances it can or must be measured.

This theme has generated an enormous debate in the literature for more than 30 years (see for example Bruzelius 1979) simply because time savings continue being the single most important benefit of transport improvement projects all over the world. However, and in spite of its importance, a consensus has not been reached about the size and nature of the values to be used in project evaluation. We will not attempt to review the subject in great detail here, but refer the reader to Jara-Díaz (2007) for a deeper discussion.

For example, in Great Britain (and other countries, such as Chile) social values of time corresponding to a fixed proportion of the average hourly rate are recommended for project appraisal. On the other hand, in the USA increasing values for three ranges of time savings (0–5 min, 5–15 min and 15 or more minutes) have been recommended (AASHTO 1977). Clearly the use of linear or non-linear valuation functions should lead to different benefits and hence to different investment priorities. For example, the British norm tends to favour schemes generating small time savings while the American norm above favours schemes generating more substantive time savings.

Most studies distinguish between *subjective* (or behavioural) and *evaluation* values of time. The first corresponds to, for example, the value of the parameter associated with in-vehicle travel time in the generalised cost functions we studied in Chapter 5 and which should have been derived by estimating, typically, a discrete choice demand model with real disaggregate data. The evaluation value is that used, as the name implies, to compare alternative schemes which produce different levels of time and other resource savings. It is argued, therefore, that the behavioural value of time reflects mostly the ability of the traveller to pay and not the intrinsic value of a particular time saving. This is why very often the value of time used for evaluation purposes is an *equity* value, taken as being the same for all travellers, independently from their age or socioeconomic group, as we will see below.

On the other hand, it may be argued that the use of different 'values of time' for evaluation and demand modelling purposes introduces inconsistencies of approach at different stages of the same exercise. This was, for example, one of the criticisms levelled at the controversial implementation of the *Transantiago* public transport system in 2007 (see Muñoz *et al.* 2009) as the low social-equity values of waiting time were confronted with normal operators earnings in a complex optimisation program used as part of the system design. There is little dispute, however, that the subjective values of time are heavily dependent on model specification and data (see Gaudry *et al.* 1989); this is an undesirable property because consistent evaluation of projects is sought over a wide range of models and areas.

15.4.2 Subjective and Social Values of Time

The utility function estimated from discrete travel choice models can be used to calculate the subjective value of time saving (SVT) or, equivalently, the willingness to pay to reduce travel time (in-vehicle, walking or waiting) by one unit. As shown in Jara-Díaz (2000), because travel utility is in fact a conditional indirect utility function, the SVT can be given a microeconomic interpretation which depends upon the arguments that are assumed to enter the utility function and the type of constraints considered; see also Bates (1987).

Time valuation analysis comes from three sources: the pure time allocation theories, the home production framework and the literature on travel demand. Everything started with Becker's (1965) approach based upon the idea of utility depending on the amount of 'final goods' (i.e. a prepared meal) consumed, each of which requires market goods and time as inputs; this was the origin of a time value equal to hourly income, because 'time can be converted into money' by spending more time at work and less in consumption. This elementary result was soon proved limited, after the successive analysis by Johnson (1966), Oort (1969), DeSerpa (1971) and Evans (1972), because work time should enter utility as an argument.

Later on, the fixed-income approach to mode choice models, introduced earlier as the *expenditure rate* approach (Jara-Díaz and Farah 1987; Jara-Díaz and Ortúzar 1989), also supported a travel time value that is not necessarily related to the wage rate. The result of this stream of papers was a framework in which the economic actions of the individual are looked at as if they maximised a utility function that depends upon all types of activities undertaken and on all goods consumed, subject to three types of constraints: a money budget, a time constraint, and a set of technical relations between goods and time (Jara-Díaz 1998; 2007).

Up to now, the SVT has been shown to reflect the value of relaxing the minimum time requirement on travel. Analytically, this is the ratio of the multiplier on that constraint over the marginal utility of income (MUI) and can be shown to be equal to the resource value of time (or, equivalently, the value of leisure) minus the value of the marginal utility of travel. The former represents the value of reassigning the travel time saved to other activities, and is analytically given by the ratio of the multiplier of the time constraint over the MUI. The latter is the lost value, in *direct utility* terms, because of less travel, and is expected to be negative. Thus, the SVT adds up the value of a gain in leisure plus the value of a reduction in an unpleasant activity (see the discussion and the extra references provided by Jara-Díaz 2007). It is important to note that if individuals choose the work schedule (hours at work) at a given wage rate, they will adjust that schedule until the value of leisure equals the value of work; this is given by the addition of the money earned (the wage rate) plus the value of the marginal utility of work (which can be positive or negative). Jara-Díaz and Guevara (2000), and Munizaga *et al.* (2006) among others have managed to estimate simultaneous models of travel and activities, obtaining not only the SVT but also its component elements.

Finally, a word on the price of travel time that should be used for social appraisal of projects (the evaluation or social value of time). There is no reason for society to value an individual's reassignment of travel time at the individual's SVT. For the analysis of discretionary travel, the state of the art is the work by Gálvez and Jara-Díaz (1998), who show that a proper social price of time (SPT), consistent with a social appraisal framework within the field of welfare economics, should be equal to the ratio of the marginal utility of time over what they call 'social utility of money'. This is given by a weighted sum of individuals' MUI, with the weights given by the proportion of marginal taxes paid by the corresponding group.

This approach advocates for potentially different SPT by group, which are generally different from each group's SVT. It is important to note that these authors show analytically that accepting the SVT as SPT is equivalent to assigning to each group a social weight that increases with income. This has important and generally undesired policy implications, but sadly it reflects the approach usually taken in practice.

15.4.3 Some Practical Results

Heggie (1983) argued that the value of time debate was more empirical than theoretical. The enormous practical difficulties associated with measuring values of time encouraged the use of indirect methods such as the discrete choice approach mentioned above. However, this method generates the usual empirical problems such as:

- how to choose an appropriate sample, i.e. one which basically contains people with a real choice among clearly defined alternatives in terms of time and cost of travel;

- how to measure the travel attributes, i.e. avoiding aggregation, perception and other sources of bias;
- which demand function to use that is consistent with the situation under study.

All these problems suggest that values derived from models estimated with revealed preference data (the large majority of cases) may be suspect.

Perhaps the most complete study ever undertaken about the value of travel time savings was performed between 1981 and 1986 by a consortium of consultants and academic experts in Britain, using a series of models estimated with revealed preference and stated preference data for various choice situations in several areas of Great Britain (Bates and Roberts 1986). Its principal recommendations (Department of Transport 1987) were:

1. The value of working time (i.e. trips made during or as part of work) is equal to the gross hourly income of the traveller, including all additional costs to the employer.
2. The trips for all other purposes, including trips to work, increased their valuation from 27% to 43% of the average hourly income of full-time employed adults (this is an increment of 85%).
3. For the majority of cases a single *equity* value of time should be used; however, in cases where the proportion of children, pensioners or employed adults is judged to differ significantly from the national average, an *ad hoc* equity value of time should be estimated using the individual values for each of these groups.
4. To update these values, information about real hourly incomes on each year should be used; to forecast, these incomes should be estimated as a function of the domestic per capita product.
5. The values of waiting and walking time should be taken as twice the value of in-vehicle travel time; bicycle users should be treated as pedestrians in this sense.
6. Small time savings should be valued equally as more significant savings.

In 1994 the UK Department of Transport commissioned a new value of time study (Accent and HCG 1996). In what follows we summarise some of their most interesting conclusions, which are broadly in line with the findings of an earlier study done in Holland (HCG 1990) using the same methodology:

1. For any level of variation around the original journey time, travel time gains are valued less than losses. For non-work-related journeys, variations up to five minutes in journey times are generally ignored. Business travellers are more sensitive to gains and losses than commuters, who in turn are more sensitive than those on non-work-related journeys.
2. There is a clear relationship between income and SVT (as was found in 1986) which is monotonically increasing but not directly proportional. At the same income levels, the 1994 SVT values are significantly lower than those recorded in 1986. This may have been caused by changes in the composition of the car-using population (those with high SVT were earlier to acquire and use cars) with the growth in usage then biased towards market segments with lower SVT.
3. SVT values under congested conditions are significantly higher than for trips done under free-flow conditions. However, the types of road mix (i.e. percentage of time travelling on motorways, trunk and other roads) were not significantly different. Finally, regular users of motorways are relatively indifferent to number of lanes, but appear to be very sensitive to travelling with lorries in the traffic, and clearly dislike roads with no shoulders (the strongest effect of all).
4. In relation to peak shifting, it was found that the disutility of departing earlier increases linearly with the time difference. This was also true for later departures up to one hour as, curiously, they found that the burden did not increase much beyond that hour; see also the discussion in Bianchi *et al.* (1998).

15.4.4   Methods of Analysis

15.4.4.1   The Revealed Preference Approach

To estimate the willingness to pay (WTP) for savings in travel time (i.e. the SVT) in the classic transport microeconomic literature, modellers need to measure the trade-offs between travel time and cost faced by a target population represented by a statistical sample (e.g. individuals commuting from certain suburbs to the CBD). The SVT corresponds to the marginal rate of substitution between perceived times t_i (in-vehicle, walking or waiting time) and costs c_i of travel at constant utility (Gaudry *et al.* 1989), yielding the following expression:

$$\text{SVT} = -\left. \frac{\mathrm{d}C_i}{\mathrm{d}t_i} \right|_v = \frac{\partial V_i/\partial t_i}{\partial V_i/\partial c_i} \tag{15.27}$$

As the representative utility function in our most classical models is assumed to be linear and additive in the (fixed) marginal utility parameters, under this assumption the SVT corresponds to the ratio between the estimated parameters, θ_t and θ_c, of the attributes travel time and cost; for example, in the case of the popular wage rate (w) specification (Train and McFadden 1978), this simply yields:

$$\text{SVT} = \frac{w\,\theta_t}{\theta_c} \tag{15.28}$$

From (15.28) one can easily see that the ratio θ_t/θ_c represents SVT as a percentage of income.

For the linear-in-parameters expenditure rate (g) specification (Jara-Díaz and Farah 1987), where g is given by (8.7), equation (15.27) yields:

$$\text{SVT} = \frac{g\,\theta_t}{\theta_c} \tag{15.29}$$

In the non-linear Box–Cox case (8.3) with wage rate specification we get, instead:

$$SVT = \frac{w\,\theta_t t_i^{\tau_i-1}}{\theta_c (C_i/w)^{\tau_c-1}} \tag{15.30}$$

which will clearly vary across alternatives if τ_k is not equal to 1. This latter formula implies that if both τ's are equal and they are less than one (i.e. as required by their micro-economic conditions), the model will necessarily yield higher value of time estimates for modes which are more expensive per minute; however, this may not be the case if the τ's differ (Gaudry *et al.* 1989).

Now, as θ_t and θ_c are estimates of the 'true' model parameters, they are not really constants but random variables with a certain probability density function (PDF). For this reason the 'SVT point estimate' (i.e. θ_t/θ_c) is also a random variable with an unknown PDF, and it is appropriate to examine the consequences of replacing this single value by the construction of a confidence interval given a certain level of confidence. A simpler but less appropriate way out consists in judging the significance of the SVT by means of a pseudo t-ratio test. Jara-Díaz *et al.* (1988) show that by making a first- order expansion of a Taylor series for the random variable θ_t/θ_c around its mean value (the ratio of the estimated coefficients), the following t-ratio may be constructed:

$$t_{tc} = \left(\frac{\sigma_t^2}{\theta_t^2} + \frac{\sigma_c^2}{\theta_c^2} - \frac{2\text{Cov}(\theta_t, \theta_c)}{\theta_t \theta_c} \right)^{-1/2} \tag{15.31}$$

where σ_t and σ_c are the standard errors of the estimated coefficients. Daly and de Jong (2006) have given a fresh look at this formula, arguing that due to the asymptotic properties of the maximum likelihood estimator, it would be an exact measure in the immediate vicinity of the maximum.

We know that the maximum likelihood parameters are asymptotically distributed multivariate Normal. Now, if a vector of random variables (in our case the parameter estimates) converges asymptotically to a joint distribution (in our case the multivariate Normal), then any continuous function of the parameters, such as the ratio, converges in distribution (to the ratio of two Normal variables), according to the *continuous mapping theorem* (see theorem 5 in Mann and Wald 1943).

Consequently, the SVT point estimate is a random variable governed by a probability distribution (i.e. that for the ratio between two Normal distributed variables) without an explicit form (Fieller 1933; Hinkley 1969; Shanmugalingam 1982) and that may turn out to be unstable (Meijer and Rouwendal 2000). In the special case of two Normal variables with mean zero the ratio follows a Cauchy PDF (Arnold and Brockett 1992), but this has an indefinite variance and its mean does not have an analytical expression.

Given these facts it is highly likely that the ratio between the parameters θ_t and θ_c, which are components of a general multivariate Normal population, will be governed by an unyielding PDF (the only exception being when the coefficient of variation of θ_c approaches zero, in which case the ratio approximates the Normal distribution). It is necessary then to find an econometric procedure to make statistical inference on this ratio without resorting to the direct use of the associated PDF.

To solve this problem, several methods have been proposed in the literature. For example, Ettema *et al.* (1997) discuss a general method to construct confidence intervals for the SVT even in cases where the parameters of travel time and cost are allowed to interact with other segmentation variables. Simulation is used to simultaneously calculate the parameters from a multivariate Normal distribution, defined by the covariance matrix of the estimated travel time and cost parameters. Values for these are generated a sufficiently large number of times and the confidence interval is constructed on the basis of the mean and variance estimates of the generated sample; it is possible to simulate values for the parameters of travel time, waiting time, walking time and cost simultaneously. Finally, by simply calculating the 0.025 and 0.975 percentiles, the limits of the confidence interval at the 95% level are obtained.

An advantage of this method is that it does not need to introduce additional assumptions (other than normality for the maximum likelihood estimators). In addition to being applicable to any type of utility function specification, it considers the variance of the parameters and the correlation between them. The results of Ettema *et al.* (1997) suggest that when the correlation increases, the size of the intervals decreases, indicating that we may obtain extreme results when correlation is not considered.

Further advances on this method and an application to an RP/SC model including interactions in the utility specification and the introduction of intangible variables, such as comfort, were done by Espino *et al.* (2006). Their results indicate that the size of the confidence interval is affected by the outliers of the simulation as well as by the magnitude of the simulated parameters. The estimated parameters should be consistent in relation to all the microeconomic principles underpinning the model, i.e. the marginal utilities of the different attributes must have a correct sign for every individual in the sample (i.e. before applying, in their case, sample enumeration to obtain the corresponding SVT); otherwise, such individuals should be removed from the calculation. They found that elimination of outliers in two steps (first, from the simulated multivariate Normal distribution and second, from the simulated distribution of the SVT), as well as the removal of individuals with inconsistent marginal utilities, was the simulation strategy that provided narrower confidence intervals. Further, in this case the amplitude of the intervals remained constant as the number of simulations (up to 100 000) was increased.

Armstrong *et al.* (2001) discuss two further methods. The first one is called the asymptotic *t*-test method and is based on the following null hypothesis:

$$H_0 : \theta_t - VT\theta_c = 0 \tag{15.32}$$

where *VT* represents the SVT point estimate (i.e. the ratio between the parameters of time and cost in a linear utility). The confidence interval is given by the set of *VT* values for which it is not possible to reject H_0 at a given level of significance. The corresponding statistic is:

$$t = \frac{\theta_t - VT\theta_c}{\sqrt{\mathrm{Var}(\theta_t - VT\theta_c)}}$$

This expression distributes Normal for linear models and asymptotically Normal for non-linear models like the MNL (Ben-Akiva and Lerman 1985). Armstrong *et al.* (2001) also derived the upper and lower bounds for the interval as follows:

$$V_{U,L} = VT \left(\frac{t_c}{t_t} \right) \frac{(t_t t_c - \rho t^2)}{(t_c^2 - t^2)} \pm VT \left(\frac{t_c}{t_t} \right) \frac{\sqrt{(\rho t^2 - t_t t_c)^2 - (t_t^2 - t^2)(t_c^2 - t^2)}}{(t_c^2 - t^2)} \tag{15.33}$$

where t_t and t_c correspond to the t-statistics for θ_t and θ_c respectively and ρ is the coefficient of correlation between both parameter estimates. Equation (15.33) is a real number only if the radical argument is non-negative; it can be shown that this condition is met when the parameters θ_t and θ_c are statistically significant (so that t_c and t_t are greater than t). This condition assures positive upper and lower bounds.

Furthermore, it can be observed that the confidence interval derived from this formulation is not symmetrical with respect to the SVT point estimate (*VT*), and that the interval's midpoint is greater than *VT* as well. Another feature is that the value of ρ has a strong influence; for example, the interval size decreases with the value of ρ and vice versa. The size of the interval also narrows as the t-statistics get more significant.

Finally, note that for large samples the following equality holds:

$$\lim_{\substack{N \to \infty \\ t_t, t_c \to \infty}} V_{U,L} = VT \tag{15.34}$$

which agrees with the intuition that the larger the sample size, the smaller should be the interval size.

The second approach proposed by Armstrong *et al.* (2001) is called the likelihood ratio test method. It is based on imposing the linear restriction (15.32) to the maximum likelihood estimation process and comparing the statistical efficiency of the estimation with respect to the unrestricted case. The procedure is to search for values of *VT* for which the linear restriction is valid given a certain significance level. The null hypothesis is still the same as in the previous case, but the test is performed according to the following statistic:

$$\mathrm{LR} = -2[l(\theta_r) - l(\theta)] \tag{15.35}$$

where $l(\theta_r)$ and $l(\theta)$ represent the logarithm of the likelihood function for the restricted and unrestricted models respectively. LR is distributed χ^2 with one degree of freedom (corresponding to the single restriction imposed).

Example 15.2 Let us consider the following systematic utility function to be estimated:

$$V_{iq} = \theta_t t_{iq} + \theta_C C_{iq} + \sum_k \theta_k x_{kiq}$$

where t_{iq} and C_{iq} are the travel time and cost for individual q; x_{kiq} are attributes (different from travel time and cost) for individual q, and θ_k are their corresponding parameters. Replacing the ratio of θ_t and θ_C by *VT* in the above equation, the following utility function is obtained:

$$V_{iq} = \theta_C (VT t_{iq} + C_{iq}) + \sum_k \theta_k x_{kiq}$$

These two equations allow us to compute the unrestricted and restricted log-likelihood functions, $l(\theta)$ and $l(\theta_r/VT)$. Clearly, if the SVT is equal to *VT* then $l(\theta) = l(\theta_r/VT)$, but different values of *VT* will yield different values of the restricted log-likelihood function. This method requires a search for the maximum and minimum values of *VT* for which the following inequality holds:

$$-2[l(\theta_r/VT) - l(\theta)] \le \chi^2_{1,1-\alpha}$$

An advantage of this method over the asymptotic t-test method (15.33) is that it is not restricted to linear utility functions. However, the process of constructing the intervals is more tedious because it requires an iterative procedure to obtain each limit. Armstrong *et al.* (2001) present the results of using all the above methods for various cases of interest.

The subjective values of the time and their confidence intervals (both bounds and size) vary strongly with model specification (i.e. they are strongly dependent on the functional form assumed for the representative utility and on the model structure). But with cross-sectional data it is not easy to give a clear rejection of any reasonable model form; see the discussion in Jara-Díaz and Ortúzar (1989).

15.4.4.2   Special Problems Brought In by the Use of More Flexible Models

If tastes are assumed to be homogeneous, as in the classical MNL or NL models, it is possible to derive a single willingness-to-pay (WTP) value for a fictitious average individual. In this case it is also straightforward to examine if the model satisfies the required micro-economic conditions. But this assumption can be too restrictive, as WTP may vary from one person to another depending not only on observable social and economic characteristics, but also on unobserved variables or attributes which are difficult to measure. For this reason it is important to study the distribution of preferences in the population to obtain more accurate measurements.

As we saw in Chapters 7 and 8, advances in the field have enabled analysts to use increasingly complex models, such as Mixed Logit (ML) that allow one to define broader behavioural patterns (Train 2009). However, these models have been infrequently applied to evaluation studies and a consensus on the correct way to interpret their results has not yet been reached (Hensher and Greene 2003; Sillano and Ortúzar 2005). Further, most applications have been limited to estimating just the mean and spread of the distribution of population parameters and not individual parameter values. Now, the estimation of WTP values involves taking ratios of stochastic variables and in this case the problem we discussed in the previous section is compounded by the fact that not only the estimates, but the parameters themselves, are random variables and this is not a trivial issue (Meijer and Rouwendal 2000).

Amador *et al.* (2005) analysed individual preference heterogeneity with different methods and compared their benefit measures. To capture heterogeneity they used two approaches discussed in Chapter 8. First, *systematic taste variations* as in equation (8.17) where each level-of-service parameter is allowed to be a function of observed socio-economic characteristics (i.e. age, sex, income, vehicle ownership). Second, capturing *random taste variations* through the specification of a ML model (see section 8.6). Both approaches can also be used in a single model allowing us to incorporate non-observed heterogeneity as well as systematic variations in preferences.

Amador *et al.* (2005) compared subjective values of time (SVT) computed from a MNL imposing preference homogeneity and from various specifications allowing for taste variations (see Table 15.3). They found that the values derived from a model with homogeneous preferences (MNL-1) were similar to those obtained when systematic variations in tastes were considered (MNL-2); however if travel time tastes were allowed to vary randomly, significant differences appeared (i.e. up to 40% increase in SVT) even when a systematic variation for gender was allowed for as in model ML-2. This suggests that using a restrictive specification may lead to an underestimation of the value of travel time savings.

However, previous experience suggests that conclusions actually depend on the nature of the data and specifications used in each study. For example, Hensher (2001a; b) also found that more

(continued)

restrictive models tend to underestimate the value of time; notwithstanding, other authors have found no significant differences between values produced by different models (Train 1998; Carlsson 2003), and in some cases even lower SVT values have been obtained when ML (Algers *et al.* 1999) or more flexible models than the MNL are specified (i.e. Box-Cox Logit, see Gaudry *et al.* 1989). Finally, Alpizar and Carlsson (2001) found that the SVT could be underestimated or overestimated depending on the chosen mode.

Table 15.3 Subjective values of travel time[1]

	Men	Women	Mean
MNL-1	–	–	14.9
			(14.3 – 15.6)
MNL-2	10.4	18.7	15.3[2]
	(10.0 – 10.8)	(17.9 – 19.4)	
ML-1	–	–	21.4
			(20.4 – 22.4)
ML-2	17.0	24.7	21.5[2]
	(16.4 – 17.6)	(23.7 – 25.9)	

[1]Following Armstrong *et al.* (2001), confidence intervals for SVT at the 95% level are presented in parenthesis;
[2]Weighted averages considering that the sample was composed of 204 men and 290 women (Amador *et al.* 2005).

One possible explanation for the empirically observed discrepancies is the re-scaling that all parameters undergo when moving from a fixed specification to one where some parameters are allowed to vary randomly (see Example 8.8). But if all parameters were re-scaled in the same proportion the SVT should not be affected by changing the specification. However, empirical evidence shows that not all parameters are re-scaled by the same magnitude. Sillano and Ortúzar (2005) suggest that an intuitive explanation for this would be that the explicit treatment of parameter variation over the population into the systematic utility component is equivalent to the incorporation of an explanatory variable previously left out in the original (MNL) model. This is analogous to one of the misspecification problems discussed in section 3.2.1.4 and would lead to the restructuring of the utility parameters to compensate for the extra explanation accounted for. Thus, depending on the variables included in the model, the functional form chosen for the indirect utility function and the nature of the data, a fixed parameters model may lead to over/under estimates of the true values of time.

In what follows, we will discuss some econometric aspects of four different methods to achieve WTP estimates from parameter distributions. These methods can be applied to jointly distributed parameters but we will assume independent distributions for simplicity. However, in many case results are coincidental (Sillano and Ortúzar 2005).

Ratios of Population Means The simplest way to derive WTP values is to take the ratio of the means of the parameter distributions involved. In other words, if

$$\theta_t \sim f(\mu_t, \sigma_t) \wedge \theta_c \sim g(\mu_c, \sigma_c) \quad \text{then} \quad \frac{\theta_t}{\theta_c} \rightarrow \frac{\mu_t}{\mu_c} \tag{15.36}$$

This is not the mean value of the WTP, but a WTP value derived from the coefficients of the 'average individual' for each parameter. Therefore, this interpretation should not be used in cost-benefit analysis, and the calculation of this index may only be used as a means of testing model specification. Also, as the method disregards the rest of the distribution it considers a unique value for the parameters neglecting all information about heterogeneity in the population. So, at the end, the model is treated almost as a MNL, making in some sense the extra estimation effort worthless.

Simulation This method has been applied to construct confidence intervals (Ettema *et al* 1997; Armstrong *et al* 2001) as we saw in the previous section, and has been also used to derive WTP values from ML models by Hensher and Greene (2003) and Espino *et al.* (2006). It is a first approach to construct a WTP distribution over the population using information neglected by the previous method. An important feature of this method is that no assumptions are needed about the resulting distribution of the of parameter ratios.

However, one problem of the method is that it can yield rather large spreads for the distributions as the simulation process may involve drawing values that are close to zero. Hensher and Greene (2003) discuss the effect of removing parts of the simulated distributions of WTP, and compare this action with constraining the distributions. But in relation to the validity of this method, the real issue is not whether or how to constrain the distribution to make it theoretically correct. Hensher and Greene (2003) acknowledge that the mere fact of applying statistic distributions – which are already analytical constructs – to behavioural parameters governed by an unknown logic, make constraining (or removing parts of) the parameters or WTP distributions neither better nor worse than an unconstrained distribution, unless there is a theoretical rationale behind.

A consistent rationale for cutting off the tails of the distributions is that there are no *real* people with such extreme values to fill in the tails we are cutting. So, when applying this method the analyst must remember that the final goal is to estimate WTP values for the sampled population, and for sample sizes smaller than infinity this is a finite set of values. Therefore, the real problem with simulating WTP distributions from sampled values is not how to constrain them in a right way, but the fact that we are simulating countless numbers of values for people who do not even exist.

Log-Normal Distribution for WTP The use of Log-Normal distributions for the parameters over the population in ML models has been proposed by many authors, as this would constrain their signs to be consistent; further, it would yield an analytical expression for the resulting WTP distribution since the ratio of two Log-Normal distributed variables is also Log-Normal.

Consider a random variable x such that $x{\sim}N(\mu_x, \sigma_x)$. Then a variable defined as $X = \exp(x)$, has a Log-Normal distribution with mean $\exp(\mu_x + \sigma_x^2/2)$, and standard deviation given by $\exp(\mu_x + \sigma_x^2/2){\cdot}\sqrt{(\exp(\sigma_x^2) - 1)}$. Now consider the ratio of two Log-Normal variables, say X/Y, then:

$$\frac{X}{Y} = \frac{\exp(x)}{\exp(y)} = \exp(x - y) = \text{WTP}$$

where

$$\text{WTP} \sim \log N \left(\exp \left(\mu_{wtp} + \frac{\sigma_{wtp}^2}{2} \right), \exp \left(\mu_{wtp} + \frac{\sigma_{wtp}^2}{2} \right) \cdot \sqrt{\exp(\sigma_{wtp}^2) - 1} \right) \quad (15.37)$$

As x and y are Normal variables, their difference is also Normal with:

$$(x - y) \sim N(\mu_x - \mu_y, \sigma_x^2 + \sigma_y^2 - 2\sigma_{xy})$$

(continued)

Since we are assuming independent parameters, in this case the covariance term disappears. Then replacing the above expression in (15.37) we get that an expression for the log-normal WTP distribution is:

$$
\text{WTP} \sim \log N \left(\exp\left((\mu_x - \mu_y) + \frac{(\sigma_x^2 + \sigma_y^2)}{2} \right), \exp\left((\mu_x - \mu_y) + \frac{(\sigma_x^2 + \sigma_y^2)}{2} \right) \cdot \sqrt{\exp(\sigma_x^2 + \sigma_y^2) - 1} \right)
$$

$$(15.38)$$

This expression can be used to calculate cumulative proportions and confidence intervals. However, both Hensher and Greene (2003), and Sillano and Ortúzar (2005) found that in the case of this distribution there are considerable differences between taking the ratios of the means and the means of the ratios; this brings in new evidence to the discussion. The ratios of the means do not yield the WTP for the mean individual household, but for a virtual one who perceives the mean marginal utility of the population for each attribute (i.e. an 'individual household' who has the mean parameter for, say, travel time and also the mean parameter for cost). The existence of this household is not a fact but a mere coincidence, and even if it existed, its WTP value would not be representative.

An analytical explanation for this difference can be easily derived. Consider two independently distributed Log-Normal structural parameters θ_t and θ_c with associated Normal means b and c and variances s_t^2 and s_c^2 respectively. The ratio of their means can be expressed as a function of the coefficients of the underlying Normal distributions:

$$
\left.\begin{array}{l} \bar{\theta}_t = \exp\left(b + \frac{s_t^2}{2} \right) \\ \bar{\theta}_c = \exp\left(c + \frac{s_c^2}{2} \right) \end{array}\right\} \quad \frac{\bar{\theta}_t}{\bar{\theta}_c} = \exp\left(b - c + \frac{s_t^2 - s_c^2}{2} \right)
$$

$$(15.39)$$

And from (15.38) we can express the mean of the WTP log-normal distribution in terms of the same coefficients:

$$
\overline{wtp} = \exp\left(b - c + \frac{s_t^2 + s_c^2}{2} \right)
$$

$$(15.40)$$

From here we can derive the relation:

$$
\overline{wtp} = \left(\frac{\bar{\theta}_t}{\bar{\theta}_c} \right) \exp\left(s_c^2 \right)
$$

$$(15.41)$$

Thus, the ratio of the means of Log-Normal parameters is equal to the mean WTP value deflated by the exponential of the variance of the Normal distribution underlying the Log-Normal cost coefficient (i.e. the parameter in the denominator of the WTP ratio). In other words, the WTP mean and the ratio of parameter means are scaled by a proportionality factor which, by the way, is fixed for the model (i.e. the three attributes considered in this example are scaled by the same factor). The logic of this effect is the following: the larger the variance of the cost coefficient, the larger the portion of the denominators' mass that will be near to zero, and hence the mean WTP will grow larger.

The use of Log-Normal distributions for valuation purposes is not recommended. Their wide tail tends to give extremely large WTP values with high probabilities yielding large portions of cumulative mass close to zero distorting the analysis. Its main appeal is that it allows constraining the parameters to be strictly positive (for negative coefficients, they enter with a negative sign in the utility formulation). However, as we saw in Example 8.8 the relative easiness of the estimation with

Normal distributions may also lead to structural parameters with correct theoretical signs. Thus, it is not worthwhile to undergo the effort to estimate the model with Log-Normal distributed parameters, since even if the individual values show a large portion of incorrectly signed people, the right course of action should be to investigate them for consistency, and perhaps remove them from the sample.

Fixing the Cost Coefficient Another method which has been used considers fixing the cost coefficient and thus letting the WTP distribution to follow the distribution of the numerator; if the parameter in the numerator follows a Normal distribution the resulting WTP distribution would be simply given by:

$$\left. \begin{array}{l} \theta_t \sim N(\mu_t, \sigma_t) \\ \theta_c \quad fixed \end{array} \right\} \quad \frac{\theta_t}{\theta_c} \sim N\left(\frac{\mu_t}{\theta_c}, \frac{\sigma_t}{\theta_c}\right) \tag{15.42}$$

Revelt and Train (2000) cite three reasons for fixing the cost coefficient:

- it effectively solves the problem under discussion;
- the ML model tends to be unstable when all coefficients vary over the population, and identification issues arise (Ruud 1996); and
- the choice of an appropriate distribution for the cost coefficient is not straightforward, since the Normal and other distributions allow for positive values, and the Log-Normal is both hard to estimate and give values close to zero, as discussed above.

Notwithstanding, there is one drawback of this method that needs attention.

Example 15.3 Table 15.4 compares estimates of WTP derived from a MNL with those of a ML model with a fixed *cost* coefficient in a residential location choice experiment (Sillano and Ortuzar 2005). As can be seen, the means of the resulting WTP distributions (for travel time to work, travel time to study and an environmental attribute, days of alert, defined as the number of days when the air quality requires additional car restraint) are considerably higher than the MNL point estimates, a result that has also been reported by Algers *et al* (1999) and Revelt and Train (1998).

Table 15.4 Mean WTP estimates for fixed cost coefficient ML and MNL

Attributes		Willingness-to-Pay	
		MNL	ML
Travel time work (Ch$/min)	Mean	36.0	51.0
	Std. Dev.		54.8
Travel time study (Ch$/min)	Mean	22.0	31.0
	Std. Dev.		47.5
Days of Alert (Ch$/DA per year)	Mean	124,362	126,160
	Std. Dev.		107,430

Hensher (2001a; b; c) have also found higher mean WTP values for heteroskedastic and autorregressive specifications; this could indicate that ML models (with any error structure) tend to overestimate WTP values. But, these works did not explore the possibilities that by constraining only

(continued)

part of the error structure they could be causing an unbalanced growth in the model coefficients, hence producing higher welfare estimates.

In Example 8.8 we explained why larger means for ML parameters, in relation to the MNL, should be expected because of the extra variance explained by the random parameters; we have also discussed above possible reasons for obtaining uneven enlargement factors. The fact is that constraining a taste coefficient to be fixed over the population, may make it grow in a less-than-average proportion (i.e. the parameters that are allowed to vary grow more than the parameters that should vary over the population, but are constrained to be fixed). Note that this is not the case of parameters which are eventually fixed because its standard deviation was originally estimated and found not significant (see the discussion by Sillano and Ortúzar 2005 on this issue).

Willingness-to-Pay Estimation from Individual Level Parameters   In section 8.6 we discussed two forms to estimate individual level parameters for ML models, both involved the use of Bayesian statistics. The estimation of individual taste parameters eliminates the issue of analysing the WTP distribution resulting from the division of two random variables over the population. Instead individual level WTP point estimates can be computed, along with their individual confidence intervals.

Example 15.4   Figure 15.10 presents frequency charts for the valuation of the two attributes the distribution of which was shown in Figure 8.5. The charts show high concentrations on each edge of the axis accounting for extremely large positive and negative WTP values. It is important to mention that notwithstanding the sign of the WTP value, all implausibly large values belong to individual households in the sample with non-significant *cost* (rent in the case of this location choice example) parameters. That is, the denominator of the WTP ratio is statistically close to zero yielding an inordinately large value.

It is also important to mention that in Figure 15.10a the only negative WTP values are also associated with extreme cases. In fact, they correspond to the few observations with an incorrect sign for the *Rent* parameter; but as it was also not significant in those cases, it caused the ratio to grow disproportionably.

This suggests paying special attention to observations with a cost parameter statistically equal to zero. In these cases the WTP ratio grows to implausibly large monetary valuations for reductions in the corresponding attribute. On the other hand, as the individual household does not place *any* weight on the cost attribute, we can debate whether those observations do not consider the cost attribute at all, or whether the weight they place on it is negligible in relation to the rest of the attributes. If the latter is the case, the interpretation of an extremely large WTP value would be correct. If not, monetary valuations cannot be computed for these observations. Further theoretical development is necessary to define criteria to help answering this question, but note that it is case specific (i.e. it depends on the survey design, the underlying microeconomic model and the characteristics of the valued attributes).

The estimation of individual level WTP values is as close as we can get to the correct method of valuation inference from ML models. However, for project evaluation and cost-benefit analysis we usually need data for different groups or strata in the population. One beauty of individual-level data is that an analysis at the level of a given stratification can simply be performed averaging the WTP values of those individuals present in each strata, along with their cluster variance. In fact, thresholds (or strata boundaries) can even be defined *ex-post* in order to minimise the variance of the WTP values across the group, and hence be able to define more homogeneous segments for project evaluation and detailed analysis. Sillano and Ortúzar (2005) discuss this and other points in more detail.

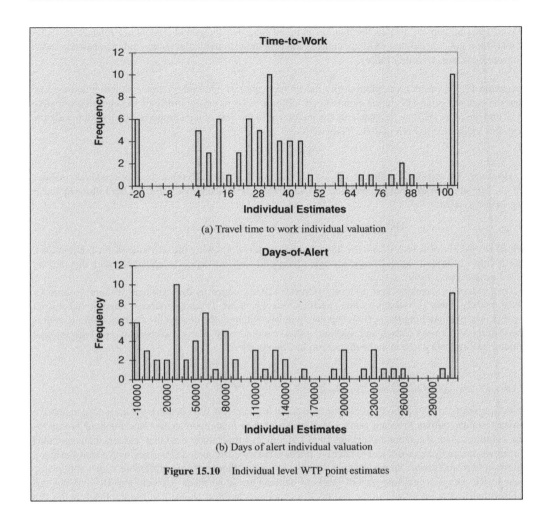

(a) Travel time to work individual valuation

(b) Days of alert individual valuation

Figure 15.10   Individual level WTP point estimates

15.4.4.3   The Transfer Price Approach

In the context of travel demand analysis, *transfer price* has been understood as the amount by which the cost of one option would have to be varied to equalise its overall attractiveness with that of another predefined option (see Bonsall 1983).

A typical application of the method involves asking individuals, for example, by how much should the fare of their currently preferred option increase to persuade them to switch to another alternative. It is clear that an important problem of the technique (in common with other forms of stated-preference analyses, and in particular contingent valuation which is the closest one) has to do with the reliability that the analyst can associate with such a data set. On the other hand, a strong advantage of the method, if it works, is that it makes it possible to know not only the direction of individual preferences but also the difference (in preference terms) among the various available options. Thus in theory, and in common with other SP studies, less data than for an RP study are required to obtain a model of similar accuracy.

We will not attempt to discuss the method in detail here, but interested readers are referred to Gunn (1984) for a good discussion of its advantages and problems, in particular its general inconsistency with conventional random utility theory.

Example 15.5 Consider a random utility model such as (7.2) in a binary-choice situation and assume that the transfer price (TP) corresponds to the difference between the utility of the chosen alternative (U_c) and the other (U_r), i.e. it represents the increment in the cost of the chosen one that would made the traveller indifferent to both options. Thus we have:

$$TP = U_c - U_r$$

However, the expected value of ($U_c - U_r$) is precisely the difference in representative utilities ($V_c - V_r$); so assuming these to be linear in the parameters, as usual, we can form the following linear regression system:

$$TP(observed) = \theta_1(X_{1c} - X_{1r}) + \theta_2(X_{2c} - X_{2r}) + \ldots$$

which should allow us to estimate the unknown parameters θ knowing the attributes **X** for both options. Furthermore, different values of time for *time savers* and *cost savers* may be calculated with this method (see Lee and Dalvi 1969).

One important problem, first noted by Hensher (1976), concerns the treatment of habit in transfer price models. Gunn (1984) shows that specifications which use TP as a dependent variable but restrict its sign (i.e. by modelling the options separately or by switching the observable characteristics to reflect the difference between chosen and rejected option) cannot easily be made consistent with conventional random utility theory (see also the discussion in Chapter 8).

15.4.4.4   *The Stated Preference Approach*

Stated preference (SP) methods, as discussed in depth in Chapter 8, have become the most used method to estimate values of time in recent years. For example, in their final report to the Department of Transport, the consultants for the ground-breaking 1994 UK value of time study note that 'evidence has amassed during the last ten years sufficient to have confidence that a well-mounted SP survey with a well-designed questionnaire and proper analysis can yield reliable results, though this is preferable with a supporting base in RP data if actual forecasts of levels of demand are to be made' (Accent and HCG 1996). An interesting discussion related to the use of SP methods in location choice and the implications for the value of time is given by Pérez *et al.* (2003).

We do not review the large number of SP-based value of time studies reported in the literature during the last few years, but we do mention some of the latest European reports on national studies. Besides the new UK study, it is interesting to have a look at those for Finland (Kurri and Pursula 1995); Holland (HCG 1990) and Sweden (Lindquist and Algers 1998). Other studies have been involved with important issues such as the estimation of randomly distributed values of time (Ben-Akiva *et al.* 1993; Gopinath and Ben-Akiva 1995), or with estimating time values using SP data allowing for interaction effects (Ortúzar *et al.* 2000c; Rizzi and Ortúzar 2003). These cover new areas and use state-of-the-art models and specifications as discussed in Chapters 7 and 8.

15.5   Valuing External Effects of Transport

15.5.1   Introduction

In many countries of the developed world, willingness-to-pay (WTP) methods have been used for the monetary valuation of a range of external effects of transport such as accidents, pollution, noise, visual intrusion and amenity loss. Several examples have been compiled by Hansson and Markham (1992), OECD (1994a), Mauch and Rothengatter (1995), Litman (1995), Maddison *et al.* (1996), Friedrich *et al.* (1998) and ECMT (1998). The thrust of this work has been to establish the full social costs of transport as a basis for efficient pricing in this sector and to extend the scope of social cost-benefit analysis (SCBA) for improved project appraisal.

Although there has been much academic enthusiasm for the monetary valuation of these non-market goods, this has been challenged on principle and practical grounds; a good expression of the nature of this dissent can be found in Adams (1992) and Whitelegg (1993). However, there is considerable force in the argument that, while empirically well-founded monetary values are difficult to achieve and may be valid only in specific contexts, their expression will help to ensure that externalities are not marginalised or understated in project and programme planning. This is particularly important in the contexts of road investment appraisal and resource allocation for accident counter-measures and pollution control strategies. Indeed, in the 1980s and 1990s, the attribution of monetary values to accidents of different severity was an important stimulus to increasing the resources towards road safety and establishing priorities over different safety measures in many countries of the world (Allsop 1999).

Also, as part of the expectation to respond to increasingly challenging environmental standards and targets, many national and local governments have been establishing, extending or refining databases relating to accidents, noise, and a variety of gaseous pollutants. These are intended for use in monitoring changes over time and for evaluating fiscal, regulatory and investment policies. In many developed countries this process is already established, while for most developing countries it is currently at a relatively early stage of development, and the scope and quality of such data vary considerably (Chesnut *et al.* 1997).

Now, although sufficient evidence was amassed over the 1990s, the economic costs of accidents, noise and pollution are all subject to considerable variation, partly due to the different sources of data and methods of measurement. For example, Quinet (1994) noted that *for all forms of transport pollution*, estimates based on WTP provided the highest numerical values of a statistical life (VOSL), a feature long known in the case of accident costing. For example, the UK government replaced the human capital approach to fatality costing by the WTP approach in 1988, and this was extended to non-fatal accidents in 1994, drawing on the national studies of Jones-Lee *et al.* (1985, 1992). However, in the case of fatalities, the government was not persuaded to accept the considerably higher values emerging from the former WTP study and instead it implemented a compromise value (Dalvi 1988), thereby exercising an element of caution in the face of a radical change of methodology (Department of Health 1999).

Until the mid 1990s, monetary valuation of environmental externalities was seldom given *official* support (OECD 1994b; Lee and Kirkpatrick 1996). However, the situation changed rapidly afterwards and surveys of 'official' transport appraisals (Bristow *et al.* 1998; DETR 1998) suggested that monetary values for noise, air pollution and (to a lesser extent) barrier effects, were increasingly used in many European countries by the new century. However, the appraisal of road investments undertaken or supported by national authorities still involves a limited cost-benefit analysis (with unit monetary values confined to savings in time, accidents and operating costs), applied in conjunction with an environmental and socio-economic impact assessment. For example, in his survey of US state agencies responsible for highway developments, Waters (1992) noted that relatively few embraced a sophisticated SCBA, preferring rather simpler needs-based or cost-effectiveness approaches.

Although several academic studies have urged the extension of the SCBA framework to embrace a wider range of impacts (Bateman *et al.* 1993; Willis *et al.* 1998), governments have remained cautious about its formal extension to pollution, noise, visual intrusion, amenity loss and ecosystem damage. This is partly because of gaps in knowledge, both in impact assessment and economic valuation (Mullen 1997), and partly because the site-specific nature of some of the impacts inhibits the use of standardised unit values. These are universal concerns.

It remains a considerable research challenge to integrate environmental impact assessment, cost-benefit and multi-criteria analysis traditions (Commission of the European Communities 1994; OECD

1994c; Lee and Kirkpatrick 1996; Nardini 1997) in a context in which environmental objectives are assuming increasing importance. Efforts include new ways of assembling and presenting qualitative and quantitative information to minimise bias against non-monetary valuation items, and the construction of appraisal frameworks which establish a 'level playing field' between different modes and allow transport problems to be addressed with less emphasis on highway solutions (Price 1999; Glaister 1999). Monetisation will increasingly be applied in multimodal settings with heavier demands on data.

15.5.2 Methods of Analysis

There are several taxonomies for valuation methods available in the literature, and a much larger economic discussion than we could attempt here (ECMT 1996; Mauch and Rothengather 1995, Nash 1997; Verhoef 1994). In this section we will just quickly review, for the sake of completeness, two methods, the *human capital* approach and the *contingent valuation* method, as it is probably fair to say that currently the method which clearly dominates the field is our old acquaintance, the stated preference approach, fairly well reviewed in Chapters 3 and 8 (see Rizzi and Ortúzar 2003 for a well-designed methodology that has been used already as far as Australia and Norway).

15.5.2.1 Human Capital Approach

It is based on the assumption that the value of an individual is what they produce, and this is usually measured by the gross salary perceived at work (i.e. before taxes in order to include the government and hence society). If the person dies this production is lost. This, almost 30-year-old approach (Landefeld and Seskin 1982) postulates that the value of preventing the death of an individual aged t is equal to the net present value (PV_t) of their expected earnings for the rest of their life:

$$PV_t = \sum_{i=1}^{T-t} \frac{\pi_{t+i} E_{t+i}}{(1+r)^i} \tag{15.43}$$

where π_{t+i} is the probability that the individual will survive from age t to age $t + i$, E_{t+i} are the expected earnings of the individual at age $t + i$, r is the discount rate and T is the retirement age.

The method has been heavily criticised as being the antithesis of the conventional premises of welfare economics. Discussion has also touched on how to value production of individuals that are not in the labour market (i.e. housewives), or what discount rate should be used to calculate PV (a sensitive issue in the case of children and young adults); classical rates ranged from 6% to 10%, but nowadays values below 5% are preferred in order to avoid punishing any age stratum in excess. Table 15.5, taken from Landefeld and Seskin (1982), shows the effects of age and discount rate on the human capital value of life.

Table 15.5   Net present value by age and discount rate

Age group	Net Present Value (US$)		
	discount = 2.5%	discount = 6.0%	discount = 10.0%
1 to 4 years	761 047	205 101	59 859
20 to 24 years	967 221	534 799	320 114
40 to 44 years	625 508	454 972	338 232
65 to 69 years	47 506	40 886	35 304

Due to difference in wages, if applied strictly, the human capital approach would yield smaller values for the life of women than for the life of men; and smaller values may also have to be assigned to non-Caucasians. Zero values would be assigned to retired individuals or to those incapacitated either by illness or for any other reason. For this reason, as in the case of the value of time, the proper methodology should be to estimate a single equity value to be used in project evaluation. Notwithstanding, it is widely accepted that as this approach does not consider pain and suffering by the victim and their relatives, its values constitute an underestimate of the true value of the social loss and therefore its use should just allow us to establish a lower bound for the value of life.

Example 15.6 Knowledge of the wages corresponding to different age and sex categories allows us to estimate net present values by sex and age given a discount rate using (15.43). Table 15.6 shows estimates of the average net present values of earnings lost by premature death by different age groups in the Santiago Metropolitan Region (Holz and Sánchez 2000).

Table 15.6   Net present value by age and sex

	Net Present Value (US$)	
	Males	**Females**
Less than 1 year	241 258	174 954
1 to 4 years	250 569	181 706
5 to 9 years	268 246	194 525
10 to 19 years	296 964	214 330
20 to 44 years	275 951	183 573
45 to 64 years	154 876	90 305
65 to 79 years	53 248	25 349
80 years and over	19 780	4 553

As can be seen the values are higher for males due to their higher wages. The net present value diminishes with more mature ages because the life horizon shortens.

In order to obtain a unit cost for mortality, Holz and Sánchez (2000) calculated the percentage of deaths for each age stratum for male and females using death statistics for 1997 in Chile (as disaggregate data by gender was not available for the first three age strata, it was assumed that mortality was evenly distributed). These percentages were multiplied by the respective net present values (Table 15.7), yielding the participation of the various age groups in the unit cost. The sum of these participations equals the average unit cost of a premature death in 1998 in Santiago, and this was estimated as US$53 224. This value assumes a measure that would affect uniformly the mortality rate of the whole population.

15.5.2.2   Contingent Valuation

As mentioned in section 3.4.1, this is a technique for eliciting values for goods which are not or cannot be bought and sold in a normal market. People are asked for their value of a good, *contingent* on a market existing for it. A hypothetical market is created and described to the respondent, who is then asked to make a market (purchase) decision. Contingent markets define the good or amenity of interest, the existing level of provision, possible increments or decrements, the institutional structure under which the good is to be provided, and the method of payment. Mitchell and Carson (1989) provide a comprehensive explanation of the theoretical foundations of the contingent valuation (CV) technique,

Table 15.7 Contributions to unit cost by age and sex

	Male deaths (%)	Unit cost contribution	Female deaths (%)	Unit cost contribution	Contribution to total unit cost
Less than 1 year	0.017	1 091	0.017	846	1 936
1 to 4 years	0.003	201	0.003	156	385
5 to 9 years	0.003	188	0.003	146	428
10 to 19 years	0.008	604	0.004	256	1 476
20 to 44 years	0.076	6 007	0.028	1 708	15 461
45 to 64 years	0.125	10 377	0.078	5 016	19 453
65 to 79 years	0.179	15 620	0.145	9 809	11 159
80 years and over	0.118	10 778	0.195	13 885	2 947
Total					**53 244**

methodological issues and practical application. Overviews are provided by Bateman and Turner (1993) and Haneman (1994).

CV questions can ask for people's willingness-to-pay (WTP) values or for their willingness-to-accept (WTA) compensation values. The WTP value is the income an individual would forego to achieve an increase in the level of a good and remain at the same level of utility, and WTA is the inverse. A problem here is property rights; WTP assumes that these belong to the consumer and WTA the contrary. However, WTP is most commonly used because it resembles familiar consumer purchase decisions (although in cases of environmental deterioration, for example, WTA should be the correct theoretical value to obtain). Thus, CV attempts to measure the change in income necessary to offset a change in amenity, while leaving utility unchanged.

There are three main methods of eliciting CV values:

- Open-ended questions, where respondents are just asked how much they would be willing to pay for a good.
- Iterative questions, where respondents are asked first whether they would be willing to pay a specified amount; if they answer yes, the question is repeated with small increments in the cost until they say no, then the cost is reduced by smaller decrements until a final figure is reached (and vice versa, if they start by saying no to the original figure).
- Referendum questions, also known as dichotomous choice questions, where respondents answer yes or no to a WTP question with a specified payment; the double-bounded dichotomous choice question has an extra question after the first.

The referendum approach is the most attractive because it presents scenarios similar to those which respondents, as consumers, encounter in day-to-day market transactions. The payment mechanisms for actually buying or selling the good can include property taxes, income or sale taxes, utility bills, community charges, fares, entry fees, subscription schemes or even an abstract instrument. Since its early application in the 1970s the CV approach has been used to value a wide range of non-market goods. Carson *et al.* (1995) provides a bibliography of CV studies containing 1400 references, indicating the wide applicability of the method.

On the other hand, a strong critical assessment is provided in a collection of conference papers edited by Hausman (1993) and Diamond and Hausman (1994), who believe that the evidence suggests that CV surveys do not measure the preferences they attempt to measure, and that changes in survey methods are unlikely to alter this. However, the method is still popular and has certainly been used in many important works related to valuing externalities in the transport sector (e.g. Jones-Lee *et al.* 1992; Feitelson *et al.* 1996).

Example 15.7 Ortúzar *et al.* (2000a) report on the use of a CV questionnaire to obtain the WTP for risk of death reductions (loosely related to environmental pollution effects) designed to overcome some problems found in typical CV studies, namely that some respondents fail to understand the basic notion of probability, attributing similar WTP to different reductions of risk, and that respondents may give zero WTP to reduce future risks of death due to lack of understanding of the commodity being valued. Their approach differed from previous CV studies of risk reduction in the following ways: (i) the timing of risk reductions and the attention given to timing of payment and (ii) the proposal of a baseline risk that has to be accepted by respondents as their own, according to age and gender.

After familiarising respondents with the concept of risk of death and its perception, their questionnaire drew attention to the main causes of death by age and gender, and about common measures to mitigate these causes and their costs. Then they introduced age- and gender-specific baseline risks (based on actual data), and checked whether respondents accepted them as their own. After this they sought WTP (using an open-ended payment method) for reductions in the risk of death in the next ten years; proposed reductions were 1 and 5 in 1000, and were presented graphically using a matrix of a 1000 circles (which presented the baseline risk as black circles), asking respondents to rub out the reductions valued. The method worked very well in the sense that surveyed individuals acquired a proper understanding of the questions asked.

After establishing the baseline risk (and having it accepted by the respondent), the fundamental question of the survey took the following form:

The measures needed to achieve a reduction in premature deaths in the next decade involve certain costs as we saw earlier in the questionnaire. Taking these into account please answer the following questions:

How much money would you be willing to pay monthly for the next ten years in order to decrease your own possibility of dying by 1 in 1000?

$/month. . . . . Nothing (why?). . . . .

How much money would you be willing to pay monthly for the next ten years in order to decrease your own possibility of dying by 5 in 1000?

$/month. . . . . Nothing (why?). . . . .

How certain are you that you would pay that amount and not another?

(a) very sure . . . (b) reasonably certain . . . (c) not very sure . . .

Table 15.8 shows results from a sample of 94 respondents. Note that the ratio of WTP for risk reductions of 5 and 1 in 1000 is close to 4, and this is consistent with expectations giving the decreasing marginal utility of risk reductions; it also suggests that people are indeed capable of distinguishing between rather small reductions in risk. And note that the VOSL is close to five times higher than that obtained with the human capital approach, consistent with findings elsewhere (e.g. Cropper and Freeman 1991).

Table 15.8 Implicit value of a statistical life per risk reduction

Risk reduction	Median WTP (US$ per month)	Net present value of WTP (US$)	Implicit value of statistical life (US$)
1 in 1000	3.0	285.1	285 113
5 in 1000	12.0	1127.0	225 400

In the case of road accidents, the two most feared outcomes are to die or to become a severely injured victim. Not surprisingly road project appraisal practice in most industrialised countries has given those two outcomes the highest economic values; fatalities being more valued than severe injuries.

Conventional practice until the end of the 90s was to elicit WTP values for preventing both fatalities and severe injuries using contingent valuation (CV) and risk-risk trade-offs (or standard gambling) methods (Jones Lee *et al.* 1993; 1995). But CV basically involved a trade-off between money and risk expressed as a tiny probability. Usually a question was posed to respondents asking for their willingness to pay to buy some special safety device designed to reduce *only* the likelihood of a particular outcome of a road crash; e.g. the likelihood of becoming a fatal victim or the likelihood of suffering – say – a head concussion.

The risk-risk trade-off, on the other hand, demanded respondents to exchange the risk of one likely trauma outcome of a road crash for another one. Usually respondents had to assume they were already a road accident victim suffering a particular trauma; then they were offered the alternative of a medical intervention that, with probability p, would return them to their health state before the crash and, with probability $1 - p$, they would end up in a health state worse than the current hypothetical one – this state was usually death. Respondents had to state the value of p that would make them undertake the medical intervention. Hence, it was possible to 'chain' different risks with the risk considered in the CV survey, allowing the researcher to monetise risks others than that considered in the CV exercise. The reader may ask why not use the CV to put a monetary value on all types of risk. The reason was that money-risk trade-offs were deemed unstable, so researchers would rather avoid the overuse of CV.

15.5.2.3 The Stated Choice Approach

Although the above methods may work as a first empirical approximation, they do not address the issue under analysis (i.e. risk of a road accident) in its proper dimension. First, the road safety schemes an authority wants to evaluate are of a public-good nature. It is about reducing a public risk; that is, a risk that displays no-rivalry in consumption since the benefits of the scheme accrue to all drivers on that particular stretch of road. The safety device considered in the CV approach is a private good, not a public one (but this could be corrected by substituting a public good for the private good, and this critique would lose substance). Second and more important, a road safety scheme is about decisions on *ex ante* risk management, in the sense of what can be done to prevent road crashes or to mitigate the impact of a road crash. However, the risk-risk trade-off is akin to a post-trauma alternative medical treatment, associated with decisions to be taken after the accident has occurred. This information should be more relevant for health insurance companies than for public road agencies.

So, if WTP values are required for appraising road safety projects stated choice (SC) methods are a superior elicitation approach (Rizzi and Ortúzar 2003; Iragüen and Ortúzar 2004; Hojman *et al.* 2005; Rizzi and Ortúzar 2006; Hensher *et al.* 2009). This technique places the respondent in the correct context, for example, having to choose between two routes with different levels-of-service (i.e. travel time, toll, number of fatalities and number of severely injured victims). This way, people implicitly reveal WTP not only for safety improvements, but also for travel time savings, probably the most important trip attribute. The quota of increased realism afforded by the SC approach is necessary to uncover the value people actually place on safer roads. It also avoids the problem of *embedding* (Sælesminde, 2003), since both the reduction of fatalities and severely injured victims are valued, together with travel time, in an integrated framework where the individual is always conscious of her budget constraint.

As a caveat, SC methods are not without problems. As with CV, the hypothetical nature of the choice scenarios is the main disadvantage of any stated preference survey, as we have discussed in Chapters 3 and 8. However, we strongly believe that SC surveys outdo conventional CV surveys with respect to increasing realism.

Example 15.8 Hojman *et al.* (2005) designed a route SC survey for car trips between the cities of Santiago and Valparaíso, and another for car trips between Santiago and Rancagua (i.e. the capital and two important Chilean cities respectively). The distance between Santiago and these two cities is around 120 km via Class A roads (Routes 68 and 5-South respectively), which are fairly safe for Chilean standards.

After a detailed experimental design phase, including focus groups and two pilots, the final survey instrument contained five parts. The first asked for the driving experience on interurban roads and on Routes 68 and 5 in particular. A question was included about the last time the respondent drove on any of these routes; if the answer was more than a year ago, the survey ended. The second part included the choice experiment itself (which varied according to the purpose of the trip and the route where the driver had more experience), and the third part different types of questions, some related to the choices themselves and others to road crash experience and attitudes. The fourth part enquired about socio-economic data and the fifth allowed respondents to give their personal definition of what constitutes a severe injury.

When respondents are asked to examine a series of choice situations it is important to set up a realistic context. According to the answer given in the first part of the survey, people were asked to consider they had to travel from Santiago to Valparaíso or to Rancagua. Invoking a recent trip to either destination was a way to reduce to a minimum the problem of not including as a third alternative the option of not doing the trip at all (see section 3.4.2.6). The trip to either city had the following characteristics (the underlined parts could vary across contexts):

- you drive your car;
- you travel during a regular weekend (without extra holiday days);
- you pay for the total cost of the trip, including the toll;
- you start the trip in the morning and is raining;
- you have to choose between two routes (both are similar to Route 68), taking into account the following four elements: 1) toll charge, 2) travel time, 3) number of fatal victims per year and 4) number of severely injured victims per year'.

A short explanation was also given on what was considered a fatal victim and a severely injured victim. In explaining the latter, several road-crash severe traumas were mentioned, so that respondents focused their attention on these types of traumas. These definitions were analysed at the focus groups and pre-test surveys. Finally, statistical data was also given about the number of fatalities and severely injured victims, and the total annual flow on Routes 68 and 5 during the previous year. Nothing was said about any accompanying member within the car; hence, a question asked whether or not the driver was considering travelling alone or with someone else.

As can be seen the context was clearly defined: the day, the time of day and the purpose of the trip were all specified; it was assumed that the person who answered the questionnaire was the driver and s/he was also assumed to pay for the toll. Many highways operate under a private toll system in Chile, thus people were already familiar with toll charges. In particular, as safety is related to a particular trip taken by the respondent there was little room for an altruistic choice.

So, as it was in the best interest of respondents to give a truthful answer, this way they managed to increase the 'realism' of the hypothetical choice context to a plausible maximum, reducing the possibility of strategic bias.

The statistical design used made it possible, in principle, to estimate different parameters for the safety variables of each alternative route. One parameter was considered for lower numbers of crashes and another for higher values. The aim was to test the *prospect theory* hypothesis (Kahneman and Tversky 1979) that increases in the level of danger are valued differently (once the sign is taken into account) than improvements in the level of safety; thus, it was expected that the modulus for higher numbers of crashes to be greater than that for lower numbers. However, as this result did not show up at the pilot study phase, they considered only one parameter in the final survey.

The survey was programmed in a web page (http://www2.ing.puc.cl/~phojman/) following the excellent results obtained in a previous experience (Iragüen and Ortúzar 2004). To recruit respondents, key officials at several institutions (both public and private) were contacted who accepted to cooperate with the study. Then, these officials sent e-mails to employees enticing them to participate. Hojman *et al.* (2005) obtained approximately 500 answers, 250 for each route, but did not calculate the response rate since (for confidentiality reasons, they did not register the e-mail of respondents and did not enquire how many people were contacted at each institution). Most individuals in the survey were middle to high-income people by Chilean standards, as car possession is low compared to European or US levels and cars are most usually owned by middle to high-income people.

Using this data, a variety of models was estimated – ranging from the simple MNL to MNL-like models allowing for systematic taste variations and ML. Hojman *et al.* (2005) concluded that the WTP values estimated from their data were between 10 to 15 times higher than the values used in social project evaluation in Chile at the time (computed from the human capital approach). Their values were also compared with values obtained in other countries using both similar and different methods, finding that – in general – the Chilean values differed from the others in more than what could be accounted for by income differences. In fact, they concluded that . . . 'our values should also be more relevant for road planners in developing nations than transferring values from industrialised nations (i.e. Miller 2000 derived a *value of risk reductions* for Chile in a range of two to three times higher), since accounting for differences in risk aversion is by no means an easy task'.

Exercises

15.1 Consider the following simple econometric model to determine car ownership as a function of income:

$$P_0/(1 - P_0) = \alpha I^\beta$$
$$P_2/[0.8(1 - P_0) - P_2] = 0.09 \ \exp(0.751)$$
$$P_0 + P_1 + P_2 = 1$$

(a) Calibrate the model using the data in the table below (*Hint*: do it graphically)

I	P_0	P_1	P_2
1	0.60	0.35	0.05
2	0.40	0.50	0.10
3	0.25	0.55	0.20
4	0.20	0.45	0.35
5	0.15	0.35	0.50

(b) Indicate what proportions with 0, 1 and 2 or more cars would the model predict for an annual income of six monetary units.

15.2 The following table presents the results of a transfer price survey made on the sample of eight individuals in Exercise 9.3; TP indicates the reported increment in the monetary cost (expressed in time units after deflating by income) of the currently chosen mode that would leave each individual indifferent to both alternatives. The study assumed that only time (t) and cost/income (c) were relevant variables.

Individual	Chosen option	TP	t_1 (min)	t_2 (min)	c_1 (min)	c_2 (min)
1	1	8.0	47.5	83.2	14.8	7.0
2	1	6.5	30.2	45.0	10.4	5.0
3	1	2.5	22.0	30.4	12.6	4.0
4	2	0.5	45.0	50.6	8.2	5.0
5	2	1.5	15.3	20.5	50.0	17.0
6	1	8.5	34.8	50.2	55.0	35.0
7	2	130.0	65.5	100.5	200.3	53.5
8	2	6.0	12.0	14.0	44.6	17.0

(a) Use the data to estimate the individuals' subjective value of time. Discuss the role, size and sign of the intercept of the transfer price linear regression equation (*Hint*: if you do not have available a calculator with a linear regression facility, do it graphically assuming the coefficient of time, θ_t, is known and equal to -0.03).

(b) If the revealed preferences parameter for the time variable is indeed -0.03 and the mode specific constant of option 1 is 1.35, estimate the subjective value of time using another method. Compare your results and discuss.

16

Pricing and Revenue

16.1 Pricing, Revenue and Forecasting

16.1.1 Background

The pricing of transport infrastructure and services is becoming a more important issue in policy development partly because of the increasing role of the private sector in their provision. This chapter will discuss some of the issues involved in this task as they put specific requirements to modelling and reporting.

Pricing transport services is not only more critical but it is also more complex than in the last century. We are now accustomed to using period or season public transport tickets and to variable pricing on air travel and in a number of other services including rail, parking, congestion-charging and tolling. The prevalence of variable pricing is likely to continue and even extend its reach in transportation. Moreover, many of these prices are converted into revenue by means of smart cards, electronic tags and even mobile phones. This poses some difficult questions in the field of modelling, namely:

(a) Is all money perceived the same by travellers? Has the money being paid for fuel or deduced as a road tax the same quality and invokes the same perception as that paid for tolls or parking?
(b) Is the de-coupling of use and payment an issue affecting behaviour?
(c) What is the best way of modelling willingness-to-pay (WTP) for transport services and how is this affected by the two questions above?
(d) How is this WTP affected by the legibility of the pricing signal?

Modelling the impact of price on demand is important as it is calculating the revenues that will result from new prices. This adds some additional considerations to our modelling effort.

We focus this chapter on the issues surrounding projects where the private sector takes some degree of revenue risk. These may be toll roads, public transport concessions or simply the opportunity to acquire or merge with an existing transport business the value of which will depend significantly on its future revenue stream. As such, this chapter focuses more on the practice of model application than on theory. Some of the concepts, however, are relevant to other type of modelling efforts; this is certainly the case of our discussion of uncertainty and revenue risk.

Modelling Transport, Fourth Edition. Juan de Dios Ortúzar and Luis G. Willumsen.
© 2011 John Wiley & Sons, Ltd. Published 2011 by John Wiley & Sons, Ltd.

16.1.2   Prices and Perceptions

Prices come in all sort of guises and sometimes they are purposely designed to be obscure, to encourage us to spend more money that we think we should. The 'crispier' concept of price should materialise when we need to take money out of our pocket or handbag to pay for a toll or to a parking attendant. From then on other versions of charging start to de-couple usage from payment: the use of credit cards, payment for period tickets, electronic payment via smart cards or tags, video tolling charged directly to your bank account and so on.

The costs of operating a car are even 'fuzzier' to the owner. It is generally believed that the nearest thing to the perceived cost of running a car is the fuel, other costs like maintenance, taxes and depreciation are mostly perceived as sunk (not variable with usage) costs. However, most of us cannot quote what this perceived fuel cost per kilometre is likely to be; we only know that filling up the tank is more or less expensive than some time ago and may adopt some change in travel behaviour as a result. The combination of crisp money (e.g. tolls) and fuzzy money (running costs per km) in assignment models is therefore a difficult task unless one uses several user classes or stochastic methods. Hensher (2010) provides an interesting discussion of this issue in the context of WTP calculations using generalised cost models.

There is evidence (see for example, Ariely 2009) that we are neither very rational nor very consistent when making decisions that involve prices, in particular for completely new services or products (i.e. the grass is greener on the other side). But we often value our 'rights' more than those of somebody else. This is why any attempt to curtail the use of my car, as opposed to their motorcycle or bus, generates such an outrage. The loss of a perceived right or service, say free parking, is not compensated by an equivalent tax break, in this case covering the cost of paid parking: the point elasticity of demand to a gain is not the same as that of a loss.

Presumably, our reaction to a price signal is influenced by how much we value it compared to a cash equivalent. If we feel that a change in electronic toll is less onerous than the actual payment in cash then our behavioural response will be different. The perception and impact of different forms of payment is an issue that will take some time to be resolved in modelling terms partly because it reflects different levels of price awareness.

16.1.3   Modelling and Forecasting

In Chapter 1 we distinguished the activities of modelling to compare alternatives and forecasting future demand. Modelling is about developing and using analytical tools that are sensitive to the policies of interest and respond logically to changes in key variables. Good modelling requires an ability to provide useful and timely information during the decision-making process, even if there may be certain caveats or limitations for that information. For example, the issues of payment media and the build up of demand may not have a critical role in the comparison of alternatives.

Forecasting requires visualising and quantifying future conditions. It normally requires projecting future travel demand and, in the case of projects involving pricing, the resulting revenue streams over time. We consider here the role of models in forecasting, but recognise that models alone are not good enough to provide sufficient evidence of future revenues to support the significant risks usually associated to these projects. Given the uncertainty about the future, it is preferable to use complementary approaches and supporting evidence to buttress any future revenue projections. The differences in outcomes must be interpreted in light of the experience of the forecaster, reasonability of the results, confidence in the model and underlying data, and the assumptions about the stability of the behaviour and trends implicit in the model. The quality of a forecast can only be objectively assessed through before and after studies.

Forecasting is, therefore, a more demanding task than modelling when comparing alternative plans or strategies. This chapter tries to address these issues and provide some guidelines to assist those embarking in producing revenue projections, either for the public or private sectors. In the case of forecasting, it is best to identify the factors most likely to affect the projections and focus on getting them right. In fact, the issues of risk and uncertainty are central to the business of revenue forecasting and we discuss them later in the chapter.

16.2   Private Sector Projects

16.2.1   Involvement of Private Sector in Transport Projects

The twenty-first century has seen a significant increase in the involvement of the private sector in the design and delivery of transport infrastructure and services. We do not discuss here the reasons for this involvement; there is plenty of literature on that topic. Our focus is on how the different tools discussed in this text should be applied to such projects. The basic tools of analysis are broadly the same as those developed for our erstwhile public sector clients. However, the questions being asked are different and the approach we need to adopt must be significantly changed to provide the advice our new clients need; there is also a stronger requirement for accountability.

Many transport projects today are implemented through some sort of concession where the private sector invest in constructing and operating a facility and then transfers it to the public sector at the end of the contract period (usually between 20 and 40 years). The sponsor of such a project, usually the government, is interested in the success of a bidding process for such a concession whilst the private investors would like to ensure they do not lose money in delivering the contract.

The overarching issue in forecasting for private sector projects is that of risk. Uncertainty has always been present in our modelling and forecasting work but the involvement of private investors and financial institutions has given a clear monetary value to the issue of risk. In terms of demand modelling, private investors and financial institutions are interested in a revenue stream, year after year for the duration of a concession, and the degree of confidence that can be associated to these figures. These risks change over time and to understand this we need to start by considering the different actors and processes that are central to private sector projects. Figure 16.1 shows a nominal and simplified profile of risks over the time of a project.

The main sources of risk during the bid preparations are:

- Is all the *right of way* required to be released to the concession in time for construction?
- Are all *construction costs* sufficiently well defined and known?
- Are the *ground* conditions sufficiently investigated, including the possible need to displace utilities?
- Are the costs of *operating and maintaining* (O&M) the future assets well known and quantifiable?
- How confident can one be about the future *traffic and revenue* (T&Revenue) streams?
- How long and steep will be the period of transition between starting operations and the time when stable traffic levels materialise? This is known as the *ramp-up* period.

As can be seen from Figure 16.1, all of these risks are higher before construction starts. During construction most of these risks are reduced so when the project starts operating the only remaining risks will be some hidden faults in construction, traffic and revenue, residual O&M and ramp up. Finally the project will reach maturity when traffic levels stabilise and the only remaining risk will be associated with the level of growth and the possibility that a competing facility is provided some time in the future.

Figure 16.1   Idealised risk profile for a private sector transport project

16.2.2   Agents and Processes

Many agents, professionals and advisors play key roles in the process of developing a transport project from conception to successful implementation by the private sector. We simplify these here into three main participants:

- the sponsor, usually the government, who identifies a project, develops it and takes it to the marketplace;
- bidders, often consortia of construction companies, operators and their advisors, who prepare offers for a concession to build, operate and eventually transfer the asset back to the sponsor;
- financial institutions, often a combination of banks, infrastructure and other funds, who would either invest in the concession or lend money under different forms of debt to the concessionaire.

We recognise the role of other agents like insurance companies, monoline insurers, rating agencies and pension funds who may take some degree of risk and/or influence the outcome of the transaction.

Figure 16.2 provides a simplification of the process and the role and concerns of the three main agents: banks and financial institutions, bidders and sponsors. We look at each stage in turn.

Project preparation   The Sponsor/Government tries to define a concession that will provide significant benefits to its people whilst offering an attractive role for the private sector. In doing this, the Sponsor will identify and assess the risks involved and allocate each to whoever has greater capacity to manage it (to do something about it). Although the acceptance of a risk costs money, those who can do something to manage and mitigate it are likely to charge less for accepting responsibility. Traffic and revenue risk is, in most cases, transferred to the concessionaire because it is in the best position, through the provision of a good service, to manage it. At this stage the Sponsor will try to provide a clear and transparent view of risks in order to get good competitive and comparable bids.

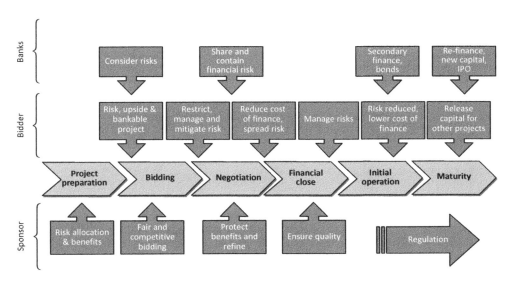

Figure 16.2   Simplified project development process

Bidding process   At this stage each consortium studies the project and its terms of reference in order to decide how much to ask for accepting the risks, obligations and compensating revenue streams on offer. The nature of the revenue risk will depend on a number of factors including the conditions of the concessions, the award criteria (on toll levels, duration, minimum subsidy/maximum payment, discounted value of revenue stream). The consortia will try to get a clear view of the risks and determine whether they have a special competitive advantage (faster construction, better finance) that could be exploited in the bid.

Negotiations   Sometimes these are very short as the conditions of the bid would have specified the project fully. More often there is a period of negotiation once a preferred bidder has been selected; this period is used to refine the Concession Contract and its conditions taking into account variations that may have not been envisaged originally. It usually deals mostly with risks other than traffic but it may involve obtaining stronger guarantees from the Government that unexpected alternative routes/services will not be provided in the future. At this stage the concession is assigned to a *Special Purpose Vehicle* or company, set up by the consortium to build and operate the project until it is transferred back to the Government. The financial institutions, in turn, will try to share and spread the risk among different banks and to press the sponsor and consortium for guarantees. Rating agencies may play a key role here in assessing project risks.

Financial close   Here all the funds needed to implement the project are finally secured and made available. This often involves obtaining a loan to cover the construction/rehabilitation period plus one or two years into operation. The Sponsor provides the rest of the finance as equity. The repayment of this loan is often structured around a lower-cost longer-term finance once the project is well into operation. Therefore, the financial institutions would like to be confident that this second stage finance is assured. There may be a grant provided by the Sponsor to strengthen the financial viability of the project. In the case of existing assets that need rehabilitation and operation over many years, the consortium may offer a payment to the government in compensation.

Second stage finance   Once the project is in operation all risks are much reduced and therefore it should be possible to obtain finance at lower rates. Often the main remaining risk is associated to the future revenue stream. A review of previous traffic and revenue projections may be needed to offer additional confidence in revenue projections. The outcome may be a lower-cost long term loan, a bond issue or some other long term financial instrument.

Third stage finance   Once the project is operating well and its long-term financial structure is in place, it is possible to offer equity participation in the market, totally or in part, depending on the conditions of the concession and the strategy of the concessionaire. This may take the form of an *Initial Public Offering* (IPO) or just a private agreed opportunity to invest in the Special Purpose Vehicle holding the concession. This injection of capital will enable the release of some capital of the original investors that they could use in another concession.

16.2.3   Some Consequences of the Process

The process just described is fairly different from the usual consideration of projects and strategic transport planning for the public sector. The number of agents or stakeholders is significant and each is concerned with risks but from different perspectives. The forecasting of traffic, patronage and revenue is central to these concerns and becomes the most important risk as the project matures.

Traffic and revenue forecasting is produced for each of these key stakeholders and in each case their different perspectives are brought to bear. It is not surprising, therefore, that what is considered important and included in the model may vary and so would the traffic and revenue projections.

Given the variety of agents and the sums of money at stake the need for transparency becomes paramount; model black boxes are not just unacceptable but are seen as a source of additional risk. The ability to explain a traffic model and deliver a compelling narrative relying on several complementary sources of evidence to support revenue projections becomes essential.

16.3   Risk

16.3.1   Uncertainty and Risk

The concepts of uncertainty and risk are obviously related but are not the same. Uncertainty may involve things that are completely unknown, whereas risks are often understood via calculable probabilities; an example, often quoted to illustrate risk, is betting on a colour at the roulette where the risk of losing is slightly above 50%.

We use here the idea of uncertainty as our failure to ascertain a present or future event with certainty. It is a reflection of our lack of knowledge and it is, in principle, impossible to quantify. Pure uncertainty is not very helpful in deciding whether to invest in a particular scheme or not; but it may be inevitable. For example, it is uncertain whether human beings will eventually abandon the idea of owning a private car and adopt the policy of renting such vehicles by the hour as and when needed. Such a change would affect car usage and traffic and would probably strengthen demand for public forms of transportation. However, the probability and timing of such a change is, at present, practically impossible to estimate.

On the other hand, we may be able to assign probabilities to the level of economic growth in the future for a particular region and from this infer future levels of traffic in a specific road section. Risk is, in this sense, quantifiable uncertainty; moreover, it may be possible to assign a monetary value to a variation over an expected future revenue stream. J P Morgan studied 14 pre-opening toll road studies in the USA and compared them with the traffic achieved after opening (Muller 1996). They found that in two cases the original studies underestimated traffic and revenue by between 10 to 30%. In four cases there were moderate over-estimations of revenue of between 12 to 25%. There were, however, 8 cases (57%) where the over-estimation of revenue was from 45 to 75%.

The main reasons for this over-estimation of revenue were considered to be:

- poor analysis of alternatives;
- poor or no analysis of willingness to pay tolls to save time;
- overoptimistic growth rates, mostly based on overestimation of generated traffic through new developments.

Similar results have been found on other markets and countries as documented by Bain (2009) in the case of toll roads and Flyvbjerg *et al.* (2005) for public works. There are many reasons for this, some of them outside the control of the traffic forecaster (for example an unforeseen economic recession that escaped even the banks financing these facilities). However, many of the criticisms levied to the craft of demand forecasting remain valid. Too often, the modelling approach adopted mirrors the classic approach employed on behalf of the public sector for the past 30 years and fails to identify and isolate the key drivers of traffic (and revenue) in a way that recognises the associated risks.

These risks relate to four main sources:

- the size of the relevant market that can possibly be attracted to the new facility (in-scope traffic);
- the estimation of capture rates for that market;
- the development of reliable growth models (including where appropriate induced traffic); and
- consideration of future new alternatives to the new facility that may affect the effective traffic and revenue capture.

16.3.2   Risk Management and Mitigation

Before discussing the implications of all of the above for modelling it is useful to consider what concessionaires can do in order to manage and mitigate risks. Bidders can, of course, request better guarantees from sponsors. This may take the form of explicit, or sometimes implicit, minimum revenue guarantees over the life of the concession. Other assurances include: automatic adjustment of tolls or fares in line with inflation or other formula, and commitment that no competing facility will be provided, at least for the initial years of the project.

During construction, the consortium can ensure minimum opposition from locals by a good communications strategy that should also help to smooth the transition into paying for the use of the new facility. The concessionaire can ensure that a good service is provided at all times and that a good relationship is developed with users and clients. The provision of complementary services (fuelling stations, food and rest facilities at toll roads, and newsstands and refreshments in the case of public transport services) is also important to attract and support customers.

Rapid response to incidents and emergencies, and quick restoration of services after a *force majeur* event are meant to be defining characteristics of private sector involvement in transport projects. Marketing can play a useful role in identifying those users that given the right information or encouragement would start using the new facility. This is important for road haulage companies that are not always fully aware of the reduction in risks and operating costs that result from using a better, if paid, road.

16.4   Demand Modelling

16.4.1   Willingness to Pay

WTP plays a key role in the estimation of patronage and revenue collection. Willingness-to-pay is usually represented through the Subjective Value of Travel Time Savings (SVTTS) ascertained through stated

(SP) and revealed preference (RP) surveys (see section 15.4). One of the problems with RP data is that this is often related to different 'types of money' and the use of SP is often unavoidable for new toll roads.

In this respect, the use of a single SVTTS is not reliable enough as will tend to exaggerate, one way or another, the real capture rate of any facility. Segmentation is very important and this can be done on the joint basis of journey purpose and income levels. Trip purpose may be important if differential growth is expected in the future. Income levels are strongly correlated with SVTTS. An additional and important segment of the travelling population are those who have their costs, including tolls and fares, covered by somebody else, usually their employer; they have a high but not unlimited WTP within the travel policies of their companies.

In the case of trucks, WTP depends on a number of factors including the size and type of goods hauled, the type of contract for each shipment (for example *just in time* arrangements), company policy and legal requirements (dangerous goods are often required to use the safest road, normally a tolled one) and, ultimately, opportunistic decisions by the driver. Some road haulage companies, in particular one-man operations, are not fully aware of their operating costs and may be more cash sensitive; these will display a lower willingness to pay for tolls to save time and operating costs than an objective evaluation would suggest.

In the case of urban schemes one must also consider that many trips will be made day after day and the impact of tolls or fares over the monthly budget may not be insignificant. Income effects will have to be considered in these cases. The attractiveness of new public transport facilities is also influenced by WTP, especially if the new mode is more expensive and better than existing services.

WTP is also influenced by the quality of the service or the road provided. One is willing to pay more to reduce the time spent under less comfortable conditions, for example heavy congestion or the need to stop at junctions as opposed to free flow on a good highway. This is sometimes referred as a 'motorway premium' or a 'standard road malus' and it is generally a difference of between 20 and 40% of the SVTTS depending on each case.

It is generally recommended to use at least 10 categories for WTP for toll roads including at least four for trucks. In the case of public transport, the level of segmentation would be less as freight is not an important component of that market.

WTP is likely to grow in line with income levels of the relevant population. This may be just the car owning population that would be the wealthier proportion of the total in an emerging country. The rate of growth of SVTTS with per capita income is somewhat uncertain and in dispute. The prevalent view is that SVTTS will grow at between 0.5 and 0.9 times the rate of growth of income per capita of the relevant population, see for example Wardman (2001) and Accent and HCG (1996).

16.4.2 Simple Projects

A small number of projects can be handled using simple models on a spreadsheet. This may be the case for some new estuarial crossings where there are only one or at most two alternatives. If the alternatives are clear and limited, it is possible to use a logit formulation to consider them and the effect of introducing a new one.

Depending on the nature of the new alternative this can be incorporated as another choice in a MNL or in a nested structure if it is, for example, a new shorter bridge to compete with a longer road and ferry crossings. Whatever the case, the alternative specific constants are always going to be contentious and should be supported by evidence complementary to SP extracted values. The need to incorporate as many service and personal attributes as possible in the choice structure should reduce the importance of these constants.

Note also that segmentation remains important even if a simple logit formulation is used. Simple cases will tend to be predominantly inter-urban ones and therefore choices might be modelled on a full day basis. The exception will be cases with significant variations in travel times during the day either because of congestion or the availability of some alternative only at certain times.

16.4.3   Complex Projects

Most transport concessions will require the development and implementation of a network model and in many cases a multi-modal one. It is very difficult to separate and distinguish the sources of revenue risk in a conventional model: the size of in-scope traffic, growth, traffic capture and the true elements of choice: willingness-to-pay and the relative advantages of each alternative. These issues are confounded in a large-scale model with less relevant material and a full range of behavioural responses. In order to de-construct the components of future demand it is useful to adopt an approach as depicted in Figure 16.3. Here we extract from a conventional transport model the main components of in-scope trips, benefits of the new facility compared to alternatives, WTP for these benefits and growth. Each of these will have risks associated with them and it is the task of the modeller to identify and reduce them to provide a more reliable forecast.

Revenue from new mode/road

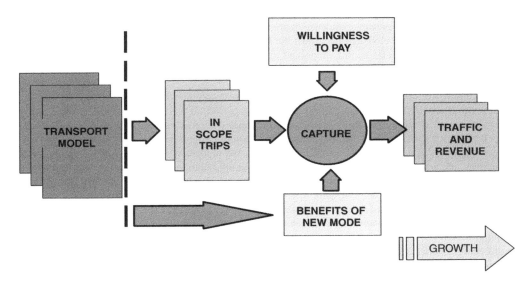

Figure 16.3   Estimation of traffic and revenue from a new facility

Traffic revenue depends on the size of the relevant travel market, its future growth, the choice mechanisms available to users and their own preferences. When faced with a new facility, users can have the following main responses:

- change their route;
- change mode;
- change their destination to one easily reachable using the new system;
- change their trip making frequency (generated/suppressed traffic);
- change the time of travel as a result of price and congestion profiles.

How many of these responses will be important depends on the new facility and on the alternatives available now and in the future. Experience with similar schemes and a bit of experimental modelling can help in deciding how many of these responses should be included in revenue projections. These will also help in identifying the scope of the traffic that could be captured by the new facility.

In-scope traffic is the traffic that *might* be attracted to the project. As such, it can be considered as the target market for the route operator. With a toll road or bridge, it can normally be considered as the traffic that would use the route if no charge were made. For public transport links, it represents an initial judgement on the traffic that could be captured both from other competing public transport services and potentially attracted from other modes under the most favourable circumstances.

The most reliable way to estimate this potential demand is to undertake a battery of transport surveys (passengers and freight). Investors are more convinced by actual *on the ground* data than by any outputs from elegant and sophisticated models. Origin-destination (O-D) surveys, traffic counts and travel time surveys, undertaken probably at different times of the year and days of the week, and for at least 16 hours per day, would be ideal. These should be combined with some permanent traffic counting methods to obtain a suitable profile of demand throughout a year.

In order to add the greatest comfort, investors should be provided with a real description of this traffic, with an understanding not merely of traffic levels but of who is travelling where and for what reason. This understanding helps them form their own judgement – in their own terms – of the function of the road, and gives them also greater confidence in predictions of growth and of capture.

Most toll roads can be modelled using just the assignment stage in a commercial package capable of handling multiple user classes. It is important that these models are handled in terms of generalised costs of travel and because of their final use it is convenient to quantify these in monetary units (see the discussion by Hensher 2010).

Three alternative approaches can be used here. The most common is to employ 10 or more user classes with equilibrium assignment during different time periods; if this is an uncongested area, it may be sufficient to model an average hour or day. A second related approach is to use fewer user classes and adopt a stochastic assignment model; the main problem with this approach is the difficulty in justifying the scaling or spread parameter(s). A third approach consists in identifying, for each O-D pair, the best two routes, one using the tolled facility and another using only untolled roads. Then, a discrete choice model is used to split demand among the alternative routes. A problem with this approach is that only two routes are identified for each O-D pair, when in practice more may be used by savvy drivers seeking to optimise the combination of tolled and untolled roads.

Some projects introduce interesting complexities that tax the ingenuity of the modeller, for example capping the toll for a facility to encourage use by long distance trips or establishing a minimum toll to discourage shorter trips. Variable pricing in high occupancy and tolled (HOT) lanes are particularly difficult to model, and one may have to rely on micro-simulation or dynamic assignment techniques.

Public transport projects, new metro, LRT, rail or BRT schemes, are more complex to model as they inevitably involve mode choice and other behavioural responses. Nevertheless, the same principles apply: the identification of in-scope traffic, transparent representation of choices, in-depth WTP analysis and consideration of present and future alternatives.

Whatever the modelling approach is adopted, the traffic advisor will need to prepare a *Base Case* (expected scenario) and *Downside* and *Optimistic* scenarios. Sometimes the Downside case will be called *Financial Case* as the debt bearing capacity of a project would be based on that revenue stream. The assumptions behind each scenario should be well documented and agreed by stakeholders in advance.

We discuss now how these different modelling approaches are influenced during the different stages in the process of implementing private sector participation in a transport project.

16.4.4   Project Preparation

The government is usually interested in offering a concession that transfers a significant element of risk to the private sector. It is also interested in tapping into the creativity and good management of the private sector to secure intelligent design, innovative financial packages and to offer a high level of service throughout the concession.

To achieve this, the sponsor requires a competitive and transparent tendering process over a well-designed *Concession Package*. This will be assisted by low bidding costs and wide international promotion of the concession programme if appropriate. The government should retain those elements of risk that it is best equipped to handle, for example securing the right of way in a timely manner.

Revenue risk is very often transferred to the concessionaire as it can handle it best through good service and pricing. Even then, in order to facilitate financial close the government may be persuaded to offer some measures to reduce revenue risk. The main instruments available to provide manageable revenue risk are the containment of future competition, minimum revenue guarantees (MRG), the choice of decision rules for awarding the concession and the provision of a well-documented database.

MRG are sometimes offered over the first few years of the concession and at a level below that of the Base Case scenario. The level of this guaranteed revenue stream is important in determining the debt/equity ratio for the concession.

If the future is very uncertain, for example when the government does not want to commit to not building alternatives in the future, some concessions have been awarded to the bidder requesting the lowest present value (LPV) of the revenue stream discounted at a pre-determined rate. In this way if the revenue stream is below expectations, the result is just an extension of the concession up to a pre-determined limit. Revenue risk is therefore reduced. Revenue projections are still needed in order to secure financial backing for the project, but they become less important than in concessions awarded on the basis of lowest toll level or minimum duration. However, LPV concessions have some undesirable side effects. In them, the focus is on reducing construction costs to the minimum and there is no incentive to offer good levels of service as any increase in O&M costs simply reduces profit.

Bid costs are generally high and naturally consortia would like to recover them through successes in their bidding programmes. The sponsors are, therefore, interested in reducing bidding costs as much as feasible without compromising the quality of the concession agreements. Traffic and revenue studies are an expensive element of bidding for a concession. There are significant advantages for the sponsor to undertake a good *Reference Study*:

- Undertaken to international standards; this means either an international company or at least a technical audit by one.
- Transparent and well documented; data should be collected with good quality assurance and provided both processed and in raw (e.g. interview records) form.
- Data should cover the relevant periods and be segmented generously; at least some traffic/person counts should be continuous over a whole year.
- If software packages are used to process the data and model demand, they should be internationally and commercially available.
- The provision of geo-coded data and the whole database on electronic format is highly desirable.

Travel surveys are expensive and time consuming. They do not fit well within the timescales and budgets available for bidding for a concession. Therefore, it is highly desirable that these are undertaken as part of the Reference Study for the sponsor in preparation for the concession. To be of use, they should be well documented and made available to all bidders on a transparent format including the processed and raw data. Geocoding these data provides an added benefit of allowing consortia to develop their own zoning systems. The bidder would like to confirm this information with its own traffic counts and other observations, seeking, at the same time, to identify opportunities to obtain a competitive advantage over other consortia.

16.4.5   Forecasting Demand and Revenue during a Bid

The viewpoints of the bidder and financial institutions are similar, although the second focuses on default risk and the first on the probability of achieving a significant surplus after debt coverage. Both benefit from looking at revenue projections in the context of the risks associated to each contribution. One way of handling this is to build different scenarios for the future: Optimistic, Base Case and Downside are commonly used and must be clearly defined.

It is very important here to adopt a multi-evidence approach. The survey-supported transport model will not provide enough evidence on its own to enable bidders and their financial advisors to estimate risks and potential upsides. The traffic advisor should be able to demonstrate a thorough understanding of the drivers behind the revenue figures: what are the main economic activities of the region and how they depend on national and international trends, what threats are posed by alternative modes or facilities, what opportunities are offered to increase revenues in the future through complementary services or pricing strategies, etc.

The financial strength of a project of this nature will depend on a number of factors. An often critical one is the *Debt Service Coverage Ratio* (DSCR). This is the ratio of revenue from operations to principal and interest obligations; that is, payments due to lenders at each period. The most critical stage will be the earlier year of the project. When a project is implemented in stages, it will be important to model them separately and to provide estimates that incorporate ramp up effects from the outset.

The ability of the traffic advisor to explain the workings of the model to non-specialists and to demonstrate in-depth understanding of the underlying drivers of its financial success, are critical to a successful bid preparation.

16.4.6   Ramp Up, Expansion, Leakage

There are a number of little issues that do not figure significantly in public works projects but have great importance in private sector projects. The ramp up, or transitional period, represents one of them as it was never considered particularly important for public works projects. However, revenue collections of the first few years of a concession play a significant role in their financial viability. This is why quick implementation and good estimation of this transitional period is essential. During ramp up, potential users learn about the new facility and the advantages it may offer to their journeys. There is often strong resistance to the introduction of a new tolled facility instead of an equivalent untolled one. This resistance may result in a slow adoption rate even if the advantages more than compensate the imposed toll. The adoption of good communication and marketing strategies should help in reducing the length of this transitional period.

Nobody has come up yet with a good theory to support the estimation of ramp up durations. We know that this will depend on issues like:

* The frequency of trip making in the area; the more frequent repeated trips are the shorter the ramp up period will be.
* The significance of the advantages offered by the new facility; a major time saving will result in shorter ramp ups.
* Information on the new facility and the advantages it will offer.
* The local tolling culture; if people are used to toll roads then it will be easier to adopt a new one.
* The provision of a short period when the new mode or facility is provided without a charge may facilitate its appreciation, but may generate a backlash when price is introduced; these periods should be short and well communicated.

In the absence of a good theory one must rely on benchmarking transitional periods with other similar facilities and contexts. Anything between six months to several years is possible depending on the characteristics above.

A second issue is the expansion from the modelled periods to a full year of operation. This is tricky for green field new facilities as there will be little evidence about the demand profile over time for a tolled road or a new rail service. One must assume that the information contained in permanent automatic traffic counters is a reliable source for considering seasonal variations along the year. This will provide limited comfort if there are real seasonal variations in the structure of the trip matrices, for example because agricultural produce movements are significant.

Whenever congestion plays a role in the capture rate of a new facility or service, it will be necessary to model different periods of the day (and sometimes of the week) to ascertain their corresponding different capture rates. The expansion task is now dependent on the number of hours a year that are represented by each modelled period.

It is generally not practical to model every year of operation using a full transport model. Common practice is to model only those years when significant changes in the network, or in prices, are expected and interpolate the other years. Years that are far in the future are sometimes extrapolated from the last year modelled with confidence. Latter years bear little influence on the financial strength of a project.

Not every penny that is collected at the toll plaza or fare box reaches the coffers of the concessionaire. There are inevitable losses in the trail from transaction to bank account, even when electronic fare and toll collection are dominant. Some losses are the result of straight avoidance on the part of users, others may be due to technical failures, misclassification of vehicles and human error. And some money reaches the wrong pockets.

In most projects the expected loss rates are reasonably well known. However, when new technology is introduced the traffic advisor must take a view on the likely levels of revenue leakage that will materialise when the concession is in operation.

Finally, some projects will offer fares or toll rates that are shared among different suppliers of services, for example metro and feeder buses. In this case, it will be necessary to perform additional calculations to correctly allocate revenue to these different agents and concessions. This is also the opportunity to account for most discounts, period tickets and concessions (free passes, exempt users) that will influence the final revenue stream figures.

As stated in previous chapters, modelling is mostly useful when benefiting from good interpretation of results. In the case of private sector projects, sound interpretation of results is of paramount importance. The ability to understand and communicate modelling results is based on the capacity to track influences from inputs to outputs. This is where good understanding of the theories underpinning the models is essential. Explanations should be delivered in the language and conceptions of the interested parties, not those of the modeller.

16.5   Risk Analysis

Although the Reference Study prepared by the sponsor will identify the main revenue risks, urgent consideration of these will only start in preparation for the bidding process. A traffic and revenue study for a bidding consortium will normally consider first the production of a comprehensive *Risk Register*. This will contain also the revenue risks and they will serve to focus the attention and the data collection effort for the traffic study.

It is difficult, even undesirable, to generalise on these risks but they are likely to include:

- poor estimation of in-scope demand;
- overestimation of willingness to pay;

- overestimation of growth prospects;
- ignoring future changes to the network;
- underestimating the importance of technology or trend changes.

Some risks are inherently difficult to identify. These are the 'black swans' of Taleb (2007), events that are almost impossible to foresee like the impact of oil prices on the price of tortillas in Mexico via biofuel production in the US. Oil prices are indeed very difficult to forecast and they do have an influence on travel behaviour in particular when they take the form of a significant shock. Experience has shown that forecasts of economic growth and recessions are also very uncertain.

There are two basic ways of handling the issue of risk in traffic and revenue projections: sensitivity analysis and stochastic simulations.

16.5.1 Sensitivity and Sources of Risk

Sensitivity analysis is performed to identify how much model outputs depend on small changes in model parameters and inputs. It is used for two reasons: first, to ensure that the model responses are reasonable and explainable; second, to identify what are the key risk sources that are most likely to affect the financial strength of a project.

Sensitivity tests are usually undertaken at least for: SVTTS, growth rates usually linked to GDP, timing of competing projects and toll or fare levels. Variations of $+/- 10$ or 20% on SVTTS are useful to assess how dependent are the estimated revenues on our evaluation of these parameters. Financial institutions linked to the project should be able to provide estimates of possible variations of future GDP growth. These will affect incomes and therefore car ownership and traffic and revenue.

Toll and fare level sensitivity tests are also important, even if these are fixed in the concession contract, because one would like to be confident that increasing them will increase revenue. This sensitivity tests may prove that toll or fares have been set too high and that more revenue (and benefits) would be collected with lower rates. An example of this type of toll sensitivity tests is shown in Figure 16.4.

The figure shows that the optimal toll rate, in terms of maximising revenue, is around 0.70 pesos per km. The sponsor would have fixed the toll at some 0.5 or 0.6 pesos per km to protect user benefits.

Different aspects of our transport models generate different levels of confidence in their outputs. We tend to believe more in the results of an assignment model because when it fails on the base years this

Figure 16.4 Toll sensitivity tests

is very easy to diagnose. Our confidence on mode, destination and frequency of travel choices is less strong because of the difficulties in performing such a diagnostic test quickly and effectively enough.

Moreover, the drivers for some component of demand capture by the new service may be different from other contributors. For example, in analysing future patronage of a high speed rail concession, capture from air travel may depend on the pricing policies of low-cost airlines that are difficult to predict; capture from other rail services or car users may be more certain as their pricing policies are better understood and predictable.

A useful way of presenting these results is to de-construct the outputs of a traffic model in a manner that enables the interested party to assign their own risk indices to different components of future demand. This is illustrated in Figure 16.5 for a hypothetical high-speed rail link. The figure shows the different contributions of *Induced* and *Redistributed Traffic* plus the traffic captured from alternative modes. A bidder who has good information about long distance bus/coach operations will be more confident of this particular component of future demand capture.

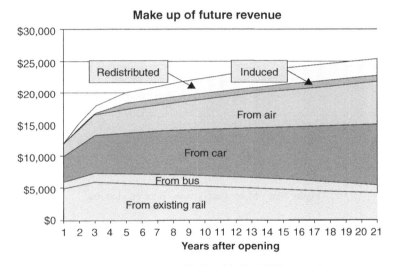

Figure 16.5 Revenue profile for a idealised HSR concession

16.5.2   Stochastic Risk Analysis

Stochastic risk analysis involves the use of Monte Carlo simulations usually implemented as an ad-on to a standard spreadsheet. In this case, the first step is to agree with stakeholders the few input or model variables that will be considered to be stochastic rather than fixed and relate the outputs from the model to them. Conventional model runs will be needed to identify, for example, how variations in GDP growth affect revenues. Most of these would have been undertaken as part of the sensitivity tests mentioned previously.

The next step would be to adopt some probabilistic distribution around the mean expected values of these variables, for example SVTTS. It is tempting to assume that these would be Gaussian, i.e. Normal distributions. However, we should be warned that the probabilistic distributions of some key variables (GDP is a good example) would have 'fat tails'; that is, they will display more extreme values more often than in a Normal distribution, see the extensive discussion on this issue by Taleb (2007). Some variables will not accept negative values that are possible in a Normal distribution. A log-normal distribution could be used in this case.

The next step is to construct a model where this handful of variables influences revenue outcomes and where their probabilistic distributions are sampled repeatedly in a Monte Carlo simulation. Note that in most cases these distributions are assumed to be independent. This is convenient but may be more difficult to accept in the case of the accepted relationship between GDP and SVTTS. Each run of the Monte Carlo simulation reflects one possible revenue path diverging from the Base Case. This is illustrated in Figure 16.6, where each path represents a diversion from the expected Base Case assumed to be unity; a revenue factor value of 0.95 in one year implies that collections in that case would be only 95% of the Base Case for that year.

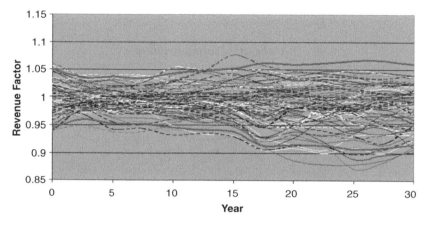

Figure 16.6   Monte Carlo revenue paths on a toll road

The end result is a distribution of revenue outcomes over the life of the project. Of these ranges, lenders would be more interested in the so called P90 or P75 revenue streams, which are the revenues that will be exceeded 90 or 75% of the time. P50 is the expected or Base Case revenue stream. Equity investors might be interested in P40, i.e. revenues that have only a 40% probability to materialise but represent a significant upside of the project. Figure 16.7 illustrates the distribution of values for a toll road with extremes of P90 (the lower band) and P10 (the upper one).

16.6   Concluding Remarks

Traffic and revenue risks have existed well before tolled facilities and public transport concessions became prevalent; given their importance in establishing the financial viability of such projects they have become explicit and more important. This type of analysis is now permeating mainstream transport modelling and will end up assisting decision making for complex and large projects. Accountability requires modellers to provide investors with results which have associated confidence intervals estimated with sensible tools as discussed in Chapter 9.

Risks are perceived differently by different agents and stakeholders of the concession process; allocating them to those who can understand and manage them best is essential, as is reducing the costs of dealing with risks.

$ Millions/yr

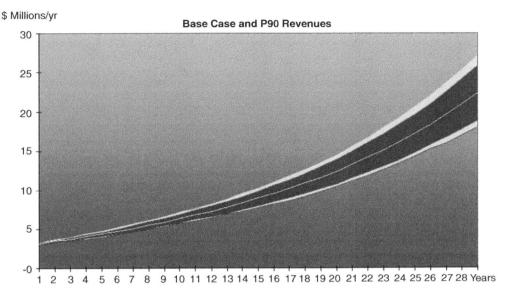

Figure 16.7   P90 Revenues after Monte Carlo risk analysis

The provision of a good, international standard, reference study is a major contribution to reducing bid costs and attaining the full benefits from a concession. The early identification of sources of revenue risk should enable to allocate modelling resources where they would add more value.

Understanding and reducing revenue risk requires transparent and traceable models with appropriate segmentation of in-scope demand. Large-scale conventional models are therefore seldom appropriate and often obscure rather than clarify risks and potential pitfalls.

Traffic and revenue projections often over-estimate economic performance because they fail to identify in-scope markets, use too coarse market segmentation coupled with inappropriate choice models and over-optimistic growth.

Willingness to pay studies based on suitable market segmentation, are key to a robust estimation of traffic capture and revenue projections. They should be supported by benchmarking against other studies and international evidence.

Ramp-up risk can be managed to some extent. There is good scope for employing and adapting marketing techniques to help price and sell tolled facilities and new transport services.

It is desirable to de-construct model results so that the level of risk associated to each contribution to total revenue can be separately ascertained. Induced and generated traffic should only be included with great caution and with a high degree of uncertainty associated to them compared with demand transferred from other routes or modes.

Risk Analysis Techniques are an element of good traffic and revenue projections; their value depends on the quality of the base modelling effort and the depth of understanding of the potential market for the facility; it is never an alternative to good traffic projections.

References

AASHTO (1977) *A Manual of User Benefit Analysis of Highway and Bus Transport Improvements.* American Association of State Highway and Transportation Officials, Washington, DC.

Abdel-Aty, M.A., Kitamura, R. and Jovanis, P.P. (1997) Using stated preference data for studying the effect of advanced traffic information on driver's route choice. *Transportation Research* 5C, 39–50.

Abraham, H. and Kavanagh, C. (1992) Modelling public transport in-vehicle congestion using EMME/2 Release 5. *Proceedings 1st European EMME/2 Users Conference*, London, April 1992, England.

Abraham, H., Shaw, N. and Willumsen, L. (1992) A micro-based incremental four stage transportation model for London. *Proceedings 20th PTRC Summer Annual Meeting*, University of Manchester Institute of Science and Technology, September 1992, England.

Abou-Zeid, M. and Ben-Akiva, M.E. (2009) A model of travel happiness and mode switching. In S. Hess and A. Daly (eds.), *Choice Modelling: The State-of-the-Art and the State-of-Practice, Proceedings from the Inaugural International Choice Modelling Conference*. Emerald, Bingley.

Accent and HCG (1996) *The Value of Travel Time in UK Roads–1994*. Final Report to the Department of Transport, Accent Marketing & Research and Hague Consulting Group, London.

Adams, J. (1992) Horse and rabbit stew. In A. Croker and C. Richards (eds), *Valuing the Environment: Economic Approaches to Environmental Valuation*. Belhaven Press, London.

Addison, J.D. and Heydecker, B.G. (1999) Dynamic traffic equilibrium with departure time choice. In A. Ceder (ed.), *Transportation and Traffic Theory*. Pergamon, Oxford.

Akçelik, R. (1991). Travel time functions for transport planning purposes: Davidson's function, its time-dependent form and an alternative travel time function. *Australian Road Research* 21, 49–59.

Algers, S., Bergström, P., Dahlberg, M. and Dillen, J.L. (1999) Mixed logit estimation of the value of travel time. *Working Paper*, Department of Economics, Upsala University.

Algers, S., Daly, A.J., Kjellman, P. and Widlert, S. (1995) Stockholm model system: application. *7th World Conference on Transport Research*, Sydney, July 1995, Australia.

Allaman, P.M., Tardiff, T.J. and Dunbar, F.C. (1982) New approaches to understanding travel behaviour. *NCHRP Report 250*, National Cooperative Highway Research Program, Transportation Research Board, Washington, DC.

Allenby, G. (1997) An introduction to hierarchical Bayesian modelling. *Tutorial Notes*, Advanced Research Techniques Forum, American Marketing Association.

Allenby, G. and Rossi, P. (1999) Marketing models for consumer heterogeneity. *Journal of Econometrics* 89, 57–78.

Allsop, R.E. (1999) Road safety strategy and targets in Great Britain. *Traffic Engineering* 34, 72–79.

Allsop, R.E. and Charlesworth, J.A. (1977) Traffic in a signal-controlled road network: an example of different signal timings inducing different routeing, *Traffic Engineering and Control* 18, 262–264.

Alonso, W. (1964) *Location and Land Use*. Harvard University Press, Cambridge.

Alonso, W. (1968) Predicting best with imperfect data. *Journal of the American Institute of Planners* 34, 248–255.

Alpizar, F and Carlsson, F. (2001) Policy implications and analysis of the determinants of travel mode choice: an application of choice experiments to metropolitan Costa Rica. *Working Paper* 56, Department of Economics, Göteborg University.

Alvarez, R. (1995) Calibración de redes de transporte: comparación de los métodos Binivel y Hookes & Jeeves. In F.J. Martínez (ed.), *Actas del Séptimo Congreso Chileno de Ingeniería de Transporte*. Sociedad Chilena de Ingeniería de Transporte, Santiago (in Spanish).

Amador, F.J., Gonzalez, R.M. and Ortúzar, J. de D. (2005) Preference heterogeneity and willingness-to-pay for travel time savings. *Transportation* **32**, 627–647.

Ampt, E.S. (2003) Respondent burden. In P.R. Stopher and P.M. Jones (eds), *Transport Survey Quality and Innovation*. Pergamon, Amsterdam.

Ampt, E.S. and Ortúzar, J. de D. (2004) On best practice in continuous large-scale mobility surveys. *Transport Reviews* **24**, 337–363.

Anas, A. (1982) *Residential Location Markets and Urban Transportation*. Academic Press, London.

Andrews, R.L., Ansari, A. and Currim, I.S. (2002) Hierarchical Bayes versus finite mixture conjoint analysis models: a comparison of fit, prediction and partworth recovery. *Journal of Marketing Research* **39**, 87–98.

Aptech Systems (1994) *GAUSS User's Manuals*. Maple Valley.

Arentze, T.A. and Timmermans, H.J.P. (2004) ALBATROSS – a learning-based transportation oriented simulation system. *Transportation Research* **38B**, 613–633.

Arezki, Y. (1986) Comparison of some algorithms for equilibrium traffic assignment with fixed demand. *Proceedings 14th PTRC Summer Annual Meeting*, University of Sussex, July 1986, England.

Arezki, Y., Chadwick, N. and Willumsen, L. (1991) Congestion, evaluation and equilibrium: some empirical results. *Proceedings 19th PTRC Summer Annual Meeting*, University of Sussex, September 1991, England.

Ariely, D. (2009) *Predictably Irrational*. Harper Collins, London.

Armoogum, J. and Madre, J.L. (1998) Weighting or imputations? The example of non responses for daily trips in the French NPTS. *Journal of Transportation and Statistics* **1**, 53–63.

Armstrong, P.M., Garrido, R.A. and Ortúzar, J. de D. (2001) Confidence intervals to bound the value of time. *Transportation Research* **37E**, 143–161.

Arnold, B. and Brockett, P. (1992) On distributions whose component ratios are Cauchy. *American Statistician* **46**, 25–26.

Arnott, R., de Palma, A. and Lindsey, R. (1993) A structural model of peak period congestion: a traffic bottleneck with elastic demand. *American Economic Review* **83**, 161–179.

Arnott, R., de Palma, A. and Lindsey, R. (1994) Welfare effects of congestion tolls and heterogeneous commuters. *Journal of Transport Economics and Policy* **XXVIII**, 139–161.

Ashley, D.J. (1978) The Regional Highway Traffic Model: the home based trip end model. *Proceedings 6th PTRC Summer Annual Meeting*, University of Warwick, July 1978, England.

Ashley, D.J., Lowe, S., Mundy, R., Stanley, R. and Baanders, A. (1985) The long distance travel model for the Netherlands: its specification and application. *Proceedings 13th PTRC Summer Annual Meeting*, University of Sussex, July 1985, England.

Ashok, K., Dillon, W. and Yuan, S. (2002) Extending discrete choice models to incorporate attitudinal and other latent variables. *Journal of Marketing Research* **39**, 31–46.

Atherton, T.J. and Ben-Akiva, M.E. (1976) Transferability and updating of disaggregate travel demand models. *Transportation Research Record* **610**, 12–18.

Axhausen, K.W., Löchl, M., Schlich, R., Buhl, T. and Widmer, P. (2007) Fatigue in long-duration travel diaries. *Transportation* **34**, 143–160.

Axhausen, K.W., Zimmermann, A., Schönfelder, S., Rindsfüser, G. and Haupt, T. (2002) Observing the rhythms of daily life: a six-week travel diary. *Transportation* **29**, 95–124.

Babin, A., Florian, M., James-Lefebre, L. and Spiess, H. (1982) EMME/2: interactive graphic method for road and transit planning. *Transportation Research Record* **866**, 1–9.

Bacharach, M. (1970) *Biproportional Matrices and Input Output Change*. Cambridge University Press, New York.

Bain, R. (2009) *Toll Road Traffic and Revenue Forecasts: an interpreter's guide*. ISBN 978-0-9561527-1-8.

Banister, D. (2003) Critical pragmatism and congestion charging in London. *International Social Science Journal* **176**, 249–264.

Barceló, J., Casas, J., Ferrer, J.L. and García, D. (1998) Modelling advanced transport telematic applications with microscopic simulators: the case of AIMSUN. In A. Bargiela and E. Kerckhoffs (eds.), *Proceedings 10th European Simulation Symposium*. Nothingham Trent University.

Bar-Gera, H. (2002) Origin-based algorithm for the traffic assignment problem. *Transportation Science* **36**, 398–417.

Bar-Gera, H. and Boyce, D.E. (2006) Solving a nonconvex combined travel forecasting model by the method of successive averages with constant step sizes. *Transportation Research* **37B**, 351–367.

Bar-Gera, H. and Luzon, A. (2007) Non-unique route flow solutions for user-equilibrium assignments. *Traffic Engineering and Control* **48**, 408–412.

Bar-Gera, H., Nie, Y. and Boyce, D.E. (2009) Practical implications of finding consistent route flows. *Transportation Planning Applications Conference*, Houston, May 2009, USA.

Bar-Gera, H., Nie, Y., Boyce, D.E., Hu, Y. and Liu, Y. (2010) Consistent route flows and the condition of proportionality. *89th Annual TRB Meeting*, Washington, DC, January 2010, USA.

Bastin, F., Cirillo, C. and Toint, P.L. (2006) Application of an adaptive Monte Carlo algorithm to mixed logit estimation. *Transportation Research* **40B**, 577–593.

Bateman, I.J. and Turner, R.K. (1993) Valuation of the environment, methods and techniques: the contingent valuation method. In R.K. Turner (ed.), *Sustainable Economics and Management: Principles and Practice*. Belhaven Press, London.

Bateman, I.J., Turner, R.K and Bateman, S. (1993) Extending cost-benefit analysis of UK highway proposals: environmental evaluation and equity. *Project Appraisal* **8**, 213–224.

Bates, J.J. (1987) Measuring travel time values with a discrete choice model: a note. *Economic Journal* **97**, 493–498.

Bates, J.J. (1988a) Econometric issues in stated preference analysis. *Journal of Transport Economics and Policy* **XXII**, 59–69.

Bates, J.J. (guest ed.) (1988b) Stated preference methods in transport research. *Journal of Transport Economics and Policy* **XXII**, 1–137.

Bates, J.J. (1996) *Time Period Choice Modelling: A Preliminary Review*. Final Report for the Department of Transport–HETA Division, John Bates Services, Oxford.

Bates, J.J., Ashley, D.J. and Hyman, G. (1987) The nested incremental logit model: theory and application to modal choice. *Proceedings 15th PTRC Summer Annual Meeting*, University of Bath, September 1987, England.

Bates, J.J., Gunn, H.F. and Roberts, M. (1978) A model of household car ownership. *Traffic Engineering and Control* **19**, 486–491, 562–566.

Bates, J.J. and Roberts, M. (1986) Value of time research: summary of methodology and findings. *Proceedings 14th PTRC Summer Annual Meeting*, University of Sussex, July 1986, England.

Battellino, H. and Peachman, J. (2003) The joys and tribulations of a continuous survey. In P.R. Stopher and P.M. Jones (eds), *Transport Survey Quality and Innovation*. Pergamon, Amsterdam.

Becker, G. (1965) A theory of the allocation of time. *Economic Journal* **75**, 493–517.

Beckman, M.J., McGuire, C.B. and Winsten, C.B. (1956) *Studies in the Economics of Transportation*. Yale University Press, New Haven.

Beckman, R.J., Baggerly, K.A. and McKay, M.D. (1995) Creating synthetic baseline populations. *LA-UR-95-1985*, Los Alamos National Laboratory, Los Alamos, New Mexico.

Bekhor, S., Ben-Akiva, M.E. and Ramming, S. (2002) Adaptation of logit kernel to route choice situation. *Transportation Research Record* **1805**, 78–85.

Bekhor, S. and Prashker, J.N. (2001 Stochastic user equilibrium formulation for the generalized nested logit model. *Transportation Research Record* **1752**, 84–90.

Bekhor, S. and Prato, C.G. (2006) Effects of choice set composition in route choice modelling. *11th International Conference on Travel Behavior Research*. Kyoto, August 2006, Japan.

Bell, M.G.H. (1983) The estimation of an origin destination matrix from traffic counts. *Transportation Science* **17**, 198–217.

Bell, M.G.H. (1984) Log-linear models for the estimation of origin-destination matrices from traffic counts: an approximation. In J. Volmüller and R. Hamerslag (eds), *Proceedings of the Ninth International Symposium on Transportation and Traffic Theory*. VNU Science Press, Utrecht.

Bell, M.G.H and Iida, Y. (1997) *Transportation Network Analysis*. John Wiley & Sons, Ltd Chichester.

Ben-Akiva, M.E. (1974) Structure of passenger travel demand models. *Transportation Research Record* **526**, 26–42.

Ben-Akiva, M.E. (1977) Choice models with simple choice set generating processes. *Working Paper*, Centre for Transportation Studies. MIT.

Ben-Akiva, M.E. (2009) Planning and action in a model of choice. In S. Hess and A. Daly (eds.), *Choice Modelling: the State-of-the-Art and the State-of-Practice, Proceedings from the Inaugural International Choice Modelling Conference*. Emerald, Bingley.

Ben-Akiva, M.E. and Bierlaire, M. (1999) Discrete choice methods and their applications in short term travel decisions. In R. Hall (ed.), *The Handbook of Transportation Science*. Kluwer, Dordrecht.

Ben-Akiva, M.E. and Bolduc, D. (1987) Approaches to model transferability and updating: the combined transfer estimator. *Transportation Research Record* **1139**, 1–7.

Ben-Akiva, M.E. and Bolduc, D. (1996) Multinomial probit with a logit kernel and a general parametric specification of the covariance structure. *Working Paper*, Department d'Economique, Université Laval.

Ben-Akiva, M., Bolduc, D. and Bradley, M. (1993) Estimation of travel choice models with randomly distributed values of time. *Transportation Research Record* **1413**, 88–97.

Ben-Akiva, M.E. and Lerman, S.R. (1979) Disaggregate travel and mobility choice models and measures of accessibility. In D.A. Hensher and P.R. Stopher (eds.), *Behavioural Travel Modelling*. Croom Helm, London.

Ben-Akiva, M.E. and Lerman, S.R. (1985) *Discrete Choice Analysis: Theory and Application to Travel Demand*. MIT Press, Cambridge, Mass.

Ben-Akiva, M. and Morikawa, T. (1990) Estimation of travel demand models from multiple data sources. *Proceedings 11th International Symposium on Transportation and Traffic Theory*, Yokohama, July 1990, Japan.

Ben-Akiva, M., Morikawa, T. and Shiroishi, F. (1992) Analysis of the reliability of preference ranking data. *Journal of Business Research* **24**, 149–164.

Ben-Akiva, M.E., Walker, J.L., Bernardino, A.T., Gopinath, D.A., Morikawa, T. and Polydoropoulou, A. (2002) Integration of choice and latent variable models. In H.S. Mahmassani (ed.), *In Perpetual Motion: Travel Behaviour Research Opportunities and Challenges*. Pergamon, Amsterdam.

Ben-Akiva, M.E. and Watanatada, T. (1980) Application of a continuous spatial choice logit model. In C.F. Manski and D. McFadden (eds), *Structural Analysis of Discrete Data: With Econometric Applications*. MIT Press, Cambridge, Mass.

Berndt, E.U., Hall, B.H., Hall, R.E. and Hausman, J.A. (1974) Estimation and inference in non-linear structural models. *Annals of Economic and Social Measurement* **3/4**, 653–655.

Bhat, C.R. (1995) A heteroskedastic extreme value model of intercity travel mode choice. *Transportation Research* **29B**, 471–483.

Bhat, C.R. (1998) Accomodating flexible substitution patterns in multidimensional choice modelling: formulation and application to travel mode and departure time choice. *Transportation Research* **32B**, 455–466.

Bhat, C.R. (2001) Quasi-random maximum simulated likelihood estimation of the mixed multinomial logit model. *Transportation Research* **35B**, 677–695.

Bhat, C.R. (2003) Simulation estimation of mixed discrete choice models using randomized and scrambled Halton sequences. *Transportation Research* **37B**, 837–855.

Bhat, C.R. (2006) Econometric choice formulations: alternative model structures, estimation techniques and emerging directions. In K. Auxhausen (ed.), *Moving Through Nets: The Physical and Social Dimensions of Travel*. Elsevier, Oxford.

Bhat, C.R. and Castelar, S. (2002) A unified mixed logit framework for modelling revealed and stated preferences: formulation and application to congestion pricing analysis in the San Francisco Bay Area. *Transportation Research* **36B**, 593–616.

Bhat, C.R. and Guo, J. (2004) A mixed spatially correlated logit model: formulation and application to residential choice modelling, *Transportation Research* **38B**, 147–168.

Bianchi, R., Jara-Díaz, S.R. and Ortúzar, J. de D. (1998) Modelling new pricing strategies for the Santiago Metro. *Transport Policy* **5**, 223–232.

Bierlaire, M. (2009) *Estimation of Discrete Choice Models with BIOGEME 1.8*, Transport and Mobility Laboratory, Ecole Polytechnique Féderale de Laussane, Zürich (http://transp-or2.epfl.ch/biogeme/doc/tutorial.pdf).

Bierlaire, M., Bolduc, D. and McFadden, D. (2008) The estimation of generalized extreme value models from choice-based samples. *Transportation Research* **42B**, 381–394.

Bierlaire, M., Hurtubia, R. and Flötteröd, G. (2009) An experimental analysis of the implicit choice set generation using the Constrained Multinomial Logit model. *Technical Report TRANSP-OR 090518*, Transport and Mobility Laboratory, Ecole Polytechnique Féderale de Laussane.

Blase, J.H. (1979) Hysteresis and catastrophe theory: a demonstration of habit and threshold effects in travel behaviour. *Proceedings 7th PTRC Summer Annual Meeting*, University of Warwick, July 1979, England.

Bliemer, M.C.J. and Rose, J.M. (2006) Designing stated choice experiments: state-of-the-art. *11th International Conference on Travel Behaviour Research*, Kyoto, August 2006, Japan.

Bliemer, M.C.J. and Rose, J.M. (2008) Construction of experimental designs for mixed logit models allowing for correlation across choice observations. *87th Annual TRB Meeting*, Washington, DC, January 2008, USA.

Bliemer, M.C.J. and Rose, J.M. (2009a) Sample optimality in the design of stated choice experiments. *European Transport Conference*, Leiden, October 2009, The Netherlands.

Bliemer, M.C.J. and Rose, J.M. (2009b) Efficiency and sample size requirements for stated choice experiments. *88th Annual TRB Meeting*, Washington, DC, January 2009, USA.

Bliemer, M.C.J. and Rose, J.M. (2010) Construction of experimental designs for mixed logit models allowing for correlation across choice observations. *Transportation Research* **44B**, 720–734.

Bliemer, M.C.J., Rose, J.M. and Hensher, D.A. (2009) Efficient stated choice experiments for estimating nested logit models. *Transportation Research* **43B**, 19–35.

Bolduc, D. and Alvarez-Daziano, R. (2009) On estimation of hybrid choice models. *International Choice Modelling Conference*, Harrogate, March 2009, England.

Bolduc, D., Boucher, N. and Alvarez-Daziano, R. (2008) Hybrid choice modelling of new technologies for car choice in Canada. *Transportation Research Record* **2082**, 63–71.

Bolduc, D. and Giroux, A. (2005) The integrated choice and latent variable (ICLV) model: handout to accompany the estimation software. Département d'économique, Université Laval.

Bollen, K.A. (1989) *Structural Equations with Latent Variables*. John Wiley & Sons, Ltd Chichester.

Bonsall, P.W. (1983) Transfer price data: its use and abuse. *Proceedings 11th PTRC Summer Annual Meeting*, University of Sussex, July 1983, England.

Börsch-Supan, A. (1990a) Recent developments in flexible discrete choice models: nested logit analysis versus simulated moments probit analysis. In M.M. Fisher, P. Nijkamp and Y.Y. Papageorgiou (eds), *Behavioural Modelling of Spatial Choices and Processes*. North Holland, Amsterdam.

Börsch-Supan, A. (1990b) On the compatibility of the nested logit model with utility maximization. *Journal of Econometrics* **43**, 373–388.

Börsch-Supan, A. and Hajivassiliou, V.A. (1993) Smooth unbiased multivariate probability simulators for maximum likelihood estimation of limited dependent variable models. *Journal of Econometrics* **58**, 347–368.

Bovy, P.H.L. and Fiorenzo-Catalano, S. (2007) Stochastic route choice set generation: behavioural and probabilistic foundations. *Transportmetrica* **3**, 173–189.

Bowman, J.L. (2009) Historical development of activity-based models: theory and practice. *Traffic Engineering and Control* **50**, 314–318.

Bowman, J.L. and Ben-Akiva, M.E. (1999) The day activity schedule approach to travel demand analysis. *78th Annual TRB Meeting*, Washington, DC, January 1999, USA.

Bowman, J.L. and Ben-Akiva, M.E. (2001) Activity-based disaggregate travel demand model system with activity schedules. *Transportation Research* **35A**, 1–28.

Bowman, J.L. and Bradley, M.A. (2008) Activity-based models: approaches used to achieve integration among trips and tours throughout the day. *2008 European Transport Conference*, Leeuwenhorst, October 2008, The Netherlands.

Boyce, D.E. (2007) Forecasting travel on congested urban transportation networks: review and prospects for network equilibrium models. *Networks and Spatial Economics* **7**, 99–128.

Boyce, D.E., Day, N.D. and McDonald, C. (1970) *Metropolitan Plan Making*. Monograph Series No. 4, Regional Science Research Institute, Philadelphia.

Boyce, D.E., O'Neill, C.R. and Scherr, W. (2008) Solving the sequential travel forecasting procedure with feedback. *Transportation Research Record* **2077**, 129–135.

Boyce, D.E., Ralevic-Dekic, B. and Bar-Gera, H. (2004) Convergence of traffic assignments: how much is enough? *Journal of Transportation Engineering of ASCE* **130**, 49–55.

Bradley, M.A., Bowman, J.L and Lawton, K. (1999) A comparison of sample enumeration and stochastic micro-simulation for application of tour-based and activity-based travel demand models. *27th European Transport Conference*, Cambridge, September 1999, England.

Bradley, M.A. and Daly, A.J. (1994) Use of the logit scaling approach to test for rank-order and fatigue effects in stated preference data. *Transportation* **21**, 167–184.

Bradley, M.A. and Daly, A.J. (1997) Estimation of logit choice models using mixed stated-preference and revealed-preference information. In P. Stopher and M. Lee-Gosselin (eds), *Understanding Travel Behaviour in an Era of Change*. Pergamon, Oxford.

Bradley, M. and Daly, A.J. (2000) New analysis issues in stated preference research. In J. de D. Ortúzar (ed.), *Stated Preference Modelling Techniques*. Perspective 4, PTRC, London.

Braess, D., Nagurney, A. and Wakolbinger, T. (2005) On a paradox of traffic planning. *Transportation Science* **39**, 446–450.

Branston, D. (1976) Link capacity functions: a review. *Transportation Research* **10**, 223–236.

Brenninger-Gothe, M., Jornsten K. and Lundgren, J. (1989) Estimation of origin-destination matrices from traffic counts using multiobjective programming formulations. *Transportation Research* **23B**, 257–269.

Bristow, A.L., Jansen, G., Mackie, P.J. and Nellthorp, J. (1998) Costs, prices and values in the appraisal of transport projects – European principles and practice. *Proceedings 8th World Conference on Transport Research*, Antwerp, July 1998, Belgium.

Bróg, W. and Ampt, E. (1982) State of the art in the collection of travel behaviour data. In *Travel Behaviour for the 1980's*, Special Report 201, National Research Council, Washington, DC.

Bróg, W. and Erl, E. (1982) Application of correction and weighting factors to obtain a representative data base. *Proceedings 10th PTRC Summer Annual Meeting*, University of Warwick, July 1982, England.

Bróg, W. and Meyburg, A.H. (1980) The non-response problem in travel surveys: an empirical investigation. *Transportation Research Record* **775**, 34–38.

Brownstone, D. (1998) Multiple imputation methodology for missing data: non-random response and panel attrition. In T. Gärling, T. Laitila and K. Westin (eds), *Theoretical Foundations of Travel Choice Modelling*. Elsevier, Oxford.

Bruton, M.J. (1985) *Introduction to Transportation Planning*. Hutchinson, London.

Bruzelius, N. (1979) *The Value of Travel Time*. Croom Helm, London.

Bunch, D.S. (1991) Estimability in the multinomial probit model. *Transportation Research* **25B**, 1–12.

Bunch, D.S., Louviere, J.J. and Anderson, D.A. (1994) A comparison of experimental design strategies for multinomial logit models: the case of generic attributes. *Working Paper*, Graduate School of Management, University of California at Davis.

Burbidge, S.K. and Goulias, K.G. (2008) Active travel behaviour. *87th Annual TRB Meeting*. Washington, DC, February 2008, USA.

Bureau of Public Roads (1964) *Traffic Assignment Manual*. Urban Planning Division, US Department of Commerce, Washington, DC.

Burgess, L. and Street, S. (2003) Optimal designs for 2k choice experiments. *Communications in Statistics: Theory and Methods* **32**, 2185–2206.

Burgess, L. and Street S. (2005) Optimal Designs for choice experiments with asymmetric attributes. *Journal of Statistical Planning and Inference* **134**, 288–301.

Burrell, J.E. (1968) Multiple route assignment and its application to capacity restraint. In W. Leutzbach and P. Baron (eds), *Beiträge zur Theorie des Verkehrsflusses*. Strassenbau und Strassenverkehrstechnik Heft, Karlsruhe.

Button, K.J., Pearman, A.D. and Fowkes, A.S. (1982) *Car Ownership Modelling and Forecasting*. Gower, Aldershot.

Caldwell, L.C. and Demetski, M.J. (1980) Transferability of trip generation models. *Transportation Research Record* **751**, 56–62.

Cantillo, V., Heydecker, B.G. and Ortúzar, J. de D. (2006) A discrete choice model incorporating thresholds for perception in attribute values. *Transportation Research* **40B**, 807–825.

Cantillo, V., Ortúzar, J. de D. and Williams, H.C.W.L. (2007) Modelling discrete choices in the presence of inertia and serial correlation. *Transportation Science* **41**, 195–205.

Cantillo, V. and Ortúzar, J. de D. (2005) A semi-compensatory discrete Choice model with explicit attribute thresholds of perception. *Transportation Research* **39B**, 641–657.

Cardell, N.S. and Reddy, B. (1977) A multinomial logit model which permits variations in tastes accross individuals. *Working Paper*, Charles River Associates, Boston.

Cardell, N.S. and Dunbar, F. (1980) Measuring the societal impacts of automobile downsizing. *Transportation Research* **14A**, 423–434.

Carlsson, F. (2003) The demand for intercity public transport: the case of business passengers. *Applied Economics* **35**, 41–50.

Carlsson, F. and Martinsson, P. (2002) Design techniques for stated preference methods in health economics. *Health Economics* **12**, 281–294.

Carrasco, J.A. (2001) *Elección Discreta de Alternativas Homocedásticas Correla-cionadas: El Modelo Logit Jerárquico en Profundidad*. M.Sc.Thesis, Department of Transport Engineering, Pontificia Universidad Católica de Chile (in Spanish).

Carrasco, J.A. and Ortúzar, J. de D. (2002) A review and assessment of the nested logit model. *Transport Reviews* **22**, 197–218.

Carson, R.T., Louviere, J.J., Anderson, D.A., Arabie, P., Bunch, D.S., Hensher, D.A., Johnson, R.M., Kuhfeld, W.F., Steinberg, D., Swait, J., Timmermans, H. and Wiley, J.B. (1994) Experimental analysis of choice. *Marketing Letters* **5**, 351–368.

Carson, R.T., Wright, J., Carson, N., Alberini, A. and Flores, N. (1995) *A Bibliography of Contingent Valuation Studies and Papers*. Natural Resource Damage Assessment, La Jolla.

Cascetta, E. (1984) Estimation of trip matrices from traffic counts and survey data: a generalised least squares approach estimator. *Transportation Research* **18B**, 289–299.

Cascetta, E. and Nguyen, S (1988) A unified framework for estimating or updating origin/destination matrices from traffic counts. *Transportation Research* **22B**, 437–455.

Cascetta, E., Nuzzolo, A. and Biggiero, L. (1992) Analysis and modelling of commuters' departure time and route choice in urban networks. *Proceedings of the Second International CAPRI Seminar in Urban Traffic Networks*, Capri, October 1992, Italy.

Cascetta, E. (1996) The Italian decision support system for transport policies and investments: general architecture and development status, *Proceedings of 7th World Conference for Transport Research*, Sydney.

Casey, H.J. (1955) Applications to traffic engineering of the law of retail gravitation. *Traffic Quarterly* **IX**, 23–35.

CASRO (1982) On the definition of response rates. *Special Report Task Force on Completion Rates*, Council of American Survey Research Organisations, New York.

Caussade, S., Ortúzar, J. de D., Rizzi, L.I. and Hensher, D.A. (2005) Assessing the influence of design dimensions on stated choice experiment estimates. *Transportation Research* **39B**, 621–640.

Chadwick, G. (1987) *Models of Urban and Regional Systems in Developing Countries*. Pergamon Press, Oxford.

Chamberlain, G. (1984) Panel data. In Z. Griliches and M. Intriligator (eds.), *Handbook of Econometrics*, Vol. 2. North-Holland, Amsterdam.

Chapman, R.G. and Staelin, R. (1982) Exploiting rank ordered choice set data within the stochastic utility model. *Journal of Marketing Research* **19**, 288–301.

Cherchi, E., Cirillo, C. and Ortúzar, J. de D. (2009) A mixed logit choice model for panel data: accounting for different correlation over time periods. *International Choice Modelling Conference*. Harrogate, March 2009, England.

Cherchi, E. y Ortúzar, J. de D. (2006) On fitting mode specific constants in the presence of new options in RP/SP models. *Transportation Research* **40A**, 1–18.

Cherchi, E. and Ortúzar, J. de D. (2008a) Predicting best with mixed logit models: understanding some confounding effects. In P.O. Inweldi (ed.), *Transportation Research Trends*. Nova Science Publishers, New York.

Cherchi, E. and Ortúzar, J. de D. (2008b) Empirical identification in the mixed logit model: analysing the effect of data richness. *Networks and Spatial Economics* **8**, 109–124.

Cherchi, E. and Ortúzar, J. de D. (2010) On the use of mixed RP/SP models in prediction: accounting for random taste heterogeneity. *Transportation Science* (in press).

Chernew, M., Gowrisankaran, G. and Scanlon, D.P. (2001) Learning and the value of information: the case of health plan report cards. *Working Paper 8589*, National Bureau of Economic Research, New York.

Chesnut, L.G., Ostro, B.D. and Vichit-Vadakan, N. (1997) Transferability of air pollution control health benefit estimates from the United States to developing countries: evidence from the Bangkok study. *American Journal of Agricultural Economics* **79**, 1630–1635.

Chib, S. and Greenberg, E. (1995) Understanding the Metropolis-Hastings algorithm. *The American Statistician* **49**, 327–335.

Chihara, T.S. (1978) *Introduction to Orthogonal Polynomials*. Gordon and Breach, New York.

Chiou, L. and Walker, J.L. (2007) Masking identification of discrete choice models under simulation methods. *Journal of Econometrics* **141**, 683–703.

Chow, A.H.F. (2009) Properties of dynamic system optimal method. *Transportation Research* **43B**, 325–344.

Chu, C. (1989) A paired combinational logit model for travel demand analysis. In World Conference on Transport Research (eds.), *Transport Policy, Management and Technology Towards 2001*. Western Periodicals Co., Ventura, California.

Cirillo, C., Daly, A.J. and Lindveld, K. (2000) Eliminating bias due to the repeated measurements problem in SP data. In J. de D. Ortúzar (ed.), *Stated Preference Modelling Techniques*. Perspectives 4, PTRC, London.

Clancy, M., Dawson, J., Catling, I., Turner, J. and Harrison, W. (1985) Electronic road pricing in Hong Kong. *Proceedings 13th PTRC Summer Annual Meeting*, University of Sussex, July 1985, England.

Commission of the European Communities (1994) *Cost-Benefit and Multi-Criteria Analysis for New Road Construction*. DGVII, Research and Development Unit, Brussels.

Cook, R.D. and Nachtsheim, C.J. (1980) A comparison of algorithms for constructing exact *D*-optimal designs. *Technometrics* **22**, 315–324.

Copley, G. and Lowe, S.R. (1981) The temporal stability of trip rates: some findings and implications. *Proceedings 9th PTRC Summer Annual Meeting*, University of Warwick, July 1981, England.

Coslett, S.R. (1981) Efficient estimation of discrete choice models. In C.F. Manski and D. McFadden (eds.), *Structural Analysis of Discrete Data: With Econometric Applications*. MIT Press, Cambridge, Mass.

Cowles, M.K. and Carlin, B.P. (1996) Markov chain Monte Carlo convergence diagnostics: a comparative review. *Journal of the American Statistical Association* **91**, 883–904.

Cropper, M.L. and Freeeman, A.M. (1991) Environmental health effects. In J.B. Braden and C.D. Kolstad (eds.), *Measuring the Demand for Environmental Quality*. North-Holland, Amsterdam.

Dafermos, S.C. (1980) Traffic equilibrium and variational inequalities. *Transportation Science* **14**, 42–54.

Daganzo, C.F. (1979) *Multinomial Probit: The Theory and its Applications to Demand Forecasting*. Academic Press, New York.

Daganzo, C.F. (1980) Optimal sampling strategies for statistical models with discrete dependent variables. *Transportation Science* **14**, 324–345.

Daganzo, C.F. and Kusnic, M. (1993) Two properties of the nested logit model. *Transportation Science* **27**, 395–400.

Daganzo, C.F. and Sheffi, Y. (1979) Estimation of choice models from panel data. *26th Annual Meeting of the Regional Science Association*, Los Angeles, November 1979, USA.

Dalal, S.R. and Klein, R.W. (1988) A flexible class of discrete choice models. *Marketing Science* **7**, 232–251.

Dalvi, Q.M. (1988) *The Value of Life and Safety: A Search for a Consensus Estimate*. Department of Transport, London.

Daly, A.J. (1982a) Estimating choice models containing attraction variables. *Transportation Research* **16B**, 5–15.

Daly, A.J. (1982b) Applicability of disaggregate models of behaviour: a question of methodology. *Transportation Research* **16A**, 363–370.

Daly, A.J. (1987) Estimating 'tree' logit models. *Transportation Research*, **21B**, 251–268.

Daly, A.J. (1998) Prototypical sample enumeration as a basis for forecasting with disaggregate models. *Proceedings 26th European Transport Conference*, University of Loughborough, September 1998, England.

Daly, A.J. (1992) *ALOGIT 3.2 User's Guide*. Hague Consulting Group, The Hague.

Daly, A.J. (1997) Improved methods for trip generation. *Proceedings 25th European Transport Forum*, Brunel University, September 1997, England.

Daly, A.J. (2001) Alternative tree logit models: comments on a paper by Koppelman and Wen. *Transportation Research* **35B**, 717–724.

Daly, A.J. and de Jong, G. (2006) Errors in functions of parameters of statistically estimated models. *European Transport Conference*, Strasbourg, October 2006, France.

Daly, A.J. and Gunn, H.F. (1986) Cost effective methods for national level demand forecasting. In A. Ruhl (ed.), *Behavioural Research for Transport Policy*. VNU Science Press, Utrecht.

Daly, A.J. and Ortúzar, J. de D. (1990) Forecasting and data aggregation: theory and practice. *Traffic Engineering and Control* **31**, 632–643.

Daly, A.J., van der Valk, J. and van Zwam, H.P.H. (1983) Application of disaggregate models for a regional transportation study in the Netherlands. In P. Baron and H. Nuppnau (eds.), *Research for Transport Policies in a Changing World*. SNV Studiengesellschaft Nahverkehr, Hamburg.

Daly, A.J. and Zachary, S. (1978) Improved multiple choice models. In D.A. Hensher and M.Q. Dalvi (eds.), *Determinants of Travel Choice*. Saxon House, Westmead.

Daor, E. (1981) The transferability of independent variables in trip generation models. *Proceedings 9th PTRC Summer Annual Meeting*, University of Warwick, July 1981, England.

Dargay, J. and Gately, D. (1999) Income's effect on car and vehicle ownership, worldwide: 1960–2015. *Transportation Research* **33A**, 101–138.

De Cea, J. and Cruz, G. (1986) ESMATUC: un modelo de estimación de matrices de viajes en transporte urbano colectivo. *Apuntes de Ingeniería* **24**, 109–125 (in Spanish).

De Cea, J., and Fernández, J.E. (1989) Transit assignment to minimal routes: an efficient new algorithm. *Traffic Engineering and Control* **30**, 491–494.

De Cea, J. and Fernández, J.E. (2000) ESTRAUS: un modelo de equilibrio oferta-demanda para redes multi-modales de transporte urbano con múltiples clases de usuarios. In J. Colomer and A. García (eds.), *Calidad e Innovación en los Transportes*. Universidad de Valencia, Valencia (in Spanish).

De Cea, J., Fernández, J.E. and de Grange, L. (2008) Combined models with hierarchical demand choices: a multi-objective entropy maximisation approach. *Transport Reviews* **28**, 415–438.

De Cea, J., Fernández, J.E., Dekock, V. and Soto, A. (2005) Solving network equilibrium problems on multimodal urban transportation networks with multiple user classes. *Transport Reviews* **35**, 293–317.

De Cea, J., Ortúzar, J. de D. and Willumsen, L.G. (1986) Evaluating marginal improvements to a transport network: an application to the Santiago underground. *Transportation* **13**, 211–233.

De Grange, L., Fernández, J.E., and de Cea, J. (2010). A consolidated model of trip distribution. *Transportation Research* **46E**, 61–75.

Dehghani, Y. and Talvitie, A.P. (1980) Model specification, model aggregation and market segmentation in mode choice models: some empirical evidence. *Transportation Research Record* **775**, 28–34.

Dehghani, Y. and Talvitie, A.P. (1983) Forecasting accuracy, transferability and updating of modal constants in disaggregate mode choice models with simple and complex specifications. *Proceedings 11th PTRC Summer Annual Meeting*, University of Sussex, July 1983, England.

De Jong, G.C. (1989) Some Joint Models of Car Ownership and Use. PhD Thesis, Faculteit der Economishe, Universiteit van Amsterdam.

De Jong, G.C. (1990) An indirect utility model of car ownership and private car use. *European Economic Review* **34**, 971–985.

De Jong, G.C. (1996) A disaggregate model system of vehicle holding duration, type choice and use. *Transportation Research* **30B**, 263–276.

De Jong, G., Daly, A.J., Pieters, M., Miller, S., Plasmeijer, R. and Hofman, F. (2007) Uncertainty in traffic forecasts: literature review and new results for The Netherlands. *Transportation* **34**, 375–395.

De la Barra, T. (1989) *Integrated Land Use and Transport Modelling: Decision Chains and Hierarchies*. Cambridge University Press, Cambridge.

Denstadli, J.M., Lines, R. and Ortúzar, J. de D. (2010) Information processing in choice-based conjoint studies: a process-tracing study. International Journal of Research in Marketing (in press).

Department of Health (1999) *Economic Appraisal of the Health Effects of Air Pollution. Ad Hoc Group on Economic Appraisal of the Health Effects of Air Pollution*. HMSO, London.

Department of Transport (1985) *Traffic Appraisal Manual (TAM)*. HMSO, London.

Department of Transport (1987) *Values for Journey Time Savings and Accident Prevention*. HMSO, London.

Department of Transport (1997) *Traffic Appraisal of Road Schemes: Design Manual for Roads and Bridges*. HMSO, London.

DeSerpa, A. (1971) A theory of the economics of time. *Economic Journal* **81**, 828–846.

DETR (1998) *A New Deal for Highways in England. Guidance to the New Appraisal Framework*. Department of the Environment, Transport and the Regions, London.

Deville, J.C., Särndal, C.E. and Sautory, O. (1993) Generalised ranking procedures in survey sampling. *Journal of the American Statistical Association* **88**, 1013–1020.

Dial, R.B. (1971) A probabilistic multipath traffic assignment model which obviates path enumeration. *Transportation Research* **5**, 83–111.

Dial, R.B. (2006) A path-based user-equilibrium traffic assignment algorithm that obviates path storage and enumeration. *Transportation Research* **40B**, 917–936.

Diamond, P. and Haussman, J.A. (1994) Contingent valuation: is some number better than no number? *Journal of Economic Perspectives* **8**, 45–64.

DICTUC (1978) *Encuesta Origen y Destino de Viajes para el Gran Santiago*. Final Report to the Ministry of Public Works, Department of Transport Engineering, Universidad Católica de Chile, Santiago (in Spanish).

DICTUC (1998) *Actualización de Encuestas Origen-Destino de Viajes*. Final Report to the Ministry of Planning, Department of Transport Engineering, Pontificia Universidad Católica de Chile, Santiago (in Spanish).

DICTUC (2003) *Encuesta Origen-Destino de Viajes del Gran Santiago 2001*. Final Report to the Ministry of Planning, Department of Transport Engineering and Logistics, Pontificia Universidad Católica de Chile, Santiago (in Spanish).

Dijkstra, E.W. (1959) Note on two problems in connection with graphs (spanning tree, shortest path). *Numerical Mathematics* **1**, 269–271.

Domencich, T.A., Kraft, G. and Valette, J.P. (1968) Estimation of urban passenger travel behaviour: an economic demand model. *Highway Research Record* **238**, 64–278.

Domencich, T. and McFadden, D. (1975) *Urban Travel Demand: A Behavioural Analysis*. North-Holland, Amsterdam.

Dong, X. and Koppelman, F.S. (2004) Comparison of continuous and discrete representations of unobserved heterogeneity in logit models. *2004 Annual Meeting of the Transportation Research Board*, Washington, DC, January 2004, USA.

Douglas, A.A. and Lewis, R.J. (1970) Trip generation techniques: (1) Introduction; (2) Zonal least squares regression analysis. *Traffic Engineering and Control* **12**, 362–365, 428–431.

Douglas, A.A. and Lewis, R.J. (1971) Trip generation techniques: (3) Household least squares regression analysis; (4) Category analysis and sumary of trip generation techniques. *Traffic Engineering and Control* **12**, 477–479, 532–535.

Downes, J.D. and Emmerson, P. (1983) *Urban transport modelling with fixed travel budgets; an evaluation of the UMOT process*. TRRL Supplementary Report SR 799, Transport and Road Research Laboratory, Crowthorne.

Downes, J.D. and Gyenes, L. (1976) *Temporal stability and forecasting ability of trip generation models in Reading*. TRRL Report LR 726, Transport and Road Research Laboratory, Crowthorne.

DRCOG (2000) *Describing and Reaching Non-Responding Populations. Denver Regional* Council of Governments (http://www.drcog.org/pub_news/about_pub_news.htm).

Duncan, G.J., Juster, F.T. and Morgan, J.N. (1987) The role of panel studies in research on economic behaviour. *Transportation Research* **21A**, 249–263.

Dunphy, R.T. (1979) Workplace interviews as an efficient source of travel survey data. *Transportation Research Record* **701**, 26–29.

Echeñique, M.H., Flowerdew, A.D.J., Hunt, J.D., Mayo, T.R., Skidmore, I.J. and Simmonds, D.C. (1990) The MEPLAN models of Bilbao, Leeds and Dortmund. *Transport Reviews* **10**, 309–322.

ECMT (1996) *The Valuation of Environmental Externalities*. European Conference of Ministers of Transport, Paris.

ECMT (1998) *Efficient Transport for Europe: Policies for Internalisation of External Costs*. European Conference of Ministers of Transport, Paris.

Economic Software, Inc. (1995) *LIMDEP, Version 7.0*. Bellport, NY.

Eilon, S. (1972) Goals and constraints in decision making. *Operations Research Quarterly* **23**, 3–15.

England, J., Hudson, K., Masters, R., Powell, K. and Shortridge, J. (eds.) (1985) *Information Systems for Policy Planning in Local Government*. Longman, Harlow.

Erlander, S. and Stewart, N.F. (1990) *The Gravity Model in Transportation Analysis: Theory and Extensions*. VSP, Utrecht.

Espino, R., Ortúzar, J. de D. and Román, C. (2006) Confidence interval for willingness-to-pay measures in mode choice models. *Networks and Spatial Economics* **6**, 81–96.

ESTRAUS (1989) *Estudio Estratégico de Transporte del Gran Santiago*. Final Report to the Executive Secretariat of the Urban Transport Commission, Consorcio SIGDO-KOPPERS/CIS, Santiago (in Spanish).

Ettema, D., Gunn, H., De Jong, G. and Lindveld, K. (1997) A simulation method for determining the confidence interval of a weighted group average value of time. *Proceedings 25th European Transport Forum*, Brunel University, September 1997, England.

Evans, A. (1972) On the theory of the valuation and allocation of time. *Scottish Journal of Political Economy* **19**, 1–17.

Evans, S.P. (1973) A relationship between the gravity model for trip distribution and the transportation problem in linear programming. *Transportation Research* **7**, 39–61.

Evans, S.P. (1976) Derivation and analysis of some models for combining trip distribution and assignment. *Transportation Research* **10**, 37–57.

Evans, S.P. and Kirby, H.R. (1974). A three dimensional Furness procedure for calibrating gravity models. *Transportation Research* **8**, 105–122.

Fang, S.C. and Tsao, S.J. (1995) Linearly-constrained entropy maximization problem with quadratic cost and its applications to transportation planning problems. *Transportation Science* **29**, 353–365.

Feil, M., Balmer, M. and Axhausen, K.W. (2009) Generating comprehensive all-day schedules: expanding activity-based travel demand modelling. *2009 European Transport Conference*, Leeuwenhorst, October 2009, The Netherlands.

Feitelson, E., Hurd, R. and Mudge, R. (1996) The impact of airport noise on willingness-to-pay for residences. *Transportation Research* **1D**, 1–14.

Fernández, J.E., Coeymans, J.E. and Ortúzar, J. de D. (1983) Evaluating extensions to the Santiago underground system. *Proceedings 11th PTRC Summer Annual Meeting*, University of Sussex, July 1983, England.

Fernández, J.E. and Friesz, T.L. (1983) Equilibrium predictions in transportation markets: the state of the art. *Transportation Research* **17B**, 155–172.

Ferrini, S. and Scarpa, R. (2007) Designs with a-priori information for nonmarket valuation with choice-experiments: a Monte Carlo study. *Journal of Environmental Economics and Management* **53**, 342–363.

FHWA (1967) *Guidelines for Trip Generation Analysis. Federal Highway Administration*, US Department of Transportation, Washington, DC.

Fieller, E. (1933) The distribution of the index in a Normal bivariate population. *Biometrika* **24**, 428–440.

Fisk, C.S. (1988) On combining maximum entropy trip matrix estimation with user optimal assignment. *Transportation Research* **22B**, 69–73.

Florian, M. and Nguyen, S. (1978) A combined trip distribution modal split and trip assignment model. *Transportation Research* **12**, 241–246.

Florian, M. and Spiess, H. (1982) The convergence of diagonalization algorithms for asymmetric network equilibrium problems. *Transportation Research* **16B**, 477–484.

Florian, M., Wu, J.H. and He, S. (1999) A multi-class multi-mode variable demand network equilibrium model with hierarchical logit structures. In J.E. Coeymans and P. Sommariva (eds.), *Actas del IX Congreso Chileno de Ingeniería de Transporte*. Sociedad Chilena de Ingeniería de Transporte, Santiago.

Flyvbjerg, B., Skamris Holm, M.K. and Buhl, S.L. (2005) How (in)accurate are demand forecasts in public works projects? *Journal of the American Planning Association* **71**, 131–146.

Foerster, J.F. (1979) Mode choice decision process models: a comparison of compensatory and non-compensatory structures. *Transportation Research* **13A**, 17–28.

Foerster, J.F. (1981) Nonlinear and non-compensatory perceptual functions of evaluations and choice. In P.R. Stopher, A.H. Meyburg and W. Brög (eds.), *New Horizons in Travel Behaviour Research*. D.C. Heath and Co., Lexington, Mass.

Foot, D. (1981) *Operational Urban Models*. Methuen, London.

Forrester, J. (1969) *Urban Dynamics*. Productivity Press, Portland.

Fosgerau, M. and Bierlaire, M. (2007) A practical test for the choice of mixing distribution in discrete choice models. *Transportation Research* **41B**, 784–794.

Fosgerau, M. and Hess, S. (2010) A comparison of methods for representing random taste heterogeneity in discrete choice models *European Transport* **42**, 1–25.

Fowkes, A.S. (2000) Recent developments in stated preference techniques in transport research. In J. de D. Ortúzar (ed.), *Stated Preference Modelling Techniques*. Perspectives 4, PTRC, London.

Fowkes, A.S. and Tweddle, G. (2000) Validation of stated preference forecasting: a case study involving anglo-continental freight. In J. de D. Ortúzar (ed.), *Stated Preference Modelling Techniques*. Perspectives 4, PTRC, London.

Fowkes, A.S. and Wardman, M. (1988) The design of stated preference travel choice experiments, with special reference to interpersonal taste variations. *Journal of Transport Economics and Policy* **22**, 27–44.

Frank, M. and Wolfe, P. (1956) An algorithm for quadratic programming, *Naval Research Logistics Quarterly* **3**, 95–110.

Friedrich, M., Haupt, T. and Nökel, K. (2003) Freight modelling: data issues, survey methods, demand and network models. *10th International Conference on Travel Behaviour Research*, Lucerne, August 2003, Switzerland.

Friedrich, M., Hofsäss, I. and Wekeck, S. (2001) Timetable-based transit assignment using branch & bound techniques. *Transportation Research Record* **1752**, 100–107.

Friedrich, R., Bickel, P. and Krewitt, W. (eds.) (1998) *External Costs of Transport*. IER, Universität Stuttgart.

Friesz, T.L., Bernstein, D., Smith, T.E., Tobin, R.L and Wie, B.V. (1993) A variational inequality formulation of the dynamic network user equilibrium problem. *Operations Research* **41**, 179–191.

Friesz, T.L., Tobin, R. and Harker, P. (1983) Predictive intercity freight network models: the state of the art. *Transportation Research* **17A**, 409–417.

Furness, K.P. (1965) Time function iteration. *Traffic Engineering and Control* **7**, 458–460.

Galbraith, R.A. and Hensher, D.A. (1982) Intra-metropolitan transferability of mode choice models. *Journal of Transport Economics and Policy* **16**, 7–29.

Galilea, P. and Ortúzar, J. de D. (2005) Valuing noise level reductions in a residential location context. *Transportation Research* **10D**, 305–322.

Gálvez, T. and Jara-Díaz, S.R. (1998) On the social valuation of travel time savings. *International Journal of Transport Economics* **25**, 205–219.

Gartner, N.H. (1980) Optimal traffic assignment with elastic demands: a review. *Transportation Science* **14**, 192–208.

Gaudry, M.J.I., Jara-Díaz, S.R. and Ortúzar, J. de D. (1989) Value of time sensitivity to model specification. *Transportation Research* **23B**, 151–158.

Gaudry, M.J.I. and Wills, M.I. (1978) Estimating the functional form of travel demand models. *Transportation Research* **12**, 257–289.

Geman, S. and Geman, D. (1984) Stochastic relaxation, Gibbs distribution and the Bayesian restoration of images. *IEEE Transaction on Pattern Analysis and Machine Intelligence* **6**, 721–741.

Gibson, J., Baeza, I. and Willumsen, L.G. (1989) Congestion, bus stops and congested bus stops. *Traffic Engineering and Control* **30**, 291–296.

Glaister, S. (1999) Observations on the new approach to the appraisal of road projects. *Journal of Transport Economics and Policy* **33**, 227–234.

Godoy, G. and Ortúzar, J. de D. (2008) On the estimation of mixed logit models. In P.O. Inweldi (ed.), *Transportation Research Trends*. Nova Science Publishers, New York.

Goldenberg, X. (1996) Choosing a household survey method: results for the Dallas-Fort Worth pretest. *Fourth International Conference on Survey Methods in Transport*, Oxford, July 1996, England.

Golledge, R.G., Kwan, M.P. and Gärling, T. (1994) Computational-process modelling of household travel decisions using a geographical information system. *Papers of the Regional Science Association* **73**, 99–117.

Golob, T.F., Kitamura, R. and Supernak, J. (1997) A panel-based evaluation of the San Diego I-15 carpool lanes project. In T.F. Golob, R. Kitamura and L. Long (eds.), *Panels for Transportation Planning: Methods and Applications*. Kluwer, Boston.

Golob, T.F. and Richardson, A.J. (1981) Non-compensatory and discontinuous constructs in travel behaviour models. In P.R. Stopher, A.H. Meyburg and W. Brög (eds.), *New Horizons in Travel Behaviour Research*. D.C. Heath and Co., Lexington, Mass.

Goodwin, P.W. (1977) Habit and hysteresis in mode choice. *Urban Studies* **14**, 95–98.

Goodman, P.R. (1973) Trip generation: a review of the category analysis and regression models. *Working Paper 9*, Institute of Transport Studies, Leeds University.

Goos, P. (2002) *The Optimal Design of Blocked and Split-Plot Experiments*. Lecture Notes in Statistics, Springer-Verlag, New York.

Gopinath, D.A. and Ben-Akiva, M. (1995) Estimation of randomly distributed values of time. *Working Paper*, Department of Civil and Environmental Engineering, MIT.

Gray, R. (1982) Behavioural approaches in freight transport modal choice. *Transport Reviews* **2**, 161–184.

Greenblat, C. and Duke, R. (1975) *Gaming-Simulation: Rationale, Design and Applications*. John Wiley & Sons, Inc. New York.

Greene, W.H. (2003) *Econometric Analysis*. Prentice Hall, Upper Saddle River, NJ.

Greene, W.H., Hensher, D.A. and Rose, J.M. (2006) Accounting for heterogeneity in the variance of the unobserved effects in mixed logit models. *Transportation Research* **40B**, 75–92.

Guevara, C.A. and Ben-Akiva, M. (2006) Endogeneity in residential location choice models. *Transportation Research Record* **1977**, 60–66.

Guevara, C.A. and Thomas, A. (2007) Multiple classification analysis in trip production models. *Transport Policy* **14**, 514–522.

Gunn, H.F. (1984) An analysis of transfer price data. *Proceedings 12th PTRC Summer Annual Meeting*, University of Sussex, July 1984, England.

Gunn, H.F. (1985a) Value of time for evaluation purposes: the state of the art. *Report No. 421–01*, Hague Consulting Group, The Hague.

Gunn, H.F. (1985b) Artificial sample applications for spatial interaction models. *Colloquium Vervoersplanologisch Speurwerk*, The Hague, November 1985, Holland.

Gunn, H.F. and Bates, J.J. (1982) Statistical aspects of travel demand modelling. *Transportation Research* **16A**, 371–382.

Gunn, H.F., Ben-Akiva, M.E. and Bradley, M.A. (1985) Tests of the scaling approach to transferring disaggregate travel demand models. *Transportation Research Record* **1037**, 21–30.

Gunn, H.F., Fisher, P., Daly, A.J. and Pol, H. (1982) Synthetic samples as a basis for enumerating disaggregate models. *Proceedings 10th PTRC Summer Annual Meeting*, University of Warwick, July 1982, England.

Gunn, H.F., Kirby, H.R., Murchland, J.D. and Whittaker, J.C. (1980) The RHTM trip distribution investigation. *Proceedings 8th PTRC Summer Annual Meeting*, University of Warwick, July 1980, England.

Gunn, H.F. and Pol, H. (1986) Model transferability: the potential for increasing cost-effectiveness. In A. Ruhl (ed.), *Behavioural Research for Transport Policy*. VNU Science Press, Utrecht.

Guo, J.Y. and Bhat, C.R. (2007) Population synthesis for microsimulating travel behaviour. *Transportation Research Record* **2014**, 92–101.

Gur, Y.J. (1982) Recalibration of disaggregate mode choice models based on on-board survey data. *Proceedings 10th PTRC Summer Annual Meeting*, University of Warwick, July 1982, England.

Hägerstrand, T. (1970) What about people in regional science? *Papers of the Regional Science Association* **24**, 7–21.

Hajivassiliou, V.A. and Ruud, P. (1994) Classical estimation methods for LDV models using simulation. In R. Engle and D. McFadden (eds.), *Handbook of Econometrics*, Vol. IV. Elsevier Science, Amsterdam.

Hall, M.D., Daly, A.J., Davies, R.F. and Russell, C.H. (1987) Modelling for an expanding city. *Proceedings 15th PTRC Summer Annual Meeting*, University of Bath, September 1987, England.

Hall, M.D., Van Vliet, D. and Willumsen, L.G. (1980) SATURN – a simulation assignment model for the evaluation of traffic management schemes. *Traffic Engineering and Control* **21**, 168–176.

Han, S.J. and Heydecker, B.G. (2006) Consistent objective and solutions of dynamic user equilibrium models. *Transportation Research* **40B**, 16–34.

Hanemann, W.M. (1994) Valuing the environment through contingent valuation. *Journal of Economic Perspectives* **8**, 19–43.

Hahn, G.J. and Shapiro, S.S. (1966) A catalogue and computer program for the design and analysis of orthogonal symmetric and asymmetric fractional factorial experiments. General Electric Research and Development Centre, Schenectady, NY.

Hansson, L. and Markham, J. (1992) *Internalisation of External Costs in Transportation*. IRU, Paris.

Harker, P.T. (1985) The state of the art in the predictive analysis of freight transport systems. *Transport Reviews* **5**, 143–164.

Harker, P.T. (1987) *Predicting Intercity Freight Flows*. VNU Science Press, Utrecht.

Hartley, T.M. and Ortúzar, J. de D. (1980) Aggregate modal split models: is current U.K. practice warranted? *Traffic Engineering and Control* **21**, 7–13.

Hauser, J.R. (1978) Testing the accuracy, usefulness and significance of probabilistic choice models: an information theoretic approach. *Operations Research* **26**, 406–421.

Hausman, J.A. (ed.) (1993) *Contingent Valuation: A Critical Assessment*, North-Holland, Amsterdam.

HCG (1990) *The Netherlands Value of Time Study*. Final Report to the Dienst Verkeerskunde, Rijkswaterstaat. Hague Consulting Group, The Hague.

HCG, Halcrow Fox and Imperial College London (2000) *User Manual for HADES 1.0*. Prepared for the Department of the Environment, Transport and the Regions London.

He, Y.P, Xu, J.P., Huang, N.J and Wu, M. (2010) Dynamic traffic network equilibrium system. *Fixed Point Theory and Applications* **2010**, Article ID 873025.

Heckman, J.J. (1981) Statistical models for discrete panel data. In C.F. Manski and D. McFadden (eds.), *Structural Analysis of Discrete Data: With Econometric Applications*. MIT Press, Cambridge, Mass.

Hedayat, A.S., Sloane, N.J.A. and Stufken, J. (1999) *Orthogonal Arrays: Theory and Applications*. Springer-Verlag, New York.

Heggie, I.G. (1983) Valueing savings in non working time: the empirical dilemma. *Transportation Research* **17A**, 13–23.

Hendrickson, C. and Kocur, G. (1981) Schedule delay and departure time decisions in a deterministic model. *Transportation Science* **15**, 62–77.

Hendrickson, C. and Plank, E. (1984) The flexibility of departure times for work trips. *Transportation Research* **18A**, 25–36.

Hensher, D.A. (1976) Valuations of commuter travel time savings: an alternative procedure. In I.G. Heggie (ed.), *Modal Choice and the Value of Time*. Clarendon Press, Oxford.

Hensher, D.A. (1987) Issues in the pre-analysis of panel data. *Transportation Research* **21A**, 265–285.

Hensher D.A. (1999) HEV choice models as a search engine for the specification of nested logit tree structures. *Marketing Letters* **10**, 339–349.

Hensher, D.A. (2001a) The valuation of commuter travel time savings for car drivers: evaluating alternative model specifications. *Transportation* **28**, 101–118.

Hensher, D.A. (2001b) Measurement of the valuation of travel time savings. *Journal of Transport Economics and Policy* **35**, 71–98.

Hensher, D.A. (2001c) The sensitivity of the valuation of travel time savings to the specification of unobserved effects. *Transportation Research* **37E**, 129–142.

Hensher, D.A. (2006) The signs of times: imposing a globally signed condition on willingness-to-pay distributions. *Transportation* **33**, 205–222.

Hensher, D.A. and Greene, W.H. (2002) Specification and estimation of the nested logit model: alternative normalizations. *Transportation Research* **36B**, 1–17.

Hensher, D.A. and Greene, W.H. (2003) The mixed logit model: the state of practice. *Transportation* **30**, 133–176.

Hensher, D.A. and Louviere, J.J. (1998) A comparison of elasticities derived from multinomial logit, nested logit and heteroskedastic extreme value SP-RP discrete choice models. *Proceedings 8th World Conference of Transportation Research*, Antwerp, July 1998, Belgium.

Hensher, D.A. (2010) A practical note on calculating generalised cost when there are two cost parameters in a utility expression. Institute of Transport and Logistics Studies, University of Sydney.

Hensher, D.A., Rose, J.M. and Greene, W.H. (2005) *Applied Choice Analysis: A Primer*. Cambridge University Press, Cambridge.

Hensher, D.A., Rose, J.M., Ortúzar, J. de D. and Rizzi, L.I. (2009) Estimating the willingness-to-pay and value of risk reduction for car occupants in the road environment. *Transportation Research* **43A**, 692–707.

Herriges, J.A. and Kling, C.L. (1995) An empirical investigation of the consistency of nested logit models with utility maximization. *Economic Letters* **50**, 33–39.

Herriges, J.A. and Kling, C.L. (1996) Testing the consistency of nested logit models with utility maximization. *American Journal of Agricultural Economics* **77**, 875–884.

Hess, S., Ben-Akiva, M.E., Gopinath, D. and Walker, J.L. (2009) Taste heterogeneity, correlation and elasticities in latent class choice models. *88th Annual TRB Meeting*, Washington, DC, January 2009, USA.

Hess, S., Bierlaire, M. and Polak, J.W. (2005a) Capturing correlation and taste heterogeneity with mixed GEV models. In R. Scarpa and A. Alberini (eds.), *Application of Simulation Methods in Environmental and Resource Economics*. Springer, Dordrecht.

Hess, S., Bierlaire, M. and Polak, J.W. (2005b) Estimation of values of travel-time savings using mixed logit models. *Transportation Research* **39A**, 221–236.

Hess, S., Bierlaire, M. and Polak, J.W. (2007) A systematic comparison of continuous and discrete mixtures models. *European Transport* **37**, 35–61.

Hess, S. and Rose, J.M. (2009) Allowing for intra-respondent variations in coefficients estimated on repeated choice data. *Transportation Research* **43B**, 708–719.

Hess, S. and Train, K.E. (2010) Approximation issues in simulation-based estimation of random coefficient models. *89th Annual TRB Meeting*, Washington, DC, January 2010, USA.

Hess, S., Train, K.E. and Polak, J.W. (2006) On the use of a Modified Latin Hypercube Sampling (MLHS) approach in the estimation of a mixed logit model for vehicle choice. *Transportation Research* **40B**, 147–163.

Heydecker, B.G. (2002) Dynamic equilibrium network design. In M.A.P. Taylor (ed.), *Transportation and Traffic Theory in the 21st Century*. Pergamon, Oxford.

Heydecker, B.G. and Addison, J.G (1996) An exact expression of dynamic traffic equilibrium. In J.B. Lesort (ed.), *Transportation and Traffic Theory*. Pergamon, Oxford.

Heydecker, B.G. and Addison, J.G (1997) Stochastic and deterministic formulations of dynamic traffic assignment. *Proceedings 25th European Transport Conference*, Vol. P415, pp. 107–120, PTRC, London.

Heydecker, B.G. and Addison, J.G (2005) Analysis of dynamic traffic equilibrium with departure time choice. *Transportation Science* **39**, 39–57.

Heydecker, B.G. and Addison, J.G. (2006) Analysis of dynamic traffic assignment. *First International Symposium on Dynamic Traffic Assignment*, Leeds, June 2006, England.

Hinkley, D. (1969) On the ratio of two correlated normal random variables. *Biometrika* **56**, 635–639.

Högberg, P. (1976) Estimation of parameters in models for traffic prediction: a non-linear-regression approach. *Transportation Research* **10**, 263–265.

Hojman, P., Ortúzar, J. de D. and Rizzi, L.I. (2005) On the joint valuation of averting fatal and serious injuries in highway accidents. *Journal of Safety Research* **36**, 377–386.

Holden, D., Fowkes, A.S. and Wardman, M. (1992) Automatic stated preference design algorithms. *Proceedings 20th PTRC Summer Annual Meeting*. University of Manchester Institute of Science and Technology, September 1992, England.

Holm, J., Jensen, T., Nielsen, S., Christensen, A. Johnsen, B. and Ronby, G. (1976) Calibrating traffic models on traffic census results only. *Traffic Engineering and Control* **17**, 137–140.

Holz, J.C. and Sánchez, J.M. (2000) Estimación de costos unitarios en morbilidad y mortalidad y su aplicación para calcular los beneficios del plan de prevención y descontaminación atmosférica de la Región Metropolitana. *Working Paper*, Department of Economics, Universidad de Chile, Santiago (in Spanish).

Horowitz, J.L. (1980) Confidence intervals for the choice probabilities of the multinomial logit model. *Transportation Research Record* **728**, 23–29.

Horowitz, J.L. (1981) Sources of error and uncertainty in behavioural travel demand models. In P.R. Stopher, A.H. Meyburg and W. Brög (eds.), *New Horizons in Travel Behaviour Research*. D.C. Heath and Co., Lexington, Mass.

Horowitz, J.L. (1982) Specification tests for probabilistic choice models. *Transportation Research* **16A**, 383–394.

Horowitz, J.L. and Manski, C.F. (1998) Censoring of outcomes and regressors due to survey nonresponse: identification and estimation using weights and imputations. *Journal of Econometrics* **84**, 37–58.

Huber, J. and Train, K.E. (2001) On the similarity of classical and Bayesian estimates of individual mean partworths. *Marketing Letters* **12**, 257–267.

Huber, J. and Zwerina, K. (1996) The importance of utility balance and efficient choice designs. *Journal of Marketing Research* **33**, 307–317.

Hunt, G.L. (2000) Alternative nested logit model structures and the special case of partial degeneracy. *Journal of Regional Science* **40**, 89–113.

Hunt, J.D., Kriger, D.S. and Miller, E.J. (2005) Current operational urban land use–transport modelling frameworks: a review. *Transport Reviews* **25**, 329–376.

Hyman, G.M. (1969) The calibration of trip distribution models. *Environment and Planning* **1**, 105–112.

Hyman, G.M. (1997) The development of operational models for time period choice. *Working Paper*, HETA Division, Department of the Environment, Transport and the Regions. London.

INRO (1996) *EMME/2 User's Manual*. INRO Inc., Montreal.

Iragüen, P. and Ortúzar, J. de D. (2004) Willingness-to-pay for reducing fatal accidents risk in urban areas: an internet-based web page stated preference survey. *Accident Analysis and Prevention* **36**, 513–524.

Iglesias, P., Godoy, F.J., Ivelic, A.M. and Ortúzar, J. de D. (2008) Un modelo de generación, distribución y partición modal conjunta para viajes interurbanos. *Proceedings XIV Panamerican Congress on Traffic and Transportation Engineering*, Cartagena de Indias, September 2008, Colombia (in Spanish).

Jansen, G.R.M. and Bovy, P.H.L. (1982) The effect of zone size and network detail on all-or-nothing and equilibrium assignment outcomes. *Traffic Engineering and Control* **23**, 311–317.

Jara-Díaz, S.R. (1998) Time and income in travel choice: towards a microeconomic activity framework. In T. Garling, T. Laitila and K. Westin (eds.), *Theoretical Foundations of Travel Choice Modelling*. Pergamon, Oxford.

Jara-Díaz, S.R. (2000) Allocation and valuation of travel time savings. In D. Hensher and K.J. Button (eds.), *Handbook of Transport Modelling*, Pergamon, Oxford.

Jara-Díaz, S.R. (2007) *Transport Economic Theory*. Elsevier Science, Amsterdam.

Jara-Díaz, S.R. and Farah, M. (1987) Transport demand and user's benefits with fixed income: the goods/leisure trade-off revisited. *Transportation Research* **21B**, 165–170.

Jara-Díaz, S.R. and Guevara, A. (2000) The contribution of work, leisure and travel to the subjective value of travel time savings. *Proceedings European Transport Conference 2000*, Cambridge, September 2000, England.

Jara-Díaz, S.R. and Ortúzar, J. de D. (1989) Introducing the expenditure rate in the estimation of mode choice models. *Journal of Transport Economics and Policy* **23**, 293–308.

Jara-Díaz, S.R. Ortúzar, J. de D. and Parra, R. (1988) Valor subjectivo del tiempo considerando efecto ingreso en la partición modal. *Actas del V Congreso Panamericano de Ingeniería de Tránsito y Transporte*, Universidad de Puerto Rico en Mayagüez, July 1988, Puerto Rico (in Spanish).

Jayakrishnan, R., Tsai, W., Prashker, J. and Rajadhyaksha, S. (1994) A faster path-based algorithm for traffic assignment. *Transportation Research Record* **1443**, 75–83.

Johnson, L.W. and Hensher, D.A. (1982) Application of multinomial probit to a two-period panel data set. *Transportation Research* **16A**, 457–464.

Johnson, M. (1966) Travel time and the price of leisure. *Western Economic Journal* **8**, 135–145.

Johnson, N.L. and Kotz, S. (1972) *Distributions in Statistics: Continuous Multivariate Distributions*. John Wiley & Sons, Inc. New York.

Jones, P.M. (1979) New approaches to understanding travel behaviour: the human activity approach. In D.A. Hensher and P.R. Stopher (eds.), *Behavioural Travel Modelling*. Croom Helm, London.

Jones-Lee, M.W., Hammerton, M. and Philips, P.R. (1985) The value of safety: results of a national sample survey. *Economic Journal* **95**, 49–72.

Jones-Lee, M.W., Loomes, G., O'Reilly, D.M. and Philips, P.R. (1992) The value of preventing non-fatal road injuries: findings of a willingness-to-pay national sample survey. *TRL Contractor Report 330*, Transport Research Laboratory, Crowthorne.

Jones-Lee, M., Loomes, G. and Philips, P. (1995) Valuing the prevention of non-fatal road injuries: contingent valuation vs. standard gamble. *Oxford Economics Papers* **47**, 676–695.

Jones Lee, M., O'Reilly, D. and Philips, P. (1993) The value of preventing non-fatal road injuries: findings of a willingness to pay national sample survey. *TRL Working Paper WPSRC2*, Transport Research Laboratory, Crowthorne.

Kahneman, D. and Tversky, A. (1979) Prospect theory: an analysis of decisions under risk. *Econometrica* **47**, 263–291.

Kam, H.B. and Morris, J. (1999) Response patterns in travel surveys: the VATS experience. *Working Paper*, Transport Research Centre, RMIT, Melbourne.

Kannel, E.J. and Heathington, K.W. (1973) Temporal stability of trip generation relations. *Highway Research Record* **472**, 17–27.

Kanninen, B.J. (2002) Optimal design for multinomial choice experiments. *Journal of Marketing Research* **39**, 214–217.

Kass, R., Carlin, B., Gelman, A. and Neal, R. (1998) Markov Chain Monte Carlo in practice: a roundtable discussion. *The American Statistician* **52**, 93–100.

Kay, J. (2010) *Obliquity: Why our Goals are Best Achieved Indirectly*. Profile Books, London.

Keeter, S., Miller, A., Kohut, A., Groves, R.M. and Presser, S. (2000) Consequences of reducing non-response in a national telephone survey. *Public Opinion Quarterly* **64**, 125–148.

Kessels, R., Goos, P. and Vandebroek, M. (2006) A comparison of criteria to design efficient choice experiments, *Journal of Marketing Research* **43**, 409–419.

Khan, A. and Willumsen, L.G. (1986) Modelling car ownership and use in developing countries. *Traffic Engineering and Control* **27**, 554–560.

Kim, H., Li, J., Roodman, S., Sen, A., Sööt, S. and Christopher, E. (1993) Factoring household travel surveys. *Transportation Research Record* **1412**, 17–22.

Kim, T.J. and Hinkle, J. (1982) Model for statewide freight transportation planning. *Transportation Research Record* **889**, 15–19.

Kimber, R.M. and Hollis, E.M. (1979) Traffic queues and delays at road junctions. *TRRL Report LR 909*, Transport and Road Research Laboratory, Crowthorne.

Kirby, H.R. (1979) Partial matrix techniques. *Traffic Engineering and Control* **20**, 422–428.

Kitamura, R. (1990a) Panel analysis in transportation planning: an overview. *Transportation Research* **24A**, 401–415.

Kitamura, R. (1990b) Longitudinal surveys. In E.S. Ampt, A.J. Richardson and A.H. Meyburg (eds), *Selected Readings in Transport Survey Methodology*. Eucalyptus Press, Melbourne.

Kitamura, R. and Bovy, P.H.L. (1987) Analysis of attrition biases and trip reporting errors for panel data. *Transportation Research* **21A**, 287–302.

Kitamura, R. and Fujii, S. (1998) Two computational process models of activity-travel behaviour. In T. Gärling, T. Laitila and K. Westin (eds.), *Theoretical Foundations of Travel Choice Modelling*, Pergamon, Oxford.

Koppelman, F.S. (1976) Guidelines for aggregate travel prediction using disaggregate choice models. *Transportation Research Record* **610**, 19–24.

Koppelman, F.S., Kuah, G-K and Rose G. (1985a) Transfer model updating with aggregate data. *64th Annual TRB Meeting*, Washington, DC, January 1985, USA.

Koppelman, F.S., Kuah, G-K and Wilmot, C.G. (1985b) Transfer model updating using disaggregate data. *Transportation Research Record* **1037**, 102–107.

Koppelman F.S., Sethi, V. and Wen, C.H. (2001) Alternative nested logit models: a response to comments by Andrew Daly on an earlier paper by Frank Koppelman and Chieh-Hua-Wen. *Transportation Research* **35B**, 725–729.

Koppelman, F.S. and Wen, C.H. (1998a) Alternative nested logit models: structure, properties and estimation. *Transportation Research* **32A**, 289–298.

Koppelman, F.S. and Wen, C.H. (1998b) Different nested logit models: which are you using? *Transportation Research Record* **1645**, 1–7.

Koppelman, F.S. and Wen, C.H. (2000) The paired combination logit model: properties, estimation and application. *Transportation Research* **34B**, 75–89.

Koppelman, F.S. and Wilmot, C.G. (1982) Transferability analysis of disaggregate choice models. *Transportation Research Record* **895**, 18–24.

Kraft, G. (1968) *Demand for Intercity Passenger Travel in the Washington-Boston Corridor*. North-East Corridor Project Report, Systems Analysis and Research Corporation, Boston, Mass.

Kresge, D.T. and Roberts, P.O. (1971) *Techniques of Transport Planning: Systems Analysis and Simulation Models*. Brookings Institution, Washington, DC.

Kroes, E.P., Daly, A.J. Gunn, H.F. and van der Hoorn, A.I.J.M. (1996) The opening of the Amsterdam ring road: a case study on short-term effects of removing a bottleneck. *Transportation* **23**, 71–82.

Kruithof, J. (1937) Calculation of telephone traffic. *Der Ingenieur* **52**, E15–E25.

Kruskal, J.B. (1965) Analysis of factorial experiments by estimating monotone transformations of the data. *Journal of the Royal Statistical Society* **27B**, 251–263.

Kumar, A. (1980) Use of incremental form of logit models in demand analysis. *Transportation Research Record* **775**, 21–27.

Kurri, J. and Pursula, M. (1995) Finish preliminary value of time studies. *Working Paper*, Laboratory of Transportation Engineering, Helsinki University of Technology.

Kurth, D.L., Coil, J.L. and Brown, M.J. (2001) Assessment of quick-refusal and no-contact non-response in household travel surveys. *Transportation Research Record* **1768**, 114–124.

Lamb, G.M. and Havers, G.E. (1970) Introduction to transportation planning: treatment of networks. *Traffic Engineering and Control* **11**, 486–489.

Lamond, B. and Stewart, N.F. (1981) Bregman's balancing method. *Transportation Research* **15B**, 239–248.

Lancaster, K.J. (1966) A new approach to consumer theory. *Journal of Political Economy* **14**, 132–157.

Landefeld, J.S. and Seskin, E.P. (1982) The economic value of life: linking theory to practice. *American Journal of Public Health* **72**, 555–566.

Langdon, M.G. (1976) Modal split models for more than two modes. *Proceedings 4th PTRC Summer Annual Meeting*, University of Warwick, July 1976, England.

Langdon, M.G. (1984) Methods of determining choice probability in utility maximising multiple alternative models. *Transportation Research* **18B**, 209–234.

Lange, K.L., Little, R.J. and Taylor, J.M. (1989) Robust statistical modelling using the t distribution. *Journal of the American Statistical Association* **84**, 881–896.

Lanzendorf, M. (2003) Mobility biographies: a new perspective for understanding travel behaviour. *10th International Conference on Travel Behaviour Research*. Lucern, August 2003, Switzerland.

Larson, R.C. and Odoni, A.R. (1981) *Urban Operations Research*. Prentice Hall, Englewood Cliffs, NJ.

Larsson, T. and Patriksson, M. (1992) Simplicial decomposition with disaggregated representation for the traffic assignment problem. *Transportation Science* **26**, 4–17.

Leamer, E. (1978) *Specification Searches: Ad-Hoc Inference with Nonexperimental Data*. John Wiley & Sons, Inc. New York.

Lee, N. and Dalvi, M.Q. (1969) Variations on the value of travel time. *Manchester School* **37**, 213–236.

Lee, N. and Kirkpatrick, C. (1996) Relevance and consistency of environmental impact assessment and cost-benefit analysis in project appraisal. *Project Appraisal* **11**, 229–236.

Leonard, D.R. and Gower, P. (1982) User guide to CONTRAM Version 4. *TRRL Supplementary Report 735*, Transport and Road Research Laboratory, Crowthorne.

Leonard, D.R. and Tough, J. (1979) Validation work on CONTRAM – a model for use in the design of traffic management schemes. *Proceedings 7th PTRC Summer Annual Meeting*, University of Warwick, July 1979, England.

Lerman, S.R. (1984) Recent advances in disaggregate demand modelling. In M. Florian (ed.), *Transportation Planning Models*. North-Holland, Amsterdam.

Lerman, S.R. and Louviere, J.J. (1978) The use of functional measurement to identify the form of utility functions in travel demand models. *Transportation Research Record* **673**, 78–85.

Lerman, S.R. and Manski, C.F. (1976) Alternative sampling procedures for calibrating disaggregate choice models. *Transportation Research Record* **592**, 24–28.

Lerman, S.R. and Manski, C.F. (1979) Sample design for discrete choice analysis of travel behaviour: the state of the art. *Transportation Research* **13B**, 29–44.

Lerman, S.R. and Manski, C.F. (1981) On the use of simulated frequencies to approximate choice probabilities. In C. Manski. and D. McFadden (eds.), *Structural Analysis of Discrete Data with Econometric Applications*. MIT Press, Cambridge, Mass.

Lerman, S.R., Manski, C.F. and Atherton, T.J. (1976) *Non-Random Sampling in the Calibration of Disaggregate Choice Models*. Final Report to the Urban Planning Division, Federal Highway Administration, US Department of Transportation, Washington, DC.

Leurent, F.M. (1998) Multicriteria assignment modelling: making explicit the determinants of mode or path choice. In P. Marcotte and S. Nguyen (eds.), *Equilibrium and Advanced Transportation Modelling*. Kluwer Academic, Boston.

Li, Z., Hensher, D.A. and Rose, J.M. (2010) Willingness to pay for travel time reliability in passenger transport: a review and some new empirical evidence. *Transportation Research* **46E**, 384–403.

Lieberman, E. (1981) Enhanced NETSIM program. *Transportation Research Board Special Report* **194**, 32–35.

Liem, T.C. and Gaudry, M.J.I. (1987) P-2: A program for the Box–Cox logit model with disaggregate data. *Publication 525*, Centre de Recherche sur les Transports, Université de Montréal.

Lindblom, C.E. (1959) The science of "muddling through". *Public Administration Review* **19**, 79–88.

Lindquist, J. and Algers, S. (1998) Further research on the Swedish national value of time study. *Proceedings 8th World Conference on Transport Research*, Antwerp, July 1998, Belgium.

Litman, T. (1995) Transportation cost analysis: techniques, estimates and implications. *Working Paper*, Victoria Transport Policy Institute, Victoria, BC.

Louviere, J.J. (1988a) Conjoint analysis modelling of stated preferences: a review of theory, methods, recent developments and external validity. *Journal of Transport Economics and Policy* **22**, 93–119.

Louviere, J.J. (1988b) *Analysing Decision Making: Metric Conjoint Analysis*. Sage Publications, Newbury Park.

Louviere, J.J., Hensher, D.A. and Swait, J.D. (2000) *Stated Choice Methods: Analysis and Application*. Cambridge University Press, Cambridge.

Louviere, J.J. and Lancsar, E. (2009) Choice experiments in health: the good, the bad, the ugly and toward a brighter future. *Health Economics, Policy and Law* **4**, 527–546.

Low, D.E. (1972) A new approach to transportation systems modelling. *Traffic Quarterly* **26**, 391–404.

Lowry, I.S. (1965) A model of a metropolis. *Technical Memorandum RM-4035-RC*, The Rand Corporation, California.

Luce, R.D. and Suppes, P. (1965) Preference, utility and subjective probability. In R.D. Luce, R.R. Bush and E. Galanter (eds.), *Handbook of Mathematical Psychology*. John Wiley & Sons, Inc. New York.

Mackett, R.L. (1985) Integrated land use-transport models. *Transport Reviews* **5**, 325–343.

Mackett, R.L. (1990) Comparative analysis of modelling land-use transport interaction at the micro and macro levels. *Environment and Planning* **22A**, 459–475.

Mackinder, I.H., and Evans, S.E. (1981) The predictive accuracy of British transport studies in urban areas. *TRRL Supplementary Report SR 699*, Transport and Road Research Laboratory, Crowthorne.

Maddison, D., Pearce, D., Johansson, O., Calthrop, E., Litman, T. and Verhoef, E. (1996) *The True Cost of Road Transport*. Earthscan Publications, London.

Maher, M.J. (1983) Inferences on trip matrices from observations on link volumes: a Bayesian statistical approach. *Transportation Research* **17B**, 435–447.

Mahmassani, H.S. (2000) Trip timing. In D. Hensher and K.J. Button (eds.), *Handbook of Transport Modelling*. Pergamon, Oxford.

Mahmassani, H.S. and Sinha, K.C. (1981) Bayesian updating of trip generation parameters. *Transportation Engineering Journal* **107**, 581–589.

Manheim, C.F. (1973) Practical implications of some fundamental properties of travel demand models. *Highway Research Record* **244**, 21–38.

Manheim, M.L. (1979) *Fundamentals of Transportation Systems Analysis*. MIT Press, Cambridge, Mass.

Mann, H.B. and Wald, A. (1943) On stochastic limit and order relationships. *The Annals of Mathematical Statistics* **14**, 217–226.

Manski, C.F. and Lerman, S.R. (1977) The estimation of choice probabilities from choice based samples. *Econometrica* **45**, 1977–1988.

Manski, C.F. and McFadden, D. (1981) Alternative estimators and sample designs for discrete choice analysis. In C.F. Manski and D. McFadden (eds.), *Structural Analysis of Discrete Data: With Econometric Applications*. MIT Press, Cambridge, Mass.

Martínez, F.J. (1987) La forma incremental del modelo logit: aplicaciones. *Actas del III Congreso Chileno de Ingeniería de Transporte*, Universidad de Concepción, November 1987, Chile (in Spanish).

Martínez, F.J. (1992) The bid-choice land-use model: an integrated economic framework. *Environment and Planning* **24A**, 871–875.

Martinez, F.J., Aguila, F. and Hurtubia, R. (2009) The constrained multinomial logit model: a semi-compensatory choice model, *Transportation Research* **43B**, 365–377.

Mauch, S.P. and Rothengatter, W. (1995) *External Effects of Transport*. Union International des Chemins de Fer, Paris.

Mayberry, J.P. (1973) Structural requirements for abstract-mode models of passenger transportation. In R.E. Quandt (ed.), *The Demand for Travel: Theory and Measurement*. D.C. Health and Co., Lexington, Mass.

McDonald, K.G. and Stopher, P.R. (1983) Some contrary indications for the use of household structure in trip generation analysis. *Transportation Research Record* **944**, 92–100.

McFadden, D. (1974) Conditional logit analysis of qualitative choice behaviour. In P. Zarembka (ed.), *Frontiers in Econometrics*. Academic Press, New York.

McFadden, D. (1978) Modelling the choice of residential location. In A. Karlquist, L. Lundquist, F. Snickars and J.W. Weibull (eds.), *Spatial Interaction Theory and Planning Models*. North-Holland, Amsterdam.

McFadden, D. (1981) Econometric models of probabilistic choice. In C. Manski and D. McFadden (eds.), *Structural Analysis of Discrete Data with Econometric Applications*. The MIT Press, Cambridge, Mass.

McFadden, D. (1986) The choice theory approach to market research. *Marketing Science* **5**, 275–297.

McFadden, D. (1989) A method of simulated moments for estimation of discrete response models without numerical integration. *Econometrica* **57**, 995–1026.

McFadden, D. and Reid, F.A. (1975) Aggregate travel demand forecasting from disaggregate behavioural models. *Transportation Research Record* **534**, 24–37.

McFadden, D. and Train, K. (2000) Mixed MNL models for discrete response. *Journal of Applied Econometrics* **15**, 447–470.

McKelvey, R.D. and Zavoina W. (1975) A statistical model for the analysis of ordinal level dependent variables. *Journal of Mathematical Sociology* **4**, 103–120.

McLeod, W.T. and Hanks, P. (eds.) (1986) *The New Collins Concise Dictionary of the English Language*. William Collins, Sons & Co., Glasgow.

McLoughlin, J. (1969) *Urban and Regional Planning: A Systems Approach*. Faber & Faber, London.

McNeil, S. and Hendrickson, C. (1985) A regression formulation of the matrix estimation problem. *Transportation Science* **19**, 278–292.

Meijer, E. and Rouwendal, J. (2000) Measuring welfare effects in models with random coefficients. *Research Report No. 00F25*, SOM Research School, University of Groningen.

Menon, A.P.G., Lam, S.H. and Fan, H.S.L. (1993) Singapore's road pricing system: its past, present and future. *ITE Journal* **63**, 44–48.

Meyer, R.J., Levin, I.P. and Louviere, J.J. (1978) Functional analysis of mode choice. *Transportation Research Record* **673**, 1–7.

Miller, T. (2000) Variations between countries in values of statistical life. *Journal of Transport Economics and Policy* **34**, 169–188.

Mirchandani, P. and Soroush, H. (1987) Generalized traffic equilibrium with probabilistic travel times and perceptions. *Transportation Science* **21**, 133–152.

Mitchell, R.B. and Rapkin, C. (1954) *Urban Traffic: A Function of Land Use*. Columbia University Press, New York.

Mitchell, R.C. and Carson, R.T. (1989) *Using Surveys to Value Public Goods: The Contingent Valuation Method*. Resources for the Future, Washington, DC.

Moavenzadeh, F., Markow, M., Brademeyer, B. and Safwat, K. (1983) A methodology for intercity transportation planning in Egypt. *Transportation Research* **17A**, 481–491.

Mogridge, M.J.H. (1983) *The Car Market*. Pion, London.

Mogridge, M.J.H. (1986) Road pricing: the right solution for the right problem? *Transportation Research* **20A**, 157–167.

Moore, E.F. (1957) The shortest path through a maze. *Proceedings International Symposium on the Theory of Switching*. Harvard University Press, Cambridge, Mass.

Morikawa, T. (1996) A hybrid probabilistic choice set model with compensatory and non-compensatory choice rules. In D.A. Hensher, J. King and T. Oum (eds.), *World Transport Research*, World Conference on Transport Research Society, Sydney.

Morikawa, T., Ben-Akiva, M. and Yamada, K. (1992) Estimation of mode choice models with serially correlated RP and SP data. *Proceedings 6th World Conference on Transport Research*, Lyon, June 1992, France.

Morikawa, T. and Sasaki, K. (1998) Discrete choice models with latent variables using subjective data. In J. de D. Ortúzar, D.A. Hensher and S.R. Jara-Diaz (eds.), *Travel Behaviour Research: Updating the State of Play*. Pergamon, Oxford.

Morley, R. (1972) *Mathematics for Modern Economics*. Fontana, London.

Morlidge, S. and Player, S. (2010) *Future Ready: How to Master Business Forecasting* John Wiley and Sons, Ltd Chichester.

Moser, C.A. and Kalton, G.K. (1985) *Survey Methods in Social Investigation*. Ashgate, Farnham.

Mullen, S. (1997) Determining monetary values of environmental impacts – a DETR perspective. Presented at *Determining Monetary Values of Environmental Impacts*, University of Westminster, October 1997, London.

Muller, R.H. (1996) Examining tollroad feasibility studies. *Municipal Market Monitor*, J.P. Morgan Securities, Inc., New York.

Munizaga, M.A. and Alvarez-Daziano, R. (2000) Modelos mixed logit: uso y potencialidades. In L.A. Lindau, J. de D. Ortúzar and O. Strambi (eds.), *Engenharia de Tráfego e Transportes 2000: Avanços para uma Era de Mudanças*. ANPET, Rio de Janeiro (in Spanish).

Munizaga, M.A. and Alvarez-Daziano, R. (2005) Testing mixed logit and probit models by simulation. *Transportation Research Record* **1921**, 53–62.

Munizaga, M.A., Correia, R., Jara-Díaz, S.R. and Ortúzar, J. de D. (2006) Valuing time with a joint mode choice-activity model. *International Journal of Transport Economics* **33**, 69–86.

Munizaga, M.A., Heydecker, B.G. and Ortúzar, J. de D. (2000) Representation of heteroskedasticity in discrete choice models. *Transportation Research* **34B**, 219–240.

Muñoz, J.C., Ortúzar, J. de D. and Gschwender, A. (2009) Transantiago: the fall and rise of a radical public transport intervention. En W. Saleh and G. Sammer (eds.), *Travel Demand Management and Road User Pricing: Success, Failure and Feasibility*. Ashgate, Farnham.

Murakami, E. and Watterson, W.T. (1990) Developing a household travel survey for the Puget Sound Region. *Transportation Research Record* **1285**, 40–48.

Murchland, J.D. (1977) The multiproportional problem. *TSG Note JDM-263*, Transport Studies Group, University College London.

Murphy, K.M. and Topel, R.H. (1985) Estimation and inference in two-step econometric models. *Journal of Business and Economic Statistics* **3**, 370–379.

MVA Systematica (1982) *TRIPS Highway Assignment Model*. MVA House, Woking.

Nash, C. (1997) Transport externalities: does monetary valuation make sense? In G. de Rus and C. Nash (eds.), *Recent Developments in Transport Economics*. Ashgate Press, London.

Nardini, A. (1997) A proposal for integrating environmental impact assessment, cost-benefit analysis and multicriteria analysis in decision making. *Project Appraisal* **12**, 173–184.

NCHRP (2010) *Advanced Practices in Travel Forecasting*. NCHRP Synthesis Report 406, US National Cooperative Highway Research Program, Washington, DC.

Newell, G.F. (1980) *Traffic Flow on Transportation Networks*. MIT Press, Cambridge, Mass.

Niederreiter, H. (1992) *Random Number Generation and Quasi Monte Carlo Methods*. Society for Industrial and Applied Mathematics, Philadelphia.

Nutt, P.C. (1981) Some guides to the selection of a decision making strategy. *Technological Forecasting and Social Change* **19**, 133–145.

OECD (1974) *Urban Traffic Models: Possibilities For Simplification*. OECD Road Research Group, Paris.

OECD (1994a) *Internalising the Social Costs of Transport*. Organisation for Economic Cooperation and Development, Paris.

OECD (1994b) *Environmental Impact Assessment of Roads*. Organisation for Economic Cooperation and Development, Paris.

OECD (1994c) *Project and Policy Appraisal: Integrating Economics and Environment*. Organisation for Economic Cooperation and Development, Paris.

Oh, J.H. (1989) Estimation of trip matrices in networks with equilibrium link flows. *Proceedings 17th PTRC Summer Annual Meeting*, University of Sussex, September 1989, England.

Oi, K.I.Y. and Shuldiner, P.W. (1962) *An Analysis of Urban Travel Demands*. Northwestern University Press, Evanston.

Olsen, G.D. and Swait, J.D. (1998) Nothing is important. *Working Paper*, Faculty of Management, University of Calgary.

Oort, O. (1969) The evaluation of travelling time. *Journal of Transport Economics and Policy* **3**, 279–286.

Oppenheim, N. (1995) *Urban Travel Demand Modelling*. John Wiley and Sons, Inc. New York.

Ortúzar, J. de D. (1980a) Mixed-mode demand forecasting techniques. *Transportation Planning and Technology* **6**, 81–95.

Ortúzar, J. de D. (1980b) Modelling park 'n' ride and kiss 'n' ride as submodal choices: a comment. *Transportation* **9**, 287–291.

Ortúzar, J. de D. (1982) Fundamentals of discrete multimodal choice modelling. *Transport Reviews* **2**, 47–78.

Ortúzar, J. de D. (1983) Nested logit models for mixed-mode travel in urban corridors. *Transportation Research* **17A**, 283–299.

Ortúzar, J. de D. (1986) The cultural and temporal transferability of discrete choice disaggregate modal split models. In T.D. Heaver (ed.), *Research for Tomorrow's Transport Requirements*, University of British Columbia, Vancouver.

Ortúzar, J. de D. (1989) Determining the preferences for frozen cargo exports. In World Conference on Transport Research (eds.), *Transport Policy, Management and Technology Towards 2001*, Western Periodicals Co, Ventura, Ca.

Ortúzar, J. de D. (ed.) (1992) *Simplified Transport Demand Modelling*. Perspectives 2, PTRC, London.

Ortúzar, J. de D. (ed.) (2000) *Stated Preference Modelling Techniques*. Perspectives 4, PTRC, London.

Ortúzar, J. de D. (2001) On the development of the nested logit model. *Transportation Research* **35B**, 213–216.

Ortúzar, J. de D., Achondo, F.J. and Espinosa, A. (1986) On the stability of logit mode choice models. *Proceedings 14th PTRC Summer Annual Meeting*, University of Sussex, July 1986, England.

Ortúzar, J. de D., Armstrong, P.M., Ivelic, A.M. and Valeze, C. (1998) Tamaño muestral y estabilidad temporal en modelos de generación de viajes. *Actas X Congreso Panamericano de Ingeniería de Tránsito y Transporte*, Santander, September 1998, Spain (in Spanish).

Ortúzar, J. de D., Cifuentes, L.A. and Williams, H.C.W.L. (2000a) Application of willingness-to-pay methods to value transport externalities in less developed countries. *Environment and Planning* **32A**, 2007–2018.

Ortúzar, J. de D. and Donoso, P.C.F. (1983) Survey design, implementation, data coding and evaluation for the estimation of disaggregate choice models in Santiago, Chile. *2nd International Conference on New Survey Methods in Transport*, Sydney, September 1983, Australia.

Ortúzar, J. de D., Donoso, P.C.F. and Hutt, G.A. (1983) The effects of measurement techniques, variable definition and model specification on demand model functions. *11th PTRC Summer Annual Meeting*, University of Sussex, July 1983, England.

Ortúzar, J. de D. and Garrido, R.A. (1994a) On the semantic scale problem in stated preference rating experiments. *Transportation* **21**, 185–201.

Ortúzar, J. de D. and Garrido, R.A. (1994b) A practical assessment of stated preference methods. *Transportation* **21**, 289–305.

Ortúzar, J. de D. and Garrido, R.A. (2002) Methodological developments: workshop report. In H.S. Mahmassani (ed.), *In Perpetual Motion: Travel Behaviour Research Opportunities and Application Challenges*. Pergamon, Oxford.

Ortúzar, J. de D. and Hutt G.A. (1984) La influencia de elementos subjetivos en funciones desagregadas de elección discreta. *Ingeniería de Sistemas* **IV**, 37–54 (in Spanish).

Ortúzar, J. de D. and Hutt, G.A. (1988) Travel diaries in Chile: the state of the art. *Proceedings 16th PTRC Summer Annual Meeting*, University of Bath, September 1988, England.

Ortúzar, J. de D. and Ivelic, A.M. (1987) Effects of using more accurately measured level-of-service variables on the specification and stability of mode choice models. *Proceedings 15th PTRC Summer Annual Meeting*, University of Bath, September 1987, England.

Ortúzar, J. de D. and Ivelic, A.M. (1988) Influencia del nivel de agregación de los datos en la estimación de modelos logit de eleción discreta. *Actas del V Congreso Panamericano de Ingeniería de Tránsito y Transporte*, Universidad de Puerto Rico en Mayagüez, July 1988, Puerto Rico (in Spanish).

Ortúzar, J. de D., Ivelic, A.M., Malbrán, H. and Thomas, A. (1993) The 1991 Great Santiago origin-destination survey: methodological design and main results. *Traffic Engineering and Control* 34, 362–368.

Ortúzar, J. de D., Martínez, F.J. and Varela, F.J. (2000b) Stated preferences in modelling accessibility. *International Planning Studies* 5, 65–85.

Ortúzar, J. de D. and Palma, A. (1992) Stated preference in refrigerated and frozen cargo exports. In J. de D. Ortúzar (ed.), *Simplified Transport Demand Modelling*. Perspectives 2, PTRC, London.

Ortúzar, J. de D., Roncagliolo, D.A. and Velarde, U.C. (2000c) Interactions and independence in stated preference modelling. In J. de D. Ortúzar (ed.), *Stated Preference Modelling Techniques*. Perspectives 4, PTRC, London.

Ortúzar, J. de D. and Rodríguez, G. (2002) Valuing reductions in environmental pollution in a residential location context. *Transportation Research* 7D, 407–427.

Ortúzar, J. de D. and Williams, H.C.W.L. (1982) Una interpretactión geométrica de los modelos de elección entre alternativas discretas basados en la teoría de la utilidad aleatoria. *Apuntes de Ingeniería* 7, 25–50 (in Spanish).

Ortúzar, J. de D. and Willumsen, L.G. (1978) Learning to manage transport systems. *Traffic Engineering and Control* 19, 239–239.

Ortúzar, J. de D. and Willumsen, L.G. (1991) Flexible long range planning using low cost information. *Transportation* 18, 151–173.

Outram, V.E. and Thompson, E. (1978) Driver route choice-behavioural and motivational studies. *Proceedings 5th PTRC Summer Annual Meeting*, University of Warwick, July 1977, England.

Ouwersloot, H. and Rietveld, P. (1996) Stated choice experiments with repeated observations. *Journal of Transport Economics and Policy* 30, 203–212.

Overgaard, K.R. (1967) Urban transportation planning: traffic estimation. *Traffic Quarterly* 21, 197–218.

Pakes, A. and Pollard, D. (1989) Simulation and the asymptotics of optimisation estimators. *Econometrica* 57, 1027–1057.

Papageorgiou, M. (ed.) (1991) *Concise Encyclopedia of Traffic and Transportation Systems*. Pergamon Press, Oxford.

Pape, U. (1974) Implementation and efficiency of Moore algorithms for the shortest route problem. *Mathematical Programming* 7, 212–222.

Patriksson, M. (1994) *The Traffic Assignment Problem: Models and Methods*. VSP, Utrecht.

Pendyala, R.M., Parashar, A. and Muthyalagari, G.R. (2001) Measuring day-to-day variability in travel characteristics using GPS data. *80th Annual TRB Meeting*, Washington, DC, January 2001, USA.

Pérez, P.E., Martínez, F.J and Ortúzar, J. de D. (2003) Microeconomic formulation and estimation of a residential location choice model: implications for the value of time. *Journal of Regional Science* 43, 771–789.

Polak, J.W. (1999) Some reflections on the application of equilibrium scheduling theory. *Working Paper*, Centre for Transport Studies, Imperial College of Science, Technology and Medicine, London.

Polak, J.W. (2002) Analysis of non-response in the LATS 2001 pilot household travel diary survey. *81st Annual Meeting of the Transportation Research Board*, Washington, D.C., January 2002, USA.

Polak, J.W., Jones, P.M., Vythoulkas, P.C., Meland, S. and Tretvik, T. (1991) The Trondheim toll ring: results of a stated preference study of travellers' responses. *Report to the European Commission DRIVE Programme*, Transport Studies Unit, University of Oxford.

Prashker J.N. and Bekhor, S. (2000) Congestion, stochastic and similarity effects in stochastic user equilibrium models. *Transportation Research Record* 1733, 80–87.

Prato, C.G. (2009) Route choice modeling: past, present and future research directions. *Journal of Choice Modelling* 2, 65–100.

Prato C.G. and Bekhor, S. (2006) Applying branch & bound techniques to route choice set generation. *Transportation Research Record* 1985, 19–28.

Press, W.H., Teukolsky, S.A., Vetterling, W.T. and Flannery, B.P. (1992) *Numerical Recipes in C: The Art of Scientific Computing*. Cambridge University Press, Cambridge.

Price, A. (1999) A new approach to the appraisal of road projects in England. *Journal of Transport Economics and Policy* 33, 221–226.

Purvis, C.L. (1989) Sample design for the 1990 Bay Area Household Travel Survey. *Working Paper 1*, Bay Area Metropolitan Transport Commission, San Francisco (http://ntl.bts.gov/DOCS/bah.html).

Quarmby, D.A. and Bates, J.J. (1970) An econometric method of car ownership forecasting in discrete areas. *MAU Note 219*. Department of the Environment, London.

Quinet, E. (1994) The social cost of transport: evaluation and links with internalisation policies. Chapter 2 in *Internalising the Social Costs of Transport*. Organisation for Economic Cooperation and Development, Paris.

Quiroga, C., Henk, R. and Jacobson, M. (2000) Innovative data collection techniques for roadside origin-destination surveys. *Transportation Research Record* **1719**, 140–146.

Raftery, A. and Lewis, S. (1992) How many iterations in the Gibbs Sampler? In J.M. Bernardo, A.F.M. Smith, A.P. David and J.O. Berger (eds.), *Bayesian Statistics 4*. Oxford University Press, New York.

RAND (2004) PRISM West Midlands tour generation modelling. *Report RED-02061-03*, Rand Europe, Cambridge.

Raveau, S., Alvarez-Daziano, R., Yáñez, M.F., Bolduc, D. and Ortúzar, J. de D. (2010) Sequential and simultaneous estimation of hybrid discrete choice models: some new findings. *Transportation Research Record* **2156**, 131–139.

Raveau, S., Ortúzar, J. de D. and Yañez, M.F. (2009) Simultaneous estimation of discrete choice models with latent variables. *XIII Euro Working Group on Transportation*, Padua, September 2009, Italy.

Regan, A.C. and Garrido, R.A. (2002) Modelling freight demand and shipper behaviour: state of the art and future directions. In D.A. Hensher (ed.), *Travel Behaviour Research: The Leading Edge*. Pergamon, Oxford.

Revelt, D. and Train, K.E. (1998) Mixed logit with repeated choices: households' choices of appliance efficiency level. *Review of Economics and Statistics* **80**, 647–657.

Revelt, D. and Train, K.E. (2000) Customer-specific taste parameters and mixed logit. *Working Paper E00-274*, Department of Economics. University of California at Berkeley.

Richardson, A.J. (1982) Search models and choice set generation. *Transportation Research* **16A**, 403–419.

Richardson A.J. Ampt, E.S. and Meyburg, A. (1995) *Survey Methods for Transport Planning*. Eucalyptus Press, Melbourne.

Richardson, A.J. and Meyburg, A.H. (2003) Definitions of unit non-response in travel surveys. In P.R. Stopher and P.M. Jones (eds), *Transport Survey Quality and Innovation*. Pergamon, Amsterdam.

Rizzi, L.I. and Ortúzar, J. de D. (2003) Stated preference in the valuation of interurban road safety. *Accident Analysis and Prevention* **35**, 9–22.

Rizzi, L.I. and Ortúzar, J. de D. (2006) Road safety valuation under a stated choice framework. *Journal of Transport Economics and Policy* **40**, 71–96.

Roberts, F.S. (1975) Weighted di-graph models for the assessment of energy use and air pollution in transportation systems. *Environment and Planning* **7A**, 703–724.

Robertson, D.I. (1969) TRANSYT: a traffic network study tool. *TRRL Report LR 253*, Transport and Road Research Laboratory, Crowthorne.

Robertson, D.I. (1974) Cyclic flow profiles. *Traffic Engineering and Control* **15**, 640–641.

Robillard, P. (1975) Estimating the O–D matrix from observed link volumes. *Transportation Research* **9**, 123–128.

Rose, G., Daskin, M. and Koppelman, F.S. (1988) An examination of convergence error in equilibrium traffic assignment models. *Transportation Research* **22B**, 261–274.

Rose, G. and Koppelman, F.S. (1984) Transferability of disaggregate trip generation models. In J. Volmüller and R. Hamerslag (eds.). *Proceedings of the Ninth International Symposium on Transportation and Traffic Theory*. VNU Science Press, Utrecht.

Rose, J.M. and Bliemer, M.C.J. (2005) Constructing efficient choice experiments. *Working Paper itls-wp-05-07*, Institute of Transport and Logistic Studies, University of Sydney.

Rose, J.M. and Bliemer, M.C.J. (2008) Stated preference experimental design strategies. In D.A. Hensher and K.J. Button (eds.), *Handbook of Transport Modelling*. Elsevier, Oxford.

Rose, J.M. and Bliemer, M.C.J. (2009) Constructing efficient stated choice experimental designs *Transport Reviews* **29**, 587–617.

Rose, J.M., Hensher, D.A., Caussade, S., Ortúzar, J. de D. and Jou, R.C. (2009a) Identifying differences in willingness to pay due to dimensionality in stated choice experiments: a cross country analysis. *Journal of Transport Geography* **17**, 21–29.

Rose, J.M., Scarpa, R. and Bliemer, M.C.J (2009b) Incorporating model uncertainty into the generation of efficient stated choice experiments: a model averaging approach. *International Choice Modelling Conference*, Harrogate, March 2009, England.

Ruijgrok, C.J. (1979) Disaggregate choice models: an evaluation. In G.R.M. Jansen, P.H.L. Bovy, J.P.J.M. van Est and F. Le Clercq (eds.), *New Developments in Modelling Travel Demand and Urban Systems*. Saxon House, Westmead.

Ruud, P. (1996) Simulation of the multinomial probit model: an analysis of covariance matrix estimation. *Working Paper*, Department of Economics, University of California at Berkeley.

Sælensminde, K. (1999) *Valuation of Non-Market Goods for Use in Cost-Benefit Analyses: Methodological Issues*. PhD. Thesis, Department of Economics and Social Sciences, Agricultural University of Norway.

Sælensminde, K. (2001) Inconsistent choices in stated choice data: use of the logit scaling approach to handle resulting variance increases. *Transportation Research* **4D**, 13–27.

Sándor, Z. and Wedel, M. (2001) Designing conjoint choice experiments using managers' prior beliefs. *Journal of Marketing Research* **38**, 430–444.

Sándor, Z. and Wedel, M. (2002) Profile construction in experimental choice designs for mixed logit models. *Marketing Science* **21**, 455–475.

Sándor, Z. and Wedel, M. (2005) Heterogeneous conjoint choice designs. *Journal of Marketing Research* **42**, 210–218.

Safwat, K.N.A. and Magnanti, T. (1988) A combined trip generation, trip distribution, modal split and trip assignment model. *Transportation Science* **22**, 14–30.

Sawtooth Software (1999) *The CBC/HB Module for Hierarchical Bayes Estimation*. Sawtooth Software Inc. (www.sawtoothsoftware.com/download/techpap/hbtech.pdf).

Scarpa, R. and Rose, J.M. (2008) Designs efficiency for non-market valuation with choice modelling: how to measure it, what to report and why. *Working Paper 21/07*, Department of Economics, University of Waikato.

Schneider, M. (1959) Gravity models and trip distribution theory. *Papers and Proceedings of the Regional Science Association* **5**, 51–56.

Shanmugalingam, S. (1982) On the analysis of the ratio of two correlated Normal variables. *The Statistician* **31**, 251–258.

Shao, J. and Tu, D. (1995) *The Jackknife and Bootstrap*. Springer, New York.

Sheffi, Y. (1985) *Urban Transportation Networks*. Prentice Hall, Englewood Cliffs, NJ.

Sheffi, Y., Hall, R. and Daganzo, C.F. (1982) On the estimation of the multinomial probit model. *Transportation Research* **16A**, 447–456.

Sikdar, P.K. and Hutchinson, B.G. (1981) Empirical studies of work trip distribution models. *Transportation Research* **15A**, 233–243.

Sillano, M. and Ortúzar, J. de D. (2005) Willingness-to-pay estimation with mixed logit models: some new evidence. *Environment and Planning* **37A**, 525–550.

Silva, M.S. (2002) *Secuencias de Baja Discrepancia para Estimación de Modelos Logit Mixto*. MSc Thesis, Department of Transport Engineering, Pontificia Universidad Católica de Chile.

Simon, H.A. (1957) *Models of Man: Social and Rational*. John Wiley & Sons, Inc. New York.

Simmonds, D.C. (2001) The objectives and design of a new land-use modelling package: DELTA. In G. Clarke and M. Madden (eds.), *Regional Science in Business*. Springer Verlag, Berlin.

Skelton, N. (1982) Determining appropriate sample sizes when two means are to be compared. *Traffic Engineering and Control* **23**, 29–37.

Small, K.A. (1982) The scheduling of consumer activities: work trips. *American Economic Review* **72**, 467–479.

Small, K.A. (1987) A discrete choice model for ordered alternatives. *Econometrica* **55**, 409–424.

Smeed, R.J. (1968) Traffic studies and urban congestion. *Journal of Transport Economics and Policy* **2**, 2–38.

Smith, M.D. (2005) State dependence and heterogeneity in fishing location choice. *Journal of Environmental Economics and Management* **50**, 319–340.

Smith, M.E. (1979) Design of small sample home interview travel surveys. *Transportation Research Record* **701**, 29–35.

Smith, M.J. (1979) Traffic control and route choice: a simple example. *Transportation Research* **13B**, 289–294.

Smith, M.J. (1979) Existence, uniqueness and stability of traffic equilibria. *Transportation Research* **13B**, 295–304.

Smith, M.J. (1981) Properties of a traffic control policy which ensures the existence of a traffic equilibrium consistent with the policy. *Transportation Research* **15B**, 453–462.

Smith, R.L. and Cleveland, D.E. (1976) Time stability analysis of trip generation and predistribution modal choice models. *Transportation Research Record* **569**, 76–86.

Smit-Kroes, N. and Nijpels, E.H.T.M. (1988) *Tweede Struktuurschema Verker en Vervoer*. State Publisher, The Hague (in Dutch).

Smock, R.J. (1962) An iterative assignment approach to capacity restraint on arterial networks. *Highway Research Board Bulletin* **156**, 1–13.

Sobel, K.L. (1980) Travel demand forecasting by using the nested multinomial logit model. *Transportation Research Record* **775**, 48–55.

Sosslau, A.B., Hassam, A., Carter, M. and Wickstrom, G. (1978) Quick response urban travel estimation techniques and transferable parameters. *NCHRP Report 817*, National Cooperative Highway Research Program, Transportation Research Board, Washington, DC.

Spear, B.D. (1977) *Applications of New Travel Demand Forecasting Techniques to Transportation: A Study of Individual Choice Models*. Final Report to the Office of Highway Planning, Federal Highway Administration, US Department of Transportation, Washington, DC.

Spiegelhalter, D.J., Thomas, A. and Best, N.G. (2001) *WinBUGS Beta Version 1.4 User Manual*. MRC Biostatistics Unit, Institute of Public Health, University of Cambridge.

Spielberg, F., Weiner, E. and Ernst, U. (1981) The shape of the 1980's: demographic, economic and travel characteristics. *Transportation Research Record* **807**, 27–34.

Spiess, H. (1983) On optimal route choice strategies in transit networks. *Publication 286*, Centre de Recherche sur les Transports, Université de Montréal, Canada.

Spiess, H. (1987) A maximum likelihood model for estimating origin-destination matrices. *Transportation Research* **21B**, 395–412.

Spiess, H. and Florian, M. (1989) Optimal strategies: a new assignment model for transit networks. *Transportation Research* **23B**, 82–102.

Steenbrink, P.A. (1974) *Optimisation of Transport Networks*. John Wiley & Sons, Inc. New York.

Steer Davies Gleave (2000) *Diseño Operacional del Sistema Transmilenio: Proyecto de Transporte Urbano para Santa Fe de Bogotá*. BIRF 4021-FONDATT-10, Bogotá (in Spanish).

Steer, J. and Willumsen, L.G. (1983) An investigation of passenger preference structures. In S. Carpenter and P.M. Jones (eds.), *Recent Advances in Travel Demand Analysis*. Gower, Aldershot.

Steinberg, R. and Zangwill, W. (1983) The prevalence of Braess's paradox. *Transportation Science* **17**, 301–318.

Stone, R. (1966) *Mathematics in the Social Sciences*. Chapman and Hall, London.

Stopher, P.R. (1975) Goodness-of-fit measures for probabilistic travel demand models. *Transportation* **4**, 67–83.

Stopher, P.R. (1982) Small-sample home-interview travel surveys: application and suggested modifications. *Transportation Research Record* **886**, 41–47.

Stopher, P.R. (1998) Household travel surveys: new perspectives and old problems. In T. Gärling, T. Laitila and K. Westin (eds.), *Theoretical Foundations of Travel Choice Modelling*. Pergamon, Oxford.

Stopher, P.R. and Greaves, S.P. (2004) Sample size requirements for measuring a change in behavior. *27th Australian Transport Research Forum*, Adelaide, September 2004, Australia.

Stopher, P.R. and Jones, P.M. (2003) Developing standards of transport survey quality. In P.R. Stopher and P.M. Jones (eds.), *Transport Survey Quality and Innovation*. Pergamon, Amsterdam.

Stopher, P.R. and Metcalf, H.M.A. (1996) Methods for household travel surveys. *NCHRP Synthesis of Highway Practice 236*, Transportation Research Board, Washington, DC.

Stopher, P.R. and Meyburg, A.H. (1979) *Survey Sampling and Multivariate Analysis for Social Scientists and Engineers*. D.C. Heath and Co., Lexington, Mass.

Stopher, P.R. and Stecher, C. (1993) Blow-up: expanding a complex random sample travel survey. *Transportation Research Record* **1412**, 10–16.

Stouffer, A. (1940) Intervening opportunities: a theory relating mobility and distance. *American Sociological Review* **5**, 845–867.

Street, D.J. and Burgess, L. (2004) Optimal and near-optimal pairs for the estimation of effects in 2-level choice experiments. *Journal of Statistical Planning and Inference* **118**, 185–199.

Street, D.J. and Burgess, L. (2007) *The Construction of Optimal Stated Choice Experiments: Theory and Methods*. John Wiley & Sons, Inc. Hoboken, NJ.

Street, D.J., Burgess, L. and Louviere, J.J. (2005) Quick and easy choice sets: constructing optimal and nearly optimal stated choice experiments. *International Journal of Research in Marketing* **22**, 459–470.

Suh, S., Park, C. and Kim, T.J. (1990) A highway capacity function in Korea: measurement and calibration. *Transportation Research* **24A**, 177–186.

Supernak, J. (1979) A behavioural approach to trip generation modelling. *Proceedings 7th PTRC Summer Annual Meeting*, University of Warwick, July 1979, England.

Supernak, J. (1981) Transferability of the person category trip generation model. *Proceedings 9th PTRC Summer Annual Meeting*, University of Warwick, July 1979, England.

Supernak, J. (1983) Transportation modelling: lessons from the past and tasks for the future. *Transportation* **12**, 79–90.

Supernak, J., Talvitie, A.P. and DeJohn, A. (1983) Person category trip generation modelling. *Transportation Research Record* **944**, 74–83.

Swait, J.D., Adamowicz, W. and Buren, M. (2004) Choice and temporal welfare impacts: incorporating history into discrete choice models. *Journal of Environmental Economics and Management* **47**, 94–116.

Swait, J.D., Louviere, J.J. and Williams, M. (1994) A sequential approach to exploiting the combined strengths of SP and RP data: application to freight shipper choice. *Transportation* **21**, 135–152.

Swanson, J. (2003) The dynamic urban model: transport and urban development. In R. Eberlein, V. Diker, R. Langer and J Rowe (eds.), *Proceedings of the 21st International Conference of the Systems Dynamics Society*. Systems Dynamic Society, New York.

Taleb, N.N. (2007) *The Black Swan*. Random House, New York.

Tamin, O.Z. and Willumsen, L.G. (1989) Transport demand model estimation from traffic counts. *Transportation* **16**, 3–26.

Tamin, O.Z. and Willumsen, L.G. (1992) Freight demand model estimation from traffic counts. In J. de D. Ortúzar (ed.), *Simplified Transport Demand Modelling*. Perspectives 2, PTRC, London.

Tanner, J.C. (1974) Forecasts of vehicles and traffic in Great Britain: 1974 revision. *TRRL Report LR 650*, Transport and Road Research Laboratory, Crowthorne.

Tanner, J.C. (1978) Long term forecasting of vehicle ownership and road traffic. *Journal of the Royal Statistical Society* **141A**, 14–63.

Tardiff, T.J. (1976) A note on goodness-of-fit statistics for probit and logit models. *Transportation* **5**, 377–388.

Tardiff, T.J. (1979) Specification analysis for quantal choice models. *Transportation Science* **13**, 179–390.

Taylor, T.L. (1971) *Instructional Planning Systems: A Gaming-Simulation Approach to Urban Problems*. Cambridge University Press, New York.

Timberlake, R.S. (1988) Traffic modelling techniques for the developing world. *67th Annual TRB Meeting*, Washington, DC, January 1988, USA.

Toner, J.P., Clark, S.D., Grant-Muller, S.M. and Fowkes, A.S. (1998) Anything you can do, we can do better: a provocative introduction to a new approach to stated preference design. *8th World Conference on Transport Research*, Antwerp, July 1998, Belgium.

Toubia, O., Hauser, J. and Garcia, R. (2007) Probabilistic polyhedral methods for adaptive choicebased conjoint analysis: theory and application. *Marketing Science* **26**, 596–610.

Train, K.E. (1977) Valuations of modal attributes in urban travel: questions of non-linearity, non-genericity and taste variations. *Working Paper*, Cambridge Systematics Inc. West, San Francisco.

Train, K.E. (1980) A structured logit model of auto ownership and mode choice. *The Review of Economic Studies* **47**, 357–370.

Train, K.E. (1986) *Qualitative Choice Analysis: Theory, Econometrics and an Application to Automobile Demand*. MIT Press, Cambridge, Mass.

Train, K.E. (1998) Recreation demand models with taste differences over people. *Land Economics* **74**, 230–239.

Train, K.E. (2001) A comparison of hierarchical Bayes and maximum simulated likelihood for mixed logit. *Working Paper*, Department of Economics, University of California at Berkeley.

Train, K.E. (2009) *Discrete Choice Methods with Simulation*. Second Edition, Cambridge University Press, Cambridge.

Train, K.E. and McFadden, D. (1978) The goods/leisure trade-off and disaggregate work trip mode choice models. *Transportation Research* **12**, 349–353.

Train, K.E., McFadden, D. and Ben-Akiva, M.E. (1987) The demand for local telephone service: a fully discrete model of residential calling patterns and service choices. *Rand Journal of Economics* **18**, 109–123.

Train, K.E. and Sonnier, G. (2005) Mixed logit with bounded distributions of correlated partworths. In R. Scarpa and A. Alberini (eds.), *Application of Simulation Methods in Environmental and Resource Economics*. Springer, Dordrecht.

Train, K.E. and Wilson, W.E. (2008) Estimation of stated-preference experiments constructed from revealed-preference choices. *Transportation Research* **40B**, 191–203.

Traugott, M.W. and Katosh, J.P. (1979) Response validity in surveys of voting behaviour. *Public Opinion Quarterly* **42**, 359–377.

Tverski, A. (1972) Elimination by aspects: a theory of choice. *Psychological Review* **79**, 281–299.

Van Es, J.V. (1982) Freight transport, an evaluation. *ECMT Round Table 58*, European Conference of Ministers of Transport, Paris.

Van Vliet, D. (1977) D'Esopo: a forgotten tree-building algorithm. *Traffic Engineering and Control* **18**, 372–375.

Van Vliet, D. (1978) Improved shortest path algorithms for transport networks. *Transportation Research* **12**, 7–20.

Van Vliet, D. (1982) SATURN: a modern assignment model. *Traffic Engineering and Control* **23**, 578–581.

Van Vliet, D. Bergman, T. and Scheltes, W. (1987) Equilibrium assignment with multiple user classes. *Proceedings 15th PTRC Summer Annual Meeting*, University of Sussex, July 1987, England.

Van Vliet, D. and Dow, P. (1979) Capacity restrained road assignment. *Traffic Engineering and Control* **20**, 296–305.

Van Wissen L.J.G and Meurs, H.J. (1989) The Dutch mobility panel: experiences and evaluation. *Transportation* **16**, 99–119.

Van Zuylen, H. and Willumsen, L.G. (1980) The most likely trip matrix estimated from traffic counts. *Transportation Research* **14B**, 281–293.

Verhoef, E. (1994) External effects and social costs of road transport. *Transportation Research* **28A**, 273–287.

Vickrey, W. (1969) Congestion theory and transport investment. *American Economic Review* **59**, 251–261.

Vovsha, P. (1997) The cross-nested logit model: application to mode choice in the Tel Aviv metropolitan area. *76th Annual TRB Meeting*, Washington, DC, January 1997, USA.

Vredin Johansson, M., Heldt, T. and Johansson, P. (2005) Latent variables in a travel mode choice model: attitudinal and behavioural indicator variables. *Working Paper*, Department of Economics, Uppsala University.

Vovsha, P., Donnelly, R. and Gupta, S. (2008) Network equilibrium with activity-based microsimulation models: the New York experience. *87th Annual TRB Meeting*, Washington, DC, February 2008, USA.

Waddell, P. (2002) UrbanSim: modeling urban development for land use, transportation and environmental planning. *Journal of the American Planning Association* **68**, 297–314.

Waddell, P., Borning, A., Noth, M., Freier, N., Becke, M. and Ulfarsson, G. (2003) UrbanSim: a simulation system for land use and transportation. *Networks and Spatial Economics* **3**, 43–67.

Walker, J.L. (2001) *Extended Discrete Choice Models: Integrated Framework, Flexible Error Structures, and Latent Variables*. PhD Thesis, Department of Civil and Environmental Engineering, MIT.

Walker, J.L. (2002) Mixed logit (or logit kernel) model: dispelling misconceptions of identification. *Transportation Research Record* **1805**, 86–98.

Walker, J.L., Ben-Akiva, M.E. and Bolduc, D. (2007) Identification of parameters in Normal error component logit-mixture (NECLM) models. *Journal of Applied Econometrics* **22**, 1095–1125.

Walker, J.L and Li, J. (2007) Latent style preferences and household location decisions. *Journal of Geographical Systems* **9**, 77–101.

Wardman, M. (2001) Inter-temporal variations in the value of time. *ITS Working Paper 566*, Institute for Transport Studies, University of Leeds.

Wardman, M., Tight, M. and Page, M. (2007) Factors influencing the propensity to cycle to work. *Transportation Research* **41A**, 339–350.

Wardrop, J.G. (1952) Some theoretical aspects of road traffic research. *Proceedings of the Institution of Civil Engineers, Part II* **1**, 325–362.

Wardrop, J.G. (1968) Journey speed and flow in central urban areas. *Traffic Engineering and Control* **9**, 528–532, 539.

Warner, S.L. (1962) *Strategic Choice of Mode in Urban Travel: A Study of Binary Choice*. Northwestern University Press, Evanston.

Waters, W.G. (1992) *The Value of Time Savings for the Economic Evaluation of Highway Investments in British Columbia*. Report to the Canadian Ministry of Transportation and Highways, Victoria, BC.

Watson, S.M., Toner, J.P., Fowkes, A.S. and Wardman, M. (2000) Efficiency properties of orthogonal stated preference designs. In J. de D. Ortúzar (ed.), *Stated Preference Modelling Techniques*. Perspectives 4, PTRC, London.

Webster, F.V., Bly, P.H. and Paulley, N.J. (eds.) (1988) *Urban Land-Use and Transport Interaction: Policies and Models*. Gower, Aldershot.

Wegener, M. (2004) Overview of land use transport models. In D.A. Hensher, K.J. Button, K.E. Haynes, and P.R. Stopher (eds.), *Handbook of Transport Geography and Spatial Systems*. Elsevier, Amsterdam.

Weintraub, A, Ortiz, C. and Gonzáles, J. (1985) Accelerating convergence of the Frank-Wolfe algorithm. *Transportation Research* **19B**, 113–122.

Wermuth, M.J. (1981) Effects of survey methods and measurement techniques on the accuracy of household travel-behaviour surveys. In P.R. Stopher, A.H. Meyburg and W. Brög (eds.), *New Horizons in Travel Behaviour Research*. D.C. Health and Co., Lexington, Mass.

Whitelegg, J. (1993) *Transport for a Sustainable Future: The Case for Europe*. Belhaven Press, London.

Williams, H.C.W.L. (1977) On the formation of travel demand models and economic evaluation measures of user benefit. *Environment and Planning* **9A**, 285–344.

Williams, H.C.W.L. (1981) Travel demand forecasting: an overview of theoretical developments. In D.J. Banister and P.G. Hall (eds.), *Transport and Public Policy Planning*. Mansell, London.

Williams, H.C.W.L. and Ortúzar, J. de D. (1982a) Behavioural theories of dispersion and the mis-specification of travel demand models. *Transportation Research* **16B**, 167–219.

Williams, H.C.W.L. and Ortúzar, J. de D. (1982b) Travel demand and response analysis-some integrating themes. *Transportation Research* **16A**, 345–362.

Williams, H.C.W.L. and Senior, M.L. (1977) Model based transport policy assessment: (2) Removing fundamental inconsistencies from the models. *Traffic Engineering and Control* **18**, 464–469.

Williams, I. (1976) A comparison of some calibration techniques for doubly constrained models with an exponential cost function. *Transportation Research* **10**, 91–104.

Willis, K.G., Garrod, G.D. and Harvey, D.R. (1998) A review of cost-benefit analysis as applied to the evaluation of new road proposals in the UK. *Transportation Research* **3D**, 141–156.

Wills, M.J. (1986) A flexible gravity-opportunities model for trip distribution. *Transportation Research* **20B**, 89–111.

Willumsen, L.G. (1978) Estimation of an O–D matrix from traffic counts: a review. *Working Paper 99*, Institute for Transport Studies, University of Leeds.

Willumsen, L.G. (1981) Simplified transport demand models based on traffic counts. *Transportation* **10**, 257–278.

Willumsen, L.G. (1982) Estimation of trip matrices from volume counts: validation of a model under congested conditions. *Proceedings 10th PTRC Summer Annual Meeting*, University of Warwick, July 1982, England.

Willumsen, L.G. (1984) Estimating time-dependent trip matrices from traffic counts. In J. Volmüller and R. Hamerslag (eds.), *Proceedings of the Ninth International Symposium on Transportation and Traffic Theory*. VNU Science Press. Utrecht.

Willumsen, L.G. (1991) Origin-destination matrix: static estimation. In M. Papageorgiou (ed.), *Concise Encyclopedia of Traffic and Transportation Systems*. Pergamon Press, Oxford.

Willumsen, L.G. (2000) Travel networks. In D.A. Hensher and K.J. Button (eds.), *Handbook of Transport Modelling*. Pergamon, Oxford.

Willumsen, L.G., Bolland, J., Arezki, Y. and Hall, M. (1993) Multi-modal modelling in congested networks: SATURN + SATCHMO. *Traffic Engineering and Control* **34**, 294–301.

Willumsen, L.G. and Hounsell, N.B. (1998) Simple models of highway reliability–supply effects. In J. de D. Ortúzar, D.A. Hensher and S.R. Jara-Díaz (eds.), *Travel Behaviour Research: Updating the State of Play*. Pergamon, Oxford.

Willumsen, L.G. and Ortúzar, J. de D. (1985) Intuition and models in transport management. *Transportation Research* **19A**, 51–58.

Willumsen, L.G. and Radovanać M. (1988) Testing the practical value of the UMOT model. *International Journal of Transport Economics* **15**, 203–23.

Wilson, A.G. (1970) *Entropy in Urban and Regional Modelling*. Pion, London.

Wilson, A.G. (1974) *Urban and Regional Models in Geography and Planning*, John Wiley & Sons, Ltd Chichester.

Wilson A.G., Hawkins, A.F., Hill, G.J. and Wagon, D.J. (1969) Calibration and testing of the SELNEC transport model. *Regional Studies* **3**, 337–350.

Wilson, A.G. and Kirby, M.J. (1980) *Mathematics for Geographers and Planners*. Clarendon Press, Oxford.

Wilson, A.G., Rees, P.H. and Leigh, C.M. (eds.) (1977) *Models of Cities and Regions: Theoretical and Empirical Developments*. John Wiley & Sons, Ltd Chichester.

Wittink, D., Krishnamurthi, L. and Nutter, J. (1982) Comparing derived importance weights accross attributes. *Journal of Consumer Research* **8**, 471–474.

Wonnacott, T.H. and Wonnacott, R.J. (1990) *Introductory Statistics for Business and Economics*. John Wiley & Sons, Inc. New York.

Wood, W., Quinn, J.M. and Kashy, D.A. (2002) Habit in everyday life: thought, emotion, and action. *Journal of Personality and Social Psychology* **83**, 1281–1297.

Wootton, H.J., Ness, M.P. and Burton, R.S. (1981) Improved direction signs and the benefits for road users. *Traffic Engineering and Control* **22**, 264–268.

Wootton, H.J. and Pick, G.W. (1967) A model for trips generated by households. *Journal of Transport Economics and Policy* **1**, 137–153.

Wootton Jeffreys and Partners (1980) *JAM User Manual*. Brookwood.

Yáñez, M.F., Cherchi, E., Heydecker, B.G. and Ortúzar, J. de D. (2010b) On the treatment of repeated observations in panel data: efficiency of mixed logit parameter estimates. *Networks and Spatial Economics* **10** (doi:10.1007/s11067-010-9143-6).

Yáñez, M.F., Cherchi, E. and Ortúzar, J. de D. (2010d) Inertia and shock effects over mode choice process: implications of the Transantiago implementation. *Transportation Science* (under review).

Yáñez, M.F., Cherchi, E. and Ortúzar, J. de D. (2010c) Defining inter-alternative error structures for joint RP-SP modeling: some new evidence. *89th Annual TRB Meeting*, Washington, DC, January 2010, USA.

Yáñez, M.F., Mansilla, P. and Ortúzar, J. de D. (2010a) The Santiago Panel: measuring the effects of implementing Transantiago. *Transportation* **37**, 125–149.

Yáñez, M.F., Raveau, S., Rojas, M. and Ortúzar, J. de D. (2009) Modelling and forecasting with latent variables in discrete choice panel models. *European Transport Conference*, Leiden, October 2009, The Netherlands.

Ye, X., Konduri, K., Pendyala, R.M., Sana, B. and Waddell, P. (2009) A methodology to match distributions of both household and person attributes in the generation of synthetic populations. *88th Annual TRB Meeting*, Washington, DC, January 2009, USA.

Yen, J., Mahmasani, H.S. and Herman, R. (1998) A model of employee participation in telecommuting programs based on stated preference data. In J. de D. Ortúzar, D.A. Hensher and S.R. Jara-Díaz (eds.), *Travel Behaviour Research: Updating the State of Play*. Pergamon, Oxford.

Youn, H., Gastner, M.T. and Jeong, H. (2008) Price of anarchy in transportation networks: efficiency and optimality control. *Physical Review Letters* **101**, 128701 (4 pages).

Young, W. and Richardson, A.J. (1980) Residential location preference models: compensatory and non-compensatory approaches. *Proceedings 8th PTRC Summer Annual Meeting*, University of Warwick, July 1980, England.

Zahavi, Y. (1979) The UMOT project. *Report No. DoT-RSPA-DPB-20-79-3*, US Department of Transportation, Washington, DC.

Zimowski, M., Tourangeau, R., Ghadialy, R and Pedlow, S. (1997) Non-response in household travel surveys. National Opinion Research Center (NORC) Report, prepared for the Federal Highway Administration, US Department of Transportation, Washington, D.C.

Index

Printed and bound by CPI Group (UK) Ltd, Croydon, CR0 4YY

12/01/2025

14624505-0002